Nonlinear Statistical Models

Nonlinear Statistical Models

A. RONALD GALLANT

Professor of Statistics and Economics
North Carolina State University
Raleigh, North Carolina

JOHN WILEY & SONS

New York · Chichester · Brisbane · Toronto · Singapore

Text Credits

SAS® is the registered trademark of SAS Institute Inc., Cary, North Carolina, USA.

Tables 1, 3, 5, 7, Chapter 1. Reprinted by permission from A. Ronald Gallant, Nonlinear regression, *The American Statistician* **29**, 73–81, © 1975 by the American Statistical Association, Washington, D.C. 20005.

Table 4, Figure 6, Chapter 1. Reprinted by permission from A. Ronald Gallant, Testing a nonlinear regression specification: A nonregular case, *The Journal of The American Statistical Association* **72**, 523–530, © 1977 by the American Statistical Association, Washington, D.C. 20005.

Table 6, 8, Chapter 1. Reprinted by permission from A. Ronald Gallant, The power of the likelihood ratio test of location in a nonlinear regression model, *The Journal of The American Statistical Association* **70**, 199–203, © 1975 by the American Statistical Association, Washington, D.C. 20005.

Table 1, Chapter 2. Courtesy of the authors: A. Ronald Gallant and J. Jeffery Goebel, Nonlinear regression with autocorrelated errors, *The Journal of The American Statistical Association* **71** (1976), 961–967.

Table 2, Figure 4, Chapter 2. Reprinted by permission from A. Ronald Gallant, Testing a nonlinear regression specification: A nonregular case, *The Journal of The American Statistical Association* **72**, 523–530, © 1977 by the American Statistical Association, Washington, D.C. 20005.

Figures 1, 2, Chapter 2. Reprinted by permission from A. Ronald Gallant and J. Jeffery Goebel, Nonlinear regression with autocorrelated errors, *The Journal of The American Statistical Association* **71**, 961–967, © 1976 by the American Statistical Association, Washington, D.C. 20005.

Tables 1*a*, 1*b*, 1*c*, Chapter 5. Courtesy of the authors: A. Ronald Gallant and Roger W. Koenker, Cost and benefits of peak-load pricing of electricity: A continuous-time econometric approach, *Journal of Econometrics* **26** (1984), 83–114.

Table 5, Chapter 5. Reprinted by permission from *Statistical Methods*, Seventh Edition, by George W. Snedecor and William G. Cochran, © 1980 by The Iowa State University Press, Ames, Iowa 50010.

Tables 6, 7, Chapter 5. Reprinted by permission from A. Ronald Gallant, On the bias in flexible functional forms and essentially unbiased form: The fourier flexible form, *The Journal of Econometrics* **15**, 211–245, © 1981 by North Holland Publishing Company, Amsterdam.

Tables 1*a*, 1*b*, Chapter 6. Courtesy of the authors: Lars Peter Hansen and Kenneth J. Singleton, Generalized instrumental variables estimators of nonlinear rational expectations models, *Econometrica* **50** (1982) 1269–1286.

Library of Congress Cataloging in Publication Data:

Gallant, A. Ronald, 1942–
    Nonlinear statistical models.

    (Wiley series in probability and mathematical statistics. Applied probability and statistics)
    Bibliography: p.
    Includes index.
    1. Regression analysis.   2. Multivariate analysis.
3. Nonlinear theories.   I. Title.   II. Series.
QA278.2.G35 1987   519.5'4   86-18955
ISBN 0-471-80260-3

Printed in the United States of America

10 9 8 7 6 5 4 3 2

To Marcia, Megan, and Drew

Preface

Any type of statistical inquiry in business, government, or academics in which principles from some body of knowledge enter seriously into the analysis is likely to lead to a nonlinear statistical model. For instance, a model obtained as the solution of a differential equation arising in engineering, chemistry, or physics is usually nonlinear. Other examples are economic models of consumer demand or of intertemporal consumption and investment.

Much applied work using linear models represents a distortion of the underlying subject matter. In the past there was little else that one could do, given the restrictions imposed by the cost of computing equipment and the lack of an adequate statistical theory. But the availability of computing resources is no longer a problem, and advances in statistical and probability theory have occurred over the last fifteen years that effectively remove the restriction of inadequate theory.

In this book, I have attempted to bring these advances together in one place, organize them, and relate them to applications, for the use of students as a text and for the use of those engaged in research as a reference. My hopes and goals in writing it will be achieved if it becomes possible for the reader to bring subject matter considerations directly to bear on data without distortion.

The coverage is comprehensive. The three major categories of statistical models relating dependent variables to explanatory variables are covered: univariate regression models, multivariate regression models, and simultaneous equations models. These models can have the classical regression structure where the independent variables are ancillary and the errors independent, or they can be dynamic, with lagged dependent variables permitted as explanatory variables and with serially correlated errors. The coverage is also comprehensive in the sense that the subject is treated at all levels: methods, theory, and computations. However, only material that I think is of practical value in making a statistical inference using a model that derives from subject matter considerations is included.

The statistical methods are accessible to anyone with a good working knowledge of the theory and methods of linear statistical models as found in a text such as Searle's *Linear Models*. It is important that Chapter 1 be read first. It lays the intuitive foundation. There the subject of univariate nonlinear regression is presented by relying on analogy with the theory and methods of linear models, on examples, and on Monte Carlo simulations. The topic lends itself to this treatment, as the role of the theory is to justify some intuitively obvious linear approximations derived from Taylor's expansions. One can get the main ideas across and save the theory for later. Generalized least squares can be applied in nonlinear regression just as in linear regression. Using this as a vehicle, the ideas, intuition, and statistical methods developed in Chapter 1 are extended to other situations, notably multivariate nonlinear regression in Chapter 5 and nonlinear simultaneous equations models in Chapter 6. These chapters include many numerical examples.

Chapter 3 is a unified theory of statistical inference for nonlinear models with regression structure, and Chapter 7 is the same for dynamic models. Some useful specialization of the general theory is possible in the case of the univariate nonlinear regression model, and this is done in Chapter 4. Notation, assumptions, and theorems are isolated and clearly identified in the theoretical chapters so that the results can be reliably applied to new situations without need for a detailed reading of the mathematics. These results should be usable by anyone who is comfortable thinking of a random variable as a function defined on an abstract probability space and understands the notion of almost sure convergence. Aside from that, application of the theory does not rise above an advanced calculus level probability course. There are examples in these chapters to provide templates.

Reading the proofs requires a good understanding of measure theoretic probability theory, as would be imparted by a course out of Tucker's *Graduate Course in Probability*, and a working knowledge of analysis, as in Royden's *Real Analysis*. For the reader's convenience, references are confined to these two books as much as possible, but this material is standard and any similar textbook will serve.

The material in Chapter 7 is at the frontier. This is the first time some of it will appear in print. As with anything new, much improvement is still possible. Regularity conditions are more onerous than need be, and there is a paucity of worked examples to determine which of them most need relaxing. I have included full details in the proofs, and have supplied the details of proofs that seemed too terse in the original source, in hopes that readers can learn the ideas and methods of proof quickly and will move the field forward.

As to computations, one must either use a programming language, with or without the aid of a scientific subroutine library, or use a statistical package. Hand calculator computations are out of the question. Using a programming language to present the ideas seems ill advised. Discussion bogs down in detail that is just tedious accounting and has nothing to do with the subject proper. For pedagogical purposes, a statistical package is the better choice. Its code should be concise and readable, even to the uninitiated. I chose SAS®, and it seems to have served well. Computational examples consist of figures displaying a few lines of SAS code and the resulting output. For those who would rather use a programming language in applications, the algorithms are in the text, and anyone accustomed to using a programming language should have no trouble implementing them; the examples will be helpful in debugging.

I have debts to acknowledge. The biggest is to my family. Hours—no, years—were spent writing that ought to have been spent with them. I owe a debt to my students Geraldo Souza and Jose Francisco Burguete. The theory for models with regression structure is their dissertation research. The theory for dynamic models was worked out while Halbert White and Jeffrey Wooldridge visited Raleigh in the summer of 1984, and much of it is theirs. I owe a special debt to my secretary, Janice Gaddy. She typed the manuscript cheerfully, promptly, and accurately. More importantly, she held every annoyance at bay.

Support while writing this book was provided by National Science Foundation Grants SES 82-07362 and SES 85-07829, North Carolina Agricultural Experiment Station Projects NC03641, NC03879, and NC05593, and the PAMS Foundation. SAS Institute Inc. let me use its computing equipment and a prerelease version of PROC SYSNLIN for the computations in Chapter 6 and has, over the years, provided generous support to the Triangle Econometrics Workshop. Many ideas in this book have come from that workshop.

A. RONALD GALLANT

December, 1986
Raleigh, North Carolina

Contents

1. Univariate Nonlinear Regression   **1**

    1. Introduction, 1
    2. Taylor's Theorem and Matters of Notation, 8
    3. Statistical Properties of Least Squares Estimators, 16
    4. Methods of Computing Least Squares Estimates, 26
    5. Hypothesis Testing, 47
    6. Confidence Intervals, 104
    7. Appendix: Distributions, 121

2. Univariate Nonlinear Regression: Special Situations   **123**

    1. Heteroscedastic Errors, 124
    2. Serially Correlated Errors, 127
    3. Testing a Nonlinear Specification, 139
    4. Measures of Nonlinearity, 146

3. A Unified Asymptotic Theory of Nonlinear Models with Regression Structure   **148**

    1. Introduction, 149
    2. The Data Generating Model and Limits of Cesaro Sums, 153
    3. Least Mean Distance Estimators, 174
    4. Method of Moments Estimators, 197
    5. Tests of Hypotheses, 217
    6. Alternative Representation of a Hypothesis, 240

7. Constrained Estimation, 242

8. Independently and Identically Distributed Regressors, 247

4. Univariate Nonlinear Regression: Asymptotic Theory   253

1. Introduction, 253

2. Regularity Conditions, 255

3. Characterizations of Least Squares Estimators and Test Statistics, 259

5. Multivariate Nonlinear Regression   267

1. Introduction, 267

2. Least Squares Estimators and Matters of Notation, 290

3. Hypothesis Testing, 320

4. Confidence Intervals, 355

5. Maximum Likelihood Estimation, 355

6. Asymptotic Theory, 379

7. An Illustration of the Bias in Inference Caused by Misspecification, 397

6. Nonlinear Simultaneous Equations Models   405

1. Introduction, 406

2. Three Stage Least Squares, 426

3. The Dynamic Case: Generalized Method of Moments, 442

4. Hypothesis Testing, 452

5. Maximum Likelihood Estimation, 465

7. A Unified Asymptotic Theory for Dynamic Nonlinear Models   487

1. Introduction, 488

2. A Uniform Strong Law and a Central Limit Theorem for Dependent, Nonstationary Random Variables, 493

3. Data Generating Process, 541

4. Least Mean Distance Estimators, 544

5. Method of Moments Estimators, 566

6. Hypothesis Testing, 584

References   595

Author Index   601

Subject Index   603

Nonlinear Statistical Models

CHAPTER 1

Univariate Nonlinear Regression

The nonlinear regression model with a univariate dependent variable is more frequently used in applications than any of the other methods discussed in this book. Moreover, these other methods are for the most part fairly straightforward extensions of the ideas of univariate nonlinear regression. Accordingly, we shall take up this topic first and consider it in some detail.

In this chapter, we shall present the theory and methods of univariate nonlinear regression by relying on analogy with the theory and methods of linear regression, on examples, and on Monte Carlo illustrations. The formal mathematical verifications are presented in subsequent chapters. The topic lends itself to this treatment because the role of the theory is to justify some intuitively obvious linear approximations derived from Taylor's expansions. Thus one can get the main ideas across first and save the theoretical details until later. This is not to say that the theory is unimportant. Intuition is not entirely reliable, and some surprises are uncovered by careful attention to regularity conditions and mathematical detail.

1. INTRODUCTION

One of the most common situations in statistical analysis is that of data which consist of observed, univariate responses y_t known to be dependent on corresponding k-dimensional inputs x_t. This situation may be represented by the regression equations

$$y_t = f(x_t, \theta^0) + e_t \qquad t = 1, 2, \ldots, n$$

1

where $f(x, \theta)$ is the known response function, θ^0 is a p-dimensional vector of unknown parameters, and the e_t represent unobservable observational or experimental errors. We write θ^0 to emphasize that it is the true, but unknown, value of the parameter vector θ that is meant; θ itself is used to denote instances when the parameter vector is treated as a variable—as, for instance, in differentiation. The errors are assumed to be independently and identically distributed with mean zero and unknown variance σ^2. The sequence of independent variables $\{x_t\}$ is treated as a fixed known sequence of constants, not random variables. If some components of the independent vectors were generated by a random process, then the analysis is conditional on that realization $\{x_t\}$ which obtained for the data at hand. See Section 2 of Chapter 3 for additional details on this point, and Section 8 of Chapter 3 for a device that allows one to consider the random regressor setup as a special case in a fixed regressor theory.

Frequently, the effect of the independent variable x_t on the dependent variable y_t is adequately approximated by a response function which is linear in the parameters

$$f(x, \theta) = x'\theta = \sum_{i=1}^{p} x_i \theta_i.$$

By exploiting various transformations of the independent and dependent variables, viz.

$$\varphi_0(y_t) = \sum_{i=1}^{p} \varphi_i(x_t)\theta_i + e_t$$

the scope of models that are linear in the parameters can be extended considerably. But there is a limit to what can be adequately approximated by a linear model. At times a plot of the data or other data analytic considerations will indicate that a model which is not linear in its parameters will better represent the data. More frequently, nonlinear models arise in instances where a specific scientific discipline specifies the form that the data ought to follow, and this form is nonlinear. For example, a response function which arises from the solution of a differential equation might assume the form

$$f(x, \theta) = \theta_1 + \theta_2 e^{x\theta_3}.$$

Another example is a set of responses that is known to be periodic in time but with an unknown period. A response function for such data is

$$f(t, \theta) = \theta_1 + \theta_2 \cos \theta_4 t + \theta_3 \sin \theta_4 t.$$

A univariate linear regression model, for our purposes, is a model that can be put in the form

$$\varphi_0(y_t) = \sum_{i=1}^{p} \varphi_i(x_t)\theta_i + e_t.$$

A univariate nonlinear regression model is of the form

$$\varphi_0(y_t) = f(x_t, \theta) + e_t$$

but since the transformation φ_0 can be absorbed into the definition of the dependent variable, the model

$$y_t = f(x_t, \theta) + e_t$$

is sufficiently general. Under these definitions a linear model is a special case of the nonlinear model in the same sense that a central chi-square distribution is a special case of the noncentral chi-square distribution. This is somewhat an abuse of language, as one ought to say regression model and linear regression model rather than nonlinear regression model and (linear) regression model to refer to these two categories. But this usage is long established and it is senseless to seek change now.

EXAMPLE 1. The example that we shall use most frequently in illustration has the response function

$$f(x, \theta) = \theta_1 x_1 + \theta_2 x_2 + \theta_4 e^{\theta_3 x_3}.$$

The vector-valued input or independent variable is

$$x = \begin{pmatrix} x_1 \\ x_2 \\ x_3 \end{pmatrix}$$

and the vector-valued parameter is

$$\theta = \begin{pmatrix} \theta_1 \\ \theta_2 \\ \theta_3 \\ \theta_4 \end{pmatrix}$$

Table 1. Data Values for Example 1.

t	Y	X1	X2	X3
1	0.98610	1	1	6.28
2	1.03848	0	1	9.86
3	0.95482	1	1	9.11
4	1.04184	0	1	8.43
5	1.02324	1	1	8.11
6	0.90475	0	1	1.82
7	0.96263	1	1	6.58
8	1.05026	0	1	5.02
9	0.98861	1	1	6.52
10	1.03437	0	1	3.75
11	0.98982	1	1	9.86
12	1.01214	0	1	7.31
13	0.66768	1	1	0.47
14	0.55107	0	1	0.07
15	0.96822	1	1	4.07
16	0.98823	0	1	4.61
17	0.59759	1	1	0.17
18	0.99418	0	1	6.99
19	1.01962	1	1	4.39
20	0.69163	0	1	0.39
21	1.04255	1	1	4.73
22	1.04343	0	1	9.42
23	0.97526	1	1	8.90
24	1.04969	0	1	3.02
25	0.80219	1	1	0.77
26	1.01046	0	1	3.31
27	0.95196	1	1	4.51
28	0.97658	0	1	2.65
29	0.50811	1	1	0.08
30	0.91840	0	1	6.11

Source: Gallant (1975d).

so that for this response function $k = 3$ and $p = 4$. A set of observed responses and inputs for this model which will be used to illustrate the computations is given in Table 1. The inputs correspond to a one way "treatment-control" design that uses experimental material whose age ($= x_3$) affects the response exponentially. That is, the first observation

$$x_1 = (1, 1, 6.28)'$$

represents experimental material with attained age $x_3 = 6.28$ months that was (randomly) allocated to the treatment group and has expected response

$$f(x_1, \theta^0) = \theta_1^0 + \theta_2^0 + \theta_4^0 e^{6.28\theta_3^0}.$$

Similarly, the second observation

$$x_2 = (0, 1, 9.86)'$$

represents an allocation of material with attained age $x_3 = 9.86$ to the control group, with expected response

$$f(x_2, \theta^0) = \theta_2^0 + \theta_4^0 e^{9.86\theta_3^0}$$

and so on. The parameter θ_1^0 is the treatment effect. The data of Table 1 are simulated. □

EXAMPLE 2. Quite often, nonlinear models arise as solutions of a system of differential equations. The following linear system has been used so often in the nonlinear regression literature (Box and Lucus, 1959; Guttman and Meeter, 1965; Gallant, 1980) that it might be called the standard pedagogical example.

Linear System

$$\frac{d}{dx} A(x) = -\theta_1 A(x)$$

$$\frac{d}{dx} B(x) = \theta_1 A(x) - \theta_2 B(x)$$

$$\frac{d}{dx} C(x) = \theta_2 B(x)$$

Boundary Conditions

$$A(x) = 1 \quad B(x) = C(x) = 0 \qquad \text{at time} \quad x = 0$$

Parameter Space

$$\theta_1 \geq \theta_2 \geq 0$$

Solution, $\theta_1 > \theta_2$

$$A(x) = e^{-\theta_1 x}$$
$$B(x) = (\theta_1 - \theta_2)^{-1}(\theta_1 e^{-\theta_2 x} - \theta_1 e^{-\theta_1 x})$$
$$C(x) = 1 - (\theta_1 - \theta_2)^{-1}(\theta_1 e^{-\theta_2 x} - \theta_2 e^{-\theta_1 x})$$

Solution, $\theta_1 = \theta_2$

$$A(x) = e^{-\theta_1 x}$$

$$B(x) = \theta_1 x e^{-\theta_1 x}$$

$$C(x) = 1 - e^{-\theta_1 x} - \theta_1 x e^{-\theta_1 x}$$

Systems such as this arise in compartment analysis where the rate of flow of a substance from compartment A into compartment B is a constant proportion θ_1 of the amount $A(x)$ present in compartment A at time x. Similarly, the rate of flow from B to C is a constant proportion θ_2 of the amount $B(x)$ present in compartment B at time x. The rate of change of the quantities within each compartment is described by the system of linear differential equations. In chemical kinetics, this model describes a reaction where substance A decomposes at a reaction rate of θ_1 to form substance B, which in turn decomposes at a rate θ_2 to form substance C. There are a great number of other instances where linear systems of differential equations such as this arise.

Following Guttman and Meeter (1965), we shall use the solutions for $B(x)$ and $C(x)$ to construct two nonlinear models (see Table 2) which they assert "represent fairly well the extremes of near linearity and extreme nonlinearity." These two models are set forth immediately below. The design points and parameter settings are those of Guttman and Meeter (1965).

Model B

$$f(x, \theta) = \begin{cases} \dfrac{\theta_1(e^{-x\theta_2} - e^{-x\theta_1})}{\theta_1 - \theta_2} & \theta_1 \neq \theta_2 \\ \theta_1 x e^{-x\theta_1} & \theta_1 = \theta_2 \end{cases}$$

$$\theta^0 = (1.4, .4)'$$

$$\{x_t\} = \{.25, .5, 1, 1.5, 2, 4, .25, .5, 1, 1.5, 2, 4\}$$

$$n = 12$$

$$\sigma^2 = (.025)^2$$

Model C

$$f(x, \theta) = \begin{cases} 1 - \dfrac{\theta_1 e^{-x\theta_2} - \theta_2 e^{-x\theta_1}}{\theta_1 - \theta_2} & \theta_1 \neq \theta_2 \\ 1 - e^{-x\theta_1} - x\theta_1 e^{-x\theta_1} & \theta_1 = \theta_2 \end{cases}$$

$$\theta^0 = (1.4, .4)'$$

$$\{x_t\} = \{1, 2, 3, 4, 5, 6, 1, 2, 3, 4, 5, 6\}$$

$$n = 12$$

$$\sigma^2 = (.025)^2 \qquad\qquad \square$$

Table 2. Data Values
for Example 2.

t	Y	X
	Model B	
1	0.316122	0.25
2	0.421297	0.50
3	0.601996	1.00
4	0.573076	1.50
5	0.545661	2.00
6	0.281509	4.00
7	0.273234	0.25
8	0.415292	0.50
9	0.603644	1.00
10	0.621614	1.50
11	0.515790	2.00
12	0.278507	4.00
	Model C	
1	0.137790	1
2	0.409262	2
3	0.639014	3
4	0.736366	4
5	0.786320	5
6	0.893237	6
7	0.163208	1
8	0.372145	2
9	0.599155	3
10	0.749201	4
11	0.835155	5
12	0.905845	6

A word regarding notation. All vectors, such as θ, are column vectors unless the contrary is indicated by θ', which is a row vector. Strict adherence to this convention in notation leads to clutter, such as

$$d = (a', b', c')'.$$

We shall usually let the primes be understood in these cases and write

$$d = (a, b, c)$$

instead. Transposition will be carefully indicated at instances where clarity seems to demand it.

2. TAYLOR'S THEOREM AND MATTERS OF NOTATION

In what follows, a matrix notation for certain concepts in differential calculus leads to a more compact and readable exposition. Suppose that $s(\theta)$ is a real valued function of a p-dimensional argument θ. The notation $(\partial/\partial\theta)s(\theta)$ denotes the gradient of $s(\theta)$,

$$\frac{\partial}{\partial\theta}s(\theta) = \begin{pmatrix} \dfrac{\partial}{\partial\theta_1}s(\theta) \\[6pt] \dfrac{\partial}{\partial\theta_2}s(\theta) \\[6pt] \vdots \\[6pt] \dfrac{\partial}{\partial\theta_p}s(\theta) \end{pmatrix}_1^{\,p}$$

a p by 1 (column) vector with typical element $(\partial/\partial\theta_i)s(\theta)$. Its transpose is denoted by

$$\frac{\partial}{\partial\theta'}s(\theta) = {}_1\!\left(\frac{\partial}{\partial\theta_1}s(\theta), \frac{\partial}{\partial\theta_2}s(\theta), \ldots, \frac{\partial}{\partial\theta_p}s(\theta) \right)_p.$$

Suppose that all second order derivatives of $s(\theta)$ exist. They can be arranged in a p by p matrix, known as the Hessian matrix of the function

$s(\theta)$,

$$\frac{\partial^2}{\partial\theta\,\partial\theta'}s(\theta) = \begin{pmatrix} \dfrac{\partial^2}{\partial\theta_1^2}s(\theta) & \dfrac{\partial^2}{\partial\theta_1\,\partial\theta_2}s(\theta) & \cdots & \dfrac{\partial^2}{\partial\theta_1\,\partial\theta_p}s(\theta) \\[3mm] \dfrac{\partial^2}{\partial\theta_2\,\partial\theta_1}s(\theta) & \dfrac{\partial^2}{\partial\theta_2^2}s(\theta) & \cdots & \dfrac{\partial^2}{\partial\theta_2\,\partial\theta_p}s(\theta) \\[2mm] \vdots & \vdots & & \vdots \\[2mm] \dfrac{\partial^2}{\partial\theta_p\,\partial\theta_1}s(\theta) & \dfrac{\partial^2}{\partial\theta_p\,\partial\theta_2}s(\theta) & \cdots & \dfrac{\partial^2}{\partial^2\theta_p}s(\theta) \end{pmatrix}_p .$$

If the second order derivatives of $s(\theta)$ are continuous functions in θ, then the Hessian matrix is symmetric (Young's theorem).

Let $f(\theta)$ be an n by 1 (column) vector valued function of a p-dimensional argument θ. The Jacobian of

$$f(\theta) = \begin{pmatrix} f_1(\theta) \\ f_2(\theta) \\ \vdots \\ f_n(\theta) \end{pmatrix}_1$$

is the n by p matrix

$$\frac{\partial}{\partial\theta'}f(\theta) = \begin{pmatrix} \dfrac{\partial}{\partial\theta_1}f_1(\theta) & \dfrac{\partial}{\partial\theta_2}f_1(\theta) & \cdots & \dfrac{\partial}{\partial\theta_p}f_1(\theta) \\[3mm] \dfrac{\partial}{\partial\theta_1}f_2(\theta) & \dfrac{\partial}{\partial\theta_2}f_2(\theta) & \cdots & \dfrac{\partial}{\partial\theta_p}f_2(\theta) \\[2mm] \vdots & \vdots & & \vdots \\[2mm] \dfrac{\partial}{\partial\theta_1}f_n(\theta) & \dfrac{\partial}{\partial\theta_2}f_n(\theta) & \cdots & \dfrac{\partial}{\partial\theta_p}f_n(\theta) \end{pmatrix}_p .$$

Let $h'(\theta)$ be a 1 by n (row) vector valued function

$$h'(\theta) = [h_1(\theta), h_2(\theta), \ldots, h_n(\theta)].$$

Then

$$
\frac{\partial}{\partial \theta} h'(\theta) = \begin{pmatrix}
\frac{\partial}{\partial \theta_1} h_1(\theta) & \frac{\partial}{\partial \theta_1} h_2(\theta) & \cdots & \frac{\partial}{\partial \theta_1} h_n(\theta) \\
\frac{\partial}{\partial \theta_2} h_1(\theta) & \frac{\partial}{\partial \theta_2} h_2(\theta) & \cdots & \frac{\partial}{\partial \theta_2} h_n(\theta) \\
\vdots & \vdots & & \vdots \\
\frac{\partial}{\partial \theta_p} h_1(\theta) & \frac{\partial}{\partial \theta_p} h_2(\theta) & \cdots & \frac{\partial}{\partial \theta_p} h_n(\theta)
\end{pmatrix}_{\substack{\\ n}}^{\substack{p}} .
$$

In this notation, the following rule governs matrix transposition:

$$
\left(\frac{\partial}{\partial \theta'} f(\theta) \right)' = \frac{\partial}{\partial \theta} f'(\theta).
$$

And the Hessian matrix of $s(\theta)$ can be obtained by successive differentiation variously as

$$
\begin{aligned}
\frac{\partial^2}{\partial \theta \, \partial \theta'} s(\theta) &= \frac{\partial}{\partial \theta} \left(\frac{\partial}{\partial \theta'} s(\theta) \right) \\
&= \frac{\partial}{\partial \theta} \left(\frac{\partial}{\partial \theta} s(\theta) \right)' \\
&= \frac{\partial}{\partial \theta'} \left(\frac{\partial}{\partial \theta} s(\theta) \right) \qquad \text{(if symmetric)} \\
&= \frac{\partial}{\partial \theta'} \left(\frac{\partial}{\partial \theta'} s(\theta) \right)' \qquad \text{(if symmetric)}.
\end{aligned}
$$

One has a product rule and a chain rule. They read as follows. If $f(\theta)$ and $h'(\theta)$ are as above, then (Problem 1)

$$
\frac{\partial}{\partial \theta'} h'(\theta) f(\theta) = \underset{1}{h'(\theta)} \underset{n}{\left(\frac{\partial}{\partial \theta'} f(\theta) \right)_p} + \underset{1}{f'(\theta)} \underset{n}{\left(\frac{\partial}{\partial \theta'} h(\theta) \right)_p}.
$$

Let $g(\rho)$ be a p by 1 (column) vector valued function of an r-dimensional argument ρ, and let $f(\theta)$ be as above: Then (Problem 2)

$$
\frac{\partial}{\partial \rho'} f[g(\rho)] = \underset{n}{\left(\frac{\partial}{\partial \theta'} f(\theta) \right)_p} \Bigg|_{\theta = g(\rho)} \frac{\partial}{\partial \rho'} g(\rho)_r .
$$

The set of nonlinear regression equations

$$y_t = f(x_t, \theta^0) + e_t \qquad t = 1, 2, \ldots, n$$

may be written in a convenient vector form

$$y = f(\theta^0) + e$$

by adopting conventions analogous to those employed in linear regression; namely

$$y = _n\!\!\left(\begin{matrix} y_1 \\ y_2 \\ \vdots \\ y_n \end{matrix}\right)_1$$

$$f(\theta) = _n\!\!\left(\begin{matrix} f(x_1, \theta) \\ f(x_2, \theta) \\ \vdots \\ f(x_n, \theta) \end{matrix}\right)_1$$

$$e = _n\!\!\left(\begin{matrix} e_1 \\ e_2 \\ \vdots \\ e_n \end{matrix}\right)_1 .$$

The sum of squared deviations

$$\mathrm{SSE}(\theta) = \sum_{t=1}^{n} [y_t - f(x_t, \theta)]^2$$

of the observed y_t from the predicted value $f(x_t, \theta)$ corresponding to a trial value of the parameter θ becomes

$$\mathrm{SSE}(\theta) = [y - f(\theta)]'[y - f(\theta)] = \|y - f(\theta)\|^2$$

in this vector notation.

The estimators employed in nonlinear regression can be characterized as linear and quadratic forms in the vector e which are similar in appearance to those that appear in linear regression to within an error of approximation that becomes negligible in large samples. Let

$$F(\theta) = \frac{\partial}{\partial \theta'} f(\theta);$$

that is, $F(\theta)$ is the matrix with typical element $(\partial/\partial \theta_j) f(x_t, \theta)$, where t is the row index and j is the column index. The matrix $F(\theta^0)$ plays the same role in these linear and quadratic forms as the design matrix X in the linear regression:

$$"y" = X\beta + e.$$

The appropriate analogy is obtained by setting $"y" = y - f(\theta^0) + F(\theta^0)\theta^0$ and setting $X = F(\theta^0)$. Malinvaud (1970a, Chapter 9) terms this equation the "linear pseudo-model." For simplicity we shall write F for the matrix $F(\theta)$ when it is evaluated at $\theta = \theta^0$:

$$F \equiv F(\theta^0).$$

Let us illustrate this notation with Example 1.

EXAMPLE 1 (Continued). Direct application of the definitions of y and $f(\theta)$ yields

$$y = \begin{pmatrix} 0.98610 \\ 1.03848 \\ 0.95482 \\ 1.04184 \\ \vdots \\ 0.50811 \\ 0.91840 \end{pmatrix}_{\substack{\\ \\ \\ \\ \\ \\ 1}}^{\substack{\\ \\ \\ \\ \\ \\ \\}}$$

$$f(\theta) = \begin{pmatrix} \theta_1 + \theta_2 + \theta_4 e^{6.28\theta_3} \\ \theta_2 + \theta_4 e^{9.86\theta_3} \\ \theta_1 + \theta_2 + \theta_4 e^{9.11\theta_3} \\ \theta_2 + \theta_4 e^{8.43\theta_3} \\ \vdots \\ \theta_1 + \theta_2 + \theta_4 e^{0.08\theta_3} \\ \theta_2 + \theta_4 e^{6.11\theta_3} \end{pmatrix}.$$

Since

$$\frac{\partial}{\partial \theta_1} f(x, \theta) = \frac{\partial}{\partial \theta_1} \left(\theta_1 x_1 + \theta_2 x_2 + \theta_4 e^{\theta_3 x_3} \right) = x_1$$

$$\frac{\partial}{\partial \theta_2} f(x, \theta) = \frac{\partial}{\partial \theta_2} \left(\theta_1 x_1 + \theta_2 x_2 + \theta_4 e^{\theta_3 x_3} \right) = x_2$$

$$\frac{\partial}{\partial \theta_3} f(x, \theta) = \frac{\partial}{\partial \theta_3} \left(\theta_1 x_1 + \theta_2 x_2 + \theta_4 e^{\theta_3 x_3} \right) = \theta_4 x_3 e^{\theta_3 x_3}$$

$$\frac{\partial}{\partial \theta_4} f(x, \theta) = \frac{\partial}{\partial \theta_4} \left(\theta_1 x_1 + \theta_2 x_2 + \theta_4 e^{\theta_3 x_3} \right) = e^{\theta_3 x_3}$$

the Jacobian of $f(\theta)$ is

$$F(\theta) = {}_{30}\begin{pmatrix} 1 & 1 & \theta_4(6.28)e^{6.28\theta_3} & e^{6.28\theta_3} \\ 0 & 1 & \theta_4(9.86)e^{9.86\theta_3} & e^{9.86\theta_3} \\ 1 & 1 & \theta_4(9.11)e^{9.11\theta_3} & e^{9.11\theta_3} \\ 0 & 1 & \theta_4(8.43)e^{8.43\theta_3} & e^{8.43\theta_3} \\ \vdots & \vdots & \vdots & \vdots \\ 1 & 1 & \theta_4(0.08)e^{0.08\theta_3} & e^{0.08\theta_3} \\ 0 & 1 & \theta_4(6.11)e^{6.11\theta_3} & e^{6.11\theta_3} \end{pmatrix}_4 . \qquad \Box$$

Taylor's theorem, as we shall use it, reads as follows:

TAYLOR'S THEOREM. Let $s(\theta)$ be a real valued function defined over Θ. Let Θ be an open, convex subset of p-dimensional Euclidean space $\mathbb{R}^p$. Let θ^0 be some point in Θ.

If $s(\theta)$ is once continuously differentiable on Θ, then

$$s(\theta) = s(\theta^0) + \sum_{i=1}^{p} \left(\frac{\partial}{\partial \theta_i} s(\bar{\theta}) \right) \left(\theta_i - \theta_i^0 \right)$$

or, in vector notation,

$$s(\theta) = s(\theta^0) + \left(\frac{\partial}{\partial \theta} s(\bar{\theta}) \right)' \left(\theta - \theta^0 \right)$$

for some $\bar{\theta} = \lambda \theta^0 + (1 - \lambda)\theta$ where $0 \leq \lambda \leq 1$.

If $s(\theta)$ is twice continuously differentiable on Θ, then

$$s(\theta) = s(\theta^0) + \sum_{i=1}^{P} \left(\frac{\partial}{\partial \theta_i} s(\theta^0) \right) (\theta_i - \theta_i^0)$$

$$+ \frac{1}{2} \sum_{i=1}^{P} \sum_{j=1}^{P} (\theta_i - \theta_i^0) \left(\frac{\partial^2}{\partial \theta_i \partial \theta_j} s(\bar{\theta}) \right) (\theta_j - \theta_j^0)$$

or, in vector notation,

$$s(\theta) = s(\theta^0) + \left(\frac{\partial}{\partial \theta} s(\theta^0) \right)' (\theta - \theta^0)$$

$$+ \tfrac{1}{2} (\theta - \theta^0)' \left(\frac{\partial^2}{\partial \theta \, \partial \theta'} s(\bar{\theta}) \right) (\theta - \theta^0)$$

for some $\bar{\theta} = \lambda \theta^0 + (1 - \lambda)\theta$ where $0 \leq \lambda \leq 1$. $\square$

Applying Taylor's theorem to $f(x, \theta)$, we have

$$f(x, \theta) = f(x, \theta^0) + \left(\frac{\partial}{\partial \theta} f(x, \theta^0) \right)' (\theta - \theta^0)$$

$$+ \tfrac{1}{2} (\theta - \theta^0)' \left(\frac{\partial^2}{\partial \theta \, \partial \theta'} f(x, \bar{\theta}) \right) (\theta - \theta^0)$$

implicitly assuming that $f(x, \theta)$ is twice continuously differentiable on some open, convex set Θ. Note that $\bar{\theta}$ is a function of both x and θ, $\bar{\theta} = \bar{\theta}(x, \theta)$. Applying this formula row by row to the vector $f(\theta)$, we have the approximation

$$f(\theta) = f(\theta^0) + \left(\frac{\partial}{\partial \theta'} f(\theta^0) \right) (\theta - \theta^0) + R(\theta - \theta^0)$$

where a typical row of R is

$$r_t' = \tfrac{1}{2} (\theta - \theta^0)' \left(\frac{\partial^2}{\partial \theta \, \partial \theta'} f(x_t, \bar{\theta}) \right) \Bigg|_{\bar{\theta} = \bar{\theta}(x_t, \theta)} ;$$

alternatively

$$f(\theta) = f(\theta^0) + F(\theta^0)(\theta - \theta^0) + R(\theta - \theta^0).$$

Using the previous formulas,

$$\frac{\partial}{\partial \theta'} \text{SSE}(\theta) = \frac{\partial}{\partial \theta'}[y - f(\theta)]'[y - f(\theta)]$$

$$= [y - f(\theta)]'\frac{\partial}{\partial \theta'}[y - f(\theta)] + [y - f(\theta)]'\frac{\partial}{\partial \theta'}[y - f(\theta)]$$

$$= 2[y - f(\theta)]'\left(-\frac{\partial}{\partial \theta'}f(\theta)\right)$$

$$= -2[y - f(\theta)]'F(\theta).$$

The least squares estimator is the value $\hat{\theta}$ that minimizes $\text{SSE}(\theta)$ over the parameter space Θ. If $\text{SSE}(\theta)$ is once continuously differentiable on some open set Θ^0 with $\theta \in \Theta^0 \subset \Theta$, then $\hat{\theta}$ satisfies the "normal equations"

$$F'(\hat{\theta})[y - f(\hat{\theta})] = 0.$$

This is because $(\partial/\partial\theta)\text{SSE}(\hat{\theta}) = 0$ at any local optimum. In linear regression,

$$y = X\beta + e$$

least squares residuals $\hat{e}$ computed as

$$\hat{e} = y - X\hat{\beta} \qquad \hat{\beta} = (X'X)^{-1}X'y$$

are orthogonal to the columns of X, viz.,

$$X'\hat{e} = 0.$$

In nonlinear regression, least squares residuals are orthogonal to the columns of the Jacobian of $f(\theta)$ evaluated at $\theta = \hat{\theta}$, viz.,

$$F'(\hat{\theta})[y - f(\hat{\theta})] = 0.$$

PROBLEMS

1. (Product rule.) Show that

$$\frac{\partial}{\partial \theta'} h'(\theta)f(\theta) = h'(\theta)\frac{\partial}{\partial \theta'}f(\theta) + f'(\theta)\frac{\partial}{\partial \theta}h(\theta)$$

by computing $(\partial/\partial\theta_i)\sum_{k=1}^{n}h_k(\theta)f_k(\theta)$ for $i = 1, 2, \ldots, p$ to obtain

$$\frac{\partial}{\partial\theta'}h'(\theta)f(\theta) = \sum_{k=1}^{n}h_k(\theta)\frac{\partial}{\partial\theta'}f_k(\theta) + \sum_{k=1}^{n}f_k(\theta)\frac{\partial}{\partial\theta'}h_k(\theta).$$

Note that $(\partial/\partial\theta')f_k(\theta)$ is the kth row of $(\partial/\partial\theta')f(\theta)$.

2. (Chain rule.) Show that

$$\frac{\partial}{\partial\rho'}f[g(\rho)] = \left(\frac{\partial}{\partial\theta'}f[g(\rho)]\right)\frac{\partial}{\partial\rho'}g(\rho)$$

by computing the (i, j) element of $(\partial/\partial\rho')f[g(\rho)], (\partial/\partial\rho_j)f_i[g(\rho)]$, and then applying the definition of matrix multiplication.

3. STATISTICAL PROPERTIES OF LEAST SQUARES ESTIMATORS

The least squares estimator of the unknown parameter θ^0 in the nonlinear model

$$y = f(\theta^0) + e$$

is the p by 1 vector $\hat{\theta}$ that minimizes

$$\text{SSE}(\theta) = [y - f(\theta)]'[y - f(\theta)] = \|y - f(\theta)\|^2.$$

The estimate of the variance of the errors e_t corresponding to the least squares estimator $\hat{\theta}$ is

$$s^2 = \frac{\text{SSE}(\hat{\theta})}{n - p}.$$

In Chapter 4 we shall show that

$$\hat{\theta} = \theta^0 + (F'F)^{-1}F'e + o_p\left(\frac{1}{\sqrt{n}}\right)$$

$$s^2 = \frac{e'\left[I - F(F'F)^{-1}F'\right]e}{n - p} + o_p\left(\frac{1}{n}\right)$$

where, recall, $F = F(\theta^0) = (\partial/\partial\theta')f(\theta^0)$ is the matrix with typical row $(\partial/\partial\theta')f(x_t, \theta^0)$. The notation $o_p(a_n)$ denotes a (possibly) matrix valued random variable $X_n = o_p(a_n)$ with the property that each element X_{ijn} satisfies

$$\lim_{n \to \infty} P\left[\left|\frac{X_{ijn}}{a_n}\right| > \epsilon\right] = 0$$

for any $\epsilon > 0$; $\{a_n\}$ is some sequence of real numbers, the most frequent choices being $a_n \equiv 1$, $a_n = 1/\sqrt{n}$, and $a_n = 1/n$.

These equations suggest that a good approximation to the joint distribution of $(\hat{\theta}, s^2)$ can be obtained by simply ignoring the terms $o_p(1/\sqrt{n})$ and $o_p(1/n)$. Noting the similarity of the equations

$$\hat{\theta} = \theta^0 + (F'F)^{-1}F'e$$

$$s^2 = \frac{e'\left[I - F(F'F)^{-1}F'\right]e}{n - p}$$

with the equations that arise in linear models theory and assuming normal errors, we have approximately that $\hat{\theta}$ has the p-dimensional multivariate normal distribution with mean θ^0 and variance-covariance matrix $\sigma^2(F'F)^{-1}$,

$$\hat{\theta} \sim N_p\left[\theta^0, \sigma^2(F'F)^{-1}\right];$$

$(n - p)s^2/\sigma^2$ has the chi-square distribution with $n - p$ degrees of freedom,

$$\frac{(n - p)s^2}{\sigma^2} \sim \chi^2(n - p);$$

and s^2 and $\hat{\theta}$ are independent, so that the joint distribution of $(\hat{\theta}, s^2)$ is the product of the marginal distributions. In applications, $(F'F)^{-1}$ must be approximated by the matrix

$$\hat{C} = \left[F'(\hat{\theta})F(\hat{\theta})\right]^{-1}.$$

The alternative to this method of obtaining an approximation to the distribution of $\hat{\theta}$—characterization coupled with a normality assumption—is to use conventional asymptotic arguments. One finds that $\hat{\theta}$ converges almost surely to θ^0, s^2 converges almost surely to σ^2, $(1/n)F'(\hat{\theta})F(\hat{\theta})$ converges almost surely to a matrix Q, and $\sqrt{n}(\hat{\theta} - \theta^0)$ is asymptotically distributed as the p-variate normal with mean zero and variance-covariance matrix $\sigma^2 Q^{-1}$,

$$\sqrt{n}\,(\hat{\theta} - \theta^0) \xrightarrow{\mathscr{L}} N_p(0, \sigma^2 Q^{-1}).$$

The normality assumption is not needed. Let

$$\hat{Q} = \frac{1}{n} F'(\hat{\theta}) F(\hat{\theta}).$$

Following the characterization-normality approach it is natural to write

$$\hat{\theta} \overset{\cdot}{\sim} N_p(\theta^0, s^2\hat{C}) \quad \left(= N_p\left[\theta^0, s^2(1/n)\hat{Q}^{-1}\right]\right)$$

Following the asymptotic normality approach, it is natural to write

$$\sqrt{n}\,(\hat{\theta} - \theta^0) \overset{\cdot}{\sim} N_p(0, s^2\hat{Q}^{-1}) \quad \left(= N_p(0, s^2 n\hat{C})\right)$$

—natural perhaps even to drop the degrees of freedom correction and use

$$\hat{\sigma}^2 = \frac{1}{n} \text{SSE}(\hat{\theta})$$

to estimate σ^2 instead of s^2. The practical difficulty with this is that one can never be sure of the scaling factors in computer output. Natural combinations to report are:

$$\hat{\theta}, s^2, \hat{C};$$
$$\hat{\theta}, s^2, s^2\hat{C};$$
$$\hat{\theta}, \hat{\sigma}^2, \hat{Q}^{-1};$$
$$\hat{\theta}, \hat{\sigma}^2, \hat{\sigma}^2\hat{Q}^{-1};$$

and so on. The documentation usually leaves some doubt in the reader's mind as to what is actually printed. Probably, the best strategy is to run the program using Example 1 and resolve the issue by comparison with the results reported in the next section.

As in linear regression, the practical importance of these distributional properties is their use to set confidence intervals on the unknown parameters θ_i^0 ($i = 1, 2, \ldots, p$) and to test hypotheses. For example, a 95% confidence interval may be found for θ_i^0 from the .025 critical value $t_{.025}$ of the t-distribution with $n - p$ degrees of freedom as

$$\hat{\theta}_i \pm t_{.025}\sqrt{s^2\hat{c}_{ii}}\,.$$

Similarly, the hypothesis $H: \theta_i^0 = \theta_i^*$ may be tested against the alternative $A: \theta_i^0 \neq \theta_i^*$ at the 5% level of significance by comparing

$$|\tilde{t}_i| = \frac{|\hat{\theta}_i - \theta_i^*|}{\sqrt{s^2 \hat{c}_{ii}}}$$

with $|t_{.025}|$ and rejecting H when $|\tilde{t}_i| > |t_{.025}|$; $\hat{c}_{ii}$ denotes the ith diagonal element of the matrix $\hat{C}$. The next few paragraphs are an attempt to convey an intuitive feel for the nature of the regularity conditions used to obtain these results; the reader is reminded once again that they are presented with complete rigor in Chapter 4.

The sequence of input vectors $\{x_t\}$ must behave properly as n tends to infinity. Proper behavior is obtained when the components x_{it} of x_t are chosen either by random sampling from some distribution or (possibly disproportionate) replication of a fixed set of points. In the latter case, some set of points $a_0, a_1, \ldots, a_{T-1}$ is chosen and the inputs assigned according to $x_{it} = a_{t \bmod T}$. Disproportionality is accomplished by allowing some of the a_j to be equal. More general schemes than these are permitted—see Section 2 of Chapter 3 for full details—but this is enough to gain a feel for the sort of stability that $\{x_t\}$ ought to exhibit. Consider, for instance, the data generating scheme of Example 1.

EXAMPLE 1 (Continued). The first two coordinates x_{1t}, x_{2t} of $x_t = (x_{1t}, x_{2t}, x_{3t})'$ consist of replication of a fixed set of design points determined by the design structure:

$$(x_1, x_2)_1 = (1, 1)$$
$$(x_1, x_2)_2 = (0, 1)$$
$$\vdots$$
$$(x_1, x_2)_t = (1, 1) \quad \text{if } t \text{ is odd}$$
$$(x_1, x_2)_t = (0, 1) \quad \text{if } t \text{ is even}$$
$$\vdots$$

That is,

$$(x_1, x_2)_t = a_{t \bmod 2}$$

with

$$a_0 = (0, 1)$$
$$a_1 = (1, 1).$$

The covariate x_{3t} is the age of the experimental material and is conceptually a random sample from the age distribution of the population due to the random allocation of experimental units to treatments. In the simulated data of Table 1, x_{3t} was generated by random selection from the uniform distribution on the interval $[0, 10]$. In a practical application one would probably not know the age distribution of the experimental material but would be prepared to assume that x_3 was distributed according to a continuous distribution function that has a density $p_3(x)$ which is positive everywhere on some known interval $[0, b]$, there being some doubt as to how much probability mass was to the right of b.   □

The response function $f(x, \theta)$ must be continuous in the argument (x, θ); that is, if $\lim_{i \to \infty}(x_i, \theta_i) = (x^*, \theta^*)$ (in Euclidean norm on $\mathbb{R}^{k+p}$) then $\lim_{i \to \infty} f(x_i, \theta_i) = f(x^*, \theta^*)$. The first partial derivatives $(\partial/\partial\theta_i)$ $f(x, \theta)$ must be continuous in (x, θ), and the second partial derivatives $(\partial^2/\partial\theta_i \, \partial\theta_j) f(x, \theta)$ must be continuous in (x, θ). These smoothness requirements are due to the heavy use of Taylor's theorem in Chapter 3. Some relaxation of the second derivative requirement is possible (Gallant, 1973). Quite probably, further relaxation is possible (Huber, 1982).

There remain two further restrictions on the limiting behavior of the response function and its derivatives which roughly correspond to estimability considerations in linear models. The first is that

$$s(\theta) = \lim_{n \to \infty} \frac{1}{n} \sum_{t=1}^{n} \left[f(x_t, \theta) - f(x_t, \theta^0) \right]^2$$

has a unique minimum at $\theta = \theta^0$, and the second is that the matrix

$$Q = \lim_{n \to \infty} \frac{1}{n} F'(\theta^0) F(\theta^0)$$

be non-singular. We term these the *identification condition* and the *rank qualification* respectively. When random sampling is involved, Kolmogorov's strong law of large numbers is used to obtain the limit, as we illustrate with Example 1 below. These two conditions are tedious to verify in applications, and few would bother to do so. However, these conditions indirectly impose restrictions on the inputs x_t and parameter θ^0 that are often easy to spot by inspection. Although θ^0 is unknown in an estimation situation, when testing hypotheses one should check whether the null hypothesis violates these assumptions. If this happens, methods to circumvent the difficulty are given in the next chapter. For Example 1, either $H: \theta_3^0 = 0$ or $H: \theta_4^0 = 0$ will violate the rank qualification and the identification condition, as we next show.

EXAMPLE 1 (Continued). We shall first consider how the problems with $H: \theta_4^0 = 0$ and $H: \theta_3^0 = 0$ can be detected by inspection, next consider how limits are to be computed, and last how one verifies that $s(\theta) = \lim_{n \to \infty}(1/n)\sum_{t=1}^{n}[f(x_t, \theta) - f(x_t, \theta^0)]^2$ has a unique minimum at $\theta = \theta^0$.

Consider the case $H: \theta_3^0 = 0$, leaving the case $H: \theta_4^0 = 0$ to Problem 1. If $\theta_3^0 = 0$ then

$$
F(\theta) = \begin{pmatrix}
1 & 1 & \theta_4 x_{31} & 1 \\
0 & 1 & \theta_4 x_{32} & 1 \\
1 & 1 & \theta_4 x_{33} & 1 \\
0 & 1 & \theta_4 x_{34} & 1 \\
\vdots & \vdots & \vdots & \vdots \\
1 & 1 & \theta_4 x_{3n-1} & 1 \\
0 & 1 & \theta_4 x_{3n} & 1
\end{pmatrix}.
$$

$F(\theta)$ has two columns of ones and is thus singular. Now this fact can be noted at sight in applications; there is no need for any analysis. It is this kind of easily checked violation of the regularity conditions that one should guard against. Let us verify that the singularity carries over to the limit. Let

$$
Q_n(\theta) = \frac{1}{n}F'(\theta)F(\theta) = \frac{1}{n}\sum_{t=1}^{n}\left(\frac{\partial}{\partial\theta}f(x_t, \theta)\right)\left(\frac{\partial}{\partial\theta}f(x_t, \theta)\right)'.
$$

The regularity conditions of Chapter 4 guarantee that $\lim_{n \to \infty}Q_n(\theta)$ exists, and we shall show it directly below. Put $\lambda' = (0, 1, 0, -1)$. Then

$$
\lambda'Q_n(\theta)|_{\theta_3=0}\lambda = \frac{1}{n}\sum_{t=1}^{n}\left(\lambda'\frac{\partial}{\partial\theta}f(x_t, \theta)\Big|_{\theta_3=0}\right)^2 = 0.
$$

Since it is zero for every n, $\lambda'[\lim_{n \to \infty}Q_n(\theta)|_{\theta_3=0}]\lambda = 0$ by continuity of $\lambda'A\lambda$ in A.

Recall that $\{x_{3t}\}$ is independently and identically distributed according to the density $p_3(x_3)$. Since it is an age distribution, there is some (possibly unknown) maximum attained age c that is biologically possible. Then for any continuous function $g(x)$ we must have $\int_0^c |g(x)| p_3(x)\, dx < \infty$, so that by Kolmogorov's strong law of large numbers (Tucker, 1967)

$$
\lim_{n \to \infty} \frac{1}{n}\sum_{t=1}^{n}g(x_{3t}) = \int_0^c g(x)p_3(x)\, dx.
$$

Applying these facts to the treatment group, we have

$$\lim_{n \to \infty} \frac{2}{n} \sum_{t \text{ odd}}^{n} \left[f(x_t, \theta) - f(x_t, \theta^0) \right]^2$$

$$= \int_0^c \left[f(x, \theta) - f(x, \theta^0) \right]^2 p_3(x_3) \, dx_3 \Big|_{(x_1, x_2) = (1, 1)}.$$

Applying them to the control group, we have

$$\lim_{n \to \infty} \frac{2}{n} \sum_{t \text{ even}}^{n} \left[f(x_t, \theta) - f(x_t, \theta^0) \right]^2$$

$$= \int_0^c \left[f(x, \theta) - f(x, \theta^0) \right]^2 p_3(x_3) \, dx_3 \Big|_{(x_1, x_2) = (0, 1)}.$$

Then

$$\lim_{n \to \infty} \frac{1}{n} \sum_{t=1}^{n} \left[f(x_t, \theta^0) - f(x_t, \theta^0) \right]^2$$

$$= \frac{1}{2} \lim_{n \to \infty} \frac{2}{n} \left\{ \sum_{t \text{ odd}}^{n} \left[f(x_t, \theta) - f(x_t, \theta^0) \right]^2 \right.$$

$$\left. + \sum_{t \text{ even}}^{n} \left[f(x_t, \theta) - f(x_t, \theta^0) \right]^2 \right\}$$

$$= \frac{1}{2} \sum_{(x_1, x_2) = (0, 1)}^{(1, 1)} \int_0^c \left[f(x, \theta) - f(x, \theta^0) \right]^2 p_3(x_3) \, dx_3.$$

Suppose we let $F_{12}(x_1, x_2)$ be the distribution function corresponding to the discrete density

$$p_{12}(x_1, x_2) = \begin{cases} \frac{1}{2} & (x_1, x_2) = (0, 1) \\ \frac{1}{2} & (x_1, x_2) = (1, 1) \end{cases}$$

and we let $F_3(x_3)$ be the distribution function corresponding to $p_3(x)$. Let $\mu(x) = F_{12}(x_1, x_2) F_3(x_3)$. Then

$$\int \left[f(x, \theta) - f(x, \theta^0) \right]^2 d\mu(x)$$

$$= \frac{1}{2} \sum_{(x_1, x_2) = (0, 1)}^{(1, 1)} \int_0^c \left[f(x, \theta) - f(x, \theta^0) \right]^2 p_3(x) \, dx$$

where the integral on the left is a Lebesgue-Stieltjes integral (Royden, 1968, Chapter 12; Tucker, 1967, Section 2.2). In this notation the limit can be given an integral representation

$$\lim_{n \to \infty} \frac{1}{n} \sum_{t=1}^{n} \left[f(x_t, \theta) - f(x_t, \theta^0) \right]^2 = \int \left[f(x, \theta) - f(x, \theta^0) \right]^2 d\mu(x).$$

These are the ideas behind Section 2 of Chapter 3. The advantage of the integral representation is that familiar results from integration theory can be used to deduce properties of limits. As an example: What is required of $f(x, \theta)$ such that

$$\frac{\partial}{\partial \theta} \lim_{n \to \infty} \sum_{t=1}^{n} f(x_t, \theta) = \lim_{n \to \infty} \sum_{t=1}^{n} \frac{\partial}{\partial \theta} f(x_t, \theta)?$$

We find later that the existence of $b(x)$ with $|(\partial/\partial\theta)f(x,\theta)| \le b(x)$ and $\int b(x)d\mu(x) < \infty$ is enough, given continuity of $(\partial/\partial\theta)f(x,\theta)$.

Our last task is to verify that

$$s(\theta) = \int \left[f(x, \theta) - f(x, \theta^0) \right]^2 d\mu(x)$$

$$= \frac{1}{2} \sum_{(x_1, x_2)=(0,1)}^{(1,1)} \int_0^c \left[f(x, \theta) - f(x, \theta^0) \right]^2 p_3(x_3) \, dx_3$$

$$= \frac{1}{2} \int_0^c \left[(\theta_2 - \theta_2^0) + \theta_4 e^{\theta_3 x} - \theta_4^0 e^{\theta_3^0 x} \right]^2 p_3(x) \, dx$$

$$+ \frac{1}{2} \int_0^c \left[(\theta_1 - \theta_1^0) + (\theta_2 - \theta_2^0) + \theta_4 e^{\theta_3 x} - \theta_4^0 e^{\theta_3^0 x} \right]^2 p_3(x) \, dx$$

has a unique minimum. Since $s(\theta) \ge 0$ in general and $s(\theta^0) = 0$, the question is: Does $s(\theta) = 0$ imply that $\theta = \theta^0$? One first notes that $\theta_3^0 = 0$ or $\theta_4^0 = 0$ must be ruled out, as in the former case any θ with $\theta_3 = 0$ and $\theta_2 + \theta_4 = \theta_2^0 + \theta_4^0$ will have $s(\theta) = 0$, and in the latter case any θ with $\theta_1 = \theta_1^0$, $\theta_2 = \theta_2^0$, $\theta_4 = 0$ will have $s(\theta) = 0$. Then assume that $\theta_3^0 \ne 0$ and $\theta_4^0 \ne 0$, and recall that $p_3(x) > 0$ on $[0, c]$. Now $s(\theta) = 0$ implies

$$\theta_2 - \theta_2^0 + \theta_4 e^{\theta_3 x} - \theta_4^0 e^{\theta_3^0 x} = 0 \qquad 0 \le x \le c.$$

Differentiating, we have

$$\theta_3 \theta_4 e^{\theta_3 x} - \theta_3^0 \theta_4^0 e^{\theta_3^0 x} = 0 \qquad 0 \le x \le c.$$

Putting $x = 0$, we have $\theta_3 \theta_4 = \theta_3^0 \theta_4^0$, whence

$$e^{(\theta_3 - \theta_3^0)x} = 1 \qquad 0 \le x \le c$$

which implies $\theta_3 = \theta_3^0$. We now have that

$$s(\theta) = 0, \quad \theta_3^0 \ne 0, \quad \theta_4^0 \ne 0 \quad \Rightarrow \quad \theta_3 = \theta_3^0, \quad \theta_4 = \theta_4^0.$$

But if $\theta_3 = \theta_3^0$, $\theta_4 = \theta_4^0$, and $s(\theta) = 0$, then

$$s(\theta) = \tfrac{1}{2}\left(\theta_2 - \theta_2^0\right)^2 + \tfrac{1}{2}\left[\left(\theta_1 - \theta_1^0\right) + \left(\theta_2 - \theta_2^0\right)\right]^2 = 0$$

which implies $\theta_1 = \theta_1^0$ and $\theta_2 = \theta_2^0$. In summary

$$s(\theta) = 0, \quad \theta_3^0 \ne 0, \quad \theta_4^0 \ne 0 \quad \Rightarrow \quad \theta = \theta^0. \qquad \square$$

As seen from Example 1, checking the identification condition and rank qualification is a tedious chore to be put to whenever one uses nonlinear methods. Uniqueness depends on the interaction of $f(x, \theta)$ and $\mu(x)$, and verification is ad hoc. Similarly for the rank qualification (Problem 2). As a practical matter, one should be on guard against obvious problems and can usually trust that numerical difficulties in computing $\hat{\theta}$ will serve as a sufficient warning against subtle problems, as seen in the next section.

An appropriate question is how accurate are probability statements based on the asymptotic properties of nonlinear least squares estimators in applications. Specifically one might ask: How accurate are probability statements obtained by using the critical points of the t-distribution with $n - p$ degrees of freedom to approximate the sampling distribution of

$$\tilde{t}_i = \frac{\theta_i - \theta_i^0}{\sqrt{s^2 \hat{c}_{ii}}}?$$

Monte Carlo evidence on this point is presented below using Example 1. We shall accumulate such information as we progress.

EXAMPLE 1 (Continued). Table 3 shows the empirical distribution of $\tilde{t}_i$ computed from 5000 Monte Carlo trials evaluated at the critical points of the t-distribution. The responses were generated using the inputs of Table 1

Table 3. Empirical Distribution of $\tilde{t}_i$ Compared with the t-Distribution.

Tabular Values		Empirical Distribution				Std.
c	$P(t \le c)$	$P(\tilde{t}_1 \le c)$	$P(\tilde{t}_2 \le c)$	$P(\tilde{t}_3 \le c)$	$P(\tilde{t}_4 \le c)$	Error
− 3.707	.0005	.0010	.0010	.0000	.0002	.0003
− 2.779	.0050	.0048	.0052	.0018	.0050	.0010
− 2.056	.0250	.0270	.0280	.0140	.0270	.0022
− 1.706	.0500	.0522	.0540	.0358	.0494	.0031
− 1.315	.1000	.1026	.1030	.0866	.0998	.0042
− 1.058	.1500	.1552	.1420	.1408	.1584	.0050
− 0.856	.2000	.2096	.1900	.1896	.2092	.0057
− 0.684	.2500	.2586	.2372	.2470	.2638	.0061
0.0	.5000	.5152	.4800	.4974	.5196	.0071
0.684	.7500	.7558	.7270	.7430	.7670	.0061
0.856	.8000	.8072	.7818	.7872	.8068	.0057
1.058	.8500	.8548	.8362	.8346	.8536	.0050
1.315	.9000	.9038	.8914	.8776	.9004	.0042
1.706	.9500	.9552	.9498	.9314	.9486	.0031
2.056	.9750	.9772	.9780	.9584	.9728	.0022
2.779	.9950	.9950	.9940	.9852	.9936	.0010
3.707	.9995	.9998	.9996	.9962	.9994	.0003

Source: Gallant (1975d).

with the parameters of the model set at

$$\theta^0 = (0, 1, -1, -.5)'$$
$$\sigma^2 = .001.$$

The standard errors shown in the table are the standard errors of an estimate of the probability $P(\tilde{t} < c)$ computed from 5000 Monte Carlo trials assuming that $\tilde{t}$ follows the t-distribution. If that assumption is correct, the Monte Carlo estimate of $P[\tilde{t} < c]$ follows the binomial distribution and has variance $P(t < c)P(t > c)/5000$.

Table 3 indicates that the critical points of the t-distribution describe the sampling behavior of $\tilde{t}_i$ reasonably well. For example, the Monte Carlo estimate of the Type I error for a two tailed test of $H: \theta_3^0 = -1$ using the tabular values ± 2.056 is .0556 with a standard error of .0031. Thus it seems that the actual level of the test is close enough to its nominal level of .05 for any practical purpose. However, in the next chapter we shall encounter instances where this is definitely not the case. □

PROBLEMS

1. Show that $H: \theta_4^0 = 0$ will violate the rank qualification in Example 1.

2. Show that $Q = \lim_{n \to \infty} (1/n) F'(\theta) F(\theta)$ has full rank in Example 1 if $\theta_3^0 \neq 0$ and $\theta_4^0 \neq 0$.

4. METHODS OF COMPUTING LEAST SQUARES ESTIMATES

The more widely used methods of computing nonlinear least squares estimators are Hartley's (1961) modified Gauss-Newton method and the Levenberg-Marquardt algorithm (Levenberg, 1944; Marquardt, 1963).

The Gauss-Newton method is based on the substitution of a first order Taylor series approximation to $f(\theta)$ about a trial parameter value θ_T in the formula for the residual sum of squares SSE(θ). The approximating sum of squares surface thus obtained is

$$\mathrm{SSE}_T(\theta) = \left\| y - f(\theta_T) - F(\theta_T)(\theta - \theta_T) \right\|^2.$$

The value of the parameter minimizing the approximating sum of squares surface is (Problem 1)

$$\theta_M = \theta_T + \left[F'(\theta_T) F(\theta_T) \right]^{-1} F'(\theta_T) \left[y - f(\theta_T) \right].$$

It would seem that θ_M should be a better approximation to the least squares estimator $\hat{\theta}$ than θ_T in the sense that SSE(θ_M) < SSE(θ_T). These ideas are displayed graphically in Figure 1 in the case that θ is univariate ($p = 1$).

As suggested by Figure 1, SSE$_T(\theta)$ is tangent to the curve SSE(θ) at the point θ_T. The approximation is first order in the sense that one can show that (Problem 2)

$$\lim_{\|\theta - \theta_T\| \to 0} \frac{\left| \mathrm{SSE}(\theta) - \mathrm{SSE}_T(\theta) \right|}{\|\theta - \theta_T\|} = 0$$

but not second order, since the best one can show in general is that (Problem 2)

$$\lim_{\delta \to 0} \sup_{\|\theta - \theta_T\| < \delta} \frac{\left| \mathrm{SSE}(\theta) - \mathrm{SSE}_T(\theta) \right|}{\|\theta - \theta_T\|^2} < \infty.$$

It is not necessarily true that θ_M is closer to $\hat{\theta}$ than θ_T in the sense that SSE(θ_M) $\leq$ SSE(θ_T). This situation is depicted in Figure 2.

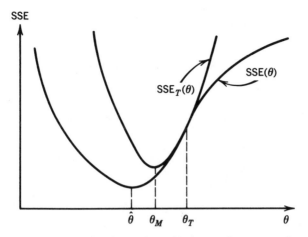

Figure 1. The linearized approximation to the residual sum of squares surface, an adequate approximation.

But as suggested by Figure 2, points on the line segment joining θ_T to θ_M that are sufficiently close to θ_T ought to lead to improvement. This is the case, and one can show (Problem 3) that there is a λ^* such that all points with

$$\theta = \theta_T + \lambda(\theta_M - \theta_T) \qquad 0 < \lambda < \lambda^*$$

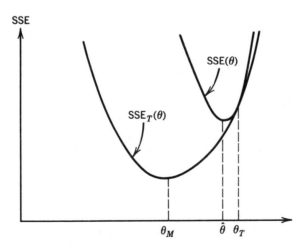

Figure 2. The linearized approximation to the residual sum of squares surface, a poor approximation.

satisfy

$$SSE(\theta) < SSE(\theta_T).$$

These are the ideas that motivate the modified Gauss-Newton algorithm which is as follows:

0. Choose a starting estimate θ_0. Compute

$$D_0 = [F'(\theta_0)F(\theta_0)]^{-1}F'(\theta_0)[y - f(\theta_0)].$$

Find a λ_0 between 0 and 1 such that

$$SSE(\theta_0 + \lambda_0 D_0) < SSE(\theta_0).$$

1. Let $\theta_1 = \theta_0 + \lambda_0 D_0$. Compute

$$D_1 = [F'(\theta_1)F(\theta_1)]^{-1}F'(\theta_1)[y - f(\theta_1)].$$

Find a λ_1 between 0 and 1 such that

$$SSE(\theta_1 + \lambda_1 D_1) < SSE(\theta_1).$$

2. Let $\theta_2 = \theta_1 + \lambda_1 D_1, \ldots$.

There are several methods for choosing the step length λ_i at each iteration, of which the simplest is to accept the first λ in the sequence

$$1, .9, .8, .7, .6, \tfrac{1}{2}, \tfrac{1}{4}, \tfrac{1}{8}, \ldots$$

for which

$$SSE(\theta_i + \lambda D_i) < SSE(\theta_i)$$

as the step length λ_i. This simple approach is nearly always adequate in applications. Hartley (1961) suggests two alternative methods in his article. Gill, Murray, and Wright (1981, Section 4.3.2.1) discuss the problem in general from a practical point of view and follow the discussion with an annotated bibliography of recent literature. Whatever rule is used, it is essential that the computer program verify that $SSE(\theta_i + \lambda_i D_i)$ is smaller than $SSE(\theta_i)$ before taking the next iterative step. This caveat is necessary when, for example, Hartley's quadratic interpolation formula is used to find λ_i.

The iterations are continued until terminated by a stopping rule such as

$$\|\theta_i - \theta_{i+1}\| < \epsilon(\|\theta_i\| + \tau)$$

and

$$|SSE(\theta_i) - SSE(\theta_{i+1})| < \epsilon[SSE(\theta_i) + \tau]$$

where $\epsilon > 0$ and $\tau > 0$ are preset tolerances. Common choices are $\epsilon = 10^{-5}$ and $\tau = 10^{-3}$. A more conservative (and costly) approach is to allow the iterations to continue until the requisite step size λ_i is so small that the fixed word length of the machine prevents differentiation between the values of $SSE(\theta_i + \lambda_i D_i)$ and $SSE(\theta_i)$. This happens sooner than one might expect and, unfortunately, sometimes before the correct answer is obtained. Gill, Murray, and Wright (1981, Section 8.2.3) discuss termination criteria in general and follow the discussion with an annotated bibliography of recent literature.

Much more difficult than deciding when to stop the iterations is determining where to start them. The choice of starting values is pretty much an *ad hoc* process. They may be obtained from prior knowledge of the situation, inspection of the data, grid search, or trial and error. A general method of finding starting values is given by Hartley and Booker (1965). Their idea is to cluster the independent variables $\{x_t\}$ into p groups

$$x_{ij} \qquad j = 1, 2, \ldots, n_i \quad i = 1, 2, \ldots, p$$

and fit the model

$$\bar{y}_i = \bar{f}_i(\theta) + \bar{e}_i$$

where

$$\bar{y}_i = \frac{1}{n_i} \sum_{j=1}^{n_i} y_{ij}$$

$$\bar{f}_i(\theta) = \frac{1}{n_i} \sum_{j=1}^{n_i} f(x_{ij}, \theta)$$

for $i = 1, 2, \ldots, p$. The hope is that one can find a value θ_0 that solves the equations

$$\bar{y}_i = \bar{f}_i(\theta) \qquad i = 1, 2, \ldots, p$$

exactly. The only reason for this hope is that one has a system of p equations in p unknowns; but as the system is not a linear system, there is no guarantee. If an exact solution cannot be found, it is hard to see why one is better off with this new problem than with the original least squares problem

$$\text{minimize} \quad \text{SSE}(\theta) = \frac{1}{n} \sum_{t=1}^{n} [y_t - f(x_t, \theta)]^2.$$

A simpler variant of their idea, and one that is much easier to use with a statistical package, is to select p representative inputs x_{t_i} with corresponding responses y_{t_i} and then solve the system of nonlinear equations

$$y_{t_i} = f(x_{t_i}, \theta) \qquad i = 1, 2, \ldots, p$$

for θ. The solution is used as the starting value. Even if iterative methods must be employed to obtain the solution, it is still a viable technique, since the correct answer can be recognized when found. This is not the case in an attempt to minimize $\text{SSE}(\theta)$ directly. As with Hartley and Booker, the method fails when there is no solution to the system of nonlinear equations. There is also a risk that this technique can place the starting value near a slight depression in the surface $\text{SSE}(\theta)$ and cause convergence to a local minimum that is not the global minimum. It is sound practice to try a few perturbations of θ_0 as starting values and see if convergence to the same point occurs each time. We illustrate these techniques with Example 1.

EXAMPLE 1 (Continued). We begin by plotting the data as shown in Figure 3. A "1" indicates the observation is in the treatment group, and a "0" that it is in the control group. Looking at the plot, the treatment effect appears to be negligible; a starting value of zero for θ_1 seems reasonable. The overall impression is that the curve is concave and increasing. That is, it appears that

$$\frac{\partial}{\partial x_3} f(x, \theta) > 0$$

and

$$\frac{\partial^2}{\partial x_3^2} f(x, \theta) < 0.$$

Since

$$\frac{\partial}{\partial x_3} f(x, \theta) = \theta_3 \theta_4 e^{\theta_3 x_3} > 0$$

SAS Statements:

```
DATA WORK01;  SET EXAMPLE1;
PX1='0';  IF X1=1 THEN PX1='1';
PROC PLOT DATA=WORK01;
PLOT Y*X3=PX1 / HAXIS = 0 TO 10 BY 2  VPOS = 34;
```

Output:

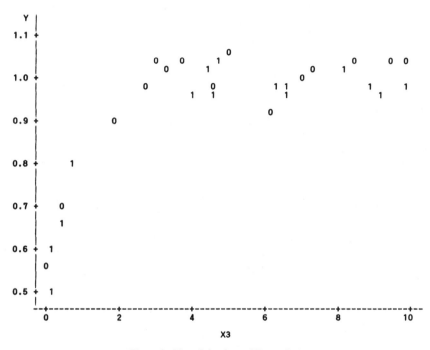

Figure 3. Plot of the data of Example 1.

and

$$\frac{\partial^2}{\partial x_3^2} f(x, \theta) = \theta_3^2 \theta_4 e^{\theta_3 x_3} < 0$$

we see that both θ_3 and θ_4 must be negative. Experience with exponential models suggests that what is important is to get the algebraic signs of the starting values of θ_3 and θ_4 correct and that, within reason, getting the correct magnitudes is not that important. Accordingly, take -1 as the starting value of both θ_3 and θ_4. Again, experience indicates that the starting values for parameters that enter the model linearly such as θ_1 and θ_2 are almost irrelevant, within reason, so take zero as the starting value of θ_2. In summary, inspection of a plot of the data suggests that

$$\theta = (0, 0, -1, -1)'$$

is a reasonable starting value.

Let us use the idea of solving the equations

$$y_{t_i} = f(x_{t_i}, \theta) \qquad i = 1, 2, \ldots, p$$

for some representative set of inputs

$$x_{t_i} \qquad i = 1, 2, \ldots, p$$

to refine these visual impressions and get better starting values. We can solve the equations by minimizing

$$\sum_{i=1}^{p} \left[y_{t_i} - f(x_{t_i}, \theta) \right]^2$$

using the modified Gauss-Newton method. If the equations have a solution, then the starting value we seek will produce a residual sum of squares of zero. The equation for observations in the control group ($x_1 = 0$) is

$$f(x, \theta) = \theta_2 + \theta_4 e^{\theta_3 x_3}.$$

If we take two extreme values of x_3 and one where the curve is bending, we should get a good fix on values for $\theta_2, \theta_3, \theta_4$. Inspecting Table 1, let us select

$$x_{14} = (0, 1, 0.07)'$$
$$x_6 = (0, 1, 1.82)'$$
$$x_2 = (0, 1, 9.86)'.$$

The equation for an observation in the treatment group ($x_1 = 1$) is

$$f(x, \theta) = \theta_1 + \theta_2 + \theta_4 e^{\theta_3 x_3}.$$

If we can find an observation in the treatment group with an x_3 near one of the x_3's that we have already chosen, then we should get a good fix on θ_1

SAS Statements:

```
DATA WORK01;   SET EXAMPLE1;
IF T=2 OR T=6 OR T=11 OR T=14 THEN OUTPUT; DELETE;
PROC NLIN DATA=WORK01 METHOD=GAUSS ITER=50 CONVERGENCE=1.0E-5;
PARMS T1=0 T2=0 T3=-1 T4=-1;
MODEL Y=T1*X1+T2*X2+T4*EXP(T3*X3);
DER.T1=X1; DER.T2=X2; DER.T3=T4*X3*EXP(T3*X3); DER.T4=EXP(T3*X3);
```

Output:

STATISTICAL   ANALYSIS   SYSTEM   1

NON-LINEAR LEAST SQUARES ITERATIVE PHASE

DEPENDENT VARIABLE: Y   METHOD: GAUSS-NEWTON

ITERATION	T1 T4	T2	T3	RESIDUAL SS
0	0.000000E+00 -1.00000000	0.000000E+00	-1.00000000	5.39707160
1	-0.04866000 -0.51074741	1.03859589	-0.82674151	0.00044694
2	-0.04866000 -0.51328803	1.03876874	-0.72975636	0.00000396
3	-0.04866000 -0.51361959	1.03883445	-0.73786415	0.00000000
4	-0.04866000 -0.51362269	1.03883544	-0.73791851	0.00000000
5	-0.04866000 -0.51362269	1.03883544	-0.73791852	0.00000000

NOTE: CONVERGENCE CRITERION MET.

Figure 4. Computation of starting values for Example 1.

that is independent of whatever blunders we make in guessing θ_2, θ_3, and θ_4. The eleventh observation is ideal:

$$x_{11} = (1, 1, 9.86)'.$$

Figure 4 displays SAS code for selecting the subsample x_2, x_6, x_{11}, x_{14} from the original data set and solving the equations

$$y_t = f(x_t, \theta) \qquad t = 2, 6, 11, 14$$

by minimizing

$$\sum_{t=2, 6, 11, 14} [y_t - f(x_t, \theta)]^2$$

using the modified Gauss-Newton method from a starting value of

$$\theta = (0,0,-1,-1)'.$$

The solution is

$$\hat{\theta} = \begin{pmatrix} -0.04866 \\ 1.03884 \\ -0.73792 \\ -0.51362 \end{pmatrix}.$$

SAS code using this as the starting value for computing the least squares estimator with the modified Gauss-Newton method is shown in Figure 5*a* together with the resulting output. The least squares estimator is

$$\hat{\theta} = \begin{pmatrix} -0.02588970 \\ 1.01567967 \\ -1.115769714 \\ -0.50490286 \end{pmatrix}.$$

The residual sum of squares is

$$\text{SSE}(\hat{\theta}) = 0.03049554$$

and the variance estimate is

$$s^2 = \frac{\text{SSE}(\hat{\theta})}{n-p} = 0.00117291.$$

As seen from Figure 5*a*, SAS prints estimated standard errors $\hat{\sigma}_i$ and correlations $\hat{\rho}_{ij}$. To recover the matrix $s^2\hat{C}$ one uses the formula

$$s^2\hat{c}_{ij} = \hat{\sigma}_i\hat{\sigma}_j\hat{\rho}_{ij}.$$

For example,

$$s^2c_{12} = (0.01262384)(0.00993793)(-0.627443)$$
$$= -0.000078716.$$

The matrices $s^2\hat{C}$ and $\hat{C}$ are shown in Figure 5*b*.   □

The obvious approach to finding starting values is grid search. When looking for starting values by a grid search, it is only necessary to search with respect to those parameters which enter the model nonlinearly. The

SAS Statements:

```
PROC NLIN DATA=EXAMPLE1 METHOD=GAUSS ITER=50 CONVERGENCE=1.0E-13;
PARMS T1=-0.04866 T2=1.03884 T3=-0.73792 T4=-0.51362;
MODEL Y=T1*X1+T2*X2+T4*EXP(T3*X3);
DER.T1=X1; DER.T2=X2; DER.T3=T4*X3*EXP(T3*X3); DER.T4=EXP(T3*X3);
```

Output:

S T A T I S T I C A L   A N A L Y S I S   S Y S T E M   1

NON-LINEAR LEAST SQUARES ITERATIVE PHASE

DEPENDENT VARIABLE: Y   METHOD: GAUSS-NEWTON

ITERATION	T1 T4	T2	T3	RESIDUAL SS
0	-0.04866000 -0.51362000	1.03884000	-0.73792000	0.05077531
1	-0.02432899 -0.49140162	1.00985922	-1.01571093	0.03235152
2	-0.02573470 -0.50457486	1.01531500	-1.11610448	0.03049761
3	-0.02588979 -0.50490158	1.01567999	-1.11568229	0.03049554
4	-0.02588969 -0.50490291	1.01567966	-1.11569767	0.03049554
5	-0.02588970 -0.50490286	1.01567967	-1.11569712	0.03049554
6	-0.02588970 -0.50490286	1.01567967	-1.11569714	0.03049554

NOTE: CONVERGENCE CRITERION MET.

S T A T I S T I C A L   A N A L Y S I S   S Y S T E M   2

NON-LINEAR LEAST SQUARES SUMMARY STATISTICS   DEPENDENT VARIABLE Y

SOURCE	DF	SUM OF SQUARES	MEAN SQUARE
REGRESSION	4	26.34594211	6.58648553
RESIDUAL	26	0.03049554	0.00117291
UNCORRECTED TOTAL	30	26.37643764	
(CORRECTED TOTAL)	29	0.71895291	

PARAMETER	ESTIMATE	ASYMPTOTIC STD. ERROR	ASYMPTOTIC 95 % CONFIDENCE INTERVAL LOWER	UPPER
T1	-0.02588970	0.01262384	-0.05183816	0.00005877
T2	1.01567967	0.00993793	0.99525213	1.03610721
T3	-1.11569714	0.16354199	-1.45185986	-0.77953442
T4	-0.50490286	0.02565721	-0.55764159	-0.45216413

ASYMPTOTIC CORRELATION MATRIX OF THE PARAMETERS

	T1	T2	T3	T4
T1	1.000000	-0.627443	-0.085786	-0.136140
T2	-0.627443	1.000000	0.373492	-0.007261
T3	-0.085786	0.373492	1.000000	0.561533
T4	-0.136140	-0.007261	0.561533	1.000000

Figure 5a. Example 1 fitted by the modified Gauss-Newton method.

35

$$s^2C$$

	COL 1	COL 2	COL 3	COL 4
ROW 1	0.00015936	-7.8716D-05	-0.00017711	-4.4095D-05
ROW 2	-7.8716D-05	9.8762D-05	0.00060702	-1.8514D-06
ROW 3	-0.00017711	0.00060702	0.026746	0.00235621
ROW 4	-4.4095D-05	-1.8514D-06	0.00235621	0.00065829

$$C$$

	COL 1	COL 2	COL 3	COL 4
ROW 1	0.13587	-0.067112	-0.15100	-0.037594
ROW 2	-0.067112	0.084203	0.51754	-0.00157848
ROW 3	-0.15100	0.51754	22.8032	2.00887
ROW 4	-0.037594	-0.00157848	2.00887	0.56125

Figure 5b. The matrices s^2C and C for Example 1.

parameters which enter the model linearly can be estimated by ordinary multiple regression methods once the nonlinear parameters are specified. For example, once θ_3 is specified, the model

$$y_t = \theta_1 x_{1t} + \theta_2 x_{2t} + \theta_4 e^{\theta_3 x_{2t}} + e_t$$

is linear in the remaining parameters $\theta_1, \theta_2, \theta_4$, and these can be estimated by linear least squares. The surface to be inspected for a minimum with respect to grid values of the parameters entering nonlinearly is the residual sum of squares after fitting for the parameters entering linearly. The trial value of the nonlinear parameters producing the minimum over the grid together with the corresponding least squares estimates of the parameters entering the model linearly is the starting value. Some examples of plots of this sort are found toward the end of this section.

The surface to be examined for a minimum is usually locally convex. This fact can be exploited in the search to eliminate the necessity of evaluating the residual sum of squares at every point in the grid. Often, a direct search with respect to the parameters entering the model nonlinearly which exploits convexity is competitive in cost and convenience with either Hartley's or Marquardt's methods. The only reason to use the latter methods in such situations would be to obtain the matrix $[F'(\hat{\theta})F(\hat{\theta})]^{-1}$, which is printed by most implementations of either algorithm.

Of course, these same ideas can be exploited in designing an algorithm. Suppose that the model is of the form

$$f(\rho, \beta) = A(\rho)\beta$$

where ρ denotes the parameters entering nonlinearly, $A(\rho)$ is an n by K matrix, and β is a K-vector denoting the parameters entering linearly. Given ρ, the minimizing value of β is

$$\hat{\beta} = [A'(\rho)A(\rho)]^{-1}A'(\rho)y.$$

The residual sum of squares surface after fitting the parameters entering linearly is

$$\mathrm{SSE}(\rho) = \left\{ y - A(\rho)[A'(\rho)A(\rho)]^{-1}A'(\rho)y \right\}'$$
$$\times \left\{ y - A(\rho)[A'(\rho)A(\rho)]^{-1}A'(\rho)y \right\}.$$

To solve this minimization problem one can simply view

$$f(\rho) = A(\rho)[A'(\rho)A(\rho)]^{-1}A'(\rho)y$$

as a nonlinear model to be fitted to y and use, say, the modified Gauss-Newton method. Of course, computing

$$\frac{\partial}{\partial\rho'} \left\{ A(\rho)[A'(\rho)A(\rho)]^{-1}A'(\rho)y \right\}$$

is not a trivial task, but it is possible. Golub and Pereyra (1973) obtain an analytic expression for $(\partial/\partial\rho')f(\rho)$ and present an algorithm exploiting it that is probably the best of its genre.

Marquardt's algorithm is similar to the Gauss-Newton method in the use of the sum of squares $\mathrm{SSE}_T(\theta)$ to approximate $\mathrm{SSE}(\theta)$. The difference between the two methods is that Marquardt's algorithm uses a ridge regression improvement of the approximating surface

$$\theta_\delta = \theta_T + [F'(\theta_T)F(\theta_T) + \delta I]^{-1}F'(\theta_T)[y - f(\theta_T)]$$

instead of the minimizing value θ_M. For all δ sufficiently large θ_δ is an improvement over θ_T [$\mathrm{SSE}(\theta_\delta)$ is smaller than $\mathrm{SSE}(\theta_T)$] under appropriate conditions (Marquardt, 1963). This fact forms the basis for Marquardt's algorithm.

The algorithm actually recommended by Marquardt differs from that suggested by this theoretical result in that a diagonal matrix S with the same diagonal elements as $F'(\theta_T)F(\theta_T)$ is substituted for the identity matrix in the expression for θ_δ. Marquardt gives the justification for this deviation in his article and also a set of rules for choosing δ at each iterative step. See Osborne (1972) for additional comments on these points.

Newton's method (Gill, Murray, and Wright, 1981, Section 4.4) is based on a second order Taylor series approximation to SSE(θ) at the point θ_T:

$$\mathrm{SSE}(\theta) \doteq \mathrm{SSE}(\theta_T) + \left(\frac{\partial}{\partial \theta'}\mathrm{SSE}(\theta_T)\right)(\theta - \theta_T)$$
$$+ \tfrac{1}{2}(\theta - \theta_T)'\left(\frac{\partial^2}{\partial \theta \, \partial \theta'}\mathrm{SSE}(\theta_T)\right)(\theta - \theta_T).$$

The value of θ that minimizes this expression is

$$\theta_M = \theta_T + \left(-\frac{\partial^2}{\partial \theta \, \partial \theta'}\mathrm{SSE}(\theta_T)\right)^{-1}\frac{\partial}{\partial \theta}\mathrm{SSE}(\theta_T).$$

As with the modified Gauss-Newton method, one finds λ_T with

$$\mathrm{SSE}\left[\theta_T + \lambda_T(\theta_M - \theta_T)\right] < \mathrm{SSE}(\theta_T)$$

and takes $\theta = \theta_T + \lambda_T(\theta_M - \theta_T)$ as the next point in the iterative sequence. Now

$$-\frac{\partial^2}{\partial \theta \, \partial \theta'}\mathrm{SSE}(\theta_T) = 2F'(\theta_T)F(\theta_T) - 2\sum_{t=1}^{n} \tilde{e}_t \frac{\partial^2}{\partial \theta \, \partial \theta'}f(x_t, \theta_T)$$

where

$$\tilde{e}_t = y_t - f(x_t, \theta_T) \qquad t = 1, 2, \ldots, n.$$

From this expression one can see that the modified Gauss-Newton method can be viewed as an approximation to the Newton method if the term

$$\sum_{t=1}^{n} \tilde{e}_t \frac{\partial^2}{\partial \theta \, \partial \theta'}f(x_t, \theta_T)$$

is negligible relative to the term $F'(\theta_T)F(\theta_T)$ for θ_T near $\hat{\theta}$—say, as a rule of thumb, when

$$\left[\sum_{t=1}^{n} \sum_{i=1}^{p} \sum_{j=1}^{p} \left(\hat{e}_t \frac{\partial^2}{\partial \theta_i \, \partial \theta_j} f(x_t, \hat{\theta}) \right)^2 \right]^{1/2}$$

is less then the smallest eigenvalue of $F'(\hat{\theta}) F(\hat{\theta})$, where $\hat{e}_t = y_t - f(x_t, \hat{\theta})$. If this is not the case, then one has what is known as the "large residual problem." In this instance it is considered sound practice to use the Newton method, or some other second order method, to compute the least squares estimator, rather than the modified Gauss-Newton method. In most instances analytic computation of $(\partial^2 / \partial \theta \, \partial \theta') f(x, \theta)$ is quite tedious and there is a considerable incentive to try and find some method to approximate

$$\sum_{t=1}^{n} \tilde{e}_t \frac{\partial^2}{\partial \theta \, \partial \theta'} f(x_t, \theta_T)$$

without being put to this bother. The best method for doing this is probably the algorithm by Dennis, Gay, and Welsch (1977).

Success, in terms of convergence to $\hat{\theta}$ from a given starting value, is not guaranteed with any of these methods. Experience indicates that failure of the iterations to converge to the correct answer depends both on the distance of the starting value from the correct answer and on the extent of overparametrization in the response function relative to the data. These problems are interrelated in that more appropriate response functions lead to greater radii of convergence. When convergence fails, one should try to find better starting values or use a similar response function with fewer parameters. A good check on the accuracy of the numerical solution is to try several reasonable starting values and see if the iterations converge to the same answer for each starting value. It is also a good idea to plot actual responses y_t against predicted responses $\hat{y}_t = f(x_t, \hat{\theta})$; if a 45° line does not obtain then the answer is probably wrong. The following example illustrates these points.

EXAMPLE 1   (Continued).   Conditional on $\rho = \theta_3$, the model

$$f(x, \theta) = \theta_1 x_1 + \theta_2 x_2 + \theta_4 e^{\theta_3 x_3}$$

has three parameters $\beta = (\theta_1, \theta_2, \theta_4)'$ that enter the model linearly. Then as

Table 4.   Performance of the Modified Gauss-Newton Method.

True value of θ_4 [a]	Least squares estimate					Modified Gauss-Newton iterations from a start of $\hat{\theta}_i - .1$
	$\hat{\theta}_1$	$\hat{\theta}_2$	$\hat{\theta}_3$	$\hat{\theta}_4$	s^2	
− .5	−.0259	1.02	−1.12	−.505	.00117	4
− .3	−.0260	1.02	−1.20	−.305	.00117	5
− .1	−.0265	1.02	−1.71	−.108	.00118	6
− .05	−.0272	1.02	−3.16	−.0641	.00117	7
− .01	−.0272	1.01	−0.0452	.00758	.00120	[b]
− .005	−.0268	1.01	−0.0971	.0106	.00119	[b]
− .001	−.0266	1.01	−0.134	.0132	.00119	202
0	−.0266	1.01	−0.142	.0139	.00119	69

Source: Gallant (1977a).
[a] Parameters other than θ_4 fixed at $\theta_1 = 0$, $\theta_2 = 1$, $\theta_3 = -1$, $\sigma^2 = .001$.
[b] Algorithm failed to converge after 500 iterations.

remarked earlier, we may write

$$f(\rho) = A(\rho)[A'(\rho)A(\rho)]^{-1}A'(\rho)y$$

where a typical row of $A(\rho)$ is

$$a'_t(\rho) = \left(x_{1t}, x_{2t}, e^{\rho x_{3t}}\right)$$

and treat this situation as a problem of fitting $f(\rho)$ to y by minimizing

$$\text{SSE}(\rho) = [y - f(\rho)]'[y - f(\rho)].$$

As ρ is univariate, $\hat{\rho}$ can easily be found simply by plotting $\text{SSE}(\rho)$ against ρ and inspecting the plot for the minimum. Once $\hat{\rho}$ is found,

$$\hat{\beta} = [A'(\hat{\rho})A(\hat{\rho})]^{-1}A'(\hat{\rho})y$$

gives the values of the remaining parameters.

Figure 6 shows the plots for data generated according to

$$y_t = \theta_1 x_{1t} + \theta_2 x_{2t} + \theta_4 e^{\theta_3 x_{3t}} + e_t$$

with normally distributed errors, input variables as in Table 1, and parameter settings as in Table 4. As θ_4 is the only parameter that is varying, it

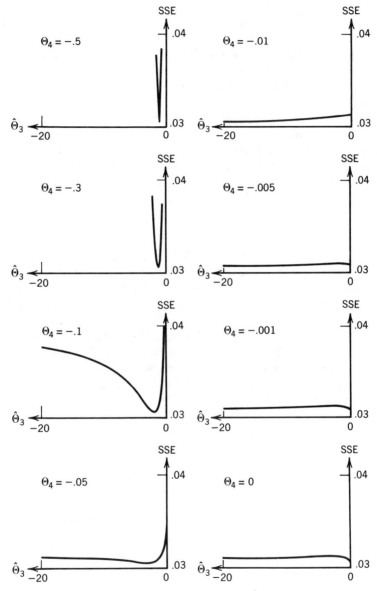

Figure 6. Residual sum of squares plotted against trial values for θ_3 for various true values of θ_4.

serves to label the plots. The 30 errors were not regenerated for each plot: the same 30 were used each time, so that θ_4 is truly all that varies in these plots.

As one sees from the various plots, the fitting of the model becomes increasingly dubious as $|\theta_4|$ decreases. Plots such as those in Figure 3 do not give any visual impression of an exponential trend in x_3 for $|\theta_4|$ smaller than 0.1.

Table 4 shows the deterioration in the performance of the modified Gauss-Newton method as the model becomes increasingly implausible—as $|\theta_4|$ decreases. The table was constructed by finding the local minimum nearest $\rho = 0$ ($\theta_3 = 0$) by grid search over the plots in Figure 6 and setting $\hat{\theta}_3 = \hat{\rho}$ and $(\hat{\theta}_1, \hat{\theta}_2, \hat{\theta}_4) = \hat{\beta}$. From the starting value

$$_{(0)}\theta_i = \hat{\theta}_i - 0.1 \qquad i = 1, 2, 3, 4$$

an attempt was made to recompute this local minimum using the modified Gauss-Newton method and the following stopping rule: Stop when two successive iterations, $_{(i)}\theta$ and $_{(i+1)}\theta$, do not differ in the fifth significant digit (properly rounded) of any component. As noted, performance deteriorates for small $|\theta_4|$.

One learns from this that problems in computing the least squares estimators will usually accompany attempts to fit models with superfluous parameters. Unfortunately one can sometimes be forced into this situation when attempting to formally test the hypothesis $H: \theta_4 = 0$. We shall return to this problem in the next chapter.   □

PROBLEMS

1. Show that

$$\text{SSE}_T(\theta) = \| y - f(\theta_T) - F(\theta_T)(\theta - \theta_T) \|^2$$

is a quadratic function of θ with minimum at

$$\theta_M = \theta_T + [F'(\theta_T)F(\theta_T)]^{-1}F'(\theta_T)[y - f(\theta_T)].$$

One can see these results at sight by applying standard linear least squares theory to the linear model "y" $= X\beta + e$ with "y" $= y - f(\theta_T) + F(\theta_T)\theta_T$, $X = F(\theta_T)$, and $\beta = \theta$.

2. Set forth regularity conditions (Taylor's theorem) such that

$$\mathrm{SSE}(\theta) = \mathrm{SSE}(\theta_T) + \left[\frac{\partial}{\partial\theta}\mathrm{SSE}(\theta_T)\right]'(\theta - \theta_T)$$

$$+ \tfrac{1}{2}(\theta - \theta_T)'\left[\frac{\partial^2}{\partial\theta\,\partial\theta'}\mathrm{SSE}(\theta_T)\right](\theta - \theta_T) + o\left(\|\theta - \theta_T\|^3\right).$$

Show that

$$\mathrm{SSE}(\theta) - \mathrm{SSE}_T(\theta) = (\theta - \theta_T)'A(\theta - \theta_T) + o\left(\|\theta - \theta_T\|^3\right)$$

where A is a symmetric matrix. Show that $|(\theta - \theta_T)'A(\theta - \theta_T)|/\|\theta - \theta_T\|^2$ is less than the largest eigenvalue of A in absolute value, $\max|\lambda_i(A)|$. Use these facts to show that

$$\lim_{\|\theta-\theta_T\|\to 0} \frac{|\mathrm{SSE}(\theta) - \mathrm{SSE}_T(\theta)|}{\|\theta - \theta_T\|} = 0$$

and

$$\lim_{\delta\to 0}\ \sup_{\|\theta-\theta_T\|<\delta} \frac{|\mathrm{SSE}(\theta) - \mathrm{SSE}_T(\theta)|}{\|\theta - \theta_T\|^2} \leq \max|\lambda_i(A)|.$$

3. Assume that θ_T is not a stationary point of $\mathrm{SSE}(\theta)$; that is, $(\partial/\partial\theta)\mathrm{SSE}(\theta_T) \neq 0$. Set forth regularity conditions (Taylor's theorem) such that

$$\mathrm{SSE}[\theta_T + \lambda(\theta_M - \theta_T)]$$

$$= \mathrm{SSE}(\theta_T) + \lambda\left[\frac{\partial}{\partial\theta}\mathrm{SSE}(\theta_T)\right]'(\theta_M - \theta_T) + o(\lambda^2).$$

Let $F_T = F(\theta_T)$, $\hat{e}_T = y - f(\theta_T)$, and show that this equation reduces to

$$\mathrm{SSE}[\theta_T + \lambda(\theta_M - \theta_T)]$$

$$= \mathrm{SSE}(\theta_T) + \left(-2\hat{e}_T'F_T(F_T'F_T)^{-1}F_T'\hat{e}_T + \frac{o(\lambda^2)}{\lambda}\right)\lambda.$$

There must be a λ^* such that

$$-2\hat{e}_T'F_T(F_T'F_T)^{-1}F_T'\hat{e}_T + o\left(\frac{\lambda^2}{\lambda}\right) < 0$$

for all λ with $0 < \lambda < \lambda^*$ (why?). Thus

$$\text{SSE}[\theta_T + \lambda(\theta_M - \theta_T)] < \text{SSE}(\theta_T)$$

for all λ with $0 < \lambda < \lambda^*$.

4. (Convergence of the modified Gauss-Newton method.) Supply the missing details in the proof of the following result.

THEOREM. Let

$$Q(\theta) = \sum_{t=1}^{n} [y_t - f(x_t, \theta)]^2.$$

Conditions: There is a convex, bounded subset S of R^p and θ_0 interior to S such that:

1. $(\partial/\partial\theta)f(x_t, \theta)$ exists and is continuous over $\bar{S}$ for $t = 1, 2, \ldots, n$;
2. $\theta \in S$ implies the rank of $F(\theta)$ is p;
3. $Q(\theta_0) < \tilde{Q} = \inf\{Q(\theta): \theta \text{ a boundary point of } S\}$;
4. there does not exist θ', θ'' in S such that

$$\frac{\partial}{\partial\theta}Q(\theta') = \frac{\partial}{\partial\theta}Q(\theta'') = 0 \quad \text{and} \quad Q(\theta') = Q(\theta'').$$

Construction: Construct a sequence $\{\theta_\alpha\}_{\alpha=1}^{\infty}$ as follows:

0. Compute $D_0 = [F'(\theta_0)F(\theta_0)]^{-1}F'(\theta_0)[y - f(\theta_0)]$. Find λ_0 which minimizes $Q(\theta_0 + \lambda D_0)$ over

$$\Lambda_0 = \{\lambda: 0 \le \lambda \le 1, \theta_0 + \lambda D_0 \in \bar{S}\}.$$

1. Set $\theta_1 = \theta_0 + \lambda_0 D_0$. Compute $D_1 = [F'(\theta_1)F(\theta_1)]^{-1}F'(\theta_1)[y - f(\theta_1)]$. Find λ_1 which minimizes $Q(\theta_1 + \lambda D_1)$ over

$$\Lambda_1 = \{\lambda: 0 \le \lambda \le 1, \theta_1 + \lambda D_1 \in \bar{S}\}.$$

2. Set $\theta_2 = \theta_1 + \lambda_1 D_1 \ldots$.
Conclusions. Then for the sequence $\{\theta_\alpha\}_{\alpha=1}^{\infty}$ it follows that:

1. θ_α is an interior point of S for $\alpha = 1, 2, \ldots$.
2. The sequence $\{\theta_\alpha\}$ converges to a limit θ^* which is interior to S.
3. $(\partial/\partial\theta)Q(\theta^*) = 0$.

Proof. We establish conclusion 1. The conclusion will follow by induction if we show that θ_α interior to S and $Q(\theta_\alpha) < \tilde{Q}$ imply that λ_α minimizing $Q(\theta_\alpha + \lambda D_\alpha)$ over Λ_α exists and $\theta_{\alpha+1}$ is an interior point of S. Let $\theta_\alpha \in S^0$ and consider the set

$$\hat{S} = \{\theta \in \bar{S} : \theta = \theta_\alpha + \lambda D_\alpha, 0 \leq \lambda \leq 1\}.$$

$\hat{S}$ is a closed, bounded line segment contained in $\bar{S}$ (why?). There is a θ' in $\hat{S}$ minimizing Q over $\hat{S}$ (why?). Hence, there is a λ_α ($\theta' = \theta_\alpha + \lambda_\alpha D_\alpha$) minimizing $Q(\theta_\alpha + \lambda D_\alpha)$ over Λ_α. Now θ' is either an interior point of $\bar{S}$ or a boundary point of S. By Lemma 2.2.1 of Blackwell and Girshick (1954, p. 32) S and $\bar{S}$ have the same interior points and boundary points. In θ' were a boundary point of S, we would have

$$\tilde{Q} \leq Q(\theta') \leq Q(\theta_\alpha) < \tilde{Q}$$

which is not possible. Then θ' is an interior point of S. Since $\theta_{\alpha+1} = \theta'$, we have established conclusion 1.

We establish conclusions 2, 3. By construction $0 \leq Q(\theta_{\alpha+1}) \leq Q(\theta_\alpha)$; hence $Q(\theta_\alpha) \to Q^*$ as $\alpha \to \infty$. The sequence $\{\theta_\alpha\}$ must have a convergent subsequence $\{\theta_\beta\}_{\beta=1}^\infty$ with limit $\theta^* \in \bar{S}$ (why?). $Q(\theta_\beta) \to Q(\theta^*)$ so $Q(\theta^*) = Q^*$ (why?). θ^* is either an interior point of $\bar{S}$ or a boundary point. The same holds for S, as we saw above. If θ^* were a boundary point of S, then $\tilde{Q} \leq Q(\theta^*) \leq Q(\theta^0)$, which is impossible because $Q(\theta_0) < \tilde{Q}$. So θ^* is an interior point of S.

The function

$$D(\theta) = [F'(\theta)F(\theta)]^{-1}F'(\theta)[y - f(\theta)]$$

is continuous over S (why?). Thus

$$\lim_{\beta \to \infty} D_\beta = \lim_{\beta \to \infty} D(\theta_\beta) = D(\theta^*) = D^*.$$

Suppose $D^* \neq 0$, and consider the function $q(\lambda) = Q(\theta^* + \lambda D^*)$ for $\lambda \in [-\eta, \eta]$, where $0 < \eta \leq 1$ and $\theta^* \pm \eta D^*$ are interior points of S. Then

$$q'(0) = \frac{\partial}{\partial \theta'} Q(\theta^* + \lambda D^*)D^* \Big|_{\lambda=0}$$
$$= (-2)[y - f(\theta^*)]'F(\theta^*)D^*$$
$$= (-2)D^{*\prime}F'(\theta^*)F(\theta^*)D^*$$
$$< 0$$

(why?). Choose $\epsilon > 0$ so that $\epsilon < -q'(0)$. By the definition of derivative, there is a $\lambda^* \in (0, \frac{1}{2}\eta)$ such that

$$Q(\theta^* + \lambda^* D^*) - Q(\theta^*) = q(\lambda^*) - q(0)$$
$$< [q'(0) + \epsilon]\lambda^*.$$

Since Q is continuous for $\theta \in S$, we may choose $\gamma > 0$ such that $-\gamma > [q'(0) + \epsilon]\lambda^*$ and there is $\delta > 0$ such that

$$\|\theta_\beta + \lambda^* D_\beta - \theta^* - \lambda^* D^*\| < \delta$$

implies

$$Q(\theta_\beta + \lambda^* D) - Q(\theta^* + \lambda^* D^*) < \gamma.$$

Then for all β sufficiently large we have

$$Q(\theta_\beta + \lambda^* D_\beta) - Q(\theta^*) < [q'(0) + \epsilon]\lambda^* + \gamma = -c^2.$$

Now for β large enough $\theta_\beta + \lambda^* D_\beta$ is interior to S, so that $\lambda^* \in \Lambda_\beta$ and we obtain

$$Q(\theta_{\beta+1}) - Q(\theta^*) < -c^2.$$

This contradicts the fact that $Q(\theta_\beta) \to Q(\theta^*) = Q^*$ as $\beta \to \infty$; thus D^* must be the zero vector. Then it follows that

$$\frac{\partial}{\partial\theta}Q(\theta^*) = (-2)F'(\theta^*)[y - f(\theta^*)]$$
$$= (-2)F'(\theta^*)F(\theta^*)D^*$$
$$= 0.$$

Given any subsequence of $\{\theta_\alpha\}$, we have by the above that there is a convergent subsequence with limit point $\theta' \in S$ such that

$$\frac{\partial}{\partial\theta}Q(\theta') = 0 = \frac{\partial}{\partial\theta}Q(\theta^*)$$

and

$$Q(\theta') = Q^* = Q(\theta^*).$$

By hypothesis 4, $\theta' = \theta^*$, so that $\theta_\alpha \to \theta'$ as $\alpha \to \infty$.

5. HYPOTHESIS TESTING

Assuming that the data follow the model

$$y = f(\theta^0) + e \qquad e \sim N_n(0, \sigma^2 I)$$

consider testing the hypothesis

$$H : h(\theta^0) = 0 \quad \text{against} \quad A : h(\theta^0) \neq 0$$

where $h(\theta)$ is a once continuously differentiable function mapping $\mathbb{R}^p$ into $\mathbb{R}^q$ with Jacobian

$$H(\theta) = \frac{\partial}{\partial \theta'} h(\theta)$$

of order q by p. When $H(\theta)$ is evaluated at $\theta = \hat{\theta}$ we shall write $\hat{H}$,

$$\hat{H} = H(\hat{\theta})$$

and at $\theta = \theta^0$ write H,

$$H = H(\theta^0)$$

In Chapter 4 we shall show that $h(\hat{\theta})$ may be characterized as

$$h(\hat{\theta}) = h(\theta^0) + H(F'F)^{-1}F'e + o_p\left(\frac{1}{\sqrt{n}}\right)$$

where, recall, $F = (\partial / \partial \theta') f(\theta^0)$. Ignoring the remainder term, we have

$$h(\hat{\theta}) \doteq N_q\left[h(\theta^0), \sigma^2 H(F'F)^{-1}H'\right]$$

whence

$$\frac{h'(\hat{\theta})\left[H(F'F)^{-1}H'\right]^{-1}h(\hat{\theta})}{\sigma^2}$$

is (approximately) distributed as the noncentral chi-square distribution (Section 7) with q degrees of freedom and noncentrality parameter

$$\lambda = \frac{h'(\theta^0)\left[H(F'F)^{-1}H'\right]^{-1}h(\theta^0)}{2\sigma^2}.$$

Recalling that to within the order of approximation $o_p(1/n)$, $(n-p)s^2/\sigma^2$ is distributed independently of $\hat{\theta}$ as the chi-square distribution with $n-p$ degrees of freedom, we have (approximately) that the ratio

$$\frac{h'(\hat{\theta})\big[H(F'F)^{-1}H'\big]^{-1}h(\hat{\theta})\big/(q\sigma^2)}{(n-p)s^2\big/\big[(n-p)\sigma^2\big]}$$

follows the noncentral F-distribution (see Appendix, Section 7 of this chapter) with q numerator degrees of freedom, $n-p$ denominator degrees of freedom, and noncentrality parameter λ; denoted as $F'(q, n-p, \lambda)$. Canceling like terms in the numerator and denominator, we have

$$\frac{h'(\hat{\theta})\big[H(F'F)^{-1}H'\big]^{-1}h(\hat{\theta})}{(qs^2)} \sim F'(q, n-p, \lambda)$$

In applications, estimates $\hat{H}$ and $\hat{C}$ must be substituted for H and $(F'F)^{-1}$, where, recall, $\hat{C} = [F'(\hat{\theta})F(\hat{\theta})]^{-1}$. The resulting statistic

$$W = \frac{h'(\hat{\theta})(\hat{H}\hat{C}\hat{H}')^{-1}h(\hat{\theta})}{qs^2}$$

is usually called the Wald test statistic.

To summarize this discussion, the Wald test rejects the hypothesis

$$H: h(\theta^0) = 0$$

when the statistic

$$W = \frac{h'(\hat{\theta})(\hat{H}\hat{C}\hat{H}')^{-1}h(\hat{\theta})}{qs^2}$$

exceeds the upper $\alpha \times 100\%$ critical point of the F-distribution with q numerator degrees of freedom and $n-p$ denominator degrees of freedom, denoted as $F^{-1}(1-\alpha; q, n-p)$. We illustrate.

EXAMPLE 1 (Continued). Recalling that

$$f(x, \theta) = \theta_1 x_1 + \theta_2 x_2 + \theta_4 e^{\theta_3 x_3}$$

consider testing the hypothesis of no treatment effect,

$$H: \theta_1 = 0 \quad \text{against} \quad A: \theta_1 \neq 0.$$

For this case

$$h(\theta) = \theta_1$$

$$H(\theta) = \frac{\partial}{\partial \theta'} h(\theta) = (1,0,0,0)$$

$$h(\hat{\theta}) = -0.02588970 \qquad \text{(from Fig. } 5a)$$

$$\hat{H} = \frac{\partial}{\partial \theta'} h(\hat{\theta}) = (1,0,0,0)$$

$$\hat{H}\hat{C}\hat{H}' = \hat{c}_{11} = 0.13587 \qquad \text{(from Fig. } 5b)$$

$$s^2 = 0.00117291 \qquad \text{(from Fig. } 5a)$$

$$q = 1$$

$$W = \frac{h'(\hat{\theta})(\hat{H}\hat{C}\hat{H}')^{-1}h(\theta)}{qs^2}$$

$$= \frac{(-0.02588970)(0.13587)^{-1}(-0.02588970)}{1 \times 0.00117291}$$

$$= 4.2060.$$

The upper 5% critical point of the F-distribution with 1 numerator degree of freedom and $26 = 30 - 4$ denominator degrees of freedom is

$$F^{-1}(.95; 1, 26) = 4.22$$

so one fails to reject the null hypothesis.

Of course, in this simple instance one can compute a t-statistic directly from the output shown in Figure 5a as

$$t = \frac{-0.025887970}{0.01262384}$$

$$= -2.0509$$

and compare the absolute value with

$$t^{-1}(.975; 26) = 2.0555. \qquad \square$$

In simple examples such as the proceeding, one can work directly from printed output such as Figure 5a. But anything more complicated requires some programming effort to compute and invert $\hat{H}\hat{C}\hat{H}'$. There are a variety of ways to do this; we shall describe a method that is useful pedagogically, as it builds on the ideas of the previous section and is easy to use with a statistical package. It also has the advantage of saving the bother of looking up the critical values of the F-distribution.

Suppose that one fits the model

$$\hat{e} = \hat{F}\beta + u$$

by least squares and tests the hypothesis

$$H: \hat{H}\beta = h(\hat{\theta}) \quad \text{against} \quad A: \hat{H}\beta \neq h(\hat{\theta}).$$

The computed F-statistic will be

$$F = \frac{[\hat{H}\beta - h(\hat{\theta})]'[\hat{H}(\hat{F}'\hat{F})^{-1}\hat{H}']^{-1}[\hat{H}\beta - h(\hat{\theta})]/q}{[\hat{e} - \hat{F}\beta]'[\hat{e} - \hat{F}\beta]/(n-p)}$$

but since

$$0 = \frac{\partial}{\partial\theta}\text{SSE}(\hat{\theta}) = -2\hat{F}'\hat{e}$$

we have

$$0 = (\hat{F}'\hat{F})^{-1}\hat{F}'\hat{e} = \hat{\beta}$$

and the computed F-statistic reduces to

$$W = \frac{h'(\hat{\theta})(\hat{H}\hat{C}\hat{H}')^{-1}h(\hat{\theta})}{qs^2}.$$

Thus, any statistical package that can compute a linear regression and test a linear hypothesis becomes a convenient tool for computing the Wald test statistic. We illustrate these ideas in the next example.

EXAMPLE 1 (Continued). Recalling that the response function is

$$f(x,\theta) = \theta_1 x_1 + \theta_2 x_2 + \theta_4 e^{\theta_3 x_3}$$

consider testing

$$H: \left.\frac{\partial}{\partial x_3}f(x,\theta)\right|_{x_3=1} = \frac{1}{5} \quad \text{against} \quad A: \left.\frac{\partial}{\partial x_3}f(x,\theta)\right|_{x_3=1} \neq \frac{1}{5}$$

or equivalently

$$H: \theta_3\theta_4 e^{\theta_3} = \tfrac{1}{5} \quad \text{against} \quad A: \theta_3\theta_4 e^{\theta_3} \neq \tfrac{1}{5}.$$

We have

$$h(\theta) = \theta_3\theta_4 e^{\theta_3} - \tfrac{1}{5}$$

$$H(\theta) = \frac{\partial}{\partial\theta'}h(\theta) = \left[0,0,\theta_4(1+\theta_3)e^{\theta_3}, \theta_3 e^{\theta_3}\right]$$

$$h(\hat\theta) = (-1.11569714)(-0.50490286)e^{-1.11569714} - 0.2$$

$$= -0.0154079303 \qquad \text{(from Fig. 5}a\text{)}$$

$$\hat H = (\partial/\partial\theta')h(\hat\theta)$$

$$= (0,0,0.0191420895, -0.365599176)$$

$$\text{(from Fig. 5}a\text{)}$$

$$\frac{h'(\hat\theta)(\hat H \hat C \hat H')^{-1}h(\hat\theta)}{1} = 0.0042964 \qquad \text{(from Fig. 7)}$$

$$s^2 = 0.001172905 \qquad \text{(from Fig. 5}a\text{ or 7)}$$

$$W = 3.6631 \qquad \text{(from Fig. 7 or}$$

$$\text{by division).}$$

Since $F^{-1}(.95; 1, 26) = 4.22$, one fails to reject at the 5% level. The p-value is 0.0667 as shown in Figure 7; that is $1 - F(3.661; 1, 26) = 0.0667$.

Also shown in Figure 7 are the computations for the previous example as well as computations for the joint hypothesis

$$H: \theta_1 = 0 \text{ and } \theta_3\theta_4 e^{\theta_3} = \tfrac{1}{5} \quad \text{against} \quad A: \theta_1 \neq 0 \text{ or } \theta_3\theta_4 e^{\theta_3} \neq \tfrac{1}{5}.$$

The joint hypothesis is included to illustrate the computations for the case $q > 1$. One rejects the joint hypothesis at the 5% level; the p-value is 0.0210.

□

We have noted in the somewhat heuristic derivation of the Wald test that W is distributed as the noncentral F-distribution. What can be shown rigorously (Chapter 4) is that

$$W = Y + o_p\left(\frac{1}{n}\right)$$

$$Y \sim F'(q, n - p, \lambda)$$

$$\lambda = \frac{h'(\theta^0)\left\{H(\theta^0)[F'(\theta^0)F(\theta^0)]^{-1}H'(\theta^0)\right\}^{-1}h(\theta^0)}{2\sigma^2}.$$

SAS Statements:

```
DATA WORK01; SET EXAMPLE1;
T1=-0.02588970; T2=1.01567967; T3=-1.11569714; T4=-0.50490286;
E=Y-(T1*X1+T2*X2+T4*EXP(T3*X3));
DER_T1=X1; DER_T2=X2; DER_T3=T4*X3*EXP(T3*X3); DER_T4=EXP(T3*X3);
PROC REG DATA=WORK01;  MODEL E = DER_T1 DER_T2 DER_T3 DER_T4 / NOINT;
FIRST: TEST DER_T1=0.02588970;
SECOND:TEST 0.0191420895*DER_T3-0.365599176*DER_T4=-0.0154079303;
JOINT: TEST DER_T1=0.02588970,
            0.0191420895*DER_T3-0.365599176*DER_T4=-0.0154079303;
```

Output:

S T A T I S T I C A L   A N A L Y S I S   S Y S T E M   1

DEP VARIABLE: E

SOURCE	DF	SUM OF SQUARES	MEAN SQUARE	F VALUE	PROB>F
MODEL	4	3.29597E-17	8.23994E-18	0.000	1.0000
ERROR	26	0.030496	0.001172905		
U TOTAL	30	0.030496			

ROOT MSE	0.034248	R-SQUARE	0.0000
DEP MEAN	4.13616E-11	ADJ R-SQ	-0.1154
C.V.	82800642118		

NOTE: NO INTERCEPT TERM IS USED. R-SQUARE IS REDEFINED.

VARIABLE	DF	PARAMETER ESTIMATE	STANDARD ERROR	T FOR H0: PARAMETER=0	PROB > |T|
DER_T1	1	1.91639E-09	0.012624	0.000	1.0000
DER_T2	1	-6.79165E-10	0.009937927	-0.000	1.0000
DER_T3	1	1.52491E-10	0.163542	0.000	1.0000
DER_T4	1	-1.50709E-09	0.025657	-0.000	1.0000

TEST: FIRST	NUMERATOR: .0049333	DF: 1	F VALUE: 4.2060
	DENOMINATOR: .0011729	DF: 26	PROB >F : 0.0505
TEST: SECOND	NUMERATOR: .0042964	DF: 1	F VALUE: 3.6631
	DENOMINATOR: .0011729	DF: 26	PROB >F : 0.0667
TEST: JOINT	NUMERATOR: .0052743	DF: 2	F VALUE: 4.4968
	DENOMINATOR: .0011729	DF: 26	PROB >F : 0.0210

Figure 7. Illustration of Wald test computations with Example 1.

That is, Y is distributed as the noncentral F-distribution with q-numerator degrees of freedom, $n - p$ denominator degrees of freedom, and non-centrality parameter λ (Section 7). The computation of power requires computation of λ and use of charts (Pearson and Hartley, 1951; Fox, 1956) of the noncentral F-distribution. One convenient source for the charts is Scheffé (1959). The computation of λ is very little different from the computation of W itself, and one can use exactly the same strategy used in the previous example to obtain

$$\frac{h'(\theta^0)\left\{H(\theta^0)\left[F'(\theta^0)F(\theta^0)\right]^{-1}H'(\theta^0)\right\}^{-1}h(\theta^0)}{q}$$

and then multiply by $q/(2\sigma^2)$ to obtain λ. Alternatively one can write code in some programming language to compute λ. To add variety to the discussion, we shall illustrate the latter approach using PROC MATRIX in SAS.

EXAMPLE 1   (Continued).   Recalling that

$$f(x,\theta) = \theta_1 x_1 + \theta_2 x_2 + \theta_4 e^{\theta_3 x_3}$$

let us approximate the probability that the Wald test rejects the following three hypotheses at the 5% level when the true values of the parameters are

$$\theta^0 = (.03, 1, -1.4, -.5)'$$
$$\sigma^2 = .001.$$

The three null hypotheses are

$$H_1: \theta_1 = 0$$
$$H_2: \theta_3 \theta_4 e^{\theta_3} = \tfrac{1}{5}$$
$$H_3: \theta_3 = 0 \text{ and } \theta_3 \theta_4 e^{\theta_3} = \tfrac{1}{5}.$$

PROC MATRIX code to compute

$$\lambda = \frac{h'(\theta^0)\left\{H(\theta^0)\left[F'(\theta^0)F(\theta^0)\right]^{-1}H'(\theta^0)\right\}^{-1}h(\theta^0)}{2\sigma^2}$$

for each of the three cases is shown in Figure 8. We obtain

$$\lambda_1 = 3.3343 \qquad \text{(from Fig. 8)}$$
$$\lambda_2 = 5.65508 \qquad \text{(from Fig. 8)}$$
$$\lambda_3 = 9.88196 \qquad \text{(from Fig. 8).}$$

Then from the Pearson-Hartley charts of the noncentral F-distribution in

SAS Statements:

```
PROC MATRIX;  FETCH X DATA=EXAMPLE1(KEEP=X1 X2 X3);
T1=.03;  T2=1;  T3=-1.4;  T4=-.5;  S=.001;  N=30;
F1=X(,1);  F2=X(,2);  F3=T4*(X(,3)#EXP(T3*X(,3)));  F4=EXP(T3*X(,3));
F=F1||F2||F3||F4;  C=INV(F'*F);
SMALL_H1=T1;  H1=1 0 0 0;
LAMBDA=SMALL_H1'*INV(H1*C*H1')*SMALL_H1#/(2*S); PRINT LAMBDA;
SMALL_H2=(T3#T4#EXP(T3)-1#/5);  H2=0||0||T4#(1+T3)#EXP(T3)||T3#EXP(T3);
LAMBDA=SMALL_H2'*INV(H2*C*H2')*SMALL_H2#/(2*S); PRINT LAMBDA;
SMALL_H3=SMALL_H1//SMALL_H2;  H3=H1//H2;
LAMBDA=SMALL_H3'*INV(H3*C*H3')*SMALL_H3#/(2*S); PRINT LAMBDA;
```

Output:

STATISTICAL ANALYSIS SYSTEM   1

	LAMBDA	COL1
	ROW1	3.3343

	LAMBDA	COL1
	ROW1	5.65508

	LAMBDA	COL1
	ROW1	9.88196

Figure 8. Illustration of Wald test power computations with Example 1.

Scheffé (1959) we obtain

$$1 - F'(4.22; 1, 26, 3.3343) = .70$$

$$1 - F'(4.22; 1, 26, 5.65508) = .90$$

$$1 - F'(3.37; 2, 26, 9.88196) = .97.$$

For the first hypothesis one approximates $P(W > F_\alpha)$ by $P(Y > F_\alpha) = .70$, where $F_\alpha = F^{-1}(.95; 1, 26) = 4.22$, and so on for the other two cases.

The natural question is: How accurate are these approximations? In this instance the Monte Carlo simulations reported in Table 5 indicate that the approximation is accurate enough for practical purposes, but later on we shall see examples showing fairly poor approximations to $P(W > F_\alpha)$ by $P(Y > F_\alpha)$. Table 5 was constructed by generating 5000 responses using the

Table 5. Monte Carlo Power Estimates for the Wald Test.

			$H_0 : \theta_1 = 0$ against $H_1 : \theta_1 \neq 0$				$H_0 : \theta_3 = -1$ against $H_1 : \theta_3 \neq -1$		
Parameters[a]				Monte Carlo				Monte Carlo	
θ_1	θ_3	λ	$P[Y > F_\alpha]$	$P[W > F_\alpha]$	Std. Err.	λ	$P[Y > F_\alpha]$	$P[W > F_\alpha]$	Std. Err.
0.0	−1.0	0.0	.050	.050	.003	0.0	.050	.056	.003
0.008	−1.1	0.2353	.101	.094	.004	0.2220	.098	.082	.004
0.015	−1.2	0.8309	.237	.231	.006	0.7332	.215	.183	.006
0.030	−1.4	3.3343	.700	.687	.006	2.1302	.511	.513	.007

Source: Gallant (1975d).

[a] $\theta_2 = 1$, $\theta_4 = -.5$, $\sigma^2 = .001$.

response function

$$f(x, \theta) = \theta_1 x_1 + \theta_2 x_2 + \theta_4^{\theta_3 x_3}$$

and the inputs shown in Table 1. The parameters used were $\theta_2 = 1$, $\theta_4 = -.5$, and $\sigma^2 = .001$ excepting θ_1 and θ_3, which were varied as shown in Table 5. The power for a test of $H : \theta_1 = 0$ and $H : \theta_3 = -1$ is computed for $P(Y > F_\alpha)$ and compared with $P(W > F_\alpha)$ estimated from the Monte Carlo trials. The standard errors in the table refer to the fact that the Monte Carlo estimate of $P(W < F_\alpha)$ is binomially distributed with $n = 5000$ and $p = P(Y > F_\alpha)$. Thus, $P(W > F_\alpha)$ is estimated with a standard of error of $\{P(Y > F_\alpha)[1 - P(Y > F_\alpha)]/5000\}^{1/2}$. These simulations are described in somewhat more detail in Gallant (1975b). □

One of the most familiar methods of testing a linear hypothesis

$$H : R\beta = r \quad \text{against} \quad A : R\beta \neq r$$

for the linear model

$$y = X\beta + e$$

is: First, fit the full model by least squares, obtaining

$$\text{SSE}_{\text{full}} = (y - X\hat{\beta})'(y - X\hat{\beta})$$
$$\hat{\beta} = (X'X)^{-1}X'y.$$

Second, refit the model subject to the null hypothesis that $R\beta = r$, obtaining

$$\text{SSE}_{\text{reduced}} = (y - X\tilde{\beta})'(y - X\tilde{\beta})$$

$$\tilde{\beta} = \hat{\beta} + (X'X)^{-1}R'\left[R(X'X)^{-1}R'\right]^{-1}(r - R\hat{\beta}).$$

Third, compute the F-statistic

$$F = \frac{(\text{SSE}_{\text{reduced}} - \text{SSE}_{\text{full}})/q}{(\text{SSE}_{\text{full}})/(n - p)}$$

where q is the number of restrictions on β (number of rows in R), p the number of columns in X, and n the number of observations—full rank matrices being assumed throughout. One rejects for large values of F. If one assumes normal errors in the nonlinear model

$$y = f(\theta) + e \qquad e \sim N_n(0, \sigma^2 I)$$

and derives the likelihood ratio test statistic for the hypothesis

$$H : h(\theta) = 0 \quad \text{against} \quad A : h(\theta) \neq 0$$

one obtains exactly the same test as just described (Problem 1). The statistic is computed as follows.

First, compute

$$\hat{\theta} \quad \text{minimizing} \quad \text{SSE}(\theta) = [y - f(\theta)]'[y - f(\theta)]$$

using the methods of the previous section, and let

$$\text{SSE}_{\text{full}} = \text{SSE}(\hat{\theta}).$$

Second, refit under the null hypothesis by computing

$$\tilde{\theta} \quad \text{minimizing} \quad \text{SSE}(\theta) \quad \text{subject to} \quad h(\theta) = 0$$

using methods discussed immediately below, and let

$$\text{SSE}_{\text{reduced}} = \text{SSE}(\tilde{\theta}).$$

Third, compute the statistic

$$L = \frac{(\text{SSE}_{\text{reduced}} - \text{SSE}_{\text{full}})/q}{(\text{SSE}_{\text{full}})/(n - p)}.$$

Recall that $h(\theta)$ maps $\mathbb{R}^p$ into $\mathbb{R}^q$, so that q is, in a sense, the number of restrictions on θ. One rejects $H: h(\theta^0) = 0$ when L exceeds the $\alpha \times 100\%$ critical point F_α of the F-distribution with q numerator degrees of freedom and $n - p$ denominator degrees of freedom; $F_\alpha = F^{-1}(1 - \alpha; q, n - p)$. Later on, we shall verify that L is distributed according to the F-distribution if $h(\theta^0) = 0$. For now, let us consider computational aspects.

General methods for minimizing SSE(θ) subject to $h(\theta) = 0$ are given in Gill, Murray, and Wright (1981). But it is almost always the case in practice that a hypothesis written as a parametric restriction

$$H: h(\theta^0) = 0 \quad \text{against} \quad A: h(\theta^0) \neq 0$$

can easily be rewritten as a functional dependence

$$H: \theta^0 = g(\rho) \text{ for some } \rho^0 \quad \text{against} \quad A: \theta^0 \neq g(\rho) \text{ for any } \rho.$$

Here ρ is an r-vector with $r = p - q$. In general one obtains $g(\rho)$ by augmenting the equations

$$h(\theta) = \tau$$

by the equations

$$\varphi(\theta) = \rho$$

which are chosen such that the system of equations

$$h(\theta) = \tau$$
$$\varphi(\theta) = \rho$$

is a one to one transformation with inverse

$$\theta = \psi(\rho, \tau).$$

Then imposing the condition

$$\theta = \psi(\rho, 0)$$

is equivalent (Problem 2) to imposing the condition

$$h(\theta) = 0$$

so that the desired functional dependence is obtained by putting

$$\theta = g(\rho) = \psi(\rho, 0).$$

But usually $g(\rho)$ can be constructed at sight on an *ad hoc* basis without resorting to these formalities, as seen in the later examples.

The null hypothesis is that the data follow the model

$$y_t = f(x_t, \theta^0) + e_t$$

and that θ^0 satisfies

$$h(\theta^0) = 0.$$

Equivalently, the null hypothesis is that the data follow the model

$$y_t = f(x_t, \theta^0) + e_t$$

and

$$\theta^0 = g(\rho) \quad \text{for some} \quad \rho^0.$$

But the latter statement can be expressed more simply as: The null hypothesis is that the data follow the model

$$y_t = f[x_t, g(\rho^0)] + e_t.$$

In vector notation,

$$y = f[g(\rho^0)] + e.$$

This is, of course, merely a nonlinear model that can be fitted by the methods described previously. One computes

$$\hat{\rho} \quad \text{minimizing} \quad \text{SSE}[g(\rho)] = \{y - f[g(\rho)]\}'\{y - f[g(\rho)]\}$$

by, say, the modified Gauss-Newton method. Then

$$\text{SSE}_{\text{reduced}} = \text{SSE}[g(\hat{\rho})]$$

because $\tilde{\theta} = g(\hat{\rho})$ (Problem 3).

The fact that $f[x, g(\rho)]$ is a composite function gives derivatives some structure that can be exploited in computations. Let

$$G(\rho) = \frac{\partial}{\partial \rho'} g(\rho)$$

that is, $G(\rho)$ is the Jacobian of $g(\rho)$, which has p rows and r columns.

Then using the differentiation rules of Section 2,

$$\frac{\partial}{\partial \rho'} f[x, g(\rho)] = \frac{\partial}{\partial \theta'} f[x, g(\rho)] \frac{\partial}{\partial \rho'} g'(\rho)$$

or

$$\frac{\partial}{\partial \rho'} f[g(\rho)] = F[g(\rho)] G(\rho).$$

These facts can be used as a labor saving device when writing code for nonlinear optimization, as seen in the examples.

EXAMPLE 1 (Continued). Recalling that the response function is

$$f(x, \theta) = \theta_1 x_1 + \theta_2 x_2 + \theta_4 e^{\theta_3 x_3},$$

reconsider the first hypothesis

$$H : \theta_1^0 = 0.$$

This is an assertion that the data follow the model

$$y_t = \theta_2 x_{2t} + \theta_4 e^{\theta_3 x_{3t}} + e_t.$$

Fitting this model to the data of Table 1 by the modified Gauss-Newton method, we have

$$\text{SSE}_{\text{reduced}} = 0.03543298 \qquad \text{(from Fig. 9a)}.$$

Previously we computed

$$\text{SSE}_{\text{full}} = 0.03049554 \qquad \text{(from Fig. 5a)}.$$

The likelihood ratio statistic is

$$
\begin{aligned}
L &= \frac{(\text{SSE}_{\text{reduced}} - \text{SSE}_{\text{full}})/q}{(\text{SSE}_{\text{full}})/(n - p)} \\[6pt]
&= \frac{(0.03543298 - 0.03049554)/1}{0.03049554/26} \\[6pt]
&= 4.210.
\end{aligned}
$$

SAS Statements:

```
PROC NLIN DATA=EXAMPLE1 METHOD=GAUSS ITER=50 CONVERGENCE=1.0E-13;
PARMS   T2=1.01567967 T3=-1.11569714 T4=-0.50490286;  T1=0;
MODEL  Y=T1*X1+T2*X2+T4*EXP(T3*X3);
DER.T2=X2; DER.T3=T4*X3*EXP(T3*X3); DER.T4=EXP(T3*X3);
```

Output:

STATISTICAL ANALYSIS SYSTEM   1

NON-LINEAR LEAST SQUARES ITERATIVE PHASE

DEPENDENT VARIABLE: Y   METHOD: GAUSS-NEWTON

ITERATION	T2	T3	T4	RESIDUAL SS
0	1.01567967	-1.11569714	-0.50490286	0.04054968
1	1.00289158	-1.14446980	-0.51206647	0.03543349
2	1.00297335	-1.14082057	-0.51178607	0.03543299
3	1.00296493	-1.14128672	-0.51182738	0.03543298
4	1.00296604	-1.14122778	-0.51182219	0.03543298
5	1.00296590	-1.14123524	-0.51182285	0.03543298
6	1.00296592	-1.14123430	-0.51182276	0.03543298
7	1.00296592	-1.14123442	-0.51182277	0.03543298

NOTE: CONVERGENCE CRITERION MET.

STATISTICAL ANALYSIS SYSTEM   2

NON-LINEAR LEAST SQUARES SUMMARY STATISTICS   DEPENDENT VARIABLE Y

SOURCE	DF	SUM OF SQUARES	MEAN SQUARE
REGRESSION	3	26.34100467	8.78033489
RESIDUAL	27	0.03543298	0.00131233
UNCORRECTED TOTAL	30	26.37643764	
(CORRECTED TOTAL)	29	0.71895291	

PARAMETER	ESTIMATE	ASYMPTOTIC STD. ERROR	ASYMPTOTIC 95 % CONFIDENCE INTERVAL LOWER	UPPER
T2	1.00296592	0.00813053	0.98628359	1.01964825
T3	-1.14123442	0.17446900	-1.49921245	-0.78325638
T4	-0.51182277	0.02718622	-0.56760385	-0.45604169

ASYMPTOTIC CORRELATION MATRIX OF THE PARAMETERS

	T2	T3	T4
T2	1.000000	0.400991	-0.120866
T3	0.400991	1.000000	0.565235
T4	-0.120866	0.565235	1.000000

Figure 9a. Illustration of likelihood ratio test computations with Example 1.

Comparing with the critical point

$$F^{-1}(.95; 1, 26) = 4.22$$

one fails to reject the null hypothesis at the 95% level.

Reconsider the second hypothesis

$$H: \theta_3\theta_4 e^{\theta_3} = \tfrac{1}{5}$$

which can be rewritten as

$$H: \theta_4 = \frac{1}{5\theta_3 e^{\theta_3}}.$$

Then writing

$$g(\rho) = \begin{pmatrix} \rho_1 \\ \rho_2 \\ \rho_3 \\ \dfrac{1}{5\rho_3 e^{\rho_3}} \end{pmatrix}$$

an equivalent form of the null hypothesis is that

$$H: \theta^0 = g(\rho) \text{ for some } \rho^0.$$

One can fit the null model in one of two ways. The first is to fit directly the model

$$y_t = \rho_1 x_{1t} + \rho_2 x_{2t} + (5\rho_3)^{-1} e^{\rho_3(x_{3t}-1)} + e_t.$$

The second is as follows:

1. Given ρ, set $\theta = g(\rho)$.
2. Use the code written previously (Figure 5a) to compute $f(x, \theta)$ and $(\partial/\partial\theta')f(x, \theta)$ given θ.
3. Use

$$\frac{\partial}{\partial\rho'}f[x, g(\rho)] = \left(\frac{\partial}{\partial\theta'}f[x, g(\rho)]\right)G(\rho)$$

to compute the partial derivatives with respect to ρ; recall that $G(\rho) = (\partial/\partial\rho')g(\rho)$.

We use this second method to fit the reduced model in Figure 9b. We have

$$G(\rho) = \begin{pmatrix} 1 & 0 & 0 \\ 0 & 1 & 0 \\ 0 & 0 & 1 \\ 0 & 0 & -(5\rho_3 e^{\rho_3})^{-2}(5e^{\rho_3} + 5\rho_3 e^{\rho_3}) \end{pmatrix}.$$

SAS Statements:

```
PROC NLIN DATA=EXAMPLE1 METHOD=GAUSS ITER=60 CONVERGENCE=1.0E-8;
PARMS R1=-0.02588970 R2=1.01567967 R3=-1.11569714;
T1=R1; T2=R2; T3=R3; T4=1/(5*R3*EXP(R3));
MODEL Y=T1*X1+T2*X2+T4*EXP(T3*X3);
DER_T1=X1; DER_T2=X2; DER_T3=T4*X3*EXP(T3*X3); DER_T4=EXP(T3*X3);
DER.R1=DER_T1; DER.R2=DER_T2;
DER.R3=DER_T3+DER_T4*(-T4**2)*(5*EXP(R3)+5*R3*EXP(R3));
```

Output:

S T A T I S T I C A L   A N A L Y S I S   S Y S T E M   1

NON-LINEAR LEAST SQUARES ITERATIVE PHASE

DEPENDENT VARIABLE: Y   METHOD: GAUSS-NEWTON

ITERATION	R1	R2	R3	RESIDUAL SS
0	-0.02588970	1.01567967	-1.11569714	0.03644046
1	-0.02286308	1.01860305	-1.19237581	0.03502362
2	-0.02314184	1.02019397	-1.13249955	0.03500414
3	-0.02291862	1.01903284	-1.18159656	0.03497186
4	-0.02309964	1.02003652	-1.14220257	0.03496229
5	-0.02295240	1.01926378	-1.17465123	0.03495011
6	-0.02307276	1.01992190	-1.14831568	0.03494536
7	-0.02297427	1.01940189	-1.17003037	0.03494040
8	-0.02305506	1.01984017	-1.15230734	0.03493808
9	-0.02298878	1.01948877	-1.16691829	0.03493597
10	-0.02304322	1.01978274	-1.15495732	0.03493486
.				
.				
.				
28	-0.02301940	1.01966026	-1.16023895	0.03493222
29	-0.02301808	1.01965320	-1.16052942	0.03493222
30	-0.02301917	1.01965901	-1.16029058	0.03493222
31	-0.02301828	1.01965423	-1.16048699	0.03493222

NOTE: CONVERGENCE CRITERION MET.

S T A T I S T I C A L   A N A L Y S I S   S Y S T E M   2

NON-LINEAR LEAST SQUARES SUMMARY STATISTICS   DEPENDENT VARIABLE Y

SOURCE	DF	SUM OF SQUARES	MEAN SQUARE
REGRESSION	3	26.34150543	8.78050181
RESIDUAL	27	0.03493222	0.00129379
UNCORRECTED TOTAL	30	26.37643764	
(CORRECTED TOTAL)	29	0.71895291	

PARAMETER	ESTIMATE	ASYMPTOTIC STD. ERROR	ASYMPTOTIC 95 % CONFIDENCE INTERVAL LOWER	UPPER
R1	-0.02301828	0.01315496	-0.05000981	0.00397326
R2	1.01965423	0.01009676	0.99893755	1.04037092
R3	-1.16048699	0.16302087	-1.49497559	-0.82599838

ASYMPTOTIC CORRELATION MATRIX OF THE PARAMETERS

	R1	R2	R3
R1	1.000000	-0.671463	-0.056283
R2	-0.671463	1.000000	0.392338
R3	-0.056283	0.392338	1.000000

Figure 9b. Illustration of likelihood ratio test computations with Example 1.

If

$$\frac{\partial}{\partial \theta'} f(x, \theta) = (\text{DER\_T1, DER\_T2, DER\_T3, DER\_T4})$$

then to compute

$$\frac{\partial}{\partial \rho'} f[x, g(\rho)] = (\text{DER.R1, DER.R2, DER.R3})$$

one codes

```
DER.R1 = DER_T1
DER.R2 = DER_T2
DER.R3 = DER_T3 + DER_T4 * (- T4**2) * (5*EXP(R3) +
         5*R3*EXP(R3))
```

where

$$T4 = \frac{1}{5\rho_3 e^{\rho_3}}$$

as shown in Figure 9b.

We have

$$\text{SSE}_{\text{reduced}} = 0.03493222 \qquad (\text{from Fig. } 9b)$$
$$\text{SSE}_{\text{full}} = 0.03049554 \qquad (\text{from Fig. } 5a)$$
$$L = \frac{(0.03493222 - 0.3049554)/1}{0.03049554/26}$$
$$= 3.783.$$

As $F^{-1}(.95; 1, 26) = 4.22$ one fails to reject the null hypothesis at the 5% level.

Reconsidering the third hypothesis

$$H: \theta_1 = 0 \text{ and } \theta_3 \theta_4 e^{\theta_3} = \tfrac{1}{5}$$

which may be rewritten as

$$H: \theta^0 = g(\rho) \text{ for some } \rho^0$$

with

$$g(\rho) = \begin{pmatrix} 0 \\ \rho_2 \\ \rho_3 \\ \dfrac{1}{5\rho_3 e^{\rho_3}} \end{pmatrix}$$

SAS Statements:

```
PROC NLIN DATA=EXAMPLE1 METHOD=GAUSS ITER=60 CONVERGENCE=1.0E-8;
PARMS R2=1.01965423 R3=-1.16048699;  R1=0;
T1=R1; T2=R2; T3=R3; T4=1/(5*R3*EXP(R3));
MODEL Y=T1*X1+T2*X2+T4*EXP(T3*X3);
DER_T1=X1; DER_T2=X2; DER_T3=T4*X3*EXP(T3*X3); DER_T4=EXP(T3*X3);
DER.R2=DER_T2; DER.R3=DER_T3+DER_T4*(-T4**2)*(5*EXP(R3)+5*R3*EXP(R3));
```

Output:

NON-LINEAR LEAST SQUARES ITERATIVE PHASE

DEPENDENT VARIABLE: Y   METHOD: GAUSS-NEWTON

ITERATION	R2	R3	RESIDUAL SS
0	1.01965423	-1.16048699	0.04287983
1	1.00779498	-1.17638081	0.03890362
2	1.00807441	-1.16332560	0.03890234
3	1.00784845	-1.17411590	0.03890127
4	1.00803764	-1.16523771	0.03890066
5	1.00788362	-1.17257272	0.03890018
6	1.00801199	-1.16653150	0.03889989
7	1.00790702	-1.17152084	0.03889967
8	1.00799423	-1.16740905	0.03889954
9	1.00792271	-1.17080393	0.03889944
10	1.00798200	-1.16800508	0.03889937
.			
.			
.			
19	1.00795079	-1.16949660	0.03889924
20	1.00795944	-1.16908756	0.03889923
21	1.00795231	-1.16942506	0.03889923
22	1.00795819	-1.16914663	0.03889923
23	1.00795334	-1.16937636	0.03889923
24	1.00795735	-1.16918683	0.03889923

NOTE: CONVERGENCE CRITERION MET.

NON-LINEAR LEAST SQUARES SUMMARY STATISTICS   DEPENDENT VARIABLE Y

SOURCE	DF	SUM OF SQUARES	MEAN SQUARE
REGRESSION	2	26.33753841	13.16876921
RESIDUAL	28	0.03889923	0.00138926
UNCORRECTED TOTAL	30	26.37643764	
(CORRECTED TOTAL)	29	0.71895291	

PARAMETER	ESTIMATE	ASYMPTOTIC STD. ERROR	ASYMPTOTIC 95 % CONFIDENCE INTERVAL LOWER	UPPER
R2	1.00795735	0.00769931	0.99218613	1.02372856
R3	-1.16918683	0.17039162	-1.51821559	-0.82015808

ASYMPTOTIC CORRELATION MATRIX OF THE PARAMETERS

	R2	R3
R2	1.000000	0.467769
R3	0.467769	1.000000

Figure 9c. Illustration of likelihood ratio computations with Example 1.

we have

$$SSE_{reduced} = 0.03889923 \quad \text{(from Fig. 9}c)$$
$$SSE_{full} = 0.03049554 \quad \text{(from Fig. 5}a)$$
$$L = \frac{(SSE_{reduced} - SSE_{full})/(p - r)}{(SSE_{full})/(n - p)}$$
$$= \frac{(0.03889923 - 0.03049554)/(4 - 2)}{(0.03049554)/(30 - 4)}$$
$$= 3.582.$$

Since $F^{-1}(.95; 2, 26) = 3.37$, one rejects the null hypothesis at the 5% level.

□

It is not always easy to convert a parametric restriction $h(\theta) = 0$ to a functional dependence $\theta = g(\rho)$ analytically. However, all that is needed is the value of θ for given ρ and the value of $(\partial/\partial\rho')g(\rho)$ for given ρ. This allows substitution of numerical methods for analytical methods in the determination of $g(\rho)$. We illustrate with the next example.

EXAMPLE 2 (Continued). Recall that the amount of substance in compartment B at time x is given by the response function

$$f(x, \theta) = \frac{\theta_1(e^{-x\theta_2} - e^{-x\theta_1})}{\theta_1 - \theta_2}.$$

By differentiating with respect to x and setting the derivative to zero one has that the time at which the maximum amount of substance present in compartment B is

$$\hat{x} = \frac{\ln \theta_1 - \ln \theta_2}{\theta_1 - \theta_2}.$$

The unconstrained fit of this model is shown in Figure 10a. Suppose that we want to test

$$H: \hat{x} = 1 \quad \text{against} \quad A: \hat{x} \neq 1.$$

This requires that

$$h(\theta) = \frac{\ln \theta_1 - \ln \theta_2}{\theta_1 - \theta_2} - 1$$

SAS Statements:

```
PROC NLIN DATA=EG2B METHOD=GAUSS ITER=50 CONVERGENCE=1.E-10;
PARMS T1=1.4 T2=.4;
MODEL Y=T1*(EXP(-T2*X)-EXP(-T1*X))/(T1-T2);
DER.T1=-T2*(EXP(-T2*X)-EXP(-T1*X))/(T1-T2)**2+T1*X*EXP(-T1*X)/(T1-T2);
DER.T2=T1*(EXP(-T2*X)-EXP(-T1*X))/(T1-T2)**2-T1*X*EXP(-T2*X)/(T1-T2);
```

Output:

S T A T I S T I C A L   A N A L Y S I S   S Y S T E M   1

NON-LINEAR LEAST SQUARES ITERATIVE PHASE

DEPENDENT VARIABLE: Y   METHOD: GAUSS-NEWTON

ITERATION	T1	T2	RESIDUAL SS
0	1.40000000	0.40000000	0.00567248
1	1.37373983	0.40266678	0.00545775
2	1.37396974	0.40265518	0.00545774
3	1.37396966	0.40265518	0.00545774

NOTE: CONVERGENCE CRITERION MET.

S T A T I S T I C A L   A N A L Y S I S   S Y S T E M   2

NON-LINEAR LEAST SQUARES SUMMARY STATISTICS   DEPENDENT VARIABLE Y

SOURCE	DF	SUM OF SQUARES	MEAN SQUARE
REGRESSION	2	2.68129496	1.34064748
RESIDUAL	10	0.00545774	0.00054577
UNCORRECTED TOTAL	12	2.68675270	
(CORRECTED TOTAL)	11	0.21359486	

PARAMETER	ESTIMATE	ASYMPTOTIC STD. ERROR	ASYMPTOTIC 95 % CONFIDENCE INTERVAL	
			LOWER	UPPER
T1	1.37396966	0.04864622	1.26557844	1.48236088
T2	0.40265518	0.01324390	0.37314574	0.43216461

ASYMPTOTIC CORRELATION MATRIX OF THE PARAMETERS

	T1	T2
T1	1.000000	0.236174
T2	0.236174	1.000000

Figure 10 a. Illustration of likelihood ratio test computations with Example 2.

be converted to a functional dependence if one is to be able to use unconstrained optimization methods. To do this numerically, set $\theta_2 = \rho$. Then the problem is to solve the equation

$$\theta_1 = \ln \theta_1 + \rho - \ln \rho$$

for θ_1. Stated differently, we are trying to find a fixed point of the equation

$$z = \ln z + const.$$

But $\ln z + const$ is a contraction mapping for $z > 1$—the derivative with respect to z is less than one—so that a fixed point can be found by successive substitution:

$$z_1 = \ln z_0 + \text{const}$$
$$z_2 = \ln z_1 + \text{const}$$
$$\vdots$$
$$z_{i+1} = \ln z_i + \text{const}$$
$$\vdots$$

This sequence $\{z_{i+1}\}$ will converge to the fixed point.

To compute $(\partial/\partial\rho)g(\rho)$ we apply the implicit function theorem to

$$\theta_1(\rho) - \ln[\theta_1(\rho)] = \rho - \ln \rho.$$

We have

$$\frac{\partial}{\partial\theta_1}\{\theta_1(\rho) - \ln[\theta_1(\rho)]\}\frac{\partial}{\partial\rho}\theta_1(\rho) = \frac{\partial}{\partial\rho}(\rho - \ln \rho)$$

or

$$\frac{\partial}{\partial\rho}\theta_1(\rho) = \frac{1 - 1/\rho}{1 - 1/\theta_1(\rho)}.$$

Then the Jacobian of $\theta = g(\rho)$ is

$$\left(\frac{\partial}{\partial\rho'}\right) = \begin{pmatrix} \dfrac{1 - 1/\rho}{1 - 1/\theta_1(\rho)} \\ 1 \end{pmatrix}$$

and

$$
\frac{\partial}{\partial \rho} f[x, g(\rho)] = \left[\frac{\left(\frac{\partial}{\partial \theta_1} f(x, \theta) \right)\left(1 - \frac{1}{\rho} \right)}{1 - \frac{1}{\theta_1}} + \frac{\partial}{\partial \theta_2} f(x, \theta_2) \right]_{\theta = g(\rho)}.
$$

SAS Statements:

```
PROC NLIN DATA=EG2B METHOD=GAUSS ITER=50 CONVERGENCE=1.E-10;
PARMS RHO=.40265518;
T2=RHO;
Z1=1.4; Z2=0; C=T2-LOG(T2);
L1: IF ABS(Z1-Z2)>1.E-13 THEN DO; Z2=Z1; Z1=LOG(Z1)+C; GO TO L1; END;
T1=Z1;
NU2=T1*(EXP(-T2*X)-EXP(-T1*X))/(T1-T2);
DER_T1=-T2*(EXP(-T2*X)-EXP(-T1*X))/(T1-T2)**2+T1*X*EXP(-T1*X)/(T1-T2);
DER_T2=T1*(EXP(-T2*X)-EXP(-T1*X))/(T1-T2)**2-T1*X*EXP(-T2*X)/(T1-T2);
DER_RHO=DER_T1*(1-1/T2)/(1-1/T1)+DER_T2;
MODEL Y=NU2; DER.RHO=DER_RHO;
```

Output:

S T A T I S T I C A L   A N A L Y S I S   S Y S T E M   1

NON-LINEAR LEAST SQUARES ITERATIVE PHASE

DEPENDENT VARIABLE: Y   METHOD: GAUSS-NEWTON

ITERATION	RHO	RESIDUAL SS
0	0.40265518	0.07004386
1	0.46811176	0.04654328
2	0.47688375	0.04621215
3	0.47750162	0.04621056
4	0.47754034	0.04621055
5	0.47754274	0.04621055
6	0.47754289	0.04621055

NOTE: CONVERGENCE CRITERION MET.

S T A T I S T I C A L   A N A L Y S I S   S Y S T E M   2

NON-LINEAR LEAST SQUARES SUMMARY STATISTICS   DEPENDENT VARIABLE Y

SOURCE	DF	SUM OF SQUARES	MEAN SQUARE
REGRESSION	1	2.64054214	2.64054214
RESIDUAL	11	0.04621055	0.00420096
UNCORRECTED TOTAL	12	2.68675270	
(CORRECTED TOTAL)	11	0.21359486	

PARAMETER	ESTIMATE	ASYMPTOTIC STD. ERROR	ASYMPTOTIC 95 % CONFIDENCE INTERVAL LOWER	UPPER
RHO	0.47754289	0.03274044	0.40548138	0.54960439

ASYMPTOTIC CORRELATION MATRIX OF THE PARAMETERS

RHO

	RHO
RHO	1.000000

Figure 10b. Illustration of likelihood ratio test computations with Example 2.

These ideas are illustrated in Figure 10b. We have

$$\text{SSE}_{\text{full}} = 0.00545774 \qquad (\text{from Fig. } 10a)$$
$$\text{SSE}_{\text{reduced}} = 0.04621055 \qquad (\text{from Fig. } 10b)$$
$$= L \frac{(\text{SSE}_{\text{reduced}} - \text{SSE}_{\text{full}})/q}{(0.00545774)/(12 - 2)}$$
$$= 74.670.$$

As $F^{-1}(.95; 1, 10) = 4.96$, one rejects H. $\qquad\qquad\qquad\qquad$ □

Now let us turn our attention to the computation of the power of the likelihood ratio test. That is, for the data that follow the model

$$y_t = f(x_t, \theta^0) + e_t$$
$$e_t \text{ iid } N(0, \sigma^2)$$
$$t = 1, 2, \ldots, n$$

we should like to compute

$$P(L > F_\alpha \mid \theta^0, \sigma^2, n)$$

the probability that the likelihood ratio test rejects at level α given θ^0, σ^2, and n, where $F_\alpha = F^{-1}(1 - \alpha; q, n - p)$. To do this, note that the test that rejects when

$$L = \frac{(\text{SSE}_{\text{reduced}} - \text{SSE}_{\text{full}})/q}{(\text{SSE}_{\text{full}})/(n - p)} > F_\alpha$$

is equivalent to the test that rejects when

$$\frac{\text{SSE}_{\text{reduced}}}{\text{SSE}_{\text{full}}} > c_\alpha$$

where

$$c_\alpha = 1 + \frac{qF_\alpha}{n - p}.$$

In Chapter 4 we shall show that

$$\frac{n}{\text{SSE}_{\text{full}}} = \frac{n}{e'P_F^\perp e} + o_p\left(\frac{1}{n}\right)$$

where

$$P_F^\perp = I - F(F'F)^{-1}F';$$

recall that $F = (\partial/\partial\theta')f(\theta^0)$. Then it remains to obtain an approximation to $(\text{SSE}_{\text{reduced}})/n$ in order to approximate $(\text{SSE}_{\text{reduced}})/(\text{SSE}_{\text{full}})$. To this end, let

$$\theta_n^* = g(\rho_n^0)$$

where

$$\rho_n^0 \quad \text{minimizes} \quad \sum_{t=1}^{n} \left\{ f(x_t, \theta^0) - f[x_t, g(\rho)] \right\}^2.$$

Recall that $g(\rho)$ is the mapping from $\mathbb{R}^r$ into $\mathbb{R}^p$ that describes the null hypothesis—$H: \theta^0 = g(\rho)$ for some ρ^0; $r = p - q$. The point θ_n^* may be interpreted as that point which is being estimated by the constrained estimator $\tilde{\theta}_n$ in the sense that $\sqrt{n}(\tilde{\theta}_n - \theta_n^*)$ converges in distribution to the multivariate normal distribution; see Chapter 3 for details. Under this interpretation,

$$\delta = f(\theta^0) - f(\theta_n^*)$$

may be interpreted as the prediction bias. We shall show later (Chapter 4) that what one's intuition would suggest is true:

$$\frac{\text{SSE}_{\text{reduced}}}{n} = \frac{(e + \delta)'P_{FG}^\perp(e + \delta)}{n} + o_p\left(\frac{1}{n}\right)$$

where

$$P_{FG}^\perp = I - FG(G'F'FG)^{-1}G'F'$$

$$F = \frac{\partial}{\partial\theta'}f(\theta^0)$$

$$G = \frac{\partial}{\partial\rho'}g(\rho_n^0).$$

It follows from the characterizations of the residual sum of squares for the full and reduced models that

$$\frac{\text{SSE}_{\text{reduced}}}{\text{SSE}_{\text{full}}} = X + o_p\left(\frac{1}{n}\right)$$

where

$$X = \frac{(e + \delta)' P_{FG}^{\perp} (e + \delta)}{e' P_F^{\perp} e}.$$

The idea, then, is to approximate the probability $P(L > F_\alpha \mid \theta^0, \sigma^2, n)$ by the probability $P(X > c_\alpha \mid \theta^0, \sigma^2, n)$. The distribution function of the random variable X is for $x > 1$ (Problem 4)

$$H(x: \nu_1, \nu_2, \lambda_1, \lambda_2)$$

$$= 1 - \int_0^\infty G\left(\frac{t}{x - 1} + \frac{2x\lambda_2}{(x - 1)^2} ; \nu_2, \frac{\lambda_2}{(x - 1)^2} \right) g(t; \nu_1, \lambda_1) \, dt$$

where $g(t; \nu, \lambda)$ denotes the noncentral chi-square density function with ν degrees of freedom and noncentrality parameter λ, and $G(t; \nu, \lambda)$ denotes the corresponding distribution function (Section 7). The two degrees of freedom entries are

$$\nu_1 = q = p - r$$
$$\nu_2 = n - p$$

and the noncentrality parameters are

$$\lambda_1 = \frac{\delta'(P_F - P_{FG})\delta}{2\sigma^2}$$

$$\lambda_2 = \frac{\delta' P_F^{\perp} \delta}{2\sigma^2}$$

where $P_F = F(F'F)^{-1}F'$, $P_{FG} = FG(G'F'FG)^{-1}G'F'$, and $P_F^{\perp} = I - P_F$. This distribution is partly tabulated in Table 6. Let us illustrate the computations necessary to use these tables and check the accuracy of the approximation of $P(L > F_\alpha)$ by $P(X > c_\alpha)$ by Monte Carlo simulation using Example 1.

EXAMPLE 1   (Continued).   Recalling that

$$f(x, \theta) = \theta_1 x_1 + \theta_2 x_2 + \theta_4 e^{\theta_3 x_3}$$

let us approximate the probability that the likelihood ratio test rejects the following three hypotheses at the 5% level when the true values of the parameters are

$$\theta^0 = (.03, 1, -1.4, -.5)'$$
$$\sigma^2 = .001.$$

Table 6. Power of the Likelihood Ratio Test at the 5% Level.

λ_2	$\lambda_1 = 0$	.5	1	2	3	4	5	6	8	10	12
					$(a)\ \nu_1 = 1,\ \nu_2 = 10$						
0.0	.050	.148	.249	.440	.599	.722	.813	.876	.949	.980	.992
.0001	.050	.148	.249	.440	.599	.722	.813	.876	.949	.980	.992
.001	.050	.148	.249	.440	.599	.723	.813	.876	.949	.980	.992
.01	.051	.150	.251	.442	.601	.724	.814	.877	.949	.980	.992
.1	.063	.168	.272	.462	.617	.735	.821	.882	.951	.980	.992
					$(b)\ \nu_1 = 1,\ \nu_2 = 20$						
0.0	.050	.159	.271	.478	.645	.768	.853	.909	.967	.989	.996
.0001	.050	.159	.271	.478	.645	.768	.853	.909	.967	.989	.996
.001	.050	.159	.271	.478	.645	.768	.853	.909	.967	.989	.996
.01	.051	.161	.273	.480	.647	.769	.853	.909	.967	.989	.996
.1	.065	.181	.296	.501	.663	.780	.860	.913	.968	.989	.996
					$(c)\ \nu_1 = 1,\ \nu_2 = 30$						
0.0	.050	.163	.278	.490	.659	.781	.864	.917	.972	.991	.997
.0001	.050	.163	.278	.490	.659	.781	.864	.917	.972	.991	.997
.001	.050	.163	.278	.491	.659	.781	.864	.917	.972	.991	.997
.01	.051	.165	.280	.493	.661	.782	.864	.918	.972	.991	.997
.1	.065	.185	.303	.514	.676	.792	.871	.921	.973	.991	.997
					$(d)\ \nu_1 = 2,\ \nu_2 = 10$						
0.0	.050	.111	.178	.318	.454	.575	.677	.759	.873	.936	.969
.0001	.050	.111	.178	.318	.454	.575	.677	.759	.873	.936	.969
.001	.050	.111	.178	.318	.454	.575	.677	.759	.873	.936	.969
.01	.051	.112	.179	.320	.456	.576	.678	.760	.873	.936	.969
.1	.058	.122	.192	.334	.469	.588	.688	.767	.877	.938	.970
					$(e)\ \nu_1 = 2,\ \nu_2 = 20$						
0.0	.050	.121	.199	.364	.517	.647	.749	.827	.922	.968	.987
.0001	.050	.121	.199	.364	.517	.647	.749	.827	.922	.968	.987
.001	.050	.121	.200	.364	.517	.647	.750	.827	.922	.968	.987
.01	.051	.122	.201	.365	.519	.648	.750	.828	.923	.968	.987
.1	.060	.135	.216	.382	.534	.660	.759	.834	.925	.969	.987
					$(f)\ \nu_1 = 2,\ \nu_2 = 30$						
0.0	.050	.124	.208	.381	.539	.671	.773	.847	.936	.975	.991
.0001	.050	.124	.208	.381	.539	.671	.773	.847	.936	.975	.991
.001	.050	.125	.208	.381	.539	.671	.773	.847	.936	.975	.991
.01	.051	.126	.210	.382	.541	.672	.774	.848	.936	.975	.991
.1	.060	.139	.226	.400	.556	.684	.782	.854	.938	.976	.991

Table 6. (*Continued*)

λ_2	$\lambda_1 = 0$	.5	1	2	3	4	5	6	8	10	12
				(g) $\nu_1 = 3, \nu_2 = 10$							
0.0	.050	.094	.145	.255	.368	.477	.576	.662	.794	.881	.933
.0001	.050	.094	.145	.255	.368	.477	.576	.662	.794	.881	.933
.001	.050	.095	.145	.255	.368	.477	.576	.662	.794	.881	.933
.01	.051	.095	.146	.256	.369	.478	.577	.662	.795	.881	.934
.1	.056	.103	.155	.267	.381	.489	.586	.670	.800	.884	.935
				(h) $\nu_1 = 3, \nu_2 = 20$							
0.0	.050	.104	.165	.300	.436	.561	.668	.755	.874	.940	.973
.0001	.050	.104	.165	.300	.436	.561	.668	.755	.874	.940	.973
.001	.050	.104	.165	.300	.437	.561	.668	.755	.874	.940	.973
.01	.051	.105	.166	.302	.438	.562	.669	.755	.875	.940	.973
.1	.057	.114	.178	.316	.452	.574	.679	.763	.878	.942	.973
				(i) $\nu_1 = 3, \nu_2 = 30$							
0.0	.050	.107	.173	.318	.462	.591	.699	.785	.897	.954	.981
.0001	.050	.107	.173	.318	.462	.591	.699	.785	.897	.954	.981
.001	.050	.107	.173	.318	.462	.592	.699	.785	.897	.954	.981
.01	.051	.108	.175	.320	.464	.593	.700	.785	.897	.954	.981
.1	.058	.119	.187	.335	.478	.605	.710	.792	.900	.956	.981

Source: Gallant (1975a).

The three null hypotheses are

$$H_1 : \theta_1 = 0$$
$$H_2 : \theta_3\theta_4 e^{\theta_3} = \tfrac{1}{5}$$
$$H_3 : \theta_3 = 0 \text{ and } \theta_3\theta_4 e^{\theta_3} = \tfrac{1}{5}.$$

The computational chore is to compute for each hypothesis

$$\rho_n^0 \quad \text{minimizing} \quad \sum_{t=1}^{n} \left\{ f(x_t, \theta^0) - f[x_t, g(\rho)] \right\}^2$$
$$\delta = f(\theta^0) - f(\theta_n^*), \qquad \theta_n^* = g(\rho_n^0)$$
$$\delta' P_F \delta, \qquad \delta' P_{FG} \delta, \quad \text{and} \quad \delta'\delta.$$

With these, the noncentrality parameters

$$\lambda_1 = \frac{\delta' P_F \delta - \delta' P_{FG} \delta}{2\sigma^2}$$

$$\lambda_2 = \frac{\delta' \delta - \delta' P_F \delta}{2\sigma^2}$$

are easily computed. As usual, there are a variety of strategies that one might employ.

To compute δ, the easiest approach is to notice that minimizing

$$\sum_{t=1}^{n} \left\{ f(x_t, \theta^0) - f[x_t, g(\rho)] \right\}^2$$

is no different than minimizing

$$\sum_{t=1}^{n} \left\{ y_t - f[x_t, g(\rho)] \right\}^2.$$

One simply replaces y_t by $f(x_t, \theta^0)$ and uses the modified Gauss-Newton method, the Levenberg-Marquardt method, or whatever.

To compute $\delta' P_F \delta$ one can either proceed directly using a programming language such as PROX MATRIX or make the following observation. If one regresses δ on F with no intercept term using a linear regression procedure, then the analysis of variance table printed by the program will have the following entries:

Source	d.f.	Sum of Squares
Regression	p	$\delta' F(F'F)^{-1} F' \delta$
Error	$n - p$	$\delta' \delta - \delta' F(F'F)^{-1} F' \delta$
Total	n	$\delta' \delta$

One can just read off

$$\delta' P_F \delta = \delta' F(F'F)^{-1} F' \delta$$

from the analysis of variance table. Similarly for a regression of δ on FG.

Figures 11a, 11b, and 11c illustrate these ideas for the hypotheses H_1, H_2, and H_3.

For the first hypothesis we have

$$\delta'\delta = 0.006668583 \qquad \text{(from Fig. 11}a\text{)}$$
$$\delta'P_F\delta = 0.006668583 \qquad \text{(from Fig. 11}a\text{)}$$
$$\delta'P_{FG}\delta = 3.25 \times 10^{-9} \qquad \text{(from Fig. 11}a\text{)}$$

whence

$$\lambda_1 = \frac{\delta'P_F\delta - \delta'P_{FG}\delta}{2\sigma^2}$$
$$= \frac{0.006668583 - 3.25 \times 10^{-9}}{2 \times 0.001}$$
$$= 3.3343$$
$$\lambda_2 = \frac{\delta'\delta - \delta'P_F\delta}{2\sigma^2}$$
$$= \frac{0.006668583 - 0.006668583}{2 \times 0.001}$$
$$= 0$$
$$c_\alpha = 1 + \frac{qF_\alpha}{n - p}$$
$$= 1 + \frac{(1)(4.22)}{26}$$
$$= 1.1623.$$

Computing $1 - H(1.1623; 1, 26, \lambda_1, \lambda_2)$ by interpolating from Table 6, we obtain

$$P(X > c_\alpha) = .700$$

as an approximation to $P(L > F_\alpha)$. Later we shall show that tables of the noncentral F will usually be accurate enough that there is no need for special tables.

For the second hypothesis we have

$$\delta'\delta = 0.01321589 \qquad \text{(from Fig. 11}b\text{)}$$
$$\delta'P_F\delta = 0.013215 \qquad \text{(from Fig. 11}b\text{)}$$
$$\delta'\delta - \delta'P_F\delta = 0.00000116542 \qquad \text{(from Fig. 11}b\text{)}$$
$$\delta'P_{FG}\delta = 0.0001894405 \qquad \text{(from Fig. 11}b\text{)}$$

SAS Statements:

```
DATA WORK01;  SET EXAMPLE1;  T1=.03;  T2=1;  T3=-1.4;  T4=-.5;
YDUMMY=T1*X1+T2*X2+T4*EXP(T3*X3);
F1=X1;  F2=X2;  F3=T4*X3*EXP(T3*X3);  F4=EXP(T3*X3);
DROP T1 T2 T3 T4;
PROC NLIN DATA=WORK01 METHOD=GAUSS ITER=50 CONVERGENCE=1.0E-13;
PARMS T2=1 T3=-1.4 T4=-.5;  T1=0;
MODEL YDUMMY=T1*X1+T2*X2+T4*EXP(T3*X3);
DER.T2=X2;  DER.T3=T4*X3*EXP(T3*X3);  DER.T4=EXP(T3*X3);
```

Output:

NON-LINEAR LEAST SQUARES ITERATIVE PHASE

DEPENDENT VARIABLE: YDUMMY   METHOD: GAUSS-NEWTON

ITERATION	T2	T3	T4	RESIDUAL SS
0	1.00000000	-1.40000000	-0.50000000	0.01350000
1	1.01422090	-1.39717572	-0.49393589	0.00666859
2	1.01422435	-1.39683401	-0.49391057	0.00666858
3	1.01422476	-1.39679638	-0.49390747	0.00666858
4	1.01422481	-1.39679223	-0.49390713	0.00666858
5	1.01422481	-1.39679178	-0.49390709	0.00666858
6	1.01422481	-1.39679173	-0.49390708	0.00666858

NOTE: CONVERGENCE CRITERION MET.

SAS Statements:

```
DATA WORK02;  SET WORK01;
T1=0;  T2=1.01422481;  T3=-1.39679173;  T4=-0.49390708;
DELTA=YDUMMY-(T1*X1+T2*X2+T4*EXP(T3*X3));
FG1=F2;  FG2=F3;  FG3=F4;  DROP T1 T2 T3 T4;
PROC REG DATA=WORK02;  MODEL DELTA=F1 F2 F3 F4 / NOINT;
PROC REG DATA=WORK02;  MODEL DELTA=FG1 FG2 FG3 / NOINT;
```

Output:

DEP VARIABLE: DELTA

SOURCE	DF	SUM OF SQUARES	MEAN SQUARE	F VALUE	PROB>F
MODEL	4	0.006668583	0.001667146	999999.990	0.0001
ERROR	26	2.89364E-13	1.11294E-14		
U TOTAL	30	0.006668583			

DEP VARIABLE: DELTA

SOURCE	DF	SUM OF SQUARES	MEAN SQUARE	F VALUE	PROB>F
MODEL	3	3.25099E-09	1.08366E-09	0.000	1.0000
ERROR	27	0.00666858	0.0002469844		
U TOTAL	30	0.006668583			

Figure 11a. Illustration of likelihood ratio test power computations with Example 1.

SAS Statements:

```
DATA WORK01;   SET EXAMPLE1;   T1=.03;   T2=1;   T3=-1.4;   T4=-.5;
YDUMMY=T1*X1+T2*X2+T4*EXP(T3*X3);
F1=X1;   F2=X2;   F3=T4*X3*EXP(T3*X3);   F4=EXP(T3*X3);
DROP T1 T2 T3 T4;
PROC NLIN DATA=WORK01 METHOD=GAUSS ITER=50 CONVERGENCE=1.0E-13;
PARMS R1=.03 R2=1 R3=-1.4;
T1=R1; T2=R2; T3=R3; T4=1/(5*R3*EXP(R3));
MODEL YDUMMY=T1*X1+T2*X2+T4*EXP(T3*X3);
DER_T1=X1; DER_T2=X2; DER_T3=T4*X3*EXP(T3*X3); DER_T4=EXP(T3*X3);
DER_R1=DER_T1; DER_R2=DER_T2;
DER.R3=DER_T3+DER_T4*(-T4**2)*(5*EXP(R3)+5*R3*EXP(R3));
```

Output:

STATISTICAL   ANALYSIS   SYSTEM   1

NON-LINEAR LEAST SQUARES ITERATIVE PHASE

DEPENDENT VARIABLE: YDUMMY   METHOD: GAUSS-NEWTON

ITERATION	R1	R2	R3	RESIDUAL SS
0	0.03000000	1.00000000	-1.40000000	0.01867856
1	0.03363136	1.01008796	-1.12533963	0.01588546
2	0.03440842	1.00692167	-1.28648656	0.01344947
.				
.				
.				
14	0.03433974	1.00978675	-1.27330943	0.01321589

NOTE: CONVERGENCE CRITERION MET.

SAS Statements:

```
DATA WORK02;   SET WORK01;
R1=0.03433974;   R2=1.00978675;   R3=-1.27330943;
T1=R1;   T2=R2;   T3=R3;   T4=1/(5*R3*EXP(R3));
DELTA=YDUMMY-(T1*X1+T2*X2+T4*EXP(T3*X3));
FG1=F1;   FG2=F2;   FG3=F3+F4*(-T4**2)*(5*EXP(R3)+5*R3*EXP(R3));
DROP T1 T2 T3 T4;
PROC REG DATA=WORK02;   MODEL DELTA=F1 F2 F3 F4 / NOINT;
PROC REG DATA=WORK02;   MODEL DELTA=FG1 FG2 FG3 / NOINT;
```

Output:

STATISTICAL   ANALYSIS   SYSTEM   1

DEP VARIABLE: DELTA

SOURCE	DF	SUM OF SQUARES	MEAN SQUARE	F VALUE	PROB>F
MODEL	4	0.013215	0.003303681	73703.561	0.0001
ERROR	26	.00000116542	4.48239E-08		
U TOTAL	30	0.013216			

STATISTICAL   ANALYSIS   SYSTEM   2

DEP VARIABLE: DELTA

SOURCE	DF	SUM OF SQUARES	MEAN SQUARE	F VALUE	PROB>F
MODEL	3	0.0001894405	.00006314682	0.131	0.9409
ERROR	27	0.013026	0.0004824611		
U TOTAL	30	0.013216			

Figure 11b. Illustration of likelihood ratio test power computations with Example 1.

SAS Statements:

```
DATA WORK01;  SET EXAMPLE1;  T1=.03;  T2=1;  T3=-1.4;  T4=-.5;
YDUMMY=T1*X1+T2*X2+T4*EXP(T3*X3);
F1=X1;  F2=X2;  F3=T4*X3*EXP(T3*X3);  F4=EXP(T3*X3);
DROP T1 T2 T3 T4;
PROC NLIN DATA=WORK01 METHOD=GAUSS ITER=50 CONVERGENCE=1.0E-13;
PARMS R2=1 R3=-1.4;  R1=0;
T1=R1;  T2=R2;  T3=R3;  T4=1/(5*R3*EXP(R3));
MODEL YDUMMY=T1*X1+T2*X2+T4*EXP(T3*X3);
DER_T1=X1;  DER_T2=X2;  DER_T3=T4*X3*EXP(T3*X3);  DER_T4=EXP(T3*X3);
DER.R2=DER_T2;  DER.R3=DER_T3+DER_T4*(-T4**2)*(5*EXP(R3)+5*R3*EXP(R3));
```

Output:

NON-LINEAR LEAST SQUARES ITERATIVE PHASE

DEPENDENT VARIABLE: YDUMMY   METHOD: GAUSS-NEWTON

ITERATION	R2	R3	RESIDUAL SS
0	1.00000000	-1.40000000	0.04431091
1	1.02698331	-1.10041642	0.02539361
2	1.02383184	-1.26840577	0.02235554
.			
.			
.			
13	1.02709006	-1.26246439	0.02204771

NOTE: CONVERGENCE CRITERION MET.

SAS Statements:

```
DATA WORK02;  SET WORK01;
R1=0;  R2=1.02709006;  R3=-1.26246439;
T1=R1;  T2=R2;  T3=R3;  T4=1/(5*R3*EXP(R3));
DELTA=YDUMMY-(T1*X1+T2*X2+T4*EXP(T3*X3));
FG1=F2;  FG2=F3+F4*(-T4**2)*(5*EXP(R3)+5*R3*EXP(R3));
DROP T1 T2 T3 T4;
PROC REG DATA=WORK02;  MODEL DELTA=F1 F2 F3 F4 / NOINT;
PROC REG DATA=WORK02;  MODEL DELTA=FG1 FG2 / NOINT;
```

Output:

DEP VARIABLE: DELTA

SOURCE	DF	SUM OF SQUARES	MEAN SQUARE	F VALUE	PROB>F
MODEL	4	0.022046	0.005511515	86947.729	0.0001
ERROR	26	.00000164811	6.33888E-08		
U TOTAL	30	0.022048			

DEP VARIABLE: DELTA

SOURCE	DF	SUM OF SQUARES	MEAN SQUARE	F VALUE	PROB>F
MODEL	2	0.0001252535	.00006262677	0.080	0.9233
ERROR	28	0.021922	0.0007829449		
U TOTAL	30	0.022048			

Figure 11c. Illustration of likelihood ratio test power computations with Example 1.

whence

$$\lambda_1 = \frac{\delta' P_F \delta - \delta' P_{FG} \delta}{2\sigma^2}$$

$$= \frac{0.013215 - 0.0001894405}{(2 \times .001)}$$

$$= 6.5128$$

$$\lambda_2 = \frac{\delta' \delta - \delta' P_F \delta}{2\sigma^2}$$

$$= \frac{0.00000116542}{2 \times .001}$$

$$= 0.0005827$$

$$c_\alpha = 1 + \frac{q F_\alpha}{n - p}$$

$$= 1 + \frac{(1)(4.22)}{26}$$

$$= 1.1623.$$

Computing $1 - H(1.1623; 1, 26, \lambda_1, \lambda_2)$ as above, we obtain

$$P(X > c_\alpha) = .935$$

as an approximation to $P(L > F_\alpha)$.

For the third hypothesis we have

$$\delta' \delta = 0.02204771 \qquad \text{(from Fig. 11}c)$$
$$\delta' P_F \delta = 0.022046 \qquad \text{(from Fig. 11}c)$$
$$\delta' \delta - \delta' P_F \delta = 0.00000164811 \qquad \text{(from Fig. 11}c)$$
$$\delta' P_{FG} \delta = 0.0001252535 \qquad \text{(from Fig. 11}c)$$

whence

$$\lambda_1 = \frac{\delta' P_F \delta - \delta' P_{FG} \delta}{2\sigma^2}$$

$$= \frac{0.022046 - 0.0001252535}{2 \times .001}$$

$$= 10.9604$$

$$\lambda_2 = \frac{\delta'\delta - \delta'P_F\delta}{2\sigma^2}$$

$$= \frac{0.00000164811}{2 \times .001}$$

$$= 0.0008241$$

$$c_\alpha = \frac{1 + qF_\alpha}{n - p}$$

$$= 1 + \frac{(2)(3.37)}{26}$$

$$= 1.2592.$$

Computing $1 - H(1.2592; 2, 26, \lambda_1, \lambda_2)$ as above, we obtain

$$P(X > c_\alpha) = .983.$$

Once again we ask: How accurate are these approximations? Table 7 indicates that the approximations are quite good, and later we shall see several more examples where this is the case. In general, Monte Carlo evidence suggests that the approximation $P(L > F_\alpha) \doteq P(X > c_\alpha)$ is very accurate over a wide range of circumstances. Table 7 was constructed exactly as Table 5. □

In most applications λ_2 will be quite small relative to λ_1, as in the three cases in the last example. This being the case, one sees by scanning the entries in Table 6 that the value of $P(X > c_\alpha)$ computed with $\lambda_2 = 0$ would be adequate to approximate $P(L > F_\alpha)$. If $\lambda_2 = 0$ then (Problem 5)

$$H(c_\alpha; \nu_1, \nu_2, \lambda_1, 0) = F'(F_\alpha; \nu_1, \nu_2, \lambda_1)$$

with

$$c_\alpha = 1 + \frac{\nu_1 F_\alpha}{\nu_2}.$$

Recall that $F'(x; \nu_1, \nu_2, \lambda)$ denotes the noncentral F-distribution with ν_1 numerator degrees of freedom, ν_2 denominator degrees of freedom, and noncentrality parameter λ (Section 7). Stated differently, the first rows of parts a through i of Table 6 are a tabulation of the power of the F-test.

Table 7. Monte Carlo Power Estimates for the Likelihood Ratio Test.

Parameters[a]				$H_0: \theta_1 = 0$ against $H_1: \theta_1 \neq 0$ Monte Carlo					$H_0: \theta_3 = -1$ against $H_1: \theta_3 \neq -1$ Monte Carlo		
θ_1	θ_3	λ_1	λ_2	$P[X > c_\alpha]$	$P[L > F_\alpha]$	Std. Err.	λ_1	λ_2	$P[X > c_\alpha]$	$P[L > F_\alpha]$	Std. Err.
0.0	−1.0	0.0	0.0	.050	.050	.003	0.0	0.0	.050	.052	.003
0.008	−1.1	0.2353	0.0000	.101	.094	.004	0.2423	0.0006	.103	.110	.004
0.015	−1.2	0.8307	0.0000	.237	.231	.006	0.8526	0.0078	.244	.248	.006
0.030	−1.4	3.3343	0.0000	.700	.687	.006	2.6928	0.0728	.622	.627	.007

Source: Gallant (1975d).

[a] $\theta_2 = 1$, $\theta_4 = -.5$, $\sigma^2 = .001$.

Thus, in most applications, an adequate approximation to the power of the likelihood ratio test is

$$P(L > F_\alpha) \doteq 1 - F'(F_\alpha; \nu_1, \nu_2, \lambda_1).$$

The next example explores the adequacy of this approximation.

EXAMPLE 3. Table 8 compares the probability $P(X > c_\alpha)$ with Monte Carlo estimates of the probability of $P(L > F_\alpha)$ for the model

$$y_t = \theta_1 e^{\theta_2 x_t} + e_t.$$

Thirty inputs $\{x_t\}_{t=1}^{30}$ were chosen by replicating the points $0(.1).7$ three times and the points $.8(.1)1$ twice. The null hypothesis is $H: \theta^0 = (\frac{1}{2}, \frac{1}{2})$. For the null hypothesis and selected departures from the null hypothesis, 5000 random samples of size 30 from the normal distribution were generated according to the model with σ^2 taken as .04. The point estimate $\hat{p}$ of $P(L > F_\alpha)$ is, of course, the ratio of the number of times L exceeded F_α to 5000. The variance of $\hat{p}$ was estimated by $\mathrm{Var}(\hat{p}) = P(X > c_\alpha)P(X \le c_\alpha)/5000$. For complete details see Gallant (1975a).

To comment on the choice of the values of $\theta^0 \ne (\frac{1}{2}, \frac{1}{2})$ shown in Table 8, the ratio λ_2/λ_1 is minimized $(= 0)$ for $\theta^0 \ne (\frac{1}{2}, \frac{1}{2})$ of the form $(\theta_1, \frac{1}{2})$ and is maximized for θ^0 of the form $(\frac{1}{2}, \frac{1}{2}) \pm r(\cos(5\pi/8), \sin(5\pi/8))$. Three points were chosen to be of the first form, and two of the latter form. Further, two sets of points were paired with respect to λ_1. This was done to evaluate the variation in power when λ_2 changes while λ_1 is held fixed.

Table 8. Monte Carlo Power Estimates for an Exponential Model.

Parameters		Noncentrality			Power	
					Monte Carlo	
θ_1	θ_2	λ_1	λ_2	$P[X > c_\alpha]$	$\hat{p}$	$SE(\hat{p})$
.5	.5	0	0	.050	.0532	.00308
.5398	.5	0.9854	0	.204	.2058	.00570
.4237	.6849	0.9853	0.00034	.204	.2114	.00570
.5856	.5	4.556	0	.727	.7140	.00630
.3473	.8697	4.556	0.00537	.728	.7312	.00629
.62	.5	8.958	0	.957	.9530	.00287

Source: Gallant (1975a).

These simulations indicate that the approximation of $P(L > F_\alpha)$ by $P(X > c_\alpha)$ is quite accurate, as is the approximation $P(X > c_\alpha) \doteq 1 - F'(F_\alpha; q, n - p, \lambda_1)$. □

EXAMPLE 2 (Continued).   As mentioned at the beginning of the chapter, the model

$$B : y_t = \frac{\theta_1(e^{-\theta_2 x_t} - e^{-\theta_1 x_t})}{\theta_1 - \theta_2} + e_t$$

was chosen by Guttman and Meeter (1965) to represent a nearly linear model as measured by measures of nonlinearity introduced by Beale (1960). The model

$$C : y_t = 1 - \frac{\theta_1 e^{-\theta_2 x_t} - \theta_2 e^{-\theta_1 x_t}}{\theta_1 - \theta_2} + e_t$$

is highly nonlinear by this same criterion. The simulations reported in Table 9 were designed to determine how the approximations

$$P(W > F_\alpha) \doteq P(Y > F_\alpha)$$

$$P(L > F_\alpha) \doteq P(X > c_\alpha)$$

hold up as we move from a nearly linear situation to more nonlinear situations. As we have hinted at all along, the approximation

$$P(W > F_\alpha) \doteq P(Y > F_\alpha)$$

deteriorates badly, while the approximation

$$P(L > F_\alpha) \doteq P(X > c_\alpha)$$

holds up quite well. The details of the simulation are as follows.

The probabilities $P(W > F_\alpha)$ and $P(L > F_\alpha)$ that the hypothesis $H : \theta^0 = (1.4, .4)$ is rejected shown in Table 9 were computed from 4000 Monte Carlo trials using the control variate method of variance reduction (Hammersley and Handscomb, 1964). The independent variables were the same as those listed in Table 2, and the simulated errors were normally distributed with mean zero and variance $\sigma^2 = (.025)^2$. The sample size in each of the 4000 trials was $n = 12$ as one sees from Table 2. An asterisk

Table 9.   Monte Carlo Estimates of Power.

$\dfrac{\theta_1 - 1.4}{\sigma_1}$	$\dfrac{\theta_2 - .4}{\sigma_2}$	Wald Test			Likelihood Ratio		
		$P[Y > F_\alpha]$	$P[W > F_\alpha]$	Std. Err.[a]	$P[X > c_\alpha]$	$P[L > c_\alpha]$	Std. Err.
			(a) Model B[b]				
−4.5	1.0	.9725	.9835	.0017*	.9889	.9893	.0020
−3.0	0.5	.6991	.7158	.0027*	.7528	.7523	.0035
−1.5	−1.5	.2943	.2738	.0023*	.3051	.3048	.0017
1.5	−0.5	.2479	.2539	.0018*	.2379	.2379	.0016
3.0	−4.0	.9938	.9948	.0008	.9955	.9948	.0006
2.0	3.0	.7127	.7122	.0017	.6829	.6800	.0028
−1.5	1.0	.3295	.3223	.0022*	.3381	.3368	.0015
0.5	−0.5	.0885	.0890	.0016	.0885	.0892	.0009
0.0	0.0	.0500	.0525	.0012*	.0500	.0501	.0008
			(b) Model C[c]				
−2.5	0.5	.9964	.9540	.0009*	1.0000	1.0000	.0000
−1.0	0.0	.5984	.4522	.0074*	.7738	.7737	.0060
2.0	−1.5	.4013	.4583	.0062*	.2807	.2782	.0071
0.5	−1.0	.2210	.2047	.0056*	.2877	.2892	.0041
4.5	−3.0	.9945	.8950	.0012*	.9736	.9752	.0025
0.0	1.0	.5984	.7127	.0054*	.5585	.5564	.0032
−2.0	3.5	.9795	.7645	.0022*	.4207	.4192	.0078
−0.5	1.0	.2210	.3710	.0055*	.1641	.1560	.0040*
0.0	0.0	.0500	.1345	.0034*	.0500	.0502	.0012

[a]Asterisk indicates difference significant at 5% level.
[b]Model B: $\sigma_1 = 0.052957$, $\sigma_2 = 0.014005$.
[c]Model C: $\sigma_1 = 0.27395$, $\sigma_2 = 0.029216$.

indicates that $P(W > F_\alpha)$ is significantly different from $P(Y > F_\alpha)$ at the 5% level; similarly for the likelihood ratio test. For complete details see Gallant (1976).   □

If the null hypothesis is written as a parametric restriction

$$H : h(\theta^0) = 0$$

and it is not convenient to rewrite it as a functional dependence $\theta = g(\rho)$, the following alternative formula (Section 6 of Chapter 3) may be used to compute P_{FG}:

$$\theta_n^* \quad \text{minimizes} \quad \sum_{t=1}^{n} \left[f(x_t, \theta^0) - f(x_t, \theta) \right]^2 \quad \text{subject to} \quad h(\theta) = 0$$

$$\overline{H} = H(\theta_n^*) = \frac{\partial}{\partial \theta'} h(\theta_n^*)$$

$$P_{FG} = P_F - F(F'F)^{-1}\overline{H}' \left[\overline{H}(F'F)^{-1}\overline{H}' \right]^{-1} \overline{H}(F'F)^{-1}F'.$$

We have discussed the Wald test and the likelihood test of

$$H : h(\theta^0) = 0 \quad \text{against} \quad A : h(\theta^0) \neq 0$$

equivalently,

$$H : \theta^0 = g(\rho) \text{ for some } \rho^0 \quad \text{against} \quad A : \theta^0 \neq g(\rho) \text{ for any } \rho.$$

There is one other test in common use, the Lagrange multiplier (Problem 6) or efficient score test. In view of the foregoing, the following motivation is likely to have the strongest intuitive appeal. Let

$$\tilde{\theta} \quad \text{minimize} \quad \text{SSE}(\theta) \quad \text{subject to} \quad h(\theta) = 0$$

or equivalently,

$$\tilde{\theta} = g(\hat{\rho}) \quad \text{where} \quad \hat{\rho} \text{ minimizes SSE}[g(\rho)].$$

Suppose that $\tilde{\theta}$ is used as a starting value; the Gauss-Newton step away from $\tilde{\theta}$ (presumably) toward $\hat{\theta}$ is

$$\tilde{D} = (\tilde{F}'\tilde{F})^{-1}\tilde{F}'[y - f(\tilde{\theta})]$$

where $\tilde{F} = F(\tilde{\theta}) = (\partial / \partial \theta') f(\tilde{\theta})$. Intuitively, if the hypothesis $h(\theta^0) = 0$ is false, then minimization of $\text{SSE}(\theta)$ subject to $h(\theta) = 0$ will cause a large displacement away from $\hat{\theta}$ and $\tilde{D}$ will be large. Conversely, if $h(\theta^0)$ is true then $\tilde{D}$ should be small. It remains to find some measure of the distance of $\tilde{D}$ from zero that will yield a convenient test statistic.

Recall that

$$\theta_n^* \quad \text{minimizes} \quad \sum_{t=1}^{n} \left[f(x_t, \theta^0) - f(x_t, \theta) \right]^2 \quad \text{subject to} \quad h(\theta) = 0$$

or equivalently,

$$\theta_n^* = g(\rho_n^0) \quad \text{where} \quad \rho_n^0 \text{ minimizes} \quad \sum_{t=1}^{n} \left\{ f(x_t, \theta^0) - f[x_t, g(\rho)] \right\}^2$$

and that

$$\delta = f(\theta^0) - f(\theta_n^*)$$
$$P_F = F(F'F)^{-1}F'$$
$$P_{FG} = FG(G'F'FG)^{-1}G'F'$$

where $G = (\partial/\partial\rho')g(\rho_n^0)$. Equivalently,

$$P_{FG} = P_F - F(F'F)^{-1}\overline{H}'\left[\overline{H}(F'F)^{-1}\overline{H}'\right]^{-1}\overline{H}(F'F)^{-1}F'$$

where $\overline{H} = (\partial/\partial\theta')h(\theta_n^*)$. We shall show in Chapter 4 that

$$\frac{\tilde{D}'(\tilde{F}'\tilde{F})\tilde{D}}{n} = \frac{(e+\delta)'(P_F - P_{FG})(e+\delta)}{n} + o_p\left(\frac{1}{n}\right),$$

$$\frac{\text{SSE}(\tilde{\theta})}{n} = \frac{(e+\delta)'(I - P_{FG})(e+\delta)}{n} + o_p\left(\frac{1}{n}\right),$$

$$\frac{\text{SSE}(\hat{\theta})}{n} = \frac{e'(I - P_F)e}{n} + o_p\left(\frac{1}{n}\right).$$

These characterizations suggest two test statistics

$$R_1 = \frac{\tilde{D}'(\tilde{F}'\tilde{F})\tilde{D}/q}{\text{SSE}(\hat{\theta})/(n-p)}$$

and

$$R_2 = \frac{n\tilde{D}'(\tilde{F}'\tilde{F})\tilde{D}}{\text{SSE}(\tilde{\theta})}.$$

The second statistic R_2 is the customary form of the Lagrange multiplier test and has the advantage that it can be computed from knowledge of $\tilde{\theta}$ alone. The first requires two minimizations, one to compute $\hat{\theta}$ and another to compute $\tilde{\theta}$. Much is gained by going to this extra bother. The distribution theory is simpler and the test has better power, as we shall see later on.

The two test statistics can be characterized as

$$R_1 = Z_1 + o_p(1)$$
$$R_2 = Z_2 + o_p(1)$$

where

$$Z_1 = \frac{(e + \delta)'(P_F - P_{FG})(e + \delta)/q}{e'(I - P_F)e/(n - p)}$$

$$Z_2 = \frac{n(e + \delta)'(P_F - P_{FG})(e + \delta)}{(e + \delta)'(I - P_{FG})(e + \delta)}.$$

The distribution function of Z_1 is (Problem 7)

$$F'(z; q, n - p, \lambda_1)$$

where

$$\lambda_1 = \frac{\delta'(P_F - P_{FG})\delta}{2\sigma^2}.$$

That is, the random variable Z_1 is distributed as the noncentral F-distribution (Section 7) with q numerator degrees of freedom, $n - p$ denominator degrees of freedom, and noncentrality parameter λ_1. Thus R_1 is approximately distributed as the (central) F-distribution under the null, and the test is: Reject H when R_1 exceeds $F_\alpha = F^{-1}(1 - \alpha; q, n - p)$.

The distribution function of Z_2 is (Problem 8) for $z < n$

$$F''\left(\frac{(n - p)z}{q(n - z)}; q, n - p, \lambda_1, \lambda_2\right)$$

where

$$\lambda_1 = \frac{\delta'(P_F - P_{FG})\delta}{2\sigma^2}$$

$$\lambda_2 = \frac{\delta'(I - P_F)\delta}{2\sigma^2}$$

and $F''(t; q, n - p, \lambda_1, \lambda_2)$ denotes the doubly noncentral F-distribution (Section 7) with q numerator degrees of freedom, $n - p$ denominator degrees of freedom, numerator noncentrality parameter λ_1, and denominator noncentrality parameter λ_2 (Section 7). If we approximate

$$P(R_2 > d) \doteq P(Z_2 > d)$$

then under the null hypothesis that $h(\theta^0) = 0$ we have $\delta = 0$, $\lambda_1 = 0$, and $\lambda_2 = 0$, whence

$$P(R_2 > d \,|\, \lambda_1 = \lambda_2 = 0) \doteq 1 - F\left(\frac{(n - p)d}{q(n - d)}; q, n - p \right).$$

Letting F_α denote the $\alpha \times 100\%$ critical point of the F-distribution, that is,

$$\alpha = 1 - F(F_\alpha; q, n - p)$$

then that value d_α of d for which

$$P(R_2 > d_\alpha \,|\, \lambda_1 = \lambda_2 = 0) = \alpha$$

is

$$F_\alpha = \frac{(n - p)d_\alpha}{q(n - d_\alpha)}$$

or

$$d_\alpha = \frac{nF_\alpha}{(n - p)/q + F_\alpha}.$$

The test is then: Reject $H: h(\theta^0) = 0$ if $R_2 > d_\alpha$. With this computation of d_α,

$$
\begin{aligned}
P(R_1 > F_\alpha) &\doteq P(Z_1 > F_\alpha) \\
&= 1 - F'(F_\alpha; q, n - p, \lambda_1) \\
&\leq 1 - F''(F_\alpha; q, n - p, \lambda_1, \lambda_2) \\
&= P(Z_2 > d_\alpha) \\
&\doteq P(R_2 > d_\alpha)
\end{aligned}
$$

and we see that to within the accuracy of these approximations, the first version of the Lagrange multiplier test always has better power than the second. Of course, as we noted earlier, in most instances λ_2 will be small relative to λ_1 and the difference in power will be negligible.

In the same vein, judging from the entries in Table 6, we have (see Problem 10)

$$1 - F'(F_\alpha; q, n - p, \lambda_1) \leq 1 - H(c_\alpha; q, n - p, \lambda_1, \lambda_2)$$

whence

$$\begin{aligned}
P(L > F_\alpha) &\doteq P(X > c_\alpha) \\
&= 1 - H(c_\alpha; q, n - p, \lambda_1, \lambda_2) \\
&\geq 1 - F'(F_\alpha; q, n - p, \lambda_1) \\
&= P(Z_1 > F_\alpha) \\
&\doteq P(R_1 > F_\alpha).
\end{aligned}$$

Thus the likelihood ratio test has better power than either of the two versions of the Lagrange multiplier test. But again, λ_2 is usually small and the difference in power negligible.

To summarize this discussion, the first version of the Lagrange multiplier test rejects the hypothesis

$$H : h(\theta^0) = 0$$

when the statistic

$$R_1 = \frac{\tilde{D}'(\tilde{F}'\tilde{F})\tilde{D}/q}{\mathrm{SSE}(\hat{\theta})/(n - p)}$$

exceeds $F_\alpha = F^{-1}(1 - \alpha; q, n - p)$. The second version rejects when the statistic

$$R_2 = \frac{n\tilde{D}'(\tilde{F}'\tilde{F})\tilde{D}}{\mathrm{SSE}(\tilde{\theta})}$$

exceeds

$$d_\alpha = \frac{nF_\alpha}{(n - p)/q + F_\alpha}.$$

As usual, there are various strategies one might employ to compute the statistics R_1 and R_2. In connection with the likelihood ratio test, we have already discussed and illustrated how one can compute $\tilde{\theta}$ by computing the unconstrained minimum $\hat{\rho}$ of the composite function $\text{SSE}[g(\rho)]$ and setting $\tilde{\theta} = g(\hat{\rho})$. Now suppose that one creates a data set with observations

$$\tilde{e}_t = y_t - f(x_t, \tilde{\theta}) \qquad t = 1, 2, \ldots, n$$

$$\tilde{f}_t' = \frac{\partial}{\partial \theta'} f(x_t, \tilde{\theta}) \qquad t = 1, 2, \ldots, n$$

or in vector notation

$$\tilde{e} = y - f(\tilde{\theta}) \qquad \tilde{F} = \frac{\partial}{\partial \theta'} f(\tilde{\theta}).$$

Note that $\tilde{F}$ is an n by p matrix; $\tilde{F}$ is *not* the n by r matrix $(\partial/\partial \rho')f[g(\tilde{\rho})]$. If one regresses $\tilde{e}$ on $\tilde{F}$ with no intercept term using a linear regression procedure, then the analysis of variance table printed by the program will have the following entries:

Source	d.f.	Sum of Squares
Regression	p	$\tilde{e}'\tilde{F}(\tilde{F}'\tilde{F})^{-1}\tilde{F}'\tilde{e}$
Error	$n - p$	$\tilde{e}'\tilde{e} - \tilde{e}'\tilde{F}(\tilde{F}'\tilde{F})^{-1}\tilde{F}'\tilde{e}$
Total	n	$\tilde{e}'\tilde{e}$

One can just read off

$$\tilde{D}'(\tilde{F}'\tilde{F})\tilde{D} = \tilde{e}'\tilde{F}(\tilde{F}'\tilde{F})^{-1}\tilde{F}'\tilde{e}$$

$$\text{SSE}(\tilde{\theta}) = \tilde{e}'\tilde{e}$$

from the analysis of variance table. Let us illustrate these ideas.

EXAMPLE 1 (Continued). Recalling that the response function is

$$f(x, \theta) = \theta_1 x_1 + \theta_2 x_2 + \theta_4 e^{\theta_3 x_3}$$

reconsider the first hypothesis

$$H : \theta_1^0 = 0.$$

SAS Statements:

```
DATA WORK01;  SET EXAMPLE1;
T1=0.0;   T2=1.00296592;   T3=-1.14123442;   T4=-0.51182277;
E=Y-(T1*X1+T2*X2+T4*EXP(T3*X3));
F1=X1;  F2=X2;  F3=T4*X3*EXP(T3*X3);  F4=EXP(T3*X3);
DROP T1 T2 T3 T4;
PROC REG DATA=WORK01;  MODEL E=F1 F2 F3 F4 / NOINT;
```

Output:

S T A T I S T I C A L   A N A L Y S I S   S Y S T E M   1

DEP VARIABLE: E

SOURCE	DF	SUM OF SQUARES	MEAN SQUARE	F VALUE	PROB>F
MODEL	4	0.004938382	0.001234596	1.053	0.3996
ERROR	26	0.030495	0.001172869		
U TOTAL	30	0.035433			

ROOT MSE	0.034247	R-SQUARE	0.1394
DEP MEAN	-5.50727E-09	ADJ R-SQ	0.0401
C.V.	-621854289		

NOTE: NO INTERCEPT TERM IS USED. R-SQUARE IS REDEFINED.

VARIABLE	DF	PARAMETER ESTIMATE	STANDARD ERROR	T FOR H0: PARAMETER=0	PROB > |T|
F1	1	-0.025888	0.012616	-2.052	0.0504
F2	1	0.012719	0.009874181	1.288	0.2091
F3	1	0.026417	0.165440	0.160	0.8744
F4	1	0.007033215	0.025929	0.271	0.7883

Figure 12a. Illustration of Lagrange multiplier test computations with Example 1.

Previously we computed

$$\tilde{\theta} = \begin{pmatrix} 0.0 \\ 1.00296592 \\ -1.14123442 \\ -0.51182277 \end{pmatrix} \quad \text{(from Fig. 9a)}$$

$$\text{SSE}(\tilde{\theta}) = 0.03543298 \quad \text{(from Fig. 9a or Fig. 12a)}$$

$$\text{SSE}(\hat{\theta}) = 0.03049554 \quad \text{(from Fig. 5a)}.$$

We implement the scheme of regressing $\tilde{e}$ on $\tilde{F}$ in Figure 12a (note the similarity with Fig. 11a) and obtain

$$\tilde{D}'(\tilde{F}'\tilde{F})\tilde{D} = 0.004938382 \qquad \text{(from Fig. 12}a\text{).}$$

The first Lagrange multiplier test statistic is

$$R_1 = \frac{\tilde{D}'(\tilde{F}'\tilde{F})\tilde{D}/q}{\text{SSE}(\hat{\theta})/(n-p)}$$

$$= \frac{(0.004938382)/(1)}{(0.03049554)/(26)}$$

$$= 4.210.$$

Comparing with the critical point

$$F^{-1}(.95; 1, 26) = 4.22$$

one fails to reject the null hypothesis at the 95% level.
The second Lagrange multiplier test statistic is

$$R_2 = \frac{n\tilde{D}'(\tilde{F}'\tilde{F})\tilde{D}}{\text{SSE}(\tilde{\theta})}$$

$$= \frac{(30)(0.004938382)}{0.03543298}$$

$$= 4.1812.$$

Comparing with the critical point

$$d_\alpha = \frac{nF_\alpha}{(n-p)/q + F_\alpha}$$

$$= \frac{(30)(4.22)}{(26/1) + 4.22}$$

$$= 4.19$$

one fails to reject the null hypothesis at the 95% level.
Reconsider the second hypothesis

$$H: \theta_3\theta_4 e^{\theta_3} = \tfrac{1}{5}$$

which can be represented equivalently as

$$H : \theta^0 = g(\rho) \text{ for some } \rho^0$$

with

$$g(\rho) = \begin{pmatrix} \rho_1 \\ \rho_2 \\ \rho_3 \\ 1 \\ \overline{5\rho_3 e^{\rho_3}} \end{pmatrix}.$$

Previously we computed

$$\tilde{\rho} = \begin{pmatrix} -0.02301828 \\ 1.01965423 \\ -1.16048699 \end{pmatrix} \quad \text{(from Fig. } 9b\text{)}$$

$$\text{SSE}(\tilde{\theta}) = 0.03493222 \quad \text{(from Fig. } 9b \text{ or Fig. } 12b\text{)}$$

$$\text{SSE}(\hat{\theta}) = 0.03049554 \quad \text{(from Fig. } 5a\text{)}.$$

Regressing $\tilde{e}$ on $\tilde{F}$, we obtain

$$\tilde{D}'(\tilde{F}'\tilde{F})\tilde{D} = 0.004439308 \quad \text{(from Fig. } 12b\text{)}.$$

The first Lagrange multiplier test statistic is

$$R_1 = \frac{\tilde{D}'(\tilde{F}'\tilde{F})\tilde{D}/q}{\text{SSE}(\hat{\theta})/(n-p)}$$

$$= \frac{0.004439308/1}{0.03049554/26}$$

$$= 3.7849.$$

Comparing with

$$F(.95; 1, 26) = 4.22$$

we fail to reject the null hypothesis at the 95% level.

SAS Statements:

```
DATA WORK01;  SET EXAMPLE1;
R1=-0.02301828;   R2=1.01965423;   R3=-1.16048699;
T1=R1;  T2=R2;  T3=R3;  T4=1/(5*R3*EXP(R3));
E=Y-(T1*X1+T2*X2+T4*EXP(T3*X3));
F1=X1;   F2=X2;   F3=T4*X3*EXP(T3*X3);   F4=EXP(T3*X3);
DROP T1 T2 T3 T4;
PROC REG DATA=WORK01;  MODEL E=F1 F2 F3 F4 / NOINT;
```

Output:

S T A T I S T I C A L   A N A L Y S I S   S Y S T E M   1

DEP VARIABLE: E

SOURCE	DF	SUM OF SQUARES	MEAN SQUARE	F VALUE	PROB>F
MODEL	4	0.004439308	0.001109827	0.946	0.4531
ERROR	26	0.030493	0.001172804		
U TOTAL	30	0.034932			

ROOT MSE	0.034246	R-SQUARE	0.1271
DEP MEAN	7.59999E-09	ADJ R-SQ	0.0264
C.V.	450609078		

NOTE: NO INTERCEPT TERM IS USED. R-SQUARE IS REDEFINED.

VARIABLE	DF	PARAMETER ESTIMATE	STANDARD ERROR	T FOR H0: PARAMETER=0	PROB > |T|
F1	1	-0.00285742	0.012611	-0.227	0.8225
F2	1	-0.00398546	0.009829362	-0.405	0.6885
F3	1	0.043503	0.156802	0.277	0.7836
F4	1	0.045362	0.026129	1.736	0.0944

Figure 12b. Illustration of Lagrange multiplier test computations with Example 1.

The second Lagrange multiplier test statistic is

$$R_2 = \frac{n\tilde{D}'(\tilde{F}'\tilde{F})\tilde{D}}{\text{SSE}(\tilde{\theta})}$$

$$= \frac{(30)(0.004439308)}{0.0349322}$$

$$= 3.8125.$$

Comparing with

$$d_\alpha = \frac{nF_\alpha}{(n-p)/q + F_\alpha}$$

$$= \frac{(30)(4.22)}{(26/1) + 4.22}$$

$$= 4.19$$

we fail to reject at the 95% level.

We reconsider the third hypothesis

$$H: \theta_1 = 0 \text{ and } \theta_3\theta_4 e^{\theta_3} = \tfrac{1}{5}$$

which may be rewritten as

$$H: \theta^0 = g(\rho) \text{ for some } \rho^0$$

with

$$g(\rho) = \begin{pmatrix} 0 \\ \rho_2 \\ \rho_3 \\ 1 \\ \dfrac{1}{5\rho_3 e^{\rho_3}} \end{pmatrix}.$$

Previously we computed

$$\tilde{\theta} = \begin{pmatrix} \tilde{\rho}_2 \\ \tilde{\rho}_3 \end{pmatrix} = \begin{pmatrix} 1.00795735 \\ -1.16918683 \end{pmatrix} \qquad \text{(from Fig. 9c)}$$

$$\text{SSE}(\tilde{\theta}) = 0.03889923 \qquad \text{(from Fig. 9c or Fig. 12c)}$$

$$\text{SSE}(\hat{\theta}) = 0.03049554 \qquad \text{(from Fig. 5a).}$$

Regressing $\tilde{e}$ on $\tilde{F}$, we obtain

$$\tilde{D}'(\tilde{F}'\tilde{F})\tilde{D} = 0.008407271 \qquad \text{(from Fig. 12c).}$$

SAS Statements:

```
DATA WORK01;  SET EXAMPLE1;
R1=0;  R2=1.00795735;  R3=-1.16918683;
T1=R1;  T2=R2;  T3=R3;  T4=1/(5*R3*EXP(R3));
E=Y-(T1*X1+T2*X2+T4*EXP(T3*X3));
F1=X1;  F2=X2;  F3=T4*X3*EXP(T3*X3);  F4=EXP(T3*X3);
DROP T1 T2 T3 T4;
PROC REG DATA=WORK01;  MODEL E=F1 F2 F3 F4 / NOINT;
```

Output:

S T A T I S T I C A L   A N A L Y S I S   S Y S T E M   1

DEP VARIABLE: E

SOURCE	DF	SUM OF SQUARES	MEAN SQUARE	F VALUE	PROB>F
MODEL	4	0.008407271	0.002101818	1.792	0.1607
ERROR	26	0.030492	0.001172768		
U TOTAL	30	0.038899			

ROOT MSE	0.034246	R-SQUARE	0.2161	
DEP MEAN	-2.83174E-09	ADJ R-SQ	0.1257	
C.V.	-1209350370			

NOTE: NO INTERCEPT TERM IS USED. R-SQUARE IS REDEFINED.

VARIABLE	DF	PARAMETER ESTIMATE	STANDARD ERROR	T FOR H0: PARAMETER=0	PROB > |T|
F1	1	-0.025868	0.012608	-2.052	0.0504
F2	1	0.007699193	0.00980999	0.785	0.4396
F3	1	0.052092	0.157889	0.330	0.7441
F4	1	0.046107	0.026218	1.759	0.0904

Figure 12c. Illustration of Lagrange multiplier test computations with Example 1.

The first Lagrange multiplier test statistic is

$$R_1 = \frac{\tilde{D}'(\tilde{F}'\tilde{F})\tilde{D}/q}{\text{SSE}(\hat{\theta})/(n-p)}$$

$$= \frac{0.008407271/2}{0.03049554/26}$$

$$= 3.5840.$$

Comparing with

$$F^{-1}(.95; 2, 26) = 3.37$$

we reject the null hypothesis at the 5% level.

The second Lagrange multiplier test statistic is

$$R_2 = \frac{n\tilde{D}'(\tilde{F}'\tilde{F})\tilde{D}}{\text{SSE}(\tilde{\theta})}$$

$$= \frac{(30)(0.008407271)}{0.03889923}$$

$$= 6.4839.$$

Comparing with

$$d_\alpha = \frac{nF_\alpha}{(n-p)/q + F_\alpha}$$

$$= \frac{(30)(3.37)}{(26/2) + 3.37}$$

$$= 6.1759$$

we reject at the 95% level.   □

As the example suggests, the approximation

$$\tilde{D}'(\tilde{F}'\tilde{F})\tilde{D} \doteq \text{SSE}(\tilde{\theta}) - \text{SSE}(\hat{\theta})$$

is quite good, so that

$$R_1 \doteq L$$

in most applications. Thus, in most instances, the likelihood ratio test and the first version of the Lagrange multiplier test will accept and reject together.

To compute power, one uses the approximations

$$P(R_1 > F_\alpha) \doteq P(Z_1 > F_\alpha)$$

and

$$P(R_2 > d_\alpha) \doteq P(Z_2 > d_\alpha).$$

The noncentrality parameters λ_1 and λ_2 appearing in the distributions of Z_1 and Z_2 are the same as those in the distribution of X. Their computation was discussed in detail during the discussion of power computations for the likelihood ratio test. We illustrate

EXAMPLE 1 (Continued).   Recalling that

$$f(x, \theta) = \theta_1 x_1 + \theta_2 x_2 + \theta_4 e^{\theta_3 x_3}$$

let us approximate the probabilities that the two versions of the Lagrange multiplier test reject the following three hypotheses at the 5% level when the true values of the parameters are

$$\theta^0 = (.03, 1, -1.4, -.5)'$$
$$\sigma^2 = .001.$$

The three hypotheses are the same as those we have used for the illustration throughout:

$$H_1 : \theta_1 = 0$$
$$H_2 : \theta_3\theta_4 e^{\theta_3} = \tfrac{1}{5}$$
$$H_3 : \theta_1 = 0 \text{ and } \theta_3\theta_4 e^{\theta_3} = \tfrac{1}{5}.$$

In connection with the illustration of power computations for the likelihood ratio test we obtained

$$H_1 : \lambda_1 = 3.3343, \lambda_2 = 0$$
$$H_2 : \lambda_1 = 6.5128, \lambda_2 = 0.0005827$$
$$H_3 : \lambda_1 = 10.9604, \lambda_2 = 0.0008241.$$

For the first hypothesis

$$
\begin{aligned}
P(R_1 > F_\alpha) &\doteq P(Z_1 > F_\alpha) \\
&= 1 - F'(F_\alpha; q, n - p, \lambda_1) \\
&= 1 - F'(4.22; 1, 26, 3.3343) \\
&= .700 \\
P(R_2 > d_\alpha) &\doteq P(Z_2 > d_\alpha) \\
&= 1 - F''(F_\alpha; q, n - p, \lambda_1, \lambda_2) \\
&= 1 - F''(4.22; 1, 26, 3.3343, 0) \\
&= .700;
\end{aligned}
$$

for the second

$$
\begin{aligned}
P(R_1 > F_\alpha) &\doteq P(Z_1 > F_\alpha) \\
&= 1 - F'(F_\alpha; q, n - p, \lambda_1) \\
&= 1 - F'(4.22; 1, 26, 6.5128) \\
&= .935 \\
P(R_2 > d_\alpha) &\doteq P(Z_2 > d_\alpha) \\
&= 1 - F''(F_\alpha; q, n - p, \lambda_1, \lambda_2) \\
&= 1 - F''(4.22; 1, 26, 6.5128, 0.0005827) \\
&= .935;
\end{aligned}
$$

Table 10a. Monte Carlo Power Estimates for Version 1 of the Lagrange Multiplier Test.

Parameters[a]				$H_0: \theta_1 = 0$ against $H_1: \theta_1 \neq 0$ Monte Carlo				$H_0: \theta_3 = -1$ against $H_1: \theta_3 \neq -1$ Monte Carlo	
θ_1	θ_3	λ_1	λ_2	$P[Z_1 > F_\alpha]$	$P[R_1 > F_\alpha]$ Std. Err.	λ_1	λ_2	$P[Z_1 > F_\alpha]$	$P[R_1 > F_\alpha]$ Std. Err.
0.0	−1.0	0.0	0.0	.050	.049 .003	0.0	0.0	.050	.051 .003
0.008	−1.1	0.2353	0.0000	.101	.094 .004	0.2423	0.0006	.103	.107 .004
0.015	−1.2	0.8307	0.0000	.237	.231 .006	0.8526	0.0078	.242	.241 .006
0.030	−1.4	3.3343	0.0000	.700	.687 .006	2.6928	0.0728	.608	.608 .007

[a] $\theta_2 = 1$, $\theta_4 = -.5$, $\sigma^2 = .001$.

Table 10b. Monte Carlo Power Estimates for Version 2 of the Lagrange Multiplier Test.

Parameters[a]				$H_0: \theta_1 = 0$ against $H_1: \theta_1 \neq 0$ Monte Carlo				$H_0: \theta_3 = -1$ against $H_1: \theta_3 \neq -1$ Monte Carlo	
θ_1	θ_3	λ_1	λ_2	$P[Z_2 > d_\alpha]$	$P[R_2 > d_\alpha]$ Std. Err.	λ_1	λ_2	$P[Z_2 > d_\alpha]$	$P[R_2 > d_\alpha]$ Std. Err.
0.0	−1.0	0.0	0.0	.050	.049 .003	0.0	0.0	.050	.050 .003
0.008	−1.1	0.2353	0.0000	.101	.094 .004	0.2423	0.0006	.103	.106 .004
0.015	−1.2	0.8307	0.0000	.237	.231 .006	0.8526	0.0078	.242	.241 .006
0.030	−1.4	3.3343	0.0000	.700	.687 .006	2.6928	0.0728	.606	.605 .007

[a] $\theta_2 = 1$, $\theta_4 = -.5$, $\sigma^2 = .001$.

and for the third

$$P(R_1 > F_\alpha) \doteq P(Z_1 > F_\alpha)$$
$$= 1 - F'(F_\alpha; q, n - p, \lambda_1)$$
$$= 1 - F'(3.37; 2, 26, 10.9604)$$
$$= .983$$
$$P(R_2 > d_\alpha) \doteq P(Z_2 > d_\alpha)$$
$$= 1 - F''(F_\alpha; q, n - p, \lambda_1, \lambda_2)$$
$$= 1 - F''(3.37; 2, 26, 10.9604, 0.0008241)$$
$$= .983.$$

Again one questions the accuracy of these approximations. Tables $10a$ and $10b$ indicate that the approximations are quite good. Also, by comparing Tables 7, $10a$, and $10b$ one can see the beginnings of the spread

$$P(L > F_\alpha) > P(R_1 > F_\alpha) > P(R_2 > d_\alpha)$$

as λ_2 increases which was predicted by the theory. Tables $10a$ and $10b$ were constructed exactly the same as Tables 5 and 7. $\square$

PROBLEMS

1. Assuming that the density of y is $p(y; \theta, \sigma) = (2\pi\sigma^2)^{-n/2} \exp\{-\frac{1}{2}[y - f(\theta)]'[y - f(\theta)]/\sigma^2\}$, show that

$$\max_{\theta, \sigma} p(y; \theta, \sigma) = [2\pi \mathrm{SSE}(\hat{\theta})/n]^{-n/2} e^{-n/2}$$
$$\max_{h(\theta)=0, \sigma} p(y; \theta, \sigma) = [2\pi \mathrm{SSE}(\tilde{\theta})/n]^{-n/2} e^{-n/2}$$

presuming, of course, that $f(\theta)$ is such that the maximum exists. The likelihood ratio test rejects when the ratio

$$\frac{\max_{h(\theta)=0, \sigma} p(y; \theta, \sigma)}{\max_{\theta, \sigma} p(y; \theta, \sigma)}$$

is small. Put this statistic in the form: Reject when

$$\frac{[\mathrm{SSE}(\tilde{\theta}) - \mathrm{SSE}(\hat{\theta})]/q}{\mathrm{SSE}(\hat{\theta})/(n - p)}$$

is large.

2. If the system of equations defined over Θ

$$h(\theta) = \tau$$
$$\varphi(\theta) = \rho$$

has an inverse

$$\theta = \psi(\rho, \tau)$$

show that

$$\{\theta \in \Theta : h(\theta) = 0\} = \{\theta : \theta = \psi(\rho, 0) \text{ for some } \rho \text{ in } R\}$$

where $R = \{\rho : \rho = \phi(\theta) \text{ for some } \theta \text{ in } \Theta \text{ with } h(\theta) = 0\}$.

3. Referring to the previous problem, show that

$$\max\{\text{SSE}(\theta) : h(\theta) = 0 \text{ and } \theta \text{ in } \Theta\} = \max\{\text{SSE}[\psi(\rho, 0)] : \rho \text{ in } R\}$$

if either maximum exists.

4. [Derivation of $H(x; \nu_1, \nu_2, \lambda_1, \lambda_2)$.] Define $H(x; \nu_1, \nu_2, \lambda_1, \lambda_2)$ to be the distribution function given by

$$0, \qquad x \le 1 \quad \lambda_2 = 0$$

$$\int_0^\infty G\left(\frac{t}{x-1} + \frac{2x\lambda_2}{(x-1)^2}; \nu_2, \frac{\lambda_2}{(x-1)^2}\right) g(t; \nu_1, \lambda_1)\, dt \qquad x < 1 \quad \lambda_2 > 0$$

$$\int_0^\infty N(-t; 2\lambda_2, 8\lambda_2) g(t; \nu_1, \lambda_1)\, dt \qquad x = 1 \quad \lambda_2 > 0$$

$$1 - \int_0^\infty G\left(\frac{t}{x-1} + \frac{2x\lambda_2}{(x-1)^2}; \nu_2, \frac{\lambda_2}{(x-1)^2}\right) g(t; \nu_1, \lambda_1)\, dt \qquad x > 1.$$

where $g(t; \nu, \lambda)$ denotes the noncentral chi-square density function with ν degrees of freedom and noncentrality parameter λ, and $G(t; \nu, \lambda)$ denotes the corresponding distribution function (Section 7). Fill in the missing steps. Set $z = (1/\sigma)e$, $\gamma = (1/\sigma)\delta_0$, and $R = P_F - P_{FG}$. The random variables $(z_1, z_2, \ldots, z_n)$ are independent with density $n(t; 0, 1)$. For an arbitrary constant b, the random variable $(z + b\gamma)'R(z + b\gamma)$ is a noncentral chi-square with q degrees of freedom and noncentrality $b^2\gamma'R\gamma/2$, since R is idempotent with rank q. Similarly, $(z + b\gamma)'P^\perp(z + b\gamma)$ is a noncentral chi-square with $n - p$ degrees of freedom and noncentrality $b^2\gamma'P^\perp\gamma/2$. These two random variables are independent because $RP^\perp = 0$ (Section 7). Let

$a > 0$. Then

$$
\begin{aligned}
P[x > a + 1] &= P\big[(z + \gamma)'P_{FG}^{\perp}(z + \gamma) > (a + 1)z'P^{\perp}z\big] \\
&= P\big[(z + \gamma)'R(z + \gamma) > az'P^{\perp}z - 2\gamma'P^{\perp}z - \gamma'P^{\perp}\gamma\big] \\
&= P\big[(z + \gamma)'R(z + \gamma) > a(z - a^{-1}\gamma)' \\
&\qquad \times P^{\perp}(z - a^{-1}\gamma) - (1 + a^{-1})\gamma'P^{\perp}\gamma\big] \\
&= \int_{0}^{\infty} P\big[t > a(z - a^{-1}\gamma)'P^{\perp}(z - a^{-1}\gamma) \\
&\qquad - (1 + a^{-1})\gamma'P^{\perp}\gamma\big]g\Big(t; q, \frac{\gamma'R\gamma}{2}\Big)\, dt \\
&= \int_{0}^{\infty} P\bigg((z - a^{-1}\gamma)'P^{\perp}(z - a^{-1}\gamma) \\
&\qquad < \frac{t + (1 + a^{-1})\gamma'P^{\perp}\gamma}{a}\bigg)g\Big(t; q, \frac{\gamma'R\gamma}{2}\Big)\, dt \\
&= \int_{0}^{\infty} G\bigg(\frac{t}{a} + \frac{(a + 1)\gamma'P^{\perp}\gamma}{a^{2}}; n - p, \frac{\gamma'P^{\perp}\gamma}{2a^{2}}\bigg) \\
&\qquad \times g\Big(t; q, \frac{\gamma'R\gamma}{2}\Big)\, dt.
\end{aligned}
$$

By substituting $x = a - 1$, $\lambda_1 = \gamma'R\gamma/2$, and $\lambda_2 = \gamma'P^{-}\gamma/2$ one obtains the form of the distribution function for $x > 1$. The derivations for the remaining cases are analogous.

5. Show that if $\lambda_2 = 0$, then

$$
P(X > c_{\alpha}) = P\bigg((n - p)\frac{(e + \delta)'(P_F - P_{FG})(e + \delta)}{qe'P_F^{\perp}e} > F_{\alpha}\bigg).
$$

Referring to Problem 4, why does this fact imply that

$$
H(c_{\alpha}; \nu_1, \nu_2, \lambda_1, 0) = F'(F_{\alpha}; \nu_1, \nu_2, \lambda_1)?
$$

6. (Alternative motivation of the Lagrange multiplier test.) Suppose that we change the sign conventions on the components of the vector valued function $h(\theta)$ in a neighborhood of $\tilde{\theta}$ so that the problem

$$
\text{minimize SSE}(\theta)
$$
$$
\text{subject to } h(\theta) \le 0
$$

is equivalent to the problem

$$
\text{minimize SSE}(\theta)
$$
$$
\text{subject to } h(\theta) = 0
$$

on that neighborhood. The vector inequality means inequality component by component.

Now consider the problem

$$\text{minimize } \text{SSE}(\theta)$$

$$\text{subject to } h(\theta) = x$$

and view the solution θ as depending on x. Under suitable regularity conditions there is a vector λ of Lagrange multipliers such that

$$\frac{\partial}{\partial \theta'}\text{SSE}(\tilde{\theta}) = \tilde{\lambda}'H(\tilde{\theta})$$

and $(\partial/\partial x')\tilde{\theta}(x)$ exists. Then

$$h[\tilde{\theta}(x)] = x$$

implies

$$H(\tilde{\theta})\frac{\partial}{\partial x'}\tilde{\theta}(x) = I$$

whence

$$\frac{\partial}{\partial x'}\text{SSE}[\tilde{\theta}(x)] = \frac{\partial}{\partial \theta'}\text{SSE}[\tilde{\theta}(x)]\frac{\partial}{\partial x'}\tilde{\theta}(x)$$

$$= \tilde{\lambda}'H[\tilde{\theta}(x)]\frac{\partial}{\partial x'}\tilde{\theta}(x)$$

$$= \tilde{\lambda}'.$$

The intuitive interpretation of this equation is that if one had one more unit of the constraint h_i, then $\text{SSE}(\theta)$ would decrease by the amount λ_i. Then one should be willing to pay $|\lambda_i|$ (in units of SSE) for one more unit of h_i. Stated differently, the absolute values of the components of λ can be viewed as the prices of the constraints. With this interpretation any reasonable measure $d(\lambda)$ of the distance of the vector $\tilde{\lambda}$ from zero could be used to test

$$H : h(\theta) = 0 \quad \text{against} \quad A : h(\theta) \neq 0.$$

One would reject for large values of $d(\tilde{\lambda})$. Show that if

$$d(\tilde{\lambda}) = \tfrac{1}{4}\tilde{\lambda}'\tilde{H}(\tilde{F}'\tilde{F})^{-1}\tilde{H}'\tilde{\lambda}$$

is chosen as the measure of distance where $\tilde{H}$ and $\tilde{F}$ denote evaluation at $\theta = \tilde{\theta}$ then

$$d(\tilde{\lambda}) = \tilde{D}'(\tilde{F}'\tilde{F})\tilde{D}$$

where, recall, $\tilde{D} = (\tilde{F}'\tilde{F})^{-1}\tilde{F}'[y - f(\tilde{\theta})]$.

7. Show that Z_1 is distributed as $F'(z; q, n - p, \lambda_1)$. Hint: $P_F(I - P_F) = 0$ and $P_{FG}(I - P_F) = 0$.

8. Fill in the missing steps: If $z < n$,

$$P(Z_2 < z) = P\big((e + \delta)'(P_F - P_{FG})(e + \delta)$$

$$< \frac{z}{n}(e + \delta)'(I - P_{FG})(e + \delta)\big)$$

$$= P\left(\frac{(e + \delta)'(P_F - P_{FG})(e + \delta)/q}{(e + \delta)'P_F^{\perp}(e + \delta)/(n - p)} < \frac{(n - p)z}{q(n - z)}\right)$$

$$= F''\left(\frac{(n - p)z}{q(n - z)}; q, n - p, \lambda_1, \lambda_2\right).$$

9. (Relaxation of the normality assumption.) The distribution of e is spherical if the distribution of Qe is the same as the distribution of e for every n by n orthogonal matrix Q. Perhaps the most useful distribution of this sort other than the normal is the multivariate Student t (Zellner, 1976). Show that the null distributions of X, Z_1, and Z_2 do not change if any spherical distribution is substituted for the normal distribution. Hint: Jensen (1981).

10. Prove that $P(X > c_\alpha) \geq P(Z_1 > F_\alpha)$. Warning: This is an open question.

6. CONFIDENCE INTERVALS

A confidence interval on any (twice continuously differentiable) parametric function $\gamma(\theta)$ can be obtained by inverting any of the tests of

$$H: h(\theta^0) = 0 \quad \text{against} \quad A: h(\theta^0) \neq 0$$

described in the previous section. That is, to construct a $100 \times (1 - \alpha)\%$ confidence interval for $\gamma(\theta)$ one lets

$$h(\theta) = \gamma(\theta) - \gamma^0$$

and puts in the interval all those γ^0 for which the hypothesis $H: h(\theta^0) = 0$ is accepted at the α level of significance (Problem 1). The same is true for confidence regions, the only difference being that $\gamma(\theta)$ and γ^0 will be q-vectors instead of being univariate.

The Wald test is easy to invert. In the univariate case $(q = 1)$, the Wald test accepts when

$$\frac{|\gamma(\hat{\theta}) - \gamma^0|}{(s^2\hat{H}\hat{C}\hat{H}')^{1/2}} \leq t_{\alpha/2}$$

where

$$\hat{H} = \frac{\partial}{\partial \theta'}[\gamma(\hat{\theta}) - \gamma^0] = \frac{\partial}{\partial \theta'}\gamma(\hat{\theta})$$

and $t_{\alpha/2} = t^{-1}(1 - \alpha/2; n - p)$; that is, $t_{\alpha/2}$ denotes the upper $\alpha/2$ critical point of the t-distribution with $n - p$ degrees of freedom. Those points γ^0 that satisfy the inequality are in the interval

$$\gamma(\hat{\theta}) \pm t_{\alpha/2}(s^2\hat{H}\hat{C}\hat{H}')^{1/2}.$$

The most common situation is when one wishes to set a confidence interval on one of the components θ_i of the parameter vector θ. In this case the interval is

$$\theta_i \pm t_{\alpha/2}\sqrt{s^2\hat{c}_{ii}}$$

where $\hat{c}_{ii}$ is the ith diagonal element of $\hat{C} = [F'(\hat{\theta})F(\hat{\theta})]^{-1}$. We illustrate with Example 1.

EXAMPLE 1 (Continued).   Recalling that

$$f(x, \theta) = \theta_1 x_1 + \theta_2 x_2 + \theta_4 e^{\theta_3 x_3}$$

let us set a confidence interval on θ_1 by inverting the Wald test. One can read off the confidence interval directly from the SAS output of Figure 5a as

$$[-0.05183816, 0.00005877]$$

or compute it as

$$\theta_1 = -0.02588970 \qquad \text{(from Fig. 5}a\text{)}$$
$$\hat{c}_{11} = .13587 \qquad \text{(from Fig. 5}b\text{)}$$
$$s^2 = 0.00117291 \qquad \text{(from Fig. 5}b\text{)}$$
$$t^{-1}(.975; 26) = 2.0555$$
$$\theta_1 \pm t_{\alpha/2}\sqrt{s^2\hat{c}_{11}} = -0.02588970 \pm (2.0555)\sqrt{(0.00117291)(.13587)}$$
$$= -0.02588970 \pm 0.0259484615$$

whence

$$[-0.051838, 0.000588].$$

SAS Statements:

```
PROC MATRIX;
C =  0.13587   -0.067112    -0.15100   -0.037594/
    -0.067112   0.084203     0.51754  -0.00157848/
    -0.15100    0.51754      22.8032    2.00887/
    -0.037594  -0.00157848   2.00887    0.56125;
H =  0  0  0.0191420895  -0.365599176;
HCH = H*C*H';  PRINT HCH;
```

Output:

S T A T I S T I C A L   A N A L Y S I S   S Y S T E M   1

$$\text{HCH} \qquad\qquad \text{COL1}$$

$$\text{ROW1} \qquad 0.0552563$$

Figure 13. Wald test confidence interval construction illustrated with Example 1.

To put a confidence interval on

$$\gamma(\theta) = \frac{\partial}{\partial x_3} f(x,\theta)\bigg|_{x_3=1} = \theta_3\theta_4 e^{\theta_3}$$

we have

$$H(\theta) = \frac{\partial}{\partial \theta'}\gamma(\theta) = \left[0, 0, \theta_4(1+\theta_3)e^{\theta_3}, \theta_3 e^{\theta_3}\right]$$

$$\gamma(\theta) = (-1.11569714)(-0.50490286)e^{-1.11569714} \quad \text{(from Fig. 5}a\text{)}$$

$$= 0.1845920697$$

$$\hat{H} = (0, 0, 0.0191420895, -0.365599176) \qquad \text{(from Fig. 5}a\text{)}$$

$$\hat{H}\hat{C}\hat{H}' = 0.0552562 \qquad\qquad\qquad\qquad\qquad \text{(from Fig. 5}b \text{ and 13)}$$

$$s^2 = 0.00117291. \qquad\qquad\qquad\qquad\qquad \text{(from Fig. 5}a\text{)}.$$

Then the confidence interval is

$$\gamma(\hat{\theta}) \pm t_{\alpha/2}(s^2\hat{H}\hat{C}\hat{H}')^{1/2}$$

$$= 0.184592 \pm (2.0555)[(0.00117291)(0.0552563)]^{1/2}$$

$$= 0.1845921 \pm 0.0165478$$

or

$$[0.168044, 0.201140]. \qquad\qquad\qquad \Box$$

In the case that $\gamma(\theta)$ is a q-vector, the Wald test accepts when

$$\frac{[\gamma(\hat{\theta}) - \gamma^0]'(\hat{H}\hat{C}\hat{H}')^{-1}[\gamma(\hat{\theta}) - \gamma^0]}{qs^2} \le F_\alpha.$$

The confidence region obtained by inverting this test is an ellipsoid with center at $\gamma(\hat{\theta})$ and the eigenvectors of $\hat{H}\hat{C}\hat{H}'$ as axes.

To construct a confidence interval for $\gamma(\theta)$ by inverting the likelihood ratio test, put

$$h(\theta) = \gamma(\theta) - \gamma^0$$

with γ^0 a q-vector, and let

$$\text{SSE}_{\gamma^0} = \min\{\text{SSE}(\theta) : \gamma(\theta) = \gamma^0\}.$$

The likelihood ratio test accepts when

$$L(\gamma^0) = \frac{(\text{SSE}_{\gamma^0} - \text{SSE}_{\text{full}})/q}{(\text{SSE}_{\text{full}})/(n - p)} \le F_\alpha$$

where, recall, $F_\alpha = F^{-1}(1 - \alpha; q, n - p)$ and $\text{SSE}_{\text{full}} = \text{SSE}(\hat{\theta}) = \min \text{SSE}(\theta)$. Thus, a likelihood ratio confidence region consists of those points γ^0 with $L(\gamma^0) \le F_\alpha$. Although it is not a frequent occurrence in applications, the likelihood ratio test can have unusual structural characteristics. It is possible that $L(\gamma^0)$ does not rise above F_α as $\|\gamma^0\|$ increases in some direction, so that the confidence region can be unbounded. Also it is possible that $L(\gamma^0)$ has local minima which can lead to confidence regions consisting of disjoint islands. But as we said, this does not happen often.

In the univariate case, the easiest way to invert the likelihood ratio test is by quadratic interpolation as follows. Take three trial values $\gamma_1^0, \gamma_2^0, \gamma_3^0$ around the lower limit of the Wald test confidence interval, and compute the corresponding values of $L(\gamma_1^0)$, $L(\gamma_2^0)$, $L(\gamma_3^0)$. Fit the quadratic equation

$$L(\gamma_i^0) = a(\gamma_i^0)^2 + b(\gamma_i^0) + c \qquad i = 1, 2, 3$$

to these three points, and let $\hat{x}$ solve the equation

$$F_\alpha = ax^2 + bx + c.$$

One can take $\hat{x}$ as the lower limit or refine the estimates by taking three trial values $\gamma_1^0, \gamma_2^0, \gamma_3^0$ around $\hat{x}$ and repeating the process. The upper confidence limit can be computed similarly. We illustrate with Example 1.

EXAMPLE 1 (Continued).   Recalling that

$$f(x, \theta) = \theta_1 x_1 + \theta_2 x_2 + \theta_4 e^{\theta_3 x_3}$$

let us set a confidence interval on θ_1. We have

$$\text{SSE}_{\text{full}} = 0.03049554 \qquad (\text{from Fig. } 5a).$$

By simply reusing the SAS code from Figure 9a and embedding it in a MACRO whose argument γ^0 is assigned to the parameter θ_1, we can easily construct the following table from Figure 14a:

γ^0	SSE_{γ^0}	$L(\gamma^0)$
$-.052$	0.03551086	4.275980
$-.051$	0.03513419	3.954837
$-.050$	0.03477221	3.646219
$-.001$	0.03505883	3.890587
$.000$	0.03543298	4.209581
$.001$	0.03582188	4.541151

Then either by hand calculator or by using PROC MATRIX as in Figure 14b, one can interpolate from this table to obtain the confidence interval

$$[-0.0518, 0.0000320].$$

Next let us set a confidence interval on the parametric function

$$\gamma(\theta) = \frac{\partial}{\partial x_3} f(x, \theta)\Big|_{x_3=1} = \theta_3 \theta_4 e^{\theta_3}.$$

As we have seen previously, the hypothesis

$$H: \theta_3 \theta_4 e^{\theta_3} = \gamma^0$$

can be rewritten as

$$H: \theta_4 = \left(\frac{\theta_3 e^{\theta_3}}{\gamma^0}\right)^{-1}.$$

Again, as we have seen previously, to compute SSE_{γ^0} let

$$g_{\gamma^0}(\rho) = \begin{pmatrix} \rho_1 \\ \rho_2 \\ \rho_3 \\ \dfrac{\rho_3 e^{\rho_3}}{\gamma^0} \end{pmatrix}$$

SAS Statements:

```
%MACRO SSE(GAMMA);
PROC NLIN DATA=EXAMPLE1 METHOD=GAUSS ITER=50 CONVERGENCE=1.0E-13;
PARMS  T2=1.01567967 T3=-1.11569714 T4=-0.50490286;  T1=&GAMMA;
MODEL Y=T1*X1+T2*X2+T4*EXP(T3*X3);
DER.T2=X2; DER.T3=T4*X3*EXP(T3*X3); DER.T4=EXP(T3*X3);
%MEND SSE;
%SSE(-.052) %SSE(-.051) %SSE(-.050) %SSE(-.001) %SSE(.000) %SSE(.001)
```

Output:

NON-LINEAR LEAST SQUARES ITERATIVE PHASE

DEPENDENT VARIABLE: Y   METHOD: GAUSS-NEWTON

ITERATION	T2	T3	T4	RESIDUAL SS
6	1.02862742	-1.08499107	-0.49757910	0.03551086
5	1.02812865	-1.08627326	-0.49786686	0.03513419
5	1.02763014	-1.08754637	-0.49815400	0.03477221
7	1.00345514	-1.14032573	-0.51156098	0.03505883
7	1.00296592	-1.14123442	-0.51182277	0.03543298
7	1.00247682	-1.14213734	-0.51208415	0.03582188

Figure 14a. Likelihood ratio test confidence interval construction illustrated with Example 1.

and SSE_{γ^0} can be computed as the unconstrained minimum of $SSE[g_{\gamma^0}(\rho)]$. Using the SAS code from Figure 9b and embedding it in a MACRO whose argument γ^0 replaces the value $\frac{1}{5}$ in the previous code, the following table can be constructed from Figure 14c:

γ^0	SSE_{γ^0}	$L(\gamma^0)$
.166	0.03591352	4.619281
.167	0.03540285	4.183892
.168	0.03491101	3.764558
.200	0.03493222	3.782641
.201	0.03553200	4.294004
.202	0.03617013	4.838063

Quadratic interpolation from this table as shown in Figure 14d yields

$$[0.1669, 0.2009].$$

□

SAS Statements:

```
PROC MATRIX;
A= 1 -.052 .002704 /
   1 -.051 .002601 /
   1 -.050 .002500 ;
TEST= 4.275980 / 3.954837 / 3.646219 ;  B=INV(A)*TEST;
ROOT=(-B(2,1)+SQRT(B(2,1)#B(2,1)-4#B(3,1)#(B(1,1)-4.22)))#/(2#B(3,1));
PRINT ROOT;
ROOT=(-B(2,1)-SQRT(B(2,1)#B(2,1)-4#B(3,1)#(B(1,1)-4.22)))#/(2#B(3,1));
PRINT ROOT;
A= 1 -.001 .000001 /
   1  .000 .000000 /
   1  .001 .000001 ;
TEST= 3.890587 / 4.209581 / 4.541151 ;  B =INV(A)*TEST;
ROOT=(-B(2,1)+SQRT(B(2,1)#B(2,1)-4#B(3,1)#(B(1,1)-4.22)))#/(2#B(3,1));
PRINT ROOT;
ROOT=(-B(2,1)-SQRT(B(2,1)#B(2,1)-4#B(3,1)#(B(1,1)-4.22)))#/(2#B(3,1));
PRINT ROOT;
```

Output:

S T A T I S T I C A L   A N A L Y S I S   S Y S T E M   1

ROOT	COL1
ROW1	0.000108776

ROOT	COL1
ROW1	-0.0518285

ROOT	COL1
ROW1	.0000320109

ROOT	COL1
ROW1	-0.0517626

Figure 14b. Likelihood ratio test confidence interval construction illustrated with Example 1.

To construct a confidence interval for $\gamma(\theta)$ by inverting the Lagrange multiplier tests, let

$$h(\theta) = \gamma(\theta) - \gamma^0$$

$\tilde{\theta}$ minimize SSE(θ) subject to $h(\theta) = 0$

$$\tilde{F} = F(\tilde{\theta}) = \frac{\partial}{\partial \theta'} f(\tilde{\theta})$$

$$\tilde{D} = (\tilde{F}'\tilde{F})^{-1}\tilde{F}'[y - f(\tilde{\theta})]$$

$$R_1(\gamma^0) = \frac{\tilde{D}'(\tilde{F}'\tilde{F})\tilde{D}/q}{\mathrm{SSE}(\tilde{\theta})/(n-p)}$$

$$R_2(\gamma^0) = \frac{n\tilde{D}'(\tilde{F}'\tilde{F})\tilde{D}}{\mathrm{SSE}(\tilde{\theta})}.$$

The first version of the Lagrange multiplier test accepts when

$$R_1(\gamma^0) \le F_\alpha$$

and the second when

$$R_2(\gamma^0) \le d_\alpha$$

SAS Statements:

```
%MACRO SSE(GAMMA);
PROC NLIN DATA=EXAMPLE1 METHOD=GAUSS ITER=60 CONVERGENCE=1.0E-8;
PARMS R1=-0.02588970 R2=1.01567967 R3=-1.11569714;  RG=1/&GAMMA;
T1=R1; T2=R2; T3=R3; T4=1/(RG*R3*EXP(R3));
MODEL Y=T1*X1+T2*X2+T4*EXP(T3*X3);
DER_T1=X1; DER_T2=X2; DER_T3=T4*X3*EXP(T3*X3); DER_T4=EXP(T3*X3);
DER.R1=DER_T1; DER.R2=DER_T2;
DER.R3=DER_T3+DER_T4*(-T4**2)*(RG*EXP(R3)+RG*R3*EXP(R3));
%MEND SSE;
%SSE(.166) %SSE(.167) %SSE(.168) %SSE(.200) %SSE(.201) %SSE(.202)
```

Output:

NON-LINEAR LEAST SQUARES ITERATIVE PHASE

DEPENDENT VARIABLE: Y   METHOD: GAUSS-NEWTON

ITERATION	R1	R2	R3	RESIDUAL SS
8	-0.03002338	1.01672014	-0.91765508	0.03591352
8	-0.02978174	1.01642383	-0.93080113	0.03540285
8	-0.02954071	1.01614385	-0.94412575	0.03491101
31	-0.02301828	1.01965423	-1.16048699	0.03493222
43	-0.02283734	1.01994671	-1.16201915	0.03553200
13	-0.02265799	1.02024775	-1.16319256	0.03617013

Figure 14c. Likelihood ratio test confidence interval construction illustrated with Example 1.

SAS Statements:

```
PROC MATRIX;
A= 1 .166 .027556 /
   1 .167 .027889 /
   1 .168 .028224 ;
TEST= 4.619281 / 4.183892 / 3.764558 ;  B=INV(A)*TEST;
ROOT=(-B(2,1)+SQRT(B(2,1)#B(2,1)-4#B(3,1)#(B(1,1)-4.22)))#/(2#B(3,1));
PRINT ROOT;
ROOT=(-B(2,1)-SQRT(B(2,1)#B(2,1)-4#B(3,1)#(B(1,1)-4.22)))#/(2#B(3,1));
PRINT ROOT;
A= 1 .200 .040000 /
   1 .201 .040401 /
   1 .202 .040804 ;
TEST= 3.782641 / 4.294004 / 4.838063 ;  B =INV(A)*TEST;
ROOT=(-B(2,1)+SQRT(B(2,1)#B(2,1)-4#B(3,1)#(B(1,1)-4.22)))#/(2#B(3,1));
PRINT ROOT;
ROOT=(-B(2,1)-SQRT(B(2,1)#B(2,1)-4#B(3,1)#(B(1,1)-4.22)))#/(2#B(3,1));
PRINT ROOT;
```

Output:

STATISTICAL   ANALYSIS   SYSTEM   1

ROOT	COL1
ROW1	0.220322

ROOT	COL1
ROW1	0.166916

ROOT	COL1
ROW1	0.200859

ROOT	COL1
ROW1	0.168861

Figure 14d. Likelihood ratio test confidence interval construction illustrated with Example 1.

where $F_\alpha = F^{-1}(1 - \alpha; q, n - p)$, $d_\alpha = nF_\alpha/[(n - p)/q + F_\alpha]$, and q is the dimension of γ^0. Confidence regions consist of those points γ^0 for which the tests accept. These confidence regions have the same structural characteristics as likelihood ratio confidence regions except that disjoint islands are much more likely with Lagrange multiplier regions (Problem 2).

In the univariate case, Lagrange multiplier tests are inverted the same as the likelihood ratio test. One constructs a table with $R_1(\gamma^0)$ and $R_2(\gamma^0)$

evaluated at three points around each of the Wald test confidence limits and then uses quadratic interpolation to find the limits. We illustrate with Example 1.

EXAMPLE 1 (Continued).   Recalling that

$$f(x, \theta) = \theta_1 x_1 + \theta_2 x_2 + \theta_4 e^{\theta_3 x_3}$$

let us set Lagrange multiplier confidence intervals on θ_1. We have

$$\text{SSE}(\hat{\theta}) = 0.03049554 \qquad \text{(from Fig. 5}a\text{)}.$$

Taking $\tilde{\theta}$ and $\text{SSE}(\tilde{\theta})$ from Figure 14a and embedding the SAS code from Figure 12a in a MACRO as shown in Figure 15a, we obtain the following table from the entries in Figure 15a:

γ^0	$\tilde{D}'(\tilde{F}'\tilde{F})\tilde{D}$	$R_1(\gamma^0)$	$R_2(\gamma^0)$
$-.052$	0.005017024	4.277433	4.238442
$-.051$	0.004640212	3.956169	3.962134
$-.050$	0.004278098	3.647437	3.690963
$-.001$	0.004564169	3.891336	3.905580
$.000$	0.004938382	4.210384	4.181174
$.001$	0.005327344	4.542001	4.461528

Interpolating as shown in Figure 15b, we obtain

$$R_1 : [-0.0518, 0.0000345]$$
$$R_2 : [-0.0518, 0.0000317].$$

In exactly the same way we construct the following table for

$$\gamma(\theta) = \theta_3 \theta_4 e^{\theta_3}$$

from the entries of Figures 14c and 15c:

γ^0	$\tilde{D}'(\tilde{F}'\tilde{F})\tilde{D}$	$R_1(\gamma^0)$	$R_2(\gamma^0)$
$.166$	0.005507692	4.695768	4.600795
$.167$	0.004986108	4.251074	4.225175
$.168$	0.004483469	3.822533	3.852770
$.200$	0.004439308	3.784882	3.812504
$.201$	0.005039249	4.296382	4.254685
$.202$	0.005677511	4.840553	4.709005

SAS Statements:

```
%MACRO DFFD(THETA1,THETA2,THETA3,THETA4,SSER);
DATA WORK01;   SET EXAMPLE1;
T1=&THETA1;   T2=&THETA2;   T3=&THETA3;   T4=&THETA4;
E=Y-(T1*X1+T2*X2+T4*EXP(T3*X3));
F1=X1;   F2=X2;   F3=T4*X3*EXP(T3*X3);   F4=EXP(T3*X3);
DROP T1 T2 T3 T4;
PROC REG DATA=WORK01;   MODEL E=F1 F2 F3 F4 / NOINT;
%MEND DFFD;
%DFFD(-.052, 1.02862742, -1.08499107, -0.49757910, 0.03551086)
%DFFD(-.051, 1.02812865, -1.08627326, -0.49786686, 0.03513419)
%DFFD(-.050, 1.02763014, -1.08754637, -0.49815400, 0.03477221)
%DFFD(-.001, 1.00345514, -1.14032573, -0.51156098, 0.03505883)
%DFFD( .000, 1.00296592, -1.14123442, -0.51182277, 0.03543298)
%DFFD( .001, 1.00247682, -1.14213734, -0.51208415, 0.03582188)
```

Output:

SOURCE	DF	SUM OF SQUARES	MEAN SQUARE	F VALUE	PROB>F
MODEL	4	0.005017024	0.001254256	1.069	0.3916
ERROR	26	0.030494	0.00117284		
U TOTAL	30	0.035511			
MODEL	4	0.004640212	0.001160053	0.989	0.4309
ERROR	26	0.030494	0.001172845		
U TOTAL	30	0.035134			
MODEL	4	0.004278098	0.001069524	0.912	0.4717
ERROR	26	0.030494	0.001172851		
U TOTAL	30	0.034772			
MODEL	4	0.004564169	0.001141042	0.973	0.4392
ERROR	26	0.030495	0.001172871		
U TOTAL	30	0.035059			
MODEL	4	0.004938382	0.001234596	1.053	0.3996
ERROR	26	0.030495	0.001172869		
U TOTAL	30	0.035433			
MODEL	4	0.005327344	0.001331836	1.136	0.3617
ERROR	26	0.030495	0.001172867		
U TOTAL	30	0.035822			

Figure 15a. Lagrange multiplier test confidence interval construction illustrated with Example 1.

Quadratic interpolation from this table as shown in Figure 15d yields

$$R_1 : [0.1671, 0.2009]$$
$$R_2 : [0.1671, 0.2009].$$

There is some risk in using quadratic interpolation around Wald test confidence limits to find likelihood ratio or Lagrange multiplier confidence

SAS Statements:

```
PROC MATRIX;
A= 1 -.052 .002704 /
   1 -.051 .002601 /
   1 -.050 .002500 ;
TEST= 4.277433 4.238442 /
      3.956169 3.962134 /
      3.647437 3.690963 ;  B=INV(A)*TEST;
ROOT1=(-B(2,1)+SQRT(B(2,1)#B(2,1)-4#B(3,1)#(B(1,1)-4.22)))#/(2#B(3,1));
ROOT2=(-B(2,1)-SQRT(B(2,1)#B(2,1)-4#B(3,1)#(B(1,1)-4.22)))#/(2#B(3,1));
ROOT3=(-B(2,2)+SQRT(B(2,2)#B(2,2)-4#B(3,2)#(B(1,2)-4.19)))#/(2#B(3,2));
ROOT4=(-B(2,2)-SQRT(B(2,2)#B(2,2)-4#B(3,2)#(B(1,2)-4.19)))#/(2#B(3,2));
PRINT ROOT1 ROOT2 ROOT3 ROOT4;
A= 1 -.001 .000001 /
   1  .000 .000000 /
   1  .001 .000001 ;
TEST= 3.891336 3.905580 /
      4.210384 4.181174 /
      4.452001 4.461528 ;  B=INV(A)*TEST;
ROOT1=(-B(2,1)+SQRT(B(2,1)#B(2,1)-4#B(3,1)#(B(1,1)-4.22)))#/(2#B(3,1));
ROOT2=(-B(2,1)-SQRT(B(2,1)#B(2,1)-4#B(3,1)#(B(1,1)-4.22)))#/(2#B(3,1));
ROOT3=(-B(2,2)+SQRT(B(2,2)#B(2,2)-4#B(3,2)#(B(1,2)-4.19)))#/(2#B(3,2));
ROOT4=(-B(2,2)-SQRT(B(2,2)#B(2,2)-4#B(3,2)#(B(1,2)-4.19)))#/(2#B(3,2));
PRINT ROOT1 ROOT2 ROOT3 ROOT4;
```

Output:

```
          S T A T I S T I C A L   A N A L Y S I S   S Y S T E M          1


              ROOT1         COL1
              ROW1     .0000950422

              ROOT2         COL1
              ROW1     -0.0518241

              ROOT3         COL1
              ROW1     0.0564016

              ROOT4         COL1
              ROW1     -0.051826

              ROOT1         COL1
              ROW1     .0000344662

              ROOT2         COL1
              ROW1     0.00720637

              ROOT3         COL1
              ROW1     .0000317425

              ROOT4         COL1
              ROW1     -0.116828
```

Figure 15b. Lagrange multiplier test confidence interval construction illustrated with Example 1.

intervals. If the confidence region is a union of disjoint intervals then the method will compute the wrong answer. To be completely safe one would have to plot $L(\gamma^0)$, $R_1(\gamma^0)$, or $R_2(\gamma^0)$ and inspect for local minima.

The usual criterion for judging the quality of a confidence procedure is the expected length, area, or volume, depending on the dimension q of $\gamma(\theta)$. Let us use volume as the generic term. If two confidence procedures have the same probability of covering $\gamma(\theta^0)$, then the one with the smallest

SAS Statements:

```
%MACRO DFFD(GAMMA,RHO1,RHO2,RHO3,SSER);
DATA WORK01;   SET EXAMPLE1;
T1=&RHO1;   T2=&RHO2;   T3=&RHO3;   T4=1/(&RHO3*EXP(&RHO3)/&GAMMA);
E=Y-(T1*X1+T2*X2+T4*EXP(T3*X3));
F1=X1;   F2=X2;   F3=T4*X3*EXP(T3*X3);   F4=EXP(T3*X3);
DROP T1 T2 T3 T4;
PROC REG DATA=WORK01;   MODEL E=F1 F2 F3 F4 / NOINT;
%MEND DFFD;
%DFFD( .166, -0.03002338, 1.01672014, -0.91765508, 0.03591352)
%DFFD( .167, -0.02978174, 1.01642383, -0.93080113, 0.03540285)
%DFFD( .168, -0.02954071, 1.01614385, -0.94412575, 0.03491101)
%DFFD( .200, -0.02301828, 1.01965423, -1.16048699, 0.03493222)
%DFFD( .201, -0.02283734, 1.01994671, -1.16201915, 0.03553200)
%DFFD( .202, -0.02265799, 1.02024775, -1.16319256, 0.03617013)
```

Output:

SOURCE	DF	SUM OF SQUARES	MEAN SQUARE	F VALUE	PROB>F
MODEL	4	0.005507692	0.001376923	1.177	0.3438
ERROR	26	0.030406	0.001169455		
U TOTAL	30	0.035914			
MODEL	4	0.004986108	0.001246527	1.066	0.3935
ERROR	26	0.030417	0.001169875		
U TOTAL	30	0.035403			
MODEL	4	0.004483469	0.001120867	0.958	0.4471
ERROR	26	0.030428	0.00117029		
U TOTAL	30	0.034911			
MODEL	4	0.004439308	0.001109827	0.946	0.4531
ERROR	26	0.030493	0.001172804		
U TOTAL	30	0.034932			
MODEL	4	0.005039249	0.001259812	1.074	0.3894
ERROR	26	0.030493	0.001172798		
U TOTAL	30	0.035532			
MODEL	4	0.005677511	0.001419378	1.210	0.3303
ERROR	26	0.030493	0.001172793		
U TOTAL	30	0.036170			

Figure 15c. Lagrange multiplier test confidence interval construction illustrated with Example 1.

expected volume is preferred. But expected volume is really just an attribute of the power curve of the test to which the confidence procedure corresponds. To see this, let a test be described by its critical function

$$\phi(y, \gamma^0) = \begin{cases} 1 & \text{reject } H : \gamma(\theta) = \gamma^0 \\ 0 & \text{accept } H : \gamma(\theta) = \gamma^0. \end{cases}$$

SAS Statements:

```
PROC MATRIX;
A= 1 .166 .027556 /
   1 .167 .027889 /
   1 .168 .028224 ;
TEST= 4.695768 4.600795 /
      4.251074 4.225175 /
      3.822533 3.852770 ;  B=INV(A)*TEST;
ROOT1=(-B(2,1)+SQRT(B(2,1)#B(2,1)-4#B(3,1)#(B(1,1)-4.22)))#/(2#B(3,1));
ROOT2=(-B(2,1)-SQRT(B(2,1)#B(2,1)-4#B(3,1)#(B(1,1)-4.22)))#/(2#B(3,1));
ROOT3=(-B(2,2)+SQRT(B(2,2)#B(2,2)-4#B(3,2)#(B(1,2)-4.19)))#/(2#B(3,2));
ROOT4=(-B(2,2)-SQRT(B(2,2)#B(2,2)-4#B(3,2)#(B(1,2)-4.19)))#/(2#B(3,2));
PRINT ROOT1 ROOT2 ROOT3 ROOT4;
A= 1 .200 .040000 /
   1 .201 .040401 /
   1 .202 .040804 ;
TEST= 3.784882 3.812504 /
      4.296382 4.254685 /
      4.840553 4.709005 ;  B=INV(A)*TEST;
ROOT1=(-B(2,1)+SQRT(B(2,1)#B(2,1)-4#B(3,1)#(B(1,1)-4.22)))#/(2#B(3,1));
ROOT2=(-B(2,1)-SQRT(B(2,1)#B(2,1)-4#B(3,1)#(B(1,1)-4.22)))#/(2#B(3,1));
ROOT3=(-B(2,2)+SQRT(B(2,2)#B(2,2)-4#B(3,2)#(B(1,2)-4.19)))#/(2#B(3,2));
ROOT4=(-B(2,2)-SQRT(B(2,2)#B(2,2)-4#B(3,2)#(B(1,2)-4.19)))#/(2#B(3,2));
PRINT ROOT1 ROOT2 ROOT3 ROOT4;
```

Output:

STATISTICAL ANALYSIS SYSTEM 1

	ROOT1 COL1
ROW1	0.220989

	ROOT2 COL1
ROW1	0.167071

	ROOT3 COL1
ROW1	0.399573

	ROOT4 COL1
ROW1	0.167094

	ROOT1 COL1
ROW1	0.200855

	ROOT2 COL1
ROW1	0.168833

	ROOT3 COL1
ROW1	0.200855

	ROOT4 COL1
ROW1	0.127292

Figure 15d. Lagrange multiplier test confidence interval construction illustrated with Example 1.

The corresponding confidence procedure is

$$R_y = \{\gamma^0 : \phi(y, \gamma^0) = 0\}.$$

The expected volume is computed as

$$\text{Expected volume}(\phi) = \int_{R^n} \int_{R_y} d\gamma \, dN_n[y; f(\theta^0), \sigma^2 I]$$

$$= \int_{R^n} \int_{R^q} [1 - \phi(y, \gamma)] \, d\gamma \, dN_n[y; f(\theta^0), \sigma^2 I].$$

As Pratt (1961) shows by interchanging the order of integration,

$$\text{Expected volume}(\phi) = \int_{\mathbf{R}^q}\int_{\mathbf{R}^n} [1 - \phi(y, \gamma)]\, dN_n[y; f(\theta^0), \sigma^2 I]\, d\gamma$$

$$= \int_{\mathbf{R}^q} P[\phi(y, \gamma) = 0 | \theta^0, \sigma^2]\, d\gamma.$$

The integrand is the probability of covering γ,

$$c_\phi(\gamma) = P[\phi(y, \gamma) = 0 | \theta^0, \sigma^2]$$

and is analogous to the operating characteristic curve of a test. The essential difference between the coverage function $c_\phi(\gamma)$ and the operating characteristic function lies in the treatment of the hypothesized value γ and the true value of the parameter θ^0. For the coverage function, θ^0 is held fixed and γ varies; the converse is true for the operating characteristic function. If a test $\phi(y, \gamma)$ has better power against $H: \gamma(\theta) = \gamma^0$ than the test $\psi(y, \gamma^0)$ for all γ^0, then we have that

$$c_\phi(\gamma^0) = P[\phi(y, \gamma^0) = 0 | \theta^0, \sigma^2]$$
$$\leq P[\psi(y, \gamma^0) = 0 | \theta^0, \sigma^2]$$
$$= c_\psi(\gamma^0)$$

which implies

$$\text{Expected volume}(\phi) \leq \text{Expected volume}(\psi).$$

In this case a confidence procedure based on ϕ is to be preferred to a confidence interval based on ψ.

If one accepts the approximations of the previous section as giving useful guidance in applications, then the confidence procedure obtained by inverting the likelihood ratio test is to be preferred to either of the Lagrange multiplier procedures. However, both the likelihood ratio and Lagrange procedures can have infinite expected volume; Example 2 is an instance (Problem 3). But for $\gamma \neq \gamma(\theta^0)$ the coverage function gives the probability that the confidence procedure covers false values of γ. Thus, even in the case of infinite expected volume, the inequality $c_\phi(\gamma) \leq c_\psi(\gamma)$ implies that the procedure obtained by inverting ϕ is preferred to that obtained by inverting ψ. Thus the likelihood ratio procedure remains preferable to the Lagrange multiplier procedures even in the case of infinite expected volume.

Again, if one accepts the approximations of the previous section, the confidence procedure obtained by inverting the Wald test has better structural characteristics than either the likelihood ratio procedure or the Lagrange multiplier procedures. Wald test confidence regions are always intervals, ellipses, or ellipsoids according to the dimension of $\gamma(\theta)$, and they are much easier to compute than likelihood ratio or Lagrange multiplier regions. The expected volume is always finite (Problem 4). It is a pity that the approximation to the probability $P(W > F_\alpha)$ by $P(Y > F_\alpha)$ of the previous section is often inaccurate. This makes the use of Wald confidence regions risky, as one cannot be sure that the actual coverage probability is accurately approximated by the nominal probability of $1 - \alpha$ short of Monte Carlo simulation in each instance. Measures of nonlinearity are intended to help remedy this defect; they are discussed in the next chapter.

PROBLEMS

1. In the notation of the last few paragraphs of this section, show that

$$P\left\{\phi[y, \gamma(\theta^0)] = 0 | \theta^0, \sigma\right\} = \int_{R_y} dN_n[y; f(\theta^0), \sigma^2 I].$$

2. (Disconnected confidence regions.) Fill in the missing details in the following argument. Consider setting a confidence region on the entire parameter vector θ. Islands in likelihood ratio confidence regions *may* occur because SSE(θ) has a local minimum at θ^* causing $L(\theta^*)$ to fall below F_α. But if θ^* is a local minimum, then $R_1(\theta^*) = R_2(\theta^*) = 0$ and a neighborhood of θ^* *must* be included in a Lagrange multiplier confidence region.

3. Referring to Model B of Example 2 and the hypothesis $H: \theta^0 = \gamma^0$, show that because $0 < f(x, \gamma) < 1$, we have $P(X > c_\alpha) < 1 - \epsilon$ for some $\epsilon > 0$ and all γ in $A = \{\gamma : 0 \leq \gamma_2 \leq \gamma_1\}$, where X and c_α are as defined in the previous section. Show also that there is an open set E such that for all e in E we have

$$\sup_{\gamma \in A} \|e + \delta(\gamma)\| \leq c_\alpha \inf_{\gamma \in A} \|e + \delta(\gamma)\|$$

where $\delta(\gamma) = f(\theta^0) - f(\gamma)$. Show that this implies that $P(L > F_\alpha) < 1 - \epsilon$ for some $\epsilon > 0$ and all γ in A. Show that these facts imply that

Table 11.   List of Distributions

Random Variable	Density Function	Distribution Function	Quantile Function
Generic: X	$p(x)$	$P(x) = \int_{-\infty}^{x} p(t)\, dt$	$P^{-1}(\alpha) = x$ that solves $P(x) = \alpha$
Univariate normal with mean μ and variance σ^2; $n(x; \mu, \sigma^2) = (2\pi\sigma^2)^{-1/2}\exp\{-\frac{1}{2}[(x-\mu)/\sigma]^2\}$	$n(x; \mu, \sigma^2)$	$N(x; \mu, \sigma^2)$ or $N(\mu, \sigma^2)$	$N^{-1}(\alpha; \mu, \sigma^2)$
Multivariate normal with dimension p, mean μ, and variance-covariance matrix Σ; $n_p(x; \mu, \Sigma) = [\det(2\pi\Sigma)]^{-1/2} \times \exp[(-\frac{1}{2})(x-\mu)'\Sigma^{-1}(x-\mu)]$	$n_p(x; \mu, \Sigma)$	$N_p(x; \mu, \Sigma)$ or $N_p(\mu, \Sigma)$	$N_p^{-1}(\alpha; \mu, \Sigma)$
Chi-square with q degrees of freedom; $X = \sum_{i=1}^{q} Z_i^2$, where the Z_i are independent, $Z_i \sim N(0,1)$.	—	$\chi^2(x; q)$ or $\chi^2(q)$	$(\chi^2)^{-1}(\alpha; q)$
Noncentral chi-square with q degrees of freedom and non-centrality parameter λ; $X = \sum_{i=1}^{q} Z_i^2$, where the Z_i are independent, $Z_i \sim N(\mu, 1)$, $\lambda = \frac{1}{2}\sum_{i=1}^{q}\mu_i^2$	—	$\chi'^2(x; q, \lambda)$ or $\chi'^2(q, \lambda)$	$(\chi'^2)^{-1}(\alpha; q, \lambda)$
F-distribution with q_1 numerator degrees of freedom and q_2 denominator degrees of freedom; $F = (X_1/q_1)/(X_2/q_2)$, where the X_i are independent, $X_i \sim \chi^2(q_i)$.	—	$F(x; q_1, q_2)$ or $F(q_1, q_2)$	$F^{-1}(\alpha; q_1, q_2)$

the expected volume of the likelihood ratio confidence region is infinite both when the approximating random variable X is used in the computation and when L itself is used.

4. Show that if $Y \sim F'[q, n-p, \lambda(\gamma^0)]$, where
$$\lambda(\gamma^0) = \frac{[\gamma(\theta^0) - \gamma^0]'\left\{ H(\theta^0)[F'(\theta^0)F(\theta^0)]^{-1}H'(\theta^0)\right\}^{-1}[\gamma(\theta^0) - \gamma^0]}{(2\sigma)^2}$$

and $c_Y(\gamma^0) = P(Y \le F_\alpha)$, then $\int_{\mathbf{R}^q} c_Y(\gamma)\, d\gamma < \infty$.

Table 11. (*Continued*)

Random Variable	Density Function	Distribution Function	Quantile Function
Noncentral F-distribution with q_1 numerator degrees of freedom, q_2 denominator degrees of freedom, and noncentrality parameter λ; $F = (X_1/q_1)/(X_2/q_2)$, where the X_i are independent, $X_1 \sim \chi'^2(q_1, \lambda)$, $X_2 \sim \chi^2(q_2)$.	—	$F'(x; q_1, q_2, \lambda)$ or $F'(q_1, q_2, \lambda)$	$F'^{-1}(\alpha; q_1, q_2, \lambda)$
Doubly noncentral F-distribution with q_1 numerator degrees of freedom, q_2 denominator degrees of freedom, and noncentrality parameters λ_1 and λ_2; $F = (X_1/q_1)/(X_2/q_2)$, where the X_i are independent, $X_i \sim \chi'^2(q_i, \lambda_i)$.	—	$F''(x; q_1, q_2, \lambda_1, \lambda_2)$ or $F''(q_1, q_2, \lambda_1, \lambda_2)$	$F''^{-1}(\alpha; q_1, q_2, \lambda_1, \lambda_2)$
t-distribution with q degrees of freedom; $t = X/\sqrt{Y/q}$, where X and Y are independent, $X \sim N(0,1)$, $Y \sim \chi^2(q)$.	—	$t(x; q)$ or $t(q)$	$t^{-1}(\alpha; q)$
Noncentral t-distribution with q degrees of freedom and noncentrality parameter μ; $t = X/\sqrt{Y/q}$, where X and Y are independent, $X \sim N(u, 1)$, $Y \sim \chi^2(q)$.	—	$t'(x; q, \mu)$ or $t'(q, \mu)$	$t'^{-1}(\alpha; q, \mu)$

7. APPENDIX: DISTRIBUTIONS

Table 11 lists the conventions used to denote various distributions that arise in linear regression analysis together with the few facts regarding them that we shall use. More details are in Section 2.4 of Searle (1971).

We shall assume familiarity with the salient facts regarding linear and quadratic forms in normally distributed random variables. In terms of the notation of the table, they are:

1. If $X \sim N_p(\mu, \Sigma)$ and A is a symmetric matrix, then $\mathscr{E}X = \mu$, $\mathscr{C}(X, X') = \mathscr{E}(X - \mu)(X - \mu)' = \Sigma$, and $\mathscr{E}X'AX = \text{trace}(A\Sigma) + \mu'A\mu$.

2. If $X \sim N_p(\mu, \Sigma)$ and H is a q by p matrix, then $HX \sim N_q(H\mu, H\Sigma H')$.

3. If $X \sim N_p(\mu, \Sigma)$ and A is a symmetric matrix, then $X'AX \sim \chi'^2[\text{rank}(A), \mu'A\mu/2]$ if and only if $A\Sigma$ is idempotent.

4. If $X \sim N_p(\mu, \Sigma)$ and A is a symmetric matrix, then $X'AX$ and HX are distributed independently if and only if $H\Sigma A = 0$.

5. If $X \sim N_p(\mu, \Sigma)$ and A, B are symmetric matrices, then $X'AX$ and $X'BX$ are distributed independently if and only if $B\Sigma A = 0$.

For proofs and additional details see Section 2.5 of Searle (1971).

CHAPTER 2

Univariate Nonlinear Regression: Special Situations

In this chapter, we shall consider some special situations that are often encountered in the analysis of univariate nonlinear models but lie outside the scope of the standard least squares methods that were discussed in the previous chapter.

The first situation considered is the problem of heteroscedastic errors. Two solutions are proposed: Either deduce the pattern of the heteroscedasticity, transform the model, and then apply standard nonlinear methods, or use least squares and substitute heteroscedastic invariant variance estimates and test statistics. The former offers efficiency gains if a suitable transformation can be found.

The second situation is the problem of serially correlated errors. The solution is much as above. If the errors appear to be covariance stationary, then a suitable transformation will reduce the model to the standard case. If the errors appear to be both serially correlated and heteroscedastic, then least squares estimators can be used with invariant variance estimates and test statistics.

The third is a testing problem involving model choice which arises quite often in applications but violates the regularity conditions needed to apply standard methods. A variant of the lack-of-fit test is proposed as a solution.

The last topic is a brief discussion of nonlinearity measures. They can be used to find transformations that will improve the performance of optimization routines and, perhaps, the accuracy of probability statements. The latter is an open question, as the measures relate to sufficient conditions, not necessary conditions, and little Monte Carlo evidence is available.

1. HETEROSCEDASTIC ERRORS

If the variance σ_t^2 of the errors in the nonlinear model

$$y_t = f(x_t, \theta^0) + e_t \qquad t = 1, 2, \ldots, n$$

is known to depend on x_t, viz.

$$\sigma_t^2 = \frac{\sigma^2}{\psi^2(x_t)}$$

then the situation can be remedied using weighted least squares—see Judge et al. (1980, Section 4.3) for various tests for heteroscedasticity. Put

$$\text{``}y_t\text{''} = \psi(x_t) y_t$$
$$\text{``}f\text{''}(x_t, \theta) = \psi(x_t) f(x_t, \theta)$$

and apply the methods of the previous chapter with "y_t" and "f"(x_t, θ) replacing y_t and $f(x_t, \theta)$ throughout. The justification for this approach is straightforward. If the errors e_t are independent, then the errors

$$\text{``}e_t\text{''} = \text{``}y_t\text{''} - \text{``}f\text{''}(x_t, \theta^0) \qquad t = 1, 2, \ldots, n$$

will be independent and have constant variance σ^2 as required.

If the transformation

$$\sigma_t^2 = \frac{\sigma^2}{\psi^2(x_t, \tau^0)}$$

depends on an unknown parameter τ, there are a variety of approaches that one might use. If one is willing to take the trouble, the best approach is to write the model as

$$q(y_t, x_t, \theta^0, \tau^0) = \psi(x_t, \tau^0)[y_t - f(x_t, \theta^0)] = \text{``}e_t\text{''}$$

and estimate the parameters $\lambda = (\theta, \tau, \sigma^2)$ jointly using maximum likelihood as discussed in Section 5 of Chapter 6. If not, and the parameters τ do not depend functionally on θ—or one is willing to forgo efficiency gains if they do—then a two step approach can be used. It is as follows.

Let $\hat{\theta}$ denote the least squares estimator computed by minimizing

$$\sum_{t=1}^{n} [y_t - f(x_t, \theta)]^2.$$

Put

$$|\hat{e}_t| = |y_t - f(x_t, \hat{\theta})|$$

and estimate τ^0 by $\hat{\tau}$ where $(\hat{\tau}, \hat{c})$ minimizes

$$\sum_{t=1}^{n} \left(|\hat{e}_t| - \frac{c}{\psi(x_t, \tau)} \right)^2;$$

$\hat{c}$ will be a consistent estimator of $\sqrt{2\sigma^2/\pi}$ if the errors are normally distributed. The methods discussed in Section 4 of Chapter 1 can be used to compute this minimum. Put

$$"y_t" = \psi(x_t, \hat{\tau}) y_t$$
$$"f"(x_t, \theta) = \psi(x_t, \hat{\tau}) f(x_t, \theta)$$

and apply the methods of the previous chapter with $"y_t"$ and $"f"(x_t, \theta)$ replacing y_t and $f(x_t, \theta)$ throughout. Section 3 of Chapter 3 provides the theoretical justification for this approach.

If one suspects that heteroscedasticity is present but cannot deduce an acceptable form for $\psi(x_t, \tau)$, another approach is to use least squares estimators and correct the variance estimate. As above, let $\hat{\theta}$ denote the least squares estimator, the value that minimizes

$$s_n(\theta) = \frac{1}{n} \sum_{t=1}^{n} [y_t - f(x_t, \theta)]^2$$

and let $\hat{e}_t$ denote residuals

$$\hat{e}_t = y_t - f(x_t, \hat{\theta}) \qquad t = 1, 2, \ldots, n.$$

Upon application of the results of Section 3 of Chapter 3,

$$\sqrt{n}(\hat{\theta} - \theta^0) \xrightarrow{\mathscr{L}} N_p(0, V)$$

with

$$V = \mathscr{J}^{-1} \mathscr{I} \mathscr{J}^{-1}.$$

$\mathscr{I}$ and $\mathscr{J}$ can be estimated using

$$\hat{\mathscr{I}} = \frac{1}{n}\sum_{t=1}^{n}\hat{e}_t^2\left(\frac{\partial}{\partial\theta}f(x_t,\hat{\theta})\right)\left(\frac{\partial}{\partial\theta}f(x_t,\hat{\theta})\right)'$$

and

$$\hat{\mathscr{J}} = \frac{1}{n}\sum_{t=1}^{n}\left(\frac{\partial}{\partial\theta}f(x_t,\hat{\theta})\right)\left(\frac{\partial}{\partial\theta}f(x_t,\hat{\theta})\right)'.$$

For testing

$$H: h(\theta^0) = 0 \quad \text{against} \quad A: h(\theta^0) \neq 0$$

where $h: \mathbb{R}^p \to \mathbb{R}^q$, the Wald test statistic is (Theorem 11, Chapter 3)

$$W = nh'(\hat{\theta})[\hat{H}\hat{V}\hat{H}']^{-1}h(\hat{\theta})$$

where $\hat{H} = (\partial/\partial\theta')h(\hat{\theta})$ and $\hat{V} = \hat{\mathscr{J}}^{-1}\hat{\mathscr{I}}\hat{\mathscr{J}}^{-1}$. The null hypothesis $H: h(\theta^0) = 0$ is rejected in favor of the alternative hypothesis $A: h(\theta^0) \neq 0$ when the test statistic exceeds the upper $\alpha \times 100$ percentage point χ_α^2 of a chi-square random variable with q degrees of freedom, $\chi_\alpha^2 = (\chi^2)^{-1}(1-\alpha; q)$.

As a consequence of this result, a 95% confidence interval on the ith element of θ^0 is computed as

$$\hat{\theta}_i \pm z_{.025}\frac{\sqrt{\hat{V}_{ii}}}{\sqrt{n}}$$

where $z_{.025} = -\sqrt{(\chi^2)^{-1}(.95; 1)} = N^{-1}(.025; 0, 1)$.

Let $\tilde{\theta}$ denote the minimizer of $s_n(\theta)$, subject to the restriction that $h(\theta) = 0$. Let $\tilde{H}$, $\tilde{\mathscr{I}}$, and $\tilde{\mathscr{J}}$ denote the formulas for $\hat{H}$, $\hat{\mathscr{I}}$, and $\hat{\mathscr{J}}$ above, but with $\tilde{\theta}$ replacing $\hat{\theta}$ throughout; put

$$\tilde{V} = \tilde{\mathscr{J}}^{-1}\tilde{\mathscr{I}}\tilde{\mathscr{J}}^{-1}.$$

The Lagrange multiplier test statistic is (Theorem 14, Chapter 3)

$$R = n\left(\frac{\partial}{\partial\theta}s_n(\tilde{\theta})\right)'\tilde{\mathscr{J}}^{-1}\tilde{H}'(\tilde{H}\tilde{V}\tilde{H}')^{-1}\tilde{H}\tilde{\mathscr{J}}^{-1}\left(\frac{\partial}{\partial\theta}s_n(\tilde{\theta})\right).$$

Again, the null hypothesis $H: h(\theta^0) = 0$ is rejected in favor of the alternative hypothesis $A : h(\theta^0) \neq 0$ when the test statistic exceeds the upper $\alpha \times 100$ percentage point χ_α^2 of a chi-square random variable with q degrees of freedom, $\chi_\alpha^2 = (\chi^2)^{-1}(1 - \alpha, q)$.

The likelihood ratio test cannot be used, because $\mathscr{I} \neq \mathscr{J}$; see Theorem 15 of Chapter 3. Formulas for computing the power of the Wald and Lagrange multiplier tests are given in Theorems 11 and 14 of Chapter 3, respectively.

2. SERIALLY CORRELATED ERRORS

In this section we shall consider estimation and inference regarding the parameter θ^0 in the univariate nonlinear model

$$y_t = f(x_t, \theta^0) + u_t \qquad t = 1, 2, \ldots, n$$

when the errors are serially correlated. In most application—methods for handling exceptions are considered at the end of the section—an assumption that the process $\{u_t\}_{t=-\infty}^{\infty}$ generating the realized disturbances $\{u_t\}_{t=1}^{n}$ is covariance stationary is plausible. This is to say that the covariances $\mathrm{cov}(u_t, u_{t+h})$ of the time series depend only on the gap h and not on the position t in time. In consequence, the variance-covariance matrix Γ_n of the disturbance vector

$$u = (u_1, u_2, \ldots, u_n)' \qquad (n \times 1)$$

will have a banded structure with typical element $\gamma_{ij} = \gamma(i - j)$, where $\gamma(h)$ is the autocovariance function of the process, viz.

$$\gamma(h) = \mathrm{cov}(u_t, u_{t+h}) \qquad h = 0, \pm 1, \pm 2, \ldots.$$

The appropriate estimator, were Γ_n known, would be the generalized nonlinear least square estimator. Specifically, one would estimate θ^0 by the value of θ that minimizes

$$[y - f(\theta)]'\Gamma_n^{-1}[y - f(\theta)]$$

where

$$y = (y_1, y_2, \ldots, y_n)' \qquad (n \times 1)$$

and

$$f(\theta) = [f(x_1, \theta), f(x_2, \theta), \ldots, f(x_n, \theta)]' \qquad (n \times 1).$$

The generalized nonlinear least squares estimator is seen to be appropriate from the following considerations. Suppose that Γ_n^{-1} can be factored as

$$\Gamma_n^{-1} = (\text{scalar}) \cdot P'P.$$

If we put

$$\text{``}y\text{''} = Py \qquad \text{``}f\text{''}(\theta) = Pf(\theta) \qquad \text{``}e\text{''} = Pu$$

then the model

$$\text{``}y\text{''} = \text{``}f\text{''}(\theta) + \text{``}e\text{''}$$

satisfies the assumptions—$\mathscr{E}(\text{``}e\text{''}) = 0$, $\mathscr{C}(\text{``}e\text{''}, \text{``}e\text{''}') = \sigma^2 I$—that justify the use of the least squares estimator and associated inference procedures. However, the least squares estimator computed from the model

$$\text{``}y\text{''} = \text{``}f\text{''}(\theta) + \text{``}e\text{''}$$

is the same as the generalized least squares estimator above. This justifies the approach. More importantly, it provides computational and inference procedures—one need only transform the model using P and then apply the methods of Chapter 1 forthwith. For this approach to be practical, the matrix P must be easy to obtain, must be representable using far fewer than n^2 storage locations, and the multiplication Pw must be convenient relative to the coding requirements of standard, nonlinear least squares statistical packages. As we shall see below, if an autoregressive assumption is justified, then P is easy to obtain, can be stored using very few storage locations, and the multiplication Pw is particularly convenient.

When Γ_n is not known, as we assume here, the obvious approach is to substitute an estimator $\hat{\Gamma}_n$ in the formulas above. Section 4 of Chapter 7 furnishes the theoretical justification for this approach provided that $\hat{\Gamma}_n$ depends on a finite length vector $\hat{\tau}_n$ of random variables with $\sqrt{n}(\hat{\tau}_n - \tau^0)$ bounded in probability for some τ^0. A proof that $\hat{\Gamma}_n$ computed as described below satisfies this restriction is given by Gallant and Goebel (1975).

An autoregressive process is a time series that can be reduced to a white noise process by using a short linear filter. Specifically, the time series $\{u_t\}_{t=-\infty}^{\infty}$ is assumed to satisfy the equations

$$u_t + a_1 u_{t-1} + a_2 u_{t-2} + \cdots + a_q u_{t-q} = e_t \qquad t = 0, \pm 1, \pm 2, \ldots$$

where $\{e_t\}_{t=-\infty}^{\infty}$ is a sequence of independently and identically distributed random variables each with mean zero and variance σ^2. In addition, we assume that the roots of the characteristic polynomial

$$m^q + a_1 m^{q-1} + a_2 m^{q-2} + \cdots + a_q$$

are less than one in absolute value. The necessity for this assumption is discussed in Fuller (1976, Chapter 2); Pantula (1985) describes a testing strategy for determining the validity of this assumption. A time series $\{u_t\}_{t=-\infty}^{\infty}$ which satisfies this assumption is called an autoregressive process of order q.

EXAMPLE 1 (Wholesale prices). The Wholesale Price Index for the years 1720 through 1973 provides an illustration. The data are listed Table 1 and plotted as Figure 1. Using least squares, an exponential growth model

$$y_t = \theta_1 e^{\theta_2 t} + u_t \qquad t = 1, 2, \ldots, n = 254$$

was fitted to the data to obtain residuals $\{\hat{u}_t\}_{t=1}^{254}$. From these residuals, the autocovariances have been estimated using

$$\hat{\gamma}(h) = \frac{1}{n} \sum_{t=1}^{n-h} \hat{u}_t \hat{u}_{t+h} \qquad h = 0, 1, \ldots, 60$$

and plotted as "autocovariance" in Figure 2. Using the methods discussed below, a second order autoregression

$$u_t + a_1 u_{t-1} + a_2 u_{t-2} = e_t$$

was fitted to the residuals $\{\hat{u}_t\}_{t=1}^{254}$ to obtain

$$\left(\hat{a}_1, \hat{a}_2, \hat{\sigma}^2 \right) = (-1.048, 0.1287, 34.09).$$

Estimates of the autocovariances can be calculated from these estimates using the Yule-Walker equations as discussed in Anderson (1971, p. 174). Doing so yields the estimates plotted "autoregressive" in Figure 2. The two plots in Figure 2—autocovariance (unrestricted estimates requiring that 60 population quantities be estimated) and autoregressive (restricted estimates requiring that only three population quantities be estimated)—are in rea-

Table 1.   U.S. Wholesale Prices.

Year	Index	Year	Index	Year	Index	Year	Index
1720	16.98	1761	25.48	1802	39.04	1843	25.06
1721	15.48	1762	28.42	1803	39.29	1844	25.20
1722	16.07	1763	27.43	1804	43.30	1845	27.47
1723	16.60	1764	24.85	1805	47.84	1846	27.62
1724	17.51	1765	24.41	1806	43.94	1847	31.99
1725	19.03	1766	25.23	1807	41.85	1848	28.02
1726	19.89	1767	25.82	1808	37.01	1849	27.79
1727	19.22	1768	24.90	1809	42.44	1850	28.17
1728	18.28	1769	25.01	1810	45.03	1851	27.77
1729	18.22	1770	25.63	1811	45.12	1852	29.37
1730	19.30	1771	26.50	1812	45.77	1853	32.71
1731	17.16	1772	29.67	1813	53.76	1854	38.34
1732	16.47	1773	28.07	1814	65.06	1855	40.81
1733	17.73	1774	26.08	1815	59.92	1856	36.38
1734	17.18	1775	25.74	·1816	53.12	1857	37.77
1735	17.29	1776	29.51	1817	53.95	1858	31.56
1736	16.47	1777	42.21	1818	50.48	1859	32.97
1737	17.94	1778	48.04	1819	41.86	1860	31.48
1738	17.94	1779	77.55	1820	34.70	1861	31.11
1739	16.19	1780	77.21	1821	32.07	1862	36.35
1740	17.20	1781	74.12	1822	35.02	1863	46.49
1741	22.18	1782	57.44	1823	34.14	1864	67.47
1742	21.33	1783	44.52	1824	31.76	1865	64.67
1743	18.83	1784	34.50	1825	32.62	1866	60.82
1744	17.90	1785	31.58	1826	31.67	1867	56.63
1745	18.26	1786	29.94	1827	31.62	1868	55.23
1746	19.64	1787	28.99	1828	31.84	1869	52.78
1747	21.78	1788	25.73	1829	32.35	1870	47.19
1748	24.56	1789	27.45	1830	29.43	1871	45.44
1749	23.93	1790	31.48	1831	31.38	1872	47.54
1750	21.69	1791	29.51	1832	31.69	1873	46.49
1751	22.57	1792	30.96	1833	32.12	1874	44.04
1752	22.65	1793	34.60	1834	30.50	1875	41.25
1753	22.28	1794	36.63	1835	33.97	1876	38.45
1754	22.20	1795	44.43	1836	39.69	1877	37.05
1755	22.19	1796	50.25	1837	41.33	1878	31.81
1756	22.43	1797	45.20	1838	38.45	1879	31.46
1757	22.00	1798	42.09	1839	38.11	1880	34.96
1758	23.10	1799	43.47	1840	31.63	1881	36.00
1759	26.21	1800	44.51	1841	29.87	1882	37.75
1760	26.28	1801	48.99	1842	26.62	1883	35.31

sonable agreement, and the autoregressive assumption seems to yield an adequate approximation to the autocovariance. Indeed, it must for large enough q if the process $\{u_t\}_{t=-\infty}^{\infty}$ is, in fact, stationary (Berk, 1974).   □

The transformation matrix $\hat{P}$ based on the autoregressive assumption is computed as follows. Write the model in vector form

$$y = f(\theta^0) + u;$$

Table 1.   (Continued).

Year	Index	Year	Index	Year	Index	Year	Index
1884	32.51	1908	32.08	1932	33.16	1956	90.70
1885	29.71	1909	34.48	1933	33.63	1957	93.30
1886	28.66	1910	35.91	1934	38.21	1958	94.60
1887	29.71	1911	33.10	1935	40.84	1959	94.80
1888	30.06	1912	35.24	1936	41.24	1960	94.90
1889	28.31	1913	35.60	1937	44.03	1961	94.50
1890	28.66	1914	34.73	1938	40.09	1962	94.80
1891	28.46	1915	35.45	1939	39.36	1963	94.50
1892	26.62	1916	43.61	1940	40.09	1964	94.70
1893	27.24	1917	59.93	1941	44.59	1965	96.60
1894	24.43	1918	66.97	1942	50.39	1966	99.80
1895	24.89	1919	70.69	1943	52.67	1967	100.00
1896	23.72	1920	78.75	1944	53.05	1968	102.50
1897	23.77	1921	49.78	1945	54.01	1969	106.50
1898	24.74	1922	49.32	1946	61.72	1970	110.40
1899	26.62	1923	51.31	1947	76.65	1971	113.90
1900	28.61	1924	50.03	1948	83.08	1972	119.10
1901	28.20	1925	52.79	1949	78.48	1973	135.50
1902	30.04	1926	51.00	1950	81.68		
1903	30.40	1927	48.66	1951	91.10		
1904	30.45	1928	49.32	1952	88.60		
1905	30.65	1929	48.60	1953	87.40		
1906	31.52	1930	44.11	1954	87.60		
1907	33.25	1931	37.23	1955	87.80		

Source: Composite derived from: Wholesale Prices for Philadelphia, 1720 to 1861, Series E82, U.S. Bureau of the Census (1960); Wholesale Prices, All Commodities, 1749 to 1890, Series E1, U.S. Bureau of the Census (1960); Wholesale Prices, All Commodities, 1890 to 1951, Series E13, U.S. Bureau of the Census (1960); Wholesale Prices, All Commodities, 1929 to 1971, Office of the President (1972); Wholesale Prices, All Commodities, 1929 to 1973, Office of the President (1974).

compute the least squares estimator $\hat{\theta}$, which minimizes

$$\mathrm{SSE}(\theta) = [y - f(\theta)]'[y - f(\theta)];$$

compute the residuals

$$\hat{u} = y - f(\hat{\theta});$$

from these, estimate the autocovariances up to lag q using

$$\hat{\gamma}(h) = \frac{1}{n} \sum_{t=1}^{n-h} \hat{u}_t \hat{u}_{t+h} \qquad h = 0, 1, \ldots, q;$$

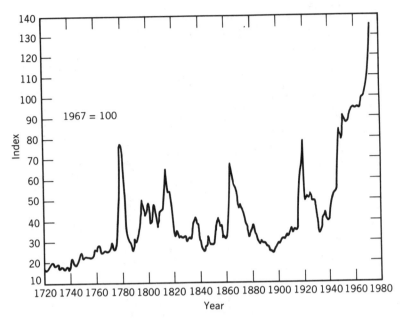

Figure 1. U.S. wholesale prices.

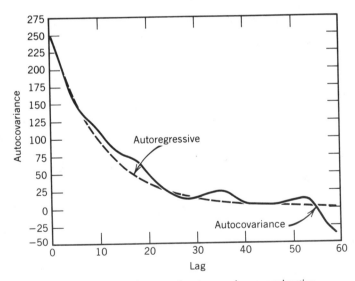

Figure 2. Autocovariances and autoregressive approximation.

put

$$\hat{\Gamma}_q = \begin{bmatrix} \hat{\gamma}(0) & \hat{\gamma}(1) & \cdots & \hat{\gamma}(q-1) \\ \hat{\gamma}(1) & \hat{\gamma}(0) & \cdots & \hat{\gamma}(q-2) \\ \vdots & \vdots & & \vdots \\ \hat{\gamma}(q-1) & \hat{\gamma}(q-2) & \cdots & \hat{\gamma}(0) \end{bmatrix} \qquad (q \times q)$$

$$\hat{\gamma}_q = [\hat{\gamma}(1), \hat{\gamma}(2), \ldots, \hat{\gamma}(q)]' \qquad (q \times 1)$$

and compute $\hat{a}$ using the Yule-Walker equations

$$\hat{a} = -\Gamma_q^{-1}\gamma_q \qquad (q \times 1)$$
$$\hat{\sigma}^2 = \gamma(0) + \hat{a}'\hat{\gamma}_q \qquad (1 \times 1);$$

factor

$$\hat{\Gamma}_q^{-1} = \hat{P}_q'\hat{P}_q$$

using, for example, Cholesky's method, and set

$$\hat{P} = \left[\begin{array}{ccccccc} \multicolumn{3}{c}{\sqrt{\hat{\sigma}^2}\,\hat{P}_q 0} \\ \hline \hat{a}_q & \hat{a}_{q-1} & \cdots & \hat{a}_1 & 1 \\ & \hat{a}_q & \hat{a}_{q-1} & \cdots & \hat{a}_1 & 1 \\ & & \ddots & & \ddots & & \ddots \\ & & & \hat{a}_q & \hat{a}_{q-1} & \cdots & \hat{a}_1 & 1 \end{array} \right] \begin{array}{l} \} q \text{ rows} \\ \\ \left.\begin{array}{l} \\ \\ \\ \end{array}\right\} n - q \text{ rows} \end{array}$$

As discussed above, $\hat{P}$ is used to transform the model $y = f(\theta) + u$ to the model "y" = "f"(θ) + "e" using

$$\text{"}y\text{"} = \hat{P}y \qquad \text{"}f\text{"}(\theta) = \hat{P}f(\theta) \qquad \text{"}e\text{"} = \hat{P}u.$$

Thereafter the methods discussed in Chapter 1 are applied to "y" and "f"(θ). This includes computation of the estimator and methods for testing hypotheses and for setting confidence intervals. The value $\hat{\theta}$ left over from the least squares estimation usually is an excellent starting value for the iterations to minimize

$$\text{"SSE"}(\theta) = [\text{"}y\text{"} - \text{"}f\text{"}(\theta)]'[\text{"}y\text{"} - \text{"}f\text{"}(\theta)].$$

Note that after the first q observations,

$$\text{"}y_t\text{"} = y_t + \hat{a}_1 y_{t-1} + \hat{a}_2 y_{t-2} + \cdots + \hat{a}_q y_{t-q}$$
$$t = q + 1, q + 2, \ldots, n$$
$$\text{"}f\text{"}(x_t, \theta) = f(x_t, \theta) + \hat{a}_1 f(x_{t-1}, \theta) + \cdots + \hat{a}_q f(x_{t-q}, \theta)$$
$$t = q + 1, q + 2, \ldots, n$$

which is particularly easy to code. We illustrate with the example.

EXAMPLE 1 (Continued). Code to compute $\hat{\theta}$ for the model

$$y_t = \theta_1 e^{\theta_2 t} + u_t, \qquad t = 1, 2, \ldots, n = 254$$
$$u_t + a_1 u_{t-1} + a_2 u_{t-2} = e_t \qquad t = 0, \pm 1, \pm 2, \ldots$$

using the data of Table 1 as shown in Figure 3.

Most of the code has to do with the organization of data sets. First, the least squares estimator $\hat{\theta}$ is computed with PROC NLIN, and the residuals from the fit are stored in WORK01. The purpose of the code which follows is to arrange these residuals so that the data set WORK03 has the form

"y"	"X"	
$\hat{u}$	0	0
0	$\hat{u}$	0
0	0	$\hat{u}$

so that standard regression formulas can be used to compute

$$\hat{\Gamma}_q = \frac{1}{n} X'X$$
$$= \begin{pmatrix} 252.32 & 234.35 \\ 234.35 & 252.32 \end{pmatrix}$$
$$\hat{a} = -(X'X)^{-1} X'y$$
$$= \begin{pmatrix} 213.20 \\ 234.35 \end{pmatrix}$$
$$\hat{\sigma}^2 = y'y - \hat{a}'X'X\hat{a}$$
$$= 34.0916$$

using PROC MATRIX. Note that the columns of "X" have been permuted to

SAS Statements:

```
PROC NLIN DATA=EG01 METHOD=GAUSS ITER=50 CONVERGENCE=1.E-10;
OUTPUT OUT=WORK01 RESIDUAL=UHAT;
PARMS T1=1 T2=.003;
MODEL Y=T1*EXP(T2*X); DER.T1=EXP(T2*X); DER.T2=T1*X*EXP(T2*X);

DATA WORK02; SET WORK01; KEEP UHAT; OUTPUT;
IF _N_=254 THEN DO; UHAT=0; OUTPUT; OUTPUT; END;
DATA WORK03; SET WORK02;
UHAT_0=UHAT; UHAT_1=LAG1(UHAT); UHAT_2=LAG2(UHAT);
IF _N_=1 THEN DO; UHAT_1=0; UHAT_2=0; END; IF _N_=2 THEN UHAT_2=0;

PROC MATRIX;
FETCH Y DATA=WORK03(KEEP=UHAT_0); FETCH X DATA=WORK03(KEEP=UHAT_2 UHAT_1);
GG=X'*X#/254; G=X'*Y#/254; A=-INV(GG)*G; SS=Y'*Y#/254-A'*GG*A;
SPQ=HALF(SS#INV(GG));
ZERO=0/0; ONE=1; P=SPQ||ZERO; ROW=A'||ONE; DO I=1 TO 252; P=P//ROW; END;
OUTPUT P OUT=WORK04;

DATA WORK05; SET EG01;
Y3=Y; Y2=LAG1(Y); Y1=LAG2(Y); X3=X; X2=LAG1(X); X1=LAG2(X);
IF _N_=3 THEN DO; OUTPUT; OUTPUT; END; IF _N_>2 THEN OUTPUT;
DATA WORK06; MERGE WORK04 WORK05; DROP ROW Y X;

PROC NLIN DATA=WORK06 METHOD=GAUSS ITER=50 CONVERGENCE=1.E-10;
PARMS T1=1 T2=.003;
Y = COL1*Y1 + COL2*Y2 + COL3*Y3;
F = COL1*T1*EXP(T2*X1) + COL2*T1*EXP(T2*X2) +COL3*T1*EXP(T2*X3);
D1 = COL1*EXP(T2*X1) + COL2*EXP(T2*X2) +COL3*EXP(T2*X3);
D2 = COL1*T1*X1*EXP(T2*X1) + COL2*T1*X2*EXP(T2*X2) +COL3*T1*X3*EXP(T2*X3);
MODEL Y = F; DER.T1 = D1; DER.T2 = D2;
```

Output:

SAS   5

NON-LINEAR LEAST SQUARES SUMMARY STATISTICS   DEPENDENT VARIABLE Y

SOURCE	DF	SUM OF SQUARES	MEAN SQUARE
REGRESSION	2	4428.02183844	2214.01091922
RESIDUAL	252	5656.56502716	22.44668662
UNCORRECTED TOTAL	254	10084.58686559	

PARAMETER	ESTIMATE	ASYMPTOTIC STD. ERROR	ASYMPTOTIC 95 % CONFIDENCE INTERVAL	
			LOWER	UPPER
T1	12.19756397	3.45880524	5.38562777	19.00950018
T2	0.00821720	0.00133383	0.00559029	0.01084410

Figure 3. Example 1 estimated using an autoregressive transformation.

permute the columns of $\hat{a}$ in this code. The transformation $\hat{P}$ is put in the data set **WORK04**, whose first two rows contains $(\sqrt{\hat{\sigma}^2}\,\hat{P}_q\,|\,0)$ and remaining rows contain $(\hat{a}_2, \hat{a}_1, 1)$. The transformation is merged with the data, and lagged values of the data, and stored in the data set **WORK06**. The appearance of the data set **WORK06** is as follows:

OBS	COL1	COL2	COL3	Y1	Y2	Y3	X1	X2	X3
1	0.991682	-0.9210	0	16.98	15.48	16.07	1	2	3
2	0.000000	0.3676	0	16.98	15.48	16.07	1	2	3
3	0.128712	-1.0483	1	16.98	15.48	16.07	1	2	3
4	0.128712	-1.0483	1	15.48	16.07	16.60	2	3	4
.									
.									
.									
254	0.128712	-1.0483	1	113.90	119.10	135.50	522	253	254

Using PROC NLIN one obtains

$$\hat{\theta} = (12.1975, 0.00821720)'$$

as shown in Figure 3.

A word of caution. This example is intended to illustrate the computations, not to give statistical guidance. Specifically, putting $x_t = t$ violates the regularity conditions of the asymptotic theory, and visual inspection of Figure 1 suggests a lack of stationarity, as the variance seems to be growing with time. □

Monte Carlo simulations reported in Gallant and Goebel (1976) suggest that the efficiency gains, relative to least squares, using this procedure can be substantial. They also suggest that the probability statements associated to hypothesis and confidence intervals are not as accurate as one might hope, but they are certainly an improvement over least squares probability statements. These statements hold true whether the series $\{u_t\}_{t=-\infty}^{\infty}$ that generates the data is actually an autoregressive process of order q or some other covariance stationary process such as a moving average process that can be approximated by an autoregression.

The order q of the autoregressive process which best approximates the error process $\{u_t\}$ is unknown in applications. One approach is to attempt to determine q from the least squares residuals $\{\hat{u}_t\}$.

This problem is very much analogous to the problem of determining the appropriate degree of polynomial to use in polynomial regression analysis. The correct analogy is obtained by viewing $\hat{\Gamma}_q a = -\hat{\gamma}_q$ as the normal equations with solution vector $\hat{a} = -\hat{\Gamma}_q\hat{\gamma}_q$ and residual mean square $s^2 = [\hat{\gamma}(0) + \hat{a}'\hat{\gamma}_q]/(n-q)$. The hypotheses $H: a_i = 0$ against $A: a_i \neq 0$, $i = 1, 2, \ldots, q$, may be tested using $t_i = |\hat{a}_i|/\sqrt{s^2\hat{\gamma}^{ii}}$, where $\hat{\gamma}^{ii}$ is the ith diagonal element of $\hat{\Gamma}_q^{-1}$, by entering tables of t with $n - q$ degrees of freedom. Standard techniques of degree determination in polynomial regression analysis may be employed, of which two are to test sequentially upward, or to start from a very high order and test downward (Anderson, 1971, Section 3.2.2).

Akaike's (1969) method is a variant on the familiar procedure of plotting the residual mean square against the degree of the fitted polynomial and terminating when the curve flattens. Akaike plots

$$\text{FPE} = \left(1 + \frac{q+p}{n}\right)\frac{1}{n-q-p}\sum_{t=1}^{n}\left(\hat{u}_t + \sum_{j=1}^{q}\hat{a}_j\hat{u}_{t-j}\right)^2$$

against q for all q less than an *a priori* upper bound; in this computation put $\hat{u}_0, \hat{u}_{-1}, \ldots, \hat{u}_{1-q} = 0$. That q at which the minimum obtains is selected as the order of the approximating autoregressive process.

The methods discussed above are appropriate if the error process $\{u_t\}$ is covariance stationary. If there is some evidence to the contrary, and a transformation such as discussed in the previous section will not induce stationarity, then an alternative approach is called for. The easiest is to make no attempt at efficiency gain as above, but simply correct the standard errors of least squares estimators and let it go at that. The method is as follows.

As above, let $\hat{\theta}$ denote the least squares estimator, the value that minimizes

$$s_n(\theta) = \frac{1}{n}\sum_{t=1}^{n}[y_t - f(x_t, \theta)]^2$$

and let $\hat{u}_t$ denote residuals

$$\hat{u}_t = y_t - f(x_t, \hat{\theta}) \qquad t = 1, 2, \ldots, n.$$

Upon application of the results of Section 4 of Chapter 7, approximately—see Theorem 6 of Chapter 7 for an exact statement—

$$\sqrt{n}(\hat{\theta} - \theta^0) \sim N_p(0, V)$$

with

$$V = \mathscr{J}^{-1}\mathscr{I}\mathscr{J}^{-1}.$$

$\mathscr{I}$ and $\mathscr{J}$ can be estimated using

$$\hat{\mathscr{J}} = \sum_{\tau=-l(n)}^{l(n)} w\left(\frac{\tau}{l(n)}\right) \hat{\mathscr{S}}_{n\tau}$$

and

$$\hat{\mathscr{I}} = \frac{1}{n} \sum_{t=1}^{n} \left(\frac{\partial}{\partial\theta} f(x_t, \hat{\theta})\right)\left(\frac{\partial}{\partial\theta} f(x_t, \hat{\theta})\right)'$$

where $l(n)$ is the integer nearest $n^{1/5}$,

$$w(x) = \begin{cases} 1 - 6|x|^2 + 6|x|^3 & 0 \le x \le \frac{1}{2} \\ 2(1 - |x|)^3 & \frac{1}{2} \le x \le 1 \end{cases}$$

and

$$\mathscr{S}_{n\tau} = \begin{cases} \frac{1}{n} \sum_{t=1+\tau}^{n} \left(\hat{u}_t \frac{\partial}{\partial\theta} f(x_t, \hat{\theta})\right)\left(\hat{u}_{t-\tau} \frac{\partial}{\partial\theta} f(x_{t-\tau}, \hat{\theta})\right)' & \tau \ge 0 \\ (\hat{\mathscr{S}}_{n, -\tau})' & \tau < 0. \end{cases}$$

For testing

$$H: h(\theta^0) = 0 \quad \text{against} \quad A: h(\theta^0) \ne 0$$

where $h: \mathbb{R}^p \to \mathbb{R}^q$, the Wald test statistic is (Theorem 12, Chapter 7)

$$W = nh'(\hat{\theta})[\hat{H}\hat{V}\hat{H}']^{-1}h(\hat{\theta})$$

where $\hat{H} = (\partial/\partial\theta')h(\hat{\theta})$ and $\hat{V} = \hat{\mathscr{J}}^{-1}\hat{\mathscr{I}}\hat{\mathscr{J}}^{-1}$. The null hypothesis $H: h(\theta^0) = 0$ is rejected in favor of the alternative hypothesis $A: h(\theta^0) \ne 0$ when the test statistic exceeds the upper $\alpha \times 100$ percentage point χ_α^2 of a chi-square random variable with q degrees of freedom, $\chi_\alpha^2 = (\chi^2)^{-1}(1 - \alpha; q)$.

As a consequence of this result, a 95% confidence interval on the ith element of θ^0 is computed as

$$\hat{\theta}_i \pm z_{.025} \frac{\sqrt{\hat{V}_{ii}}}{\sqrt{n}}$$

where $z_{.025} = -\sqrt{(\chi^2)^{-1}(.95; 1)} = N^{-1}(.025; 0, 1)$.

Let $\tilde{\theta}$ denote the minimizer of $s_n(\theta)$, subject to the restriction that $h(\theta) = 0$. Let $\tilde{H}$, $\tilde{\mathscr{I}}$, and $\tilde{\mathscr{J}}$ denote the formulas for $\hat{H}$, $\hat{\mathscr{I}}$, and $\hat{\mathscr{J}}$ above but with $\tilde{\theta}$ replacing $\hat{\theta}$ throughout; put

$$\tilde{V} = \tilde{\mathscr{J}}^{-1}\tilde{\mathscr{I}}\tilde{\mathscr{J}}^{-1}.$$

The Lagrange multiplier test statistic is (Theorem 16, Chapter 7)

$$R = n\left(\frac{\partial}{\partial\theta}s_n(\tilde{\theta})\right)'\tilde{\mathscr{J}}^{-1}\tilde{H}'(\tilde{H}\tilde{V}\tilde{H}')^{-1}\tilde{H}\tilde{\mathscr{J}}^{-1}\left(\frac{\partial}{\partial\theta}s_n(\tilde{\theta})\right).$$

Again, the null hypothesis $H: h(\theta^0) = 0$ is rejected in favor of the alternative hypothesis $A: h(\theta^0) \neq 0$ when the test statistic exceeds the upper $\alpha \times 100$ percentage point χ^2_α of a chi-square random variable with q degrees of freedom $\chi^2_\alpha = (\chi^2)^{-1}(1 - \alpha; q)$.

The likelihood ratio test cannot be used, because $\mathscr{I} \neq \mathscr{J}$; see Theorem 17 of Chapter 7. Formulas for computing the power of the Wald and Lagrange multiplier tests are given in Theorems 14 and 16 of Chapter 7, respectively.

3. TESTING A NONLINEAR SPECIFICATION

Often, it is helpful to be able to choose between two model specifications:

$$H: y_t = g(x_t, \psi) + e_t$$

and

$$A: y_t = g(x_t, \psi) + \tau h(x_t, \omega) + e_t.$$

The unknown parameters are ψ, τ, and ω of dimension u, 1, and v, respectively. The functional forms $g(x, \psi)$ and $h(x, \omega)$ are known. The errors e_t are normally and independently distributed with mean zero and unknown variance σ^2. Parametrically, the situation is equivalent to testing

$$H: \tau = 0 \quad \text{against} \quad A: \tau \neq 0$$

regarding ψ, ω, and σ^2 as nuisance parameters.

It would be natural to employ one of the tests discussed in Section 5 of the previous chapter. In the formal sense, the Lagrange multiplier test is undefined because ω cannot be estimated if $\tau = 0$. The likelihood ratio test is defined in the sense that the residual sum of squares from the model

$$H: y_t = g(x_t, \psi) + e_t$$

can be computed and used as SSE($\tilde{\theta}$). But one must also compute the unconstrained estimate of

$$\theta = (\psi, \omega, \tau)$$

to obtain SSE($\hat{\theta}$) in order to compute the likelihood ratio test statistic; the Wald test statistic also requires computation of $\hat{\theta}$. When H is true, this dependence on $\hat{\theta}$ causes two difficulties:

1. It is likely that the attempt to fit the full model will fail or, at best, converge very slowly, as seen in Figure 6 and Table 4 of Chapter 1.
2. The regularity conditions used to obtain the asymptotic properties of the unconstrained least squares estimator $\hat{\theta}$—and also of test statistics that depend on $\hat{\theta}$—are violated, as neither the identification condition or the rank condition discussed in Section 3 of the previous chapter is satisfied.

It is useful to consider when the situation of testing H against A using data which support H is likely to arise. It is improbable that one will attempt to fit a nonlinear model which is not supported by the data if one is merely attempting to represent data parametrically without reference to a substantive problem. For example, in the cases considered in Table 4 of Chapter 1, plots of the observed response y_t against the input x_{3t} failed to give any visual impression of exponential growth for values of $|\theta_4|$ less than .1. Consequently, substantive rather than data analytic considerations will likely have suggested A. As we shall see, it will be helpful if these same substantive considerations also imply probable values for ω.

The lack-of-fit test has been discussed by several authors (Beale, 1960; Halperin, 1963; Hartley, 1964; Turner, Monroe, and Lucas, 1961; Williams, 1962) in the context of finding exact tests or confidence regions in nonlinear regression analysis. Here the same idea is employed, but an asymptotic theory is substituted for an exact small sample theory. The basic idea is straightforward: If $\tau^0 = 0$, then the least squares estimator of the parameter δ in the model

$$\hat{A} : y_t = g(x_t, \psi) + z_t'\delta + e_t$$

where the w-vector z_t does not depend on any unknown parameters, is estimating the zero vector. Thus any (asymptotically) level α test of

$$H : y_t = g(x_t, \psi) + e_t \quad \text{against} \quad \hat{A} : y_t = g(x_t, \psi) + z_t'\delta + e_t$$

is an (asymptotically) level α test of

$$H: y_t = g(x_t, \psi) + e_t \quad \text{against} \quad A: y_t = g(x_t, \psi) + \tau h(x_t, \omega) + e_t.$$

Note that since z_t does not depend on any unknown parameters, the computational problems that arise when trying to fit A by least squares will not arise when fitting $\hat{A}$.

When H is true, any of the tests considered in Section 5 of Chapter 1 are asymptotically level α. Regularity conditions such that the Wald and likelihood ratio test statistics for H against $\hat{A}$ follow the noncentral F-distribution plus an asymptotically negligible remainder term when A is true are in Gallant (1977a). Simulations reported in Gallant (1977a) suggest that the problem of inaccurate probability statements associated with the Wald test statistic are exacerbated in the present circumstance. The simulations support the use of the likelihood ratio test; the Lagrange multiplier test was not considered. The likelihood ratio test is computed as follows.

Let $(\hat{\psi}, \hat{\delta})$ denote the least squares estimator for the model $\hat{A}$, and define

$$\text{SSE}(\hat{\psi}, \hat{\delta}) = \sum_{t=1}^{n} \left[y_t - g(x_t, \hat{\psi}) + z_t' \hat{\delta} \right]^2$$

$$G(\psi) = \text{the } n \text{ by } u \text{ matrix with } t\text{th } row \ \frac{\partial}{\partial \psi'} g(x_t, \psi)$$

$$Z = \text{the } n \text{ by } w \text{ matrix with } t\text{th } row \ z_t'$$

$$Q_G = I - G(\psi^0) \left[G'(\psi^0) G(\psi^0) \right]^{-1} G'(\psi^0).$$

Let $\tilde{\psi}$ denote the least squares estimator for the model H, and define

$$\text{SSW}(\tilde{\psi}) = \sum_{t=1}^{n} \left[y_t - g(x_t, \tilde{\psi}) \right]^2.$$

The likelihood ratio test for H against $\hat{A}$ rejects when

$$L = \frac{\left[\text{SSE}(\tilde{\psi}) - \text{SSE}(\hat{\psi}, \hat{\delta}) \right]/w}{\text{SSE}(\hat{\psi}, \hat{\delta})/(n - u - w)}$$

exceeds F_α, the upper $\alpha \times 100$ percentage point of an F-random variable with w numerator degrees of freedom and $n - u - w$ denominator degrees of freedom; $F_\alpha = F^{-1}(1 - \alpha; w, n - u - w)$.

The objective governing the choice of the vector z_t of additional regressors is to find those which will maximize the power of the test of H against

$\hat{A}$ when A is true. The asymptotic power of the likelihood ratio test is given by the probability that a doubly noncentral F-statistic exceeds F_α (Gallant, 1977a). The noncentrality parameters of this statistic are

$$\lambda_1 = (\tau^0)^2 \frac{h'Q_G Z(Z'Q'_G Q_G Z)^{-1} Z'Q'_G h}{2\sigma^2}$$

for the numerator, and

$$\lambda_2 = (\tau^0)^2 \frac{h'Q_G h}{2\sigma^2} - \lambda_1$$

for the denominator, where

$$h = \left[h(x_1, \omega^0), h(x_2, \omega^0), \ldots, h(x_n, \omega^0) \right]'.$$

Thus one should attempt to find those z_t which best approximate h in the sense of maximizing the ratio

$$\frac{h'Q_G Z(Z'Q'_G Q_G Z)^{-1} Z'Q'_G h}{h'Q_G h}$$

while attempting, simultaneously, to keep the number of columns of Z as small as possible. We consider, next, how this might be done in applications.

In a situation where substantive considerations or previous experimental evidence suggest a single point estimate $\hat{\omega}$ for ω^0, the natural choice is $z_t = h(x_t, \hat{\omega})$.

If, instead of a point estimate, ranges of plausible values for the components of ω are available then a representative selection of values of ω,

$$\{ \hat{\omega}_i : i = 1, 2, \ldots, K \}$$

whose components fall within these ranges can be chosen—either deterministically or by random sampling from a distribution defined on the plausible values—and the vectors $h(\hat{\omega}_i)$ made the columns of Z. If, following this procedure, the number of columns of Z is unreasonably large, it may be reduced as follows. Decompose the matrix

$$B = \left[h(\hat{\omega}_1) | \cdots | h(\hat{\omega}_K) \right]$$

into its principal component vectors, and choose the first few of these to

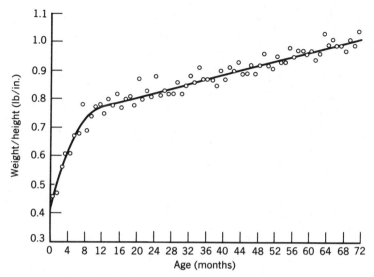

Figure 4. Preschool boys' weight/height ratio.

make up Z; equivalently, obtain the singular value decomposition (Bussinger and Golub, 1969) $B = USV'$ where $U'U = V'V = VV' = I$, and S is diagonal with nonnegative entries, and choose the first few columns of U to make up Z. We illustrate with an example.

EXAMPLE 2 (Preschool boys' weight/height ratio). The data shown in Figure 4 are preschool boys' weight/height ratios plotted against age and were obtained from Eppright et al. (1972); the tabular values are shown in Table 2. The question is whether the data support the choice of a three segment quadratic-quadratic-linear polynomial response function as opposed to a two segment quadratic-linear response function. In both cases, the response function is required to be continuous in x (age) and to have a continuous first derivative in x. Formally,

$$H: y_t = \theta_1 + \theta_2 x_t + \theta_3 T_2(\theta_4 - x_t) + e_t$$

and

$$A: y_t = \theta_1 + \theta_2 x_t + \theta_3 T_2(\theta_4 - x_t) + \theta_5 T_2(\theta_6 - x_t) + e_t$$

where

$$T_k(z) = \begin{cases} z^k & \text{when} \quad z \geq 0 \\ 0 & \text{when} \quad z \leq 0; \end{cases}$$

Table 2. Boys' Weight / Height Versus Age.

W/H	Age	W/H	Age	W/H	Age
0.46	0.5	0.88	24.5	0.92	48.5
0.47	1.5	0.81	25.5	0.96	49.5
0.56	2.5	0.83	26.5	0.92	50.5
0.61	3.5	0.82	27.5	0.91	51.5
0.61	4.5	0.82	28.5	0.95	52.5
0.67	5.5	0.86	29.5	0.93	53.5
0.68	6.5	0.82	30.5	0.93	54.5
0.78	7.5	0.85	31.5	0.98	55.5
0.69	8.5	0.88	32.5	0.95	56.5
0.74	9.5	0.86	33.5	0.97	57.5
0.77	10.5	0.91	34.5	0.97	58.5
0.78	11.5	0.87	35.5	0.96	59.5
0.75	12.5	0.87	36.5	0.97	60.5
0.80	13.5	0.87	37.5	0.94	61.5
0.78	14.5	0.85	38.5	0.96	62.5
0.82	15.5	0.90	39.5	1.03	63.5
0.77	16.5	0.87	40.5	0.99	64.5
0.80	17.5	0.91	41.5	1.01	65.5
0.81	18.5	0.90	42.5	0.99	66.5
0.78	19.5	0.93	43.5	0.99	67.5
0.87	20.5	0.89	44.5	0.97	68.5
0.80	21.5	0.89	45.5	1.01	69.5
0.83	22.5	0.92	46.5	0.99	70.5
0.81	23.5	0.89	47.5	1.04	71.5

Source: Gallant (1977a).

see Gallant and Fuller (1973) for a discussion of the derivation and fitting of grafted polynomial models.

The correspondence with the notation above is

$$\psi = (\theta_1, \theta_2, \theta_3, \theta_4)'$$
$$\tau = \theta_5$$
$$\omega = \theta_6$$
$$g(x, \psi) = \psi_1 + \psi_2 x + \psi_3 T_2(\psi_4 - x)$$
$$h(x, \omega) = T_2(\omega - x).$$

The parameter ω is the join point associated with the quadratic term whose omission is proposed.

Suppose plausible values for ω are $\hat{\omega}_1 = 4$, $\hat{\omega}_2 = 8$, and $\hat{\omega}_3 = 12$. The matrix B, described above, has typical row

$$B_t = [T_2(4 - x_t), T_2(8 - x_t), T_2(12 - x_t)].$$

The first principal component vector of B, with elements

$$z_t = [(2.08)T_2(4 - x_t) + (14.07)T_2(8 - x_t) + (39.9)T_2(12 - x_t)] \times 10^{-4}$$

```
SAS Statements:

DATA B; SET EG02;
Z1=(4-AGE>0)*(4-AGE)**2; Z2=(8-AGE>0)*(8-AGE)**2; Z3=(12-AGE>0)*(12-AGE)**2;
PROC MATRIX; FETCH B DATA=B(KEEP=Z1 Z2 Z3); SVD U Q V B; OUTPUT U OUT=WORK01;
DATA WORK02; MERGE EG02 WORK01;  KEEP AGE WH Z; Z=COL1;

PROC NLIN DATA=WORK02 METHOD=GAUSS ITER=50 CONVERGENCE=1.E-10;
PARMS T1=1 T2=.004 T3=-.002 T4=12; X=(T4-AGE>0)*(T4-AGE);
MODEL WH = T1+T2*AGE+T3*X**2;
DER.T1=1; DER.T2=AGE; DER.T3=X**2; DER.T4=2*T3*X;

PROC NLIN DATA=WORK02 METHOD=GAUSS ITER=50 CONVERGENCE=1.E-10;
PARMS T1=.73 T2=.004 T3=-5.E-5 T4=21.181 D=-.4; X=(T4-AGE>0)*(T4-AGE);
MODEL WH = T1+T2*AGE+T3*X**2+Z*D;
DER.T1=1; DER.T2=AGE; DER.T3=X**2; DER.T4=2*T3*X; DER.D=Z;
```

Output:

SAS   3

NON-LINEAR LEAST SQUARES SUMMARY STATISTICS   DEPENDENT VARIABLE WH

SOURCE	DF	SUM OF SQUARES	MEAN SQUARE
REGRESSION	4	53.67750135	13.41937534
RESIDUAL	68	0.03789865	0.00055733
UNCORRECTED TOTAL	72	53.71540000	

SAS   5

NON-LINEAR LEAST SQUARES SUMMARY STATISTICS   DEPENDENT VARIABLE WH

SOURCE	DF	SUM OF SQUARES	MEAN SQUARE
REGRESSION	5	53.67770969	10.73554194
RESIDUAL	67	0.03769031	0.00056254
UNCORRECTED TOTAL	72	53.71540000	

Figure 5. Lack-of-fit test illustrated using Example 2.

was chosen as the additional regressor. This choice yields

$$\text{SSE}(\tilde{\psi}) = 0.03789865 \quad \text{(from Fig. 5)}$$

$$\text{SSE}(\hat{\psi}, \hat{\delta}) = 0.03769031 \quad \text{(from Fig. 5)}$$

$$L = \frac{(0.03789865 - 0.03769031)/1}{0.03769031/(72 - 4 - 1)}$$

$$= 0.370$$

$$P\left[F(1, 67) > 0.370\right] \doteq .485.$$

These data give little support to A.

Simulations reported by Gallant (1977a) indicate that the best choice for z_t in this example is to take as Z the first principal component of B. For practical purposes, the power of the test is as good as if the true value ω^0 were known.

4. MEASURES OF NONLINEARITY

Consider the nonlinear model

$$y = f(\theta^0) + e$$

with normality distributed errors. The lack-of-fit test statistic for $H : \theta^0 = \theta*$ against $A : \theta^0 \neq \theta*$,

$$R = \frac{[y - f(\theta*)]'F(\theta*)[F'(\theta*)F(\theta*)]^{-1}F'(\theta*)[y - f(\theta*)]/p}{[y - f(\theta*)]'[I - F(\theta*)[F'(\theta*)F(\theta*)]^{-1}F'(\theta*)][y - f(\theta*)]/(n - p)}$$

is distributed exactly as an F with p numerator degrees of freedom and $n - p$ denominator degrees of freedom when H is true. Beale (1960) studied the extent to which confidence contours constructed using the lack-of-fit test statistic coincide with contours constructed using the likelihood ratio test statistic

$$L = \frac{[SSE(\theta*) - SSE(\hat{\theta})]/p}{SSE(\hat{\theta})/(n - p)}.$$

On the basis of this study, he constructed measures of nonlinearity that measure the extent of the coincidence and suggested corrections to critical points based on these measures to improve the accuracy of confidence statements. Coincidence is a sufficient condition for accurate probability statements, not a necessary condition. Thus a large value of Beale's nonlinearity measure does not imply inaccurate probability statements, and it is possible for Beale's corrections actually to be counterproductive. Simulations reported by Gallant (1976) indicate that there are such instances.

Bates and Watts (1980) take a geometric approach in deriving their measures of nonlinearity, an approach somewhat related in spirit to Efron's (1975); Ratkowsky (1983) summarizes their work and contains FORTRAN code to compute these measures. The most interesting aspect of their work is that they break their measure into two pieces, one a measure of intrinsic curvature and the other a measure of parameter effects curvature. The latter can be reduced by reparametrization; the former cannot. In their examples, which are rather extensive, they find the parameter effects curvature is more important than the intrinsic in each case.

What is interesting about this decomposition is that it sheds some intuitive light on the question of why the likelihood ratio statistic leads to more accurate probability statements regarding the size of tests and level of confidence regions than the Wald statistic. We have seen that the Wald test is not invariant to nonlinear transformation, which means that there may exist a transformation that would make the total nonlinearity nearly equal to the intrinsic nonlinearity and so improve accuracy. The Bates-Watts measure provides some guidance in finding it; see Bates and Watts (1981). The likelihood ratio test is invariant to reparametrization, which means that it can be regarded as a statistic where this transformation has been found automatically, so that it is only the intrinsic nonlinearity that is operative. This is, of course, rather intuitive and speculative, and suffers from the same defect that was noted above: the measures, like Beale's, represent sufficient conditions, not necessary conditions; see Cook and Witmer (1985) in this regard.

The advice given in Chapter 1 was to avoid the whole issue as regards inference and simply use the likelihood ratio statistic in preference to the Wald statistic. A reparametrization will usually destroy the principal advantage of the Wald statistic, which is that it provides ellipsoidal confidence regions on model parameters. After a reparametrization, the ellipsoid will correspond to new parameters that will not necessarily be naturally associated to the problem, and one is no better off in this regard than with the likelihood ratio statistic. As regards computations, reparametrization can be helpful; see Ross (1970).

CHAPTER 3

A Unified Asymptotic Theory
of Nonlinear Models
with Regression Structure

Models have a regression structure if the predictor or explanatory variables either are design variables, (variables subject to experimental control) or are ancillary (variables that have a joint marginal distribution that does not depend on model parameters). With this type of structure, the analysis can be made conditional on the explanatory variables, and it is customary to do so. Models with lagged dependent variables as explanatory variables are excluded by this definition, and as a matter of convenience we shall also exclude models with serially correlated errors. Models with either lagged dependent variables or serially correlated errors, or both, are classified as dynamic models. An asymptotic theory for them is developed in Chapter 7.

The estimators customarily employed with models having a regression structure are defined as solutions of an optimization problem. For instance, the least squares estimator is defined as the parameter value that minimizes the residual sum of squares. This fact provides the unifying concept. In this chapter, an asymptotic theory is obtained by borrowing from the classical theory of maximum likelihood estimation, treating the objective function of the optimization problem as the analog of the log-likelihood. A theory of inference is obtained in the same way. The objective function is treated as if it were the log-likelihood to derive a Wald test statistic, a "likelihood ratio" test statistic, and a Lagrange multiplier test statistic. Their asymptotic null and nonnull distributions are found using arguments fairly similar to the classical maximum likelihood arguments. The differences from the classical theory are caused by conditioning the analysis on the explanatory variables. Observations are independently distributed, as in the classical theory, but

148

they are not identically distributed, since, at the minimum, location parameters are being shifted about by the explanatory variables.

The model that actually generates the data need not be the same as the model that was presumed to generate the data when the optimization problem was set forth. Thus, the results of this chapter can be used to obtain the asymptotic behavior of estimation and inference procedures under specification error. For example, it is not necessary to resort to Monte Carlo simulation to determine if inferences based on an exponential fit would be much affected if some other plausible growth model were to generate the data. The asymptotic approximations we give here will provide an analytical answer to the question that is sufficiently accurate for most purposes. An analytical solution also provides a qualitative understanding of the effect of misspecification that cannot be obtained otherwise.

An early version of this chapter appeared as Burguete, Gallant, and Souza (1982) together with comment by Huber (1982), Phillips (1982a), and White (1982). This chapter differs from the earlier work in that the Pitman drift assumption is isolated from the results on estimation. See especially Phillips's Comment and the Reply in this regard.

1. INTRODUCTION

An estimator is the solution of an optimization problem. It is convenient to divide these optimization problems into two groups and study these groups separately. Afterwards, one can ignore this classification and study inference in unified fashion. These two groups are least mean distance estimators and method of moments estimators. We shall define these in turn.

Multivariate nonlinear least squares is an example of a least mean distance estimator. The estimator for the model

$$y_t = f(x_t, \theta) + e_t \qquad t = 1, 2, \ldots, n$$

where y_t is an M-vector, is computed as follows. Least squares residuals $\hat{e}_{it}$ are obtained by fitting the univariate models

$$y_{it} = f_i(x_t, \theta) + e_{it} \qquad i = 1, 2, \ldots, M \quad t = 1, 2, \ldots, n$$

individually by least squares. Let $\hat{e}_t = (\hat{e}_{1t}, \hat{e}_{2t}, \ldots, \hat{e}_{Mt})'$, and put

$$\hat{\tau} = \frac{1}{n} \sum_{t=1}^{n} \hat{e}_t \hat{e}_t'.$$

The multivariate nonlinear least squares estimator is that value $\hat{\theta}$ which minimizes

$$\frac{1}{n} \sum_{t=1}^{n} \frac{1}{2} [y_t - f(x_t, \theta)]' (\hat{\tau})^{-1} [y_t - f(x_t, \theta)].$$

A general description of estimators of this type is: A least mean distance estimator is that value $\hat{\lambda}_n$ which minimizes an objective function of the form

$$s_n(\lambda) = \frac{1}{n} \sum_{t=1}^{n} s(y_t, x_t, \hat{\tau}_n, \lambda).$$

The literature subsumed by this definition is:

> Single equation nonlinear least squares—Jennrich (1969), Malinvaud (1970b), Gallant (1973, 1975a, 1975b).
>
> Multivariate nonlinear least squares—Malinvaud (1970a), Gallant (1975c), Holly (1978).
>
> Single equation and multivariate maximum likelihood—Malinvaud (1970a), Barnett (1976), Holly (1978).
>
> Maximum likelihood for simultaneous systems—Amemiya (1977), Gallant and Holly (1980).
>
> M-estimators—Balet-Lawrence (1975), Grossman (1976), Ruskin (1978).
>
> Iteratively rescaled M-estimators—Souza (1979).

Two stage nonlinear least squares is an example of a method of moments estimator. The estimator for the αth equation,

$$q_\alpha(y_t, x_t, \theta) = e_{\alpha t} \qquad t = 1, 2, \ldots, n$$

of a simultaneous system of M such equations—y_t is an M-vector—is computed as follows. One chooses instrumental variables z_t as functions of the exogenous variables x_t. Theoretical discussions of this choice consume much of the literature, but the most frequent choice in applications is low order monomials in x_t, viz.

$$z_t = \left(x_1, x_1^2, x_2, x_2^2, x_1 x_2, x_3, \ldots \right)'_t.$$

The moment equations are

$$m_n(\theta) = \frac{1}{n} \sum_{t=1}^{n} z_t q_\alpha(y_t, x_t, \theta)$$

and the true value of θ is presumed to satisfy $\mathscr{E}m_n(\theta^0) = 0$. [Note that $q_\alpha(y_t, x_t, \theta)$ is a scalar and z_t is a vector.] The two stage least squares estimator is defined as the value $\hat{\theta}$ which minimizes

$$s_n(\theta) = \tfrac{1}{2}m_n'(\theta)\left(\frac{1}{n}\sum_{t=1}^{n}z_tz_t'\right)^{-1}m_n(\theta).$$

A general description of estimators of this type is as follows. Define moment equations

$$m_n(\lambda) = \frac{1}{n}\sum_{t=1}^{n}m(y_t, x_t, \hat{\tau}_n, \lambda)$$

and a notion of distance

$$d(m, \hat{\tau}_n)$$

where we permit a dependence on a random variable $\hat{\tau}_n$ via the argument τ in $m(y, x, \tau, \lambda)$ and $d(m, \tau)$ so as to allow preliminary estimates of nuisance parameters as in three stage least squares. The estimator is that $\hat{\lambda}_n$ which minimizes

$$s_n(\lambda) = d[m_n(\lambda), \hat{\tau}_n].$$

Estimators which are properly thought of as method of moment estimators, in the sense that they can be posed no other way, are:

The Hartley-Booker estimator—Hartley and Booker (1965).

Scale invariate M-estimators—Ruskin (1978).

Two stage nonlinear least squares estimators—Amemiya (1974).

Three stage nonlinear least squares estimators—Jorgenson and Laffont (1974), Gallant (1977b), Amemiya (1977), Gallant and Jorgenson (1979).

In both least mean distance estimation and method of moments estimation, one is led to regard an estimator as the value $\hat{\lambda}_n$ which minimizes an objective function $s_n(\lambda)$. This objective function depends on the sample $\{(y_t, x_t): t = 1, 2, \ldots, n\}$ and possibly on a preliminary estimator $\hat{\tau}_n$ of some nuisance parameters. Now the negative of $s_n(\lambda)$ may be treated as if it were a likelihood function, and the Wald test statistic W_n, the "likelihood ratio" test statistic L_n, and the Lagrange multiplier test statistic R_n may be derived for a null hypothesis $H: h(\lambda) = 0$ against its alternative $A: h(\lambda)$

$\neq 0$. Almost all of the inference procedures used in the analysis of nonlinear statistical models can be derived in this way. It is only a matter of finding the appropriate objective function $s_n(\lambda)$.

We emerge from this discussion with an interest in four statistics—$\hat{\lambda}_n$, W_n, L_n, R_n—all of which depend on $s_n(\lambda)$. We should like to find their asymptotic distribution in three cases: the null case where the model is correctly specified and the null hypothesis $h(\lambda) = 0$ holds, the nonnull case where the model is correctly specified and the null hypothesis is violated, and in the case where the model is misspecified. By misspecification, one has in mind the following. The definition of an objective function $s_n(\lambda)$ which defines the four statistics of interest is motivated by a model and assumptions on the error distribution. For example, the multivariate nonlinear least squares estimator is predicated on the assumption that the data follow the model

$$y_t = f(x_t, \theta) + e_t \qquad t = 1, 2, \ldots, n$$

and that the errors have mean zero. Misspecification means that either the model assumption or the error assumption or both are violated. We find that we can obtain an asymptotic theory for all three cases at once by presuming that the data actually follow the multivariate implicit model

$$q\left(y_t, x_t, \gamma_n^0\right) = e_t \qquad t = 1, 2, \ldots, n$$

where y, q, and e are M-vectors and the parameter γ may be infinite dimensional. That is, we obtain our results with misspecification and violation of the null hypothesis presumed throughout, and then specialize to consider correctly specified null and nonnull situations. The following results are obtained.

The least mean distance estimator $\hat{\lambda}_n$, the estimator which minimizes

$$s_n(\lambda) = \frac{1}{n} \sum_{t=1}^{n} s(y_t, x_t, \hat{\tau}_n, \lambda)$$

is shown to be asymptotically normally distributed with a limiting variance-covariance matrix of the form $\mathcal{J}^{-1}\mathcal{I}\mathcal{J}^{-1}$. Consistent estimators $\hat{\mathcal{I}}_n$ and $\hat{\mathcal{J}}_n$ are set forth. Two examples—an M-estimator and an iteratively rescaled M-estimator—are carried throughout the development to illustrate the regularity conditions and results as they are introduced.

Next, method of moments estimation is taken up. The method of moments estimator $\hat{\lambda}_n$, the estimator that minimizes

$$s_n(\lambda) = d\left[m_n(\lambda), \hat{\tau}_n\right]$$

is shown to be asymptotically normally distributed with a limiting variance-covariance matrix of the form $\mathscr{J}^{-1}\mathscr{I}\mathscr{J}^{-1}$. Again, consistent estimators $\hat{\mathscr{I}}_n$ and $\hat{\mathscr{J}}_n$ are set forth. The example carried throughout the discussion is a scale invariant M-estimator.

Both analyses—least mean distance estimation and method of moments estimation—terminate with the same conclusion: $\hat{\lambda}_n$ minimizing $s_n(\lambda)$ is asymptotically normally distributed with a limiting variance-covariance matrix that may be estimated consistently by using $\hat{\mathscr{J}}_n$ and $\hat{\mathscr{I}}_n$ as intermediate statistics. As a result, an asymptotic theory for the test statistics W_n, L_n, and R_n can be developed in a single section, Section 5, without regard to whether the source of the objective function $s_n(\lambda)$ was least mean distance estimation or method of moments estimation. The discussion is illustrated with a misspecified nonlinear regression model fitted by least squares.

Observe that a least mean distance estimator may be cast into the form of a method of moments estimator by putting

$$m_n(\lambda) = \frac{1}{n} \sum_{t=1}^{n} \frac{\partial}{\partial \lambda} s(y_t, x_t, \hat{\tau}_n, \lambda)$$

and $d(m, \tau) = m'm$, because $\hat{\lambda}$ which minimizes

$$\frac{1}{n} \sum_{t=1}^{n} s(y_t, x_t, \hat{\tau}_n, \lambda)$$

solves

$$\frac{1}{n} \sum_{t=1}^{n} \frac{\partial}{\partial \lambda} s(y_t, x_t, \hat{\tau}_n, \lambda) = 0.$$

If one's only interest is the asymptotic distribution of $\hat{\lambda}_n$, then posing the problem as a method of moments estimator is the more convenient approach, as algebraic simplifications of the equations $m_n(\lambda) = 0$ prior to analysis can materially simplify the computation of the parameters of the asymptotic distribution. However, one pays two penalties for this convenience: The problem is no longer posed in a way that permits the use of the statistic L_n, and consistency results are weaker.

2. THE DATA GENERATING MODEL AND LIMITS OF CESARO SUMS

The objective is to find asymptotic approximations in situations such as the following. An analysis is predicted on the assumption that the data were generated according to the model

$$y_t = f(x_t, \lambda) + e_t \qquad t = 1, 2, \ldots, n$$

when actually they were generated according to

$$y_t = g(x_t) + e_t \qquad t = 1, 2, \ldots, n$$

where the errors have mean zero and variance σ^2. One estimates λ by $\hat{\lambda}_n$ that minimizes

$$s_n(\lambda) = \frac{1}{n} \sum_{t=1}^{n} [y_t - f(x_t, \lambda)]^2$$

and tests $H: \lambda = \lambda^*$ by rejecting when the test statistic

$$W_n = n(\hat{\lambda}_n - \lambda^*)'(\hat{\mathscr{F}}^{-1}\hat{\mathscr{I}}\hat{\mathscr{F}}^{-1})^{-1}(\hat{\lambda}_n - \lambda^*)$$

exceeds some critical value. Later we shall show that $\hat{\lambda}_n$ is estimating λ^0 that minimizes

$$s_n^0(\lambda) = \sigma^2 + \frac{1}{n} \sum_{t=1}^{n} [g(x_t) - f(x_t, \lambda)]^2.$$

Thus, one is actually testing the null hypothesis $H: \lambda^0 = \lambda^*$. Depending on the context, a test of $H: \lambda^0 = \lambda^*$ when the data are generated according to

$$y_t = g(x_t) + e_t \qquad t = 1, 2, \ldots, n$$

and not according to

$$y_t = f(x_t, \lambda) + e_t \qquad t = 1, 2, \ldots, n$$

may or may not make sense. In order to make a judgement as to whether the inference procedure is sensible, it is necessary to have the (asymptotic approximation to the) sampling distribution of W_n.

A problem in deriving asymptotic approximations to the sampling distribution of W_n is that if $\lambda^0 \neq \lambda^*$, then W_n will reject the null hypothesis with probability one as n tends to infinity, whence its limiting distribution is degenerate. The classical solution to this problem is to index the parameter as λ_n^0 and let it drift at a rate such that $\sqrt{n}(\lambda_n^0 - \lambda^*)$ converges to a finite limit, called a Pitman drift. Thus, we need some mechanism for subjecting the true model $g(x)$ to drift so as to induce the requisite drift on λ_n^0.

One possible drift mechanism is the following. Suppose that the independent variable x is univariate, that $\mathscr{X} = [0, 1]$, and that $f(x, \lambda^*)$ is continuous on $\mathscr{X}$. Then $f(x, \lambda^*)$ has a polynomial expansion

$$f(x, \lambda^*) = \sum_{i=0}^{\infty} \gamma_i^* x^i$$

by the Stone-Weierstrass theorem. If the data are generated according to the sequence of models

$$g_1(x) = \gamma_0^* + \gamma_1^* x \qquad\qquad n = 1$$
$$g_2(x) = \gamma_0^* + \gamma_1^* x + \gamma_2^* x^2 \qquad\qquad n = 2$$
$$g_3(x) = \gamma_0^* + \gamma_1^* x + \gamma_2^* x^2 + \gamma_3^* x^3 \qquad n = 3$$
$$\vdots$$

then λ_n^0 will converge to that λ^* specified by $H: \lambda = \lambda^*$ (Problem 2). Convergence can be accelerated so that $\lim \sqrt{n}\,(\lambda_n^0 - \lambda^*)$ is finite by changing a few details (Problem 2). The natural representation of this scheme is to put

$$\gamma_1^0 = (\gamma_0^*, \gamma_1^*, 0, \dots)$$
$$\gamma_2^0 = (\gamma_0^*, \gamma_1^*, \gamma_2^*, 0, \dots)$$
$$\gamma_3^0 = (\gamma_0^*, \gamma_1^*, \gamma_2^*, \gamma_3^*, 0, \dots)$$
$$\vdots$$

and let

$$g_n(x) = g(x, \gamma_n^0) = \sum_{i=0}^{\infty} \gamma_{in}^0 x^i.$$

We see from this discussion that the theory should at least be general enough to accommodate data generating models with an infinite dimensional parameter space. Rather than working directly with an infinite dimensional parameter space, it is easier to let the parameter space be an abstract metric space (Γ, ρ). To specialize to the infinite dimensional case, let Γ be the collection of infinite dimensional vectors and put $\rho(\gamma, \gamma^0) = \sum_{i=0}^{\infty} |\gamma_i - \gamma_i^0|$ or some other convenient metric (Problem 2). To specialize further to the finite dimensional case, let $\Gamma = \mathbb{R}^s$ and put $\rho(\gamma, \gamma^0) = (\sum_{i=1}^{s} |\gamma_i - \gamma_i^0|^2)^{1/2}$.

Moving on to the formal assumptions, we assume that the observed data

$$(y_1, x_1), (y_2, x_2), (y_3, x_3), \ldots$$

are generated according to the model

$$q(y_t, x_t, \gamma_n^0) = e_t \qquad t = 1, 2, \ldots, n$$

with $x_t \in \mathcal{X}$, $y_t \in \mathcal{Y}$, $e_t \in \mathcal{E}$, and $\gamma_n^0 \in \Gamma$. The dimensions are: x_t is a k-vector, y_t and e_t are M-vectors, and (Γ, ρ) is an abstract metric space with γ_n^0 some point in Γ. The observed values of y_t are actually doubly indexed and form a triangular array

$$
\begin{array}{ll}
y_{11} & n = 1 \\
y_{12} \, y_{22} & n = 2 \\
y_{13} \, y_{23} \, y_{33} & n = 3 \\
\quad \vdots &
\end{array}
$$

due to the dependence of γ_n^0 on the sample size n. This second index will simply be understood throughout.

ASSUMPTION 1. The errors are independently and identically distributed with common distribution $P(e)$.

Obviously, for the model to make sense, some measure of central tendency of $P(e)$ ought to be zero, but no formal use is made of such an assumption. If $P(e)$ is indexed by parameters, they cannot drift with the sample size as γ_n^0 may.

The assumption appears to rule out heteroscedastic errors. Actually it does not if one is willing to presume that the error variance-covariance matrix depends on the independent variable x_t,

$$\mathscr{E}(e_t, e_t') = \Sigma(x_t).$$

Factor $\Sigma^{-1}(x_t)$ as $R'(x_t) R(x_t)$ and write

$$R(x_t) q(y_t, x_t, \gamma_n^0) = R(x_t) e_t.$$

Then $R(x_t) e_t$ is homoscedastic. If one is willing to assume a common distribution for $R(x_t) e_t$ as well, then Assumption 1 is satisfied. Note that the actual construction of $R(x_t)$ is not required in applications, as the estimation is based only on the known function $s_n(\lambda)$. Similarly, many other apparent departures from Assumption 1 can be accommodated by presuming the existence of a transformation $\psi[q(y, x, \gamma_{(1)}), x, \gamma_{(2)}]$ that will yield residuals that satisfy Assumption 1.

The model is supposed to describe the behavior of some physical, biological, economic, or social system. If so, to each value of (e, x, γ^0) there should correspond one, and only one, outcome y. This condition and continuity are imposed.

ASSUMPTION 2. For each $(x, \gamma) \in \mathscr{X} \times \Gamma$ the equation $q(y, x, \gamma) = e$ defines a one to one mapping of $\mathscr{E}$ onto $\mathscr{Y}$, denoted as $Y(e, x, \gamma)$. Moreover, $Y(e, x, \gamma)$ is continuous on $\mathscr{E} \times \mathscr{X} \times \Gamma$ and Γ is compact.

It should be emphasized that it is not necessary to have a closed form expression for $Y(e, x, \gamma)$, or even to be able to compute it using numerical methods, in order to use the statistical methods set forth here. Inference is based only on the known function $s_n(\lambda)$. The existence of $Y(e, x, \gamma)$ is needed, but its construction is not required. This point is largely irrelevant to standard regression models, but it is essential to nonlinear simultaneous equation models, where $Y(e, x, \gamma)$ is often difficult to compute. Since Γ may be taken as $\{\gamma^*, \gamma_1, \gamma_2, \ldots\}$ if desired, no generality is lost by assuming that Γ is compact.

Repeatedly in the analysis of nonlinear models a Cesaro sum such as

$$\frac{1}{n} \sum_{t=1}^{n} f(y_t, x_t, \lambda) = \frac{1}{n} \sum_{t=1}^{n} f[Y(e_t, x_t, \gamma^0), x_t, \lambda]$$

must converge uniformly in (γ^0, λ) to obtain a desired result. If the results are to be useful in applications, the conditions imposed to insure this uniform convergence should be plausible and easily recognized as obtaining or not obtaining in an application. The conditions imposed here have evolved in Jennrich (1969), Malinvaud (1970b), Gallant (1977b), Gallant and Holly (1980), and Burguete, Gallant, and Souza (1982).

As motivation for these conditions, consider the sequence of independent variables resulting from a treatment-control experiment where the response depends on the age of the experimental material. Suppose subjects are randomly selected from a population whose age distribution is $F_A(\cdot)$ and then subjected to either the treatment or the control. The observed sequence of independent variables is

$$x_1 = (1, a_1) \qquad \text{treatment}$$
$$x_2 = (0, a_2) \qquad \text{control}$$
$$x_3 = (1, a_3) \qquad \text{treatment}$$
$$x_4 = (0, a_4) \qquad \text{control}$$
$$\vdots$$

Let $F_p(\cdot)$ denote the point binomial distribution with $p = \frac{1}{2}$, and set

$$d\mu(x) = dF_p(x_1) \times dF_A(x_2).$$

Then for any continuous function $f(x)$ whose expectation exists,

$$\lim_{n \to \infty} \frac{1}{n} \sum_{t=1}^{n} f(x_t) = \sum_{i=0}^{1} \frac{1}{2} \int f(i, a)\, dF_A(a) = \int_{\mathscr{X}} f(x)\, d\mu(x)$$

for almost every realization of $\{x_t\}$, by the strong law of large numbers. The null set depends on the function $f(x)$, which would be an annoyance, as the discussion flows more naturally if one has the freedom to hold a realization of $\{x_t\}$ fixed while permitting $f(x)$ to vary over a possibly uncountable collection of functions. Fortunately, the collection of functions considered later is dominated, and we can take advantage of that fact now to eliminate this dependence of the null set on $f(x)$. Consider the following consequence of the generalized Glivenko-Cantelli theorem.

PROPOSITION 1 (Gallant and Holly, 1980). Let V_t, $t = 1, 2, \ldots$, be a sequence of independent and identically distributed s-dimensional random variables defined on a complete probability space $(\Omega, \mathscr{A}_0, P_0)$ with common distribution ν. Let ν be absolutely continuous with respect to some product measure on $\mathbb{R}^s$, and let b be a nonnegative function with $\int b\, d\nu < \infty$. Then there exists E with $P_0(E) = 0$ such that if $\omega \notin E$,

$$\lim_{n \to \infty} \frac{1}{n} \sum_{t=1}^{n} f[V_t(\omega)] = \int f(v)\, d\nu(v)$$

for every continuous function with $|f(v)| \le b(v)$.

The conclusion of this proposition describes the behavior that is required of a sequence $v_t = x_t$ or $v_t = (e_t, x_t)$. As terminology for it, such a sequence is called a Cesaro sum generator.

DEFINITION (Cesaro sum generator: Gallant and Holly, 1980). A sequence $\{v_t\}$ of points from a Borel set $\mathcal{V}$ is said to be a Cesaro sum generator with respect to a probability measure ν defined on the Borel subsets of $\mathcal{V}$ and a dominating function $b(v)$ with $\int b \, d\nu < \infty$ if

$$\lim_{n \to \infty} \frac{1}{n} \sum_{t=1}^{n} f(v_t) = \int f(v) \, d\nu(v)$$

for every real valued, continuous function f with $|f(v)| \le b(v)$.

We have seen that independent variables generated according to an experimental design or by random sampling satisfy this definition. Many other situations such as stratified or cluster sampling will satisfy the definition as well. We shall assume, below, that the sequence $\{x_t\}$ upon which the results are conditioned is a Cesaro sum generator as is almost every joint realization $\{(e_t, x_t)\}$. Then we derive the uniform strong law of large numbers.

ASSUMPTION 3 (Gallant and Holly, 1980). Almost every realization of $\{v_t\}$ with $v_t = (e_t, x_t)$ is a Cesaro sum generator with respect to the product measure

$$\nu(A) = \int_{\mathcal{X}} \int_{\mathcal{E}} I_A(e, x) \, dP(e) \, d\mu(x)$$

and dominating function $b(e, x)$. The sequence $\{x_t\}$ is a Cesaro sum generator with respect to μ and $b(x) = \int_{\mathcal{E}} b(e, x) \, dP(e)$. For each $x \in \mathcal{X}$ there is a neighborhood N_x such that $\int_{\mathcal{E}} \sup_{N_x} b(e, x) \, dP(e) < \infty$. [$I_A(e, x) = 1$ if $(e, x) \in A$, 0 otherwise.]

THEOREM 1 (Uniform strong law of large numbers). Let Assumptions 1 through 3 hold. Let $\langle B, \sigma \rangle$ and $\langle \Gamma, \rho \rangle$ be compact metric spaces, and let $f(y, x, \beta)$ be continuous on $\mathcal{Y} \times \mathcal{X} \times B$. Let

$$|f(y, x, \beta)| \le |b[q(y, x, \gamma), x]|$$

or equivalently

$$|f[Y(e, x, \gamma), x, \beta]| \le b(e, x)$$

for all $(y, x) \in \mathcal{Y} \times \mathcal{X}$ and all $(\beta, \gamma) \in B \times \Gamma$, where $b(e, x)$ is given by

Assumption 3. Then both

$$\frac{1}{n} \sum_{t=1}^{n} f(y_t, x_t, \beta)$$

and

$$\frac{1}{n} \sum_{t=1}^{n} \int_{\mathscr{E}} f[Y(e, x_t, \gamma), x_t, \beta] \, dP(e)$$

converge uniformly to

$$\int_{\mathscr{X}} \int_{\mathscr{E}} f[Y(e, x, \gamma), x, \beta] \, dP(e) \, d\mu(x)$$

over $B \times \Gamma$ except on the event E with $P_0(E) = 0$ given by Assumption 3.

Recall that the uniform limit of continuous functions is continuous.

Proof. (Jennrich, 1969). Let $v = (e, x)$ denote a typical element of $\mathscr{V} = \mathscr{E} \times \mathscr{X}$, let $\alpha = (\beta, \gamma)$ denote a typical element of $A = B \times \Lambda$, and let $\{v_t\}$ be a Cesaro sum generator. The idea of the proof is to use the dominated convergence theorem and Cesaro summability to show that

$$h_n(\alpha) = \frac{1}{n} \sum_{t=1}^{n} h(v_t, \alpha)$$

where

$$h(v, \alpha) = f[Y(e, x, \gamma), x, \beta]$$

is an equicontinuous sequence on A. An equicontinuous sequence that has a pointwise limit on a compact set converges uniformly; see, for example, Chapter 9 of Royden (1968).

In order to apply Cesaro summability, we show that $\sup_{\alpha \in O} h(v, \alpha)$ and $\inf_{\alpha \in O} h(v, \alpha)$ are continuous for any $O \subset A$; they are obviously dominated by $b(e, x)$. Put $\tau(\alpha, \alpha^0) = [\sigma^2(\beta, \beta^0) + \rho^2(\gamma, \gamma^0)]^{1/2}$, whence $\langle A, \tau \rangle$ is a compact metric space. Let v^0 in $\mathscr{V}$ and $\epsilon > 0$ be given. Let V be a compact neighborhood of v^0, and let $\overline{O}$ be the closure of O in $\langle A, \tau \rangle$, whence $\langle \overline{O}, \tau \rangle$ is compact. By assumption, $h(v, \alpha)$ is continuous on $\mathscr{V} \times A$, so it is uniformly continuous on $\overline{V} \times \overline{O}$. Then there is a $\delta > 0$ such that for all $|v - v^0| < \delta$ and $\alpha \in \overline{O}$

$$h(v^0, \alpha) - \epsilon < h(v, \alpha) < h(v^0, \alpha) + \epsilon.$$

This establishes continuity (Problem 4).

A sequence is equicontinuous if for each $\epsilon > 0$ and α^0 in A there is a $\delta > 0$ such that $\tau(\alpha, \alpha^0) < \delta$ implies $\sup_n |h_n(\alpha) - h_n(\alpha^0)| < \epsilon$. When each $h_n(\alpha)$ is continuous over A, it suffices to show that $\sup_{n > N} |h_n(\alpha) - h_n(\alpha^0)| < \epsilon$ for some finite N. Let $\epsilon > 0$ and $\delta > 0$ be given and let $O_\delta = \{\alpha : \tau(\alpha, \alpha^0) < \delta\}$. By the dominated convergence theorem and continuity

$$\lim_{\delta \to 0} \int_{\mathcal{Y}} \sup_{O_\delta} h(v, \alpha) - h(v, \alpha^0) \, dv(v)$$

$$= \int_{\mathcal{Y}} \lim_{\delta \to 0} \sup_{O_\delta} h(v, \alpha) - h(v, \alpha^0) \, dv(v)$$

$$= 0.$$

Then there is a $\delta > 0$ such that $\tau(\alpha, \alpha^0) < \delta$ implies

$$\int_{\mathcal{Y}} \sup_{O_\delta} h(v, \alpha) - h(v, \alpha^0) \, dv(v) < \frac{\epsilon}{2}.$$

By Cesaro summability, there is an N such that $n > N$ implies

$$\sup_{O_\delta} h_n(\alpha) - h_n(\alpha^0) - \int_{\mathcal{Y}} \sup_{O_\delta} h(v, \alpha) - h(v, \alpha^0) \, dv(v) < \frac{\epsilon}{2}$$

whence

$$h_n(\alpha) - h_n(\alpha^0) \leq \sup_{O_\delta} h_n(\alpha) - h_n(\alpha^0) < \epsilon$$

for all $n > N$ and all $\tau(\alpha, \alpha^0) < \delta$. A similar argument applied to $\inf_{O_\delta} h_n(\alpha)$ yields

$$-\epsilon < h_n(\alpha) - h_n(\alpha^0) < \epsilon$$

for all $n > N$ and all $\tau(\alpha, \alpha^0) < \delta$. This establishes equicontinuity.

To show that

$$\bar{h}_n(\alpha) = \frac{1}{n} \sum_{t=1}^{n} \bar{h}(x_t, \alpha)$$

where

$$\bar{h}(x, \alpha) = \int_{\mathscr{E}} f[Y(e, x, \gamma), x, \beta] \, dP(e)$$

is an equicontinuous sequence, the same argument can be applied. It is only necessary to show that $\bar{h}(x, \alpha)$ is continuous on $\mathscr{X} \times A$ and dominated by $b(x)$. Now

$$|\bar{h}(x, \alpha)| \le \int_{\mathscr{E}} |h(v, \alpha)| \, dP(e) \le \int_{\mathscr{E}} b(e, x) \, dP(e) = b(x)$$

which establishes domination. By continuity on $\mathscr{V} \times A$ and the dominated convergence theorem with $\sup_{N_{x^0}} b(e, x)$ of Assumption 3 as the dominating function,

$$\lim_{(x, \alpha) \to (x^0, \alpha^0)} \bar{h}(x, \alpha) = \int_{\mathscr{E}} \lim_{(x, \alpha) \to (x^0, \alpha^0)} h(e, x, \alpha) \, dP(e)$$

$$= \int_{\mathscr{E}} h(e, x^0, \alpha^0) \, dP(e)$$

$$= \bar{h}(x^0, \alpha^0).$$

This establishes continuity.   □

In typical applications, an error density $p(e)$ and a Jacobian

$$J(y, x, \gamma^0) = \frac{\partial}{\partial y'} q(y, x, \gamma^0)$$

are available. With these in hand, the conditional density

$$p(y \mid x, \gamma^0) = |\det J(y, x, \gamma^0)| p[q(y, x, \gamma^0)]$$

may be used for computing limits of Cesaro sums, since

$$\int_{\mathscr{X}} \int_{\mathscr{E}} f[Y(e, x, \gamma^0), x, \gamma] \, dP(e) \, d\mu(x)$$

$$= \int_{\mathscr{X}} \int_{\mathscr{Y}} f(y, x, \gamma) p(y \mid x, \gamma^0) \, dy \, d\mu(x).$$

The choice of integration formulas is dictated by convenience.

The main use of the uniform strong law is in the following type of argument:

$$\lim_{n \to \infty} \hat{\lambda}_n = \lambda^*$$

$$\lim_{n \to \infty} \sup_{\Lambda} |s_n(\lambda) - s^*(\lambda)| = 0$$

$$s^*(\lambda) \text{ continuous}$$

implies

$$\lim_{n \to \infty} s_n(\hat{\lambda}_n) = s^*(\lambda^*)$$

because

$$|s_n(\hat{\lambda}_n) - s^*(\lambda^*)| = |s_n(\hat{\lambda}_n) - s^*(\hat{\lambda}_n) + s^*(\hat{\lambda}_n) - s^*(\lambda^*)|$$

$$\leq \sup_{\Lambda} |s_n(\lambda) - s^*(\lambda)| + |s^*(\hat{\lambda}_n) - s^*(\lambda^*)|.$$

We could get by with a weaker result that merely stated

$$\lim_{n \to \infty} s_n(\hat{\lambda}_n) = s^*(\lambda^*)$$

for any sequence with

$$\lim_{n \to \infty} \hat{\lambda}_n = \lambda^*.$$

For the central limit theorem, we shall make do with this weaker notion of convergence, called continuous convergence.

THEOREM 2 (Central limit theorem). Let Assumptions 1 through 3 hold. Let $\langle \Gamma, \rho \rangle$ be a compact metric space; let T be a closed ball in a Euclidean space centered at τ^* with finite, nonzero radius; and let Λ be a compact subset of a Euclidean space. Let $\{\gamma_n^0\}$ be a sequence from Γ that converges to γ^*; let $\{\hat{\tau}_n\}$ be a sequence of random variables with range in T that converges almost surely to τ^*; let $\{\tau_n^0\}$ be a sequence from T with $\sqrt{n}(\hat{\tau}_n - \tau_n^0)$ bounded in probability; let $\{\lambda_n^0\}$ be a sequence from Λ that converges to λ^*. Let $f(y, x, \tau, \lambda)$ be a vector valued function such that each element of $f(y, x, \tau, \lambda)$, $f(y, x, \tau, \lambda)f'(y, x, \tau, \lambda)$, and $(\partial/\partial\tau')f(y, x, \tau, \lambda)$ is continuous on $\mathscr{Y} \times \mathscr{X} \times T \times \Lambda$ and dominated by $b[q(y, x, \gamma), x]$ for all $(y, x) \in \mathscr{Y} \times \mathscr{X}$ and all $(\gamma, \tau, \lambda) \in \Gamma \times T \times \Lambda$; $b(e, x)$ is given by Assumption 3. If

$$\frac{\partial}{\partial \tau} \int_{\mathscr{X}} \int_{\mathscr{E}} f[Y(e, x, \gamma^*), x, \tau^*, \lambda^*] \, dP(e) \, d\mu(x) = 0$$

then

$$\frac{1}{\sqrt{n}} \sum_{t=1}^{n} \left[f(y_t, x_t, \hat{\tau}_n, \lambda_n^0) - \mu(x_t, \gamma_n^0, \tau_n^0, \lambda_n^0) \right] \xrightarrow{\mathscr{L}} N_p(0, I^*)$$

where

$$\mu(x, \gamma, \tau, \lambda) = \int_{\mathscr{E}} f[Y(e, x, \gamma), x, \tau, \lambda] \, dP(e)$$

$$I^* = \int_{\mathscr{X}} \int_{\mathscr{E}} f[Y(e, x, \gamma^*), x, \tau^*, \lambda^*]$$

$$\times f'[Y(e, x, \gamma^*), x, \tau^*, \lambda^*] \, dP(e) \, d\mu(x) - \mathscr{U}^*$$

$$\mathscr{U}^* = \int_{\mathscr{X}} \mu(x, \gamma^*, \tau^*, \lambda^*) \mu'(x, \gamma^*, \tau^*, \lambda^*) \, d\mu(x).$$

I^* may be singular.

Proof.   Let

$$Z(e, x, \gamma, \tau, \lambda) = f[Y(e, x, \gamma), x, \tau, \lambda]$$

$$- \int_{\mathscr{E}} f[Y(e, x, \gamma), x, \tau, \lambda] \, dP(e).$$

Given l with $\|l\| = 1$, consider the triangular array of random variables

$$Z_{tn} = l'Z(e_t, x_t, \gamma_n^0, \tau_n^0, \lambda_n^0) \qquad t = 1, 2, \ldots, n \quad n = 1, 2, \ldots \ .$$

Each Z_{tn} has mean zero and variance

$$\sigma_{tn}^2 = l' \int_{\mathscr{E}} Z(e, x_t, \gamma_n^0, \tau_n^0, \lambda_n^0) Z'(e, x_t, \gamma_n^0, \tau_n^0, \lambda_n^0) \, dP(e) \, l.$$

Putting $V_n = \sum_{t=1}^n \sigma_{tn}^2$, by Theorem 1 and the assumption that $\lim_{n \to \infty} (\gamma_n^0, \tau_n^0, \lambda_n^0) = (\gamma^*, \tau^*, \lambda^*)$ it follows that $\lim_{n \to \infty} (1/n)V_n = l'I^*l$ (Problem 5). Now $(1/n)V_n$ is the variance of $(1/\sqrt{n})\sum_{t=1}^n Z_{tn}$, and if $l'I^*l = 0$, then $(1/\sqrt{n})\sum_{t=1}^n Z_{tn}$ converges in distribution to $N(0, l'I^*l)$ by Chebyshev's inequality. Suppose, then, that $l'I^*l > 0$. If it is shown that for every $\epsilon > 0$ one has $\lim_{n \to \infty} B_n = 0$, where

$$B_n = \frac{1}{n} \sum_{t=1}^n \int_{\mathscr{E}} I_{[|z| > \epsilon \sqrt{V_n}]}[l'Z(e, x_t, \gamma_n^0, \tau_n^0, \lambda_n^0)]$$

$$\times [l'Z(e, x_t, \gamma_n^0, \tau_n^0, \lambda_n^0)]^2 \, dP(e)$$

then $\lim_{n \to \infty} (n/V_n)B_n = 0$. This is the Lindeberg-Feller condition (Chung,

1974); it implies that $(1/\sqrt{n})\Sigma_{t=1}^{n}Z_{tn}$ converges in distribution to $N(0, l'I^*l)$.

Let $\eta > 0$ and $\epsilon > 0$ be given. Choose $a > 0$ such that $\bar{B}(\gamma^*, \tau^*, \lambda^*) < \eta/2$, where

$$\bar{B}(\gamma^*, \tau^*, \lambda^*) = \int_{\mathscr{X}}\int_{\mathscr{E}} I_{[|z|>\epsilon a]}[l'Z(e, x, \gamma^*, \tau^*, \lambda^*)]$$

$$\times [l'Z(e, x, \gamma^*, \tau^*, \lambda^*)]^2 \, dP(e) \, d\mu(x).$$

This is possible because $\bar{B}(\gamma^*, \tau^*, \gamma^*)$ exists when $a = 0$. Choose a continuous function $\varphi(z)$ and an N_1 such that, for all $n > N_1$,

$$I_{[|z|>\epsilon\sqrt{V_n}]}(z) \le \varphi(z) \le I_{[|z|>\epsilon a]}(z)$$

and set

$$\tilde{B}_n(\gamma, \tau, \lambda) = \frac{1}{n}\sum_{t=1}^{n}\int_{\mathscr{E}}\varphi[l'Z(e, x_t, \gamma, \tau, \lambda)][l'Z(e, x_t, \gamma, \tau, \lambda)]^2 \, dP(e).$$

By Theorem 1, $\tilde{B}_n(\gamma, \tau, \lambda)$ conveges uniformly on $\Gamma \times T \times \Lambda$ to, say, $\tilde{B}(\gamma, \tau, \lambda)$. By assumption $\lim_{n\to\infty}(\gamma_n^0, \tau_n^0, \lambda_n^0) = (\gamma^*, \tau^*, \lambda^*)$, whence $\lim_{n\to\infty}\tilde{B}_n(\gamma_{n_2}^0, \tau_n^0, \lambda_n^0) = \tilde{B}(\gamma^*, \tau^*, \lambda^*)$. Then there is an N_2 such that, for all $n > N_2$, $\tilde{B}_n(\gamma_n^0, \tau_n^0, \lambda_n^0) < \tilde{B}(\gamma^*, \tau^*, \lambda^*) + \eta/2$. But for all $n > N = \max\{N_1, N_2\}$, $B_n \le \tilde{B}_n(\gamma_n^0, \tau_n^0, \lambda_n^0)$, whence

$$B_n \le \tilde{B}_n\left(\gamma_n^0, \tau_n^0, \lambda_n^0\right) < \tilde{B}(\gamma^*, \tau^*, \lambda^*) + \frac{\eta}{2} \le \bar{B}(\gamma^*, \tau^*, \lambda^*) + \frac{\eta}{2} < \eta.$$

By Taylor's theorem, expanding about τ_n^0,

$$\frac{1}{\sqrt{n}}\sum_{t=1}^{n}l'\left[f(y_t, x_t, \hat{\tau}_n, \lambda_n^0) - \mu(x_t, \gamma_n^0, \tau_n^0, \lambda_n^0)\right]$$

$$= \frac{1}{\sqrt{n}}\sum_{t=1}^{n}Z_{tn}$$

$$+ \left(\frac{1}{n}\sum_{t=1}^{n}l'\frac{\partial}{\partial\tau'}f(y_t, x_t, \bar{\tau}_n, \lambda_n^0)\right)\sqrt{n}\left(\hat{\tau}_n - \tau_n^0\right)$$

where $\bar{\tau}_n$ lies on the line segment joining $\hat{\tau}_n$ to τ_n^0; thus $\bar{\tau}_n$ converges almost surely to τ^*. The almost sure convergence of $(\gamma_n^0, \bar{\tau}_n, \lambda_n^0)$ to $(\gamma^*, \tau^*, \lambda^*)$ and the uniform almost sure convergence of

$$\frac{1}{n}\sum_{t=1}^{n}l'\frac{\partial}{\partial\tau'}f(y_t, x_t, \tau, \lambda)$$

over $\Gamma \times T \times \Lambda$ given by Theorem 1 imply that

$$\frac{1}{n}\sum_{t=1}^{n} l'\frac{\partial}{\partial\tau'}f\left(y_t, x_t, \bar{\tau}_n, \lambda_n^0\right)$$

converges almost surely to

$$\int_{\mathscr{X}}\int_{\mathscr{E}} l'\frac{\partial}{\partial\tau'}f\left[Y(e, x, \gamma^*), x, \tau^*, \lambda^*\right] dP(e)\, d\mu(x) = 0.$$

Since $\sqrt{n}\,(\hat{\tau}_n - \tau_n^0)$ is bounded in probability, we have that

$$\frac{1}{\sqrt{n}}\sum_{t=1}^{n}\left[f\left(y_t, x_t, \hat{\tau}_n, \lambda_n^0\right) - \mu\left(x_t, \gamma_n^0, \tau_n^0, \lambda_n^0\right)\right]$$

$$=\frac{1}{\sqrt{n}}\sum_{t=1}^{n} Z_{tn} + o_p(1)$$

$$\xrightarrow{\mathscr{L}} N(0, l'I^*l).$$

This holds for every l with $\|l\| = 1$, whence the desired result obtains. □

In the main, small sample regression analysis is conditional. With a model such as

$$y_t = f(x_t, \theta) + e_t \qquad t = 1, 2, \ldots, n$$

the independent variables are held fixed and the sampling variation enters via the errors $e_1, e_2, \ldots, e_n$. The principal of ancillarity seems to provide the strongest theoretical support for a conditional analysis of regression situations (Cox and Hinkley, 1974, Section 2.2.viii). It seems appropriate to maintain this conditioning when passing to the limit. This is what we shall do in the sequel, excepting dynamic models. In a conditional analysis, one fixes an infinite sequence

$$x_\infty = (x_1, x_2, \ldots)$$

that satisfies the Cesaro summability property, and all sampling variation enters via the random variables $\{e_t\}_{t=1}^\infty$. To give an unambiguous description of this conditioning, it is necessary to spell out the probability structure in detail. The reader who has no patience with details of this sort is invited to skip to the fourth from the last paragraph of this section at this point.

We begin with an abstract probability space $(\Omega, \mathcal{A}_0, P_0)$ on which are defined random variables $\{E_t\}_{t=1}^{\infty}$ and $\{X_t\}_{t=1}^{\infty}$ which represent the errors and independent variables respectively. Nonrandom independent variables are represented in this scheme by random variables that take on a single value with probability one. A realization of the errors can be denoted by an infinite dimensional sequence

$$e_{\infty} = (e_1, e_2, \dots)$$

where $e_t = E_t(\omega)$ for some ω in Ω. Similarly for the independent variables

$$x_{\infty} = (x_1, x_2, \dots).$$

Let $\mathcal{E}_{\infty} = \times_{t=1}^{\infty} \mathcal{E}$ and $\mathcal{X}_{\infty} = \times_{t=1}^{\infty} \mathcal{X}$ so that all joint realizations of the errors and independent variables take their values in $\mathcal{E}_{\infty} \times \mathcal{X}_{\infty}$ and all realizations of the independent variables take their values in $\mathcal{X}_{\infty}$.

Using the Daniell-Kolmogorov construction (Tucker, 1967, Section 2.3), this is enough to define a joint probability space

$$(\mathcal{E}_{\infty} \times \mathcal{X}_{\infty}, \mathcal{A}_{e,x}, \nu_{\infty})$$

such that if a random variable is a function of (e_{∞}, x_{∞}), one can perform all computations with respect to the more structured space $(\mathcal{E}_{\infty} \times \mathcal{X}_{\infty}, \mathcal{A}_{e,x}, \nu_{\infty})$, and one is spared the trouble of tracing preimages back to the space $(\Omega, \mathcal{A}_0, P_0)$. Similarly one can construct the marginal probability space

$$(\mathcal{X}_{\infty}, \mathcal{A}_x, \mu_{\infty}).$$

Assumption 3 imposes structure on both of these probability spaces. The set on which Cesaro summability fails jointly,

$$F_{e,x} = \bigcup_{\epsilon>0} \bigcap_{j=0}^{\infty} \bigcup_{n=j}^{\infty} \left\{ (e_{\infty}, x_{\infty}) : \exists | f(e,x)| < b(e,x) \right.$$

$$\left. \times \text{s.t.} \left| \frac{1}{n} \sum_{t=1}^{n} f(e_t, x_t) - \int\int f dP\, d\mu \right| > \epsilon \right\}$$

has ν_{∞}-measure zero. And the set on which Cesaro summability fails marginally,

$$F_x = \bigcup_{\epsilon>0} \bigcap_{j=0}^{\infty} \bigcup_{n=j}^{\infty} \left\{ x_\infty : \exists |f(x)| < b(x) \right.$$

$$\left. \times \text{ s.t. } \left| \frac{1}{n} \sum_{t=1}^{n} f(x_t) - \int f \, d\mu \right| > \epsilon \right\}$$

has μ_∞-measure zero.

By virtue of the construction of $(\mathscr{E}_\infty \times \mathscr{X}_\infty, \mathscr{A}_{e,x}, \nu_\infty)$ from countable families of random variables, there exists (Loeve, 1963, Section 27.2, Regularity Theorem) a regular conditional probability $P(A \mid x_\infty)$ connecting the joint and the marginal spaces by

$$\nu_\infty(A) = \int_{\mathscr{X}} P(A \mid x_\infty) \, d\mu_\infty(x_\infty).$$

Recall that a regular conditional probability is a mapping of $\mathscr{A}_{e,x} \times \mathscr{X}_\infty$ into $[0,1]$ such that $P(A \mid x_\infty)$ is a probability measure on $(\mathscr{E}_\infty \times \mathscr{X}_\infty, \mathscr{A}_{e,x})$ for each fixed x_∞, such that $P(A \mid x_\infty)$ is a measurable function over $(\mathscr{X}_\infty, \mathscr{A}_x)$ for each fixed A, and such that $\int_B P(A \mid x_\infty) \, d\mu_\infty(x_\infty) = \nu_\infty[A \cap (\mathscr{E}_\infty \times B)]$ for every B in $\mathscr{A}_x$. The simplest example that comes to mind is to assume that $\{E_t\}_{t=1}^\infty$ and $\{X_t\}_{t=1}^\infty$ are independent families of random variables, to construct $(\mathscr{E}_\infty, \mathscr{A}_e, P_e)$, and to put

$$P(A \mid x_\infty) = \int_{\mathscr{E}_\infty} I_A(e_\infty, x_\infty) \, dP_e(e_\infty).$$

Define the marginal conditional distribution on $(\mathscr{E}_\infty, \mathscr{A}_e)$ by

$$P_{e|x}(E \mid x_\infty) = P(E \times \mathscr{X}_\infty \mid x_\infty).$$

All probability statements in the sequel are with respect to $P_{e|x}(E \mid x_\infty)$. Assumption 1 puts additional structure on $P_{e|x}(E \mid x_\infty)$. It states that $P_{e|x}(E \mid x_\infty)$ is a product measure corresponding to a sequence of independent random variables each having common distribution $P(e)$ defined over measurable subsets of $\mathscr{E}$. This distribution can depend on x_∞. For example, $\{e_t\}_{t=1}^\infty$ could be a sequence of independently and normally distributed random variables each with mean zero and variance-covariance matrix $\lim_{n \to \infty} (1/n) \sum_{t=1}^\infty T(x_t) T'(x_t)$. But as indicated by the discussion following Assumption 1, this dependence on x_∞ is very restricted. So restricted, in fact, that we do not bother to reflect it in our notation; we do not index P of Assumption 1 by x_∞.

If all probability statements are with respect to $P_{e|x}(E \mid x_\infty)$, then the critical question becomes: Does the set where Cesaro summability fails conditionally at $x_\infty = x_\infty^0$,

$$F_{e|x}^0 = \bigcup_{\epsilon > 0} \bigcap_{j=0}^{\infty} \bigcup_{n=j}^{\infty} \left\{ e_\infty : \exists |f(e, x)| < b(e, x) \right.$$

$$\left. \times \text{s.t.} \left| \frac{1}{n} \sum_{t=1}^{n} f(e_t, x_t^0) - \iint f \, dP \, d\mu \right| > \epsilon \right\}$$

have conditional measure zero? The following computation shows that the answer is yes for almost every choice of x_∞^0:

$$P_{e|x}\left(F_{e|x}^0 \mid x_\infty^0 \right)$$

$$= \int_{\mathcal{E}_\infty} I_{F_{e|x}^0}(e_\infty) \, dP_{e|x}\left(e_\infty \mid x_\infty^0 \right) \qquad \text{(marginal} \mid x^0)$$

$$= \int_{\mathcal{E}_\infty} I_{F_{e|x}^0 \times \{x_\infty^0\}}(e_\infty, x_\infty^0) \, dP_{e|x}\left(e_\infty \mid x_\infty^0 \right) \qquad \text{(marginal} \mid x^0)$$

$$= \int_{\mathcal{E}_\infty \times \mathcal{X}_\infty} I_{F_{e|x}^0 \times \{x_\infty^0\}}(e_\infty, x_\infty) \, dP\left[(e_\infty, x_\infty) \mid x_\infty^0 \right] \qquad \text{(joint} \mid x^0)$$

$$= \int_{\mathcal{E}_\infty \times \mathcal{X}_\infty} I_{F_{e|x}^0 \times \{x_\infty^0\}}(e_\infty, x_\infty) \, dP\left[(e_\infty, x_\infty) \middle| x_\infty^0 \right] \qquad \text{(joint} \mid x^0)$$

$$\leq \int_{\mathcal{E}_\infty \times \mathcal{X}_\infty} I_{F_{e,x}}(e_\infty, x_\infty) \, dP\left[(e_\infty, x_\infty) \middle| x_\infty^0 \right] \qquad \text{(joint} \mid x^0)$$

$$= P\left(F_{e,x} \mid x_\infty^0 \right).$$

Since

$$\nu_\infty(F_{e,x}) = \int_{\mathcal{X}} P\left(F_{e,x} \mid x_\infty^0 \right) d\mu_\infty\left(x_\infty^0 \right) = 0$$

we have

$$P_{e|x}\left(F_{e|x}^0 \mid x_\infty^0 \right) = 0 \qquad \text{a.e. } (\mathcal{X}_\infty, \mathcal{A}_\infty, \mu_\infty).$$

Since the parameter γ_n^0 is subject to drift, it is as well to spell out a few additional details. For each n, the conditional distribution of the dependent variables $\{y_t\}_{t=1}^n$ given x_∞ and γ_n^0 is defined by

$$P_n\left(A \mid x_\infty, \gamma_n^0\right) = P_{e\mid x}\left\{e_\infty \in \mathscr{E}_\infty : \left[Y(e_t, x_1, \gamma_n^0), \dots,\right.\right.$$
$$\left.\left. Y(e_n, x_n, \gamma_n^0)\right] \in A \middle| x_\infty \right\}$$

for each measurable subset A of $\times_{i=1}^n \mathscr{Y}$. A statement such as $\hat{\lambda}_n$ converges almost surely to λ^* means that $\hat{\lambda}_n$ is a random variable with argument $(y_1, \dots, y_n, x_1, \dots, x_n)$, and that $P_{e\mid x}(E \mid x_\infty) = 0$ where

$$E = \bigcup_{\epsilon > 0} \bigcap_{j=1}^\infty \bigcup_{n=j}^\infty \left\{e_\infty : |\hat{\lambda}_n - \lambda^*| > \epsilon\right\}\Bigg|_{(y_t, x_t) = [Y(e_t, x_t, \gamma_n^0), x_t]} .$$

A statement that $\sqrt{n}(\hat{\lambda}_n - \lambda^*)$ converges in distribution to a multivariate normal distribution $N_p(\cdot \mid \delta, V)$ means that for A of the form

$$A = (-\infty, \lambda_1] \times (-\infty, \lambda_2] \times \cdots \times (-\infty, \lambda_p]$$

it is true that

$$\lim_{n \to \infty} P_n\left(\sqrt{n}(\hat{\lambda}_n - \lambda^*) \in A \middle| x_\infty, \gamma_n^0\right) = \int_A dN_p(z \mid \delta, V).$$

There is very little qualitative difference between an analysis that conditions on $\{x_t\}$ and an analysis that takes $\{x_t\}$ to be a nonstationary, random process, as can be seen by a comparison of the results of this chapter with the results of Chapter 7. However, one can be seriously misled if one assumes that $\{x_t\}$ is a stationary process, particularly if one assumes that $\{x_t\}$ is a sequence of independently and identically distributed random variables. The details are spelled out in Section 8.

We shall assume that the estimation space Λ is compact. Our defense of this assumption is that it does not cause problems in applications as a general rule and it can be circumvented on an *ad hoc* basis as necessary without affecting the results. We explain.

One does not wander haphazardly into nonlinear estimation. As a rule, one has need of a considerable knowledge of the situation in order to construct the model. In the computations, a fairly complete knowledge of admissible values of λ is required in order to be able to find starting values for nonlinear optimization algorithms. Thus, a statistical theory which presumes this same knowledge is not limited in its scope of applications. Most authors apparently take this position, as the assumption of a compact estimation space is more often encountered than not.

One may be reluctant to impose bounds on scale parameters and parameters that enter the model linearly. Frequently these are regarded as nuisance parameters in an application, and one has little feel for what values they ought to have. Scale parameter estimates are often computed from residuals, so start values are unnecessary, and, at least for least squares, linear parameters need no start values either (Golub and Pereyra, 1973). Here, then, a compact parameter space is an annoyance.

These situations can be accommodated without disturbing the results obtained here as follows. Our results are asymptotic, so if there is a compact set Λ' such that for each realization of $\{e_t\}$ there is an N where $n > N$ implies

$$\sup_{\Lambda'} s_n(\lambda) = \sup_{\Lambda} s_n(\lambda)$$

then the asymptotic properties of $\hat{\lambda}_n$ are the same whether the estimation space is Λ or Λ'. For examples using this device to accommodate parameters entering linearly, see Gallant (1973). See Gallant and Holly (1980) for application to scale parameters. Other devices, such as the use of an initial consistent estimator as the start value for an algorithm which is guaranteed to converge to a local minimum of $s_n(\lambda)$, are effective as well.

PROBLEMS

1. Referring to the discussion following Theorem 2, show that if $\{X_t\}$ and $\{E_t\}$ are independent sequences of random variables, then $P_{e|x}(E \mid x_\infty)$ does not depend on x_∞.

2. (Construction of a Pitman drift.) Consider the example of the first few paragraphs of this section where the fitted model is

$$y_t = f(x_t, \lambda) + u_t \qquad t = 1, 2, \ldots, n$$

    but the data actually follow

$$y_t = g(x_t, \gamma_n^0) + e_t \qquad t = 1, 2, \ldots, n$$

    where

$$g(x, \gamma) = \sum_{j=0}^{\infty} \gamma_j x^j.$$

The equality is with respect to uniform convergence. That is, one restricts attention to the set Γ^* of $\gamma = (\gamma_0, \gamma_1, \ldots)$ with

$$\lim_{J \to \infty} \; \sup_{x \in [0,1]} \left| \sum_{j=0}^{J} \gamma_j x^j \right| < \infty$$

and $g(x, \gamma)$ denotes that continuous function on $[0, 1]$ with

$$\lim_{J \to \infty} \; \sup_{x \in [0,1]} \left| g(x, \gamma) - \sum_{j=0}^{J} \gamma_j x^j \right| = 0.$$

Take γ as equivalent to γ^0, and write $\gamma = \gamma^0$ if $g(x, \gamma) = g(x, \gamma^0)$ for all x in $[0, 1]$. Define

$$\rho(\gamma, \gamma^0) = \lim_{J \to \infty} \; \sup_{x \in [0,1]} \left| \sum_{j=0}^{J} (\gamma_j - \gamma_j^0) x^j \right|.$$

Show that (Γ^*, ρ) is a metric space on these equivalence classes (Royden, 1968, Section 7.1). If the model is fitted by least squares, if $f(x, \lambda)$ is continuous over $[0, 1] \times \Lambda$, and if the estimation space Λ is compact, we shall show later that

$$\lambda_n^0 \quad \text{minimizes} \quad s_n^0(\lambda) = \sigma^2 + \frac{1}{n} \sum_{t=1}^{n} \left[g(x_t, \gamma_n^0) - f(x_t, \lambda) \right]^2.$$

Assume that $f(x, \lambda)$ and $\{x_t\}$ are such that $s_n^0(\lambda)$ has a unique minimum for n larger than some N. By the Stone-Weierstrass theorem (Royden, 1968, Section 9.7) we can find a γ^0 in Γ^* with

$$g(x, \gamma^0) = f\left(x, \lambda^* + \frac{\Delta}{\sqrt{n}}\right).$$

That is, $\lim_{J \to \infty} \sup_{x \in [0,1]} |\sum_{j=0}^{J} \gamma_j^0 x^j - f(x, \lambda^* + \Delta/\sqrt{n})| = 0$. Show that it is possible to truncate γ^0 at some point m_n such that if

$$\gamma_n^0 = \left(\gamma_0, \gamma_1, \ldots, \gamma_{m_n}, 0, \ldots\right)$$

then

$$\left| \lambda_n^0 - \lambda^* - \frac{\Delta}{\sqrt{n}} \right| < \frac{1}{n}.$$

for $n > N$. Hint: See the proof of Theorem 3. Show that

$$\lim_{n \to \infty} \rho(\gamma_n^0, \gamma^*) = 0.$$

Let $\Gamma = \{\gamma_n^0\}_{n=N}^{\infty}$. Show that (Γ, ρ) is a compact metric space.

3. (Construction of a Pitman drift.) Let $g(x)$ be once continuously differentiable on a bounded, open, convex set in $\mathbb{R}^k$ containing $\mathscr{X}$. By rescaling the data, we may assume that $\mathscr{X} \subset \times_{i=1}^{k}[0, 2\pi]$ without loss of generality. Then $g(x)$ can be expanded in a multivariate Fourier series. Letting r denote a multiindex—that is, a vector with integer (positive, negative, or zero) components—and letting $|r| = \sum_{i=1}^{k}|r_i|$, a multivariate Fourier series of order R is written $\sum_{|r| \le R} \gamma_r e^{ir'x}$ with $e^{ir'x} = \cos(r'x) + i \sin(r'x)$ and $i = \sqrt{-1}$. The restriction $\gamma_r = \bar{\gamma}_{-r}$, where the overbar denotes complex conjugation, will cause the Fourier series to be real valued. We have (Edmunds and Moscatelli, 1977)

$$\lim_{R \to \infty} \sup_{\mathscr{X}} \left| g(x) - \sum_{|r| \le R} \gamma_r e^{ir'x} \right| = 0.$$

Construct a Pitman drift using a multivariate Fourier series expansion along the same lines as in Problem 2.

4. Show that if for any $\epsilon > 0$ there is a $\delta > 0$ such that $|v - v^0| < \delta$ implies that

$$h(v^0, \alpha) - \epsilon < h(v, \alpha) < h(v^0, \alpha) + \epsilon$$

for all α in $\mathcal{O}$, then $\sup_{\alpha \in \mathcal{O}} h(v, \alpha)$ and $\inf_{\alpha \in \mathcal{O}} h(v, \alpha)$ are continuous.

5. Referring to the proof of Theorem 2, show that

$$\{l'f[Y(e, x, \gamma), x, \tau, \lambda]\}^2 \le b(e, x)$$

implies that

$$\left\{ \int_{\mathscr{E}} l'f[Y(e, x, \gamma), x, \tau, \lambda] \, dP(e) \right\}^2$$

$$\le \int_{\mathscr{E}} \{l'f[Y(e, x, \gamma), x, \tau, \lambda]\}^2 \, dP(e) \le b(x).$$

Show that $\lim_{n \to \infty} (1/n)V_n = l'I*l$.

6. Show that if $\hat{\tau}_n$ converges almost surely to $\tau*$ and $\sqrt{n}(\hat{\tau}_n - \tau_n^0)$ is bounded in probability, then $\lim_{n \to \infty} \tau_n^0 = \tau*$.

3. LEAST MEAN DISTANCE ESTIMATORS

Recall that a least mean distance estimator $\hat{\lambda}_n$ is defined as the solution of the optimization problem

$$\text{minimize}\quad s_n(\lambda) = \frac{1}{n}\sum_{t=1}^{n} s(y_t, x_t, \hat{\tau}_n, \lambda)$$

where $\hat{\tau}_n$ is a random variable which corresponds conceptually to estimators of nuisance parameters. A constrained least mean distance estimator $\tilde{\lambda}_n$ is the solution of the optimization problem

$$\text{minimize}\quad s_n(\lambda)\quad \text{subject to}\quad h(\lambda) = 0$$

where $h(\lambda)$ maps $\mathbb{R}^p$ into $\mathbb{R}^q$.

The objective of this section is to find the almost sure limit and the asymptotic distribution of the unconstrained estimator $\hat{\lambda}_n$ under regularity conditions that do not rule out specification error. Some ancillary facts regarding the asymptotic distribution of the constrained estimator $\tilde{\lambda}_n$ under a Pitman drift are also derived for use in later sections on hypothesis testing. In order to permit this Pitman drift, and to allow generality that may be useful in other contexts, the parameter γ_n^0 of the data generating model is permitted to depend on the sample size n throughout. A more conventional asymptotic theory regarding the unconstrained estimator $\hat{\lambda}_n$ is obtained by applying these results with γ_n^0 held fixed at a point γ^* for all n. These results are due to Souza (1979) in the main, with some refinements made here to center $\hat{\lambda}_n$ about a point λ_n^0 so as to isolate results regarding $\hat{\lambda}_n$ from the Pitman drift assumption.

An example, a correctly specified iteratively rescaled M-estimator, is carried throughout the discussion to serve as a template in applications.

EXAMPLE 1 (Iteratively rescaled M-estimator). The data generating model is

$$y_t = f(x_t, \gamma_n^0) + e_t \qquad t = 1, 2, \ldots, n.$$

An estimate of scale is obtained by first minimizing

$$\frac{1}{n}\sum_{t=1}^{n} \rho[y_t - f(x_t, \theta)]$$

with respect to θ to obtain $\hat{\theta}_n$, where

$$\rho(u) = \ln\cosh\left(\frac{u}{2}\right)$$

and then solving

$$\frac{1}{n} \sum_{t=1}^{n} \Psi^2\left(\frac{y_t - f(x_t, \hat{\theta}_n)}{\tau} \right) = \int \Psi^2(e) \, d\Phi(e)$$

with respect to τ to obtain $\hat{\tau}_n$, where

$$\Psi(u) = \frac{d}{du}\rho(u) = \tfrac{1}{2}\tanh\left(\frac{u}{2} \right)$$

and Φ is the standard normal distribution function. The parameters of the model are estimated by minimizing

$$s_n(\lambda) = \frac{1}{n} \sum_{t=1}^{n} \rho\left(\frac{y_t - f(x_t, \lambda)}{\hat{\tau}_n} \right),$$

whence

$$s(y, x, \tau, \lambda) = \rho\left(\frac{y - f(x, \lambda)}{\tau} \right).$$

The error distribution $P(e)$ is symmetric, puts positive probability on every open interval of the real line, and has finite first and second moments. See Huber (1964) for the motivation.   □

The first question one must address is: What is $\hat{\lambda}_n$ to be regarded as estimating in a finite sample? Ordinarily, in an asymptotic estimation theory, the parameter γ^0 of the data generating model is held fixed, and $\hat{\lambda}_n$ would be regarded as estimating the almost sure limit λ^* of $\hat{\lambda}_n$ in each finite sample. But we are in a conditional setting and have both misspecification and a parameter γ_n^0 that is subject to drift. In a conditional setting, either of these situations is enough to make that answer unsatisfactory, since if we regarded $\hat{\lambda}_n$ as centered about its almost sure limit λ^* (Theorem 3), we would find it necessary to impose a Pitman drift, accelerate the rate of convergence of Cesaro sums generated from $\{x_t\}_{t=1}^{\infty}$, or impose other regularity conditions to show that $\sqrt{n}(\hat{\lambda}_n - \lambda^*)$ is asymptotically normally distributed. Such conditions are unnatural in an estimation setting. A more satisfactory answer to the question is obtained if one regards $\hat{\lambda}_n$ as estimating λ_n^0 that is the solution to

$$\text{minimize} \quad s_n^0(\lambda) = \frac{1}{n} \sum_{t=1}^{n} \int_{\mathscr{e}} s\left[Y(e, x_t, \gamma_n^0), x_t, \tau_n^0, \lambda \right] dP(e);$$

τ_n^0 is defined later (Assumption 4). With this choice, one can show that $\sqrt{n}(\hat{\lambda}_n - \lambda_n^0)$ is asymptotically normally distributed without unusual regularity conditions. Moreover, in analytically tractable situations such as a linear model fitted by least squares to data that actually follow a nonlinear model, it turns out that λ_n^0 is indeed the mean of $\hat{\lambda}_n$ in finite samples.

We call the reader's attention to some heavily used notation and then state the identification condition:

NOTATION 1.

$$s_n(\lambda) = \frac{1}{n} \sum_{t=1}^{n} s(y_t, x_t, \hat{\tau}_n, \lambda)$$

$$s_n^0(\lambda) = \frac{1}{n} \sum_{t=1}^{n} \int_{\mathscr{E}} s\left[Y(e, x_t, \gamma_n^0), x_t, \tau_n^0, \lambda\right] dP(e)$$

$$s^*(\lambda) = \int_{\mathscr{X}} \int_{\mathscr{E}} s\left[Y(e, x, \gamma^*), x, \tau^*, \lambda\right] dP(e) \, d\mu(x)$$

$\hat{\lambda}_n$ minimizes $s_n(\lambda)$
$\tilde{\lambda}_n$ minimizes $s_n(\lambda)$ subject to $h(\lambda) = 0$
λ_n^0 minimizes $s_n^0(\lambda)$
λ_n^* minimizes $s_n^0(\lambda)$ subject to $h(\lambda) = 0$
λ^* minimizes $s^*(\lambda)$.

ASSUMPTION 4 (Identification). The parameter γ^0 is indexed by n, and the sequence $\{\gamma_n^0\}$ converges to a point γ^*. The sequence of nuisance parameter estimators is centered at τ_n^0 in the sense that $\sqrt{n}(\hat{\tau}_n - \tau_n^0)$ is bounded in probability; the sequence $\{\tau_n^0\}$ converges to a point τ^*, and $\{\hat{\tau}_n\}$ converges almost surely to τ^*. The function $s^*(\lambda)$ has a unique minimum over the estimation space Λ^* at λ^*.

The critical condition imposed by Assumption 4 is that $s^*(\lambda)$ must have a unique minimum over Λ^*. In a correctly specified situation, the usual approach to verification is to commence with an obviously minimal identification condition. Then known results for the simple location problem that motivated the choice of distance function $s(y, x, \tau, \lambda)$ are exploited to verify a unique association of λ^* to γ^* over Λ^*. We illustrate with the example:

EXAMPLE 1 (Continued). We are trapped in a bit of circularity in that we need the results of this section and the next in order to compute the center τ_n^0 of the nuisance parameter estimator $\hat{\tau}_n$ and show that $\sqrt{n}(\hat{\tau}_n - \tau_n^0)$ is bounded in probability. So we must defer verification until the end of Section 4. At that time we shall find that $\tau_n^0 > 0$ and $\tau^* > 0$, which facts we shall use now.

To verify that $s^*(\lambda)$ has a unique minimum one first notes that it will be impossible to determine λ by observing $\{y_t, x_t\}$ if $f(x, \lambda) = f(x, \gamma)$ for $\lambda \neq \gamma$ at each x in $\mathscr{X}$ that is given weight by the measure μ. Then a minimal identification condition is

$$\lambda \neq \gamma \quad \Rightarrow \quad \mu\{x : f(x, \lambda) \neq f(x, \gamma)\} > 0.$$

This is a condition on both the function $f(x, \lambda)$ and the infinite sequence $\{x_t\}_{t=1}^{\infty}$.

Now for $\tau > 0$

$$\varphi(\delta) \int_{\mathscr{E}} \rho\left(\frac{e + \delta}{\tau}\right) dP(e)$$

is known (Problem 9) to have a unique minimum at $\delta = 0$ when $P(e)$ is symmetric about zero, has finite first moment, and assigns positive probability to every nonempty, open interval. Let

$$\delta(x) = f(x, \gamma) - f(x, \lambda).$$

If $\lambda \neq \gamma$ then $\varphi[\delta(x)] \geq \varphi(0)$ for every x. Again, if $\lambda \neq \gamma$ the identification condition implies that $\varphi[\delta(x)] > \varphi(0)$ on some set A of positive μ measure. Consequently, if $\lambda \neq \gamma$,

$$s^*(\gamma, \tau, \lambda) = \int_{\mathscr{X}} \varphi[\delta(x)] d\mu(x) > \int_{\mathscr{X}} \varphi(0) d\mu(x) = \varphi(0).$$

Now $s^*(\lambda) = s^*(\gamma^*, \tau^*, \lambda)$, so that $s^*(\lambda) > \varphi(0)$ if $\lambda \neq \gamma^*$ and $s^*(\lambda) = \varphi(0)$ if $\lambda = \gamma^*$, which shows that $s^*(\lambda)$ has a unique minimum at $\lambda = \gamma^*$.

A similar argument can be used to compute λ_n^0. It runs as follows. Let

$$s_n^0(\gamma, \tau, \lambda) = \frac{1}{n} \sum_{t=1}^{n} \varphi[\delta(x_t)] \geq \frac{1}{n} \sum_{t=1}^{n} \varphi(0) = \varphi(0).$$

Since $s_n^0(\lambda) = s_n^0(\gamma_n^0, \tau_n^0, \lambda)$, $s_n^0(\lambda)$ has a minimum at $\lambda = \gamma_n^0$. It is not necessary to the theory which follows that λ_n^0 be unique. Existence is all that is required. Similarly for τ_n^0. □

We shall adjoin some technical conditions. To comment, note that the almost sure convergence of $\hat{\tau}_n$ imposed in Assumption 4 implies that there is a sequence which takes its values in a neighborhood of τ^* and is tail equivalent (Lemma 2) to $\hat{\tau}_n$. Consequently, without loss of generality, it may be assumed that $\hat{\tau}_n$ takes its values in a compact ball T for which τ^* is an interior point. Thus, the effective conditions of the next assumption are domination of the objective function and a compact estimation space Λ^*. As noted in the previous section, a compact estimation space is not a serious restriction in applications.

ASSUMPTION 5. The estimation space Λ^* is compact; $\{\hat{\tau}_n\}$ and $\{\tau_n^0\}$ are contained in T, which is a closed ball centered at τ^* with finite, nonzero radius. The distance function $s(y, x, \tau, \lambda)$ is continuous on $\mathscr{Y} \times \mathscr{X} \times T \times \Lambda^*$, and $|s(y, x, \tau, \lambda)| \leq b[q(y, x, \gamma), x]$ on $\mathscr{Y} \times \mathscr{X} \times T \times \Lambda^* \times \Gamma$; $b(e, x)$ is that of Assumption 3.

The exhibition of the requisite dominating function $b(e, x)$ is an *ad hoc* process, and one exploits the special characteristics of an application. We illustrate with Example 1:

EXAMPLE 1 (Continued).   Now $\rho(u) \leq \frac{1}{2}|u|$ (Problem 9), so that

$$
\begin{aligned}
|s(y, x, \tau, \lambda)| &= \rho\left(\frac{e + f(x, \gamma) - f(x, \lambda)}{\tau}\right) \\
&\leq \frac{\frac{1}{2}|e + f(x, \gamma) - f(x, \lambda)|}{\tau} \\
&\leq \frac{|e| + \sup_\Gamma |f(x, \gamma)| + \sup_{\Lambda*}|f(x, \lambda)|}{\min T}.
\end{aligned}
$$

Suppose that $\Gamma = \Lambda^*$ and that $\sup_\Gamma |f(x, \gamma)|$ is μ-integrable. Then

$$
b_1(e, x) = \frac{|e| + 2\sup_\Gamma |f(x, \gamma)|}{\min T}
$$

will serve to dominate $s(y, x, \tau, \lambda)$. If $\mathscr{X}$ is compact, then $b_1(e, x)$ is integrable for any μ. To see this observe that $f(x, \gamma)$ must be continuous over $\mathscr{X} \times \Gamma$ to satisfy Assumption 2. A continuous function over a compact set is bounded, so $\sup_\Gamma f(x, \gamma)$ is a bounded function.

Later (Assumption 6) we shall need to dominate

$$
\left\|\frac{\partial}{\partial\lambda}s(y, x, \tau, \lambda)\right\| = \left\|\Psi\left(\frac{e + f(x, \gamma) - f(x, \lambda)}{\tau}\right)\frac{(\partial/\partial\lambda)f(x, \lambda)}{\tau}\right\|
$$

$$
\leq \frac{\sup_\Gamma\|(\partial/\partial\lambda)f(x, \lambda)\|}{\min T}
$$

since $|\Psi(u)| = |(1/2)\tanh(u/2)| \leq 1/2$. Thus

$$
b_2(e, x) = \frac{\sup_\Gamma\|(\partial/\partial\lambda)f(x, \lambda)\|}{\min T}
$$

serves as a dominating function.

One continues the construction of suitable $b_1(e, x), b_2(e, x), \ldots$ to dominate each of the functions listed in Assumptions 4 and 6. Then the overall dominating function of Assumption 3 is

$$
b(e, x) = \sum_i b_i(e, x).
$$

This construction will satisfy the formal logical requirements of the theory. In many applications $\mathscr{X}$ can be taken as compact and $P(e)$ to possess enough moments, so that the domination requirements of the general theory obtain trivially.   □

We can now prove that $\hat{\lambda}_n$ is a strongly consistent estimator of λ^*. First a lemma, then the proof:

LEMMA 1. Let Assumptions 1 through 5 hold. Then $s_n(\lambda)$ converges almost surely to $s^*(\lambda)$ uniformly on Λ^*, and $s_n^0(\lambda)$ converges to $s^*(\lambda)$ uniformly on Λ^*.

Proof. We shall prove the result for $s_n(\lambda)$. The argument for $s_n^0(\lambda)$ is much the same (Problem 1). Now

$$\sup_{\Lambda^*} |s_n(\lambda) - s^*(\lambda)|$$

$$\leq \sup_{\Lambda^*} \left| \frac{1}{n} \sum_{t=1}^{n} s\left[Y(e_t, x_t, \gamma_n^0), x_t, \hat{\tau}_n, \lambda \right] \right.$$

$$\left. - \int_{\mathscr{X}} \int_{\mathscr{E}} s\left[Y(e, x, \gamma_n^0), x, \hat{\tau}_n, \lambda \right] dP(e)\, d\mu(x) \right|$$

$$+ \sup_{\Lambda^*} \left| \int_{\mathscr{X}} \int_{\mathscr{E}} s\left[Y(e, x, \gamma_n^0), x, \hat{\tau}_n, \lambda \right] dP(e)\, d\mu(x) \right.$$

$$\left. - \int_{\mathscr{X}} \int_{\mathscr{E}} s\left[Y(e, x, \gamma^*), x, \tau^*, \lambda \right] dP(e)\, d\mu(x) \right|$$

$$\leq \sup_{\Gamma \times T \times \Lambda^*} \left| \frac{1}{n} \sum_{t=1}^{n} s\left[Y(e_t, x_t, \gamma), x_t, \tau, \lambda \right] \right.$$

$$\left. - \int_{\mathscr{X}} \int_{\mathscr{E}} s\left[Y(e, x, \gamma), x, \tau, \lambda \right] dP(e)\, d\mu(x) \right|$$

$$+ \sup_{\Lambda^*} \int_{\mathscr{X}} \int_{\mathscr{E}} \left| s\left[Y(e, x, \gamma_n^0), x, \hat{\tau}_n, \lambda \right] \right.$$

$$\left. - s\left[Y(e, x, \gamma^*), x, \tau^*, \lambda \right] \right| dP(e)\, d\mu(x)$$

$$= \sup_{\Gamma \times T \times \Lambda^*} f_n(\gamma, \tau, \lambda) + \sup_{\Lambda^*} g(\gamma_n^0, \hat{\tau}_n, \lambda).$$

Since $\Gamma \times T \times \Lambda^*$ is compact, and $s(y, x, \tau, \lambda)$ is continuous on $\mathcal{Y} \times \mathcal{X} \times T \times \Lambda^*$ with $|s(y, x, \tau, \lambda)| \leq b[q(y, x, \gamma), x]$ for all (y, x) in $\mathcal{Y} \times \mathcal{X}$ and all (γ, τ, λ) in $\Gamma \times T \times \Lambda^*$, we have, by Theorem 1, that $\sup_{\Gamma \times T \times \Lambda^*} f_n(\gamma, \tau, \lambda)$ converges almost surely to zero. Given any sequence $\{(\gamma_n, \tau_n, \lambda_n)\}$ that converges to, say, $(\gamma^0, \tau^0, \lambda^0)$ we have, by the dominated convergence theorem with $2b(e, x)$ as the dominating function, that $\lim_{n \to \infty} g(\gamma_n, \tau_n, \lambda_n) = g(\gamma^0, \tau^0, \lambda^0)$. This shows that $g(\gamma, \tau, \lambda)$ is continuous in (γ, τ, λ). Moreover, $\sup_{\Lambda^*} g(\gamma, \tau, \lambda)$ is continuous in (γ, τ); see the proof of Theorem 1 for details. Then, since $(\gamma_n^0, \hat{\tau}_n)$ converges almost surely to (γ^*, τ^*), $\sup_{\Lambda^*} g(\gamma_n^0, \hat{\tau}_n^0, \lambda)$ converges almost surely to zero. $\square$

THEOREM 3 (Strong consistency). Let Assumptions 1 through 5 hold. Then $\hat{\lambda}_n$ converges almost surely to λ^*, and λ_n^0 converges to λ^*.

Proof. If a realization $\{e_t\}$ of the errors is held fixed, then $\{\hat{\lambda}_n\}$ becomes a fixed, vector valued sequence and $\{s_n(\lambda)\}$ becomes a fixed sequence of functions. We shall hold fixed a realization $\{e_t\}$ with the attribute that $s_n(\lambda)$ converges uniformly to $s^*(\lambda)$ on Λ^*; almost every realization is such by Lemma 1. If we can show that the corresponding sequence $\{\hat{\lambda}_n\}$ converges to λ^*, then we have the first result. This is the plan.

Now $\hat{\lambda}_n$ lies in the compact set Λ^*. Thus the sequence $\{\hat{\lambda}_n\}$ has at least one limit point $\hat{\lambda}$ and one subsequence $\{\hat{\lambda}_{n_m}\}$ with $\lim_{m \to \infty} \hat{\lambda}_{n_m} = \hat{\lambda}$. Now, by uniform convergence (see Problem 2),

$$s^*(\hat{\lambda}) = \lim_{m \to \infty} s_{n_m}(\hat{\lambda}_{n_m})$$

$$\leq \lim_{m \to \infty} s_{n_m}(\lambda^*)$$

$$= s^*(\lambda^*)$$

where the inequality is due to the fact that $s_n(\hat{\lambda}_n) \leq s_n(\lambda^*)$ for every n as $\hat{\lambda}_n$ is a minimizing value. The assumption of a unique minimum, Assumption 4, implies $\hat{\lambda} = \lambda^*$. Then $\{\hat{\lambda}_n\}$ has only the one limit point λ^*.

An analogous argument implies that λ_n^0 converges to λ^* (Problem 3). $\square$

The following notation defines the parameters of the asymptotic distribution of $\hat{\lambda}_n$.

NOTATION 2.

$$\bar{\bar{\mathscr{U}}}(\lambda) = \int_{\mathscr{X}} \left(\int_{\mathscr{E}} \frac{\partial}{\partial \lambda} s[Y(e, x, \gamma^*), x, \tau^*, \lambda] \, dP(e) \right)$$
$$\times \left(\int_{\mathscr{E}} \frac{\partial}{\partial \lambda} s[Y(e, x, \gamma^*), x, \tau^*, \lambda] \, dP(e) \right)' d\mu(x)$$

$$\bar{\bar{\mathscr{I}}}(\lambda) = \int_{\mathscr{X}} \int_{\mathscr{E}} \left(\frac{\partial}{\partial \lambda} s[Y(e, x, \gamma^*), x, \tau^*, \lambda] \right)$$
$$\times \left(\frac{\partial}{\partial \lambda} s[Y(e, x, \gamma^*), x, \tau^*, \lambda] \right)' dP(e) \, d\mu(x) - \bar{\bar{\mathscr{U}}}(\lambda)$$

$$\bar{\bar{\mathscr{J}}}(\lambda) = \int_{\mathscr{X}} \int_{\mathscr{E}} \frac{\partial^2}{\partial \lambda \, \partial \lambda'} s[Y(e, x, \gamma^*), x, \tau^*, \lambda] \, dP(e) \, d\mu(x)$$

$$\mathscr{I}^* = \bar{\bar{\mathscr{I}}}(\lambda^*), \qquad \mathscr{J}^* = \bar{\bar{\mathscr{J}}}(\lambda^*), \quad \mathscr{U}^* = \bar{\bar{\mathscr{U}}}(\lambda^*).$$

If this were maximum likelihood estimation with $s(y, x, \tau, \lambda) = -\ln p(y \mid x, \lambda)$, then $\mathscr{I}^*$ would be the information matrix and $\mathscr{J}^*$ the expectation of the Hessian of the log-likelihood. Under correct specification one would have $\mathscr{U}^* = 0$ and $\mathscr{I}^* = \mathscr{J}^*$ (Section 7).

We illustrate the computations with the example.

EXAMPLE 1 (Continued). The first and second derivatives of $s(y, x, \tau, \lambda)$ are

$$\frac{\partial}{\partial \lambda} s(y, x, \tau, \lambda) = \frac{\partial}{\partial \lambda} \rho \left(\frac{y - f(x, \lambda)}{\tau} \right)$$
$$= -\frac{1}{\tau} \Psi \left(\frac{y - f(x, \lambda)}{\tau} \right) \frac{\partial}{\partial \lambda} f(x, \lambda)$$

$$\frac{\partial^2}{\partial \lambda \, \partial \lambda'} s(y, x, \tau, \lambda) = \frac{\partial}{\partial \lambda} \left(-\frac{1}{\tau} \right) \Psi \left(\frac{y - f(x, \lambda)}{\tau} \right) \frac{\partial}{\partial \lambda'} f(x, \lambda)$$
$$= \frac{1}{\tau^2} \Psi' \left(\frac{y - f(x, \lambda)}{\tau} \right) \left(\frac{\partial}{\partial \lambda} f(x, \lambda) \right) \left(\frac{\partial}{\partial \lambda} f(x, \lambda) \right)'$$
$$- \frac{1}{\tau} \Psi \left(\frac{y - f(x, \lambda)}{\tau} \right) \frac{\partial^2}{\partial \lambda \, \partial \lambda'} f(x, \lambda).$$

Evaluating the first derivative at $y = f(x, \gamma) + e$, $\tau = \tau^*$, and $\lambda = \gamma$, we have

$$\int_{\mathscr{E}} \frac{\partial}{\partial \lambda} s[Y(e, x, \gamma), x, \tau^*, \lambda] \, dP(e) \Big|_{\gamma = \lambda}$$

$$= -\frac{1}{\tau^*} \int_{\mathscr{E}} \Psi\left(\frac{e}{\tau^*}\right) dP(e) \frac{\partial}{\partial \lambda} f(x, \lambda)$$

$$= -\frac{1}{\tau^*}(0) \frac{\partial}{\partial \lambda} f(x, \lambda)$$

$$= 0$$

because $\Psi(e/\tau)$ is an odd function [i.e., $\Psi(u) = \Psi(-u)$], and an odd function integrates to zero against a symmetric error distribution. Thus, $\mathscr{U}^* = 0$. In fact, $\mathscr{U}^*$ is always zero in a correctly specified situation when using a sensible estimation procedure. To continue, writing $\mathscr{E}\Psi^2(e/\tau^*)$ for $\int_{\mathscr{E}} \Psi^2(e/\tau^*) \, dP(e)$ and $\mathscr{E}\Psi'(e/\tau^*)$ for $\int_{\mathscr{E}} (d/du)\Psi(u)|_{u = e/\tau^*} \, dP(e)$, we have

$$\int_{\mathscr{E}} \left(\frac{\partial}{\partial \lambda} s[Y(e, x, \gamma), x, \tau^*, \lambda] \right) \left(\frac{\partial}{\partial \lambda} s[Y(e, x, \gamma), x, \tau^*, \lambda] \right)' dP(e) \Big|_{\gamma = \lambda}$$

$$= \left(\frac{1}{\tau^*} \right)^2 \mathscr{E}\Psi^2\left(\frac{e}{\tau^*}\right) \left(\frac{\partial}{\partial \lambda} f(x, \lambda) \right) \left(\frac{\partial}{\partial \lambda} f(x, \lambda) \right)'$$

and

$$\int_{\mathscr{E}} \frac{\partial^2}{\partial \lambda \, \partial \lambda'} s[Y(e, x, \gamma), x, \tau^*, \lambda] \, dP(e) \Big|_{\gamma = \lambda}$$

$$= \left(\frac{1}{\tau^*} \right)^2 \mathscr{E}\Psi'\left(\frac{e}{\tau^*}\right) \left(\frac{\partial}{\partial \lambda} f(x, \lambda) \right) \left(\frac{\partial}{\partial \lambda} f(x, \lambda) \right)'.$$

Thus,

$$\mathscr{I}^* = \left(\frac{1}{\tau^*} \right)^2 \mathscr{E}\Psi^2\left(\frac{e}{\tau^*}\right) \int_{\mathscr{X}} \left(\frac{\partial}{\partial \lambda} f(x, \lambda^*) \right) \left(\frac{\partial}{\partial \lambda} f(x, \lambda^*) \right)' d\mu(x)$$

and

$$\mathscr{J}^* = \left(\frac{1}{\tau^*} \right)^2 \mathscr{E}\Psi'\left(\frac{e}{\tau^*}\right) \int_{\mathscr{X}} \left(\frac{\partial}{\partial \lambda} f(x, \lambda^*) \right) \left(\frac{\partial}{\partial \lambda} f(x, \lambda^*) \right)' d\mu(x). \quad \square$$

In Section 5, the distributions of test statistics are characterized in terms of the following quantities:

NOTATION 3.

$$\bar{\mathcal{U}}_n(\lambda) = \frac{1}{n} \sum_{t=1}^{n} \left(\int_{\mathcal{E}} \frac{\partial}{\partial \lambda} s \left[Y(e, x_t, \gamma_n^0), x_t, \tau_n^0, \lambda \right] dP(e) \right)$$

$$\times \left(\int_{\mathcal{E}} \frac{\partial}{\partial \lambda} s \left[Y(e, x_t, \gamma_n^0), x_t, \tau_n^0, \lambda \right] dP(e) \right)'$$

$$\bar{\mathcal{I}}_n(\lambda) = \frac{1}{n} \sum_{t=1}^{n} \int_{\mathcal{E}} \left(\frac{\partial}{\partial \lambda} s \left[Y(e, x_t, \gamma_n^0), x_t, \tau_n^0, \lambda \right] \right)$$

$$\times \left(\frac{\partial}{\partial \lambda} s \left[Y(e, x_t, \gamma_n^0), x_t, \tau_n^0, \lambda \right] \right)' dP(e) - \bar{\mathcal{U}}_n(\lambda)$$

$$\bar{\mathcal{J}}_n(\lambda) = \frac{1}{n} \sum_{t=1}^{n} \int_{\mathcal{E}} \frac{\partial^2}{\partial \lambda \, \partial \lambda'} s \left[Y(e, x_t, \gamma_n^0), x_t, \tau_n^0, \lambda \right] dP(e)$$

$$\mathcal{J}_n^0 = \bar{\mathcal{J}}_n(\lambda_n^0), \qquad \mathcal{I}_n^0 = \bar{\mathcal{I}}_n(\lambda_n^0), \qquad \mathcal{U}_n^0 = \bar{\mathcal{U}}_n(\lambda_n^0)$$

$$\mathcal{J}_n^* = \bar{\mathcal{J}}_n(\lambda_n^*), \qquad \mathcal{I}_n^* = \bar{\mathcal{I}}_n(\lambda_n^*), \qquad \mathcal{U}_n^* = \bar{\mathcal{U}}_n(\lambda_n^*).$$

We illustrate their computation with Example 1:

EXAMPLE 1 (Continued). Let

$$\mu_t(\lambda) = \int_{\mathcal{E}} \Psi \left(\frac{e + f(x, \gamma_n^0) - f(x, \lambda)}{\tau_n^0} \right) dP(e)$$

$$\sigma_t^2(\lambda) = \int_{\mathcal{E}} \Psi^2 \left(\frac{e + f(x, \gamma_n^0) - f(x, \lambda)}{\tau_n^0} \right) dP(e) - \mu_t^2(\lambda)$$

$$\beta_t(\lambda) = \int_{\mathcal{E}} \Psi' \left(\frac{e + f(x, \gamma_n^0) - f(x, \lambda)}{\tau_n^0} \right) dP(e).$$

Note that if one evaluates at $\lambda = \lambda_n^0$, then $\mu_t(\lambda_n^0) = 0$, $\sigma_t^2(\lambda_n^0) = \mathcal{E} \Psi^2(e/\tau_n^0)$, and $\beta_t(\lambda_n^0) = \mathcal{E} \Psi'(e/\tau_n^0)$, which eliminates the variation with t; but if one evaluates at $\lambda = \lambda_n^*$, then the variation with t remains. We have by direct computation that

$$\bar{\mathscr{U}}(\lambda) = \left(\frac{1}{\tau_n^0}\right)^2 \frac{1}{n} \sum_{t=1}^{n} u_t^2(\lambda)\left(\frac{\partial}{\partial\lambda}f(x_t, \lambda)\right)\left(\frac{\partial}{\partial\lambda}f(x_t, \lambda)\right)'$$

$$\bar{\mathscr{I}}(\lambda) = \left(\frac{1}{\tau_n^0}\right)^2 \frac{1}{n} \sum_{t=1}^{n} \sigma_t^2(\lambda)\left(\frac{\partial}{\partial\lambda}f(x_t, \lambda)\right)\left(\frac{\partial}{\partial\lambda}f(x_t, \lambda)\right)'$$

$$\bar{\mathscr{J}}(\lambda) = \left(\frac{1}{\tau_n^0}\right)^2 \frac{1}{n} \sum_{t=1}^{n} \beta_t(\lambda)\left(\frac{\partial}{\partial\lambda}f(x_t, \lambda)\right)\left(\frac{\partial}{\partial\lambda}f(x_t, \lambda)\right)'$$

$$- \frac{1}{\tau_n^0}\frac{1}{n} \sum_{t=1}^{n} \mu_t(\lambda)\frac{\partial^2}{\partial\lambda\,\partial\lambda'}f(x_t, \lambda). \qquad\qquad \square$$

Some plausible estimators of $\mathscr{I}^*$ and $\mathscr{J}^*$—or of $(\mathscr{I}_n^0, \mathscr{J}_n^0)$ and $(\mathscr{I}_n^*, \mathscr{J}_n^*)$ respectively, depending on one's point of view—are as follows:

NOTATION 4.

$$\mathscr{I}_n(\lambda) = \frac{1}{n} \sum_{t=1}^{n} \left(\frac{\partial}{\partial\lambda}s(y_t, x_t, \hat{\tau}_n, \lambda)\right)\left(\frac{\partial}{\partial\lambda}s(y_t, x_t, \hat{\tau}_n, \lambda)\right)'$$

$$\mathscr{J}_n(\lambda) = \frac{1}{n} \sum_{t=1}^{n} \frac{\partial^2}{\partial\lambda\,\partial\lambda'}s(y_t, x_t, \hat{\tau}_n, \lambda)$$

$$\hat{\mathscr{I}} = \mathscr{I}_n(\hat{\lambda}_n), \qquad \hat{\mathscr{J}} = \mathscr{J}_n(\hat{\lambda}_n), \qquad \tilde{\mathscr{I}} = \mathscr{I}_n(\tilde{\lambda}_n), \qquad \tilde{\mathscr{J}} = \mathscr{J}_n(\tilde{\lambda}_n).$$

We illustrate the computations and point out some alternatives using Example 1.

EXAMPLE 1 (Continued). Let

$$\hat{\Psi}_t = \Psi\left(\frac{y_t - f(x_t, \hat{\lambda}_n)}{\hat{\tau}_n}\right)$$

$$\hat{\Psi}_t' = \Psi'\left(\frac{y_t - f(x_t, \hat{\lambda}_n)}{\hat{\tau}_n}\right).$$

Then

$$\hat{\mathscr{I}}_n = \left(\frac{1}{\hat{\tau}_n}\right)^2 \frac{1}{n} \sum_{t=1}^{n} \hat{\Psi}_t^2\left(\frac{\partial}{\partial\lambda}f(x_t, \hat{\lambda}_n)\right)\left(\frac{\partial}{\partial\lambda}f(x_t, \hat{\lambda}_n)\right)'$$

$$\hat{\mathscr{J}}_n = \left(\frac{1}{\hat{\tau}_n}\right)^2 \frac{1}{n} \sum_{t=1}^{n} \hat{\Psi}_t'\left(\frac{\partial}{\partial\lambda}f(x_t, \hat{\lambda}_n)\right)\left(\frac{\partial}{\partial\lambda}f(x_t, \hat{\lambda}_n)\right)'$$

$$- \frac{1}{\hat{\tau}_n}\frac{1}{n} \sum_{t=1}^{n} \hat{\Psi}_t\frac{\partial^2}{\partial\lambda\,\partial\lambda'}f(x_t, \hat{\lambda}_n).$$

Alternatives that are similar to the forms used in least squares are

$$\hat{I}_n = \frac{\hat{\sigma}_n^2}{\hat{\tau}_n^2} \frac{1}{n} \sum_{t=1}^{n} \left(\frac{\partial}{\partial \lambda} f(x_t, \hat{\lambda}_n) \right) \left(\frac{\partial}{\partial \lambda} f(x_t, \hat{\lambda}_n) \right)'$$

$$\hat{J}_n = \frac{\hat{\beta}_n}{\hat{\tau}_n^2} \frac{1}{n} \sum_{t=1}^{n} \left(\frac{\partial}{\partial \lambda} f(x_t, \hat{\lambda}_n) \right) \left(\frac{\partial}{\partial \lambda} f(x_t, \hat{\lambda}_n) \right)'$$

with

$$\hat{\sigma}_n^2 = \frac{1}{n} \sum_{t=1}^{n} \hat{\Psi}_t^2$$

$$\hat{\beta}_n = \frac{1}{n} \sum_{t=1}^{n} \hat{\Psi}_t'.$$

The former are heteroscedastic invariant, the latter are not.   □

Some additional, technical restrictions needed to prove asymptotic normality are:

ASSUMPTION 6. The estimation space Λ^* contains a closed ball Λ centered at λ^* with finite, nonzero radius such that the elements of $(\partial/\partial\lambda)s(y, x, \tau, \lambda)$, $(\partial^2/\partial\lambda\,\partial\lambda')s(y, x, \tau, \lambda)$, $(\partial^2/\partial\tau\partial\lambda')s(y, x, \tau, \lambda)$, and $[(\partial/\partial\lambda)s(y, x, \tau, \lambda)][(\partial/\partial\lambda)s(y, x, \tau, \lambda)]'$ are continuous and dominated by $b[q(y, x, \gamma), x]$ on $\mathcal{Y} \times \mathcal{X} \times \Gamma \times T \times \Lambda$. Moreover, $\mathcal{J}^*$ is nonsingular and

$$\int_{\mathcal{X}} \int_{\mathcal{E}} \frac{\partial^2}{\partial\tau\partial\lambda'} s[Y(e, x, \gamma^*), x, \tau^*, \lambda^*]\, dP(e)\, d\mu(x) = 0.$$

The integral condition is sometimes encountered in the theory of maximum likelihood estimation; see Durbin (1970) for a detailed discussion. It validates the application of maximum likelihood theory to a subset of the parameters when the remainder are treated as if known in the derivations but are subsequently estimated. The assumption plays the same role here. It can be avoided in maximum likelihood estimation at a cost of additional complexity in the results; see Gallant and Holly (1980) for details. It can be avoided here as well, but there is no reason to further complicate the results in view of the intended applications. In an application where the condition is not satisfied, the simplest solution is to estimate λ and τ jointly and not use a two step estimator. We illustrate with the example:

EXAMPLE 1   (Continued).

$$\frac{\partial^2}{\partial\tau\partial\lambda'}s\big[Y(e,x,\gamma^*),x,\tau,\lambda^*)\big]\bigg|_{\gamma^*=\lambda^*}$$

$$=\frac{1}{\tau^2}\Big[\Psi\Big(\frac{e}{\tau}\Big)+\Psi'\Big(\frac{e}{\tau}\Big)\frac{e}{\tau}\Big]\frac{\partial}{\partial\lambda}f(x,\lambda^*).$$

Both $\Psi(e/\tau)$ and $\Psi'(e/\tau)(e/\tau)$ are odd functions and will integrate to zero for symmetric $P(e)$.   □

The derivative of the distance function plays the same role here as does the derivative of the log density function or score in maximum likelihood estimation. Hence, we use the same terminology here. As with the scores in maximum likelihood estimation, their normalized sum is asymptotically normally distributed:

THEOREM 4   (Asymptotic normality of the scores).   Under Assumptions 1 through 6

$$\sqrt{n}\,\frac{\partial}{\partial\lambda}s_n\big(\lambda_n^0\big)\xrightarrow{\mathscr{L}}N_p(0,\mathscr{I}^*).$$

$\mathscr{I}^*$ may be singular.

Proof.   By Theorem 2

$$\frac{1}{\sqrt{n}}\sum_{t=1}^{n}\Bigg(\frac{\partial}{\partial\lambda}s\big(y_t,x_t,\hat{\tau}_n,\lambda_n^0\big)$$

$$-\int_{\mathscr{E}}\frac{\partial}{\partial\lambda}s\big[Y(e,x_t,\gamma_n^0),x_t,\tau_n^0,\lambda_n^0\big]\,dP(e)\Bigg)$$

$$\xrightarrow{\mathscr{L}}N_p(0,\mathscr{I}^*).$$

Domination permits the interchange of differentiation and integration (Problem 11), and λ_n^0 is defined as a minimizing value, whence

$$\sum_{t=1}^{n}\int_{\mathscr{E}}\frac{\partial}{\partial\lambda}s\big[Y(e,x_t,\gamma_n^0),x_t,\tau_n^0,\lambda_n^0\big]\,dP(e)$$

$$=\sum_{t=1}^{n}\frac{\partial}{\partial\lambda}\int_{\mathscr{E}}s\big[Y(e,x_t,\gamma_n^0),x_t,\tau_n^0,\lambda_n^0\big]\,dP(e)$$

$$=0.$$ □

We can now show that $\hat{\lambda}_n$ is asymptotically normally distributed. First we prove two lemmas:

LEMMA 2 (Tail equivalence). Let $\{\lambda_n\}$ be a sequence of vector valued random variables that take their values in $\Lambda^* \subset \mathbb{R}^p$ and that converge almost surely to a point λ^* in Λ^*. Let $\{s_n(\lambda)\}$ be a sequence of real valued random functions defined on Λ^*. Let $g(\lambda)$ be a vector valued function defined on Λ^*. Let Λ^0 be an open subset of $\mathbb{R}^p$ with $\lambda^* \in \Lambda^0 \subset \Lambda^*$. Then there is a sequence $\{\bar{\lambda}_n\}$ of random variables that take their values in Λ^0, that satisfy

$$g(\lambda_n) = g(\bar{\lambda}_n) + o_s(n^{-\alpha})$$

for every $\alpha > 0$, and such that:

1. If $(\partial/\partial\lambda)s_n(\lambda)$ is continuous on Λ^0, and λ_n minimizes $s_n(\lambda)$ over Λ^*, then

$$\frac{\partial}{\partial\lambda}s_n(\bar{\lambda}_n) = o_s(n^{-\alpha})$$

for every $\alpha > 0$.

2. If $(\partial/\partial\lambda)s_n(\lambda)$ and $(\partial/\partial\lambda')h(\lambda)$ are continuous on Λ^0, if $(\partial/\partial\lambda')h(\lambda)$ has full rank at $\lambda = \lambda^*$, and if λ_n minimizes $s_n(\lambda)$ over Λ^* subject to $h(\lambda) = 0$, then there is a vector $\bar{\theta}_n$ of (random) Lagrange multipliers such that

$$\frac{\partial}{\partial\lambda'}[s_n(\bar{\lambda}_n) + \bar{\theta}_n'h(\bar{\lambda}_n)] = o_s(n^{-\alpha})$$
$$h(\bar{\lambda}_n) = o_s(n^{-\alpha})$$

for every $\alpha > 0$.

Proof. The idea of the proof is that eventually λ_n is in Λ^0 and itself has the desired properties, due to the almost sure convergence of λ_n to λ^*. Stating that the residual random variables are of almost sure order $o_s(n^{-\alpha})$ is a way of expressing the fact that the requisite large n depends on the realization $\{e_t\}$ that obtains; that is, the convergence is not uniform in $\{e_t\}$.

We shall prove part 2. By Problem 5, $(\partial/\partial\lambda')h(\lambda)$ has full rank on some open set $\mathcal{O}$ with $\lambda^* \in \mathcal{O} \subset \Lambda^0$. Define

$$\bar{\lambda}_n = \begin{cases} \lambda^* & \text{if } \lambda_n \notin \mathcal{O} \\ \lambda_n & \text{if } \lambda_n \in \mathcal{O}. \end{cases}$$

Fix a realization $\{e_t\}$ for which $\lim_{n \to \infty} \lambda_n = \lambda^*$; almost every realization is such. There is an N such that $n > N$ implies $\lambda_n \in \mathcal{O}$ for all $n > N$. Since $\mathcal{O}$ is open and λ_n is the constrained optimum, we have that $\bar{\theta}_n$ exists and that

$$\bar{\lambda}_n = \lambda_n$$

$$\frac{\partial}{\partial \lambda'} [s_n(\lambda_n) + \bar{\theta}'_n h(\lambda_n)] = 0$$

$$h(\lambda_n) = 0$$

(Bartle, 1964, Section 21). Then, trivially,

$$\lim_{n \to \infty} n^\alpha \| g(\bar{\lambda}_n) - g(\lambda_n) \| = 0$$

$$\lim_{n \to \infty} n^\alpha \left\| \frac{\partial}{\partial \lambda'} [s_n(\bar{\lambda}_n) + \bar{\theta}'_n h(\bar{\lambda}_n)] \right\| = 0,$$

$$\lim_{n \to \infty} n^\alpha \| h(\bar{\lambda}_n) \| = 0. \qquad \square$$

LEMMA 3. Under Assumptions 1 though 6, interchange of differentiation and integration is permitted in these instances:

$$\frac{\partial}{\partial \lambda} s^*(\lambda) = \int_{\mathscr{X}} \int_{\mathscr{E}} \frac{\partial}{\partial \lambda} s[Y(e, x, \gamma^*), x, \tau^*, \lambda] \, dP(e) \, d\mu(x)$$

$$\frac{\partial^2}{\partial \lambda \, \partial \lambda'} s^*(\lambda) = \int_{\mathscr{X}} \int_{\mathscr{E}} \frac{\partial^2}{\partial \lambda \, \partial \lambda'} s[Y(e, x, \gamma^*), x, \tau^*, \lambda] \, dP(e) \, d\mu(x)$$

$$\frac{\partial}{\partial \lambda} s_n^0(\lambda) = \frac{1}{n} \sum_{t=1}^{n} \int_{\mathscr{E}} \frac{\partial}{\partial \lambda} s[Y(e, x_t, \gamma_n^0), x_t, \tau_n^0, \lambda] \, dP(e)$$

$$\frac{\partial^2}{\partial \lambda \, \partial \lambda'} s_n^0(\lambda) = \frac{1}{n} \sum_{t=1}^{n} \int_{\mathscr{E}} \frac{\partial^2}{\partial \lambda \, \partial \lambda'} s[Y(e, x_t, \gamma_n^0), x_t, \tau_n^0, \lambda] \, dP(e).$$

Moreover,

$$\lim_{n \to \infty} \frac{\partial}{\partial \lambda} s_n^0(\lambda) = \frac{\partial}{\partial \lambda} s^*(\lambda) \qquad \text{uniformly on } \Lambda$$

$$\lim_{n \to \infty} \frac{\partial^2}{\partial \lambda \, \partial \lambda'} s_n^0(\lambda) = \frac{\partial^2}{\partial \lambda \, \partial \lambda'} s^*(\lambda) \qquad \text{uniformly on } \Lambda$$

$$\lim_{n \to \infty} \frac{\partial}{\partial \lambda} s_n(\lambda) = \frac{\partial}{\partial \lambda} s^*(\lambda) \qquad \text{almost surely, uniformly on } \Lambda$$

$$\lim_{n \to \infty} \frac{\partial^2}{\partial \lambda \, \partial \lambda'} s_n(\lambda) = \frac{\partial^2}{\partial \lambda \, \partial \lambda'} s^*(\lambda) \qquad \text{almost surely, uniformly on } \Lambda.$$

Proof. *Interchange*: We shall prove the result for $(\partial/\partial\lambda)s^*(\lambda)$, the argument for the other three cases being much the same. Let λ be in Λ, and $\{h_m\}$ be any sequence with $\lim_{m\to\infty}h_m = 0$ and $\lambda - h_m\xi_i$ in Λ, where ξ_i is the ith elementary vector. By the mean value theorem,

$$\frac{s[Y(e, x, \gamma^*), x, \tau^*, \lambda] - s[Y(e, x, \gamma^*), x, \tau^*, \lambda - h_m\xi_i]}{h_m}$$

$$= \frac{\partial}{\partial\lambda_i}s[Y(e, x, \gamma^*), x, \tau^*, \lambda - \bar{h}_m(e, x)\xi_i]$$

where $|\bar{h}_m(e, x)| \le h_m$. [One can show that $\bar{h}_m(e, x)$ is measurable, but it is not necessary for the validity of the proof, as the composite function on the right hand side is measurable by virtue of being equal to the left.] Thus

$$\frac{s^*(\lambda) - s^*(\lambda - h_m\xi_i)}{h_m}$$

$$= \int_{\mathcal{X}}\int_{\mathcal{E}}\frac{\partial}{\partial\lambda_i}s[Y(e, x, \gamma^*), x, \tau^*, \lambda - \bar{h}_m(e, x)\xi_i]\, dP(e)\, d\mu(x).$$

By the dominated convergence theorem, with $b(e, x)$ as the dominating function, and continuity,

$$\frac{\partial}{\partial\lambda_i}s^*(\lambda) = \lim_{m\to\infty}\frac{s^*(\lambda) - s^*(\lambda - h_m\xi_i)}{h_m}$$

$$= \int_{\mathcal{X}}\int_{\mathcal{E}}\lim_{m\to\infty}\frac{\partial}{\partial\lambda_i}s[Y(e, x, \gamma^*), x, \tau^*,$$

$$\lambda - \bar{h}_m(e, x)\xi_i]\, dP(e)\, d\mu(x)$$

$$= \int_{\mathcal{X}}\int_{\mathcal{E}}\frac{\partial}{\partial\lambda_i}s[Y(e, x, \gamma^*), x, \tau^*, \lambda]\, dP(e)\, d\mu(x).$$

Uniform convergence: The argument is the same as that used in the proof of Lemma 1.   □

THEOREM 5 (Asymptotic normality).   Let Assumptions 1 through 6 hold. Then:

$$\sqrt{n}\left(\hat{\lambda}_n - \lambda_n^0\right) \xrightarrow{\mathscr{L}} N_p\left[0, (\mathscr{J}^*)^{-1}\mathscr{I}^*(\mathscr{J}^*)^{-1}\right]$$

$\hat{\mathscr{I}}$ converges almost surely to $\mathscr{I}^* + \mathscr{U}^*$,

$\mathscr{I}_n^0$ converges to $\mathscr{I}^*$,

$\hat{\mathscr{J}}$ converges almost surely to $\mathscr{J}^*$,

$\mathscr{J}_n^0$ converges to $\mathscr{J}^*$.

Proof. By Lemma 2, we may assume without loss of generality that $\hat{\lambda}_n, \lambda_n^0 \in \Lambda$ and that $(\partial/\partial\lambda)s_n(\hat{\lambda}_n) = o_s(n^{-1/2})$, $(\partial/\partial\lambda)s_n^0(\lambda_n^0) = o(n^{-1/2})$: see Problem 6.

By Taylor's theorem

$$\sqrt{n}\,\frac{\partial}{\partial\lambda}s_n(\lambda_n^0) = \sqrt{n}\,\frac{\partial}{\partial\lambda}s_n(\hat{\lambda}_n) + \bar{\mathscr{J}}\sqrt{n}\left(\lambda_n^0 - \hat{\lambda}_n\right)$$

where $\bar{\mathscr{J}}$ has rows

$$\frac{\partial}{\partial\lambda'}\frac{\partial}{\partial\lambda_i}s_n(\bar{\lambda}_{in})$$

and $\bar{\lambda}_{in}$ lies on the line segment joining λ_n^0 to $\hat{\lambda}_n$. Now both λ_n^0 and $\hat{\lambda}_n$ converge almost surely to λ^* by Theorem 3, so that $\bar{\lambda}_{in}$ converges almost surely to λ^*. Also, $(\partial/\partial\lambda')(\partial/\partial\lambda_i)s_n(\lambda)$ converges almost surely to $(\partial/\partial\lambda')(\partial/\partial\lambda_i)s^*(\lambda)$ uniformly on Λ by Lemma 3. Taking these two facts together (Problem 2), $\bar{\mathscr{J}}$ converges almost surely to $(\partial^2/\partial\lambda\,\partial\lambda')s^*(\lambda^*)$. By interchanging integration and differentiation as permitted by Lemma 3, $\mathscr{J}^* = (\partial^2/\partial\lambda\,\partial\lambda')s^*(\lambda^*)$. Thus we may write $\bar{\mathscr{J}} = \mathscr{J}^* + o_s(1)$ and, as $(\partial/\partial\lambda)s_n(\tilde{\lambda}_n) = o_s(n^{-1/2})$, we may write

$$[\mathscr{J}^* + o_s(1)]\sqrt{n}\left(\hat{\lambda}_n - \lambda_n^0\right) = -\sqrt{n}\,\frac{\partial}{\partial\lambda}s_n(\lambda_n^0) + o_s(1).$$

The first result follows at once from Slutsky's theorem (Serfling, 1980, Section 1.5.4, or Rao, 1973, Section 2c.4).

By Theorem 1, with Assumption 6 providing the dominating function, and the almost sure convergence of $(\gamma_n^0, \hat{\tau}_n, \hat{\lambda}_n)$ to $(\gamma^*, \tau^*, \lambda^*)$ it follows that $\lim_{n\to\infty}[\mathscr{I}_n(\hat{\lambda}_n), \mathscr{J}_n(\hat{\lambda}_n)] = (\mathscr{I}^* + \mathscr{U}^*, \mathscr{J}^*)$ almost surely (Problem 7). Similar arguments apply to $\mathscr{I}_n^0$ and $\mathscr{J}_n^0$. $\qquad\square$

As illustrated by Example 1, the usual consequence of a correctly specified model and a sensible estimation procedure is:

$$\gamma_n^0 = \gamma^* \quad \text{for all } n \quad \text{implies} \quad \lambda_n^0 = \lambda^* \quad \text{for all } n.$$

If $\lambda_n^0 = \lambda^*$ for all n, then we have

$$\sqrt{n}\left(\hat{\lambda}_n - \lambda^*\right) \xrightarrow{\mathscr{L}} N_p\left[0, (\mathscr{J}^*)^{-1}\mathscr{I}^*(\mathscr{J}^*)^{-1}\right].$$

But, in general, even if $\gamma_n^0 \equiv \gamma^*$ for all n, it is not true that

$$\sqrt{n}\left(\hat{\lambda}_n - \lambda^*\right) \xrightarrow{\mathscr{L}} N_p\left[\Delta, (\mathscr{J}^*)^{-1}\mathscr{I}^*(\mathscr{J}^*)^{-1}\right]$$

for some finite Δ. To reach the conclusion that $\sqrt{n}\,(\hat{\lambda}_n - \lambda^*)$ is asymptotically normally distributed, one must append additional regularity conditions. There are three options.

The first is to impose a Pitman drift. Estimation methods of the usual sort are designed with some class of models in mind. The idea is to embed this intended class in a larger class $Y(e, x, \gamma)$ so that any member of the intended class is given by $Y(e, x, \gamma^*)$ for some choice of γ^*. For this choice one has

$$\gamma_n^0 \equiv \gamma^* \quad \text{for all } n \quad \text{implies} \quad \lambda_n^0 = \lambda^* \quad \text{for all } n.$$

A misspecified model would correspond to some $\gamma^\#$ such that $Y(e, x, \gamma^\#)$ is outside the intended class of models. Starting with $\gamma_1^0 = \gamma^\#$, one chooses a sequence $\gamma_2^0, \gamma_3^0, \ldots$ that converges to γ^* fast enough that $\lim_{n \to \infty} \sqrt{n}\,(\lambda_n^0 - \lambda^*) = \Delta$ for some finite Δ; the most natural choice would seem to be $\Delta = 0$. See Problem 14 for the details of this construction. Since γ can be infinite dimensional, one has considerable latitude in the choice of $Y(e, x, \gamma^\#)$.

The second is to hold $\gamma_n^0 \equiv \gamma^*$ and speed up the rate of convergence of Cesaro sums. If the sequence $\{x_t\}_{t=1}^\infty$ is chosen such that

$$\lim_{n \to \infty} \sqrt{n}\left(\frac{\partial}{\partial \lambda} s_n^0(\lambda) - \frac{\partial}{\partial \lambda} s^*(\lambda) \right) = K(\lambda) \quad \text{uniformly on } \Lambda$$

where $K(\lambda)$ is some finite valued function, then (Problem 15)

$$\lim_{n \to \infty} \sqrt{n}\,\left(\lambda_n^0 - \lambda^* \right) = \Delta$$

for some finite Δ. For example, suppose that $\tau_n^0 = \tau^*$ and the sequence $\{x_t\}_{t=1}^\infty$ consists of replicates of T points—that is, one puts $x_t = a_{t \bmod T}$ for some set of points $a_0, a_1, \ldots, a_{T-1}$. Then for $i = 1, 2, \ldots, p$

$$\sup_\Lambda \sqrt{n}\left| \frac{\partial}{\partial \lambda_i} s_n^0(\lambda) - \frac{\partial}{\partial \lambda_i} s^*(\lambda) \right|$$

$$\leq \frac{\sqrt{n}}{n} \sup_\Lambda \sum_{j=0}^{T-2} \left| \int_{\mathscr{E}} \frac{\partial}{\partial \lambda_i} s\left[Y(e, a_j, \gamma^*), a_j, \tau^*, \lambda \right] dP(e) \right|$$

whence $K(\lambda) \equiv 0$. See Berger and Naftali (1984) for additional discussion of this technique and applications to experimental design.

The third is to hold $\gamma_n^0 \equiv \gamma^*$ for all n and assume that the x_t are iid random variables. This has the effect of imposing $\lambda_n^0 \equiv \lambda^*$ for all n. See Section 8 for details.

Next we establish some ancillary facts regarding the constrained estimator for use in Section 5 under the assumption of a Pitman drift. Due to the Pitman drift, these results are not to be taken as an adequate theory of constrained estimation. See Section 7 for that.

ASSUMPTION 7 (Pitman drift).   The sequence $\{\gamma_n^0\}$ is chosen such that $\lim_{n \to \infty} \sqrt{n}\,(\lambda_n^0 - \lambda_n^*) = \Delta$. Moreover, $h(\lambda^*) = 0$.

THEOREM 6.   Let Assumptions 1 through 7 hold. Then:

$\tilde{\lambda}_n$ converges almost surely to λ^*,

λ_n^* converges to λ^*,

$\tilde{\mathscr{I}}$ converges almost surely to $\mathscr{I}^* + \mathscr{U}^*$,

$\mathscr{I}_n^*$ converges to $\mathscr{I}^*$,

$\tilde{\mathscr{J}}$ converges almost surely to $\mathscr{J}^*$,

$\mathscr{J}_n^*$ converges to $\mathscr{J}^*$,

$\sqrt{n}\,(\partial/\partial\lambda)s_n(\lambda_n^*) - \sqrt{n}\,(\partial/\partial\lambda)s_n^0(\lambda_n^*) \overset{\mathscr{L}}{\to} N_p(0, \mathscr{I}^*)$,

$\sqrt{n}\,(\partial/\partial\lambda)s_n^0(\lambda_n^*)$ converges to $-\mathscr{J}^*\Delta$.

Proof.   The proof that $\tilde{\lambda}_n$ converges almost surely to λ^* is nearly word for word the same as the proof of Theorem 3. The critical inequality

$$\lim_{m \to \infty} s_{n_m}\!\left(\tilde{\lambda}_{n_m}\right) \le \lim_{m \to \infty} s_{n_m}(\lambda^*)$$

obtains by realizing that both $h(\tilde{\lambda}_{n_m}) = 0$ and $h(\lambda^*) = 0$ under the Pitman drift assumption.

The convergence properties of $\tilde{\mathscr{I}}, \mathscr{I}_n^*, \tilde{\mathscr{J}}, \mathscr{J}_n^*$ follow directly from the convergence of $\tilde{\lambda}_n$ and λ_n^* using the argument of the proof of Theorem 5.

Since domination implies that (Problem 11)

$$\frac{\partial}{\partial\lambda}s_n^0(\lambda_n^*) = \frac{1}{n}\sum_{t=1}^{n}\int_{\mathscr{E}}\frac{\partial}{\partial\lambda}s\big[Y(e, x_t, \gamma_n^0), x_t, \tau_n^0, \lambda_n^*\big]\,dP(e).$$

We have from Theorem 2 that

$$\sqrt{n}\,\frac{\partial}{\partial\lambda}s_n(\lambda_n^*) - \sqrt{n}\,\frac{\partial}{\partial\lambda}s_n^0(\lambda_n^*) \overset{\mathscr{L}}{\to} N_p(0, \mathscr{I}^*).$$

Note that convergence of $\{\lambda_n^*\}$ to λ^* is all that is needed here; the rate $\lim_{n \to \infty}\sqrt{n}\,(\lambda_n^0 - \lambda_n^*)$ is not required up to this point in the proof.

By Taylor's theorem, recalling that $(\partial/\partial\lambda)s_n^0(\lambda_n^0) = o(n^{-1/2})$,

$$\sqrt{n}\,\frac{\partial}{\partial\lambda}s_n^0(\lambda_n^*) = o(1) + \bar{\bar{\mathscr{J}}}\sqrt{n}\,\big(\lambda_n^* - \lambda_n^0\big)$$

where $\bar{\bar{\mathcal{I}}}$ is similar to $\bar{\mathcal{I}}$ in the proof of Theorem 5 and converges to $\mathcal{I}^*$ for similar reasons. Since $\sqrt{n}\,(\lambda_n^* - \lambda_n^0)$ converges to $-\Delta$ by Assumption 7, the last result follows. □

PROBLEMS

1. Prove that $s_n^0(\lambda)$ converges uniformly to $s^*(\lambda)$ on Λ^*.

2. Hold an $\{e_t\}$ fixed for which $\lim_{n\to\infty} \sup_{\Lambda^*} \|g_n(\lambda) - g^*(\lambda)\| = 0$ and $\lim_{n\to\infty}\hat{\lambda}_n = \lambda^*$. Show that $\lim_{n\to\infty} g_n(\hat{\lambda}_n) = g^*(\lambda^*)$ if $g^*(\lambda)$ is continuous.

3. Prove that λ_n^0 converges to λ^*.

4. Prove Part 1 of Lemma 2.

5. Let $(\partial/\partial\lambda')h(\lambda)$ be a matrix of order $q \times p$ with $q < p$ such that each element of $(\partial/\partial\lambda')h(\lambda)$ is continuous on an open set Λ^0 containing λ^*. Let $(\partial/\partial\lambda')h(\lambda)$ have rank q at $\lambda = \lambda^*$. Prove that there is an open set containing λ^* such that $\operatorname{rank}[(\partial/\partial\lambda')h(\lambda)] = q$ for every λ in $\mathcal{O}$. Hint: There is a matrix K' of order $(p - q) \times p$ and of rank $p - q$ such that

$$A(\lambda) = \begin{bmatrix} (\partial/\partial\lambda')h(\lambda) \\ K' \end{bmatrix}$$

has rank $A(\lambda^*) = p$ (why?). Also, $\det A(\lambda)$ is continuous and $\mathcal{O} = \{\lambda : |\det A(\lambda)| > 0\}$ is the requisite set (why?).

6. Verify the claim of the first line of the proof of Theorem 5. The essence of the argument is that one could prove Theorem 5 for a set of random variables $\bar{\lambda}_n$, $\bar{\bar{\lambda}}_n$, and so on given by Lemma 2, and then $\sqrt{n}\,\hat{\lambda}_n = \sqrt{n}\,\bar{\lambda}_n + o_s(1)$, $\sqrt{n}\,(\partial/\partial\lambda)s_n(\lambda_n^0) = \sqrt{n}\,(\partial/\partial\lambda)s_n(\bar{\bar{\lambda}}_n) + o_s(1)$, and so on. Make this argument rigorous.

7. Use Theorem 1 to prove that $[\mathcal{I}_n(\lambda), \mathcal{J}_n(\lambda)]$ converges almost surely, uniformly on Λ, and compute the uniform limit. Why does $(\gamma_n^0, \hat{\tau}_n, \hat{\lambda}_n)$ converge almost surely to $(\gamma^*, \tau^*, \lambda^*)$? Show that $[\mathcal{I}_n(\hat{\lambda}_n), \mathcal{J}_n(\hat{\lambda}_n)]$ converges almost surely to $(\mathcal{I}^*, \mathcal{J}^*)$.

8. Show that Assumption 6 suffices to dominate the elements of

$$\int_{\mathcal{E}} \frac{\partial}{\partial\lambda} s[Y(e, x, \gamma), x, \tau, \lambda]\, dP(e) \int_{\mathcal{E}} \frac{\partial}{\partial\lambda'} s[Y(e, x, \gamma), x, \tau, \lambda]\, dP(e)$$

by $b(x)$. Then apply Theorem 1 to show that $\mathcal{U}_n^0$ converges to $\mathcal{U}^*$.

9. Show that if $\rho(u) = \ln\cosh(u/2)$ and $P(e)$ is symmetric, has finite first moment, and assigns positive probability to every nonempty,

open interval, then $\varphi(\delta) = \int_{\mathscr{E}} \rho(e + \delta) \, dP(e)$ exists and has a unique minimum at $\delta = 0$. Hint: Rewrite $\rho(u)$ in terms of exponentials and show that $\rho(u) \le \frac{1}{2}|u|$. Use the mean value theorem and the dominated convergence theorem to show that $\varphi'(\delta) = \int_{\mathscr{E}} \Psi(e + \delta) \, dP(e)$. Then show that $\varphi'(0) = 0$, $\varphi'(\delta) < 0$ if $\delta < 0$, and $\varphi'(\delta) > 0$ if $\delta > 0$.

10. Suppose that $\hat{\lambda}_n$ is computed by minimizing

$$s_n(\lambda) = \frac{1}{n} \sum_{t=1}^{n} \rho \left(\frac{y_t - f(x_t, \lambda)}{\tau^*} \right)$$

where $\tau^* > 0$ is known, but that the data are actually generated according to

$$y_t = g(x_t, \gamma_n^0) + e_t.$$

Assuming that $s_n^0(\lambda)$ has a unique minimum λ_n^0 which converges to some point λ^*, compute $\mathscr{U}_n^0$, $\mathscr{I}_n^0$, and $\mathscr{J}_n^0$.

11. Prove that under Assumptions 1 through 6,

$$\int_{\mathscr{E}} \frac{\partial}{\partial \lambda} s[Y(e, x, \gamma), x, \tau, \lambda] \, dP(e)$$

$$= \frac{\partial}{\partial \lambda} \int_{\mathscr{E}} s[Y(e, x, \gamma), x, \tau, \lambda] \, dP(e).$$

Hint: See the proof of Lemma 3.

12. Suppose that G_n is a matrix with $(\partial/\partial \lambda')h(\lambda_n^*)G_n = 0$ and $\lim_{n \to \infty} G_n = G$. Show that under Assumptions 1 through 6

$$\sqrt{n}\, G_n' \frac{\partial}{\partial \lambda} s_n(\lambda_n^*) \xrightarrow{\mathscr{L}} N_p(0, G'\mathscr{I}^*G);$$

Assumption 7 is not needed. Hint: There are Lagrange multipliers θ_n such that $(\partial/\partial \lambda')[s_n(\lambda_n^*) + \theta_n' h(\lambda_n^*)] = 0$.

13. Suppose that there is a function $\varphi(\lambda)$ such that

$$\tau = h(\lambda)$$
$$\rho = \varphi(\lambda)$$

is a once continuously differentiable mapping with a once continuously differentiable inverse

$$\lambda = \Psi(\tau, \rho).$$

Put

$$g(\rho) = \Psi(0, \rho)$$

$$\rho_n^0 = \varphi(\lambda_n^*)$$

$$H_n = \frac{\partial}{\partial \lambda'} h(\lambda_n^*)$$

$$G_n = \frac{\partial}{\partial \rho'} g(\rho_n^0).$$

Show that G_n is the matrix required in Problem 12. Show also that

$$\text{rank}\begin{pmatrix} G_n' \\ H_n \end{pmatrix} = p.$$

14. (Construction of a Pitman drift.) Fill in the missing steps and supply the necessary regularity conditions. Let

$\lambda_n^0(\gamma)$ minimize

$$s_n^0(\gamma, \lambda) = \frac{1}{n} \sum_{t=1}^n \int_{\mathscr{E}} s\left[Y(e, x_t, \gamma), x_t, \tau_n^0, \lambda\right] dP(e)$$

and let

$\lambda^*(\gamma)$ minimize

$$s^*(\gamma, \lambda) = \int_{\mathscr{X}} \int_{\mathscr{E}} s\left[Y(e, x, \gamma), x, \tau^*, \lambda\right] dP(e) \, d\mu(x).$$

Suppose that there is a point γ^* in Γ such that

$$\gamma_n^0 \equiv \gamma^* \quad \text{for all } n \quad \text{implies} \quad \lambda_n^0(\gamma^*) = \lambda^*(\gamma^*) \quad \text{for all } n.$$

Suppose also that Γ is a linear space and that $(\partial/\partial\alpha)Y(e, x, \gamma^* + \alpha\gamma^\#)$ exists for $0 \le \alpha \le 1$ and for some point $\gamma^\#$ in Γ. Note that Γ can be an infinite dimensional space; a directional derivative of this sort on a normed, linear space is called a Gateau derivative (Luenberger, 1969, Section 7.2, or Wouk, 1979, Section 12.1). Let

$$\gamma(\alpha) = \gamma^* + \alpha\gamma^\#$$

$$\lambda_n^0(\alpha) = \lambda_n^0[\gamma^* + \alpha\gamma^\#]$$

and

$$\lambda^*(\alpha) = \lambda^*\left[\gamma^* + \alpha\gamma^\#\right].$$

Under appropriate regularity conditions, $(\partial/\partial\alpha)\lambda_n^0(\alpha)$ exists and can be computed from

$$0 = \frac{1}{n}\sum_{t=1}^{n}\int_{\mathscr{E}}\frac{\partial^2}{\partial\lambda\,\partial y'}s\{Y[e, x_t, \gamma(\alpha)], x_t, \tau_n^0, \lambda_n^0(\alpha)\}$$

$$\times \frac{\partial}{\partial\alpha}Y[e, x_t, \gamma(\alpha)]\,dP(e)$$

$$+ \frac{\partial^2}{\partial\lambda\,\partial\lambda'}s_n^0\left[\gamma(\alpha), \lambda_n^0(\alpha)\right]\frac{\partial}{\partial\alpha}\lambda(\alpha).$$

Again under appropriate regularity conditions,

$$\lim_{n\to\infty}\sup_{0\le\alpha\le 1}\left\|\frac{\partial}{\partial\alpha}\lambda_n^0(\alpha) - \frac{\partial}{\partial\alpha}\lambda^*(\alpha)\right\| = 0.$$

Then by Taylor's theorem, for $i = 1, 2, \ldots, p$,

$$\sqrt{n}\left[\lambda_{in}^0(\alpha) - \lambda_{in}^0(0)\right] = \sqrt{n}\,\alpha\frac{\partial}{\partial\alpha}\lambda_{in}^0(\bar{\alpha}_i)$$

where $0 \le \bar{\alpha}_i \le \alpha$. Let $\{\alpha_n\}_{n=1}^{\infty}$ be any sequence such that $\lim_{n\to\infty}\sqrt{n}\,\alpha_n = \delta$ with δ finite. Since $\lambda_n^0(0) = \lambda^*(0)$ for all n and $(\partial/\partial\alpha)\lambda_n^0(\alpha)$ converges uniformly to $(\partial/\partial\alpha)\lambda^*(\alpha)$, we have

$$\lim_{n\to\infty}\sqrt{n}\left[\lambda_n^0(\alpha_n) - \lambda^*(0)\right] = \delta\frac{\partial}{\partial\alpha}\lambda^*(0).$$

If the parameters of the data generating model are set to $\gamma_n^0 = \gamma^* + \alpha_n\gamma^\#$, then

$$\lim_{n\to\infty}\sqrt{n}\left(\lambda_n^0 - \lambda^*\right) = \Delta$$

for some finite Δ as required. Note that α_n can be chosen so that $\Delta = 0$.

Suppose that the parametric constraint $h(\lambda) = 0$ can be equivalently represented as a functional dependence $\lambda = g(\rho)$; see Problem 13 or Section 6 for the construction. What is required of $g(\rho)$ so that $\lim_{n\to\infty}\sqrt{n}\,(\rho_n^0 - \rho^*) = \beta$? Put $\lambda_n^* = g(\rho_n^0)$. What is required of $g(\rho)$ so that $\lim_{n\to\infty}\sqrt{n}\,(\lambda_n^* - \lambda^*) = \Delta^*$? Note that $\lim_{n\to\infty}\sqrt{n}\,(\lambda_n^0 - \lambda_n^*) = \Delta - \Delta^*$ in this case.

15. Use Taylor's theorem twice to write

$$\sqrt{n}\left[\frac{\partial}{\partial\lambda}s^*(\lambda_n^0) - \frac{\partial}{\partial\lambda}s_n^0(\lambda^*)\right] = [2\mathscr{I}^* + o(1)]\sqrt{n}\left(\lambda_n^0 - \lambda^*\right);$$

recall that $(\partial/\partial\lambda)s^*(\lambda^*) = (\partial/\partial\lambda)s_n^0(\lambda_n^0) = 0$. Referring to the comments following Theorem 5, verify that speeding up the rate at which Cesaro sums converge will cause $\sqrt{n}\,(\hat{\lambda}_n - \lambda^*)$ to be asymptotically normally distributed.

4. METHOD OF MOMENTS ESTIMATORS

Recall that a method of moments estimator $\hat{\lambda}_n$ is defined as the solution of the optimization problem

$$\text{minimize}\quad s_n(\lambda) = d\big[m_n(\lambda), \hat{\tau}_n\big]$$

where $d[m, \tau]$ is a measure of the distance of m from zero, $\hat{\tau}_n$ is an estimator of nuisance parameters, and

$$m_n(\lambda) = \frac{1}{n}\sum_{t=1}^{n} m(y_t, x_t, \hat{\tau}_n, \lambda).$$

The constrained method of moments estimator $\tilde{\lambda}_n$ is the solution of the optimization problem

$$\text{minimize}\quad s_n(\lambda)\quad \text{subject to}\quad h(\lambda) = 0.$$

The objective of this section is to find the almost sure limit and the asymptotic distribution of the unconstrained estimator $\hat{\lambda}_n$ under regularity conditions that do not rule out specification error. Some ancillary facts regarding the asymptotic distribution of the constrained estimator $\tilde{\lambda}_n$ under a Pitman drift are also derived for use in the later sections on hypothesis testing. This section differs from the previous section in detail, but the general pattern is much the same. Accordingly the comments on motivations, regularity conditions, and results will be abbreviated. These results are due to Burguete (1980) in the main, with some refinements made here to isolate the Pitman drift assumption.

As before, an example—a correctly specified scale invariant M-estimator—is carried throughout the discussion to illustrate how the regularity conditions may be satisfied in applications.

EXAMPLE 2 (Scale invariant M-estimator). The data generating model is

$$y_t = f(x_t, \gamma_n^0) + e_t \qquad t = 1, 2, \ldots, n.$$

Proposal 2 of Huber (1964) leads to the moment equations

$$m_n(\lambda) = \frac{1}{n} \sum_{t=1}^{n} \begin{pmatrix} \Psi\left(\dfrac{y_t - f(x_t, \theta)}{\sigma}\right) \dfrac{\partial}{\partial \theta} f(x_t, \theta) \\ \Psi^2\left(\dfrac{y_t - f(x_t, \theta)}{\sigma}\right) - \beta \end{pmatrix}$$

with $\lambda = (\theta', \sigma)'$. For specificity let

$$\Psi(u) = \tfrac{1}{2}\tanh\left(\frac{u}{2}\right),$$

a bounded odd function with bounded even derivative, and let

$$\beta = \int \Psi^2(e) \, d\Phi(e)$$

where Φ is the standard normal distribution function. There is no preliminary estimator $\hat{\tau}_n$ with this example, so the argument τ of $m(y, x, \tau, \lambda)$ is suppressed to obtain

$$m(y, x, \lambda) = \begin{pmatrix} \Psi\left(\dfrac{y - f(x, \theta)}{\sigma}\right) \dfrac{\partial}{\partial \theta} f(x, \theta) \\ \Psi^2\left(\dfrac{y - f(x, \theta)}{\sigma}\right) - \beta \end{pmatrix}.$$

The distance function is

$$d(m) = \tfrac{1}{2} m'm,$$

again suppressing the argument τ, whence the estimator $\hat{\lambda}_n$ is defined as that value of λ which minimizes

$$s_n(\lambda) = \tfrac{1}{2} m'_n(\lambda) m_n(\lambda).$$

The error distribution $P(e)$ is symmetric and puts positive probability on every open interval of the real line.   □

We call the reader's attention to some heavily used notation and then state the identification condition.

NOTATION 5.

$$m_n(\lambda) = \frac{1}{n} \sum_{t=1}^{n} m(y_t, x_t, \hat{\tau}_n, \lambda)$$

$$m_n^0(\lambda) = \frac{1}{n} \sum_{t=1}^{n} \int_{\mathscr{E}} m\left[Y(e, x_t, \gamma_n^0), x_t, \tau_n^0, \lambda\right] dP(e)$$

$$m^*(\lambda) = \int_{\mathscr{X}}\int_{\mathscr{E}} m[Y(e, x, \gamma^*), x, \tau^*, \lambda]\, dP(e)\, d\mu(x)$$

$$s_n(\lambda) = d[m_n(\lambda), \hat{\tau}_n]$$

$$s_n^0(\lambda) = d[m_n^0(\lambda), \tau_n^0]$$

$$s^*(\lambda) = d[m^*(\lambda), \tau^*]$$

$\hat{\lambda}_n$ minimizes $s_n(\lambda)$

$\tilde{\lambda}_n$ minimizes $s_n(\lambda)$ subject to $h(\lambda) = 0$

λ_n^0 minimizes $s_n^0(\lambda)$

λ_n^* minimizes $s_n^0(\lambda)$ subject to $h(\lambda) = 0$

λ^* minimizes $s^*(\lambda)$.

ASSUMPTION 8 (Identification). The parameter γ^0 is indexed by n, and the sequence $\{\gamma_n^0\}$ converges to a point γ^*. The sequence of nuisance parameter estimators is centered at a point τ_n^0 in the sense that $\sqrt{n}\,(\hat{\tau}_n - \tau_n^0)$ is bounded in probability; the sequence $\{\tau_n^0\}$ converges to a point τ^*, and $\{\hat{\tau}_n\}$ converges almost surely to τ^*. Either the solution λ^* of the equations $m^*(\lambda) = 0$ is unique or there is one solution λ^* that can be regarded as being naturally associated to γ^*. Further, $(\partial/\partial\lambda')m^*(\lambda^*)$ has full column rank $(= p)$.

The assumption that $m^*(\lambda^*) = 0$ is somewhat implausible in those misspecified situations where the range of $m_n(\lambda)$ is in a higher dimension than the domain. As a sensible estimation procedure will have $m^*(\lambda^*) = 0$ if $Y(e, x, \gamma^*)$ falls into the class of models for which it was designed, one could have both $m^*(\lambda^*) = 0$ and misspecification with a Pitman drift: Problem 14 of Section 3 spells out the details; see also Problems 2 and 3 of Section 2. But this is not really satisfactory. One would rather have the freedom to hold $\gamma_n^0 \equiv \gamma^*$ for all n at some point γ^* for which $m^*(\lambda^*) \neq 0$. Such a theory is not beyond reach, but it is more complicated than for the case $m^*(\lambda^*) = 0$. As we have no need of the case $m^*(\lambda^*) \neq 0$ in the sequel, we shall spare the reader these complications in the text; the more general result is given in Problem 6.

For the example, $m^*(\lambda^*) = 0$:

EXAMPLE 2 (Continued). Let σ^* solve $\int_{\mathscr{E}}\Psi^2(e/\sigma)\, dP(e) = \beta$; a solution exists, since $G(\sigma) = \int_{\mathscr{E}}\Psi^2(e/\sigma)\, dP(e)$ is a continuous, decreasing function with $G(0) = 1$ and $G(\infty) = 0$. Consider putting $\lambda = (\gamma', \sigma^*)'$.

With this choice

$$\int_{\mathcal{E}} m[Y(e, x, \gamma), x, (\gamma', \sigma^*)] \, dP(e) = \int_{\mathcal{E}} m[e + f(x, \gamma), x, (\gamma', \sigma^*)] \, dP(e)$$

$$= \begin{pmatrix} \int_{\mathcal{E}} \Psi\left(\dfrac{e}{\sigma^*}\right) dP(e) \dfrac{\partial}{\partial \theta} f(x, \gamma) \\[2mm] \int_{\mathcal{E}} \Psi^2\left(\dfrac{e}{\sigma^*}\right) dP(e) - \beta \end{pmatrix}$$

$$= \begin{pmatrix} 0 \\ 0 \end{pmatrix}.$$

As the integral is zero for every x, it follows that $m^*(\lambda^*) = 0$ at $\lambda^* = (\gamma^{*\prime}, \sigma^*)'$. Similarly $m_n^0(\lambda_n^0) = 0$ at $\lambda_n^0 = (\gamma_n^{0\prime}, \sigma^*)'$. $\square$

The following notation defines the parameters of the asymptotic distribution of $\hat{\lambda}_n$. The notation is not as formidable as it looks; it merely consists of breaking a variance computation down into its component parts.

NOTATION 6.

$$\bar{\bar{K}}(\lambda) = \int_{\mathcal{X}} \int_{\mathcal{E}} m[Y(e, x, \gamma^*), x, \tau^*, \lambda] \, dP(e)$$

$$\times \int_{\mathcal{E}} m'[Y(e, x, \gamma), x, \tau^*, \lambda] \, dP(e) \, d\mu(x)$$

$$\bar{\bar{S}}(\lambda) = \int_{\mathcal{X}} \int_{\mathcal{E}} m[Y(e, x, \gamma^*), x, \tau^*, \lambda]$$

$$\times m'[Y(e, x, \gamma^*), x, \tau^*, \lambda] \, dP(e) \, d\mu(x) - \bar{\bar{K}}(\lambda)$$

$$\bar{\bar{M}}(\lambda) = \int_{\mathcal{X}} \int_{\mathcal{E}} \frac{\partial}{\partial \lambda'} m[Y(e, x, \gamma^*), x, \tau^*, \lambda] \, dP(e) \, d\mu(x)$$

$$\bar{\bar{D}}(\lambda) = \frac{\partial^2}{\partial m \, \partial m'} d[m^*(\lambda), \tau^*]$$

$$\bar{\bar{\mathcal{J}}}(\lambda) = \bar{\bar{M}}'(\lambda) \bar{\bar{D}}(\lambda) \bar{\bar{S}}(\lambda) \bar{\bar{D}}(\lambda) \bar{\bar{M}}(\lambda)$$

$$\bar{\bar{\mathcal{J}}}(\lambda) = \bar{\bar{M}}'(\lambda) \bar{\bar{D}}(\lambda) \bar{\bar{M}}(\lambda)$$

$$\bar{\bar{\mathcal{U}}}(\lambda) = \bar{\bar{M}}'(\lambda) \bar{\bar{D}}(\lambda) \bar{\bar{K}}(\lambda) \bar{\bar{D}}(\lambda) \bar{\bar{M}}(\lambda)$$

$$\mathcal{J}^* = \bar{\bar{\mathcal{J}}}(\lambda^*) \qquad \mathcal{J}^* = \bar{\bar{\mathcal{J}}}(\lambda^*) \qquad \mathcal{U}^* = \bar{\bar{\mathcal{U}}}(\lambda^*)$$

$$S^* = \bar{\bar{S}}(\lambda^*) \qquad M^* = \bar{\bar{M}}(\lambda^*) \qquad D^* = \bar{\bar{D}}(\lambda^*) \qquad K^* = \bar{\bar{K}}(\lambda^*).$$

We illustrate the computations with the example:

EXAMPLE 2 (Continued). For $\lambda = (\gamma', \sigma^*)'$ we have

$$\int_{\mathscr{E}} m[Y(e, x, \gamma), x, \lambda]\, dP(e)\Big|_{\lambda = (\gamma', \sigma^*)'} = \begin{pmatrix} \int_{\mathscr{E}} \Psi\left(\frac{e}{\sigma^*}\right) dP(e)\, \frac{\partial}{\partial\theta} f(x, \gamma) \\ \int_{\mathscr{E}} \Psi^2\left(\frac{e}{\sigma^*}\right) dP(e) - \beta \end{pmatrix}$$

$$= \begin{pmatrix} 0 \\ 0 \end{pmatrix}.$$

Thus $\overline{\overline{K}}(\lambda^*) = 0$, whence $\mathscr{U}^* = 0$. Further computation yields

$$\overline{\overline{S}}(\lambda^*) = \begin{pmatrix} \int_{\mathscr{E}} \Psi^2\left(\frac{e}{\sigma^*}\right) dP(e)\, \mathscr{F}'\mathscr{F} & 0 \\ 0' & \int_{\mathscr{E}}\left[\Psi^2\left(\frac{e}{\sigma^*}\right) - \beta\right]^2 dP(e) \end{pmatrix}$$

$$\overline{\overline{M}}(\lambda^*) = \begin{pmatrix} \dfrac{1}{\sigma^*} \int_{\mathscr{E}} \Psi'\left(\frac{e}{\sigma^*}\right) dP(e)\, \mathscr{F}'\mathscr{F} & 0 \\ 0' & -2\left(\dfrac{1}{\sigma^*}\right)^2 \int_{\mathscr{E}} \Psi\left(\frac{e}{\sigma^*}\right)\Psi'\left(\frac{e}{\sigma^*}\right) e\, dP(e) \end{pmatrix}$$

$$\overline{\overline{D}}(\lambda^*) = I$$

where

$$\mathscr{F}'\mathscr{F} = \int_{\mathscr{X}} \left(\frac{\partial}{\partial\theta} f(x, \theta)\right)\left(\frac{\partial}{\partial\theta} f(x, \theta)\right)' d\mu(x)\Big|_{\theta = \gamma^*}.$$

As will be seen later, it is only $V^* = (\mathscr{J}^*)^{-1}\mathscr{I}^*(\mathscr{J}^*)^{-1}$ that is needed. Observing that $\overline{M}(\lambda)$ is invertible, we have

$$V^* = (\mathscr{J}^*)^{-1}\mathscr{I}^*(\mathscr{J}^*)^{-1}$$

$$= \left[\overline{\overline{M}}(\lambda^*)\right]^{-1}\left[\overline{\overline{S}}(\lambda^*)\right]\left[\overline{\overline{M}}'(\lambda^*)\right]^{-1}$$

$$= \begin{pmatrix} \dfrac{(\sigma^*)^2 \mathscr{E}\Psi^2(e/\sigma^*)}{[\mathscr{E}\Psi'(e/\sigma^*)]^2}(\mathscr{F}'\mathscr{F})^{-1} & 0 \\ 0' & \dfrac{(\sigma^*)^4 \mathscr{E}[\Psi^2(e/\sigma^*) - \beta]^2}{4[\mathscr{E}e\Psi(e/\sigma^*)\Psi'(e/\sigma^*)]^2} \end{pmatrix}. \quad \Box$$

In Section 5, the distributions of test statistics are characterized in terms of the following quantities:

NOTATION 7.

$$\bar{K}_n(\lambda) = \frac{1}{n} \sum_{t=1}^{n} \int_{\mathscr{E}} m\left[Y(e, x_t, \gamma_n^0), x_t, \tau_n^0, \lambda\right] dP(e)$$

$$\times \int_{\mathscr{E}} m'\left[Y(e, x_t, \gamma_n^0), x_t, \tau_n^0, \lambda\right] dP(e)$$

$$\bar{S}_n(\lambda) = \frac{1}{n} \sum_{t=1}^{n} \int_{\mathscr{E}} m\left[Y(e, x_t, \gamma_n^0), x_t, \tau_n^0, \lambda\right]$$

$$\times m'\left[Y(e, x_t, \gamma_n^0), x_t, \tau_n^0, \lambda\right] dP(e) - \bar{K}_n(\lambda)$$

$$\bar{M}_n(\lambda) = \frac{1}{n} \sum_{t=1}^{n} \int_{\mathscr{E}} \frac{\partial}{\partial \lambda'} m\left[Y(e, x_t, \gamma_n^0), x_t, \tau_n^0, \lambda\right] dP(e)$$

$$\bar{D}_n(\lambda) = \frac{\partial^2}{\partial m \, \partial m'} d\left[m_n^0(\lambda), \tau_n^0\right]$$

$$\tilde{\mathscr{I}}_n(\lambda) = \tilde{M}_n'(\lambda) \bar{D}_n(\lambda) \bar{S}_n(\lambda) \bar{D}_n(\lambda) \bar{M}_n(\lambda)$$

$$\bar{\mathscr{I}}_n(\lambda) = \bar{M}_n'(\lambda) \bar{D}_n(\lambda) \bar{M}_n(\lambda)$$

$$\bar{\mathscr{U}}_n(\lambda) = \bar{M}_n'(\lambda) \bar{D}_n(\lambda) \bar{K}_n(\lambda) \bar{D}_n(\lambda) \bar{M}_n(\lambda)$$

$$\mathscr{I}_n^0 = \tilde{\mathscr{I}}_n(\lambda_n^0) \qquad \mathscr{J}_n^0 = \bar{\mathscr{I}}_n(\lambda_n^0) \qquad \mathscr{U}_n^0 = \bar{\mathscr{U}}_n(\lambda_n^0)$$

$$\mathscr{I}_n^* = \tilde{\mathscr{I}}_n(\lambda_n^*) \qquad \mathscr{J}_n^* = \bar{\mathscr{I}}_n(\lambda_n^*) \qquad \mathscr{U}_n^* = \bar{\mathscr{U}}_n(\lambda_n^*).$$

We illustrate the computations with the example:

EXAMPLE 2 (Continued). Computations similar to those for $\mathscr{I}^*$ and $\mathscr{J}^*$ yield

$$V_n^0 = \left(\mathscr{J}_n^0\right)^{-1}\left(\mathscr{I}_n^0\right)\left(\mathscr{J}_n^0\right)^{-1}$$

$$= \begin{vmatrix} \dfrac{(\sigma^*)^2 \mathscr{E}\Psi^2(e/\sigma^*)}{[\mathscr{E}\Psi'(e/\sigma^*)]^2}(F'F)^{-1} & 0 \\[4ex] 0' & \dfrac{(\sigma^*)^4 \mathscr{E}\left[\Psi^2(e/\sigma^*) - \beta\right]^2}{4[\mathscr{E}e\Psi(e/\sigma^*)\Psi'(e/\sigma^*)]^2} \end{vmatrix}$$

where

$$F'F = \frac{1}{n} \sum_{t=1}^{n} \left(\frac{\partial}{\partial \theta} f(x_t, \theta_n^0) \right) \left(\frac{\partial}{\partial \theta} f(x_t, \theta_n^0) \right)'.$$

☐

Some estimators of $\mathscr{I}^*$ and $\mathscr{J}^*$ are:

NOTATION 8.

$$S_n(\lambda) = \frac{1}{n} \sum_{t=1}^{n} m(y_t, x_t, \hat{\tau}_n, \lambda) m'(y_t, x_t, \hat{\tau}_n, \lambda)$$

$$M_n(\lambda) = \frac{1}{n} \sum_{t=1}^{n} \frac{\partial}{\partial \lambda'} m(y_t, x_t, \hat{\tau}_n, \lambda)$$

$$D_n(\lambda) = \frac{\partial^2}{\partial m \, \partial m'} d[m_n(\lambda), \hat{\tau}_n]$$

$$\mathscr{I}_n(\lambda) = M_n'(\lambda) D_n(\lambda) S_n(\lambda) D_n(\lambda) M_n(\lambda)$$

$$\mathscr{J}_n(\lambda) = M_n'(\lambda) D_n(\lambda) M_n(\lambda)$$

$$\hat{\mathscr{I}} = \mathscr{I}_n(\hat{\lambda}_n) \qquad \tilde{\mathscr{I}} = \mathscr{I}_n(\tilde{\lambda}_n)$$

$$\hat{\mathscr{J}} = \mathscr{J}_n(\hat{\lambda}_n) \qquad \tilde{\mathscr{J}} = \mathscr{J}_n(\tilde{\lambda}_n).$$

For Example 2, there are alternative choices:

EXAMPLE 2 (Continued). Reasoning by analogy with the forms that obtain from Notation 6, most would probably substitute the following estimators for those given by Notation 8:

$$\hat{S}_n = \begin{pmatrix} \frac{1}{n} \sum_{t=1}^{n} \Psi^2\left(\frac{\hat{e}_t}{\hat{\sigma}}\right) \hat{F}'\hat{F} & 0 \\ 0' & \frac{1}{n} \sum_{t=1}^{n} \left[\Psi^2\left(\frac{\hat{e}_t}{\hat{\sigma}}\right) - \beta\right]^2 \end{pmatrix}$$

$$\hat{M}_n = \begin{pmatrix} \frac{1}{\hat{\sigma}^2} \frac{1}{n} \sum_{t=1}^{n} \Psi'\left(\frac{\hat{e}_t}{\hat{\sigma}}\right) \hat{F}'\hat{F} & 0 \\ 0' & -\frac{2}{\hat{\sigma}^2} \frac{1}{n} \sum_{t=1}^{n} \Psi\left(\frac{\hat{e}_t}{\hat{\sigma}}\right) \Psi'\left(\frac{\hat{e}_t}{\hat{\sigma}}\right) \hat{e}_t \end{pmatrix}$$

$$\hat{D}_n = I$$

where

$$\hat{e}_t = y_t - f(x_t, \hat{\theta}_n)$$

$$\hat{F}'\hat{F} = \frac{1}{n} \sum_{t=1}^{n} \left(\frac{\partial}{\partial \theta} f(x_t, \hat{\theta}_n) \right) \left(\frac{\partial}{\partial \theta} f(x_t, \hat{\theta}_n) \right)'. \qquad \square$$

We shall adjoin some technical assumptions. As before, one may assume that $\hat{\tau}_n$ takes its values in a compact ball T for which τ^* is an interior point without loss of generality. Similarly for the parameter space Γ. This leaves domination as the essential condition. We have commented previously (Section 2) on the implications of a compact estimation space Λ^*. In the previous section we commented on the construction of the requisite dominating function $b(e, x)$.

ASSUMPTION 9. There are closed balls Λ^* and T centered at λ^* and τ^* respectively with finite, nonzero radii for which the elements of $m(y, x, \tau, \lambda), (\partial/\partial\lambda_i)m_\alpha(y, x, \tau, \lambda), (\partial^2/\partial\lambda_i \partial\lambda_j)m_\alpha(y, x, \tau, \lambda)$ are continuous and dominated by $b[q(y, x, \gamma), x]$ on $\mathcal{Y} \times \mathcal{X} \times T \times \Lambda^* \times \Gamma$; $b(e, x)$ is that of Assumption 3. The distance function $d(m, \tau)$ and derivatives $(\partial/\partial m) d(m, \tau), (\partial^2/\partial m \partial \tau') d(m, \tau), (\partial^2/\partial m \partial m') d(m, \tau)$ are continuous on $\mathcal{F} \times T$ where $\mathcal{F}$ is some closed ball centered at the zero vector with finite, nonzero radius.

The only distance functions that we shall ever consider have the form

$$d(m, \tau) = m'\Psi(\tau)m$$

with $\Psi(\tau)$ positive definite over T. There seems to be no reason to abstract beyond the essential properties of distance functions of this form, so we impose:

ASSUMPTION 10. The distance function satisfies: $(\partial/\partial m) d(0, \tau) = 0$ for all τ in T [which implies $(\partial^2/\partial m \partial \tau') d(0, \tau) = 0$ for all τ in T], and $(\partial^2/\partial m \partial m') d(0, \tau)$ is positive definite for all τ in T.

If the point λ^* that satisfies $m^*(\lambda) = 0$ is unique over Λ^*, then $s^*(\lambda)$ will have a unique minimum over Λ^* for any distance function that increases with $\|m\|$. In this case the same argument used to prove Theorem 3 can be used to conclude that $\hat{\lambda}_n$ converges almost surely to λ^*. But in many applications, the moment equations are the first order conditions of an optimization problem. In these applications it is unreasonable to expect

$m^*(\lambda)$ to have a unique root over some natural estimation space Λ^*. To illustrate, consider posing Example 1 as a method of moments problem:

EXAMPLE 1 (Continued). The optimization problem

$$\text{Minimize} \quad s_n(\lambda) = \frac{1}{n} \sum_{t=1}^{n} \rho\left(\frac{y_t - f(x_t, \lambda)}{\hat{\tau}_n}\right)$$

has first order conditions $m_n(\hat{\lambda}_n) = 0$ with

$$m_n(\lambda) = \frac{1}{n} \sum_{t=1}^{n} \Psi\left(\frac{y_t - f(x_t, \lambda)}{\hat{\tau}_n}\right) \frac{\partial}{\partial \lambda} f(x, \lambda).$$

We have seen that it is quite reasonable to expect that the almost sure limit $s^*(\lambda)$ of $s_n(\lambda)$ will have a unique minimum λ^* over Λ^*. But, depending on the choice of $f(x, \theta)$, $s^*(\lambda)$ can have local minima and saddle points over Λ^* as well. In this case $m^*(\lambda)$ will have a root at λ^*, but $m^*(\lambda)$ will also have roots at each local minimum and each saddle point. Thus, if Example 1 is recast as the problem

$$\text{Minimize} \quad s_n(\lambda) = \tfrac{1}{2} m'_n(\lambda) m_n(\lambda)$$

we cannot reasonably assume that $s^*(\lambda)$ will have a unique minimum.   □

Without the assumption that $m^*(\lambda)$ has a unique root, the best consistency result that we can obtain is that $s_n(\lambda)$ will eventually have a local minimum near λ^*. We collect together a list of facts needed throughout this section as a lemma and then prove the result:

LEMMA 4. Under Assumptions 1 through 3 and 8 through 10, interchange of differentiation and integration is permitted in these instances:

$$\frac{\partial}{\partial \lambda_i} m^*_\alpha(\lambda) = \int_{\mathcal{X}} \int_{\mathcal{E}} \frac{\partial}{\partial \lambda_i} m_\alpha[Y(e, x, \gamma^*), x, \tau^*, \lambda] \, dP(e) \, d\mu(x)$$

$$\frac{\partial^2}{\partial \lambda_i \, \partial \lambda_j} m^*_\alpha(\lambda) = \int_{\mathcal{X}} \int_{\mathcal{E}} \frac{\partial^2}{\partial \lambda_i \, \partial \lambda_j} m_\alpha[Y(e, x, \gamma^*), x, \tau^*, \lambda] \, dP(e) \, d\mu(x)$$

$$\frac{\partial}{\partial \lambda_i} m^0_{\alpha n}(\lambda) = \frac{1}{n} \sum_{t=1}^{n} \int_{\mathcal{E}} \frac{\partial}{\partial \lambda_i} m_\alpha[Y(e, x, \gamma^0_n), x, \tau^0_n, \lambda] \, dP(e)$$

$$\frac{\partial^2}{\partial \lambda_i \, \partial \lambda_j} m^0_{\alpha n}(\lambda) = \frac{1}{n} \sum_{t=1}^{n} \int_{\mathcal{E}} \frac{\partial^2}{\partial \lambda_i \, \partial \lambda_j} m_\alpha[Y(e, x, \gamma^0_n), x, \tau^0_n, \lambda] \, dP(e).$$

There is a closed ball Λ centered at λ^* with finite nonzero radius such that

$$\lim_{n \to \infty} m_n^0(\lambda) = m^*(\lambda) \qquad \text{uniformly on } \Lambda$$

$$\lim_{n \to \infty} \frac{\partial}{\partial \lambda_i} m_n^0(\lambda) = \frac{\partial}{\partial \lambda_i} m^*(\lambda) \qquad \text{uniformly on } \Lambda$$

$$\lim_{n \to \infty} \frac{\partial^2}{\partial \lambda_i \, \partial \lambda_j} m_n^0(\lambda) = \frac{\partial^2}{\partial \lambda_i \, \partial \lambda_j} m^*(\lambda) \qquad \text{uniformly on } \Lambda$$

$$\lim_{n \to \infty} m_n(\lambda) = m^*(\lambda) \qquad \text{almost surely, uniformly on } \Lambda$$

$$\lim_{n \to \infty} \frac{\partial}{\partial \lambda_i} m_n(\lambda) = \frac{\partial}{\partial \lambda_i} m^*(\lambda) \qquad \text{almost surely, uniformly on } \Lambda$$

$$\lim_{n \to \infty} \frac{\partial^2}{\partial \lambda_i \, \partial \lambda_j} m_n(\lambda) = \frac{\partial^2}{\partial \lambda_i \, \partial \lambda_j} m^*(\lambda) \qquad \text{almost surely, uniformly on } \Lambda$$

$$\lim_{n \to \infty} s_n^0(\lambda) = s^*(\lambda) \qquad \text{uniformly on } \Lambda$$

$$\lim_{n \to \infty} \frac{\partial}{\partial \lambda} s_n^0(\lambda) = \frac{\partial}{\partial \lambda} s^*(\lambda) \qquad \text{uniformly on } \Lambda$$

$$\lim_{n \to \infty} \frac{\partial^2}{\partial \lambda \, \partial \lambda'} s_n^0(\lambda) = \frac{\partial^2}{\partial \lambda \, \partial \lambda'} s^*(\lambda) \qquad \text{uniformly on } \Lambda$$

$$\lim_{n \to \infty} s_n(\lambda) = s^*(\lambda) \qquad \text{almost surely, uniformly on } \Lambda$$

$$\lim_{n \to \infty} \frac{\partial}{\partial \lambda} s_n(\lambda) = \frac{\partial}{\partial \lambda} s^*(\lambda) \qquad \text{almost surely, uniformly on } \Lambda$$

$$\lim_{n \to \infty} \frac{\partial^2}{\partial \lambda \, \partial \lambda'} s_n(\lambda) = \frac{\partial^2}{\partial \lambda \, \partial \lambda'} s^*(\lambda) \qquad \text{almost surely, uniformly on } \Lambda$$

and

$$M^* = \frac{\partial}{\partial \lambda'} m^*(\lambda^*)$$

$$\frac{\partial}{\partial \lambda} s^*(\lambda^*) = 0$$

$$\frac{\partial^2}{\partial \lambda \, \partial \lambda'} s^*(\lambda^*) = \mathscr{I}^*.$$

Proof. The arguments used in the proof of Lemma 3 may be repeated to show that interchange of differentiation and integration is permitted on Λ^*

and that the sequences involving $m_n^0(\lambda)$ and $m_n(\lambda)$ converge uniformly on Λ^*. So let us turn our attention to $s_n(\lambda) = d[m_n(\lambda), \hat{\tau}_n]$.

Differentiating, we have for $m_n(\lambda) \in \mathscr{F}$ that

$$\frac{\partial}{\partial \lambda_i} s_n(\lambda) = \sum_\alpha \frac{\partial}{\partial m_\alpha} d[m_n(\lambda), \hat{\tau}_n] \frac{\partial}{\partial \lambda_i} m_{\alpha n}(\lambda)$$

and

$$\frac{\partial^2}{\partial \lambda_i \partial \lambda_j} s_n(\lambda) = \sum_\alpha \sum_\beta \frac{\partial^2}{\partial m_\alpha \partial m_\beta} d[m_n(\lambda), \hat{\tau}_n] \frac{\partial}{\partial \lambda_i} m_{\alpha n}(\lambda) \frac{\partial}{\partial \lambda_j} m_{\beta n}(\lambda)$$

$$+ \sum_\alpha \frac{\partial}{\partial m_\alpha} d[m_n(\lambda), \hat{\tau}_n] \frac{\partial^2}{\partial \lambda_i \partial \lambda_j} m_{\alpha n}(\lambda).$$

Fix a sequence $\{e_t\}$ for which $\hat{\tau}_n$ converges to τ^* and for which $m_n(\lambda)$ converges uniformly to $m^*(\lambda)$ on Λ^*; almost every $\{e_t\}$ is such. Now $m^*(\lambda^*) = 0$ by assumption, and $m^*(\lambda)$ is continuous on the compact set Λ^*, as it is the uniform limit of continuous functions. Thus there is a $\delta > 0$ such that

$$\|\lambda - \lambda^*\| \leq \delta \quad \Rightarrow \quad \|m^*(\lambda)\| < \eta$$

where η is the radius of the closed ball $\mathscr{F}$ given by Assumption 9. Then there is an N such that

$$n > N, \quad \|\lambda - \lambda^*\| \leq \delta \quad \Rightarrow \quad \|m_n(\lambda)\| < \eta.$$

Set $\Lambda = \{\lambda : \|\lambda - \lambda^*\| \leq \delta\}$.

Now $(\partial/\partial m_\alpha) d(m, \tau)$ is a continuous function on the compact set $\mathscr{F} \times T$, so it is uniformly continuous on $\mathscr{F} \times T$: see Problem 1. Then since $m_n(\lambda)$ converges uniformly to $m^*(\lambda)$ and $\hat{\tau}_n$ converges to τ^*, it follows that $(\partial/\partial m_\alpha) d[m_n(\lambda), \hat{\tau}_n]$ converges uniformly to $(\partial/\partial m_\alpha) d[m^*(\lambda), \tau^*]$; similarly for $d[m_n(\lambda), \hat{\tau}_n]$ and $(\partial^2/\partial m_\alpha \partial m_\beta) d[m_n(\lambda), \hat{\tau}_n]$. The uniform convergence of $s_n(\lambda), (\partial/\partial \lambda_i) s_n(\lambda)$, and $(\partial^2/\partial \lambda_i \partial \lambda_j) s_n(\lambda)$ follows at once. Since the convergence is uniform for almost every $\{e_t\}$, it is uniform almost surely. Similar arguments apply to $s_n^0(\lambda)$.

By the interchange result $M^* = (\partial/\partial \lambda') m^*(\lambda^*)$. Differentiating,

$$\frac{\partial}{\partial \lambda_i} s^*(\lambda^*) = \sum_\alpha \frac{\partial}{\partial m_\alpha} d[m^*(\lambda^*), \tau^*] \frac{\partial}{\partial \lambda_i} m_\alpha^*(\lambda^*).$$

As $m^*(\lambda^*) = 0$ and $(\partial/\partial m)d(0, \tau^*) = 0$, we have $(\partial/\partial\lambda)s^*(\lambda^*) = 0$. Differentiating once more,

$$\frac{\partial^2}{\partial\lambda_i\,\partial\lambda_j}s^*(\lambda^*) = \sum_\alpha\sum_\beta\frac{\partial^2}{\partial m_\alpha\,\partial m_\beta}d(0, \tau^*)\frac{\partial}{\partial\lambda_i}m_\alpha^*(\lambda^*)\frac{\partial}{\partial\lambda_j}m_\beta^*(\lambda^*)$$

$$+ \sum_\alpha\frac{\partial}{\partial m_\alpha}d(0, \tau^*)\frac{\partial^2}{\partial\lambda_i\,\partial\lambda_j}m_\alpha^*(\lambda^*).$$

The second term is zero as $(\partial/\partial m)d(0, \tau^*) = 0$ whence

$$\frac{\partial^2}{\partial\lambda\,\partial\lambda'}s^*(\lambda) = \left(\frac{\partial}{\partial\lambda'}m^*(\lambda^*)\right)'\frac{\partial^2}{\partial m\,\partial m'}d(0, \tau^*)\frac{\partial}{\partial\lambda'}m^*(\lambda^*)$$

$$= (M^*)'D^*M^* = \mathscr{J}^*. \qquad \square$$

THEOREM 7 (Existence of consistent local minima). Let Assumptions 1 through 3 and 8 through 10 hold. Then there is a closed ball Λ centered at λ^* with finite, nonzero radius such that the sequence $\{\hat{\lambda}_n\}$ of $\hat{\lambda}_n$ that minimize $s_n(\lambda)$ over Λ converges almost surely to λ^* and the sequence $\{\lambda_n^0\}$ of λ_n^0 that minimize $s_n^0(\lambda)$ over Λ converges to λ^*.

Proof. By Lemma 4 and by assumption, $(\partial/\partial\lambda)s^*(\lambda^*) = 0$ and $(\partial^2/\partial\lambda\,\partial\lambda')s^*(\lambda^*)$ is positive definite. Then there is a closed ball Λ' centered at λ^* with finite, nonzero radius on which $s^*(\lambda)$ has a unique minimum at $\lambda = \lambda^*$ (Bartle, 1964, Section 21). Let Λ'' be the set given by Lemma 4, and put $\Lambda = \Lambda' \cap \Lambda''$. Then $s^*(\lambda)$ has a unique minimum on Λ, and both $s_n(\lambda)$ and $s_n^0(\lambda)$ converge almost surely to $s^*(\lambda)$ uniformly on Λ. The argument used to prove Theorem 3 may be repeated here word for word to obtain the conclusions of the theorem. $\qquad \square$

The following additional regularity conditions are needed to obtain asymptotic normality. The integral condition is similar to that in Assumption 6; the comments following Assumption 6 apply here as well.

ASSUMPTION 11. The elements of $m(y, x, \tau, \lambda)m'(y, x, \tau, \lambda)$ and $(\partial/\partial\tau')m(y, x, \tau, \lambda)$ are continuous and dominated by $b[q(y, x, \gamma), x]$ on $\mathscr{Y}\times\mathscr{X}\times T\times\Lambda^*\times\Gamma$; $b(e, x)$ is that of Assumption 3. The elements of

$(\partial^2/\partial\tau\partial m')\,d(m,\tau)$ are continuous on $\mathscr{F}\times T$, and

$$\int_{\mathscr{X}}\int_{\mathscr{E}}\frac{\partial}{\partial\tau'}m[Y(e,x,\gamma^*),x,\tau^*,\lambda^*]\,dP(e)\,d\mu(x)=0.$$

Next we show that the "scores" $(\partial/\partial\lambda)s_n(\lambda_n^0)$ are asymptotically normally distributed. As noted earlier, we rely heavily on the assumption that $m^*(\lambda^*)=0$. To remove it, see Problem 6.

THEOREM 8 (Asymptotic normality of the scores). Under Assumptions 1 through 3 and 8 through 11

$$\sqrt{n}\,\frac{\partial}{\partial\lambda}s_n(\lambda_n^0)\xrightarrow{\mathscr{L}}N_p(0,\mathscr{I}^*).$$

$\mathscr{I}^*$ may be singular.

Proof. By Lemma 2, we may assume without loss of generality that $\hat{\lambda}_n$ and λ_n^0 lie in the smallest of the closed balls given by Assumptions 9 and 11, Lemma 4, and Theorem 7 and that $(\partial/\partial\lambda)s_n(\hat{\lambda}_n)=o_s(n^{-1/2})$ and $(\partial/\partial\lambda)s_n^0(\lambda_n^0)=o(n^{-1/2})$.

A typical element of the vector $\sqrt{n}\,(\partial/\partial m)d[m_n(\lambda_n^0),\hat{\tau}_n]$ can be expanded about $[m_n^0(\lambda_n^0),\tau_n^0]$ to obtain

$$\sqrt{n}\,\frac{\partial}{\partial m_\alpha}d\big[m_n(\lambda_n^0),\hat{\tau}_n\big]=\sqrt{n}\,\frac{\partial}{\partial m_\alpha}d\big[m_n^0(\lambda_n^0),\tau_n^0\big]$$

$$+\frac{\partial}{\partial\tau'}\frac{\partial}{\partial m_\alpha}d(\bar{m},\bar{\tau})\sqrt{n}\,\big(\hat{\tau}_n-\tau_n^0\big)$$

$$+\frac{\partial}{\partial m'}\frac{\partial}{\partial m_\alpha}d(\bar{m},\bar{\tau})\sqrt{n}\,\big[m_n(\lambda_n^0)-m_n^0(\lambda_n^0)\big]$$

where $(\bar{m},\bar{\tau})$ is on the line segment joining $[m_n(\lambda_n^0),\hat{\tau}_n]$ to $[m_n^0(\lambda_n^0),\tau_n^0]$. Thus $(\bar{m},\bar{\tau})$ converges almost surely to (m^*,τ^*) where $m^*=m^*(\lambda^*)$. Noting that $\sqrt{n}\,(\hat{\tau}_n-\tau_n^0)$ is bounded in probability by Assumption 8 and that

$$\sqrt{n}\,\big[m_n(\lambda_n^0)-m_n^0(\lambda_n^0)\big]\xrightarrow{\mathscr{L}}N(0,S^*)$$

by Theorem 2, we may write (Problem 3)

$$\sqrt{n}\,\frac{\partial}{\partial m}d\big[m_n(\lambda_n^0),\hat{\tau}_n\big]$$

$$=\sqrt{n}\,\frac{\partial}{\partial m}d\big[m_n^0(\lambda_n^0),\tau_n^0\big]+\frac{\partial^2}{\partial m\,\partial\tau'}d(m^*,\tau^*)\sqrt{n}\,\big(\hat{\tau}_n-\tau_n^0\big)$$

$$+\frac{\partial^2}{\partial m\,\partial m'}d(m^*,\tau^*)\sqrt{n}\,\big[m_n(\lambda_n^0)-m_n^0(\lambda_n^0)\big]+o_p(1).$$

Then

$$\sqrt{n}\,\frac{\partial}{\partial\lambda}s_n(\lambda_n^0)=\sqrt{n}\,\frac{\partial}{\partial\lambda}s_n(\lambda_n^0)-\sqrt{n}\,\frac{\partial}{\partial\lambda}s_n^0(\lambda_n^0)+o(1)$$

$$=\sqrt{n}\,M_n'(\lambda_n^0)\frac{\partial}{\partial m}d\big[m_n(\lambda_n^0),\hat{\tau}_n\big]$$

$$-\sqrt{n}\,\overline{M}_n'(\lambda_n^0)\frac{\partial}{\partial m}d\big[m_n^0(\lambda_n^0),\tau_n^0\big]+o_s(1)$$

$$=\sqrt{n}\,\big[M_n(\lambda_n^0)-\overline{M}_n(\lambda_n^0)\big]'\frac{\partial}{\partial m}d\big[m_n^0(\lambda_n^0),\tau_n^0\big]$$

$$+M_n'(\lambda_n^0)\Big(\frac{\partial^2}{\partial m\,\partial\tau'}d(m^*,\tau^*)\Big)\sqrt{n}\,\big(\hat{\tau}_n-\tau_n^0\big)$$

$$+M_n'(\lambda_n^0)\Big(\frac{\partial^2}{\partial m\,\partial m'}d(m^*,\tau^*)\Big)\sqrt{n}\,\big[m_n(\lambda_n^0)-m_n^0(\lambda_n^0)\big]$$

$$+o_p(1).$$

Note that by Theorem 2, $\sqrt{n}\,[M_n(\lambda_n^0)-\overline{M}_n(\lambda_n^0)]$ is also bounded in probability, so that we have (Problem 3) the critical equation of the proof:

$$\sqrt{n}\,\frac{\partial}{\partial\lambda}s_n(\lambda_n^0)=\sqrt{n}\,\big[M_n(\lambda_n^0)-\overline{M}_n(\lambda_n^0)\big]'\frac{\partial}{\partial m}d(m^*,\tau^*)$$

$$+(M^*)'\Big(\frac{\partial^2}{\partial m\,\partial\tau'}d(m^*,\tau^*)\Big)\sqrt{n}\,\big(\hat{\tau}_n-\tau_n^0\big)$$

$$+(M^*)'D^*\sqrt{n}\,\big[m_n(\lambda_n^0)-m_n^0(\lambda_n^0)\big]+o_p(1).$$

We assumed that $m^*=0$, so that the first two terms on the right hand side drop out by Assumption 10. Inspecting the third term, we can conclude at once that

$$\sqrt{n}\,\frac{\partial}{\partial\lambda}s_n(\lambda_n^0) \xrightarrow{\mathscr{L}} N_p[0,(M^*)'D^*S^*D^*M^*].$$

In general the first two terms must be taken into account (Problem 6).   □

Asymptotic normality of the unconstrained method of moments estimator follows at once:

THEOREM 9. Let Assumptions 1 through 3 and 8 through 11 hold. Then:

$$\sqrt{n}\,(\hat{\lambda}_n - \lambda_n^0) \xrightarrow{\mathscr{L}} N_p[0,(\mathscr{I}^*)^{-1}\mathscr{I}^*(\mathscr{I}^*)^{-1}],$$

$\hat{\mathscr{I}}$ converges almost surely to $\mathscr{I}^* + \mathscr{U}^*$,

$\mathscr{I}_n^0$ converges to $\mathscr{I}^*$,

$\hat{\mathscr{J}}$ converges almost surely to $\mathscr{J}^*$,

$\mathscr{J}_n^0$ converges to $\mathscr{J}^*$.

$\mathscr{I}^*$ may be singular.

Proof. By Lemma 2, we may assume without loss of generality that $\hat{\lambda}_n$ and λ_n^0 lie in the smallest of the closed balls given by Assumptions 9 and 11, Lemma 4, and Theorem 7 and that $(\partial/\partial\lambda)s_n(\hat{\lambda}_n) = o_s(n^{-1/2})$, $(\partial/\partial\lambda)s_n^0(\lambda_n^0) = o(n^{-1/2})$.

By Taylor's theorem and arguments similar to the previous proof,

$$\sqrt{n}\,\frac{\partial}{\partial\lambda}s_n(\lambda_n^0) = \sqrt{n}\,\frac{\partial}{\partial\lambda}s_n(\hat{\lambda}_n) + [\mathscr{I}^* + o_s(1)]\sqrt{n}\,(\lambda_n^0 - \hat{\lambda}_n)$$

$$= o_s(1) + [\mathscr{I}^* + o_s(1)]\sqrt{n}\,(\lambda_n^0 - \hat{\lambda}_n).$$

Then by Slutsky's theorem (Serfling, 1980, Section 1.5.4, or Rao, 1973, Section 2c.4)

$$\sqrt{n}\,(\lambda_n^0 - \hat{\lambda}_n) \xrightarrow{\mathscr{L}} N_p[0,(\mathscr{I}^*)^{-1}\mathscr{I}^*(\mathscr{I}^*)^{-1}].$$

This establishes the first result.

We shall show that $\hat{\mathscr{I}}$ converges almost surely to $\mathscr{I}^* + \mathscr{U}^*$. The arguments for $\mathscr{I}_n^0$, $\hat{\mathscr{J}}$, and $\mathscr{J}_n^0$ are similar. Now $\hat{\mathscr{I}}$ is defined as

$$\hat{\mathscr{I}} = M_n'(\hat{\lambda}_n)D_n(\hat{\lambda}_n)S_n(\hat{\lambda}_n)D_n(\hat{\lambda}_n)M_n(\hat{\lambda}_n).$$

Since the Cesaro sum

$$\frac{1}{n} \sum_{t=1}^{n} m\big[Y(e_t, x_t, \gamma), x_t, \tau, \lambda\big] m'\big[Y(e_t, x_t, \gamma), x_t, \tau, \lambda\big]$$

converges almost surely to the integral

$$\int_{\mathscr{X}}\int_{\mathscr{E}} m\big[Y(e, x, \gamma), x, \tau, \lambda\big] m'\big[Y(e, x, \gamma), x, \tau, \lambda\big] \, dP(e) \, d\mu(x)$$

uniformly on $\Gamma \times T \times \Lambda$ by Theorem 1, with Assumption 11 providing the dominating function, and since $(\gamma_n^0, \hat{\tau}_n, \hat{\lambda}_n)$ converges almost surely to $(\gamma^*, \tau^*, \lambda^*)$, we have that

$$\lim_{n \to \infty} S_n(\hat{\lambda}_n) = \int_{\mathscr{X}}\int_{\mathscr{E}} m\big[Y(e, x, \gamma^*), x, \tau^*, \lambda^*\big]$$

$$\times m'\big[Y(e, x, \gamma^*), x, \tau^*, \lambda^*\big] \, dP(e) \, d\mu(x)$$

$$= S^* + K^*$$

almost surely. A similar argument shows that $M_n(\hat{\lambda}_n)$ converges almost surely to M^*. Since $(\partial^2/\partial m \, \partial m') d(m, \tau)$ is continuous in (m, τ) by Assumption 9 and $[m_n(\hat{\lambda}_n), \hat{\tau}_n]$ converges almost surely to $(0, \tau^*)$ by Lemma 4, Theorem 7, and Assumption 8, we have that $D_n(\hat{\lambda}_n)$ converges almost surely to D^*. Thus

$$\lim_{n \to \infty} \hat{\mathscr{I}} = (M^*)'D^*(S^* + K^*)D^*M^*$$

$$= \mathscr{I}^* + \mathscr{U}^*$$

almost surely.   □

The variance formula

$$\mathscr{I}^{-1}\mathscr{I}\mathscr{I}^{-1} = (M'DM)^{-1}(M'DSDM)(M'DM)^{-1}$$

is the same as that which would result if the generalized least squares estimator

$$\hat{\beta} = (M'DM)^{-1}M'Dy$$

were employed for the linear model

$$y = M\beta + e, \qquad e \sim (0, S).$$

Thus, the greatest efficiency for given moment equations results when $D^* = (S^*)^{-1}$.

A construction of τ_n^0 for Example 1 was promised:

EXAMPLE 1 (Continued). Assume that f, μ, and P are such that Assumptions 1 through 6 are satisfied for the preliminary estimator $\hat{\theta}_n$. Then $\hat{\theta}_n$ has a center θ_n^0 such that $\sqrt{n}\,(\hat{\theta}_n - \theta_n^0)$ is bounded in probability and $\lim_{n \to \infty} \theta_n^0 = \gamma^*$. Let

$$m(y, x, \theta, \tau) = \Psi^2\left(\frac{y - f(x, \theta)}{\tau}\right) - \int \Psi^2(e)\, d\Phi(e)$$

and

$$m_n(\tau) = \frac{1}{n} \sum_{t=1}^{n} m(y_t, x_t, \hat{\theta}_n, \tau).$$

The almost sure limit of $m_n(\tau)$ is

$$m^*(\tau) = \int_{\mathscr{X}} \int_{\mathscr{E}} m[Y(e, x, \gamma^*), x, \gamma^*, \tau]\, dP(e)\, d\mu(x)$$

$$= \int_{\mathscr{E}} \Psi^2\left(\frac{e}{\tau}\right) dP(e) - \int \Psi^2(e)\, d\Phi(e).$$

Since $0 < \int \Psi^2(e)\, d\Phi(e) < 1$ and $G(\tau) = \int_{\mathscr{E}} \Psi^2(e/\tau)\, dP(e)$ is a continuous, decreasing function with $G(0) = 1$ and $G(\infty) = 0$, there is a τ^* with $m^*(\tau^*) = 0$. Assume that f, μ, and P are such that Assumption 8 through 11 are satisfied for $s_n(\tau) = \frac{1}{2} m_n^2(\tau)$. Then by Theorem 7 and 9, $\hat{\tau}_n$ has a center τ_n^0 such that $\sqrt{n}\,(\hat{\tau}_n - \tau_n^0)$ is bounded in probability and $\lim_{n \to \infty} \tau_n^0 = \tau^*$. □

The argument used in the example is a fairly general approach for verifying the regularity conditions regarding nuisance parameter estimators. Typically, a nuisance parameter estimator solves an equation of the form

$$m_n(\tau) = \frac{1}{n} \sum_{t=1}^{n} m(y_t, x_t, \hat{\theta}_n, \tau) = 0$$

where $\hat{\theta}_n$ minimizes an $s_n(\theta)$ that is free of nuisance parameters. Thus $\hat{\theta}_n$ comes equipped with a center θ_n^0 as defined in either Section 3 or 4. Let

$$m_n^0(\tau) = \frac{1}{n} \sum_{t=1}^{n} \int_{\mathscr{E}} m[Y(e, x_t, \gamma_n^0), x_t, \theta_n^0, \tau]\, dP(e)$$

and let $d(m) = m'm/2$; then the appropriate center

$$\tau_n^0 \quad \text{minimizes} \quad s_n^0(\tau) = d[m_n^0(\tau)].$$

Next we establish some ancillary facts regarding the constrained estimator under a Pitman drift for use in Section 5. As noted previously, these results are not to be taken as an adequate theory of constrained estimation; that is found in Section 7.

ASSUMPTION 12 (Pitman drift).   The sequence $\{\gamma_n^0\}$ is chosen such that $\lim_{n \to \infty} \sqrt{n}\,(\lambda_n^0 - \lambda_n^*) = \Delta$. Moreover, $h(\lambda^*) = 0$.

THEOREM 10.   Let Assumptions 1 through 3 and 8 through 12 hold. Then there is a closed ball Λ centered at λ^* with finite, nonzero radius such that the constrained estimator $\tilde{\lambda}_n$ converges almost surely to λ^* and λ_n^* converges to λ^*. Moreover:

$\tilde{\mathscr{I}}$ converges almost surely to $\mathscr{I}^* + \mathscr{U}^*$,

$\mathscr{I}_n^*$ converges to $\mathscr{I}^*$,

$\tilde{\mathscr{J}}$ converges almost surely to $\mathscr{J}^*$,

$\mathscr{J}_n^*$ converges to $\mathscr{J}^*$,

$\sqrt{n}\,(\partial/\partial\lambda)s_n(\lambda_n^*) - \sqrt{n}\,(\partial/\partial\lambda)s_n^0(\lambda_n^*) \xrightarrow{\mathscr{L}} N_p(0,\,\mathscr{I}^*)$,

$\sqrt{n}\,(\partial/\partial\lambda)s_n^0(\lambda_n^*)$ converges to $-\mathscr{J}^*\Delta$.

Proof.   The argument showing the convergence of $\{\tilde{\lambda}_n\}$ and $\{\lambda_n^*\}$ is the same as the proof of Theorem 7 with the argument modified as per the proof of Theorem 6. The argument showing the convergence of $\tilde{\mathscr{I}}$, $\mathscr{I}_n^*$, $\tilde{\mathscr{J}}$, and $\mathscr{J}_n^*$ is the same as in the proof of Theorem 9. The same argument used in the proof of Theorem 8 may be used to derive the equation

$$\sqrt{n}\,\frac{\partial}{\partial\lambda}s_n(\lambda_n^*)$$

$$-\sqrt{n}\,\frac{\partial}{\partial\lambda}s_n^0(\lambda_n^*) = \sqrt{n}\,\left[M_n(\lambda_n^*) - \overline{M}_n(\lambda_n^*)\right]'\frac{\partial}{\partial m}d(m^*,\tau^*)$$

$$+ (M^*)'\left(\frac{\partial^2}{\partial m\,\partial\tau'}d(m^*,\tau^*)\right)\sqrt{n}\,(\hat{\tau}_n - \tau_n^0)$$

$$+ (M^*)'D^*\sqrt{n}\,\left[m_n(\lambda_n^*) - m_n^0(\lambda_n^*)\right] + o_p(1).$$

We assumed that $m^* = 0$ so that the first two terms on the right hand side drop out. By Theorem 2,

$$\sqrt{n}\,\left[m_n(\lambda_n^*) - m_n^0(\lambda_n^*)\right] \xrightarrow{\mathscr{L}} N.(0,\,S^*)$$

whence

$$\sqrt{n}\,\frac{\partial}{\partial\lambda}s_n(\lambda_n^*) - \sqrt{n}\,\frac{\partial}{\partial\lambda}s_n^0(\lambda_n^*) \overset{\mathscr{L}}{\to} N_p[0,(M^*)'D^*S^*D^*M^*]$$

and the first result follows.

The argument that $\sqrt{n}\,(\partial/\partial\lambda)s_n^0(\lambda_n^*)$ converges to $-\mathscr{J}^*\Delta$ is the same as in the proof of Theorem 6.   □

PROBLEMS

1. A vector valued function $f(x)$ is said to be uniformly continuous on X if given $\epsilon > 0$ there is a $\delta > 0$ such that for all x, x' in X with $\|x - x'\| < \delta$ we have $\|f(x) - f(x')\| < \epsilon$. Suppose $f(x)$ is a continuous function and X is compact; $f(x)$ is uniformly continuous on X (Royden, 1968, Chapter 9). Let $g_n(t)$ take its values in X, and let $\{g_n(t)\}$ converge uniformly to $g(t)$ on T. Show that $\{f[g_n(t)]\}$ converges uniformly to $f[g(t)]$ on T.

2. Prove that $\lim_{n\to\infty}m_n^0(\lambda) = m^*(\lambda)$ uniformly on Λ^*. Prove that $\lim_{n\to\infty}s_n^0(\lambda) = s^*(\lambda)$ uniformly on Λ^*.

3. A (vector valued) random variable Y_n is bounded in probability if given any $\epsilon > 0$ and $\delta > 0$ there is an M and an N such that $P(\|Y_n\| > M) < \delta$ for all $n > N$. Show that if $Y_n \overset{\mathscr{L}}{\to} N.(\mu, V)$ then Y_n is bounded in probability. Show that if X_n is a random matrix each element of which converges in probability to zero and Y_n is bounded in probability, then X_nY_n converges in probability to the zero vector. Hint: See Rao (1973, Section 2c.4).

4. Prove that $\mathscr{J}_n^0$ converges to $\mathscr{J}^*$ and that $\hat{\mathscr{J}}$ converges almost surely to $\mathscr{J}^*$.

5. Compute $\overline{K}_n(\lambda)$, $\overline{M}_n(\lambda)$, and $\overline{S}_n(\lambda)$ for Example 2 in the case $\lambda \neq \lambda_n^0$.

6. Let Assumptions 1 through 3 and 8 through 11 hold except that $m^*(\lambda^*) \neq 0$; also, $(\partial/\partial m)d(0, \tau)$ and $(\partial^2/\partial m\,\partial\lambda')d(0, \tau)$ can be nonzero. Suppose that the nuisance parameter estimator can be written as

$$\sqrt{n}\,(\hat{\tau}_n - \tau_n^0) = A_n\frac{1}{\sqrt{n}}\sum_{t=1}^n f(y_t, x_t, \theta_n^0) + o_p(1)$$

where $\lim_{n\to\infty}\theta_n^0 = \theta^*$, $\lim_{n\to\infty}A_n = A^*$ almost surely, and $f(y, x, \theta)$

satisfies the hypotheses of Theorem 2. Let $m^* = m^*(\lambda^*)$, and define

$$Z(e, x) = \begin{pmatrix} m[Y(e, x, \gamma^*), x, \tau^*, \lambda^*] \\ \text{vec } \frac{\partial}{\partial \lambda} m'[Y(e, x, \gamma^*), x, \tau^*, \lambda^*] \\ f[Y(e, x, \gamma^*), x, \theta^*] \end{pmatrix}$$

$$\mathcal{X}^* = \int_{\mathcal{X}} \left(\int_{\mathcal{E}} Z(e, x) \, dP(e) \right) \left(\int_{\mathcal{E}} Z(e, x) \, dP(e) \right)' d\mu(x)$$

$$\mathcal{S}^* = \int_{\mathcal{X}} \int_{\mathcal{E}} Z(e, x) Z'(e, x) \, dP(e) \, d\mu(x) - \mathcal{X}^*$$

$$\mathcal{A}^* = \left[(M^*)'D^* \;\vdots\; \frac{\partial}{\partial m'} d(m^*, \tau^*) \otimes I_p \;\vdots\; (M^*)' \frac{\partial^2}{\partial m \, \partial \tau'} d(m^*, \tau^*) A^* \right]$$

$$\mathcal{I}^* = \mathcal{A}^* \mathcal{S}^* (\mathcal{A}^*)'$$

$$\mathcal{J}^* = \frac{\partial^2}{\partial \lambda \, \partial \lambda'} s^*(\lambda^*).$$

Show that

$$\sqrt{n} \, \frac{\partial}{\partial \lambda} s_n(\lambda_n^0) \xrightarrow{\mathcal{L}} N_p(0, \mathcal{I}^*)$$

$$\sqrt{n} \, (\hat{\lambda}_n - \lambda_n^0) \xrightarrow{\mathcal{L}} N_p\left[0, (\mathcal{J}^*)^{-1} \mathcal{I}^* (\mathcal{J}^*)^{-1} \right].$$

Hint: Recall that if A of order r by c is partitioned as $A = [a_1 \;\vdots\; a_2 \;\vdots\; \cdots \;\vdots\; a_c]$, then

$$\text{vec } A = \begin{pmatrix} a_1 \\ a_2 \\ \vdots \\ a_c \end{pmatrix}$$

and vec $AB = (B' \otimes I_r)$vec A, where $\otimes$ denotes the Kronecker product of two matrices. See the proofs of Theorems 8 and 9.

7. Under the same assumptions as in Problem 6, show that

$$\sqrt{n} \, \frac{\partial}{\partial \lambda} s_n(\lambda_n^*) - \sqrt{n} \, \frac{\partial}{\partial \lambda} s_n^0(\lambda_n^*) \xrightarrow{\mathcal{L}} N_p(0, \mathcal{I}^*)$$

where $\mathcal{I}^*$ is defined as in Problem 6.

5. TESTS OF HYPOTHESES

One arrives at the same destination following either the least mean distance or the method of moments path. The starting point, which is the description of the data generating process given in Assumptions 1 through 3, is the same. Then the road forks. One can follow the least mean distance path with Notations 1 through 4 defining the quantities:

$\hat{\lambda}_n$, $\tilde{\lambda}_n$, λ_n^0, and λ_n^*;

$s_n(\lambda)$ and $s_n^0(\lambda)$;

$\hat{\mathscr{I}}$, $\tilde{\mathscr{I}}$, $\mathscr{I}_n^0$, and $\mathscr{I}_n^*$;

$\hat{\mathscr{J}}$, $\tilde{\mathscr{J}}$, $\mathscr{J}_n^0$, and $\mathscr{J}_n^*$;

$\mathscr{U}_n^0$ and $\mathscr{U}_n^*$.

Or one can follow the method of moments path with Notations 5 through 8 defining these quantities. In either case the results are the same and may be summarized as follows:

SUMMARY. Let Assumptions 1 through 3 hold, and let either Assumptions 4 through 7 or 8 through 12 hold. Then on a closed ball Λ centered at λ^* with finite, nonzero radius:

$s_n(\lambda)$ and $s_n^0(\lambda)$ converge almost surely and uniformly on Λ to $s^*(\lambda)$;

$(\partial/\partial\lambda)s_n(\lambda)$ and $(\partial/\partial\lambda)s_n^0(\lambda)$ converge almost surely and uniformly on Λ to $(\partial/\partial\lambda)s^*(\lambda)$;

$(\partial^2/\partial\lambda\,\partial\lambda')s_n(\lambda)$ and $(\partial^2/\partial\lambda\,\partial\lambda')s_n^0(\lambda)$ converge almost surely and uniformly on Λ to $(\partial^2/\partial\lambda\,\partial\lambda')s^*(\lambda)$, and $(\partial^2/\partial\lambda\,\partial\lambda')s^*(\lambda^*) = \mathscr{J}^*$;

$\sqrt{n}\,(\partial/\partial\lambda)s_n(\lambda_n^*) - \sqrt{n}\,(\partial/\partial\lambda)s_n^0(\lambda_n^*) \xrightarrow{\mathscr{L}} N_p(0, \mathscr{I}^*)$;

$\sqrt{n}\,(\hat{\lambda}_n - \lambda_n^0) \xrightarrow{\mathscr{L}} N_p[0, (\mathscr{J}^*)^{-1}\mathscr{I}^*(\mathscr{J}^*)^{-1}]$;

$\sqrt{n}\,(\lambda_n^0 - \lambda_n^*)$ converges to Δ, and $\sqrt{n}\,(\partial/\partial\lambda)s_n^0(\lambda_n^*)$ converges to $-\mathscr{J}^*\Delta$;

$\hat{\lambda}_n$ and $\tilde{\lambda}_n$ converge almost surely to λ^* and $h(\lambda^*) = 0$;

$\hat{\mathscr{I}}$ and $\tilde{\mathscr{I}}$ converge almost surely to $\mathscr{I}^* + \mathscr{U}^*$, and $\mathscr{I}_n^0$ and $\mathscr{I}_n^*$ converge to $\mathscr{I}^*$;

$\hat{\mathscr{I}}$ and $\tilde{\mathscr{I}}$ converge almost surely to $\mathscr{I}^*$, and $\mathscr{I}_n^0$ and $\mathscr{I}_n^*$ converge to $\mathscr{I}^*$;

$\mathscr{U}_n^0$ and $\mathscr{U}_n^*$ converge to $\mathscr{U}^*$.

Taking the Summary as the point of departure, consider testing

$$H : h(\lambda_n^0) = 0 \quad \text{against} \quad A : h(\lambda_n^0) \neq 0.$$

Three tests for this hypothesis will be studied: the Wald test, the Lagrange multiplier test (Rao's efficient score test), and an analog of the likelihood ratio test. The test statistics to be studied are defined in terms of the following notation.

NOTATION 9.

$$V_n^0 = (\mathscr{I}_n^0)^{-1} \mathscr{I}_n^0 (\mathscr{I}_n^0)^{-1}, \qquad V_n^* = (\mathscr{I}_n^*)^{-1} \mathscr{I}_n^* (\mathscr{I}_n^*)^{-1}$$

$$\hat{V} = \hat{\mathscr{I}}^{-1} \hat{\mathscr{I}} \hat{\mathscr{I}}^{-1}, \qquad \tilde{V} = \tilde{\mathscr{I}}^{-1} \tilde{\mathscr{I}} \tilde{\mathscr{I}}^{-1}$$

$$\hat{h} = h(\hat{\lambda}_n), \qquad \tilde{h} = h(\tilde{\lambda}_n)$$

$$H(\lambda) = \frac{\partial}{\partial \lambda'} h(\lambda)$$

$$\hat{H} = H(\hat{\lambda}_n), \qquad \tilde{H} = H(\tilde{\lambda}_n).$$

In Theorem 11,

$$V = V_n^0, \quad \mathscr{I} = \mathscr{I}_n^0, \quad \mathscr{I} = \mathscr{I}_n^0, \quad \mathscr{U} = \mathscr{U}_n^0, \quad H = H_n^0.$$

In Theorems 12, 13, 14, and 15

$$V = V_n^*, \quad \mathscr{I} = \mathscr{I}_n^*, \quad \mathscr{I} = \mathscr{I}_n^*, \quad \mathscr{U} = \mathscr{U}_n^*, \quad H = H_n^*.$$

In some applications it is more convenient to take

$$V = V_n^0, \qquad \mathscr{I} = \mathscr{I}_n^0, \qquad \mathscr{I} = \mathscr{I}_n^0, \qquad \mathscr{U} = \mathscr{U}_n^0$$

in Theorems 12, 13, 14, and 15. The asymptotics remain valid with these substitutions.

The following assumption imposes full rank on the matrices H and V. This assumption is not strictly necessary, but the less than full rank case does not appear to be of any practical importance, and a full rank assumption does eliminate much clutter from the theorems and proofs.

ASSUMPTION 13. The function $h(\lambda)$ that defines the null hypothesis $H: h(\lambda_n^0) = 0$ is a once continuously differentiable mapping of the estimation space into $\mathbb{R}^q$. Its Jacobian $H(\lambda) = (\partial/\partial\lambda')h(\lambda)$ has full rank $(= q)$ at $\lambda = \lambda^*$. The matrix $V = \mathscr{J}^{-1}\mathscr{I}\mathscr{J}^{-1}$ has full rank. The statement "the null hypothesis is true" means that $h(\lambda_n^0) = 0$ for all n or, equivalently, that $\lambda_n^0 = \lambda_n^*$ for all n sufficiently large.

The first statistic considered is the Wald test statistic

$$W = n\hat{h}'(\hat{H}\hat{V}\hat{H}')^{-1}\hat{h}$$

which is the same idea as division of an estimator by its standard error or studentization. The statistic is simple to compute and may be computed solely from the results of an unconstrained optimization of $s_n(\lambda)$. It has two disadvantages. First, its asymptotic distribution is a poorer approximation to its small sample distribution than for the next two statistics if Monte Carlo simulations are any guide (Chapter 1). Second, it is not invariant to reparametrization. With the same data and an equivalent model and hypothesis, two investigators could obtain different values of the test statistic (Problem 6).

The second statistic considered is the Lagrange multiplier test statistic

$$R = n\left(\frac{\partial}{\partial\lambda}s_n(\tilde{\lambda}_n)\right)'\tilde{\mathscr{J}}^{-1}\tilde{H}'(\tilde{H}\tilde{V}\tilde{H}')^{-1}\tilde{H}\tilde{\mathscr{J}}^{-1}\left(\frac{\partial}{\partial\lambda}s_n(\tilde{\lambda}_n)\right).$$

Since $(\partial/\partial\lambda)[s_n(\tilde{\lambda}_n) + \tilde{\theta}_n'h(\tilde{\lambda}_n)] = 0$ for large n, an alternative form is

$$R = n\tilde{\theta}_n'\tilde{H}\tilde{\mathscr{J}}^{-1}\tilde{H}'(\tilde{H}\tilde{V}\tilde{H}')^{-1}\tilde{H}\tilde{\mathscr{J}}^{-1}\tilde{H}'\tilde{\theta}_n$$

which gives rise to the term Lagrange multiplier test. Quite often $V = \mathscr{J}^{-1}$ $= \mathscr{I}^{-1}$ so that $\tilde{\mathscr{I}}^{-1}$ could be substituted for $\tilde{V}$ and $\tilde{\mathscr{J}}^{-1}$ in these formulas resulting in a material simplification. The statistic may be computed solely from a constrained optimization of $s_n(\lambda)$. Often, the minimization of $s_n(\lambda)$ subject to $h(\lambda) = 0$ is considerably easier than an unconstrained minimization; $H: \lambda_n^0 = 0$ for example. In these cases R is easier to compute than W. There are several motivations for the statistic R, of which the simplest is probably the following. Suppose that the quadratic surface

$$q(\lambda) = s_n(\tilde{\lambda}_n) + \frac{\partial}{\partial\lambda'}s_n(\tilde{\lambda}_n)(\lambda - \tilde{\lambda}_n) + \tfrac{1}{2}(\lambda - \tilde{\lambda}_n)'\tilde{\mathscr{J}}(\lambda - \tilde{\lambda}_n)$$

is an accurate approximation to the surface $s_n(\lambda)$ over a region that includes $\hat{\lambda}_n$. The quadratic surface is minimized at

$$\lambda = \tilde{\lambda}_n - \mathscr{J}^{-1} \frac{\partial}{\partial \lambda} s_n(\tilde{\lambda}_n)$$

so that

$$\tilde{\lambda}_n - \hat{\lambda}_n \doteq \mathscr{J}^{-1} \frac{\partial}{\partial \lambda} s_n(\tilde{\lambda}_n).$$

Thus, $\mathscr{J}^{-1}(\partial/\partial\lambda)s_n(\tilde{\lambda}_n)$ is the difference between $\tilde{\lambda}_n$ and $\hat{\lambda}_n$ induced by the constraint $h(\lambda) = 0$, and R is a measure of the squared length of this difference. Stated differently, $\mathscr{J}^{-1}(\partial/\partial\lambda)s_n(\tilde{\lambda}_n)$ is a full Newton iterative step from $\tilde{\lambda}_n$ (presumably) toward $\hat{\lambda}_n$, and R is a measure of the step length.

The third test statistic considered is an analog of the likelihood ratio test

$$L = 2n\left[s_n(\tilde{\lambda}_n) - s_n(\hat{\lambda}_n)\right].$$

The statistic measures the increase in the objective function due to the constraint $h(\tilde{\lambda}_n) = 0$; one rejects for large values of L. The statistic is derived by treating $s_n(\lambda)$ as if it were the negative of the log-likelihood and applying the definition of the likelihood ratio test.

Our plan is to derive approximations to the sampling distributions of these three statistics that are reasonably accurate in applications. To illustrate the ideas as we progress, we shall carry along a misspecified model as an example:

EXAMPLE 3. One fits the nonlinear model

$$y_t = f(x_t, \lambda) + u_t \qquad t = 1, 2, \ldots, n$$

by least squares to data that actually follow the model

$$y_t = g(x_t, \gamma_n^0) + e_t \qquad t = 1, 2, \ldots, n$$

where the errors e_t are independently distributed with mean zero and variance σ^2. The hypothesis of interest is

$$H: \tau_n^0 = \tau^* \quad \text{against} \quad A: \tau_n^0 \neq \tau^*$$

where

$$\lambda = (\rho', \tau')'$$

p is an r-vector, and τ is a q-vector with $p = r + q$. As in Chapter 1, we can put the model in a vector form:

$$y = f(\lambda) + u \qquad y = g(\gamma_n^0) + e \qquad F(\lambda) = \frac{\partial}{\partial \lambda'} f(\lambda).$$

We shall presume throughout that this model satisfies Assumptions 1 through 7, and 13. Direct computation yields

$$s(y_t, x_t, \lambda) = [y_t - f(x_t, \lambda)]^2$$

$$s_n(\lambda) = \frac{1}{n} \|y - f(\lambda)\|^2 = \frac{1}{n} [y - f(\lambda)]'[y - f(\lambda)]$$

$$\frac{\partial}{\partial \lambda} s_n(\lambda) = -\frac{2}{n} F'(\lambda)[y - f(\lambda)]$$

$$s_n^0(\lambda) = \sigma^2 + \frac{1}{n} \|g(\gamma_n^0) - f(\lambda)\|^2$$

$$\frac{\partial}{\partial \lambda} s_n^0(\lambda) = -\frac{2}{n} F'(\lambda)[g(\gamma_n^0) - f(\lambda)]$$

$$\lambda_n^0 \quad \text{minimizes} \quad \frac{1}{n} \|g(\gamma_n^0) - f(\lambda)\|^2$$

$$\tilde{\rho}_n \quad \text{minimizes} \quad \frac{1}{n} \|y - f(\rho, \tau^*)\|^2$$

$$\tilde{\lambda}_n = (\tilde{\rho}_n', \tau^{*'})'$$

$$\rho_n^* \quad \text{minimizes} \quad \frac{1}{n} \|g(\gamma_n^0) - f(\rho, \tau^*)\|^2$$

$$\lambda_n^* = (\rho_n^{*'}, \tau^{*'})'$$

$$F_n^0 = F(\lambda_n^0), \qquad F_n^* = F(\lambda_n^*), \qquad \hat{F} = F(\hat{\lambda}_n), \qquad \tilde{F} = F(\tilde{\lambda}_n)$$

$$\bar{\mathcal{J}}_n(\lambda) = \frac{4\sigma^2}{n} F'(\lambda) F(\lambda)$$

$$\bar{\mathcal{I}}_n(\lambda) = \frac{2}{n} F'(\lambda) F(\lambda) - \frac{2}{n} \sum_{t=1}^{n} [g(x_t, \gamma_n^0) - f(x_t, \lambda)] \frac{\partial^2}{\partial \lambda \, \partial \lambda'} f(x_t, \lambda)$$

$$\bar{\mathcal{U}}_n(\lambda) = \frac{4}{n} \sum_{t=1}^{n} [g(x_t, \gamma_n^0) - f(x_t, \lambda)]^2 \left(\frac{\partial}{\partial \lambda} f(x_t, \lambda) \right) \left(\frac{\partial}{\partial \lambda} f(x_t, \lambda) \right)'$$

$$\hat{\mathcal{J}} = \frac{4}{n} \sum_{t=1}^{n} [y_t - f(x_t, \hat{\lambda}_n)]^2 \left(\frac{\partial}{\partial \lambda} f(x_t, \hat{\lambda}_n) \right) \left(\frac{\partial}{\partial \lambda} f(x_t, \hat{\lambda}_n) \right)'$$

$$\hat{\mathcal{I}} = \frac{2}{n} \hat{F}' \hat{F} - \frac{2}{n} \sum_{t=1}^{n} [y_t - f(x_t, \hat{\lambda}_n)] \frac{\partial^2}{\partial \lambda \, \partial \lambda'} f(x_t, \hat{\lambda}_n)$$

$$H = [0 \vdots I_q]$$

where I_q is the identity matrix of order q.

The estimator

$$\hat{V} = (\hat{\mathscr{J}})^{-1}\hat{\mathscr{J}}(\hat{\mathscr{J}})^{-1}$$

obtained according to the general theory is not that customarily used in nonlinear regression analysis as we have seen in Chapter 1. It has an interesting property in that if the model is correctly specified—that is, γ and λ have the same dimension and $g(x, \gamma) = f(x, \gamma)$—then $\hat{V}$ will yield the correct standard errors for $\hat{\lambda}_n$ even if $\mathrm{Var}(e_t) = \sigma^2(x_t)$. For this reason, White (1980) terms $\hat{V}$ the heteroscedastic invariant estimator of the variance-covariance matrix of $\hat{\lambda}_n$.

The estimator customarily employed is

$$\hat{\Omega} = ns^2(\hat{F}'\hat{F})^{-1}$$

with

$$s^2 = (n - p)^{-1}\|y - f(\hat{\lambda}_n)\|^2.$$

We shall substitute $\hat{\Omega}$ for $\hat{V}$ in what follows, mainly to illustrate how the general theory is to be modified to accommodate special situations. □

The limiting distributions that have been derived thus far have been stated in terms of the parameters $\mathscr{I}^*$, $\mathscr{J}^*$, and $\mathscr{U}^*$. To use these results, it is necessary to compute $\mathscr{I}^*$, $\mathscr{J}^*$, and $\mathscr{U}^*$ and to compute them it is necessary to specify the limit of $\hat{\lambda}_n$ and $\hat{\tau}_n$ and to specify the limiting measure μ on $\mathscr{X}$. Most would prefer to avoid the arbitrariness resulting from having to specify what is effectively unknowable in any finite sample. More appealing is to center $\hat{\lambda}_n$ at λ_n^0 rather than at λ^*, center $\hat{\tau}_n$ at τ_n^0, and use the empirical distribution function computed from $\{x_t\}_{t=1}^n$ to approximate μ. What results is $\mathscr{I}_n^0$, $\mathscr{J}_n^0$, and $\mathscr{U}_n^0$ as approximations to $\mathscr{I}^*$, $\mathscr{J}^*$, and $\mathscr{U}^*$. The next theorem uses a Skorokhod representation to lend some formality to this approach in approximating the finite sample distribution of W. For the example we need an approximation to the limit of $\hat{\Omega}$:

EXAMPLE 3 (Continued). The almost sure limit of $\hat{\Omega}$ is

$$\Omega^* = \left(\sigma^2 + \int_{\mathscr{X}}[g(x, \gamma^*) - f(x, \lambda^*)]^2 \, d\mu(x)\right)$$
$$\times \left[\int_{\mathscr{X}}\left(\frac{\partial}{\partial\lambda}f(x, \lambda^*)\right)\left(\frac{\partial}{\partial\lambda}f(x, \lambda^*)\right)' d\mu(x)\right]^{-1}.$$

Following the same logic that leads to the approximation of $\mathscr{I}^*$, $\mathscr{J}^*$, and

$\mathscr{U}^*$ by $\mathscr{I}_n^0$, $\mathscr{J}_n^0$, and $\mathscr{U}_n^0$, we obtain

$$\bar{\Omega}_n(\lambda) = n\left(\sigma^2 + \frac{1}{n}\|g(\gamma_n^0) - f(\lambda)\|^2\right)[F'(\lambda)F(\lambda)]^{-1}$$

and

$$\Omega_n^0 = n\left(\sigma^2 + \frac{1}{n}\|g(\gamma_n^0) - f(\lambda_n^0)\|^2\right)(F_n^{0\prime}F_n^0)^{-1}$$

where $F_n^0 = F(\lambda_n^0)$. $\square$

THEOREM 11. Let Assumptions 1 through 3 hold, and let either Assumptions 4 through 7 or 8 through 12 hold. Let

$$W = n\hat{h}'(\hat{H}\hat{V}\hat{H}')^{-1}\hat{h}.$$

Under Assumption 13,

$$W \sim Y + o_p(1)$$

where

$$Y = Z'\left[H\mathscr{J}^{-1}(\mathscr{I}+\mathscr{U})\mathscr{J}^{-1}H'\right]^{-1}Z$$

and

$$Z \sim N_q\left[\sqrt{n}\,h(\lambda_n^0), HVH'\right].$$

Recall: $V = V_n^0$, $\mathscr{I} = \mathscr{I}_n^0$, $\mathscr{J} = \mathscr{J}_n^0$, $\mathscr{U} = \mathscr{U}_n^0$, and $H = H_n^0$. If $\mathscr{U} = 0$, then Y has the noncentral, chi-square distribution with q degrees of freedom and noncentrality parameter $\alpha = nh'(\lambda_n^0)(HVH')^{-1}h(\lambda_n^0)/2$. Under the null hypothesis $\alpha = 0$.

Proof. By Lemma 2, we may assume without loss of generality that $\hat{\lambda}_n$, $\lambda_n^0 \in \Lambda$ and that $(\partial/\partial\lambda)s_n(\hat{\lambda}_n) = o_s(n^{-1/2})$, $(\partial/\partial\lambda)s_n^0(\lambda_n^0) = o(n^{1/2})$. By Taylor's theorem

$$\sqrt{n}\left[h_i(\hat{\lambda}_n) - h_i(\lambda_n^0)\right] = \frac{\partial}{\partial\lambda}h_i(\bar{\lambda}_{in})\sqrt{n}\left(\hat{\lambda}_n - \lambda_n^0\right) \qquad i = 1,2,\ldots,q$$

where $\|\bar{\lambda}_{in} - \lambda_n^0\| \le \|\hat{\lambda}_n - \lambda_n^0\|$. By the almost sure convergence of λ_n^0 and $\hat{\lambda}_n$ to λ^*, $\lim_{n\to\infty}\|\bar{\lambda}_{in} - \lambda^*\| = 0$ almost surely, whence $\lim_{n\to\infty}(\partial/\partial\lambda)h_i(\bar{\lambda}_{in}) = (\partial/\partial\lambda)h_i(\lambda^*)$ almost surely. Thus we may write

$$\sqrt{n}\left[h(\hat{\lambda}_n) - h(\lambda_n^0)\right] = [H^* + o_s(1)]\sqrt{n}\left(\hat{\lambda}_n - \lambda_n^0\right).$$

Since $\sqrt{n}\,(\hat{\lambda} - \lambda_n^0) \overset{\mathscr{L}}{\to} N_p(0, V*)$, we have

$$\sqrt{n}\left[h(\hat{\lambda}_n) - h(\lambda_n^0) \right] \overset{\mathscr{L}}{\to} N_q(0, H*V*H*').$$

By Problem 3, $\lim_{n \to \infty} \sqrt{n}\, h(\lambda_n^0) = H\Delta$ so that $\sqrt{n}\, h(\hat{\lambda}_n)$ is bounded in probability. Now $\hat{H}\hat{V}\hat{H}'$ converges almost surely to $H*(\mathscr{J}*)^{-1}$ $(\mathscr{J}* + \mathscr{U}*)(\mathscr{J}*)^{-1}H*'$ which is nonsingular, whence

$$(\hat{H}\hat{V}\hat{H}')^{-1} = \left[H\mathscr{J}^{-1}(\mathscr{J} + \mathscr{U})\mathscr{J}^{-1}H' \right]^{-1} + o_s(1).$$

Then

$$W = n h'(\hat{\lambda}_n)\left[H\mathscr{J}^{-1}(\mathscr{J} + \mathscr{U})\mathscr{J}^{-1}H' \right]^{-1} h(\hat{\lambda}_n) + o_p(1).$$

By the Skorokhod representation theorem (Serfling, 1980, Section 1.6), there are random variables Y_n with the same distribution as $\sqrt{n}\, h(\hat{\lambda}_n)$ such that $Y_n - \sqrt{n}\, h(\lambda_n^0) = Y + o_s(1)$, where $Y \sim N_q(0, H*V*H*')$. Factor $H*V*H*'$ as $H*V*H*' = P*P*'$, and for large n factor $HVH' = QQ'$ (Problem 1). Then

$$Y_n = \sqrt{n}\, h(\lambda_n^0) + Q(P*)^{-1}Y + \left[I - Q(P*)^{-1} \right]Y + o_s(1).$$

Since Y is bounded in probability and $I - Q(P*)^{-1} = o_s(1)$ (Problem 1), we have

$$Y_n = \sqrt{n}\, h(\lambda_n^0) + Q(P*)^{-1}Y + o_p(1)$$

where $Q(P*)^{-1}Y \sim N_q(0, HVH')$. Let $Z = \sqrt{n}\, h(\lambda_n^0) + Q(P*)^{-1}\,Y$ and the result follows. $\qquad\square$

Occasionally in the literature one sees an alternative form of the Wald test statistic

$$W = n(\hat{\lambda}_n - \tilde{\lambda}_n)'\hat{H}'(\hat{H}\hat{V}\hat{H}')^{-1}\hat{H}(\hat{\lambda}_n - \tilde{\lambda}_n).$$

The alternative form is obtained from the approximation $\hat{h} \doteq \hat{H}(\hat{\lambda}_n - \tilde{\lambda}_n)$, which is derived as follows. By Taylor's theorem

$$h(\hat{\lambda}_n) = h(\tilde{\lambda}_n) + \bar{H}(\hat{\lambda}_n - \tilde{\lambda}_n)$$

where $\bar{H}$ has rows $(\partial/\partial\lambda')h_i(\bar{\lambda})$ and $\bar{\lambda}$ is on the line segment joining $\tilde{\lambda}_n$ to

$\hat{\lambda}_n$. By noting that $h(\tilde{\lambda}_n) = 0$ and approximating $\bar{H}$ by $\hat{H}$, one has that $\hat{h} \doteq \hat{H}(\hat{\lambda}_n - \tilde{\lambda}_n)$. Any solution of $h(\lambda) = 0$ with $(\partial/\partial \lambda')H(\lambda) \doteq \hat{H}$ would serve as well as $\tilde{\lambda}_n$ by this logic, and one sees other choices at times.

As seen from Theorem 11, an asymptotically level α test in a correctly specified situation is to reject $H : h(\lambda_n^0) = 0$ when W exceeds the upper $\alpha \times 100\%$ critical point of a chi-square random variable with q degrees of freedom. In a conditional analysis of an incorrectly specified situation, $\mathscr{U}$, $h(\lambda_n^0)$, and α will usually be nonzero, so nothing can be said in general. One has a quadratic form in normally distributed random variables. Direct computation for a specified $q(y, x, \gamma_n^0)$ is required. We illustrate with the example.

EXAMPLE 3 (Continued). The hypothesis of interest is

$$H : \tau_n^0 = \tau^* \quad \text{against} \quad A : \tau_n^0 \neq \tau^*$$

where

$$\lambda = (\rho', \tau')'.$$

Substituting $\hat{\Omega}$ for $\hat{V}$ the Wald statistic is

$$W = \frac{(\hat{\tau}_n - \tau^*)'\left[H(\hat{F}'\hat{F})^{-1}H'\right]^{-1}(\hat{\tau}_n - \tau^*)}{s^2}$$

where $H = [0 : I_q]$. Thus $H(\hat{F}'\hat{F})^{-1}H'$ is the submatrix of $(\hat{F}'\hat{F})^{-1}$ formed by deleting the first r rows and columns of $(\hat{F}'\hat{F})^{-1}$. W is distributed as

$$W \sim Y + o_p(1)$$

where

$$Y = \frac{Z'\left[H\left(\frac{1}{n}F_n^{0\prime}F_n^0\right)^{-1}H'\right]^{-1}Z}{\sigma^2 + \frac{1}{n}\left\|g(\gamma_n^0) - f(\lambda_n^0)\right\|^2},$$

$$Z \sim N_q\left[\sqrt{n}\left(\tau_n^0 - \tau^*\right), HVH'\right]$$

$$V = \mathscr{J}^{-1}\mathscr{I}\mathscr{J}^{-1}$$

$$\mathscr{I} = \frac{4\sigma^2}{n}F_n^{0\prime}F_n^0$$

$$\mathscr{J} = \frac{2}{n}F_n^{0\prime}F_n^0 - \frac{2}{n}\sum_{t=1}^n \left[g(x_t, \gamma_n^0) - f(x_t, \lambda_n^0)\right]\frac{\partial^2}{\partial\lambda\,\partial\lambda'}f(x_t, \lambda_n^0).$$

If the model is correctly specified, then $g(x_t, \gamma_n^0) = f(x_t, \lambda_n^0)$ and these equations simplify to

$$
Y = \frac{Z'\left[H\left(\frac{1}{n}F_n^{0\prime}F_n^0\right)^{-1}H'\right]^{-1}Z}{\sigma^2}
$$

$$
Z \sim N_q\left[\sqrt{n}\left(\tau_n^0 - \tau^*\right),\, \sigma^2 H\left(\frac{1}{2}F_n^{0\prime}F_n^0\right)^{-1}H'\right]
$$

whence Y is distributed as a noncentral chi-square random variable with q degrees of freedom and noncentrality parameter

$$
\alpha = \frac{\left(\tau_n^0 - \tau^*\right)'\left[H(F'F)^{-1}H'\right]^{-1}\left(\tau_n^0 - \tau^*\right)}{\sigma^2}. \qquad \square
$$

The statistic R is a quadratic form in $(\partial/\partial\lambda)s_n(\tilde{\lambda}_n)$, and for n large enough that $\mathscr{I}_n(\lambda)$ can be inverted in a neighborhood of $\tilde{\lambda}_n$, the statistic L is also a quadratic form in $(\partial/\partial\lambda)s_n(\tilde{\lambda}_n)$ (Problem 8). Thus, a characterization of the distribution of $(\partial/\partial\lambda)s_n(\tilde{\lambda}_n)$ is needed. We shall divide this derivation into two steps. First (Theorem 12), a characterization of $(\partial/\partial\lambda)s_n(\lambda_n^*)$ is obtained. Second (Theorem 13), $(\partial/\partial\lambda)s_n(\tilde{\lambda}_n)$ is characterized as a projection of $(\partial/\partial\lambda)s_n(\lambda_n^*)$ into the column space of $H_n^{*\prime}$.

THEOREM 12. Let Assumptions 1 through 3 hold, and let either Assumptions 4 through 7 or 8 through 12 hold. Then

$$
\sqrt{n}\,\frac{\partial}{\partial\lambda}s_n(\lambda_n^*) \sim X + o_p(1)
$$

where

$$
X \sim N_p\left(\sqrt{n}\,\frac{\partial}{\partial\lambda}s_n^0(\lambda_n^*),\, \mathscr{I}_n^*\right).
$$

Proof. By either Theorem 6 or Theorem 10,

$$
\sqrt{n}\,\frac{\partial}{\partial\lambda}s_n(\lambda_n^*) - \sqrt{n}\,\frac{\partial}{\partial\lambda}s_n^0(\lambda_n^*) \xrightarrow{\mathscr{L}} N_p(0, \mathscr{I}^*).
$$

By the Skorokhod representation theorem (Serfling, 1980, Section 1.6), there are random variables Y_n with the same distribution as $\sqrt{n}\,(\partial/\partial\lambda)s_n(\lambda_n^*)$ such that $Y_n - \sqrt{n}\,(\partial/\partial\lambda)s_n^0(\lambda_n^*) = Y + o_s(1)$, where $Y \sim N_p(0, \mathscr{I}^*)$. Then factor $\mathscr{I}^*$ as $\mathscr{I}^* = P^*(P^*)'$ and for large n factor $\mathscr{I}_n^*$ as $\mathscr{I}_n^* = QQ'$ (Problem 1). Then

$$
Y_n = \sqrt{n}\,\frac{\partial}{\partial\lambda}s_n^0(\lambda_n^*) + Q(P^*)^{-1}Y + \left[I - Q(P^*)^{-1}\right]Y + o_s(1).
$$

Let $X = \sqrt{n}(\partial/\partial\lambda)s_n^0(\lambda_n^*) + Q(P^*)^{-1}Y$, whence

$$X \sim N_p\left[\sqrt{n}(\partial/\partial\lambda)s_n^0(\lambda_n^*), \mathscr{I}_n^*\right]$$

and, since $\lim_{n \to \infty} Q = P^*$ (Problem 1), $[I - Q(P^*)^{-1}]Y = o_p(1)$. □

In Theorem 12, any matrix that is equal to $\mathscr{I}_n^*$ to within $o(1)$ can be substituted for $\mathscr{I}_n^*$ and the result will be corrected. In most applications, $\mathscr{I}_n^0$ is a far more convenient choice because it is much easier to compute and its use in derivations permits many helpful algebraic simplifications; the derivations in Chapter 4 provide an illustration. Nonetheless, $\mathscr{I}_n^*$ seems the more natural choice, because it is the exact small sample variance of $\sqrt{n}(\partial/\partial\lambda)s_n(\lambda_n^*)$ when $s_n(\lambda)$ does not depend on any estimators of nuisance parameters. Moreover, in the nonstationary case where one has rather less flexibility, $\mathscr{I}_n^*$ arises naturally from the theory; see Theorem 13 of Chapter 7. The same remarks apply to $\mathscr{J}_n^*$ and $\mathscr{U}_n^*$ in the theorems that follow. $\mathscr{J}_n^*$ and $\mathscr{U}_n^*$ arise naturally, but the theorems remain true if $\mathscr{J}_n^0$ and $\mathscr{U}_n^0$ are substituted, and these choices are more convenient in applications. Actually, these matrices will only differ by $O(1/\sqrt{n})$ in most applications, as in the example:

EXAMPLE 3 (Continued). In a correctly specified situation we have $g(x, \gamma_n^0) = f(x, \lambda_n^0)$, and from previous computations we have

$$\bar{\mathscr{I}}_n(\lambda) = \frac{4\sigma^2}{n} F'(\lambda)F(\lambda)$$

$$\bar{\mathscr{J}}_n(\lambda) = \frac{2}{n} F'(\lambda)F(\lambda)$$

$$- \frac{2}{n} \sum_{t=1}^{n} \left[f\left(x_t, \lambda_n^0\right) - f(x_t, \lambda)\right]^2 \frac{\partial^2}{\partial\lambda\,\partial\lambda'}f(x_t, \lambda)$$

$$\bar{\mathscr{U}}_n(\lambda) = \frac{4}{n} \sum_{t=1}^{n} \left[f\left(x_t, \lambda_n^0\right) - f(x_t, \lambda)\right]^2 \left(\frac{\partial}{\partial\lambda}f(x_t, \lambda)\right)\left(\frac{\partial}{\partial\lambda}f(x_t, \lambda)\right)'$$

$$\bar{\Omega}_n(\lambda) = n\left(\sigma^2 + \frac{1}{n}\|f(\lambda_n^0) - f(\lambda)\|^2\right)[F'(\lambda)F(\lambda)]^{-1}.$$

A verification that $\bar{\mathscr{I}}_n(\lambda_n^*)$, $\bar{\mathscr{J}}_n(\lambda_n^*)$, and $\bar{\mathscr{U}}_n(\lambda_n^*)$ differ from $\bar{\mathscr{I}}_n(\lambda_n^0)$, $\bar{\mathscr{J}}_n(\lambda_n^0)$, and $\bar{\mathscr{U}}_n(\lambda_n^0)$ by terms of order $O(1/\sqrt{n})$ is a fairly straightforward application of Taylor's theorem. Using $\bar{\mathscr{J}}_n(\lambda_n^*)$ to illustrate, an application of Taylor's theorem, with $\bar{\lambda}_n$ denoting a point on the line joining λ_n^0 to λ_n^*, together with Theorems 1, 3, 6 yields

$$\frac{2}{n} \sum_{t=1}^{n} \left[f(x_t, \lambda_n^0) - f(x_t, \lambda_n^*) \right] \frac{\partial^2}{\partial \lambda_i \, \partial \lambda_j} f(x_t, \lambda_n^*)$$

$$= \frac{1}{\sqrt{n}} \frac{2}{n} \sum_{t=1}^{n} \frac{\partial^2}{\partial \lambda_i \, \partial \lambda_j} f(x_t, \lambda_n^*) \frac{\partial}{\partial \lambda} f(x_t, \bar{\lambda}_n) \sqrt{n} (\lambda_n^0 - \lambda_n^*)$$

$$= \frac{1}{\sqrt{n}} \left(2 \int_{\mathscr{X}} \frac{\partial^2}{\partial \lambda_i \, \partial \lambda_j} f(x, \lambda^*) \frac{\partial}{\partial \lambda} f(x, \lambda^*) \Delta \, d\mu(x) + o(1) \right)$$

$$\frac{2}{n} \sum_{t=1}^{n} \frac{\partial}{\partial \lambda_i} f(x_t, \lambda_n^0) \frac{\partial}{\partial \lambda_j} f(x_t, \lambda_n^0) - \frac{\partial}{\partial \lambda_i} f(x_t, \lambda_n^*) \frac{\partial}{\partial \lambda_j} f(x_t, \lambda_n^*)$$

$$= \frac{1}{\sqrt{n}} \frac{2}{n} \sum_{t=1}^{n} \left(\frac{\partial}{\partial \lambda_i} f(x_t, \bar{\lambda}_n) \frac{\partial}{\partial \lambda'} \frac{\partial}{\partial \lambda_j} f(x_t, \bar{\lambda}_n) \right) \sqrt{n} (\lambda_n^0 - \lambda_n^*)$$

$$+ \frac{1}{\sqrt{n}} \frac{2}{n} \sum_{t=1}^{n} \left(\frac{\partial}{\partial \lambda_j} f(x_t, \bar{\lambda}_n) \frac{\partial}{\partial \lambda'} \frac{\partial}{\partial \lambda_i} f(x_t, \bar{\lambda}_n) \right) \sqrt{n} (\lambda_n^0 - \lambda_n^*)$$

$$= \frac{1}{\sqrt{n}} \left(2 \int_{\mathscr{X}} \frac{\partial}{\partial \lambda_i} f(x, \lambda^*) \frac{\partial}{\partial \lambda'} \frac{\partial}{\partial \lambda_j} f(x, \lambda^*) \Delta \, d\mu(x) + o(1) \right)$$

$$+ \frac{1}{\sqrt{n}} \left(2 \int_{\mathscr{X}} \frac{\partial}{\partial \lambda_j} f(x, \lambda^*) \frac{\partial}{\partial \lambda'} \frac{\partial}{\partial \lambda_i} f(x, \lambda^*) \Delta \, d\mu(x) + o(1) \right)$$

whence

$$\bar{\mathscr{I}}_n(\lambda_n^0) - \bar{\mathscr{I}}_n(\lambda_n^*) = O\left(\frac{1}{\sqrt{n}} \right). \qquad \square$$

THEOREM 13. Let Assumptions 1 through 3 hold, and let either Assumptions 4 through 7 or 8 through 12 hold. Under Assumption 13,

$$\sqrt{n} \, \frac{\partial}{\partial \lambda} s_n(\tilde{\lambda}_n) = H'(H \mathscr{I}^{-1} H')^{-1} H \mathscr{I}^{-1} \sqrt{n} \, \frac{\partial}{\partial \lambda} s_n(\lambda_n^*) + o_p(1)$$

where $\mathscr{I} = \mathscr{I}_n^*$ and $H = H_n^*$.

Proof. By Lemma 2, we may assume without loss of generality that $\hat{\lambda}_n, \tilde{\lambda}_n, \lambda_n^0, \lambda_n^* \in \Lambda$. By Taylor's theorem

$$\sqrt{n} \, \frac{\partial}{\partial \lambda} s_n(\tilde{\lambda}_n) = \sqrt{n} \, \frac{\partial}{\partial \lambda} s_n(\lambda_n^*) + \mathscr{I} \sqrt{n} (\tilde{\lambda}_n - \lambda_n^*)$$

$$\sqrt{n} \, h(\tilde{\lambda}_n) = \sqrt{n} \, h(\lambda_n^*) + \bar{H} \sqrt{n} (\tilde{\lambda}_n - \lambda_n^*)$$

where $\mathscr{J}$ has rows

$$\frac{\partial}{\partial \lambda'}\frac{\partial}{\partial \lambda_i}s_n(\bar{\lambda}_{in}) \qquad i = 1, 2, \ldots, p$$

and $\bar{H}$ has rows

$$\frac{\partial}{\partial \lambda'}h_j(\bar{\lambda}_{jn}) \qquad j = 1, 2, \ldots, q$$

with $\bar{\lambda}_{in}$ and $\bar{\lambda}_{jn}$ on the line segment joining $\tilde{\lambda}_n$ to λ_n^*. By Lemma 2, $\sqrt{n}\,h(\tilde{\lambda}_n) = o_s(1)$. Recalling that $\sqrt{n}\,h(\lambda_n^*) \equiv 0$, we have $\bar{H}\sqrt{n}\,(\tilde{\lambda}_n - \lambda_n^*) = o_s(1)$. Since $\tilde{\lambda}_n$ and λ_n^* converge almost surely to λ^*, each $\bar{\lambda}_{in}$, $\bar{\lambda}_{jn}$ converges almost surely to λ^*, whence $\mathscr{J}$ converges almost surely to $\mathscr{J}^*$ by the uniform almost sure convergence of $(\partial^2/\partial\lambda\,\partial\lambda')s_n(\lambda)$; $\bar{H}$ converges almost surely to H^* by the continuity of $H(\lambda)$. Thus $\mathscr{J} = \mathscr{J} + o_s(1)$ and $\bar{H} = H + o_s(1)$. Moreover, there is an N corresponding to almost every realization of $\{e_t\}$ such that $\det(\mathscr{J}) > 0$ for all $n > N$. Defining $\mathscr{J}^{-1}$ arbitrarily when $\det(\mathscr{J}) = 0$, we have

$$\mathscr{J}^{-1}\mathscr{J}\sqrt{n}\,(\tilde{\lambda}_n - \lambda_n^*) \equiv \sqrt{n}\,(\tilde{\lambda}_n - \lambda_n^*)$$

for all $n > N$. Thus, $\mathscr{J}^{-1}\mathscr{J}\sqrt{n}\,(\tilde{\lambda}_n - \lambda_n^*) = \sqrt{n}\,(\tilde{\lambda}_n - \lambda_n^*) + o_s(1)$. Combining these observations, we may write

$$\bar{H}\sqrt{n}\,(\tilde{\lambda}_n - \lambda_n^*) = o_s(1)$$

$$\sqrt{n}\,(\tilde{\lambda}_n - \lambda_n^*) = \mathscr{J}^{-1}\sqrt{n}\,\frac{\partial}{\partial \lambda}s_n(\tilde{\lambda}_n) - \mathscr{J}^{-1}\sqrt{n}\,\frac{\partial}{\partial \lambda}s_n(\lambda_n^*) + o_s(1)$$

whence

$$\bar{H}\mathscr{J}^{-1}\sqrt{n}\,\frac{\partial}{\partial \lambda}s_n(\tilde{\lambda}_n) = \bar{H}\mathscr{J}^{-1}\sqrt{n}\,\frac{\partial}{\partial \lambda}s_n(\lambda_n^*) + o_s(1).$$

Since $\sqrt{n}\,(\partial/\partial\lambda)s_n(\lambda_n^*)$ converges in distribution, it is bounded in probability, whence

$$\bar{H}\mathscr{J}^{-1}\sqrt{n}\,\frac{\partial}{\partial \lambda}s_n(\tilde{\lambda}_n) = H\mathscr{J}^{-1}\sqrt{n}\,\frac{\partial}{\partial \lambda}s_n(\lambda_n^*) + o_p(1).$$

By Lemma 2, there is a sequence of Lagrange multipliers $\tilde{\theta}_n$ such that

$$\sqrt{n}\,\frac{\partial}{\partial \lambda}s_n(\tilde{\lambda}_n) + \tilde{H}'\sqrt{n}\,\tilde{\theta}_n = o_s(1).$$

Substituting into the previous equation, we have

$$\bar{H}\bar{\mathscr{J}}^{-1}\tilde{H}\sqrt{n}\,\tilde{\theta}_n = H\mathscr{J}^{-1}\sqrt{n}\,\frac{\partial}{\partial\lambda}s_n(\lambda_n^*) + o_p(1).$$

By Slutsky's theorem (Serfling, 1980, Section 1.5.4, or Rao, 1973, Section 2c.4), $\sqrt{n}\,\tilde{\theta}_n$ converges in distribution. In consequence, both $\sqrt{n}\,\tilde{\theta}_n$ and $\sqrt{n}\,(\partial/\partial\lambda)s_n(\tilde{\lambda}_n)$ are bounded in probability and we have

$$H'(H\mathscr{J}^{-1}H')^{-1}H\mathscr{J}^{-1}\sqrt{n}\,\frac{\partial}{\partial\lambda}s_n(\lambda_n^*)$$

$$= H'(H\mathscr{J}^{-1}H')^{-1}H\mathscr{J}^{-1}\sqrt{n}\,\frac{\partial}{\partial\lambda}s_n(\tilde{\lambda}_n) + o_p(1)$$

$$= -H'(H\mathscr{J}^{-1}H')^{-1}H\mathscr{J}^{-1}H'\sqrt{n}\,\tilde{\theta}_n + o_p(1)$$

$$= -H'\sqrt{n}\,\tilde{\theta}_n + o_p(1)$$

$$= \sqrt{n}\,\frac{\partial}{\partial\lambda}s_n(\tilde{\lambda}_n) + o_p(1). \qquad \square$$

A characterization of the distribution of the statistic R follows immediately from Theorem 13:

THEOREM 14.   Let Assumptions 1 through 3 hold and let either Assumptions 4 through 7 or 8 through 12 hold. Let

$$R = n\left(\frac{\partial}{\partial\lambda}s_n(\tilde{\lambda}_n)\right)'\tilde{\mathscr{J}}^{-1}\tilde{H}'(\tilde{H}\tilde{V}\tilde{H}')^{-1}\tilde{H}\tilde{\mathscr{J}}^{-1}\left(\frac{\partial}{\partial\lambda}s_n(\tilde{\lambda}_n)\right).$$

Under Assumption 13

$$R \sim Y + o_p(1)$$

where

$$Y = Z'\mathscr{J}^{-1}H'\left[H\mathscr{J}^{-1}(\mathscr{I} + \mathscr{U})\mathscr{J}^{-1}H'\right]^{-1}H\mathscr{J}^{-1}Z$$

and

$$Z \sim N_p\left(\sqrt{n}\,\frac{\partial}{\partial\lambda}s_n^0(\lambda_n^*), \mathscr{I}\right).$$

Recall: $V = V_n^*$, $\mathscr{I} = \mathscr{I}_n^*$, $\mathscr{J} = \mathscr{J}_n^*$, $\mathscr{U} = \mathscr{U}_n^*$, and $H = H_n^*$. Alternatively, $V = V_n^0$, $\mathscr{I} = \mathscr{I}_n^0$, $\mathscr{J} = \mathscr{J}_n^0$, and $\mathscr{U} = \mathscr{U}_n^0$. If $\mathscr{U} = 0$, then Y has the noncentral chi-square distribution with q degrees of freedom and noncentrality parameter

$$\alpha = n\big[(\partial/\partial\lambda)s_n^0(\lambda_N^*)\big]'\mathscr{J}^{-1}H'(HVH')^{-1}H\mathscr{J}^{-1}\big[(\partial/\partial\lambda)s_n^0(\lambda_N^*)\big]/2.$$

Under the null hypothesis, $\alpha = 0$.

Proof. By Lemma 2, we may assume without loss of generality that $\tilde{\lambda}_n \in \Lambda$. By the Summary,

$$\tilde{\mathscr{J}}^{-1}\tilde{H}'(\tilde{H}\tilde{V}\tilde{H}')^{-1}\tilde{H}\tilde{\mathscr{J}}^{-1} = \mathscr{J}^{-1}H'\big[H\mathscr{J}^{-1}(\mathscr{J}+\mathscr{U})\mathscr{J}^{-1}H'\big]^{-1}H\mathscr{J}^{-1} + o_s(1).$$

By Theorem 13, $\sqrt{n}\,(\partial/\partial\lambda)s_n(\tilde{\lambda}_n)$ is bounded in probability, whence we have

$$\begin{aligned}
R &= n\left(\frac{\partial}{\partial\lambda}s_n(\tilde{\lambda}_n)\right)'\mathscr{J}^{-1}H'\big[H\mathscr{J}^{-1}(\mathscr{J}+\mathscr{U})\mathscr{J}^{-1}H'\big]^{-1}H\mathscr{J}^{-1} \\
&\quad \times\left(\frac{\partial}{\partial\lambda}s_n(\tilde{\lambda}_n)\right) + o_p(1) \\
&= n\left(\frac{\partial}{\partial\lambda}s_n(\lambda_n^*)\right)'\mathscr{J}^{-1}H'\big[H\mathscr{J}^{-1}(\mathscr{J}+\mathscr{U})\mathscr{J}^{-1}H'\big]^{-1}H\mathscr{J}^{-1} \\
&\quad \times\left(\frac{\partial}{\partial\lambda}s_n(\lambda_n^*)\right) + o_p(1).
\end{aligned}$$

The distributional result follows by Theorem 13. The matrix $\mathscr{J}^{-1}H'[H\mathscr{J}^{-1}\mathscr{J}\mathscr{J}^{-1}H']^{-1}H\mathscr{J}^{-1}\mathscr{J}$ is idempotent, so Y follows the noncentral chi-square distribution of $\mathscr{U} = 0$.   □

The remarks following Theorem 11 apply here as well. In a correctly specified situation one rejects $H: h(\lambda_n^0) = 0$ when R exceeds the upper $\alpha \times 100\%$ critical point of a chi-square random variable with q degrees of freedom. Under correct specification and $A: h(\lambda_n^0) \neq 0$ one approximates the distribution of R with the noncentral chi-square distribution. Under misspecification one must approximate with a quadratic form in normally distributed random variables.

In many applications $\tilde{\mathscr{J}}^{-1} = a\tilde{V}$ for some scalar multiple a. In this event the statistic R can be put in a simpler form as follows. Since $\mathrm{rank}(\tilde{H}) = q$ and $\tilde{H}$ is q by p, one can always find a matrix $\tilde{G}$ of order p by r with $\mathrm{rank}(\tilde{G}) = r = p - q$ and $\tilde{H}\tilde{G} = 0$. For such $\tilde{G}$ we shall show in the next section that

$$\tilde{H}'(\tilde{H}\tilde{V}\tilde{H}')^{-1}\tilde{H} = \tilde{V}^{-1} - \tilde{V}^{-1}\tilde{G}(\tilde{G}'\tilde{V}^{-1}\tilde{G})^{-1}\tilde{G}'\tilde{V}^{-1}.$$

Recalling that there are Lagrange multipliers $\tilde{\theta}_n$ such that

$$\frac{\partial}{\partial\lambda}s_n(\tilde{\lambda}_n) = \tilde{H}'\tilde{\theta}_n$$

we have

$$\tilde{G}'\tilde{V}^{-1}\tilde{\mathscr{J}}^{-1}\frac{\partial}{\partial\lambda}s_n(\tilde{\lambda}_n) = a\tilde{G}'\tilde{H}'\tilde{\theta}_n = 0.$$

Consequently we may substitute $\tilde{V}^{-1}$ for $\tilde{H}'(\tilde{H}\tilde{V}\tilde{H}')^{-1}\tilde{H}$ in the formula for R to obtain the simpler form

$$R = a^2 n\left(\frac{\partial}{\partial\lambda}s_n(\tilde{\lambda}_n)\right)'\tilde{V}\left(\frac{\partial}{\partial\lambda}s_n(\tilde{\lambda}_n)\right).$$

We illustrate with Example 3:

EXAMPLE 3 (Continued).   Substituting

$$\tilde{\Omega} = n\tilde{s}^2(\tilde{F}'\tilde{F})^{-1}$$

with

$$\tilde{s}^2 = (n - p + q)^{-1}\|y - f(\tilde{\lambda}_n)\|^2$$

for $\tilde{V}$, and substituting

$$\tilde{J} = \frac{2}{n}(\tilde{F}'\tilde{F})$$

for $\tilde{\mathscr{J}}$, we have

$$\tilde{J}^{-1} = (2\tilde{s}^2)^{-1}\tilde{\Omega}$$

whence

$$R = \frac{1}{\tilde{s}^2}[y - f(\tilde{\lambda}_n)]'\tilde{F}(\tilde{F}'\tilde{F})^{-1}\tilde{F}'[y - f(\tilde{\lambda}_n)].$$

Putting

$$F_n^0 = F(\lambda_n^0) \qquad F_n^* = F(\lambda_n^*)$$

R is distributed as

$$R \sim Y + o_p(1)$$

where

$$Y = Z'J^{-1}H'(H\Omega H')^{-1}HJ^{-1}Z$$

$$J^{-1} = \frac{n}{2}(F_n^{0\prime}F_n^0)^{-1}$$

$$\Omega = n\left(\sigma^2 + \frac{1}{n}\|g(\gamma_n^0) - f(\lambda_n^0)\|^2\right)(F_n^{0\prime}F_n^0)^{-1}$$

$$H = [0 : I_q] \qquad (I_q \text{ is the identity matrix of order } q)$$

$$Z \sim N_p\left(\frac{-2}{\sqrt{n}}F_n^{*\prime}[g(\gamma_n^0) - f(\lambda_n^0)], \mathscr{I}_n^0\right)$$

$$\mathscr{I}_n^0 = \frac{4\sigma^2}{n}F_n^{0\prime}F_n^0.$$

When the model is correctly specified, these equations reduce to

$$Y \sim \frac{n}{4\sigma^2}Z'(F_n^{0\prime}F_n^0)^{-1}H'\left[H(F_n^{0\prime}F_n^0)^{-1}H'\right]^{-1}H(F_n^{0\prime}F_n^0)^{-1}Z$$

$$Z \sim N_p\left(\frac{-2}{\sqrt{n}}F_n^{*\prime}[f(\lambda_n^0) - f(\lambda_n^*)], \frac{4\sigma^2}{n}(F_n^{0\prime}F_n^0)\right)$$

Y is distributed as a noncentral chi-square random variable with q degrees of freedom and noncentrality parameter

$$\alpha = (2\sigma^2)^{-1}[f(\lambda_n^0) - f(\lambda_n^*)]'F_n^*(F_n^{0\prime}F_n^0)^{-1}$$
$$\times H'\left[H(F_n^{0\prime}F_n^0)^{-1}H'\right]^{-1}H(F_n^{0\prime}F_n^0)^{-1}F_n^{*\prime}[f(\lambda_n^0) - f(\lambda_n^*)].$$

The noncentrality parameter may be put in the form (Problem 9)

$$\alpha = \frac{[f(\lambda_n^0) - f(\lambda_n^*)]'F_n^*(F_n^{0\prime}F_n^0)^{-1}F_n^{*\prime}[f(\lambda_n^0) - f(\lambda_n^*)]}{2\sigma^2}. \qquad \square$$

THEOREM 15. Let Assumptions 1 through 3 hold, and let either Assumptions 4 through 7 or 8 through 12 hold. Let

$$L = 2n[s_n(\tilde{\lambda}_n) - s_n(\hat{\lambda}_n)].$$

Under Assumption 13,

$$L \sim Y + o_p(1)$$

where

$$Y = Z' \mathscr{I}^{-1} H' \left(H \mathscr{I}^{-1} H' \right)^{-1} H \mathscr{I}^{-1} Z$$

and

$$Z \sim N_p \left(\sqrt{n} \, \frac{\partial}{\partial \lambda} s_n^0(\lambda_n^*), \, \mathscr{I} \right).$$

Recall: $V = V_n^*$, $\mathscr{I} = \mathscr{I}_n^*$, $\mathscr{J} = \mathscr{J}_n^*$, $\mathscr{U} = \mathscr{U}_n^*$, and $H = H_n^*$. Alternatively: $V = V_n^0$, $\mathscr{I} = \mathscr{I}_n^0$, $\mathscr{J} = \mathscr{J}_n^0$, and $\mathscr{U} = \mathscr{U}_n^0$. If $HVH' = H \mathscr{I}^{-1} H'$, then Y has the noncentral chi-square distribution with q degrees of freedom and noncentrality parameter

$$\alpha = n(\partial/\partial\lambda) s_n^0(\lambda_n^*) \mathscr{I}^{-1} H' \left(H \mathscr{I}^{-1} H' \right)^{-1} H \mathscr{I}^{-1} (\partial/\partial\lambda) s_n^0(\lambda_n^*)/2.$$

Under the null hypothesis, $\alpha = 0$.

Proof. By Lemma 2 we may assume without loss of generality that $\hat{\lambda}_n$, $\tilde{\lambda}_n \in \Lambda$. By Taylor's theorem

$$2n\left[s_n(\tilde{\lambda}_n) - s_n(\hat{\lambda}_n) \right] = 2n \left(\frac{\partial}{\partial\lambda} s_n(\hat{\lambda}_n) \right)' (\tilde{\lambda}_n - \hat{\lambda}_n)$$

$$+ n(\tilde{\lambda}_n - \hat{\lambda}_n)' \left(\frac{\partial^2}{\partial\lambda\,\partial\lambda'} s_n(\bar{\lambda}_n) \right) (\tilde{\lambda}_n - \hat{\lambda}_n)$$

where $\|\bar{\lambda}_n - \hat{\lambda}_n\| \le \|\tilde{\lambda}_n - \hat{\lambda}_n\|$. By the Summary, $(\tilde{\lambda}_n, \hat{\lambda}_n)$ converges almost surely to (λ^*, λ^*), and $(\partial^2/\partial\lambda\,\partial\lambda') s_n(\lambda)$ converges almost surely uniformly to $(\partial^2/\partial\lambda\,\partial\lambda') s^*(\lambda)$ uniformly on Λ which implies $(\partial^2/\partial\lambda\,\partial\lambda') s_n(\bar{\lambda}_n) = \mathscr{I} + o_s(1)$. By Lemma 2, $2n[(\partial/\partial\lambda) s_n(\hat{\lambda}_n)]'(\tilde{\lambda}_n - \hat{\lambda}_n) = o_s(1)$, whence

$$2n\left[s_n(\tilde{\lambda}_n) - s_n(\hat{\lambda}_n) \right] = n(\tilde{\lambda}_n - \hat{\lambda}_n)' [\mathscr{I} + o_s(1)](\tilde{\lambda}_n - \hat{\lambda}_n) + o_s(1).$$

Again by Taylor's theorem

$$[\mathscr{I} + o_s(1)] \sqrt{n} (\tilde{\lambda}_n - \hat{\lambda}_n) = \sqrt{n} \, \frac{\partial}{\partial\lambda} s_n(\tilde{\lambda}_n).$$

Then by Slutsky's theorem (Serfling, 1980, Section 1.5.4, or Rao, 1973, Section 2c.4), $\sqrt{n}(\tilde{\lambda}_n - \hat{\lambda}_n)$ converges in distribution and is therefore bounded. Thus

$$2n\left[s_n(\tilde{\lambda}_n) - s_n(\hat{\lambda}_n)\right] = n(\tilde{\lambda}_n - \hat{\lambda}_n)'\mathscr{I}(\tilde{\lambda}_n - \hat{\lambda}_n) + o_p(1)$$

$$\sqrt{n}\,(\tilde{\lambda}_n - \hat{\lambda}_n) = \mathscr{I}^{-1}\sqrt{n}\,\frac{\partial}{\partial\lambda}s_n(\tilde{\lambda}_n) + o_p(1)$$

whence

$$2n\left[s_n(\tilde{\lambda}_n) - s_n(\hat{\lambda}_n)\right] = n\left(\frac{\partial}{\partial\lambda}s_n(\tilde{\lambda}_n)\right)'\mathscr{I}^{-1}\left(\frac{\partial}{\partial\lambda}s_n(\tilde{\lambda}_n)\right) + o_p(1)$$

and the distributional result follows at once from Theorem 13. To see that Y is distributed as the noncentral chi-square when $HVH' = H\mathscr{I}^{-1}H'$ note that $\mathscr{I}^{-1}H'(H\mathscr{I}^{-1}H')^{-1}H\mathscr{I}^{-1}\mathscr{I}$ is idempotent under this condition. □

The remarks immediately following Theorems 11 and 14 apply here as well. One rejects when L exceeds the upper $\alpha \times 100\%$ critical point of a chi-square with q degrees of freedom, and so on.

In the event that $\mathscr{I} = a\mathscr{I} + o(1)$ for some scalar multiple a, the "likelihood ratio test statistic" can be modified as follows. Let $\hat{a}_n$ be a random variable that converges either almost surely or in probability to a. Then

$$\hat{a}_n L \sim aY + o_p(1)$$

where

$$aY = Z'\mathscr{I}^{-1}H'(H\mathscr{I}^{-1}H')^{-1}H\mathscr{I}^{-1}Z.$$

Since $\mathscr{I}^{-1}H'(H\mathscr{I}^{-1}H')H\mathscr{I}^{-1}\mathscr{I}$ is an idempotent matrix, aY is distributed as the noncentral chi-square distribution with q degrees of freedom and noncentrality parameter

$$\alpha = \frac{n(\partial/\partial\lambda)s_n^0(\lambda_n^*)\mathscr{I}^{-1}H'(H\mathscr{I}^{-1}H')^{-1}H\mathscr{I}^{-1}(\partial/\partial\lambda)s_n^0(\lambda_n^*)}{2}.$$

We illustrate with the example:

EXAMPLE 3 (Continued).   Assuming that the model is correctly specified,

$$\mathscr{I} = \frac{4\sigma^2}{n}F_n^{0\prime}F_n^0$$

$$\mathscr{I} = \frac{2}{n}F_n^{0\prime}F_n^0.$$

Thus we have

$$\mathscr{I} = (2\sigma^2)^{-1}\mathscr{J}.$$

An estimator of σ^2 is

$$s^2 = (n - p)^{-1}\|y - f(\hat{\lambda}_n)\|^2.$$

The modified "likelihood ratio test statistic" is

$$(2s^2)^{-1}L = (2s^2)^{-1}(2n)\left(\frac{1}{n}\|y - f(\tilde{\lambda}_n)\|^2 - \frac{1}{n}\|y - f(\hat{\lambda}_n)\|^2\right)$$

$$= \frac{\|y - f(\tilde{\lambda}_n)\|^2 - \|y - f(\hat{\lambda}_n)\|^2}{s^2}.$$

A further division by q would convert $(2s^2)^{-1}L$ to the F-statistic of Chapter 1. Assuming correct specification, $(2s^2)^{-1}L$ is distributed to within $o_p(1)$ as the noncentral chi-square distribution with q degrees of freedom and noncentrality parameter (Problem 9)

$$\alpha = \frac{\left[f(\lambda_n^0) - f(\lambda_n^*)\right]' F_n^*\left(F_n^{0\prime}F_n^0\right)^{-1}F_n^{*\prime}\left[f(\lambda_n^0) - f(\lambda_n^*)\right]}{2\sigma^2}.$$

Under specification error

$$(2s^2)^{-1}L \sim aY + o_p(1)$$

where

$$aY = \frac{Z'\mathscr{J}^{-1}H'\left(H\mathscr{J}^{-1}H'\right)^{-1}H\mathscr{J}^{-1}Z}{2\sigma^2 + (2/n)\|g(\gamma_n^0) - f(\lambda_n^0)\|^2},$$

$$\mathscr{J} = \frac{2}{n}F_n^{0\prime}F_n^0 - \frac{2}{n}\sum_{t=1}^{n}\left[g(x_t, \gamma_n^0) - f(x_t, \lambda_n^0)\right]^2 \frac{\partial^2}{\partial\lambda\,\partial\lambda'}f(x_t, \lambda_n^0)$$

$$Z \sim N_p\left(\frac{-2}{\sqrt{n}}F_n^{*\prime}\left[g(\gamma_n^0) - f(\lambda_n^*)\right], \frac{4\sigma^2}{n}F_n^{0\prime}F_n^0\right). \qquad \square$$

PROBLEMS

1. (Cholesky factorization.) The validity of the argument in the proof of Theorems 11 and 12 depends on the fact that it is possible to factor a

symmetric, positive definite matrix A as $A = R'R$ in such a way that R is a continuous function of the elements of the matrix A. To see that this is so, observe that

$$A = \begin{bmatrix} a_{11} & a'_{(1)} \\ \hline a_{(1)} & A_{22} \\ & \\ & \end{bmatrix}$$

$$= \begin{bmatrix} r_{11} & 0 \\ \hline r_{12} & \\ \vdots & I \\ r'_{1p} & \end{bmatrix} \begin{bmatrix} 1 & 0 \\ \hline 0 & D_1 \\ & \end{bmatrix} \begin{bmatrix} r_{11} & r_{12} & \cdots & r_{1p} \\ \hline 0 & & I & \end{bmatrix}$$

where

$$r_{11} = \sqrt{a_{11}}$$

$$r_{1k} = \frac{a_{1k}}{\sqrt{a_{11}}} \qquad k = 2, \ldots, p$$

$$D_1 = A_{22} - \frac{1}{a_{11}} a_{(1)} a'_{(1)}.$$

The r_{1k} are the continuous elements of A and D_1 is a symmetric, positive definite matrix whose elements are continuous functions of the elements of A, why? This same argument can be applied to D_1 to obtain

$$A = \begin{bmatrix} r_{11} & 0 & & \\ & & & 0 \\ r_{12} & r_{22} & & \\ \hline r_{13} & r_{23} & & \\ \vdots & \vdots & & I \\ r_{1p} & r_{2p} & & \end{bmatrix} \begin{bmatrix} 1 & 0 & & \\ & & & 0 \\ 0 & 1 & & \\ \hline & & & \\ 0 & & D_2 & \end{bmatrix} \begin{bmatrix} r_{11} & r_{12} & r_{13} & \cdots & r_{1p} \\ 0 & r_{22} & r_{23} & \cdots & r_{2p} \\ \hline 0 & & & I & \end{bmatrix}$$

with continuity preserved. Supply the missing steps. This argument can be repeated a finite number of times to obtain the result. The recursion formula for the Cholesky square root method is

$$r_{1k} = \frac{a_{1k}}{r_{11}} \qquad\qquad k = 1, 2, 3, \ldots, p$$

$$r_{jk} = \frac{1}{r_{jj}}\left(a_{jk} - \sum_{i=1}^{j-1} r_{ij} r_{ik} \right) \qquad j = 2, 3, \ldots, p \quad k = j, j+1, \ldots, p.$$

Observe that on a computer A can be factored in place, using only the upper triangle of A with this recursion.

2.   Suppose that θ_n^0 converges to θ^*, and $\hat\theta_n$ converges almost surely to θ^*. Let $g(\theta)$ be defined on an open set Θ, and let $g(\theta)$ be continuous at $\theta^* \in \Theta$. Define $g(\theta)$ arbitrarily off Θ. Show that

$$g(\hat\theta_n) = g(\theta_n^0) + o_s(1).$$

Let θ be a square matrix and $g(\theta)$ a matrix valued function giving the inverse of θ when it exists. If θ^* is nonsingular, show that there is an open neighborhood Θ about θ^* where each $\theta \in \Theta$ is nonsingular and show that $g(\theta)$ is continuous at θ^*. Hint: Use the facts that $\|\theta\| = [\sum_{ij}\theta_{ij}^2]^{1/2}$, $\|g\| = [\sum_{ij}g_{ij}^2]^{1/2}$, the determinant of a matrix is continuous, and an inverse is the product of the adjoint matrix and the reciprocal of the determinant. Show that if $\sqrt n (\partial/\partial\lambda)s_n(\lambda_n^*)$ converges in distribution, then $\sqrt n (\partial/\partial\lambda)s_n(\lambda_n^*)$ is bounded in probability. Show that

$$[H + o_s(1)][\mathscr{I} + o_s(1)]^{-1}\sqrt n \,\frac{\partial}{\partial\lambda}s_n(\lambda_n^*)$$

$$= H\mathscr{I}^{-1}\sqrt n \,\frac{\partial}{\partial\lambda}s_n(\lambda_n^*) + o_p(1).$$

3.   Expand $\sqrt n\, h(\lambda_n^0)$ in a Taylor's series and show that $\lim_{n\to\infty}\sqrt n\, h(\lambda_n^0) = H\Delta$.

4.   Verify that if the linear model

$$y_t = x_t'\beta + u_t$$

is estimated by least squares from data that actually follow

$$y_t = g(x_t, \gamma_n^0) + e_t$$

with e_t independently and normally distributed, and if one tests the linear hypothesis

$$H : R\beta = r \quad \text{against} \quad A : R\beta \neq r$$

then

$$\sqrt{n}\, \frac{\partial}{\partial \lambda} s_n(\lambda_n^*) \sim N_p\!\left(\sqrt{n}\, \frac{\partial}{\partial \lambda} s_n^0(\lambda_n^*),\ \mathscr{I}_n^*\right).$$

5. Verify that $\alpha = 0$ when the null hypothesis is true in Theorem 14.
6. (Invariance.) Consider a least mean distance estimator

$$\hat{\lambda} \quad \text{minimizes} \quad s_n(\lambda) = \frac{1}{n}\sum_{t=1}^{n} s(y_t, x_t, \hat{\tau}_n, \lambda)$$

and the hypothesis

$$H : \lambda_n^0 = \lambda^* \quad \text{against} \quad A : \lambda_n^0 \neq \lambda^*.$$

Let $g(\rho)$ be a twice differentiable function with twice differentiable inverse $\rho = \varphi(\lambda)$. Then an equivalent formulation of the problem is

$$\hat{\rho} \quad \text{minimizes} \quad s_n[g(\rho)] = \frac{1}{n}\sum_{t=1}^{n} s[y_t, x_t, \hat{\tau}_n, g(\rho)]$$

with the hypothesis

$$H : \rho_n^0 = \varphi(\lambda^*) \quad \text{against} \quad A : \rho_n^0 \neq \varphi(\lambda^*).$$

Show that the computed value of the Wald test statistic can be different for these two equivalent problems. Show that the computed value of the Lagrange multiplier and "likelihood ratio" test statistics are invariant to this reparametrization.

7. (Equivalent local power.) Suppose that $\mathscr{I}^* = \mathscr{J}^*$ and that $\mathscr{U}^* = 0$, so that each of the three test statistics— W, R, L—is distributed as a noncentral chi-square with noncentrality parameter α_n. Show that $\lim_{n\to\infty}\alpha_n = \Delta' H'(HVH')^{-1}H\Delta/2$ in each case, with $H = (\partial/\partial\lambda')h(\lambda^*)$ and $V = (\mathscr{I}^*)^{-1}\mathscr{I}^*(\mathscr{I}^*)^{-1}$.

8. Fix a realization of the errors. For large enough n, $\hat{\lambda}_n$ and $\tilde{\lambda}_n$ must be in an open neighborhood of λ^* on which $(\partial^2/\partial\lambda\,\partial\lambda')s_n(\lambda)$ is invertible. (Why?) Use Taylor's theorem to show that for large enough n, L is exactly given as a quadratic form in $(\partial/\partial\lambda)s_n(\tilde{\lambda}_n)$.

9. Using the identity derived in Section 6, verify the alternative form for α given in the examples following Theorems 14 and 15.

10. Verify the claim in Assumption 13 that $h(\lambda_n^0) = 0$ for all n implies that there is an N with $\lambda_n^0 = \lambda_n^*$ for all $n > N$.

6. ALTERNATIVE REPRESENTATION OF A HYPOTHESIS

The results of the previous section presume that the hypothesis is stated as a parametric restriction

$$H : h(\lambda_n^0) = 0 \quad \text{against} \quad A : h(\lambda_n^0) \neq 0.$$

As we have seen, at times it is much more natural to express a hypothesis as a functional dependence

$$H : \lambda_n^0 = g(\rho_n^0) \quad \text{for some } \rho_n^0 \text{ in } \mathscr{R}$$

$$\text{against} \quad A : \lambda_n^0 \neq g(\rho) \quad \text{for any } \rho \text{ in } \mathscr{R}.$$

We assume that these hypotheses are equivalent in the sense that

$$\{\lambda : h(\lambda) = 0, \lambda \text{ in } \Lambda^*\} = \{\lambda : \lambda = g(\rho), \rho \text{ in } \mathscr{R}\}$$

where Λ^* is the set over which $s_n(\lambda)$ is to be minimized when computing $\hat{\lambda}_n$, and append the following regularity conditions: $h : \mathbb{R}^p \to \mathbb{R}^q$, $g : \mathbb{R}^r \to \mathbb{R}^p$, $p = r + q$, $H(\lambda) = (\partial/\partial\lambda')h(\lambda)$ is continuous and has rank q on Λ^*, $G(\rho) = (\partial/\partial\rho')g(\rho)$ is continuous and has rank r on $\mathscr{R}$, and $H[g(\rho)]G(\rho) = 0$.

Given a once continuously differentiable function $h(\lambda)$ mapping $\Lambda^* \subset \mathbb{R}^p$ into $\mathbb{R}^q$, a construction of the function $g(\rho)$ and domain $\mathscr{R}$ can be obtained as follows. Suppose that there is a once differentiable function $\varphi(\lambda)$ defined on Λ^* such that the transformation

$$\tau = h(\lambda)$$
$$\rho = \varphi(\lambda)$$

has a once differentiable inverse

$$\lambda = \Psi(\rho, \tau).$$

Put

$$g(\rho) = \Psi(\rho, 0)$$

and

$$\mathscr{R} = \{\rho : \rho = \varphi(\lambda), \, h(\lambda) = 0, \, \lambda \text{ in } \Lambda^*\}.$$

We have immediately that

$$\{\lambda : h(\lambda) = 0, \, \lambda \text{ in } \Lambda^*\} = \{\lambda : \lambda = g(\rho), \, \rho \text{ in } \mathscr{R}\}$$

and by differentiating the equations

$$0 = h[g(\rho)]$$
$$\rho = \varphi[g(\rho)]$$

we have

$$_q 0_r = H[g(\rho)] G(\rho)$$

$$_r I_r = \frac{\partial}{\partial \lambda'} \varphi[g(\rho)] G(\rho)$$

which implies that the rank of $G(\rho)$ is r.

Let us now consider how $g(\rho)$ can be used instead of $h(\lambda)$ in implementing Theorems 14 and 15. $\tilde{\lambda}_n$ can be computed by minimizing the composite function $s_n[g(\rho)]$ over $\mathscr{R}$ to obtain $\hat{\rho}_n$ and putting $\tilde{\lambda}_n = g(\hat{\rho}_n)$. Similarly, λ_n^* can be computed by minimizing $s_n^0[g(\rho)]$ over $\mathscr{R}$ to obtain ρ_n^0 and putting $\lambda_n^* = g(\rho_n^0)$. The statistics R and L, the vector $(\partial/\partial\lambda)s_n^0(\lambda_n^*)$, and the matrices $\mathscr{I}$, $\mathscr{J}$, $\mathscr{U}$, and V can now be computed directly. What remains is to compute matrices of the form $H'(HAH')^{-1}H$ where A is a computable, positive definite, symmetric matrix and $H = (\partial/\partial\lambda')h(\lambda_n^*)$. Let

$$G = \frac{\partial}{\partial\rho'} g(\rho_n^0).$$

We shall show that

$$H'(HAH')^{-1}H = A^{-1} - A^{-1}G(G'A^{-1}G)^{-1}G'A^{-1}$$

for any positive definite symmetric A. Factor A as $A = PP'$ (Problem 1 of Section 5). Trivially $HPP^{-1}G = 0$, which implies that there is a nonsingular matrix B of order q and there is nonsingular matrix C of order r such that $\mathcal{O}_1 = P'H'B$ has orthonormal columns, $\mathcal{O}_2 = P^{-1}GC$ has orthonormal columns, and the matrix $\mathcal{O} = [\mathcal{O}_1 \vdots \mathcal{O}_2]$ is orthogonal. Then

$$
\begin{aligned}
I &= \left[\mathcal{O}_1 \vdots \mathcal{O}_2 \right] \begin{bmatrix} \mathcal{O}_1' \\ \mathcal{O}_2' \end{bmatrix} \\
&= \mathcal{O}_1 \mathcal{O}_1' + \mathcal{O}_2 \mathcal{O}_2' \\
&= \mathcal{O}_1 (\mathcal{O}_1' \mathcal{O}_1)^{-1} \mathcal{O}_1' + \mathcal{O}_2 (\mathcal{O}_2' \mathcal{O}_2)^{-1} \mathcal{O}_2' \\
&= P'H'B \left(B'HPP'H'B \right)^{-1} B'HP \\
&\quad + P^{-1}GC \left[C'G'(P^{-1})'P^{-1}GC \right]^{-1} C'G'(P^{-1})' \\
&= P'H'(HAH')^{-1}HP + P^{-1}G(G'A^{-1}G)^{-1}G'(P^{-1})'
\end{aligned}
$$

whence

$$
\begin{aligned}
A^{-1} &= (P^{-1})'I(P^{-1}) \\
&= H'(HAH')^{-1}H + A^{-1}G(G'A^{-1}G)^{-1}G'A^{-1}.
\end{aligned}
$$

To illustrate, suppose that $\mathcal{I} = \mathcal{J}$ in Theorem 15. Then the noncentrality parameter is

$$
\begin{aligned}
\alpha &= n \frac{\partial}{\partial \lambda'} s_n^0(\lambda_n^*) \mathcal{J}^{-1} H' \left(H \mathcal{J}^{-1} H' \right)^{-1} H \mathcal{J}^{-1} \frac{\partial}{\partial \lambda} s_n^0(\lambda_n^*) \\
&= n \frac{\partial}{\partial \lambda'} s_n^0(\lambda_n^*) \left[\mathcal{J}^{-1} - G(G'\mathcal{J}G)^{-1}G' \right] \frac{\partial}{\partial \lambda} s_n^0(\lambda_n^*) \\
&= n \frac{\partial}{\partial \lambda'} s_n^0(\lambda_n^*) \mathcal{J}^{-1} \frac{\partial}{\partial \lambda} s_n^0(\lambda_n^*)
\end{aligned}
$$

since $(\partial / \partial \lambda) s_n^0(\lambda_n^*) = H'\theta$, where θ is the Lagrange multiplier.

7. CONSTRAINED ESTIMATION

Throughout this section we shall assume that the constraint has two equivalent representations:

parametric restriction: $h(\lambda) = 0$, λ in Λ^*,
functional dependence: $\lambda = g(\rho)$, ρ in $\mathcal{R}$,

where $h: \mathbb{R}^p \to \mathbb{R}^q$, $g: \mathbb{R}^r \to \mathbb{R}^p$, and $r + q = p$. They are equivalent in the sense that the null space of $h(\lambda)$ is the range space of $g(\rho)$:

$$\Lambda_H = \{\lambda : h(\lambda) = 0, \lambda \text{ in } \Lambda^*\} = \{\lambda : \lambda = g(\rho), \rho \text{ in } \mathscr{R}\}.$$

We also assume that both $g(\rho)$ and $h(\lambda)$ are twice continuously differentiable. From

$$h[g(\rho)] = 0$$

we have

$$\frac{\partial}{\partial \lambda'} h[g(\rho)] \frac{\partial}{\partial \rho'} g(\rho) = HG = 0.$$

If $\text{rank}[H' \vdots G] = p$, we have from Section 6 that for any symmetric, positive definite matrix $\mathscr{L}$

$$G(G'\mathscr{L}G)^{-1}G' = \mathscr{L}^{-1} - \mathscr{L}^{-1}H'(H\mathscr{L}^{-1}H')^{-1}H\mathscr{L}^{-1}.$$

Section 6 gives a construction which lends plausibility to these assumptions.

Let the data generating model satisfy Assumptions 1 through 3. Let the objective function $s_n[g(\rho)]$ satisfy either Assumptions 4 through 6 or Assumptions 8 through 11. Let

$$\hat{\rho}_n \text{ minimize } s_n[g(\rho)],$$
$$\rho_n^0 \text{ minimize } s_n^0[g(\rho)],$$
$$\rho^* \text{ minimize } s^*[g(\rho)].$$

Then from either Theorem 3 or Theorem 7 we have that

$$\lim_{n \to \infty} \hat{\rho}_n = \rho^* \qquad \text{almost surely}$$

$$\lim_{n \to \infty} \rho_n^0 = \rho^* \qquad \text{almost surely}$$

and from either Theorem 5 or Theorem 9 that

$$\sqrt{n}\left(\hat{\rho}_n - \rho_n^0\right) \xrightarrow{\mathscr{L}} N_r\left[0, (\mathscr{J}_\rho^*)^{-1}\mathscr{J}_\rho^*(\mathscr{J}_\rho^*)^{-1}\right].$$

The matrices $\mathscr{J}_\rho^*$ and $\mathscr{J}_\rho^*$ are of order r by r and can be computed by direct application of Notation 2 or Notation 6. In these computations one is working in an r-dimensional space, not in a p-dimensional space. We emphasize this point with the subscript ρ: $\mathscr{J}_\rho^*$, $\mathscr{J}_\rho^*$, and $\mathscr{U}_\rho^*$. To illustrate, computing according to Notation 2, one has

$$\mathcal{U}_\rho^* = \int_{\mathscr{X}} \left(\int_{\mathscr{E}} \frac{\partial}{\partial \rho} s[Y(e, x, \gamma^*), x, \tau^*, g(\rho^*)] \, dP(e) \right)$$

$$\times \left(\int_{\mathscr{E}} \frac{\partial}{\partial \rho} s[Y(e, x, \gamma^*), x, \tau^*, g(\rho^*)] \, dP(e) \right)' d\mu(x)$$

$$\mathcal{I}_\rho^* = \int_{\mathscr{X}} \int_{\mathscr{E}} \left(\frac{\partial}{\partial \rho} s[Y(e, x, \gamma^*), x, \tau^*, g(\rho^*)] \right)$$

$$\times \left(\frac{\partial}{\partial \rho} s[Y(e, x, \gamma^*), x, \tau^*, g(\rho^*)] \right)' dP(e) \, d\mu(x) - \mathcal{U}_\rho^*$$

$$\mathcal{I}_\rho^* = \frac{\partial^2}{\partial \rho \, \partial \rho'} s^*[g(\rho^*)].$$

Estimators of $\mathcal{I}_\rho^*$ and $\mathcal{I}_\rho^*$ are computed according to Notation 4 or Notation 9. To illustrate, computing according to Notation 4, one has

$$\hat{\mathcal{I}}_\rho = \frac{1}{n} \sum_{t=1}^{n} \left(\frac{\partial}{\partial \rho} s[y_t, x_t, \hat{\tau}_n, g(\hat{\rho}_n)] \right) \left(\frac{\partial}{\partial \rho} s[y_t, x_t, \hat{\tau}_n, g(\hat{\rho}_n)] \right)'$$

$$\hat{\mathcal{I}}_\rho = \frac{1}{n} \sum_{t=1}^{n} \frac{\partial^2}{\partial \rho \, \partial \rho'} s[y_t, x_t, \hat{\tau}_n, g(\hat{\rho}_n)].$$

As to testing hypotheses, the theory of Section 5 applies directly. The computations according to Notation 3 or Notation 8 are similar to those illustrated above.

Often results reported in terms of

$$\tilde{\lambda}_n = g(\hat{\rho}_n)$$

are more meaningful than results reported in terms of $\hat{\rho}_n$. As an instance, one wants to show the effect of a restriction by presenting $\hat{\lambda}_n$ and its (estimated) standard errors together with $\tilde{\lambda}_n$ and its (estimated) standard errors in a tabular display. To do this, let

$$\lambda_n^* = g(\rho_n^0)$$

$$\lambda^\# = g(\rho^*).$$

By continuity of $g(\rho)$

$$\lim_{n \to \infty} \tilde{\lambda}_n = \lambda^\# \qquad \text{almost surely}$$

$$\lim_{n \to \infty} \lambda_n^* = \lambda^\#.$$

Note that $\lambda^{\#}$ is not equal to λ^* of either Section 3 or Section 4 unless the Pitman drift assumption is imposed. From the Taylor series expansion

$$g(\hat{\rho}_n) - g(\rho_n^0) = [G^* + o_s(1)]\sqrt{n}(\hat{\rho}_n - \rho_n^0)$$

where $G^* = (\partial/\partial\rho')g(\rho^*)$, we have that

$$\sqrt{n}(\tilde{\lambda}_n - \lambda_n^*) \xrightarrow{\mathcal{L}} N_r[0, G^*(\mathcal{I}_\rho^*)^{-1}\mathcal{I}_\rho^*(\mathcal{I}_\rho^*)^{-1}G^{*\prime}].$$

The variance-covariance matrix is estimated by

$$\hat{G}(\hat{\mathcal{I}}_\rho)^{-1}\hat{\mathcal{I}}_\rho(\hat{\mathcal{I}}_\rho)^{-1}\hat{G}'$$

where $\hat{G} = (\partial/\partial\rho')g(\hat{\rho}_n)$.

To use these results given the parametric restriction $h(\lambda) = 0$, one would actually have to construct the equivalent functional dependence $\lambda = g(\rho)$. This construction can be avoided as follows.

Let $\hat{\theta}_n$ be the Lagrange multiplier for the minimization of $s_n(\lambda)$ subject to $h(\lambda) = 0$, and let

$$\mathscr{L} = \frac{\partial^2}{\partial\lambda\,\partial\lambda'}[s_n(\tilde{\lambda}_n) + \hat{\theta}_n'h(\tilde{\lambda}_n)].$$

One can show that (Problem 1)

$$\hat{\mathcal{I}}_\rho = \hat{G}'\mathscr{L}\hat{G}$$

and using the chain rule with either Notation 4 or Notation 9, one finds that

$$\hat{\mathcal{I}}_\rho = \hat{G}'\tilde{\mathcal{I}}\hat{G}$$

where $\tilde{\mathcal{I}} = \mathcal{I}_n(\tilde{\lambda}_n)$. Thus

$$\hat{G}(\hat{\mathcal{I}}_\rho)^{-1}\hat{\mathcal{I}}_\rho(\hat{\mathcal{I}}_\rho)^{-1}\hat{G}' = \hat{G}(\hat{G}'\mathscr{L}\hat{G})^{-1}\hat{G}'\tilde{\mathcal{I}}\hat{G}(\hat{G}'\mathscr{L}\hat{G})^{-1}\hat{G}'.$$

Using the identity given earlier on, one has

$$\hat{G}(\hat{\mathcal{I}}_\rho)^{-1}\hat{\mathcal{I}}_\rho(\hat{\mathcal{I}}_\rho)^{-1}\hat{G}'$$
$$= \mathscr{L}^{-1}[I - \tilde{H}'(\tilde{H}\mathscr{L}^{-1}\tilde{H}')^{-1}\tilde{H}\mathscr{L}^{-1}]\tilde{\mathcal{I}}[I - \mathscr{L}^{-1}\tilde{H}'(\tilde{H}\mathscr{L}^{-1}\tilde{H}')\tilde{H}]\mathscr{L}^{-1}$$

where $\tilde{H} = (\partial/\partial\lambda')h(\tilde{\lambda}_n)$. The right hand side of this expression can be computed from knowledge of $s_n(\lambda)$ and $h(\lambda)$ alone.

Similarly, if

$$\lambda^{\#} \quad \text{minimizes} \quad s^*(\lambda) \quad \text{subject to} \quad h(\lambda) = 0$$

with Lagrange multipliers $\theta^{\#}$

$$G^*(\mathcal{J}_\rho^*)^{-1}\mathcal{J}_\rho^*(\mathcal{J}_\rho^*)^{-1}G^{*\prime}$$
$$= \mathcal{L}^{-1}\left[I - H'(H\mathcal{L}^{-1}H')^{-1}H\mathcal{L}^{-1}\right]\mathcal{J}\left[I - \mathcal{L}^{-1}H'(H\mathcal{L}^{-1}H')H\right]\mathcal{L}^{-1}$$

where

$$H = \frac{\partial}{\partial\lambda'}h(\lambda^{\#})$$

$$\mathcal{L} = \frac{\partial^2}{\partial\lambda\,\partial\lambda'}\left[s^*(\lambda^{\#}) + \theta^{\#\prime}h(\lambda^{\#})\right]$$

$$\mathcal{J} = \bar{\bar{\mathcal{F}}}(\lambda^{\#}).$$

Under a Pitman drift, $\theta^{\#} = 0$, and the expression that one might expect from the proof of Theorem 13 obtains.

PROBLEM

1. Show that the equation $h[g(\rho)] = 0$ implies

$$\sum_{l=1}^{p}\frac{\partial}{\partial\lambda_l}h_u[g(\rho)]\frac{\partial^2}{\partial\rho_i\,\partial\rho_j}g_l(\rho)$$
$$= -\sum_{l=1}^{p}\sum_{k=1}^{p}\frac{\partial^2}{\partial\lambda_k\,\partial\lambda_l}h_u[g(\rho)]\frac{\partial}{\partial\rho_j}g_k(\rho)\frac{\partial}{\partial\rho_i}g_l(\rho).$$

Suppose that $\tilde{\lambda} = g(\hat{\rho})$ minimizes $s(\lambda)$ subject to $h(\lambda) = 0$ and that $\tilde{\theta}$ is the corresponding vector of Lagrange multipliers. Show that

$$\sum_{l=1}^{p}\frac{\partial}{\partial\lambda_l}s[g(\hat{\rho})]\frac{\partial^2}{\partial\rho_i\,\partial\rho_j}g_l(\hat{\rho})$$
$$= \sum_{l=1}^{p}\sum_{k=1}^{p}\sum_{u=1}^{q}\tilde{\theta}_u\frac{\partial^2}{\partial\lambda_k\,\partial\lambda_l}h_u(\tilde{\lambda})\frac{\partial}{\partial\rho_j}g_k(\hat{\rho})\frac{\partial}{\partial\rho_i}g_l(\hat{\rho}).$$

Compute $(\partial^2/\partial\rho_i\,\partial\rho_j)s[g(\hat{\rho})]$, and substitute the expression above to obtain

$$\frac{\partial^2}{\partial\rho\,\partial\rho'}s[g(\hat{\rho})] = \left(\frac{\partial}{\partial\rho'}g(\hat{\rho})\right)'\frac{\partial^2}{\partial\lambda\,\partial\lambda'}[s(\tilde{\lambda}) + \tilde{\theta}'h(\tilde{\lambda})]\left(\frac{\partial}{\partial\rho'}g(\hat{\rho})\right).$$

8. INDEPENDENTLY AND IDENTICALLY DISTRIBUTED REGRESSORS

As noted earlier, the standard assumption in regression analysis is that the observed independent variables $\{x_t\}_{t=1}^n$ are fixed. With a model such as

$$y_t = g(x_t, \gamma_n^0) + e_t \qquad t = 1, 2, \ldots, n$$

the independent variables $\{x_t\}_{t=1}^n$ are held fixed and the sampling variation enters via sampling variation in the errors $\{e_t\}_{t=1}^n$. If the independent variables are random variables, then the analysis is conditional on that realization $\{x_t\}_{t=1}^n$ that obtains. Stated differently, the model

$$y_t = g(x_t, \gamma_n^0) + e_t \qquad t = 1, 2, \ldots, n$$

defines the conditional distribution of $\{y_t\}_{t=1}^n$ given $\{x_t\}_{t=1}^n$, and the analysis is based on the conditional distribution.

An alternative approach is to assume that the independent variables $\{x_t\}_{t=1}^n$ are random and to allow sampling variation to enter both through the errors $\{e_t\}_{t=1}^n$ and the independent variables $\{x_t\}_{t=1}^n$. We shall see that the theory developed thus far is general enough to accommodate an assumption of independently and identically distributed (iid) regressors and that the results are little changed save in one instance, namely the misspecified model. Therefore we shall focus the discussion on this case.

We have seen that under the fixed regressor setup the principal consequence of misspecification is the inability to estimate the matrix $\mathscr{I}^*$ from sample information because the obvious estimator $\hat{\mathscr{I}}$ converges almost surely to $\mathscr{I}^* + \mathscr{U}^*$ rather than to $\mathscr{I}^*$. As a result, test statistics are distributed asymptotically as general quadratic forms in normal random variables rather than as noncentral chi-square random variables. In contrast, a consequence of the assumption of iid regressors is that $\mathscr{U}^* = 0$. With iid regressors test statistics are distributed asymptotically as the noncentral chi-square. Considering least mean difference estimators, let us trace through the details as to why this is so. Throughout, $\mathscr{I} = \mathscr{I}_n^0$, $\mathscr{J} = \mathscr{J}_n^0$, and $\mathscr{U} = \mathscr{U}_n^0$.

With least mean distance estimators, the problem of nonzero $\mathscr{U}^*$ originates with the variables

$$\frac{\partial}{\partial\lambda} s\left[Y(e_t, x_t, \gamma_n^0), x_t, \tau_n^0, \lambda \right] \qquad t = 1, 2, \ldots, n$$

that appear in the proof of Theorem 4. In a correctly specified situation, sensible estimation procedures will have the property that at each x the minimum of

$$\int_{\mathcal{E}} s\left[Y\left(e, x, \gamma_n^0\right), x, \tau_n^0, \lambda\right] dP(e)$$

will occur at $\lambda = \lambda_n^0$. Under the regularity conditions, this implies that

$$0 = \frac{\partial}{\partial \lambda} \int_{\mathcal{E}} s\left[Y\left(e, x, \gamma_n^0\right), x, \tau_n^0, \lambda_n^0\right] dP(e)$$

$$= \int_{\mathcal{E}} \frac{\partial}{\partial \lambda} s\left[Y\left(e, x, \gamma_n^0\right), x, \tau_n^0, \lambda_n^0\right] dP(e).$$

Thus, the random variables

$$Z_t(e_t) = \frac{\partial}{\partial \lambda} s\left[Y\left(e_t, x_t, \gamma_n^0\right), x_t, \tau_n^0, \lambda_n^0\right]$$

have mean zero and their normalized sum

$$\frac{1}{\sqrt{n}} \sum_{t=1}^{n} Z_t(e_t)$$

has variance-covariance matrix

$$\mathscr{I} = \frac{1}{n} \sum_{t=1}^{n} \int_{\mathcal{E}} Z_t(e) Z_t'(e) \, dP(e)$$

which can be estimated by $\hat{\mathscr{I}}$. In an incorrectly specified situation the mean of $Z_t(e_t)$ is

$$\mu_t = \int_{\mathcal{E}} \frac{\partial}{\partial \lambda} s\left[Y\left(e, x_t, \gamma_n^0\right), x_t, \tau_n^0, \lambda_n^0\right] dP(e)$$

with, as a rule, $\mu_t \neq 0$, so that μ_t varies systematically with x_t. Under misspecification, the normalized sum

$$\frac{1}{\sqrt{n}} \sum_{t=1}^{n} Z_t$$

has variance-covariance matrix

$$\mathscr{I} = \frac{1}{n} \sum_{t=1}^{n} \int_{\mathcal{E}} \left[Z_t(e) - \mu_t\right]\left[Z_t(e) - \mu_t\right]' dP(e)$$

as before, but $\hat{\mathcal{J}}$ is, in essence, estimating

$$\mathcal{J} + \frac{1}{n} \sum_{t=1}^{n} \mu_t \mu_t'.$$

Short of assuming replicates at each point x_t, there seems to be no way to form an estimate of

$$\mathcal{U} = \frac{1}{n} \sum_{t=1}^{n} \mu_t \mu_t'.$$

Without being able to estimate $\mathcal{U}$, one cannot estimate $\mathcal{J}$.

The effect of an assumption of iid regressors is to convert the deterministic variation in μ_t to random variation. The μ_t become independently distributed, each having mean zero. From the point of view of the fixed regressors theory, one could argue that the independent variables have all been set to a constant value so that each observation is now a replicate. We illustrate with Example 3 and then return to the general discussion.

EXAMPLE 3 (Continued). To put the model into the form of an iid regressors model within the framework of the general theory, let the data be generated according to the model

$$y_{(1)t} = g\left(y_{(2)t}, \gamma_n^0\right) + e_{(1)t}$$
$$y_{(2)t} = \mu_{(2)} + e_{(2)t}$$

which we presume satisfies Assumptions 1 through 3 with $x_t \equiv 1$ and μ the measure putting all its mass at $x = 1$; in other words, x_t enters the model trivially. The $y_{(2)t}$ are the random regressors. Convention has it, in this type of analysis, that $y_{(2)t}$ and $e_{(1)t}$ are independent, whence $P(e)$ is a product measure

$$dP(e) = dP_{(1)}(e_{(1)}) \times dP_{(2)}(e_{(2)}).$$

The fitted model is

$$y_{(1)t} = f\left(y_{(2)t}, \lambda\right) + \mu_t \qquad t = 1, 2, \ldots, n$$

and λ is estimated by $\hat{\lambda}_n$ that minimizes

$$s_n(\lambda) = \frac{1}{n} \sum_{t=1}^{n} \left[y_{(1)t} - f\left(y_{(2)t}, \lambda\right)\right]^2.$$

Let ν be the measure defined by

$$\int_{\mathcal{Y}_{(2)}} g(y_{(2)}) \, d\nu(y_{(2)}) = \int_{\mathcal{E}_{(2)}} g(\mu_{(2)} + e_{(2)}) \, dP_{(2)}(e_{(2)})$$

where $\mathcal{Y}_{(2)}$ is the set of admissible values for the random variable $y_{(2)}$. We have

$$s(y, x, \lambda) = [y_{(1)} - f(y_{(2)}, \lambda)]^2$$

$$s_n^0(\lambda) = \sigma_{(1)}^2 + \int_{\mathcal{Y}_{(2)}} [g(y_{(2)}, \gamma_n^0) - f(y_{(2)}, \lambda)]^2 \, d\nu(y_{(2)})$$

$$(\partial/\partial\lambda)s_n^0(\lambda) = -2\int_{\mathcal{Y}_{(2)}} [g(y_{(2)}, \gamma_n^0) - f(y_{(2)}, \lambda)] \frac{\partial}{\partial\lambda} f(y_{(2)}, \lambda) \, d\nu(y_{(2)})$$

$$\lambda_n^0 \quad \text{minimizes} \quad \int_{\mathcal{Y}_{(2)}} [g(y_{(2)}, \gamma_n^0) - f(y_{(2)}, \lambda)]^2 \, d\nu(y_{(2)}).$$

The critical change from the fixed regressor case occurs in the computation of

$$\int_{\mathcal{E}} Z_t(e) \, dP(e) = \int_{\mathcal{E}} \frac{\partial}{\partial\lambda} s[Y(e, x_t, \gamma_n^0), x_t, \lambda_n^0] \, dP(e) \Big|_{x_t \equiv 1}.$$

Let us decompose the computation into two steps. First compute the conditional mean of $Z_t(e)$ given that $y_{(2)} = y_{(2)t}$:

$$\mu_t = \int_{\mathcal{E}_{(1)}} Z_t(e_{(1)}, e_{(2)}) \, dP_{(1)}(e_{(1)})$$

$$= -2[g(y_{(2)t}, \gamma_n^0) - f(y_{(2)t}, \lambda_n^0)] \frac{\partial}{\partial\lambda} f(y_{(2)t}, \lambda_n^0).$$

Second, compute the mean of μ_t:

$$\int_{\mathcal{Y}_{(2)}} \mu_t \, d\nu(y_{(2)}) = \int_{\mathcal{Y}_{(2)}} -2[g(y_{(2)}, \gamma_n^0) - f(y_{(2)}, \lambda_n^0)] \frac{\partial}{\partial\lambda} f(y_{(2)}, \lambda_n^0) \, d\nu(y_{(2)})$$

$$= \frac{\partial}{\partial\lambda} s_n^0(\lambda_n^0)$$

$$= 0$$

because λ_n^0 minimizes $s_n^0(\lambda)$. Consequently

$$\int_{\mathcal{E}} Z_t(e) \, dP(e) = 0$$

and $\mathcal{U} = \mathcal{U}^* = 0$. One can see that in the fixed regressor case the conditional mean μ_t of $(\partial/\partial\lambda)s[Y(e, x, \gamma_n^0), x, \lambda_n^0]$ given the regressor is treated as a deterministic quantity, whereas in the iid regressor case the conditional mean μ_t is treated as a random variable having mean zero.

Further computations yield

$$\mathcal{I} = 4\sigma_{(1)}^2 \int_{\mathcal{Y}_{(2)}} \frac{\partial}{\partial\lambda} f\left(y_{(2)}, \lambda_n^0\right) \frac{\partial}{\partial\lambda'} f\left(y_{(2)}, \lambda_n^0\right) d\nu\left(y_{(2)}\right)$$

$$\mathcal{J} = 2\int_{\mathcal{Y}_{(2)}} \frac{\partial}{\partial\lambda} f\left(y_{(2)}, \lambda_n^0\right) \frac{\partial}{\partial\lambda'} f\left(y_{(2)}, \lambda_n^0\right) d\nu\left(y_{(2)}\right)$$

$$- 2\int_{\mathcal{Y}_{(2)}} \left[g\left(y_{(2)}, \gamma_n^0\right) - f\left(y_{(2)}, \lambda_n^0\right)\right] \frac{\partial^2}{\partial\lambda\,\partial\lambda'} f\left(y_{(2)}, \lambda_n^0\right) d\nu\left(y_{(2)}\right). \quad \square$$

Returning to the general case, use the same strategy employed in the example to write

$$q\left(y, x, \gamma_n^0\right) = \begin{pmatrix} e_{(1)} \\ e_{(2)} \end{pmatrix} = \begin{pmatrix} q_{(1)}\left(y_{(1)}, y_{(2)}, \gamma_n^0\right) \\ y_{(2)} - \mu_{(2)} \end{pmatrix}$$

with $x \equiv 1$ and

$$dP(e) = dP_{(1)}(e_{(1)}) \times dP_{(2)}(e_{(2)});$$

$y_{(2)}$ is the iid regressor. The reduced form can be written as

$$y_{(1)} = Y_{(1)}\left(e_{(1)}, y_{(2)}, \gamma_n^0\right)$$

$$y_{(2)} = \mu_{(2)} + e_{(2)}.$$

Let ν be the measure such that

$$\int_{\mathcal{Y}_{(2)}} g(y_{(2)}) \, d\nu(y_{(2)}) = \int_{\mathcal{E}_{(2)}} g(\mu_{(2)} + e_{(2)}) \, dP_{(2)}(e_{(2)})$$

where $\mathcal{Y}_{(2)}$ is the set of admissible values of the iid regressor $y_{(2)}$.

The distance function for the least mean distance estimator will have the form

$$s\left(y_{(1)}, y_{(2)}, \tau, \lambda\right).$$

Since the distance function depends trivially on x_t, we have

$$s_n^0(\lambda_n^0) = \int_{\mathscr{E}} s\left[Y_{(1)}\left(e_{(1)}, \mu_{(2)} + e_{(2)}, \gamma_n^0\right), \mu_{(2)} + e_{(2)}, \tau_n^0, \lambda_n^0\right] dP(e)$$

$$= \int_{\mathscr{Y}_{(2)}} \int_{\mathscr{E}_{(1)}} s\left[Y_{(1)}\left(e_{(1)}, y_{(2)}, \gamma_n^0\right), y_{(2)}, \tau_n^0, \lambda_n^0\right] dP_{(1)}(e_{(1)}) \, d\nu(y_{(2)}).$$

Since $(\partial/\partial\lambda)s_n^0(\lambda_n^0) = 0$ and the regularity conditions permit interchange of differentiation and integration, we have

$$\int_{\mathscr{Y}_{(2)}} \int_{\mathscr{E}_{(1)}} \frac{\partial}{\partial\lambda} s\left[Y_{(1)}\left(e_{(1)}, y_{(2)}, \gamma_n^0\right), y_{(2)}, \tau_n^0, \lambda_n^0\right] dP_{(1)}(e_{(1)}) \, d\nu(y_{(2)}) = 0$$

whence $\mathscr{U} = \mathscr{U}^* = 0$. Other computations assume a similar form, for example

$$\mathscr{I} = \int_{\mathscr{Y}_{(2)}} \int_{\mathscr{E}_{(1)}} \frac{\partial^2}{\partial\lambda \, \partial\lambda'} s\left[Y_{(1)}\left(e_{(1)}, y_{(2)}, \gamma_n^0\right), y_{(2)}, \tau_n^0, \lambda_n^0\right] dP_{(1)}(e_{(1)}) \, d\nu(y_{(2)}).$$

Sample quantities retain their previous form, for example

$$\hat{\mathscr{I}} = \frac{1}{n} \sum_{t=1}^{n} \frac{\partial^2}{\partial\lambda \, \partial\lambda'} s\left(y_{(1)t}, y_{(2)t}, \hat{\tau}_n, \hat{\lambda}_n\right).$$

For a method of moments estimator, in typical cases one can exploit the structure of the problem and show directly that

$$\int_{\mathscr{Y}_{(2)}} \int_{\mathscr{E}_{(2)}} m\left[Y_{(1)}\left(e_{(1)}, y_{(2)}, \gamma_n^0\right), y_{(2)}, \tau_n^0, \lambda_n^0\right] dP_{(1)}(e_{(1)}) \, d\nu(y_{(2)}) = 0.$$

This implies that $K = K^* = 0$ whence $\mathscr{U} = \mathscr{U}^* = 0$. The remaining computations are modified similarly to the foregoing.

The critical operative in the above discussion is the word identically. A sequence of fixed regressors $\{x_t\}$ is a sequence of independent random variables, but not a sequence of identically distributed random variables except in the trivial case $x_t = \text{const}$. Thus, a theory that is general enough to permit nonidentically distributed errors would be general enough to include the fixed regressors model as a special case, and the inability to estimate $\mathscr{U}^*$ under specification error ought to be a characteristic of such a theory. We shall see that this is indeed the case in Chapter 7.

CHAPTER 4

Univariate Nonlinear Regression: Asymptotic Theory

In this chapter, the results of Chapter 3 are specialized to the case of a correctly specified univariate nonlinear regression model estimated by least squares. The specialization is basically a matter of restating Assumptions 1 through 7 of Chapter 3 in context. This done, the asymptotic theory follows immediately. The characterizations used in Chapter 1 are established using probability bounds that follow from the asymptotic theory.

1. INTRODUCTION

Let us review some notation. The univariate nonlinear model is written as

$$y_t = f(x_t, \theta^0) + e_t \qquad t = 1, 2, \ldots, n$$

with θ^0 known to lie in some compact set Θ^*. The functional form of $f(x, \theta)$ is known, x is k-dimensional, θ is p-dimensional, and the model is assumed to be correctly specified. Following the conventions of Chapter 1, the model can be written in a vector notation as

$$y = f(\theta^0) + e$$

with the Jacobian of $f(\theta)$ written as $F(\theta) = (\partial/\partial\theta')f(\theta)$. The parameter θ is estimated by $\hat{\theta}$ that minimizes

$$s_n(\theta) = \frac{1}{n}\|y - f(\theta)\|^2 = \frac{1}{n}\sum_{t=1}^{n}[y_t - f(x_t, \theta)]^2.$$

We are interested in testing the hypothesis

$$H: h(\theta^0) = 0 \quad \text{against} \quad A: h(\theta^0) \neq 0$$

253

which we assume can be given the equivalent representation

$$H: \theta^0 = g(\rho^0) \text{ for some } \rho^0 \quad \text{against} \quad A: \theta^0 \neq g(\rho) \text{ for any } \rho$$

where $h: \mathbb{R}^p \to \mathbb{R}^q$, $g: \mathbb{R}^r \to \mathbb{R}^p$, and $p = r + q$. The correspondence with the notation of Chapter 3 is given in Notation 1.

NOTATION 1.

General (Chapter 3)	Specific (Chapter 4)
$e_t = q(y_t, x_t, \gamma_n^0)$	$e_t = y_t - f(x_t, \theta_n^0)$
$\gamma \in \Gamma$	$\theta \in \Theta^*$
$y = Y(e, x, \gamma)$	$y = f(x, \theta) + e$
$s(y_t, x_t, \hat{\tau}_n, \lambda)$	$[y_t - f(x_t, \theta)]^2$
$\lambda \in \Lambda^*$	$\theta \in \Theta^*$
$s_n(\lambda)$ $= (1/n)\sum_{t=1}^n s(y_t, x_t, \hat{\tau}_n, \lambda)$	$s_n(\theta)$ $= (1/n)\sum_{t=1}^n [y_t - f(x_t, \theta)]^2$
$s_n^0(\lambda)$ $= (1/n)\sum_{t=1}^n \int_{\mathscr{E}} s[Y(e, x_t, \gamma_n^0), x_t, \tau_n^0, \lambda]$ $\times dP(e)$	$s_n^0(\theta) = \sigma^2$ $+ (1/n)\sum_{t=1}^n [f(x_t, \theta_n^0) - f(x_t, \theta)]^2$
$s^*(\lambda)$ $= \int_{\mathscr{X}} \int_{\mathscr{E}} s[Y(e, x, \gamma^*), x, \tau^*, \lambda]$ $\times dP(e)\, d\mu(x)$	$s^*(\theta)$ $= \sigma^2 + \int_{\mathscr{X}} [f(x, \theta^*) - f(x, \theta)]^2$ $\times d\mu(x)$
$\hat{\lambda}_n$ minimizes $s_n(\lambda)$	$\hat{\theta}_n$ minimizes $s_n(\theta)$
$\tilde{\lambda}_n$ minimizes $s_n(\lambda)$ subject to $h(\lambda) = 0$	$\tilde{\theta}_n = g(\hat{\rho}_n)$ minimizes $s_n(\theta)$ subject to $h(\theta) = 0$
λ_n^0 minimizes $s_n^0(\lambda)$	θ_n^0 minimizes $s_n^0(\theta)$
λ_n^* minimizes $s_n^0(\lambda)$ subject to $h(\lambda) = 0$	$\theta_n^* = g(\rho_n^0)$ minimizes $s_n^0(\theta)$ subject to $h(\theta) = 0$
λ^* minimizes $s^*(\lambda)$	θ^* minimizes $s^*(\theta)$

2. REGULARITY CONDITIONS

Application of the general theory to a correctly specified univariate nonlinear regression is just a matter of restating Assumptions 1 through 7 of Chapter 3 in terms of Notation 1. As the data are presumed to be generated according to

$$y_t = f(x_t, \theta_n^0) + e_t \qquad t = 1, 2, \ldots, n$$

Assumptions 1 through 5 of Chapter 3 read as follows.

ASSUMPTION 1'. The errors are independently and identically distributed with common distribution $P(e)$.

ASSUMPTION 2'. $f(x, \theta)$ is continuous on $\mathcal{X} \times \Theta^*$, and Θ^* is compact.

ASSUMPTION 3' (Gallant and Holly, 1980). Almost every realization of $\{v_t\}$ with $v_t = (e_t, x_t)$ is a Cesaro sum generator with respect to the product measure

$$\nu(A) = \int_{\mathcal{X}} \int_{\mathcal{E}} I_A(e, x) \, dP(e) \, d\mu(x)$$

and dominating function $b(e, x)$. The sequence $\{x_t\}$ is a Cesaro sum generator with respect to μ and $b(x) = \int_{\mathcal{E}} b(e, x) \, dP(e)$. For each $x \in \mathcal{X}$ there is a neighborhood N_x such that $\int_{\mathcal{E}} \sup_{N_x} b(e, x) \, dP(e) < \infty$.

ASSUMPTION 4' (Identification). The parameter θ^0 is indexed by n, and the sequence $\{\theta_n^0\}$ converges to θ^*.

$$s^*(\theta) = \sigma^2 + \int_{\mathcal{X}} [f(x, \theta^*) - f(x, \theta)]^2 \, d\mu(x)$$

has a unique minimum over Θ^* at θ^*.

ASSUMPTION 5'. Θ^* is compact, $[e + f(x, \theta^0) - f(x, \theta)]^2$ is dominated by $b(e, x)$; $b(e, x)$ is that of Assumption 3.

The sample objective function is

$$s_n(\theta) = \frac{1}{n} \| y - f(\theta) \|^2$$

with expectation

$$s_n^0(\theta) = \sigma^2 + \frac{1}{n} \| f(\theta_n^0) - f(\theta) \|^2.$$

By Lemma 1 of Chapter 3, both $s_n(\theta)$ and $s_n^0(\theta)$ have uniform, almost sure limit

$$s^*(\theta) = \sigma^2 + \int_{\mathscr{X}} [f(x, \theta^*) - f(x, \theta)]^2 \, d\mu(x).$$

Note that the true value θ_n^0 of the unknown parameter is also a minimizer of $s_n^0(\theta)$, so that our use of θ_n^0 to denote them both is not ambiguous. We may apply Theorem 3 of Chapter 3 and conclude that

$$\lim_{n \to \infty} \theta_n^0 = \theta^*$$

$$\lim_{n \to \infty} \hat{\theta}_n = \theta^* \qquad \text{almost surely.}$$

Assumption 6 of Chapter 3 may be restated as follows.

ASSUMPTION 6'. Θ^* contains a closed ball Θ centered at θ^* with finite, nonzero radius such that

$$\frac{\partial}{\partial \theta_i} s[Y(e, x, \theta^0), x, \theta] = -2[e + f(x, \theta^0) - f(x, \theta)] \frac{\partial}{\partial \theta_i} f(x, \theta)$$

$$\frac{\partial^2}{\partial \theta_i \, \partial \theta_j} s[Y(e, x, \theta^0), x, \theta] = 2\left(\frac{\partial}{\partial \theta_i} f(x, \theta)\right)\left(\frac{\partial}{\partial \theta_j} f(x, \theta)\right)$$

$$-2[e + f(x, \theta^0) - f(x, \theta)] \frac{\partial^2}{\partial \theta_i \, \partial \theta_j} f(x, \theta)$$

$$\left(\frac{\partial}{\partial \theta_i} s[Y(e, x, \theta^0), x, \theta]\right)\left(\frac{\partial}{\partial \theta_j} s[Y(e, x, \theta^0), x, \theta]\right)$$

$$= 4[e + f(x, \theta^0) - f(x, \theta)]^2 \left(\frac{\partial}{\partial \theta_i} f(x, \theta)\right)\left(\frac{\partial}{\partial \theta_j} f(x, \theta)\right)$$

are continuous and dominated by $b(e, x)$ on $\mathscr{E} \times \mathscr{X} \times \Theta^* \times \Theta$ for $i, j = 1, 2, \ldots, p$. Moreover,

$$\mathscr{J}^* = 2\int_{\mathscr{X}} \left(\frac{\partial}{\partial \theta} f(x, \theta^*)\right)\left(\frac{\partial}{\partial \theta} f(x, \theta^*)\right)' d\mu(x)$$

is nonsingular.

Define

NOTATION 2.

$$Q = \int_{\mathscr{X}} \left(\frac{\partial}{\partial \theta} f(x, \theta^*) \right) \left(\frac{\partial}{\partial \theta} f(x, \theta^*) \right)' d\mu(x)$$

$$Q_n^0 = \frac{1}{n} F'(\theta_n^0) F(\theta_n^0)$$

$$Q_n^* = \frac{1}{n} F'(\theta_n^*) F(\theta_n^*).$$

Direct computation according to Notations 2 and 3 of Chapter 3 yields (Problem 1).

$$\mathscr{I}^* = 4\sigma^2 Q$$

$$\mathscr{J}^* = 2Q$$

$$\mathscr{U}^* = 0$$

$$\mathscr{I}_n^0 = 4\sigma^2 Q_n^0$$

$$\mathscr{J}_n^0 = 2Q_n^0$$

$$\mathscr{U}_n^0 = 0$$

$$\mathscr{I}_n^* = 4\sigma^2 Q_n^*$$

$$\mathscr{J}_n^* = 2Q_n^* - \frac{2}{n} \sum_{t=1}^{n} \left[f(x_t, \theta_n^0) - f(x_t, \theta_n^*) \right] \frac{\partial^2}{\partial \theta \, \partial \theta'} f(x_t, \theta_n^*)$$

$$\mathscr{U}_n^* = \frac{4}{n} \sum_{t=1}^{n} \left[f(x_t, \theta_n^0) - f(x_t, \theta_n^*) \right]^2 \left(\frac{\partial}{\partial \theta} f(x_t, \theta_n^*) \right) \left(\frac{\partial}{\partial \theta} f(x_t, \theta_n^*) \right)'.$$

Noting that

$$\frac{\partial}{\partial \theta} s_n(\theta) = -\frac{2}{n} F'(\theta) \left[e + f(\theta_n^0) - f(\theta) \right]$$

we have from Theorem 4 of Chapter 3 that

$$\frac{1}{\sqrt{n}} F'(\theta_n^0) e \xrightarrow{\mathscr{L}} N_p(0, \sigma^2 Q)$$

and from Theorem 5 that

$$\sqrt{n}\left(\hat{\theta}_n - \theta_n^0\right) \xrightarrow{\mathscr{L}} N_p\left(0, \sigma^2 Q^{-1}\right)$$

$$\lim_{n \to \infty} Q_n^0 = Q.$$

The Pitman drift assumption is restated as follows.

ASSUMPTION 7′ (Pitman drift). The sequence θ_n^0 is chosen such that $\lim_{n \to \infty} \sqrt{n}\,(\theta_n^0 - \theta_n^*) = \Delta$. Moreover, $h(\theta^*) = 0$.

Noting that

$$\frac{\partial}{\partial \theta} s_n^0(\theta) = -\frac{2}{n} F'(\theta)\left[f(\theta_n^0) - f(\theta) \right]$$

we have from Theorem 6 that

$$\lim_{n \to \infty} \tilde{\theta}_n = \theta* \qquad \text{almost surely}$$

$$\lim_{n \to \infty} \theta_n^* = \theta*$$

$$\lim_{n \to \infty} Q_n^* = Q$$

$$\frac{1}{\sqrt{n}} F'(\theta_n^*) e \xrightarrow{\mathscr{L}} N_p\left(0, \sigma^2 Q\right)$$

$$\lim_{n \to \infty} \frac{1}{\sqrt{n}} F'(\theta_n^*)\left[f(\theta_n^0) - f(\theta_n^*) \right] = Q\Delta.$$

Assumption 13 of Chapter 3 is restated as follows.

ASSUMPTION 13′. The function $h(\theta)$ is a once continuously differentiable mapping of Θ into $\mathbb{R}^q$. Its Jacobian $H(\theta) = (\partial/\partial\theta')h(\theta)$ has full rank $(= q)$ at $\theta = \theta*$.

PROBLEM

1. Use the derivatives given in Assumption 6′ to compute $\bar{\bar{\mathscr{F}}}(\theta)$, $\bar{\mathscr{F}}(\theta)$, $\bar{\bar{\mathscr{U}}}(\theta)$ and $\bar{\mathscr{F}}(\theta)$, $\bar{\mathscr{F}}(\theta)$, $\bar{\mathscr{U}}(\theta)$ as defined in Notations 2 and 3 of Chapter 3.

3. CHARACTERIZATIONS OF LEAST SQUARES ESTIMATORS AND TEST STATISTICS

The first of the characterizations appearing in Chapter 1 is

$$\hat{\theta}_n = \theta_n^0 + \left[F'(\theta_n^0) F(\theta_n^0) \right]^{-1} F'(\theta_n^0) e + o_p\left(\frac{1}{\sqrt{n}} \right).$$

It is derived using the same sort of arguments as used in the proof of Theorem 5 of Chapter 3, so we shall be brief here; one can look at Theorem 5 for details. By Lemma 2 of Chapter 3 we may assume without loss of generality that $\hat{\theta}_n$ and θ_n^0 are in Θ and that $(\partial/\partial\theta)s_n(\hat{\theta}_n) = o_p(1/\sqrt{n})$. Recall that $Q_n^0 = Q + o(1)$, whence $\mathscr{I}_n^0 = \mathscr{I}^* + o(1)$. By Taylor's theorem

$$\sqrt{n}\, \frac{\partial}{\partial\theta} s_n(\theta_n^0) = \sqrt{n}\, \frac{\partial}{\partial\theta} s_n(\hat{\theta}_n) + \bar{\mathscr{I}}\sqrt{n}\left(\theta_n^0 - \hat{\theta}_n \right)$$

where $\bar{\mathscr{I}} = \mathscr{I}^* + o_s(1)$. Then

$$\left[\mathscr{I}^* + o_s(1) \right]\sqrt{n}\left(\hat{\theta}_n - \theta_n^0 \right) = -\sqrt{n}\, \frac{\partial}{\partial\theta} s_n(\theta_n^0) + o_s(1)$$

which can be rewritten as

$$\mathscr{I}_n^0 \sqrt{n}\left(\hat{\theta}_n - \theta_n^0 \right) = -\sqrt{n}\, \frac{\partial}{\partial\theta} s_n(\theta_n^0) - \left[\mathscr{I}^* - \mathscr{I}_n^0 + o_s(1) \right]\sqrt{n}\left(\hat{\theta}_n - \theta_n^0 \right)$$
$$+ o_s(1).$$

Now $\mathscr{I}^* - \mathscr{I}_n^0 + o_s(1) = o_s(1)$ and $\sqrt{n}\,(\hat{\theta}_n - \theta_n^0) \overset{\mathscr{L}}{\to} N_p(0, \sigma^2 Q)$, which implies that $\sqrt{n}\,(\hat{\theta}_n - \theta_n^0) = O_p(1)$, whence $[\mathscr{I}^* - \mathscr{I}_n^0 + o_s(1)]\sqrt{n}\,(\hat{\theta}_n - \theta_n^0) = o_p(1)$. Thus we have that

$$\mathscr{I}_n^0 \sqrt{n}\left(\hat{\theta}_n - \theta_n^0 \right) = \sqrt{n}\, \frac{\partial}{\partial\theta} s_n(\theta_n^0) + o_p(1).$$

There is an N such that for $n > N$ the inverse of $\mathscr{I}_n^0$ exists, whence

$$\sqrt{n}\left(\hat{\theta}_n - \theta_n^0 \right) = -\sqrt{n}\left(\mathscr{I}_n^0 \right)^{-1} \frac{\partial}{\partial\theta} s_n(\theta_n^0) + o_p(1)$$

or

$$\hat{\theta}_n = \theta_n^0 - \left(\mathscr{I}_n^0 \right)^{-1} \frac{\partial}{\partial\theta} s_n(\theta_n^0) + o_p\left(\frac{1}{\sqrt{n}} \right).$$

Finally, $-(\mathscr{I}_n^0)^{-1}(\partial/\partial\theta)s_n(\theta_n^0) = [F'(\theta_n^0)F(\theta_n^0)]^{-1}F'(\theta_n^0)e$, which completes the argument.

The next characterization that needs justification is

$$s^2 = \frac{e'\left\{I - F(\theta_n^0)[F'(\theta_n^0)F(\theta_n^0)]^{-1}F'(\theta_n^0)\right\}e}{n-p} + o_p(1/n).$$

The derivation is similar to the arguments used in the proof of Theorem 15 of Chapter 3; again we shall be brief, and one can look at the proof of Theorem 15 for details. By Taylor's theorem

$$n[s_n(\theta_n^0) - s_n(\hat{\theta}_n)] = n\left(\frac{\partial}{\partial\theta}s_n(\hat{\theta}_n)\right)'(\hat{\theta}_n - \theta_n^0)$$

$$+ \frac{n}{2}(\hat{\theta}_n - \theta_n^0)'\left(\frac{\partial^2}{\partial\theta\,\partial\theta'}s_n(\bar{\theta}_n)\right)(\hat{\theta}_n - \theta_n^0)$$

$$= no_s\left(\frac{1}{\sqrt{n}}\right)(\hat{\theta}_n - \theta_n^0)$$

$$+ \frac{n}{2}(\hat{\theta}_n - \theta_n^0)'[\mathscr{I}_n^0 + o_s(1)](\hat{\theta}_n - \theta_n^0)$$

$$= \frac{n}{2}(\hat{\theta}_n - \theta_n^0)'\mathscr{I}_n^0(\hat{\theta}_n - \theta_n^0) + o_p(1).$$

From the proceeding result we have

$$\hat{\theta}_n - \theta_n^0 = [F'(\theta_n^0)F(\theta_n^0)]^{-1}F'(\theta_n^0)e + o_p\left(\frac{1}{\sqrt{n}}\right)$$

whence

$$n[s_n(\theta_n^0) - s_n(\hat{\theta}_n)] = ne'F(\theta_n^0)[F'(\theta_n^0)F(\theta_n^0)]^{-1}F'(\theta_n^0)e + o_p(1).$$

This equation reduces to

$$\|y - f(\hat{\theta})\|^2 = e'\left\{I - F(\theta_n^0)[F'(\theta_n^0)F(\theta_n^0)]^{-1}F'(\theta_n^0)\right\}e + o_p\left(\frac{1}{n}\right)$$

which completes the argument.

Next we show that

$$h(\hat{\theta}_n) = h(\theta_n^0) + H(\theta_n^0)[F'(\theta_n^0)F(\theta_n^0)]^{-1}F'(\theta_n^0)e + o_p\left(\frac{1}{\sqrt{n}}\right).$$

A straightforward argument using Taylor's theorem yields

$$\sqrt{n}\,h(\hat{\theta}_n) = \sqrt{n}\,h(\theta_n^0) + \bar{H}\sqrt{n}\,(\hat{\theta}_n - \theta_n^0)$$

where $\bar{H}$ has rows $(\partial/\partial\theta')h(\bar{\theta}_i)$ with $\bar{\theta}_i = \lambda_i\hat{\theta}_n + (1 - \lambda_i)\theta_n^0$ for some λ_i with $0 \le \lambda_i \le 1$, whence

$$\sqrt{n}\,h(\hat{\theta}_n) = \sqrt{n}\,h(\theta_n^0) + \left[H(\theta_n^0) + o_s(1)\right]\sqrt{n}\,(\hat{\theta}_n - \theta_n^0).$$

Since $\sqrt{n}\,(\hat{\theta}_n - \theta_n^0)$ is bounded in probability, we have

$$\sqrt{n}\,h(\hat{\theta}_n) = \sqrt{n}\,h(\theta_n^0) + \sqrt{n}\,H(\theta_n^0)(\hat{\theta}_n - \theta_n^0) + o_s(1)$$

$$= \sqrt{n}\,h(\theta_n^0) + H(\theta_n^0)\sqrt{n}\left\{\left[F'(\theta_n^0)F(\theta_n^0)\right]^{-1}F'(\theta_n^0)e\right.$$

$$\left. + o_p\!\left(\frac{1}{\sqrt{n}}\right)\right\} + o_s(1)$$

$$= \sqrt{n}\,h(\theta_n^0) + \sqrt{n}\,H(\theta_n^0)\left[F'(\theta_n^0)F(\theta_n^0)\right]^{-1}F'(\theta_n^0)e + o_p(1).$$

We next show that

$$\frac{1}{s^2} = \frac{n-p}{e'(I - P_F)e} + o_p\!\left(\frac{1}{n}\right)$$

where

$$P_F = F(\theta_n^0)\left[F'(\theta_n^0)F(\theta_n^0)\right]^{-1}F'(\theta_n^0).$$

Fix a realization of the errors $\{e_t\}$ for which $\lim_{n\to\infty}s^2 = \sigma^2$ and $\lim_{n\to\infty}e'(I - P_F)e/(n - p) = \sigma^2$; almost every realization is such (Problem 2). Choose N so that if $n > N$ then $s^2 > 0$ and $e'(I - P_F)e > 0$. Using

$$s^2 = \frac{e'(I - P_F)e}{n - p} + o_p(1/n)$$

and Taylor's theorem, we have

$$\frac{1}{s^2} = \frac{n-p}{e'(I - P_F)e} - \left(\frac{n-p}{e'(I - P_F)e}\right)^2 o_p\!\left(\frac{1}{n}\right).$$

The term $[(n - p)/e'(I - P_F)e]^2$ is bounded for $n > N$ because

$\lim_{n \to \infty}[(n - p)/e'(I - P_F)e]^2 = 1/\sigma^4$. One concludes that $1/s^2 = (n - p)/e'(I - P_F)e + o_p(1/n)$, which completes the argument.

The next task is to show that if the errors are normally distributed, then

$$W = Y + o_p(1)$$

where

$$Y \sim F'(q, n - p, \lambda)$$

$$\lambda = \frac{h'(\theta_n^0)\left\{ H(\theta_n^0)\left[F'(\theta_n^0)F(\theta_n^0)\right]^{-1}H'(\theta_n^0)\right\}^{-1}h(\theta_n^0)}{2\sigma^2}.$$

Now

$$W = n\frac{h'(\hat{\theta}_n)\left\{ \hat{H}\left[(1/n)F'(\hat{\theta}_n)F(\hat{\theta}_n)\right]^{-1}\hat{H}'\right\}^{-1}h(\hat{\theta}_n)}{qs^2}$$

and as notation write

$$\sqrt{n}\,h(\hat{\theta}_n) = \sqrt{n}\,h(\theta_n^0)$$
$$+ \sqrt{n}\,H(\theta_n^0)\left[F'(\theta_n^0)F(\theta_n^0)\right]^{-1}F'(\theta_n^0)e$$
$$+ o_p(1) = \mu + U + o_p(1)$$

$$\left[\hat{H}\left(\frac{1}{n}F'(\hat{\theta}_n)F(\hat{\theta}_n)\right)^{-1}\hat{H}'\right]^{-1} = \left[H(\theta_n^0)\left(\frac{1}{n}F'(\theta_n^0)F(\theta_n^0)\right)^{-1}H'(\theta_n^0)\right]^{-1}$$
$$+ o_p(1)$$
$$= A^{-1} + o_p(1)$$

whence

$$W = \left[\mu + U + o_p(1)\right]'A^{-1}\left[\mu + U + o_p(1)\right]\left[\frac{(n - p)}{e'(I - P_F)e} + o_p(1)\right]q^{-1}$$

$$= \frac{(\mu + U)'A^{-1}(\mu + U)/(q\sigma^2)}{e'(I - P_F)e/[\sigma^2(n - p)]} + o_p(1)$$

$$= Y + o_p(1).$$

Assuming normal errors, then

$$U \sim N_q(0, \sigma^2 A)$$

which implies that (Appendix to Chapter 1)

$$\frac{(\mu + U)'A^{-1}(\mu + U)}{\sigma^2} \sim \chi^{2'}(q, \lambda)$$

with

$$\lambda = \frac{\mu'A^{-1}\mu}{2\sigma^2}$$

$$= n \frac{h'(\theta_n^0)\left\{H(\theta_n^0)\left[(1/n)F'(\theta_n^0)F(\theta_n^0)\right]^{-1}H'(\theta_n^0)\right\}^{-1}h(\theta_n^0)}{2\sigma^2}.$$

Since $A(I - P_F) = 0$, U and $(I - P_F)e$ are independently distributed, whence $(\mu + U)'A^{-1}(\mu + U)$ and $e'(I - P_F)e = e'(I - P_F)'(I - P_F)e$ are independently distributed. This implies that $Y \sim F'(q, n - p, \lambda)$, which completes the argument.

Simply by rescaling s^2 in the foregoing we have that

$$\frac{\text{SSE}_{\text{full}}}{n} = \frac{e'P_F^\perp e}{n} + o_p\left(\frac{1}{n}\right)$$

$$\frac{n}{\text{SSE}_{\text{full}}} = \frac{n}{e'P_F^\perp e} + o_p\left(\frac{1}{n}\right)$$

where

$$P_F^\perp = I - P_F = I - F(\theta_n^0)\left[F'(\theta_n^0)F(\theta_n^0)\right]^{-1}F'(\theta_n^0);$$

recall that

$$\text{SSE}_{\text{full}} = \|y - f(\hat{\theta}_n)\|^2$$

$$\text{SSE}_{\text{reduced}} = \|y - f(\tilde{\theta}_n)\|^2 = \|y - f[g(\hat{\rho}_n)]\|^2.$$

The claim that

$$\frac{\text{SSE}_{\text{reduced}}}{n} = \frac{(e + \delta)'P_{FG}^\perp(e + \delta)}{n} + o_p\left(\frac{1}{n}\right)$$

with

$$\delta = f(\theta_n^0) - f(\theta_n^*) = f(\theta_n^0) - f[g(\rho_n^0)]$$

$$P_{FG}^\perp = I - P_{FG}$$

$$= I - F(\theta_n^0)G(\rho_n^0)\left[G'(\rho_n^0)F'(\theta_n^0)F(\theta_n^0)G(\rho_n^0)\right]^{-1}G'(\rho_n^0)F'(\theta_n^0)$$

comes fairly close to being a restatement of a few lines of the proof of Theorem 13 of Chapter 3. In that proof we find the equations

$$\bar{H}\sqrt{n}\,(\tilde{\theta}_n - \theta_n^*) = o_s(1)$$

$$\sqrt{n}\,(\tilde{\theta}_n - \theta_n^*) = \bar{\mathscr{J}}^{-1}\sqrt{n}\,\frac{\partial}{\partial\theta}s_n(\tilde{\theta}_n) - \bar{\mathscr{J}}^{-1}\sqrt{n}\,\frac{\partial}{\partial\theta}s_n(\theta_n^*) + o_s(1)$$

which, using arguments that have become repetitive at this point, can be rewritten as

$$H\sqrt{n}\,(\tilde{\theta}_n - \theta_n^*) = o_s(1)$$

$$\sqrt{n}\,(\tilde{\theta}_n - \theta_n^*) = \mathscr{J}^{-1}\!\left(\sqrt{n}\,\frac{\partial}{\partial\theta}s_n(\tilde{\theta}_n) - \sqrt{n}\,\frac{\partial}{\partial\theta}s_n(\theta_n^*)\right) + o_p(1)$$

with $\mathscr{J} = \mathscr{J}_n^0$ and $H = H(\theta_n^*)$. Using the conclusion of Theorem 13 of Chapter 3, one can substitute for $\sqrt{n}\,(\partial/\partial\theta)s_n(\tilde{\theta}_n)$ to obtain

$$\sqrt{n}\left[\frac{\partial}{\partial\theta}s_n(\tilde{\theta}_n)\right]'\sqrt{n}\,(\tilde{\theta}_n - \theta_n^*) = o_p(1)$$

$$\sqrt{n}\,(\tilde{\theta}_n - \theta_n^*) = -\mathscr{J}^{-1}\left[\mathscr{J} - H'(H\mathscr{J}^{-1}H')^{-1}H\right]\mathscr{J}^{-1}\sqrt{n}\,\frac{\partial}{\partial\theta}s_n(\theta_n^*) + o_p(1).$$

Then using Taylor's theorem

$$n\left[s_n(\tilde{\theta}_n) - s_n(\theta_n^*)\right]$$

$$= -n\left(\frac{\partial}{\partial\theta}s_n(\tilde{\theta}_n)\right)'(\tilde{\theta}_n - \theta_n^*) - \frac{n}{2}(\tilde{\theta}_n - \theta_n^*)'[\mathscr{J} + o_s(1)](\tilde{\theta}_n - \theta_n^*)$$

$$= -\frac{n}{2}(\tilde{\theta}_n - \theta_n^*)'\mathscr{J}(\tilde{\theta}_n - \theta_n^*) + o_p(1)$$

$$= -\frac{n}{2}\left(\frac{\partial}{\partial\theta}s_n(\theta_n^*)\right)'\left[\mathscr{J}^{-1} - \mathscr{J}^{-1}H'(H\mathscr{J}^{-1}H')^{-1}H\mathscr{J}^{-1}\right]\left(\frac{\partial}{\partial\theta}s_n(\theta_n^*)\right).$$

Using the identity obtained in Section 6 of Chapter 3, we have

$$\mathscr{J}^{-1} - \mathscr{J}^{-1}H'(H\mathscr{J}^{-1}H')^{-1}H\mathscr{J}^{-1} = G(G'\mathscr{J}G)^{-1}G'$$

whence

$$ns_n(\tilde{\theta}_n) = ns_n(\theta_n^*) - \frac{n}{2}\left(\frac{\partial}{\partial\theta}s_n(\theta_n^*)\right)'G(G'\mathscr{J}G)^{-1}G'\left(\frac{\partial}{\partial\theta}s_n(\theta_n^*)\right) + o_p(1).$$

Using Taylor's theorem, the uniform strong law, and the Pitman drift assumption, we have

$$\frac{\partial}{\partial \theta} s_n(\theta_n^*) = -\frac{2}{n} \sum_{t=1}^{n} \left[e_t + f(x_t, \theta_n^0) - f(x_t, \theta_n^*) \right] \frac{\partial}{\partial \theta} f(x_t, \theta_n^*)$$

$$= -\frac{2}{n} \sum_{t=1}^{n} \left[e_t + f(x_t, \theta_n^0) - f(x_t, \theta_n^*) \right] \frac{\partial}{\partial \theta} f(x_t, \theta_n^0)$$

$$+ \frac{1}{\sqrt{n}} \frac{-2}{n} \sum_{t=1}^{n} \left[e_t + f(x_t, \theta_n^0) - f(x_t, \theta_n^*) \right]$$

$$\times \begin{pmatrix} \frac{\partial}{\partial \theta'} \frac{\partial}{\partial \theta_1} f(x_t, \bar{\theta}_{1n}) \\ \vdots \\ \frac{\partial}{\partial \theta'} \frac{\partial}{\partial \theta_p} f(x_t, \bar{\theta}_{pn}) \end{pmatrix} \sqrt{n} \left(\theta_n^0 - \theta_n^* \right)$$

$$= -\frac{2}{n} F'(\theta_n^0)(e + \delta) + o_p\left(\frac{1}{\sqrt{n}} \right).$$

Substitution and algebraic reduction yields (Problem 3)

$$n s_n(\tilde{\theta}_n) = (e + \delta)'(e + \delta) - (e + \delta)' P_{FG}(e + \delta) + o_p(1)$$

which proves the claim.

The following are the characterizations used in Chapter 1 that have not yet been verified:

$$\frac{\text{SSE}_{\text{reduced}}}{\text{SSE}_{\text{full}}} = \frac{(e + \delta)' P_{FG}^{\perp}(e + \delta)}{e' P_F^{\perp} e} + o_p\left(\frac{1}{n} \right)$$

$$\frac{\tilde{D}'(\tilde{F}'\tilde{F})^{-1}\tilde{D}}{n} = \frac{(e + \delta)'(P_F - P_{FG})(e + \delta)}{n} + o_p\left(\frac{1}{n} \right)$$

$$\frac{\tilde{D}'(\tilde{F}'\tilde{F})\tilde{D}/q}{\text{SSE}(\hat{\theta})/(n - p)} = \frac{(e + \delta)'(P_F - P_{FG})(e + \delta)/q}{e'(I - P_F)e/(n - p)} + o_p(1)$$

$$\frac{n\tilde{D}'(\tilde{F}'\tilde{F})\tilde{D}}{\text{SSE}(\hat{\theta})} = \frac{n(e + \delta)'(P_F - P_{FG})(e + \delta)}{(e + \delta)'(I - P_{FG})(e + \delta)} + o_p(1).$$

Except for the second, these are obvious at sight. Let us sketch the verification of the second characterization:

$$\tilde{D}'\tilde{F}'\tilde{F}\tilde{D} = [y - f(\tilde{\theta}_n)]'\tilde{F}(\tilde{F}'\tilde{F})^{-1}\tilde{F}'[y - f(\tilde{\theta}_n)]$$

$$= \frac{n}{4}\left(\frac{\partial}{\partial\theta}s_n(\tilde{\theta}_n)\right)'\left(\frac{1}{n}\tilde{F}'\tilde{F}\right)^{-1}\left(\frac{\partial}{\partial\theta}s_n(\tilde{\theta}_n)\right)$$

$$= \frac{n}{2}\left(\frac{\partial}{\partial\theta}s_n(\tilde{\theta}_n)\right)'[\mathcal{J} + o_s(1)]^{-1}\left(\frac{\partial}{\partial\theta}s_n(\tilde{\theta}_n)\right)$$

$$= \frac{n}{2}\left(\frac{\partial}{\partial\theta}s_n(\theta_n^*)\right)'\mathcal{J}^{-1}H'(H\mathcal{J}^{-1}H')^{-1}H\mathcal{J}^{-1}\left(\frac{\partial}{\partial\theta}s_n(\theta_n^*)\right) + o_p(1)$$

$$= \frac{n}{2}\left(\frac{\partial}{\partial\theta}s_n(\theta_n^*)\right)'[\mathcal{J}^{-1} - G(G'\mathcal{J}G)^{-1}G']\left(\frac{\partial}{\partial\theta}s_n(\theta_n^*)\right) + o_p(1)$$

$$= \frac{1}{n}(e + \delta)'F(\theta_n^0)\left[(Q_n^0)^{-1} - G(G'Q_n^0G)^{-1}G'\right]$$
$$\times F'(\theta_n^0)(e + \delta) + o_p(1)$$

$$= \frac{1}{n}(e + \delta)'F(\theta_n^0)\left[(Q_n^0)^{-1} - G(G'Q_n^0G)^{-1}G'\right]$$
$$\times F'(\theta_n^0)(e + \delta) + o_p(1)$$

$$= (e + \delta)'(P_F - P_{FG})(e + \delta) + o_p(1).$$

PROBLEMS

1. Give a detailed derivation of the four characterizations listed in the preceding paragraph.

2. Cite the theorem which permits one to claim that $\lim_{n\to\infty} s^2 = \sigma^2$ almost surely, and prove directly that $\lim_{n\to\infty} e'(I - P_F)e/(n - p) = \sigma^2$ almost surely.

3. Show in detail that $(\partial/\partial\theta)s_n(\theta_n^*) = (-2/n)F'(\theta_n^0)(e + \delta) + o_p(1/\sqrt{n})$ suffices to reduce

$$\frac{n}{2}\left(\frac{\partial}{\partial\theta}s_n(\theta_n^*)\right)'G(G'\mathcal{J}G)^{-1}G'\left(\frac{\partial}{\partial\theta}s_n(\theta_n^*)\right)$$

to $(e + \delta)'P_{FG}(e + \delta)$.

CHAPTER 5

Multivariate Nonlinear Regression

All that separates multivariate regression from univariate regression is a linear transformation. Accordingly, the main thrust of this chapter is to identify the transformation, to estimate it, and then to apply the ideas of Chapter 1.

1. INTRODUCTION

In Chapter 1 we considered the univariate nonlinear model

$$y_t = f(x_t, \theta^0) + e_t, \qquad t = 1, 2, \dots, n.$$

Here we consider the case where there are M such regressions

$$y_{\alpha t} = f_\alpha(x_t, \theta_\alpha^0) + e_{\alpha t} \qquad t = 1, 2, \dots, n \quad \alpha = 1, 2, \dots, M$$

that are related in one of two ways. The first arises most naturally when

$$y_{\alpha t} \qquad \alpha = 1, 2, \dots, M$$

represent repeated measures on the same subject—height and weight measurements on the same individual for instance. In this case one would expect the observations with the same t index to be correlated, viz.

$$\mathscr{C}(y_{\alpha t}, y_{\beta t}) = \sigma_{\alpha\beta}$$

267

often called contemporaneous correlation. The second way these regressions can be related is through shared parameters. Stacking the parameter vectors and writing

$$\theta = \begin{pmatrix} \theta_1 \\ \theta_2 \\ \vdots \\ \theta_M \end{pmatrix}$$

one can have

$$\theta = g(\rho)$$

where ρ has smaller dimension than θ. If either or both of these relationships obtain (contemporaneous correlation or shared parameters), estimators with improved efficiency can be obtained—improved in the sense of better efficiency than that which obtains by applying the methods of Chapter 1 M times (Problem 12, Section 3). An example that exhibits these characteristics that we shall use heavily for illustration is the following.

EXAMPLE 1 (Consumer demand).   The data shown in Tables 1*a* and 1*b* is to be transformed as follows:

$y_1 = \ln(\text{peak expenditure share}) - \ln(\text{base expenditure share})$

$y_2 = \ln(\text{intermediate expenditure share}) - \ln(\text{base expenditure share})$

$x_1 = \ln(\text{peak price/expenditure})$

$x_2 = \ln(\text{intermediate price/expenditure})$

$x_2 = \ln(\text{base price/expenditure}).$

As notation, set

$$y = \begin{pmatrix} y_1 \\ y_2 \end{pmatrix} \qquad x = \begin{pmatrix} x_1 \\ x_2 \\ x_3 \end{pmatrix}$$

$$y_t = \begin{pmatrix} y_{1t} \\ y_{2t} \end{pmatrix} \qquad x_t = \begin{pmatrix} x_{1t} \\ x_{2t} \\ x_{3t} \end{pmatrix} \qquad t = 1, 2, \ldots, 224$$

These data are presumed to follow the model

$$y_{1t} = \ln \frac{a_1 + x_t' b_{(1)}}{a_3 + x_t' b_{(3)}} + e_{1t}$$

$$y_{2t} = \ln \frac{a_2 + x_t' b_{(2)}}{a_3 + x_t' b_{(3)}} + e_{2t}$$

where

$$a = \begin{pmatrix} a_1 \\ a_2 \\ a_3 \end{pmatrix}$$

$$B = \begin{pmatrix} b_{11} & b_{12} & b_{13} \\ b_{21} & b_{22} & b_{23} \\ b_{31} & b_{32} & b_{33} \end{pmatrix}$$

and $b_{(i)}'$ denotes the ith row of B, viz.

$$b_{(i)}' = (b_{i1}, b_{i2}, b_{i3}).$$

The errors

$$e_t = \begin{pmatrix} e_{1t} \\ e_{2t} \end{pmatrix}$$

are assumed to be independently and identically distributed, each with mean zero and variance-covariance matrix Σ.

There are various hypotheses that one might impose on the model. Two are of the nature of maintained hypotheses that follow directly from the theory of demand and ought to be satisfied. These are:

H_1: a_3 and $b_{(3)}$ are the same in both equations, as the notation suggests.
H_2: B is a symmetric matrix.

There is a third hypothesis that would be a considerable convenience if it were true:

$$H_3: \sum_{i=1}^{3} a_i = -1, \ \sum_{j=1}^{3} b_{ij} = 0 \text{ for } i = 1, 2, 3.$$

The theory supporting this model specification follows; the reader who has no interest in the theory can skip over the rest of the example.

The theory of consumer demand is fairly straightforward. Given an income Y which can be spent on N different goods which sell at prices $p_1, p_2, \ldots, p_N$, the consumer's problem is to decide what quantities $q_1, q_2, \ldots, q_N$ of each good to purchase. One assumes that the consumer has the ability to rank various bundles of goods in order of preference. Denoting a bundle by the vector

$$q = (q_1, q_2, \ldots, q_N)'$$

the assumption of an ability to rank bundles is equivalent to the assumption that there is a (utility) function $u(q)$ such that $u(q^0) > u(q^*)$ means bundle q^0 is preferred to bundle q^*. Since a bundle costs $p'q$ with $p' = (p_1, p_2, \ldots, p_N)$, the consumer's problem is

$$\text{maximize} \quad u(q)$$
$$\text{subject to} \quad p'q = Y.$$

This is the same problem as

$$\text{maximize} \quad u(q)$$
$$\text{subject to} \quad (p/Y)'q = 1$$

which means that the solution must be of the form

$$q = q(v)$$

with $v = p/Y$. The function $q(v)$ mapping the positive orthant of $\mathbb{R}^N$ into the positive orthant of $\mathbb{R}^N$ is called the consumer's demand system. It is usually assumed in applied work that all prices are positive and that a bundle with some $q_i = 0$ is never chosen.

If one substitutes the demand system $q(v)$ back into the utility function, one obtains the function

$$g(v) = u[q(v)]$$

which gives the maximum utility that a consumer can achieve at the price/income point v. The function $g(v)$ is called the indirect utility function. A property of the indirect utility function that makes it extremely useful in applied work is that the demand system is proportional to the gradient of the indirect utility function (Deaton and Muellbauer, 1980), viz.

$$q(v) = \frac{(\partial/\partial v)g(v)}{v'(\partial/\partial v)g(v)}.$$

This relationship is called Roy's identity. Thus, to implement the theory of consumer demand one need only specify a parametric form $g(v|\theta)$ and then fit the system

$$q = \frac{(\partial/\partial v)g(v|\theta)}{v'(\partial/\partial v)g(v|\theta)}$$

to observed values of (q, v) in order to estimate θ. The theory asserts that the fitted function $g(v|\theta)$ should be decreasing in each argument, $(\partial/\partial v_i)g(v|\theta) < 0$, and should be quasiconvex, $v'(\partial^2/\partial v\,\partial v')g(v|\theta)v > 0$ for every v with $v'(\partial/\partial v)g(v|\theta) = 0$ (Deaton and Muellbauer, 1980). If $g(v|\theta)$ has these properties, then there exists a corresponding $u(q)$. Thus, in applied work, there is no need to bother with $u(q)$; $g(v|\theta)$ is enough.

It is easier to arrive at a stochastic model if we reexpress the demand system in terms of expenditure shares. Accordingly let diag(v) denote a diagonal matrix with the components of the vector v along the diagonal, and set

$$s = \text{diag}(v)\, q$$

$$s(v|\theta) = \text{diag}(v)\, \frac{(\partial/\partial v)g(v|\theta)}{v'(\partial/\partial v)g(v|\theta)}.$$

Observe that

$$s_i = v_i q_i = \frac{p_i q_i}{Y}$$

so that s_i denotes that proportion of total expenditure Y spent on the ith good. As such, $1's = \sum_{i=1}^{N} s_i = 1$ and $1's(v|\theta) = 1$.

The deterministic model suggests that the distribution of the shares has a location parameter that depends on $s(v|\theta)$ in a simple way. What seems to be the case with this sort of data (Rossi, 1983) is that observed shares follow the logistic-normal distribution (Aitchison and Shen, 1980) with location parameter

$$\mu = \ln s(v|\theta)$$

where $\ln s(v|\theta)$ denotes the N-vector with components $\ln s_i(v|\theta)$ for $i = 1, 2, \ldots, N$. The logistic-normal distribution is characterized as follows. Let w be normally distributed with mean vector μ and a variance-covariance matrix $\mathscr{C}(w, w')$ that satisfies $1'\mathscr{C}(w, w')1 = 0$. Then s has the logistic-normal distribution if

$$s = \frac{e^w}{\sum_{i=1}^{N} e^{w_i}}$$

where e^w denotes the vector with components e^{w_i} for $i = 1, 2, \ldots, N$. A logarithmic transform yields

$$\ln s = w - \ln\left(\sum_{i=1}^{N} e^{w_i}\right) 1$$

whence

$$\ln s_i - \ln s_N = w_i - w_N \qquad i = 1, 2, \ldots, N - 1.$$

Writing $w_i - w_N = \mu_i - \mu_N + e_i$ for $i = 1, 2, \ldots, N - 1$, we have equations that can be fitted to data

$$\ln s_i - \ln s_N = \ln \frac{\partial}{\partial v_i} g(v|\theta) - \ln \frac{\partial}{\partial v_N} g(v|\theta) + e_i \qquad i = 1, 2, \ldots, N - 1.$$

The last step in implementing this model is the specification of a functional form for $g(v|\theta)$. Theory implies a strong preference for a low order multivariate Fourier series expansion (Gallant, 1981, 1982; Elbadawi, Gallant, and Souza, 1983) but since our purpose is illustrative, the choice will be governed by simplicity and manipulative convenience. Accordingly, let $g(v|\theta)$ be specified as the translog (Christensen, Jorgenson, and Lau, 1975)

$$g(v|\theta) = \sum_{i=1}^{N} a_i \ln(v_i) + \frac{1}{2} \sum_{i=1}^{N} \sum_{j=1}^{N} b_{ij} \ln(v_i) \ln(v_j)$$

or

$$g(v|\theta) = a'x + \tfrac{1}{2} x'Bx$$

with $x = \ln v$ and

$$a' = (a_1, a_2, a_3)$$

$$B = \begin{pmatrix} b_{11} & b_{12} & b_{13} \\ b_{21} & b_{22} & b_{23} \\ b_{31} & b_{32} & b_{33} \end{pmatrix}.$$

Differentiation yields

$$\frac{\partial}{\partial v} g(v|\theta) = [\text{diag}(v)]^{-1}[a + \tfrac{1}{2}(B + B')x].$$

One can see from this expression that B can be taken to be symmetric without loss of generality. With this assumption we have

$$\frac{\partial}{\partial v} g(v|\theta) = [\text{diag}(v)]^{-1}(a + Bx).$$

Recall that in general shares are computed as

$$s(v|\theta) = \text{diag}(v) \frac{(\partial/\partial v)g(v|\theta)}{v'(\partial/\partial v)g(v|\theta)}$$

which reduces to

$$s(v|\theta) = \frac{a + Bx}{1'(a + Bx)}$$

in this instance. The differenced logarithmic shares are

$$\ln s_i(v|\theta) - \ln s_N(v|\theta) = \ln \frac{a_i + x'b_{(i)}}{a_N + x'b_{(N)}}.$$

The model set forth in the beginning paragraphs of this discussion follows from the above equation. The origins of hypotheses H_1 and H_2 are apparent as well.

One notes, however, that we applied this model not to all goods $q_1, q_2, \ldots, q_N$ and income Y but rather to three categories of electricity expenditure—peak $= q_1$, intermediate $= q_2$, base $= q_3$—and to the total electricity expenditure E. A (necessary and sufficient) condition that permits one to apply the theory of demand essentially intact to the electricity subsystem, as we have done, is that the utility function is of the form (Blackorby, Primont, and Russell, 1978, Chapter 5)

$$u[u_{(1)}(q_1, q_2, q_3), q_4, \ldots, q_N].$$

If the utility function is of this form and E is known, it is fairly easy to see

Table 1a. Household Electricity Expenditures by Time of Use, North Carolina, Average over Weekdays in July 1978.

t	Treatment	Base	Intermediate	Peak	Expenditure ($ per day)
			Expenditure Share		
1	1	0.056731	0.280382	0.662888	0.46931
2	1	0.103444	0.252128	0.644427	0.79539
3	1	0.158353	0.270089	0.571558	0.45756
4	1	0.108075	0.305072	0.586853	0.94713
5	1	0.083921	0.211656	0.704423	1.22054
6	1	0.112165	0.290532	0.597302	0.93181
7	1	0.071274	0.240518	0.688208	1.79152
8	1	0.076510	0.210503	0.712987	0.51442
9	1	0.066173	0.202999	0.730828	0.78407
10	1	0.094836	0.270281	0.634883	1.01354
11	1	0.078501	0.293953	0.627546	0.83854
12	1	0.059530	0.228752	0.711718	1.53957
13	1	0.208982	0.328053	0.462965	1.06694
14	1	0.083702	0.297272	0.619027	0.82437
15	1	0.138705	0.358329	0.502966	0.80712
16	1	0.111378	0.322564	0.566058	0.53169
17	1	0.092919	0.259633	0.647448	0.85439
18	1	0.039353	0.158205	0.802442	1.93326
19	1	0.066577	0.247454	0.685970	1.37160
20	2	0.102844	0.244335	0.652821	0.92766
21	2	0.125485	0.230305	0.644210	1.80934
22	2	0.154316	0.235135	0.610549	2.41501
23	2	0.165714	0.276980	0.557305	0.84658
24	2	0.145370	0.173112	0.681518	1.60788
25	2	0.184467	0.268865	0.546668	0.73838
26	2	0.162269	0.280939	0.556792	0.81116
27	2	0.112016	0.220850	0.667133	2.01503
28	2	0.226863	0.257833	0.515304	2.32035
29	2	0.118028	0.219830	0.662142	2.40172
30	2	0.137761	0.345117	0.517122	0.57141
31	2	0.079115	0.257319	0.663566	0.94474
32	2	0.185022	0.265051	0.549928	1.63778
33	2	0.144524	0.276133	0.579343	0.75816
34	2	0.201734	0.241966	0.556300	1.00136
35	2	0.094890	0.227651	0.677459	1.11384
36	2	0.102843	0.264515	0.632642	1.07185
37	2	0.107760	0.214232	0.678009	1.53659
38	2	0.156552	0.236422	0.607026	0.24099
39	2	0.088431	0.222746	0.688822	0.58066
40	2	0.146236	0.301884	0.551880	2.52983
41	3	0.080802	0.199005	0.720192	1.14741
42	3	0.100711	0.387758	0.511531	0.97934
43	3	0.073483	0.335280	0.591237	1.09361
44	3	0.059455	0.259823	0.680722	2.19468
45	3	0.076195	0.378371	0.545434	1.98221

Table 1*a*. (Continued).

t	Treatment	Expenditure Share			Expenditure ($ per day)
		Base	Intermediate	Peak	
46	3	0.076926	0.325032	0.598042	1.78194
47	3	0.086052	0.339653	0.574295	3.24274
48	3	0.069359	0.278369	0.652272	0.47593
49	3	0.071265	0.273866	0.654869	1.38369
50	3	0.100562	0.306247	0.593191	1.57831
51	3	0.050203	0.294285	0.655513	2.16900
52	3	0.059627	0.311932	0.628442	2.11575
53	3	0.081433	0.328604	0.589962	0.35681
54	3	0.075762	0.285972	0.638265	1.55275
55	3	0.042910	0.372337	0.584754	1.06305
56	3	0.086846	0.340184	0.572970	4.02013
57	3	0.102537	0.335535	0.561928	0.60712
58	3	0.068766	0.310782	0.620452	1.15334
59	3	0.058405	0.307111	0.634485	2.43797
60	4	0.055227	0.300839	0.643934	0.10082
61	4	0.107435	0.273937	0.618628	0.69302
62	4	0.105958	0.291205	0.602837	1.12592
63	4	0.132278	0.279429	0.588293	1.84425
64	4	0.094195	0.328866	0.576940	1.57972
65	4	0.115259	0.401079	0.483663	1.27034
66	4	0.150229	0.317866	0.531905	0.56330
67	4	0.168780	0.307669	0.523551	3.43139
68	4	0.118222	0.318080	0.563698	1.00979
69	4	0.103394	0.307671	0.588936	2.08458
70	4	0.124007	0.362115	0.513879	1.30410
71	4	0.197987	0.280130	0.521884	3.48146
72	4	0.108083	0.337004	0.554913	0.53206
73	5	0.088798	0.232568	0.678634	3.28987
74	5	0.100508	0.272139	0.627353	0.32678
75	5	0.127303	0.298519	0.574178	0.52452
76	5	0.109718	0.228172	0.662109	0.36622
77	5	0.130080	0.231037	0.638883	0.63788
78	5	0.148562	0.323579	0.527859	1.42239
79	5	0.106306	0.252137	0.641556	0.93535
80	5	0.080877	0.214172	0.704951	1.26243
81	5	0.081810	0.135665	0.782525	1.51472
82	5	0.131749	0.278338	0.589913	2.07858
83	5	0.059180	0.254533	0.686287	1.60681
84	5	0.078620	0.267252	0.654128	1.54706
85	5	0.090220	0.293831	0.615949	2.61162
86	5	0.086916	0.193967	0.719117	2.96418
87	5	0.132383	0.230489	0.637127	0.26912
88	5	0.085560	0.252321	0.662120	0.42554
89	5	0.071368	0.276238	0.652393	1.01926
90	5	0.061196	0.245025	0.693780	1.53807

Table 1a. (Continued).

t	Treatment	Expenditure Share			Expenditure ($ per day)
		Base	Intermediate	Peak	
91	5	0.086608	0.233981	0.679411	0.75711
92	5	0.105628	0.305471	0.588901	0.83647
93	5	0.078158	0.202536	0.719307	1.92096
94	5	0.048632	0.216807	0.734560	1.57795
95	5	0.094527	0.224344	0.681128	0.83216
96	5	0.092809	0.209154	0.698037	1.39364
97	5	0.035751	0.166231	0.798018	1.72697
98	5	0.065205	0.205058	0.729736	2.04120
99	5	0.092561	0.193848	0.713591	2.04708
100	5	0.063119	0.234114	0.702767	3.43969
101	5	0.091186	0.224488	0.684326	2.66918
102	5	0.047291	0.262623	0.690086	2.71072
103	5	0.081575	0.206400	0.712025	3.36803
104	5	0.108165	0.243650	0.648185	0.65682
105	5	0.079534	0.320450	0.600017	0.95523
106	5	0.084828	0.247189	0.667984	0.61441
107	5	0.063747	0.210343	0.725910	1.85034
108	5	0.081108	0.249960	0.668932	2.11274
109	5	0.089942	0.206601	0.703457	1.54120
110	5	0.046717	0.224784	0.728499	3.54351
111	5	0.114925	0.272279	0.612796	2.61769
112	5	0.115055	0.264415	0.620530	3.00236
113	5	0.081511	0.223870	0.694618	1.74166
114	5	0.109658	0.343593	0.546750	1.17640
115	5	0.114263	0.304761	0.580976	0.74566
116	5	0.115089	0.226412	0.658499	1.30392
117	5	0.040622	0.198986	0.760392	2.13339
118	5	0.073245	0.238522	0.688234	2.83039
119	5	0.087954	0.287450	0.624596	1.62179
120	5	0.091967	0.206131	0.701902	2.18534
121	5	0.142746	0.302939	0.554315	0.26503
122	5	0.117972	0.253811	0.628217	0.05082
123	5	0.071573	0.248324	0.680103	0.42740
124	5	0.073628	0.290586	0.635786	0.47979
125	5	0.121075	0.350781	0.528145	0.59551
126	5	0.077335	0.339358	0.583307	0.47506
127	5	0.074766	0.167202	0.758032	2.11867
128	5	0.208580	0.331363	0.460058	1.13621
129	5	0.080195	0.210619	0.709185	2.61204
130	5	0.066156	0.204118	0.729726	1.45227
131	5	0.112282	0.252638	0.635080	0.79071
132	5	0.041310	0.093106	0.865584	1.30697
133	5	0.102675	0.297009	0.600316	0.93691
134	5	0.102902	0.270832	0.626266	0.98718
135	5	0.118932	0.250104	0.630964	1.40085

Table 1a. (Continued).

t	Treatment	Expenditure Share			Expenditure ($ per day)
		Base	Intermediate	Peak	
136	5	0.139760	0.322394	0.537846	1.78710
137	5	0.121616	0.214626	0.663758	8.46237
138	5	0.065701	0.263818	0.670481	1.58663
139	5	0.034029	0.175181	0.790790	2.62535
140	5	0.074476	0.194744	0.730780	4.29430
141	5	0.059568	0.229705	0.710727	0.65404
142	5	0.088128	0.295546	0.616326	0.41292
143	5	0.075522	0.213622	0.710856	2.02370
144	5	0.057089	0.195720	0.747190	1.76998
145	5	0.096331	0.301692	0.601977	0.99891
146	5	0.120824	0.250280	0.628896	0.27942
147	6	0.034529	0.193456	0.772015	0.91673
148	6	0.026971	0.180848	0.792181	1.15617
149	6	0.045271	0.141894	0.812835	1.57107
150	6	0.067708	0.219302	0.712990	1.24515
151	6	0.079335	0.230693	0.689972	1.70748
152	6	0.022703	0.178896	0.798401	1.79959
153	6	0.043053	0.157142	0.799805	4.61665
154	6	0.057157	0.245931	0.696912	0.59504
155	6	0.063229	0.136192	0.800579	1.42499
156	6	0.076873	0.214209	0.708918	1.34371
157	6	0.027353	0.124894	0.847753	2.74908
158	6	0.067823	0.146994	0.785183	1.84628
159	6	0.056388	0.189185	0.754428	3.82472
160	6	0.036841	0.194994	0.768165	1.18199
161	6	0.059160	0.138681	0.802158	2.07338
162	6	0.051980	0.215700	0.732320	0.80376
163	6	0.027300	0.145072	0.827628	1.52316
164	6	0.014790	0.179619	0.805591	3.17526
165	6	0.047865	0.167561	0.784574	3.30794
166	6	0.115629	0.231381	0.652990	0.72456
167	7	0.104970	0.147525	0.747505	0.50274
168	7	0.119254	0.187409	0.693337	1.22571
169	7	0.042564	0.112839	0.844596	2.13534
170	7	0.096756	0.150178	0.753066	5.56011
171	7	0.063013	0.168422	0.768565	3.11725
172	7	0.080060	0.143934	0.776006	0.99796
173	7	0.097493	0.173391	0.729116	0.67859
174	7	0.102526	0.220954	0.676520	0.79027
175	7	0.085538	0.195686	0.718776	2.24498
176	7	0.068733	0.166248	0.765019	2.01993
177	7	0.094915	0.140119	0.764966	4.07330
178	7	0.076163	0.132046	0.791792	3.66432
179	7	0.099943	0.176885	0.723172	0.40768
180	7	0.081494	0.175082	0.743425	1.09065

Table 1a. (Continued).

| t | Treatment | Expenditure Share | | | Expenditure ($ per day) |
		Base	Intermediate	Peak	
181	7	0.196026	0.299348	0.504626	1.35008
182	7	0.093173	0.235816	0.671011	1.06138
183	7	0.172293	0.173032	0.654675	0.99219
184	7	0.067736	0.159600	0.772663	3.69199
185	7	0.102033	0.171697	0.726271	2.36676
186	7	0.067977	0.151109	0.780914	1.84563
187	8	0.071073	0.238985	0.689942	0.18316
188	8	0.049453	0.286788	0.663759	2.23986
189	8	0.062748	0.255129	0.682123	3.48084
190	8	0.032376	0.154905	0.812719	7.26135
191	8	0.055055	0.225296	0.719648	1.68814
192	8	0.037829	0.179051	0.783120	1.13804
193	8	0.020102	0.172396	0.807502	1.40894
194	8	0.021917	0.149092	0.828992	3.47472
195	8	0.047590	0.174735	0.777675	3.37689
196	8	0.063446	0.235823	0.700731	3.14810
197	8	0.034719	0.159398	0.805883	3.21710
198	8	0.055428	0.200488	0.744084	1.13941
199	8	0.058074	0.254823	0.687103	2.55414
200	8	0.060719	0.209763	0.729518	0.29071
201	8	0.045681	0.206177	0.748142	1.21336
202	8	0.040151	0.263161	0.696688	1.02370
203	8	0.072230	0.281460	0.646310	1.40580
204	8	0.064366	0.269816	0.665819	0.97704
205	8	0.035993	0.191422	0.772585	2.09909
206	9	0.091638	0.215290	0.693073	1.03679
207	9	0.072171	0.236658	0.691171	2.36788
208	9	0.056187	0.195345	0.748468	3.45908
209	9	0.095888	0.229586	0.674526	3.63796
210	9	0.069809	0.219558	0.710633	2.56887
211	9	0.142920	0.223801	0.633279	2.00319
212	9	0.087323	0.196401	0.716276	2.40644
213	9	0.064517	0.218711	0.716772	2.58552
214	9	0.086882	0.194778	0.718341	8.94023
215	9	0.067463	0.219228	0.713309	3.75275
216	9	0.105610	0.230661	0.663730	0.34082
217	9	0.138992	0.283123	0.577885	1.62649
218	9	0.081364	0.186967	0.731670	2.31678
219	9	0.114535	0.221751	0.663714	1.77709
220	9	0.069940	0.280622	0.649438	1.38765
221	9	0.073137	0.143219	0.783643	3.46442
222	9	0.096326	0.243241	0.660434	1.74696
223	9	0.083284	0.202951	0.713765	1.28613
224	9	0.179133	0.299403	0.521465	1.15897

Source: Courtesy of the authors, Gallant and Koenker (1984).

that optimal allocation of E to q_1, q_2, and q_3 can be computed by solving

$$\text{maximize} \quad u_{(1)}(q_1, q_2, q_3)$$

$$\text{subject to} \quad \sum_{i=1}^{3} p_i q_i = E.$$

Since this problem has exactly the same structure as the original problem, one just applies the previous theory with $N = 3$ and $Y = E$.

There is a problem in passing from the deterministic version of the subsystem to the stochastic specification. One usually prefers to regard prices and income as independent variables and condition the analysis on p and Y. Expenditure in the subsystem, from this point of view, is to be regarded as stochastic with a location parameter depending on p, on Y, and possibly on demographic characteristics, viz.

$$E = f(p, Y, \text{etc.}) + \text{error.}$$

For now, we shall ignore this problem, implicitly treating it as an errors in variables problem of negligible consequence. That is, we assume that in observing E we are actually observing $f(p, Y, \text{etc.})$ with negligible error, so that an analysis conditioned on E will be adequate. In Chapter 6 we shall present methods that take formal account of this problem.

Table 1b. Experimental Rates in Effect on a Weekday in July 1978.

	Price (cents per kwh)		
Treatment	Base	Intermediate	Peak
1	1.06	2.86	3.90
2	1.78	2.86	3.90
3	1.06	3.90	3.90
4	1.78	3.90	3.90
5	1.37	3.34	5.06
6	1.06	2.86	6.56
7	1.78	2.86	6.56
8	1.06	3.90	6.56
9	1.78	3.90	6.56

Table 1c. Consumer Demographic Characteristics.

t	Family Size	Income ($ per yr)	Residence		Elec. Range (1=yes)	Washer (1=yes)	Dryer (1=yes)	Air Condition.	
			Size (SqFt)	Heat Loss (Btuh)				Central (1=yes)	Window (Btuh)
1	2	17000	600	4305	0	1	0	0	13000
2	6	13500	900	9931	1	1	0	0	0
3	2	7000	1248	18878	1	1	0	0	0
4	3	11000	1787	17377	1	1	0	0	0
5	4	27500	2900	24894	1	0	0	1	5000
6	3	13500	2000	22526	1	1	1	0	24000
7	4	22500	3800	17335	1	1	1	1	0
8	7	3060	216	4496	1	0	0	0	0
9	3	7000	1000	8792	0	1	1	0	18000
10	1	6793	1200	14663	0	0	0	0	.
11	5	11000	1000	14480	1	1	0	0	0
12	5	17000	704	3192	1	1	1	1	24000
13	3	5500	2100	8631	1	1	0	1	0
14	2	13500	1400	19720	1	1	1	0	19000
15	4	22500	1252	7386	1	1	1	0	24000
16	7	17000	916	7194	0	1	0	0	0
17	2	11000	1800	17957	1	1	1	1	0
18	2	13500	780	4641	1	1	0	1	0
19	3	6570	960	11396	1	1	0	0	24000
20	4	9000	768	8195	1	1	1	0	0
21	2	11000	1200	7812	1	1	1	1	10000
22	4	13500	900	8878	1	1	1	1	0
23	3	40000	2200	15098	1	1	1	0	0
24	5	7000	1000	7041	1	1	0	0	10000
25	3	13500	720	5130	0	1	1	0	0
26	2	13500	550	7532	1	1	0	0	12000
27	4	17000	1600	9674	1	1	1	1	0
28	4	27500	2300	13706	1	1	0	1	0
29	6	15797	1000	10372	1	1	1	0	10000
30	2	11000	880	7477	0	1	1	0	19000
31	4	9000	1200	14013	1	1	1	0	0
32	4	17052	2200	15230	1	1	0	0	0
33	2	14812	1080	13170	1	0	0	0	0
34	3	27500	870	10843	1	1	1	0	18500
35	2	4562	800	9393	1	1	1	0	6000
36	2	7000	1200	11395	1	1	0	0	0
37	3	9000	900	6175	1	1	0	0	23000
38	2	4711	1500	17655	1	0	0	0	0
39	5	14652	1500	11916	1	1	1	0	0
40	4	70711	2152	16552	1	1	1	1	0
41	2	7000	832	4316	1	1	1	1	0
42	3	22500	1700	9209	1	1	1	1	0
43	11	4500	1248	9607	1	1	0	0	0
44	5	11000	1808	19400	1	1	1	0	28000
45	6	22500	1800	19981	1	1	1	1	0

Table 1c. (Continued).

t	Family Size	Income ($ per yr)	Size (SqFt)	Heat Loss (Btuh)	Elec. Range (1=yes)	Washer (1=yes)	Dryer (1=yes)	Central (1=yes)	Window (Btuh)
								Air Condition.	
46	4	22500	1800	18573	0	0	0	1	0
47	3	40000	4200	16264	1	1	1	1	0
48	2	9000	1400	10541	1	1	1	0	24000
49	2	13500	2500	29231	1	1	0	0	16000
50	6	17000	1300	5805	1	1	1	0	21000
51	3	11000	780	5894	1	1	1	1	0
52	1	4500	1000	13714	0	0	0	0	6000
53	2	11267	960	7863	1	1	0	0	0
54	3	2500	1000	12973	1	1	0	0	0
55	1	7430	1170	9361	1	1	1	0	0
56	4	17000	2900	12203	1	1	1	1	0
57	1	22500	1000	10131	0	1	0	0	0
58	3	22500	1250	12773	1	1	1	0	12000
59	3	7000	1400	11011	1	1	1	0	29000
60	1	2500	835	12730	1	0	0	0	0
61	1	13500	1300	7196	1	1	0	0	32000
62	7	11000	540	7798	1	1	0	0	0
63	4	14381	1100	8700	1	1	1	0	30000
64	2	9000	900	5726	1	0	0	0	12000
65	3	11000	720	3854	1	1	1	1	0
66	5	5500	780	6236	1	1	0	1	0
67	4	40000	1450	8160	1	1	1	0	28000
68	2	3500	1100	10102	1	1	0	0	12000
69	2	17000	3000	36124	1	1	0	1	0
70	4	11000	1534	15711	1	0	0	0	0
71	2	40000	2000	11250	1	1	1	1	0
72	2	2500	1400	15040	0	0	0	0	6000
73	4	17000	1400	13544	1	0	1	1	0
74	2	1500	656	7383	1	0	0	0	0
75	3	9000	772	13229	1	0	0	0	1800
76	1	9000	600	4035	1	1	0	0	0
77	5	5500	500	6110	1	0	0	0	0
78	3	13500	1200	11097	1	1	1	0	10000
79	2	13590	1300	12869	1	0	0	0	24000
80	4	11000	1045	11224	1	1	0	0	0
81	2	9687	768	7565	1	1	1	0	10000
82	2	17000	1100	9159	0	1	1	0	10000
83	11	4500	480	6099	1	1	0	0	0
84	5	13500	1976	12498	1	1	1	0	0
85	4	40000	2500	23213	1	1	1	0	.
86	5	22500	2100	12314	1	1	1	1	0
87	3	3500	1196	14725	0	0	0	0	0
88	3	12100	950	11174	0	0	0	0	0
89	3	3500	1080	12186	1	0	0	0	0
90	2	7000	1400	10050	1	1	0	0	28000

Table 1c.   (Continued).

t	Family Size	Income ($ per yr)	Size (SqFt)	Heat Loss (Btuh)	Elec. Range (1=yes)	Washer (1=yes)	Dryer (1=yes)	Central (1=yes)	Window (Btuh)
			Residence					Air Condition.	
91	2	3500	1800	16493	1	1	1	0	2000
92	2	7000	1456	17469	0	1	0	0	18000
93	4	9000	1100	6177	1	1	1	0	23000
94	2	3500	1500	21659	1	1	1	0	18000
95	4	9894	720	6133	1	1	1	0	6000
96	1	22500	1500	7952	1	0	0	1	0
97	4	13500	1500	10759	1	0	1	1	0
98	4	17000	1900	10176	1	1	1	1	0
99	2	17000	1100	10869	1	1	1	0	23000
100	5	27500	2300	16610	1	1	1	1	0
101	3	13500	1500	11304	1	1	1	1	0
102	2	27500	3000	23727	1	1	1	1	0
103	4	24970	2280	18602	1	1	1	1	0
104	2	3500	970	10065	1	1	0	0	0
105	2	17000	1169	10810	1	1	0	0	30000
106	2	13500	1800	20614	1	1	1	0	0
107	2	13500	728	4841	1	1	1	1	0
108	2	11000	1500	11235	1	1	1	1	0
109	3	17000	1500	9774	1	1	0	1	0
110	5	5500	900	12085	1	1	0	0	23000
111	3	17000	1500	17859	1	1	1	1	0
112	1	70711	2600	16661	1	1	1	1	0
113	3	7000	780	5692	1	1	1	0	20000
114	4	22500	1600	8191	1	1	1	1	0
115	2	13500	600	5086	0	1	1	0	2000
116	3	4500	1200	14178	1	1	1	0	1000
117	5	17000	900	8966	1	1	1	0	18000
118	4	13500	1500	11142	1	1	1	1	0
119	5	17000	2000	19555	1	1	1	1	0
120	3	23067	1740	10183	1	1	1	0	42000
121	1	17000	696	5974	1	0	0	0	0
122	1	2500	900	10111	1	1	0	0	0
123	2	7265	970	20437	1	1	0	0	0
124	2	10415	1500	9619	1	0	0	0	0
125	3	5500	750	16955	0	0	1	0	18000
126	2	4500	824	11647	1	1	0	0	0
127	1	22500	1900	11401	1	0	1	1	0
128	4	40000	2500	15205	1	1	1	1	0
129	2	4500	840	5984	1	1	1	1	0
130	1	22500	1800	18012	1	1	1	1	0
131	2	5500	1200	8447	1	1	0	0	1000
132	1	3689	576	12207	0	0	0	0	0
133	3	16356	1600	16227	0	1	1	0	28500
134	4	11000	1360	17045	1	1	0	0	0
135	3	5500	600	4644	0	1	0	0	9000

Table 1c. (Continued).

t	Family Size	Income ($ per yr)	Residence Size (SqFt)	Heat Loss (Btuh)	Elec. Range (1=yes)	Washer (1=yes)	Dryer (1=yes)	Air Condition. Central (1=yes)	Window (Btuh)
136	3	17000	2000	16731	1	1	1	1	2300
137	2	32070	6000	61737	1	1	1	1	0
138	2	27500	1250	7397	1	1	1	1	0
139	4	17000	840	5426	1	1	1	1	0
140	4	27500	3300	11023	1	1	1	1	0
141	2	11000	1200	10888	1	0	0	0	18000
142	1	.	1000	5446	1	0	0	0	0
143	3	36919	1200	8860	1	1	1	1	0
144	5	9000	720	5882	1	1	1	0	10000
145	5	21400	1300	6273	1	1	1	0	0
146	1	1500	375	6727	0	0	0	0	0
147	2	5063	1008	7195	1	0	0	0	0
148	1	3500	1650	13164	1	0	0	1	0
149	1	9488	850	9830	0	0	1	0	10000
150	1	27500	1200	8469	1	1	1	1	0
151	5	17000	1000	8006	0	1	1	0	16000
152	3	11000	2000	12608	1	1	1	1	0
153	7	22500	1225	11505	1	0	0	1	0
154	6	3500	1200	16682	1	1	0	0	0
155	3	9273	600	5078	1	1	0	0	15000
156	8	17000	1100	17912	1	0	0	0	0
157	3	17459	980	7984	0	1	1	1	0
158	5	11000	1200	14113	1	1	1	0	18000
159	3	9000	1600	21529	1	1	1	0	6000
160	2	11000	899	5731	0	1	1	0	28000
161	3	12068	1350	16331	1	1	1	0	6000
162	2	7000	672	8875	1	1	0	0	0
163	3	22500	1200	10424	1	1	0	0	23000
164	2	5500	1300	8636	1	1	1	1	0
165	2	12519	1000	24210	1	1	1	0	37000
166	2	29391	1400	12837	1	1	1	1	0
167	2	9000	400	4519	1	0	0	0	0
168	3	4664	1235	14274	1	1	0	0	6000
169	4	11000	720	6393	0	1	1	0	23000
170	.	.	.	.	.	.	.	.	.
171	3	18125	2300	16926	1	1	0	1	0
172	.	.	.	.	.	.	.	.	.
173	5	9000	720	6439	1	1	1	0	0
174	6	5500	1000	13651	1	1	0	0	0
175	5	14085	1400	14563	1	1	0	0	15000
176	2	9000	720	6540	0	1	1	1	0
177	6	17000	1470	8439	1	1	1	1	0
178	4	27500	1900	12345	1	1	1	1	18500
179	3	7000	480	3796	0	0	0	0	10000
180	3	13500	1300	7352	1	1	0	0	23000

Table 1c. (Continued).

t	Family Size	Income ($ per yr)	Size (SqFt)	Heat Loss (Btuh)	Elec. Range (1=yes)	Washer (1=yes)	Dryer (1=yes)	Central (1=yes)	Window (Btuh)
181	3	13437	1200	9502	1	1	1	1	0
182	3	14150	1300	8334	1	1	0	0	0
183	1	7000	1200	11941	1	1	0	0	21000
184	4	27500	1350	7585	1	1	1	1	0
185	2	32444	2900	15158	1	1	0	1	0
186	1	4274	400	7859	1	0	0	0	0
187	1	3500	600	14441	0	0	0	0	0
188	4	27500	2000	15462	1	1	1	1	0
189	4	40000	2900	13478	1	1	0	1	0
190	6	17000	5000	24132	1	0	1	1	0
191	1	2500	1400	17016	1	1	0	0	2000
192	7	9000	1400	13293	1	1	0	0	0
193	.	.	.	.	0	0	0	.	0
194	4	13500	780	5629	1	1	0	1	0
195	5	13500	1000	7281	1	1	1	1	0
196	2	13500	1169	11273	1	1	0	0	12000
197	2	40000	2400	13515	1	1	0	1	0
198	4	27500	1320	9865	1	1	1	0	29000
199	4	27500	1250	5759	1	1	1	1	0
200	1	3449	1200	18358	0	0	0	0	0
201	2	3500	425	4554	1	0	0	0	.
202	2	27500	1400	13496	1	0	0	1	0
203	4	7000	1300	11555	1	1	1	0	14000
204	2	3500	1800	23271	1	1	0	0	0
205	4	11000	720	5879	1	1	1	0	16000
206	7	9000	680	11528	1	0	0	0	0
207	4	14077	780	4829	1	1	1	0	10000
208	3	13500	2200	22223	1	1	1	0	24000
209	4	17000	1342	12050	1	1	1	1	0
210	4	3500	628	5369	1	1	1	0	24000
211	2	11000	920	5590	1	1	1	1	0
212	5	9000	1300	11510	1	1	1	0	19000
213	3	5500	1400	18584	1	1	1	0	23000
214	5	27500	2300	15480	1	1	1	1	0
215	3	20144	1700	11212	1	1	1	1	0
216	5	3500	1080	13857	0	0	0	0	0
217	2	22500	1800	17588	1	1	0	0	23000
218	6	22500	1900	15115	1	1	1	0	22000
219	5	6758	1200	16868	1	0	0	0	0
220	6	11000	2200	21884	1	1	1	1	0
221	3	17000	1500	11504	1	1	1	1	0
222	2	9000	600	5825	1	0	1	1	0
223	2	15100	1932	15760	1	1	1	0	0
224	1	7000	979	11700	1	1	1	0	1000

284

Table 1c. (Continued).

t	Type of Residence			Elec. Water Heater (1=yes)	Freezer (kw)	Refrigerator (kw)
	Detached (1=yes)	Duplex or Apartment (1=yes)	Mobile Home (1=yes)			
1	0	0	1	1	0	0.700
2	1	0	0	1	1.320	0.700
3	1	0	0	1	1.320	0.700
4	1	0	0	1	1.320	2.495
5	1	0	0	0	1.320	3.590
6	1	0	0	1	0	1.795
7	1	0	0	1	0	1.795
8	1	0	0	0	1.320	0.700
9	1	0	0	1	1.320	0.700
10	1	0	0	0	1.985	1.795
11	1	0	0	1	2.640	0.700
12	0	0	1	1	1.985	1.795
13	1	0	0	1	1.320	1.795
14	1	0	0	1	1.320	1.795
15	1	0	0	1	1.320	1.795
16	0	0	1	1	0	1.795
17	1	0	0	1	1.985	1.795
18	0	0	1	1	0	0.700
19	1	0	0	1	1.320	1.795
20	1	0	0	1	0	1.795
21	1	0	0	1	1.320	0.700
22	1	0	0	1	1.320	0.700
23	1	0	0	1	3.305	1.795
24	0	1	0	1	0	0.700
25	0	0	1	1	1.320	0.700
26	1	0	0	1	1.985	0.700
27	1	0	0	1	1.320	1.795
28	1	0	0	1	1.320	1.795
29	1	0	0	1	0	1.795
30	0	0	1	1	0	0.700
31	1	0	0	1	1.320	0.700
32	1	0	0	1	0	1.795
33	1	0	0	1	1.320	3.590
34	1	0	0	1	0	1.795
35	1	0	0	1	1.320	1.795
36	1	0	0	1	1.320	0.700
37	1	0	0	1	1.320	1.795
38	1	0	0	1	0	0.700
39	1	0	0	1	0	1.795
40	1	0	0	1	1.320	1.400
41	1	0	0	1	1.985	1.795
42	1	0	0	1	2.640	0.700
43	1	0	0	1	1.320	1.795
44	1	0	0	1	3.970	1.795
45	1	0	0	1	1.985	1.795

Table 1c. (Continued).

t	Type of Residence Detached (1=yes)	Duplex or Apartment (1=yes)	Mobile Home (1=yes)	Elec. Water Heater (1=yes)	Freezer (kw)	Refrigerator (kw)
46	1	0	0	0	0	1.795
47	1	0	0	1	0	1.795
48	1	0	0	1	0	0.700
49	1	0	0	1	0	2.495
50	1	0	0	1	1.985	1.795
51	0	0	1	1	1.985	0.700
52	1	0	0	0	1.320	1.795
53	0	0	1	1	0	0.700
54	1	0	0	1	1.320	1.795
55	1	0	0	1	1.320	1.795
56	1	0	0	1	1.320	1.795
57	1	0	0	1	1.320	1.795
58	1	0	0	1	1.320	1.795
59	1	0	0	1	1.320	1.795
60	1	0	0	0	0	0.700
61	1	0	0	1	1.320	0.700
62	1	0	0	1	1.320	0.700
63	1	0	0	1	1.320	0.700
64	0	1	0	1	0	1.795
65	0	0	1	1	0	0.700
66	0	0	1	1	0	1.795
67	1	0	0	1	1.320	1.795
68	1	0	0	1	1.320	0.700
69	1	0	0	1	0	1.400
70	1	0	0	1	1.320	1.795
71	1	0	0	1	1.985	1.795
72	1	0	0	0	1.320	1.795
73	1	0	0	0	0	1.795
74	1	0	0	0	0	0.700
75	1	0	0	0	1.320	0.700
76	1	0	0	1	0	0.700
77	0	1	0	0	1.320	0.700
78	1	0	0	1	0	1.795
79	1	0	0	0	1.320	1.795
80	1	0	0	1	1.320	1.795
81	0	0	1	1	1.320	2.495
82	1	0	0	1	1.985	1.795
83	1	0	0	1	1.320	0.700
84	1	0	0	1	1.985	1.795
85	1	0	0	1	0	1.795
86	1	0	0	1	1.320	2.495
87	1	0	0	0	0	1.795
88	1	0	0	0	3.305	0.700
89	1	0	0	1	1.985	0.700
90	1	0	0	1	1.985	1.795

t	Type of Residence			Elec. Water Heater (1=yes)	Freezer (kw)	Refrigerator (kw)
	Detached (1=yes)	Duplex or Apartment (1=yes)	Mobile Home (1=yes)			
91	1	0	0	1	0	1.795
92	1	0	0	1	1.985	1.795
93	1	0	0	1	1.320	1.795
94	1	0	0	1	0	1.795
95	0	0	1	1	0	1.795
96	1	0	0	1	1.985	0.700
97	1	0	0	0	1.320	0.700
98	1	0	0	1	1.320	0.700
99	1	0	0	1	2.640	1.795
100	1	0	0	1	1.320	1.795
101	1	0	0	1	1.320	1.795
102	1	0	0	1	0	2.495
103	1	0	0	1	1.320	1.795
104	1	0	0	1	1.320	0.700
105	1	0	0	1	1.320	0.700
106	1	0	0	1	0	1.795
107	1	0	0	1	1.320	0.700
108	1	0	0	1	1.320	0.700
109	1	0	0	1	0	1.795
110	1	0	0	1	1.320	0.700
111	1	0	0	1	1.320	1.795
112	1	0	0	1	3.970	1.795
113	0	0	1	1	0	1.795
114	1	0	0	1	1.320	1.795
115	0	0	1	1	0	0.700
116	1	0	0	1	0	1.795
117	1	0	0	1	1.320	1.795
118	1	0	0	1	1.985	0.700
119	1	0	0	1	0	0.700
120	0	0	1	1	1.320	1.795
121	0	0	1	1	0	1.795
122	1	0	0	1	0	0.700
123	1	0	0	1	0	0.700
124	1	0	0	1	0	1.795
125	1	0	0	0	1.320	1.795
126	1	0	0	1	0	0.700
127	1	0	0	0	1.320	1.795
128	1	0	0	1	0	2.495
129	0	0	1	1	1.320	1.795
130	1	0	0	1	0	1.795
131	1	0	0	1	0	0.700
132	1	0	0	0	1.320	1.795
133	1	0	0	1	1.320	0.700
134	1	0	0	1	1.320	0.700
135	0	0	1	1	0	0.700

Table 1c. (Continued).

t	Type of Residence Detached (1=yes)	Duplex or Apartment (1=yes)	Mobile Home (1=yes)	Elec. Water Heater (1=yes)	Freezer (kw)	Refrigerator (kw)
136	1	0	0	1	1.985	1.795
137	1	0	0	1	1.985	1.400
138	0	1	0	1	0	1.795
139	0	0	1	1	0	0.700
140	0	1	0	1	7.265	1.795
141	1	0	0	1	1.320	0.700
142	0	1	0	1	0	1.795
143	0	1	0	1	0	0.700
144	0	0	1	1	0	0.700
145	0	0	1	1	0	1.795
146	1	0	0	0	0	1.795
147	1	0	0	0	0	0.700
148	1	0	0	0	0	1.795
149	1	0	0	0	1.320	1.795
150	0	1	0	1	0	1.795
151	1	0	0	1	2.640	1.795
152	1	0	0	1	0	1.795
153	1	0	0	0	1.320	1.795
154	1	0	0	1	1.320	0.700
155	0	0	1	1	0	0.700
156	1	0	0	1	1.320	0.700
157	0	0	1	1	1.320	1.795
158	1	0	0	1	3.970	2.495
159	1	0	0	1	0	0.700
160	1	0	0	1	1.320	1.795
161	1	0	0	1	1.320	1.795
162	1	0	0	1	1.320	0.700
163	1	0	0	1	0	1.795
164	1	0	0	1	0	1.795
165	1	0	0	1	2.640	3.590
166	1	0	0	1	1.320	0.700
167	0	1	0	1	0	0.700
168	1	0	0	1	2.640	0.700
169	0	0	1	1	0	1.795
170	.	.	.	.	.	.
171	1	0	0	1	1.320	1.795
172	.	.	.	.	.	.
173	0	0	1	1	1.320	0.700
174	1	0	0	1	1.320	0.700
175	1	0	0	1	1.985	1.795
176	0	0	1	1	0	0.700
177	1	0	0	1	3.970	1.795
178	1	0	0	1	1.985	1.795
179	0	0	1	0	0	0.700
180	1	0	0	1	1.320	1.795

Table 1c. (Continued).

t	Type of Residence Detached (1=yes)	Duplex or Apartment (1=yes)	Mobile Home (1=yes)	Elec. Water Heater (1=yes)	Freezer (kw)	Refrigerator (kw)
181	1	0	0	1	1.985	0.700
182	1	0	0	1	1.985	1.795
183	1	0	0	1	1.320	1.795
184	1	0	0	1	1.320	1.795
185	1	0	0	1	0	2.495
186	1	0	0	1	0	1.795
187	1	0	0	0	1.320	0
188	1	0	0	1	1.320	1.795
189	1	0	0	1	0	1.795
190	1	0	0	0	1.985	2.495
191	1	0	0	1	0	1.795
192	1	0	0	1	0	1.795
193	0	0	1	0	0	0
194	0	0	1	1	1.320	1.795
195	1	0	0	1	1.320	0.700
196	1	0	0	1	1.985	1.795
197	1	0	0	1	1.985	1.795
198	0	0	1	1	0	1.795
199	1	0	0	1	1.320	1.795
200	1	0	0	0	0	0.700
201	0	1	0	1	1.985	0
202	0	1	0	1	0	1.795
203	1	0	0	1	1.320	0.700
204	1	0	0	1	0	0.700
205	0	0	1	1	1.320	0.700
206	1	0	0	0	1.320	0.700
207	0	0	1	1	0	0.700
208	1	0	0	1	1.320	2.495
209	1	0	0	1	1.320	1.795
210	1	0	0	1	0	0.700
211	0	1	0	1	0	0.700
212	1	0	0	1	0	1.795
213	1	0	0	1	1.320	1.795
214	1	0	0	1	1.985	1.795
215	1	0	0	1	1.320	1.795
216	1	0	0	0	0	0.700
217	1	0	0	1	1.320	1.795
218	1	0	0	1	1.985	1.795
219	1	0	0	0	0	0.700
220	1	0	0	1	1.320	1.795
221	1	0	0	1	1.320	1.795
222	0	1	0	1	0	1.795
223	1	0	0	1	1.320	1.795
224	1	0	0	1	0	1.795

Source: Courtesy of the authors, Gallant and Koenker (1984).

In this connection, hypothesis H_3 implies that $g(v|\theta)$ is homogeneous of degree one in v, which in turn implies that the first stage allocation function has the form

$$f(p, Y, \text{etc.}) = f\left[\Pi(p_1, p_2, p_3), p_4, \ldots, p_N, Y, \text{etc.}\right]$$

where $\Pi(p_1, p_2, p_3)$ is a price index for electricity which must itself be homogeneous of degree one in p_1, p_2, p_3 (Blackorby, Primont, and Russell, 1978, Chapter 5). This leads to major simplifications in the interpretation of results, for which see Caves and Christensen (1980).

One word of warning regarding Table 1c, all data are constructed following the protocol described in Gallant and Koenker (1984) except income. Some income values have been imputed by prediction from a regression equation. These values can be identified as those not equal to one of the values 500, 1500, 2500, 3500, 4500, 5500, 7000, 9000, 11,000, 13,500, 17,00, 22,500, 27,500, 40,000, 70,711. The listed values are the midpoints of the questionnaire's class boundaries except the last, which is the mean of an open ended interval assuming that income follows the Pareto distribution. The prediction equation includes variables not shown in Table 1c, namely age and years of education of a member of the household—the respondent or head in most instances.   □

2. LEAST SQUARES ESTIMATORS AND MATTERS OF NOTATION

Univariate responses $y_{\alpha t}$ for $t = 1, 2, \ldots, n$ and $\alpha = 1, 2, \ldots, M$ are presumed to be related to k-dimensional input vectors x_t as follows:

$$y_{\alpha t} = f_\alpha\left(x_t, \theta_\alpha^0\right) + e_{t\alpha} \qquad \alpha = 1, 2, \ldots, M \quad t = 1, 2, \ldots, n$$

where each $f_\alpha(x, \theta_\alpha)$ is a known function, each θ_α^0 is a p_α-dimensional vector of unknown parameters, and the $e_{\alpha t}$ represent unobservable observational or experimental errors. As previously, we write θ_α^0 to emphasize that it is the true, but unknown, value of the parameter vector θ_α that is meant; θ_α itself is used to denote instances when the parameter vector is treated as a variable. Writing

$$e_t = \begin{pmatrix} e_{1t} \\ e_{2t} \\ \vdots \\ e_{Mt} \end{pmatrix}$$

the error vectors e_t are assumed to be independently and identically distributed with mean zero and unknown variance-covariance matrix Σ,

$$\Sigma = \mathscr{C}(e_t, e_t') \qquad t = 1, 2, \ldots, n$$

whence

$$\mathscr{C}(e_{\alpha t}, e_{\beta s}) = \begin{cases} \sigma_{\alpha\beta} & t = s \\ 0 & t \neq s \end{cases}$$

with $\sigma_{\alpha\beta}$ denoting the elements of Σ.

In the literature one finds two conventions for writing this model in a vector form. One emphasizes the fact that the model consists of M separate univariate nonlinear regression equations

$$y_\alpha = f_\alpha(\theta_\alpha^0) + e_\alpha \qquad \alpha = 1, 2, \ldots, M$$

with y_α an n-vector as described below, and the other emphasizes the fact that the data consists of repeated observations on the same subject

$$y_t = f(x_t, \theta^0) + e_t \qquad t = 1, 2, \ldots, n$$

with y_t an M-vector. To have labels to distinguish the two, we shall refer to the first arrangement of the data as grouped by equation, and the second as grouped by subject.

The grouped by equation arrangement follows the same notational convention used in Chapter 1. Write

$$y_\alpha = \underset{n}{\left(\begin{matrix} y_{\alpha 1} \\ y_{\alpha 2} \\ \vdots \\ y_{\alpha n} \end{matrix}\right)_1}$$

$$f_\alpha(\theta_\alpha) = \underset{n}{\left(\begin{matrix} f_\alpha(x_1, \theta_\alpha) \\ f_\alpha(x_2, \theta_\alpha) \\ \vdots \\ f_\alpha(x_n, \theta_\alpha) \end{matrix}\right)_1}$$

$$e_\alpha = \underset{n}{\left(\begin{matrix} e_{\alpha 1} \\ e_{\alpha 2} \\ \vdots \\ e_{\alpha n} \end{matrix}\right)_1}.$$

In this notation, each regression is written as

$$y_\alpha = f_\alpha(\theta_\alpha^0) + e_\alpha \qquad \alpha = 1, 2, \ldots, M$$

with (Problem 1)

$$\mathscr{C}(e_\alpha, e_\beta') = \sigma_{\alpha\beta} I$$

of order n by n. Denote the Jacobian of $f_\alpha(\theta_\alpha)$ by

$$F_\alpha(\theta_\alpha) = \frac{\partial}{\partial \theta_\alpha'} f_\alpha(\theta_\alpha)$$

which is of order n by p_α. Illustrating with Example 1, we have:

EXAMPLE 1 (Continued). The independent variables are the logarithms of expenditure normalized prices. From Tables 1a and 1b we obtain a few instances

$$x_1 = \ln\left(\frac{(3.90, 2.86, 1.06)}{0.46931}\right)' = (2.11747, 1.80731, 0.81476)'$$

$$x_2 = \ln\left(\frac{(3.90, 2.86, 1.06)}{0.79539}\right)' = (1.58990, 1.27974, 0.28719)'$$

$$\vdots$$

$$x_{19} = \ln\left(\frac{(3.90, 2.86, 1.06)}{1.37160}\right)' = (1.04500, 0.73484, -0.25771)'$$

$$x_{20} = \ln\left(\frac{(3.90, 2.86, 1.78)}{0.92766}\right)' = (1.43607, 1.12591, 0.65170)'$$

$$\vdots$$

$$x_{40} = \ln\left(\frac{(3.90, 2.86, 1.78)}{2.52983}\right)' = (0.43282, 0.12267, -0.35154)'$$

$$x_{41} = \ln\left(\frac{(3.90, 3.90, 1.06)}{1.14741}\right)' = (1.22347, 1.22347, -0.079238)'$$

$$\vdots$$

$$x_{224} = \ln\left(\frac{(6.56, 3.90, 1.78)}{1.15897}\right)' = (1.73346, 1.21344, 0.42908)'.$$

The vectors of dependent variables are for $\alpha = 1$

$$y_\alpha = \begin{pmatrix} \ln(0.662888/0.056731) \\ \ln(0.644427/0.103444) \\ \vdots \\ _{224}\left(\ln(0.521465/0.179133)\right) \end{pmatrix}_1 = \begin{pmatrix} 2.45829 \\ 1.82933 \\ \vdots \\ _{224}\left(1.06851\right) \end{pmatrix}_1$$

and for $\alpha = 2$

$$y_\alpha = \begin{pmatrix} \ln(0.280382/0.056731) \\ \ln(0.252128/0.103444) \\ \vdots \\ _{224}\left(\ln(0.299403/0.179133)\right) \end{pmatrix}_1 = \begin{pmatrix} 1.59783 \\ 0.89091 \\ \vdots \\ _{223}\left(0.51366\right) \end{pmatrix}_1.$$

Recall that

$$y_{1t} = \ln\frac{a_1 + x_t'b_{(1)}}{a_3 + x_t'b_{(3)}} + e_{1t}$$

$$y_{2t} = \ln\frac{a_2 + x_t'b_{(2)}}{a_3 + x_3'b_{(3)}} + e_{2t}$$

with $b_{(\alpha)}$ denoting the αth row of

$$B = \begin{pmatrix} b_{11} & b_{12} & b_{13} \\ b_{21} & b_{22} & b_{23} \\ b_{31} & b_{32} & b_{33} \end{pmatrix}$$

and with $a' = (a_1, a_2, a_3)$. Note that if both a and B are multiplied by some common factor δ to obtain $\bar{a} = \delta a$ and $\bar{B} = \delta B$, we shall have

$$\frac{a_\alpha + x'b_{(\alpha)}}{a_3 + x'b_{(3)}} = \frac{\bar{a}_\alpha + x'\bar{b}_{(\alpha)}}{\bar{a}_3 + x'\bar{b}_{(3)}}.$$

Thus the parameters of the model can only be determined to within a scalar multiple. In order to estimate the model it is necessary to impose a normalization rule. Our choice is to set $a_3 = -1$. With this choice we write the model as

$$y_{1t} = f_1(x_t, \theta_1^0) + e_{1t}$$

$$y_{2t} = f_2(x_t, \theta_2^0) + e_{2t}$$

with

$$f_\alpha(x, \theta_\alpha) = \ln \frac{a_\alpha + b'_{(\alpha)}x}{-1 + b'_{(3)}x} \qquad \alpha = 1, 2$$

$$\theta_1' = (a_1, b_{11}, b_{12}, b_{13}, b_{31}, b_{32}, b_{33})$$

$$\theta_2' = (a_2, b_{21}, b_{22}, b_{23}, b_{31}, b_{32}, b_{33}). \qquad \qquad \square$$

Recognizing that what we have is M instances of the univariate nonlinear regression model of Chapter 1, we can apply our previous results and estimate the parameters θ_α^0 of each model by computing $\hat{\theta}_\alpha^\#$ to minimize

$$\text{SSE}_\alpha(\theta_\alpha) = [y_\alpha - f_\alpha(\theta_\alpha)]'[y_\alpha - f_\alpha(\theta_\alpha)]$$

for $\alpha = 1, 2, \ldots, M$. This done, the elements $\sigma_{\alpha\beta}$ of Σ can be estimated by

$$\hat{\sigma}_{\alpha\beta} = \frac{[y_\alpha - f_\alpha(\hat{\theta}_\alpha^\#)]'[y_\beta - f_\beta(\hat{\theta}_\beta^\#)]}{n} \qquad \alpha, \beta = 1, 2, \ldots, M.$$

Let $\hat{\Sigma}$ denote the M by M matrix with typical element $\hat{\sigma}_{\alpha\beta}$. Equivalently, if we write

$$\hat{e}_\alpha = y_\alpha - f_\alpha(\hat{\theta}_\alpha^\#) \qquad \alpha = 1, 2, \ldots, M$$

$$\hat{E} = [\hat{e}_1 \vdots \hat{e}_2 \vdots \cdots \vdots \hat{e}_M]$$

then

$$\hat{\Sigma} = \frac{1}{n} \hat{E}'\hat{E}.$$

We illustrate with Example 1.

EXAMPLE 1 (Continued). Fitting

$$y_1 = f_1(\theta_1) + e_1$$

SAS Statements:

```
PROC NLIN DATA=EXAMPLE1 METHOD=GAUSS ITER=50 CONVERGENCE=1.E-13;
PARMS B11=0 B12=0 B13=0 B31=0 B32=0 B33=0 A1=-9; A3=-1;
PEAK=A1+B11*X1+B12*X2+B13*X3;    BASE=A3+B31*X1+B32*X2+B33*X3;
MODEL Y1=LOG(PEAK/BASE);
DER.A1 =1/PEAK;
DER.B11=1/PEAK*X1;   DER.B31=-1/BASE*X1;
DER.B12=1/PEAK*X2;   DER.B32=-1/BASE*X2;
DER.B13=1/PEAK*X3;   DER.B33=-1/BASE*X3;
OUTPUT OUT=WORKO2 RESIDUAL=E1;
```

Output:

<pre>
              S T A T I S T I C A L   A N A L Y S I S   S Y S T E M   1

                   NON-LINEAR LEAST SQUARES ITERATIVE PHASE

              DEPENDENT VARIABLE: Y1   METHOD: GAUSS-NEWTON

   ITERATION   B11   B12   B13   RESIDUAL SS
                      B31   B32   B33
                      A1

       0   0.000000E+00   0.000000E+00   0.000000E+00   72.21326991
                0.000000E+00   0.000000E+00   0.000000E+00
               -9.00000000

       .
       .
       .

      16   -0.83862780   -1.44241315   2.01535561   36.50071896
                0.46865734   -0.19468166   -0.38299626
               -1.98254583

NOTE: CONVERGENCE CRITERION MET.

              S T A T I S T I C A L   A N A L Y S I S   S Y S T E M   3

        NON-LINEAR LEAST SQUARES SUMMARY STATISTICS   DEPENDENT VARIABLE Y1

           SOURCE   DF   SUM OF SQUARES   MEAN SQUARE

           REGRESSION   7   1019.72335676   145.67476525
           RESIDUAL   217   36.50071896   0.16820608
           UNCORRECTED TOTAL 224   1056.22407572

           (CORRECTED TOTAL) 223   70.01946051

PARAMETER   ESTIMATE   ASYMPTOTIC   ASYMPTOTIC 95 %
                                 STD. ERROR   CONFIDENCE INTERVAL
                                                     LOWER   UPPER
B11   -0.83862780   1.37155782   -3.54194099   1.86468538
B12   -1.44241315   1.87671707   -5.14138517   2.25655887
B13   2.01535561   1.44501283   -0.83273595   4.86344716
B31   0.46865734   0.12655505   0.21921985   0.71809482
B32   -0.19468166   0.21864114   -0.62561901   0.23625569
B33   -0.38299626   0.09376286   -0.56780098   -0.19819153
A1   -1.98254583   1.03138455   -4.01538427   0.05029260
</pre>

Figure 1a. First equation of Example 1 fitted by the modified Gauss-Newton method.

SAS Statements:

```
PROC NLIN DATA=EXAMPLE1 METHOD=GAUSS ITER=50 CONVERGENCE=1.E-13;
PARMS B21=0 B22=0 B23=0 B31=0 B32=0 B33=0 A2=-3; A3=-1;
INTER=A2+B21*X1+B22*X2+B23*X3;   BASE=A3+B31*X1+B32*X2+B33*X3;
MODEL Y2=LOG(INTER/BASE);
DER.A2 =1/INTER;
DER.B21=1/INTER*X1;   DER.B31=-1/BASE*X1;
DER.B22=1/INTER*X2;   DER.B32=-1/BASE*X2;
DER.B23=1/INTER*X3;   DER.B33=-1/BASE*X3;
OUTPUT OUT=WORK03 RESIDUAL=E2;
```

Output:

S T A T I S T I C A L   A N A L Y S I S   S Y S T E M   4

NON-LINEAR LEAST SQUARES ITERATIVE PHASE

DEPENDENT VARIABLE: Y2   METHOD: GAUSS-NEWTON

ITERATION	B21 B31 A2	B22 B32	B23 B33	RESIDUAL SS
0	0.000000E+00 0.000000E+00 -3.00000000	0.000000E+00 0.000000E+00	0.000000E+00 0.000000E+00	37.16988980
.				
.				
.				
16	0.41684196 0.24777391 -1.11401781	-1.30951752 0.07675306	0.73956410 -0.39514717	19.70439405

NOTE: CONVERGENCE CRITERION MET.

S T A T I S T I C A L   A N A L Y S I S   S Y S T E M   6

NON-LINEAR LEAST SQUARES SUMMARY STATISTICS   DEPENDENT VARIABLE Y2

SOURCE	DF	SUM OF SQUARES	MEAN SQUARE
REGRESSION	7	265.36865902	37.90980843
RESIDUAL	217	19.70439405	0.09080366
UNCORRECTED TOTAL	224	285.07305307	
(CORRECTED TOTAL)	223	36.70369496	

PARAMETER	ESTIMATE	ASYMPTOTIC STD. ERROR	ASYMPTOTIC 95 % CONFIDENCE INTERVAL	
			LOWER	UPPER
B21	0.41684196	0.44396622	-0.45820663	1.29189056
B22	-1.30951752	0.60897020	-2.50978567	-0.10924936
B23	0.73956410	0.54937638	-0.34324582	1.82237401
B31	0.24777391	0.13857700	-0.02535860	0.52090642
B32	0.07675306	0.18207332	-0.28210983	0.43561595
B33	-0.39514717	0.08932410	-0.57120320	-0.21909114
A2	-1.11401781	0.34304923	-1.79016103	-0.43787460

Figure 1b. Second equation of Example 1 fitted by the modified Gauss-Newton method.

by the methods of Chapter 1 we have from Figure 1a that

$$
\hat{\theta}_1^{\#} = \begin{pmatrix} \hat{a}_1 \\ \hat{b}_{11} \\ \hat{b}_{12} \\ \hat{b}_{13} \\ \hat{b}_{31} \\ \hat{b}_{32} \\ \hat{b}_{33} \end{pmatrix} = \begin{pmatrix} -1.9825483 \\ -0.83862780 \\ -1.44241315 \\ 2.01535561 \\ 0.46865734 \\ -0.19468166 \\ -0.38299626 \end{pmatrix}
$$

and from Figure 1b that

$$
\hat{\theta}_2^{\#} = \begin{pmatrix} \hat{a}_2 \\ \hat{b}_{21} \\ \hat{b}_{22} \\ \hat{b}_{23} \\ \hat{b}_{31} \\ \hat{b}_{32} \\ \hat{b}_{33} \end{pmatrix} = \begin{pmatrix} -1.11401781 \\ 0.41684196 \\ -1.30951752 \\ 0.73956410 \\ 0.24777391 \\ 0.07675306 \\ -0.39514717 \end{pmatrix}.
$$

Some aspects of these computations deserve comment. In this instance, the convergence of the modified Gauss-Newton method is fairly robust to the choice of starting values so we have taken the simple expedient of starting with a value $_0\theta_\alpha$ with $f_\alpha(x, {}_0\theta_\alpha) \doteq \bar{y}_\alpha$. The first full step away from $_0\theta_\alpha$,

$$
_1\theta_\alpha = {}_0\theta_\alpha + \left[F_\alpha'({}_0\theta_\alpha) F_\alpha({}_0\theta_\alpha) \right]^{-1} F_\alpha'({}_0\theta_\alpha) \left[y_\alpha - f_\alpha({}_0\theta_\alpha) \right]
$$

is such that

$$
\frac{_1a_\alpha + {}_1b_{(\alpha)}x}{-1 + {}_1b_{(3)}'x}
$$

is negative for some of the x_i; this results in an error condition when taking logarithms. Obviously one need only take care to choose a step length $_0\lambda_\alpha$ small enough that

$$
_1\theta_\alpha = {}_0\theta_\alpha + {}_0\lambda_\alpha \left[F_\alpha'({}_0\theta_\alpha) F_\alpha({}_0\theta_\alpha) \right]^{-1} F'({}_0\theta_\alpha) \left[y_\alpha - f_\alpha({}_0\theta_\alpha) \right]
$$

SAS Statements:

```
DATA WORK04; MERGE WORK02 WORK03; KEEP T E1 E2;
PROC MATRIX FW=20; FETCH E DATA=WORK04(KEEP=E1 E2);
SIGMA=E'*E#/224; PRINT SIGMA; P=HALF(INV(SIGMA)); PRINT P;
```

Output:

S T A T I S T I C A L   A N A L Y S I S   S Y S T E M   7

SIGMA	COL1	COL2
ROW1	0.1629496382006	0.09015433203941
ROW2	0.09015433203941	0.08796604486025

P	COL1	COL2
ROW1	3.764814163903	-3.85846955764
ROW2	0	3.371649857133

Figure 1c. Contemporaneous variance-covariance matrix of Example 1 estimated from single equation residuals.

is in range to avoid this difficulty. Thus, this situation is not a problem for properly written code. Other than cluttering up the output (suppressed in the figures), the SAS code seems to behave reasonably well. See Problem 7 for another approach to this problem.

Lastly, we compute

$$\hat{\Sigma} = \begin{pmatrix} 0.1629496382006 & 0.09015433203941 \\ 0.09015433203941 & 0.08796604486025 \end{pmatrix}$$

as shown in Figure 1c. For later use we compute

$$\hat{P} = \begin{pmatrix} 3.764814163903 & -3.85846955764 \\ 0 & 3.371659857133 \end{pmatrix}$$

with $\hat{\Sigma}^{-1} = \hat{P}'\hat{P}$. □

The set of M regressions can be arranged in a single regression

$$y = f(\theta^0) + e$$

by writing

$$
y = \begin{matrix} & \begin{pmatrix} y_1 \\ y_2 \\ \vdots \\ {}_{nM}\!\begin{pmatrix} y_M \end{pmatrix}_1 \end{pmatrix} \end{matrix}
$$

$$
f(\theta) = {}_{nM}\begin{pmatrix} f_1(\theta_1) \\ f_2(\theta_2) \\ \vdots \\ f_M(\theta_M) \end{pmatrix}_1
$$

$$
e = {}_{nM}\begin{pmatrix} e_1 \\ e_2 \\ \vdots \\ e_M \end{pmatrix}_1
$$

$$
\theta = {}_{p}\begin{pmatrix} \theta_1 \\ \theta_2 \\ \vdots \\ \theta_M \end{pmatrix}_1
$$

with $p = \sum_{\alpha=1}^{M} p_\alpha$. In order to work out the variance-covariance matrix of e, let us review Kronecker product notation.

If A is a k by l matrix and B is m by n, then their Kronecker product, denoted as $A \otimes B$, is the km by ln matrix

$$
A \otimes B = \begin{pmatrix} a_{11}B & a_{12}B & \cdots & a_{1l}B \\ a_{21}B & a_{22}B & \cdots & a_{2l}B \\ \vdots & \vdots & & \vdots \\ a_{k1}B & a_{k2}B & \cdots & a_{kl}B \end{pmatrix}.
$$

The operations of matrix transposition and Kronecker product formation commute; viz.

$$
(A \otimes B)' = A' \otimes B'.
$$

If A and C are conformable for multiplication, that is, C has as many rows as A has columns, and B and D are conformable as well, then

$$(A \otimes B)(C \otimes D) = AC \otimes BD.$$

It follows immediately that if both A and B are square and invertible, then

$$(A \otimes B)^{-1} = A^{-1} \otimes B^{-1}$$

that is, inversion and Kronecker product formation commute.

In this notation, the variance-covariance matrix of the errors is

$$
\mathscr{C}(e, e') = \begin{vmatrix}
\mathscr{C}(e_1, e_1') & \mathscr{C}(e_1, e_2') & \cdots & \mathscr{C}(e_1, e_M') \\
\mathscr{C}(e_2, e_1') & \mathscr{C}(e_2, e_2') & \cdots & \mathscr{C}(e_2, e_M') \\
\vdots & & & \\
\mathscr{C}(e_M, e_1') & \mathscr{C}(e_M, e_2') & \cdots & \mathscr{C}(e_M, e_M')
\end{vmatrix}
$$

$$
= \begin{vmatrix}
\sigma_{11}I & \sigma_{12}I & \cdots & \sigma_{1M}I \\
\sigma_{21}I & \sigma_{22}I & \cdots & \sigma_{2M}I \\
\vdots & \vdots & & \vdots \\
\sigma_{M1}I & \sigma_{M2}I & \cdots & \sigma_{MM}I
\end{vmatrix}
$$

$$= \Sigma \otimes I;$$

the identity is n by n, while Σ is M by M, so the resultant $\Sigma \otimes I$ is nM by nM.

Factor Σ^{-1} as $\Sigma^{-1} = P'P$, and consider the rotated model

$$(P \otimes I)y = (P \otimes I)f(\theta) + (P \otimes I)e$$

or

$$\text{``}y\text{''} = \text{``}f\text{''}(\theta) + \text{``}e\text{''}.$$

Since

$$
\begin{aligned}
\mathscr{C}(\text{``}e\text{''}, \text{``}e\text{''}') &= (P \otimes I)(\Sigma \otimes I)(P \otimes I)' \\
&= (P\Sigma P') \otimes I \\
&= \left[PP^{-1}(P')^{-1}P' \right] \otimes I \\
&= {}_M I_M \otimes {}_n I_n \\
&= {}_{nM} I_{nM}
\end{aligned}
$$

the model

$$\text{``}y\text{''} = \text{``}f\text{''}(\theta^0) + \text{``}e\text{''}$$

is simply a univariate nonlinear model, and θ^0 can be estimated by minimizing

$$S(\theta, \Sigma) = [``y" - ``f"(\theta)]'[``y" - ``f"(\theta)]$$
$$= [y - f(\theta)]'(P \otimes I)'(P \otimes I)[y - f(\theta)]$$
$$= [y - f(\theta)]'(\Sigma^{-1} \otimes I)[y - f(\theta)].$$

Of course Σ is unknown, so one adopts the obvious expedient (Problem 4) of replacing Σ by $\hat{\Sigma}$ and estimating θ^0 by

$$\hat{\theta} \text{ minimizing } S(\theta, \hat{\Sigma}).$$

These ideas are easier to implement if we adopt the grouped by subject data arrangement rather than the grouped by equation arrangement. Accordingly, let

$$y_t = \underset{M}{\begin{pmatrix} y_{1t} \\ y_{2t} \\ \vdots \\ y_{Mt} \end{pmatrix}_1} \quad t = 1, 2, \ldots, n$$

$$f(x, \theta) = \underset{M}{\begin{pmatrix} f_1(x, \theta_1) \\ f_2(x, \theta_2) \\ \vdots \\ f_M(x, \theta_M) \end{pmatrix}_1}$$

$$e_t = \underset{M}{\begin{pmatrix} e_{1t} \\ e_{2t} \\ \vdots \\ e_{Mt} \end{pmatrix}_1} \quad t = 1, 2, \ldots, n$$

$$\theta = \underset{p}{\begin{pmatrix} \theta_1 \\ \theta_2 \\ \vdots \\ \theta_M \end{pmatrix}_1} \quad p = \sum_{\alpha=1}^{M} p_\alpha$$

whence the model may be written as the multivariate nonlinear regression

$$y_t = f(x_t, \theta^0) + e_t \quad t = 1, 2, \ldots, n.$$

In this scheme,

$$S(\theta, \Sigma) = \sum_{t=1}^{n} [y_t - f(x_t, \theta)]' \Sigma^{-1} [y_t - f(x_t, \theta)].$$

To see that this is so, let $\sigma^{\alpha\beta}$ denote the elements of Σ^{-1} and write

$$
\begin{aligned}
S(\theta, \Sigma) &= \sum_{t=1}^{n} [y_t - f(x_t, \theta)]' \Sigma^{-1} [y_t - f(x_t, \theta)] \\
&= \sum_{t=1}^{n} \sum_{\alpha=1}^{M} \sum_{\beta=1}^{M} \sigma^{\alpha\beta} [y_{\alpha t} - f_\alpha(x_t, \theta_\alpha)][y_{\beta t} - f_\beta(x_t, \theta_\beta)] \\
&= \sum_{\alpha=1}^{M} \sum_{\beta=1}^{M} \sigma^{\alpha\beta} [y_\alpha - f(\theta_\alpha)]' [y_\beta - f(\theta_\beta)] \\
&= [y - f(\theta)]' (\Sigma^{-1} \otimes I)[y - f(\theta)].
\end{aligned}
$$

In writing code, the grouped by subject arrangement coupled with a summation notation is more convenient because it is natural to group observations (y_t, x_t) on the same subject together and process them serially for $t = 1, 2, \ldots, n$. With $S(\theta, \Sigma)$ written as

$$S(\theta, \Sigma) = \sum_{t=1}^{n} [y_t - f(x_t, \theta)]' \Sigma^{-1} [y_t - f(x_t, \theta)]$$

one can see at sight that it suffices to fetch (y_t, x_t), compute $[y_t - f(x_t, \theta)]' \Sigma^{-1} [y_t - f(x_t, \theta)]$, add the result to an accumulator, and continue.

The notation is also suggestive of a transformation that permits the use of univariate nonlinear regression programs for multivariate computations. Observe that if Σ^{-1} factors as $\Sigma^{-1} = P'P$ then

$$S(\theta, \Sigma) = \sum_{t=1}^{n} [Py_t - Pf(x_t, \theta)]' [Py_t - Pf(x_t, \theta)].$$

Writing $p'_{(\alpha)}$ to denote the αth row of P, we have

$$S(\theta, \Sigma) = \sum_{t=1}^{n} \sum_{\alpha=1}^{M} [p'_{(\alpha)} y_t - p'_{(\alpha)} f(x_t, \theta)]^2.$$

One now has $S(\theta, \Sigma)$ expressed as the sum of squares of univariate entities; what remains is to find a notational scheme to remove the double summation. To this end, put

$$s = M(t - 1) + \alpha$$
$$\text{``}y_s\text{''} = p'_{(\alpha)} y_t$$
$$\text{``}x_s\text{''} = (p'_\alpha, x'_t)'$$
$$\text{``}f_s\text{''}(\text{``}x_s\text{''}, \theta) = p'_{(\alpha)} f(x_t, \theta)$$

for $\alpha = 1, 2, \ldots, M$ and $t = 1, 2, \ldots, n$, whence

$$S(\theta, \Sigma) = \sum_{s=1}^{nM} [\text{``}y_s\text{''} - \text{``}f\text{''}(\text{``}x_s\text{''}, \theta)]^2.$$

We illustrate these ideas with the example.

EXAMPLE 1 (Continued). Recall that the model is

$$y_{1t} = f_1(x_t, \theta_1) + e_{t1}$$
$$y_{2t} = f_2(x_t, \theta_2) + e_{t2}$$

with

$$f_\alpha(x, \theta_\alpha) = \ln \frac{a_\alpha + b'_{(\alpha)}}{-1 + b'_{(3)} x} \qquad \alpha = 1, 2$$

$$\theta'_1 = (a_1, b_{11}, b_{12}, b_{13}, b_{31}, b_{32}, b_{33})$$
$$\theta'_2 = (a_2, b_{21}, b_{22}, b_{23}, b_{31}, b_{32}, b_{33}).$$

As the model is written, the notation suggests that $b_{(3)}$ is the same for both $\alpha = 1$ and $\alpha = 2$, which up to now has not been the case. To have a notation that reflects this fact, write

$$f_\alpha(x, \theta_\alpha) = \ln \frac{a_\alpha + b'_{(\alpha)}}{-1 + b'_{\alpha(3)} x} \qquad \alpha = 1, 2$$

$$\theta'_1 = (a_1, b_{11}, b_{12}, b_{13}, b_{131}, b_{132}, b_{133})$$
$$\theta'_2 = (a_2, b_{21}, b_{22}, b_{23}, b_{231}, b_{232}, b_{233})$$

to emphasize the fact that the equality constraint is not imposed. The multivariate model is then

$$y_t = f(x_t, \theta) + e_t$$

with

$$f(x_t, \theta) = \begin{pmatrix} \ln \dfrac{a_1 + b_{11}x_{1t} + b_{12}x_{2t} + b_{13}x_{3t}}{-1 + b_{131}x_{1t} + b_{132}x_{2t} + b_{133}x_{3t}} \\[3mm] \ln \dfrac{a_2 + b_{21}x_{2t} + b_{22}x_{2t} + b_{23}x_{3t}}{-1 + b_{231}x_{1t} + b_{232}x_{2t} + b_{233}x_{3t}} \end{pmatrix}$$

$$\theta = \begin{pmatrix} a_1 \\ b_{11} \\ b_{12} \\ b_{13} \\ b_{131} \\ b_{132} \\ b_{133} \\ a_2 \\ b_{21} \\ b_{22} \\ b_{23} \\ b_{231} \\ b_{232} \\ b_{233} \end{pmatrix}$$

and

$$y_t = \begin{pmatrix} y_{1t} \\ y_{2t} \end{pmatrix} \qquad e_t = \begin{pmatrix} e_{1t} \\ e_{2t} \end{pmatrix};$$

x_t as before. To illustrate, from Table 1a for $t = 1$ we have

$$y_t = \begin{pmatrix} \ln(0.662888/0.056731) \\ \ln(0.280382/0.056731) \end{pmatrix} = \begin{pmatrix} 2.45829 \\ 1.59783 \end{pmatrix}$$

and for $t = 2$

$$y_t = \begin{pmatrix} \ln(0.644427/0.103444) \\ \ln(0.252128/0.013444) \end{pmatrix} = \begin{pmatrix} 1.82933 \\ 0.89091 \end{pmatrix};$$

as previously, from Tables 1a and 1b we have

$$x_1 = \begin{pmatrix} 2.11747 \\ 1.80731 \\ 0.81476 \end{pmatrix}, \qquad x_2 = \begin{pmatrix} 1.58990 \\ 1.27974 \\ 0.28719 \end{pmatrix}.$$

To illustrate the scheme for minimizing $S(\theta, \hat{\Sigma})$ using a univariate nonlinear program, recall that

$$\hat{P} = \begin{pmatrix} 3.7648 & -3.8585 \\ 0 & 3.3716 \end{pmatrix} \qquad \text{(from Figure } 1c)$$

whence

$$\text{“}y_1\text{”} = (3.7648, -3.8585)\begin{pmatrix} 2.45829 \\ 1.59783 \end{pmatrix} = 3.08980$$

$$\text{“}y_2\text{”} = (0 \quad , \quad 3.3716)\begin{pmatrix} 2.45829 \\ 1.59783 \end{pmatrix} = 5.38733$$

$$\text{“}y_3\text{”} = (3.7648, -3.8585)\begin{pmatrix} 1.82933 \\ 0.89091 \end{pmatrix} = 3.44956$$

$$\text{“}y_4\text{”} = (0 \quad , \quad 3.3716)\begin{pmatrix} 1.82933 \\ 0.89091 \end{pmatrix} = 3.00382$$

$$\text{“}x_1\text{”} = (3.7648, -3.8585, 2.11747, 1.80731, 0.81476)'$$

$$\text{“}x_2\text{”} = (0 \quad , \quad 3.3716, 2.11747, 1.80731, 0.81476)'$$

$$\text{“}x_3\text{”} = (3.7648, -3.8585, 1.58990, 1.27974, 0.28719)'$$

$$\text{“}x_4\text{”} = (0 \quad , \quad 3.3716, 1.58990, 1.27974, 0.28719)'$$

$$\text{“}f\text{”}(\text{“}x_1\text{”}, \theta) = 3.7648 \ln \frac{a_1 + x_1'b_{(1)}}{-1 + x_1'b_{1(3)}}$$

$$- 3.8585 \ln \frac{a_2 + x_1'b_{(2)}}{-1 + x_1'b_{2(3)}}$$

$$\text{“}f\text{”}(\text{“}x_2\text{”}, \theta) = 3.3716 \ln \frac{a_2 + x_1'b_{(2)}}{-1 + x_1'b_{2(3)}}.$$

SAS Statements:

```
DATA WORK01;  SET EXAMPLE1;
P1=3.764814163903;  P2=-3.85846955764;  Y=P1*Y1+P2*Y2;  OUTPUT;
P1=0;  P2=3.371649857133;  Y=P1*Y1+P2*Y2;  OUTPUT;  DELETE;
PROC NLIN DATA=WORK01 METHOD=GAUSS ITER=50 CONVERGENCE=1.E-8;
PARMS B11=-.8 B12=-1.4 B13=2 B131=.5 B132=-.2 B133=-.4
      B21=.4  B22=-1.3 B23=.7 B231=.2 B232=.1  B233=-.4
      A1=-2  A2=-1;  A3=-1;
PEAK =A1+B11*X1+B12*X2+B13*X3;  BASE1=A3+B131*X1+B132*X2+B133*X3;
INTER=A2+B21*X1+B22*X2+B23*X3;  BASE2=A3+B231*X1+B232*X2+B233*X3;
MODEL Y=P1*LOG(PEAK/BASE1)+P2*LOG(INTER/BASE2);
DER.A1 =P1/PEAK;        DER.A2 =P2/INTER;
DER.B11=P1/PEAK*X1;     DER.B21=P2/INTER*X1;
DER.B12=P1/PEAK*X2;     DER.B22=P2/INTER*X2;
DER.B13=P1/PEAK*X3;     DER.B23=P2/INTER*X3;
DER.B131=-P1/BASE1*X1;  DER.B231=-P2/BASE2*X1;
DER.B132=-P1/BASE1*X2;  DER.B232=-P2/BASE2*X2;
DER.B133=-P1/BASE1*X3;  DER.B233=-P2/BASE2*X3;
OUTPUT OUT=WORK02 RESIDUAL=EHAT;
PROC UNIVARIATE DATA=WORK02 PLOT NORMAL;  VAR EHAT;  ID T;
```

Output:

S T A T I S T I C A L   A N A L Y S I S   S Y S T E M   1

NON-LINEAR LEAST SQUARES ITERATIVE PHASE

DEPENDENT VARIABLE: Y   METHOD: GAUSS-NEWTON

ITERATION	B11	B12	B13	RESIDUAL SS
	B131	B132	B133	
	B21	B22	B23	
	B231	B232	B233	
	A1	A2		
0	-0.80000000	-1.40000000	2.00000000	631.16222217
	0.50000000	-0.20000000	-0.40000000	
	0.40000000	-1.30000000	0.70000000	
	0.20000000	0.10000000	-0.40000000	
	-2.00000000	-1.00000000		
⋮				
6	-2.98669756	0.90158533	1.66353998	442.65919896
	0.26718356	0.07113302	-0.47013242	
	0.20848925	-1.33081849	0.85048354	
	0.18931302	0.10756268	-0.40539911	
	-1.52573841	-0.96432128		

Figure 2a. Example 1 fitted by multivariate least squares, unconstrained.

S T A T I S T I C A L   A N A L Y S I S   S Y S T E M   2

NON-LINEAR LEAST SQUARES SUMMARY STATISTICS   DEPENDENT VARIABLE Y

SOURCE	DF	SUM OF SQUARES	MEAN SQUARE
REGRESSION	14	6540.63880955	467.18848640
RESIDUAL	434	442.65919896	1.01995207
UNCORRECTED TOTAL	448	6983.29800851	
(CORRECTED TOTAL)	447	871.79801949	

PARAMETER	ESTIMATE	ASYMPTOTIC STD. ERROR	ASYMPTOTIC 95 % CONFIDENCE INTERVAL LOWER	UPPER
B11	-2.98669756	1.27777798	-5.49813789	-0.47525724
B12	0.90158533	1.41306196	-1.87575225	3.67892291
B13	1.66353998	1.31692369	-0.92484026	4.25192022
B131	0.26718356	0.10864198	0.05365048	0.48071663
B132	0.07113302	0.17067332	-0.26432109	0.40658712
B133	-0.47013242	0.07443325	-0.61642910	-0.32383574
B21	0.20848925	0.41968687	-0.61639469	1.03337319
B22	-1.33081849	0.48055515	-2.27533750	-0.38629949
B23	0.85048354	0.54542139	-0.22152841	1.92249550
B231	0.18931302	0.12899074	-0.06421501	0.44284105
B232	0.10756268	0.14251811	-0.17255306	0.38767842
B233	-0.40539911	0.07932153	-0.56130357	-0.24949465
A1	-1.52573841	0.98851033	-3.46863048	0.41715366
A2	-0.96432128	0.34907493	-1.65041924	-0.27822332

Figure 2a. (Continued).

SAS code to implement this scheme is shown in Figure 2a together with the resulting output.

Least squares methods lean rather heavily on normality for their validity. Accordingly, it is a sensible precaution to check residuals for evidence of severe departures from normality. Figure 2a includes a residual analysis of the unconstrained fit. There does not appear to be a gross departure from normality. Notably, the Kolmogorov-Smirnov test does not reject normality.

Consider, now, fitting the model subject to the restriction that $b_{(3)}$ is the same in both equations, viz

$$H_1 : b_{131} = b_{231}, \ b_{132} = b_{232}, \ b_{133} = b_{233}.$$

As we have seen before, there are two approaches. The first is to leave the

UNIVARIATE

VARIABLE=EHAT

	MOMENTS				QUANTILES(DEF=4)		
N	448	SUM WGTS	448	100% MAX	4.62368	99%	2.68474
MEAN	-1.213E-09	SUM	-5.434E-07	75% Q3	0.648524	95%	1.63032
STD DEV	0.995133	VARIANCE	0.990289	50% MED	-0.057296	90%	1.18026
SKEWNESS	0.378159	KURTOSIS	1.30039	25% Q1	-0.665501	10%	-1.1508
USS	442.659	CSS	442.659	0% MIN	-3.31487	5%	-1.60095
CV	-8.205E+10	STD MEAN	0.0470156			1%	-2.15196
T:MEAN=0	-2.580E-08	PROB>|T|	1	RANGE	7.93855		
SGN RANK	-1017	PROB>|S|	0.710841	Q3-Q1	1.31402		
NUM ^= 0	448			MODE	-3.31487		
D:NORMAL	0.0310503	PROB>D	>0.15				

EXTREMES

LOWEST	ID		HIGHEST	ID
-3.31487(	181)		2.65142(	152)
-2.44543(	128)		2.71678(	60)
-2.31033(	13)		2.7925(	81)
-2.21378(	183)		3.96674(	164)
-2.08762(	13)		4.62368(	132)

```
        BAR CHART                                         #    BOXPLOT
   4.75+*                                                 1       *
      .*                                                  1
      .                                                           0
      .                                                           0
      .**                                                 4
      .***                                                5
      .********                                          17       |
      .********************                              39       |
   0.75+****************************************------*   80    +-----+
      .*********************************                 63     |     |
      .***********************************************   93     *--+--*
      .*****************************************         74     +-----+
      .*********************                             42       |
      .*********                                         18       |
      .*****                                             10
      .                                                          0
  -3.25+*                                                 1       0
     ----+----+----+----+----+----+----+----+----+--
     * MAY REPRESENT UP TO 2 COUNTS
```

VARIABLE=EHAT

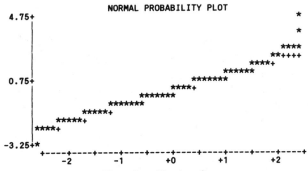

NORMAL PROBABILITY PLOT

Figure 2a. (Continued).

model as written and impose the constraint using the functional dependence

$$
\theta = \begin{pmatrix} a_1 \\ b_{11} \\ b_{12} \\ b_{13} \\ b_{131} \\ b_{132} \\ b_{133} \\ a_2 \\ b_{21} \\ b_{22} \\ b_{23} \\ b_{231} \\ b_{232} \\ b_{233} \end{pmatrix} = \begin{pmatrix} \rho_1 \\ \rho_2 \\ \rho_3 \\ \rho_4 \\ \rho_5 \\ \rho_6 \\ \rho_7 \\ \rho_8 \\ \rho_9 \\ \rho_{10} \\ \rho_{11} \\ \rho_5 \\ \rho_6 \\ \rho_7 \end{pmatrix} = g(\rho).
$$

One fits the model

$$
y_t = f\left[x_t, g(\rho^0)\right] + e_t
$$

by minimizing $S[g(\rho), \hat{\Sigma}]$; derivatives are computed using the chain rule

$$
\frac{\partial}{\partial \rho'} \text{``} f \text{''} \left[ \text{``} x_s \text{''}, g(\rho) \right] = \frac{\partial}{\partial \rho'} p'_{(\alpha)} f\left[x_t, g(\rho)\right]
$$

$$
= p'_{(\alpha)} \frac{\partial}{\partial \theta'} f(x_t, \theta) \Big|_{\theta = g(\rho)} \frac{\partial}{\partial \rho'} g(\rho).
$$

These ideas were illustrated in Figure 9b of Chapter 1 and will be seen again in Figure 2d below.

The second approach is to simply rewrite the model with the constraint imposed. We adopt the second alternative, viz.

$$
f(x_t, \theta) = \begin{pmatrix} \ln \dfrac{a_1 + b_{11}x_{1t} + b_{12}x_{2t} + b_{13}x_{3t}}{-1 + b_{31}x_{1t} + b_{32}x_{2t} + b_{33}x_{3t}} \\[2ex] \ln \dfrac{a_2 + b_{21}x_{1t} + b_{22}x_{2t} + b_{23}x_{3t}}{-1 + b_{31}x_{1t} + b_{32}x_{2t} + b_{33}x_{3t}} \end{pmatrix}
$$

$$
\theta' = \left(a_1, b_{11}, b_{12}, b_{13}, a_2, b_{21}, b_{22}, b_{23}, b_{31}, b_{32}, b_{33}\right).
$$

SAS code to fit this model is shown in Figure 2b.

SAS Statements:

```
DATA WORK01;  SET EXAMPLE1;
P1=3.764814163903;  P2=-3.85846955764;  Y=P1*Y1+P2*Y2;  OUTPUT;
P1=0;  P2=3.371649857133;  Y=P1*Y1+P2*Y2;  OUTPUT;  DELETE;
PROC NLIN DATA=WORK01 METHOD=GAUSS ITER=50 CONVERGENCE=1.E-8;
PARMS B11=-.8 B12=-1.4 B13=2 B21=.4 B22=-1.3 B23=.7 B31=.5 B32=-.2 B33=-.4
      A1=-2 A2=-1;  A3=-1;
PEAK=A1+B11*X1+B12*X2+B13*X3;  INTER=A2+B21*X1+B22*X2+B23*X3;
BASE=A3+B31*X1+B32*X2+B33*X3;
MODEL Y=P1*LOG(PEAK/BASE)+P2*LOG(INTER/BASE);
DER.A1 =P1/PEAK;      DER.A2 =P2/INTER;
DER.B11=P1/PEAK*X1;   DER.B21=P2/INTER*X1;   DER.B31=(-P1-P2)/BASE*X1;
DER.B12=P1/PEAK*X2;   DER.B22=P2/INTER*X2;   DER.B32=(-P1-P2)/BASE*X2;
DER.B13=P1/PEAK*X3;   DER.B23=P2/INTER*X3;   DER.B33=(-P1-P2)/BASE*X3;
```

Output:

NON-LINEAR LEAST SQUARES ITERATIVE PHASE

DEPENDENT VARIABLE: Y   METHOD: GAUSS-NEWTON

ITERATION	B11 B21 B31 A1	B12 B22 B32 A2	B13 B23 B33	RESIDUAL SS
0	-0.80000000 0.40000000 0.50000000 -2.00000000	-1.40000000 -1.30000000 -0.20000000 -1.00000000	2.00000000 0.70000000 -0.40000000	641.48045300
. . .				
8	-3.27643190 0.40180449 0.23944183 -1.58236942	1.30488351 -1.11931853 0.10626154 -1.20266408	1.66561680 0.41058766 -0.45982238	447.31829119

NOTE: CONVERGENCE CRITERION MET.

NON-LINEAR LEAST SQUARES SUMMARY STATISTICS   DEPENDENT VARIABLE Y

SOURCE	DF	SUM OF SQUARES	MEAN SQUARE
REGRESSION	11	6535.97971731	594.17997430
RESIDUAL	437	447.31829119	1.02361165
UNCORRECTED TOTAL	448	6983.29800851	
(CORRECTED TOTAL)	447	871.79801949	

PARAMETER	ESTIMATE	ASYMPTOTIC STD. ERROR	ASYMPTOTIC 95 % CONFIDENCE INTERVAL LOWER	UPPER
B11	-3.27643190	1.27198559	-5.77643950	-0.77642431
B12	1.30488351	0.95321400	-0.56859860	3.17836562
B13	1.66561680	1.01449051	-0.32830042	3.65953403
B21	0.40180449	0.29689462	-0.18172319	0.98533217
B22	-1.11931853	0.36601761	-1.83870310	-0.39993395
B23	0.41058766	0.33172431	-0.24139558	1.06257090
B31	0.23944183	0.09393101	0.05482635	0.42405730
B32	0.10626154	0.11605620	-0.12183961	0.33436270
B33	-0.45982238	0.05409256	-0.56613790	-0.35350687
A1	-1.58236942	0.85859333	-3.26988056	0.10514171
A2	-1.20266408	0.23172071	-1.65809655	-0.74723161

Figure 2b. Example 1 fitted by multivariate least squares, $H1$ imposed.

Following these, same ideas we impose the additional constraint of symmetry,

$$H_2 : b_{12} = b_{21}, \ b_{13} = b_{31}, \ b_{23} = b_{32}$$

by writing

$$f(x_t, \theta) = \begin{pmatrix} \ln \dfrac{a_1 + b_{11}x_{1t} + b_{12}x_{2t} + b_{13}x_{3t}}{-1 + b_{13}x_{1t} + b_{23}x_{2t} + b_{33}x_{3t}} \\ \ln \dfrac{a_2 + b_{12}x_{1t} + b_{22}x_{2t} + b_{23}x_{3t}}{-1 + b_{13}x_{1t} + b_{23}x_{2t} + b_{33}x_{3t}} \end{pmatrix}$$

$$\theta' = (a_1, b_{11}, b_{12}, b_{13}, a_2, b_{22}, b_{23}, b_{33}).$$

SAS code is shown in Figure 2c.

The last restriction to be imposed, in addition to H_1 and H_2, is the homogeneity restriction

$$\sum_{i=1}^{3} a_i = -1 \qquad \sum_{j=1}^{3} b_{ij} = 0 \quad \text{for } i = 1, 2, 3.$$

As we have noted, the scaling convention is irrelevant as far as the data are concerned. The restriction $a_1 + a_2 + a_3 = -1$ is just a scaling convention, and, other than asthetics, there is no reason to prefer it to the convention $a_3 = -1$ that we have imposed thus far.* Retaining $a_3 = -1$, the hypothesis of homogeneity can be rewritten as the parametric restriction

$$H_3 : \sum_{i=1}^{3} b_{ij} = 0 \text{ for } j = 1, 2, 3.$$

Equivalently, H_3 can be written as the functional dependence

$$\theta = \begin{pmatrix} a_1 \\ b_{11} \\ b_{12} \\ b_{13} \\ a_2 \\ b_{22} \\ b_{23} \\ b_{33} \end{pmatrix} = \begin{pmatrix} a_1 \\ -b_{12} - b_{13} \\ b_{12} \\ b_{13} \\ a_2 \\ -b_{23} - b_{12} \\ b_{23} \\ -b_{23} - b_{13} \end{pmatrix} = \begin{pmatrix} \rho_1 \\ -\rho_2 - \rho_3 \\ \rho_2 \\ \rho_3 \\ \rho_4 \\ -\rho_5 - \rho_2 \\ \rho_5 \\ -\rho_5 - \rho_3 \end{pmatrix} = g(\rho)$$

*In economic parlance, it is impossible to tell the difference between a linear homogeneous and a homothetic indirect utility function by looking at a demand system.

SAS Statements:

```
DATA WORK01;  SET EXAMPLE1;
P1=3.764814163903;  P2=-3.85846955764;  Y=P1*Y1+P2*Y2;  OUTPUT;
P1=0;  P2=3.371649857133;  Y=P1*Y1+P2*Y2;  OUTPUT;  DELETE;
PROC NLIN DATA=WORK01 METHOD=GAUSS ITER=50 CONVERGENCE=1.E-8;
PARMS B11=0 B12=0 B13=0 B22=0 B23=0 B33=0 A1=-1 A2=-1;  A3=-1;
PEAK=A1+B11*X1+B12*X2+B13*X3;  INTER=A2+B12*X1+B22*X2+B23*X3;
BASE=A3+B13*X1+B23*X2+B33*X3;
MODEL Y=P1*LOG(PEAK/BASE)+P2*LOG(INTER/BASE);
DER.A1 =P1/PEAK;  DER.A2 =P2/INTER;  DER.B11=P1/PEAK*X1;
DER.B12=P1/PEAK*X2+P2/INTER*X1;  DER.B22=P2/INTER*X2;
DER.B13=P1/PEAK*X3+(-P1-P2)/BASE*X1;  DER.B23=P2/INTER*X3+(-P1-P2)/BASE*X2;
DER.B33=(-P1-P2)/BASE*X3;
```

Output:

S T A T I S T I C A L   A N A L Y S I S   S Y S T E M   1

NON-LINEAR LEAST SQUARES ITERATIVE PHASE

DEPENDENT VARIABLE: Y   METHOD: GAUSS-NEWTON

ITERATION	B11 B22 A1	B12 B23 A2	B13 B33	RESIDUAL SS
0	0.000000E+00 0.000000E+00 -1.00000000	0.000000E+00 0.000000E+00 -1.00000000	0.000000E+00 0.000000E+00	6983.29800851
· · ·				
11	-1.28362479 -1.04835591 -2.92727122	0.81889299 0.03049767 -1.53786463	0.36106759 -0.46735947	450.95423403

NOTE: CONVERGENCE CRITERION MET.

S T A T I S T I C A L   A N A L Y S I S   S Y S T E M .   3

NON-LINEAR LEAST SQUARES SUMMARY STATISTICS   DEPENDENT VARIABLE Y

SOURCE	DF	SUM OF SQUARES	MEAN SQUARE
REGRESSION	8	6532.34377448	816.54297181
RESIDUAL	440	450.95423403	1.02489599
UNCORRECTED TOTAL	448	6983.29800851	
(CORRECTED TOTAL)	447	871.79801949	

PARAMETER	ESTIMATE	ASYMPTOTIC STD. ERROR	ASYMPTOTIC 95 % CONFIDENCE INTERVAL LOWER	UPPER
B11	-1.28362479	0.22679435	-1.72936637	-0.83788321
B12	0.81889299	0.08096691	0.65976063	0.97802535
B13	0.36106759	0.03024703	0.30162008	0.42051510
B22	-1.04835591	0.08359301	-1.21264961	-0.88406221
B23	0.03049767	0.03608943	-0.04043249	0.10142783
B33	-0.46735947	0.01923198	-0.50515801	-0.42956093
A1	-2.92727122	0.27778075	-3.47322147	-2.38132098
A2	-1.53786463	0.09167461	-1.71804189	-1.35768737

Figure 2c. Example 1 fitted by multivariate least squares, $H1$ and $H2$ imposed.

312

SAS Statements:

```
DATA WORK01;  SET EXAMPLE1;
P1=3.764814163903;   P2=-3.85846955764;   Y=P1*Y1+P2*Y2;   OUTPUT;
P1=0;   P2=3.371649857133;   Y=P1*Y1+P2*Y2;   OUTPUT;   DELETE;
PROC NLIN DATA=WORK01 METHOD=GAUSS ITER=50 CONVERGENCE=1.E-8;
PARMS R1=-3 R2=.8 R3=.4 R4=-1.5 R5=.03;   A3=-1;
A1=R1;  B11=-R2-R3;  B12=R2;  B13=R3;  A2=R4;  B22=-R5-R2;  B23=R5;  B33=-R5-R3;
PEAK=A1+B11*X1+B12*X2+B13*X3;   INTER=A2+B12*X1+B22*X2+B23*X3;
BASE=A3+B13*X1+B23*X2+B33*X3;
MODEL Y=P1*LOG(PEAK/BASE)+P2*LOG(INTER/BASE);
DER_A1 =P1/PEAK;   DER_A2 =P2/INTER;   DER_B11=P1/PEAK*X1;
DER_B12=P1/PEAK*X2+P2/INTER*X1;   DER_B22=P2/INTER*X2;
DER_B13=P1/PEAK*X3+(-P1-P2)/BASE*X1;   DER_B23=P2/INTER*X3+(-P1-P2)/BASE*X2;
DER_B33=(-P1-P2)/BASE*X3;
DER.R1=DER_A1;  DER.R2=-DER_B11+DER_B12-DER_B22;  DER.R3=-DER_B11+DER_B13-DER_B33;
DER.R4=DER_A2;  DER.R5=-DER_B22+DER_B23-DER_B33;
```

Output:

STATISTICAL ANALYSIS SYSTEM 1

NON-LINEAR LEAST SQUARES ITERATIVE PHASE

DEPENDENT VARIABLE: Y   METHOD: GAUSS-NEWTON

ITERATION	R1 R4	R2 R5	R3	RESIDUAL SS
0	-3.00000000 -1.50000000	0.80000000 0.03000000	0.40000000	560.95959664
1	-2.70479542 -1.59049135	0.85805995 0.05440110	0.37705186	478.82185398
2	-2.72429979 -1.59215078	0.85764215 0.05770545	0.37433263	478.79661265
3	-2.72517893 -1.59211417	0.85757097 0.05794017	0.37413074	478.79654696
4	-2.72523507 -1.59210976	0.85756494 0.05795637	0.37411703	478.79654666

NOTE: CONVERGENCE CRITERION MET.

STATISTICAL ANALYSIS SYSTEM 2

NON-LINEAR LEAST SQUARES SUMMARY STATISTICS   DEPENDENT VARIABLE Y

SOURCE	DF	SUM OF SQUARES	MEAN SQUARE
REGRESSION	5	6504.50146185	1300.90029237
RESIDUAL	443	478.79654666	1.08080485
UNCORRECTED TOTAL	448	6983.29800851	
(CORRECTED TOTAL)	447	871.79801949	

PARAMETER	ESTIMATE	ASYMPTOTIC STD. ERROR	ASYMPTOTIC 95 % CONFIDENCE INTERVAL LOWER	UPPER
R1	-2.72523507	0.17799072	-3.07505148	-2.37541867
R2	0.85756494	0.06718212	0.72552768	0.98960220
R3	0.37411703	0.02709873	0.32085818	0.42737587
R4	-1.59210976	0.07719388	-1.74382378	-1.44039575
R5	0.05795637	0.03403316	-0.00893116	0.12484390

Figure 2d. Example 1 fitted by multivariate least squares, $H1$, $H2$, and $H3$ imposed.

with Jacobian

$$G(\rho) = \frac{\partial}{\partial \rho'} g(\rho) = \begin{pmatrix} 1 & 0 & 0 & 0 & 0 \\ 0 & -1 & -1 & 0 & 0 \\ 0 & 1 & 0 & 0 & 0 \\ 0 & 0 & 1 & 0 & 0 \\ 0 & 0 & 0 & 1 & 0 \\ 0 & -1 & 0 & 0 & -1 \\ 0 & 0 & 0 & 0 & 1 \\ 0 & 0 & -1 & 0 & -1 \end{pmatrix}.$$

SAS code implementing this restriction is shown in Figure 2d.

The results of Figures 2a through 2b are summarized in Table 2.

As was seen in Chapter 1, these regressions can be assessed using the likelihood ratio test statistic

$$L = \frac{(\mathrm{SSE}_{reduced} - \mathrm{SSE}_{full})/q}{(\mathrm{SSE}_{full})/(n - p)}.$$

As with linear regression, when one has a number of such tests to perform, it is best to organize them into an analysis of variance table as shown in Table 3. For each hypothesis listed under source in Table 3, the entry listed under d.f. is q, as above, and that listed under Sum of Squares is $(\mathrm{SSE}_{reduced} - \mathrm{SSE}_{full})$, as above. As an instance, to test $H1$, $H2$, and $H3$ jointly one has

$$\mathrm{SSE}_{reduced} = 478.79654666 \quad \text{(from Fig. } 2d\text{)}$$
$$\mathrm{SSE}_{full} = 442.65919896 \quad \text{(from Fig. } 2a\text{)}$$

with 443 and 434 degrees of freedom respectively, which yields

$$\mathrm{SSE}_{reduced} - \mathrm{SSE}_{full} = 36.1374$$
$$q = 443 - 434 = 9$$

as shown in Table 3. In general, the mean sum of squares cannot be split from the total regression sum of squares, but in this instance it would be possible to fit a mean to the data as a special case of the nonlinear model by setting $B = 0$ and choosing

$$a = -\exp\left[\mu_3\left(\begin{matrix} P^{-1} & 1 \\ 0 \end{matrix}\right)_1\right].$$

Table 2. Parameter Estimates and Standard Errors[a] for Example 1.

Parameter	Unconstrained	Subject to		
		$H1$	$H1$ & $H2$	$H1$, $H2$ & $H3$
a_1	−1.5257	−1.5824	−2.9273	−2.7252
	(0.9885)	(0.8586)	(0.2778)	(0.1780)
b_{11}	−2.9867	−3.2764	−1.2836	−1.2317
	(1.2778)	(1.2720)	(0.2268)	
b_{12}	0.9016	1.3049	0.8189	0.8576
	(1.4131)	(0.9532)	(0.0810)	(0.0672)
b_{13}	1.6635	1.6656	0.3611	0.3741
	(1.3169)	(1.0145)	(0.0302)	(0.0271)
a_2	−0.9643	−1.2027	−1.5379	−1.5921
	(0.3491)	(0.2317)	(0.0917)	(0.0772)
b_{21}	0.2085	0.4018	0.8189	0.8576
	(0.4197)	(0.2969)		
b_{22}	−1.3308	−1.1193	−1.0484	−0.9155
	(0.4806)	(0.3660)	(0.0836)	
b_{23}	0.8505	0.4106	0.0305	0.0580
	(0.5454)	(0.3317)	(0.0361)	(0.0340)
a_3	−1.0	−1.0	−1.0	−1.0
	(0.0)	(0.0)	(0.0)	(0.0)
b_{131}/b_{31}	0.2672	0.2394	0.3611	0.3741
	(0.1086)	(0.0939)		
b_{132}/b_{32}	0.0711	0.1063	0.0305	0.0580
	(0.1707)	(0.1161)		
b_{133}/b_{33}	−0.4701	−0.4598	−0.4674	−0.4321
	(0.0744)	(0.0541)	(0.0192)	
b_{231}	0.1893			
	(0.1290)			
b_{232}	0.1076			
	(0.1425)			
b_{233}	−0.4054			
	(0.0793)			

[a] Standard errors shown in parentheses.

Table 3.   Analysis of Variance.

Source	d.f.	Sum of Squares	Mean Square	F	$P > F$
Mean	1	6111.5000	6111.5000		
Regression	4	393.0015	98.2504	96.324	0.00000
$H1, H2, H3$	9	36.1374	4.0153	3.937	0.0001
$H1$	3	4.6591	1.5530	1.523	0.206
$H2$ after $H1$	3	3.6360	1.2120	1.188	0.313
$H3$ after $H1, H2$	3	27.8423	9.2808	9.099	0.00001
Error	434	442.6592	1.0200		
Total	448	6983.2980			

The existence of a parametric restriction that will produce the model

$$"y_s" = \mu \begin{pmatrix} 1 \\ 1 \end{pmatrix} + "e_s"$$

justifies the split. The sum of squares for the mean is computed from

$$SSE_{reduced} = 6983.29800851 \quad \text{(from Fig. } 2d\text{)}$$
$$SSE_{full} = 871.798001949 \quad \text{(from Fig. } 2d\text{)}$$

with 448 and 447 degrees of freedom respectively, yielding

$$SSE_{reduced} - SSE_{full} = 6111.5000$$
$$q = 448 - 447 = 1$$

which is subtracted from

$$SSE_{regression} = 6504.50146185 \quad \text{(from Fig. } 2d\text{)}$$

with 5 degrees of freedom to yield the entry shown in Table 3.

From Table 3 one sees that the model of Figure 2c is reasonably well supported by the data and the model of Figure 2d is not. Accordingly, we shall accept it as adequate throughout most of the rest of the book, realizing that there are potential specification errors of at least two sorts. The first are omitted variables of which those listed in Table 1c are prime candidates and the second is an erroneous specification of functional form. But our purpose is illustrative and we shall not dwell on the matter. The model of Figure 2c will serve. □

As suggested by the preceding analysis, in the sequel we shall accept the information provided by $\hat{\Sigma}^{-1} = \hat{P}'\hat{P}$ regarding the rotation $\hat{P}$ that will reduce the multivariate model to a univariate model, as we must to make any progress, but we shall disregard the scale information and shall handle scaling in accordance with standard practice for univariate models. To state this differently, in using Table 3 we could have entered a table of the chi-square distribution using 27.8423 with 3 degrees of freedom, but instead we entered a table of the F-distribution using 9.099 with 3 and 434 degrees of freedom.

The idea of rewriting the multivariate model

$$y_t = f(x_t, \theta) + e_t \qquad t = 1, 2, \ldots, n$$

in the form

$$\text{``}y_s\text{''} = \text{``}f\text{''}(\text{``}x_s\text{''}, \theta) + \text{``}e_s\text{''} \qquad s = 1, 2, \ldots, nM$$

using the transformation

$$\text{``}y_s\text{''} = p'_{(\alpha)} y_t \qquad s = M(t - 1) + \alpha$$

in order to be able to use univariate nonlinear regression methods is useful pedagogically and is even convenient for small values of M. In general, however, one needs to be able to minimize $S(\theta, \Sigma)$ directly. To do this note that the Gauss-Newton correction vector is, from Section 4 of Chapter 1,

$$D(\theta, \Sigma) = \left[\sum_{s=1}^{nM} \left(\frac{\partial}{\partial \theta} \text{``}f\text{''}(\text{``}x_s\text{''}, \theta) \right) \left(\frac{\partial}{\partial \theta} \text{``}f\text{''}(\text{``}x_s\text{''}, \theta) \right)' \right]^{-1}$$

$$\times \sum_{s=1}^{nM} \left(\frac{\partial}{\partial \theta} \text{``}f\text{''}(\text{``}x_s\text{''}, \theta) \right) \left[\text{``}y_s\text{''} - \text{``}f\text{''}(\text{``}x_s\text{''}, \theta) \right]$$

$$= \left[\sum_{t=1}^{n} \sum_{\alpha=1}^{M} \left(\frac{\partial}{\partial \theta} f'(x_t, \theta) \right) p_{(\alpha)} p'_{(\alpha)} \left(\frac{\partial}{\partial \theta'} f(x_t, \theta) \right) \right]^{-1}$$

$$\times \sum_{t=1}^{n} \sum_{\alpha=1}^{M} \left(\frac{\partial}{\partial \theta} f'(x_t, \theta) \right) p_{(\alpha)} p'_{(\alpha)} \left[y_t - f(x_t, \theta) \right]$$

$$= \left[\sum_{t=1}^{n} \left(\frac{\partial}{\partial \theta} f'(x_t, \theta) \right) \Sigma^{-1} \left(\frac{\partial}{\partial \theta'} f(x_t, \theta) \right) \right]^{-1}$$

$$\times \sum_{t=1}^{n} \left(\frac{\partial}{\partial \theta} f'(x_t, \theta) \right) \Sigma^{-1} \left[y_t - f(x_t, \theta) \right].$$

The modified Gauss-Newton algorithm for minimizing $S(\theta, \Sigma)$ is then:

0. Choose a starting estimate θ_0. Compute $D_0 = D(\theta_0, \Sigma)$, and find a λ_0 between zero and one such that

$$S(\theta_0 + \lambda_0 D_0, \Sigma) < S(\theta_0, \Sigma).$$

1. Let $\theta_1 = \theta_0 + \lambda_0 D_0$. Compute $D_1 = D(\theta_1, \Sigma)$, and find a λ_1 between zero and one such that

$$S(\theta_1 + \lambda_1 D_1, \Sigma) < S(\theta_1, \Sigma).$$

2. Let $\theta_2 = \theta_1 + \lambda_1 D_1$. ...

The comments in Section 4 of Chapter 1 regarding starting rules, stopping rules, and alternative algorithms apply directly.

PROBLEMS

1. Show that if e_α is an n-vector with typical element $e_{\alpha t}$ for $t = 1, 2, \ldots, n$ and $\alpha = 1, 2, \ldots, M$, and $\mathscr{C}(e_{\alpha t}, e_{\beta s}) = \sigma_{\alpha\beta}$ if $t = s$ and is zero otherwise, then

$$\mathscr{C}(e_\alpha, e_\beta') = \sigma_{\alpha\beta} I.$$

2. Reestimate Example 1 (in unconstrained form, subject to H_1, subject to H_1 and H_2, and subject to H_1, H_2, and H_3) using the normalizing convention $a_1 + a_2 + a_3 = -1$ (instead of $a_3 = -1$ as used in Figs. 2a, 2b, 2c, 2d).

3. Using the grouped by equation data arrangement, show that the Gauss-Newton correction vector can be written as

$$D(\theta, \Sigma) = \left[F'(\theta)(\Sigma^{-1} \otimes I)F(\theta)\right]^{-1} F'(\theta)(\Sigma^{-1} \otimes I)[y - f(\theta)]$$

where $F(\theta) = (\partial/\partial\theta')f(\theta) = \mathrm{diag}[F_1(\theta_1), \ldots, F_M(\theta_M)]$.

4. Show that $S(\theta, \Sigma)$ satisfies the integral condition of Assumption 6 of Chapter 3, which justifies the expedient of replacing Σ by $\hat{\Sigma}$ and subsequently acting as if $\hat{\Sigma}$ were the true value of Σ.

5. If the model used in Example 1 is misspecified as to choice of functional form, then theory suggests (Gallant, 1981, 1982; Elbadawi, Gallant, and Souza, 1983) that the misspecification must take the form

of omission of additive terms of the form

$$a_{\alpha j}\cos(jk'_\alpha x) - b_{\alpha j}\sin(jk'_\alpha x)$$

from the indirect utility function

$$g(x|\theta) = a'x + \tfrac{1}{2}x'Bx;$$

recall that $x = \ln(p/E)$. Test for the joint omission of these terms for

$$k_\alpha = \begin{pmatrix} 1 \\ 0 \\ 0 \end{pmatrix}, \begin{pmatrix} 0 \\ 1 \\ 0 \end{pmatrix}, \begin{pmatrix} 0 \\ 0 \\ 1 \end{pmatrix}, \begin{pmatrix} 1 \\ 1 \\ 0 \end{pmatrix}, \begin{pmatrix} 1 \\ 0 \\ 1 \end{pmatrix}, \begin{pmatrix} 0 \\ 1 \\ 1 \end{pmatrix}$$

and $j = 1, 2$, a total of 24 additional parameters.

6. Instead of splitting out one degree of freedom for the model

$$``y_s" = \mu\begin{pmatrix} 1 \\ 1 \end{pmatrix} + ``e_s"$$

from the five degree of freedom regression sum of squares of Figure 2d, as was done in Table 3, split out two degrees of freedom for the model

$$``y_s" = \begin{pmatrix} \mu_1 \\ \mu_2 \end{pmatrix} + ``e_s".$$

7. (Out of range argument.) Show that the constants $t_1, t_2, a, b, c, \alpha, \beta$ can be chosen so that the function

$$slog(x) = \begin{cases} \alpha + \beta x & -\infty < x \le t_1 \\ a + bx + cx^2 & t_1 \le x \le t_2 \\ \ln x & t_2 \le x < \infty \end{cases}$$

is continuous with continuous first derivative

$$\frac{d}{dx}slog(x) = \begin{cases} \beta & -\infty < x \le t_1 \\ b + 2cx & t_1 \le x \le t_2 \\ 1/x & t_2 \le x < \infty \,. \end{cases}$$

Verify that $slog(x)$ is once continuously differentiable if the constants are chosen as

$$t_2 = 10^{-7}$$

$$t_1 = 0$$

$$\alpha = -299.999999999999886$$

$$\beta = 5667638086.9808321$$

$$a = -299.999999999999886$$

$$b = 5667638086.9808321$$

$$c = -28288190434904165.$$

Use slog(x) in place of ln(x) in Figures 1a through 2d, and observe that the same numerical results obtain.

3. HYPOTHESIS TESTING

We shall derive our results on hypothesis testing in a summation notation using the grouped by subject data arrangement. We do so mainly for pedagogical reasons, although, as remarked earlier, this form is more natural for translating results to machine code.

The prevalent form in the literature is a vector notation using the grouped by equation data arrangement. To provide convenient access to the literature, the results are restated in this notation at the end of the section. There are no new ideas or different results involved in this restatement, just an algebraic rearrangement of terms.

The data follow the model

$$y_t = f(x_t, \theta^0) + e_t \qquad t = 1, 2, \ldots, n$$

with the functional form $f(x, \theta)$ known, x_t a k-vector, θ a p-vector, y_t an M-vector, and e_t an M-vector. Assume that the errors $\{e_t\}$ are independently and normally distributed, each with mean zero and variance-covariance matrix Σ. The unknown parameters are θ^0 and Σ.

Consider testing a hypothesis that can be expressed either as a parametric restriction

$$H : h(\theta^0) = 0 \quad \text{against} \quad A : h(\theta^0) \neq 0$$

or as a functional dependence

$$H : \theta^0 = g(\rho^0) \text{ for some } \rho^0 \quad \text{against} \quad A : \theta^0 \neq g(\rho) \text{ for any } \rho.$$

Here, $h(\theta)$ maps $\mathbb{R}^p$ into $\mathbb{R}^q$ with Jacobian

$$H(\theta) = \frac{\partial}{\partial \theta'} h(\theta)$$

which we assume is continuous and has rank q over the parameter space; $g(\rho)$ maps $\mathbb{R}^r$ into $\mathbb{R}^p$ and has Jacobian

$$G(\rho) = \frac{\partial}{\partial \rho'} g(\rho).$$

The Jacobians are of order q by p for $H(\theta)$ and p by r for $G(\rho)$; we assume that $p = r + q$, and from $h[g(\rho)] = 0$ we have $H[g(\rho)]G(\rho) = 0$. For complete details see Section 6 of Chapter 3. Let us illustrate with the example.

EXAMPLE 1　(Continued).　Recall that the model

$$y_t = f\left(x_t, \theta^0\right) + e_t \qquad t = 1, 2, \ldots, 224$$

with

$$f(x, \theta) = \begin{pmatrix} \ln \dfrac{a_1 + b_{11}x_1 + b_{12}x_2 + b_{13}x_3}{-1 + b_{13}x_1 + b_{23}x_2 + b_{33}x_3} \\[2mm] \ln \dfrac{a_2 + b_{12}x_1 + b_{22}x_2 + b_{23}x_3}{-1 + b_{13}x_1 + b_{23}x_2 + b_{33}x_3} \end{pmatrix}$$

$$\theta' = \left(a_1, b_{11}, b_{12}, b_{13}, a_2, b_{22}, b_{23}, b_{33}\right)$$

was chosen as a reasonable representation of the data of Table 1 on the basis of the computations reported in Table 3. Since we have settled on a model specification, let us henceforth adopt the simpler subscripting scheme

$$f(x, \theta) = \begin{pmatrix} \ln \dfrac{\theta_1 + \theta_2 x_1 + \theta_3 x_2 + \theta_4 x_3}{-1 + \theta_4 x_1 + \theta_7 x_2 + \theta_8 x_3} \\[2mm] \ln \dfrac{\theta_5 + \theta_3 x_1 + \theta_6 x_2 + \theta_7 x_3}{-1 + \theta_4 x_1 + \theta_7 x_2 + \theta_8 x_3} \end{pmatrix}$$

$$\theta = (\theta_1, \theta_2, \theta_3, \theta_4, \theta_5, \theta_6, \theta_7, \theta_8)'.$$

In this notation, the hypothesis of homogeneity may be written as the parametric restriction

$$h(\theta) = \begin{pmatrix} \theta_2 + \theta_3 + \theta_4 \\ \theta_3 + \theta_6 + \theta_7 \\ \theta_4 + \theta_7 + \theta_8 \end{pmatrix} = 0$$

with Jacobian

$$H(\theta) = \begin{pmatrix} 0 & 1 & 1 & 1 & 0 & 0 & 0 & 0 \\ 0 & 0 & 1 & 0 & 0 & 1 & 1 & 0 \\ 0 & 0 & 0 & 1 & 0 & 0 & 1 & 1 \end{pmatrix}.$$

The hypothesis may also be written as a functional dependence

$$\theta = \begin{pmatrix} \theta_1 \\ \theta_2 \\ \theta_3 \\ \theta_4 \\ \theta_5 \\ \theta_6 \\ \theta_7 \\ \theta_8 \end{pmatrix} = \begin{pmatrix} \theta_1 \\ -\theta_3 - \theta_4 \\ \theta_3 \\ \theta_4 \\ \theta_5 \\ -\theta_7 - \theta_3 \\ \theta_7 \\ -\theta_4 - \theta_7 \end{pmatrix} = \begin{pmatrix} \rho_1 \\ -\rho_2 - \rho_3 \\ \rho_2 \\ \rho_3 \\ \rho_4 \\ -\rho_5 - \rho_2 \\ \rho_5 \\ -\rho_5 - \rho_3 \end{pmatrix} = g(\rho)$$

with Jacobian

$$G(\rho) = \begin{pmatrix} 1 & 0 & 0 & 0 & 0 \\ 0 & -1 & -1 & 0 & 0 \\ 0 & 1 & 0 & 0 & 0 \\ 0 & 0 & 1 & 0 & 0 \\ 0 & 0 & 0 & 1 & 0 \\ 0 & -1 & 0 & 0 & -1 \\ 0 & 0 & 0 & 0 & 1 \\ 0 & 0 & -1 & 0 & -1 \end{pmatrix}$$

which is, of course, the same as was obtained in Section 2. In passing, observe that $H[g(\rho)]G(\rho) = 0$. □

Throughout this section we shall take $\hat{\Sigma}$ to be any random variable that converges almost surely to Σ and has $\sqrt{n}(\hat{\Sigma} - \Sigma)$ bounded in probability. To obtain a level α test this condition on $\hat{\Sigma}$ need only hold when $H : h(\theta^0) = 0$ is true; but in order to use the approximations to power derived below, the condition must hold when $A : h(\theta^0) \neq 0$ as well.

There are two commonly used estimators of Σ that satisfy the condition under both the null and alternative hypotheses. We illustrated one of them in Section 2. There one fitted each equation in the model separately in the style of Chapter 1 and then estimated Σ from single equation residuals. Recalling that

$$S(\theta, \Sigma) = \sum_{t=1}^{n} [y_t - f(x_t, \theta)]' \Sigma^{-1} [y_t - f(x_t, \theta)]$$

an alternative approach is to put $\Sigma = I$, minimize $S(\theta, I)$ with respect to θ to obtain $\hat{\theta}^{\#}$, and estimate Σ by

$$\hat{\Sigma} = \frac{1}{n} \sum_{t=1}^{n} [y_t - f(x_t, \hat{\theta}^{\#})][y_t - f(x_t, \hat{\theta}^{\#})]'$$

$$= \frac{1}{n} \sum_{t=1}^{n} \hat{e}_t \hat{e}_t'.$$

If there are no across equation restrictions on the model, these two estimators will be the same. When there are across equation restrictions, there is a tendency to incorporate them directly into the model specification when using the grouped by subject data arrangement as we have just done with the example. (The restrictions that θ_3, θ_4, θ_7, and θ_8 must be the same in both equations are the across equation restrictions, a total of four. The restriction that θ_4 must be the same in the numerator and denominator of the first equation is called a within equation restriction.) This tendency to incorporate across equation restrictions in the model specification causes the two estimators of Σ to be different in most instances. Simply for variety's sake, we shall use the estimator computed from the fit that minimizes $S(\theta, I)$ in this section.

We illustrate these ideas with the example. In reading what follows, recall the ideas used to write a multivariate model in a univariate notation. Factor $\hat{\Sigma}^{-1}$ as $\hat{\Sigma}^{-1} = \hat{P}'\hat{P}$, let $\hat{p}'_{(\alpha)}$ denote a typical row of $\hat{P}$, and put

$$s = M(t - 1) + \alpha$$
$$\text{``}y_s\text{''} = \hat{p}'_{(\alpha)} y_t$$
$$\text{``}x_s\text{''} = (\hat{p}'_{(\alpha)}, x_t')'$$
$$\text{``}f\text{''}(\text{``}x_s\text{''}, \theta) = \hat{p}'_{(\alpha)} f(x_t, \theta).$$

In this notation

$$S(\theta, \hat{\Sigma}) = \sum_{s=1}^{nM} [\text{``}y_s\text{''} - \text{``}f\text{''}(\text{``}x_s\text{''}, \theta)]^2.$$

SAS Statements:

```
DATA WORK01;  SET EXAMPLE1;
P1=1.0;  P2=0.0;  Y=P1*Y1+P2*Y2;  OUTPUT;
P1=0.0;  P2=1.0;  Y=P1*Y1+P2*Y2;  OUTPUT;  DELETE;
PROC NLIN DATA=WORK01 METHOD=GAUSS ITER=50 CONVERGENCE=1.E-13;
PARMS T1=-2.9 T2=-1.3 T3=.82 T4=.36 T5=-1.5 T6=-1.  T7=-.03 T8=-.47;
PEAK=T1+T2*X1+T3*X2+T4*X3;   INTER=T5+T3*X1+T6*X2+T7*X3;
BASE=-1+T4*X1+T7*X2+T8*X3;
MODEL Y=P1*LOG(PEAK/BASE)+P2*LOG(INTER/BASE);
DER.T1=P1/PEAK;  DER.T2=P1/PEAK*X1;  DER.T3=P1/PEAK*X2+P2/INTER*X1;
DER.T4=P1/PEAK*X3+(-P1-P2)/BASE*X1;  DER.T5=P2/INTER;
DER.T6=P2/INTER*X2;  DER.T7=P2/INTER*X3+(-P1-P2)/BASE*X2;
DER.T8=(-P1-P2)/BASE*X3;
OUTPUT OUT=WORK02 RESIDUAL=E;
```

Output:

SAS   1

NON-LINEAR LEAST SQUARES ITERATIVE PHASE

DEPENDENT VARIABLE: Y   METHOD: GAUSS-NEWTON

ITERATION	T1	T2	T3	RESIDUAL SS
	T4	T5	T6	
	T7	T8		
0	-2.90000000	-1.30000000	0.82000000	68.32779625
	0.36000000	-1.50000000	-1.00000000	
	-0.03000000	-0.47000000		
⋮				
14	-2.98025942	-1.16088895	0.78692676	57.02306899
	0.35309087	-1.50604388	-0.99985707	
	0.05407441	-0.47436347		

Figure 3a. Example 1 fitted by least squares, across equation constraints imposed.

EXAMPLE 1 (Continued). SAS code to minimize $S(\theta, I)$ for

$$f(x, \theta) = \left(\begin{array}{c} \ln \dfrac{\theta_1 + \theta_2 x_1 + \theta_3 x_2 + \theta_4 x_3}{-1 + \theta_4 x_1 + \theta_7 x_2 + \theta_8 x_3} \\ \ln \dfrac{\theta_5 + \theta_3 x_1 + \theta_6 x_2 + \theta_7 x_3}{-1 + \theta_4 x_1 + \theta_7 x_2 + \theta_8 x_3} \end{array} \right)$$

is shown in Figure 3a. A detailed discussion of the ideas is found in connection with Figure 2a, briefly they are as follows.

Trivially the identity factors as $I = P'P$ with $P = I$. The multivariate observations y_t, x_t for $t = 1, 2, \ldots, 224 = n$ are transformed to the univariate entities

$$\text{``}y_s\text{''} = p'_{(\alpha)} y_t \qquad \text{``}x_s\text{''} = (p'_{(\alpha)}, x'_t)'$$

SAS Statements:

```
DATA WORK03;  SET WORK02;   E1=E;  IF MOD(_N_,2)=0 THEN DELETE;
DATA WORK04;  SET WORK02;   E2=E;  IF MOD(_N_,2)=1 THEN DELETE;
DATA WORK05;  MERGE WORK03 WORK04;  KEEP E1 E2;
PROC MATRIX FW=20; FETCH E DATA=WORK05(KEEP=E1 E2);
SIGMA=E'*E#/224;  PRINT SIGMA;  P=HALF(INV(SIGMA));  PRINT P;
```

Output:

SAS   4

SIGMA	COL1	COL2
ROW1	0.1649246288351	0.09200572942276
ROW2	0.09200572942276	0.08964264342294

P	COL1	COL2
ROW1	3.76639099219	-3.865677509
ROW2	0	3.339970820524

Figure 3b. Contemporaneous variance-covariance matrix of Example 1 estimated from least squares residuals, across equation constraints imposed.

for $s = 1, 2, \ldots, 448 = nM$, which are then stored in the data set WORK01 as shown in Figure 3a. The univariate nonlinear model

$$\text{``}y_s\text{''} = \text{``}f\text{''}(\text{``}x_s\text{''}, \theta) + \text{``}e_s\text{''} \qquad s = 1, 2, \ldots, 448 = nM$$

with

$$\text{``}f\text{''}(\text{``}x_s\text{''}, \theta) = p'_{(\alpha)} f(x_t, \theta)$$

$$s = M(t - 1) + \alpha$$

is fitted to these data using PROC NLIN, and the residuals "$\hat{e}_s$" for $s = 1, 2, \ldots, 448 = nM$ are stored in the data set named WORK02.

In Figure 3b the univariate residuals stored in WORK02 are regrouped into the residuals $\hat{e}_t$ for $t = 1, 2, \ldots, 224 = n$ and stored in a data set named WORK05; here we are exploiting the fact that $P = I$. From the residuals stored in WORK05, $\hat{\Sigma}$ and $\hat{P}$ with $\hat{\Sigma}^{-1} = \hat{P}'\hat{P}$ are computed using PROC MATRIX. Compare this estimate of Σ with the one obtained in Figure 1c. Imposing the across equation restrictions results in a slight difference between the two estimates.

Using $\hat{P}$ as computed in Figure 3b, $S(\theta, \hat{\Sigma})$ is minimized to obtain

$$\hat{\theta} = \begin{pmatrix} -2.92458126 \\ -1.28674630 \\ 0.81856986 \\ 0.36115784 \\ -1.53758854 \\ -1.04895916 \\ 0.03008670 \\ -0.46742014 \end{pmatrix} \quad \text{(from Fig. 3}c\text{)}$$

as shown in Figure 3c; the ideas are the same as for Figure 3a. The difference between $\hat{\Sigma}$ in Figures 1c and 3b results in a slight difference between the estimate of θ computed in Figure 2c and $\hat{\theta}$ above. ☐

The theory in Section 6 would lead one to test $H : h(\theta) = 0$ by computing, for instance,

$$L' = S(\tilde{\theta}, \hat{\Sigma}) - S(\hat{\theta}, \hat{\Sigma})$$

and rejecting if L' exceeds the α-level critical point of the χ^2-distribution with q degrees of freedom, recall that $\tilde{\theta}$ minimizes $S(\theta, \hat{\Sigma})$ subject to $h(\theta) = 0$ and that $\hat{\theta}$ is the unconstrained minimum of $S(\theta, \hat{\Sigma})$. This is what one usually encounters in the applied literature. We shall not use that approach here. In this instance we shall compute

$$L = \frac{[S(\tilde{\theta}, \hat{\Sigma}) - S(\hat{\theta}, \hat{\Sigma})]/q}{S(\hat{\theta}, \hat{\Sigma})/(nM - p)}$$

and reject if L exceeds the α-level critical point of the F-distribution with q numerator degrees of freedom and $nM - p$ denominator degrees of freedom. There are two reasons for doing so. One is pedagogical: we wish to transfer the ideas in Chapter 1 intact to the multivariate setting. The other is to make some attempt to compensate for the sampling variation due to having to estimate Σ. We might note that $S(\hat{\theta}^{\#}, \hat{\Sigma}) \equiv nM$ (Problem 1), so that in typical instances $S(\hat{\theta}, \hat{\Sigma}) \doteq nM$. If nM is larger than 100, the difference between what we recommend here and what one usually encounters in the applied literature is slight.

In notation used here, the matrix $\hat{C}$ of Chapter 1 is written as (Problem 2)

$$\hat{C} = \left[\sum_{t=1}^{n} \left(\frac{\partial}{\partial \theta'} f(x_t, \theta) \right)' \hat{\Sigma}^{-1} \left(\frac{\partial}{\partial \theta'} f(x_t, \theta) \right) \right]^{-1}$$

SAS Statements:

```
DATA WORK01;  SET EXAMPLE1;
P1=3.76639099219;  P2=-3.865677509;  Y=P1*Y1+P2*Y2;  OUTPUT;
P1=0.0;  P2=3.339970820524;  Y=P1*Y1+P2*Y2;  OUTPUT;  DELETE;
PROC NLIN DATA=WORK01 METHOD=GAUSS ITER=50 CONVERGENCE=1.E-13;
PARMS T1=-2.9 T2=-1.3 T3=.82 T4=.36 T5=-1.5 T6=-1. T7=-.03 T8=-.47;
PEAK=T1+T2*X1+T3*X2+T4*X3;  INTER=T5+T3*X1+T6*X2+T7*X3;
BASE=-1+T4*X1+T7*X2+T8*X3;
MODEL Y=P1*LOG(PEAK/BASE)+P2*LOG(INTER/BASE);
DER.T1=P1/PEAK;  DER.T2=P1/PEAK*X1;  DER.T3=P1/PEAK*X2+P2/INTER*X1;
DER.T4=P1/PEAK*X3+(-P1-P2)/BASE*X1;  DER.T5=P2/INTER;
DER.T6=P2/INTER*X2;  DER.T7=P2/INTER*X3+(-P1-P2)/BASE*X2;
DER.T8=(-P1-P2)/BASE*X3;
```

Output:

SAS   5

NON-LINEAR LEAST SQUARES ITERATIVE PHASE

DEPENDENT VARIABLE: Y   METHOD: GAUSS-NEWTON

ITERATION	T1 T4 T7	T2 T5 T8	T3 T6	RESIDUAL SS
0	-2.90000000 0.36000000 -0.03000000	-1.30000000 -1.50000000 -0.47000000	0.82000000 -1.00000000	543.55788176
⋮				
14	-2.92458126 0.36115784 0.03008670	-1.28674630 -1.53758854 -0.46742014	0.81856986 -1.04895916	446.85695247

NOTE: CONVERGENCE CRITERION MET.

SAS   7

NON-LINEAR LEAST SQUARES SUMMARY STATISTICS   DEPENDENT VARIABLE Y

SOURCE	DF	SUM OF SQUARES	MEAN SQUARE
REGRESSION	8	6468.84819992	808.60602499
RESIDUAL	440	446.85695247	1.01558398
UNCORRECTED TOTAL	448	6915.70515239	
(CORRECTED TOTAL)	447	866.32697265	

PARAMETER	ESTIMATE	ASYMPTOTIC STD. ERROR	ASYMPTOTIC 95 % CONFIDENCE INTERVAL LOWER	UPPER
T1	-2.92458126	0.27790948	-3.47078451	-2.37837801
T2	-1.28674630	0.22671234	-1.73232670	-0.84116589
T3	0.81856986	0.08088226	0.65960389	0.97753584
T4	0.36115784	0.03029057	0.30162474	0.42069093
T5	-1.53758854	0.09192958	-1.71826692	-1.35691016
T6	-1.04895916	0.08367724	-1.21341839	-0.88449993
T7	0.03008670	0.03614145	-0.04094570	0.10111909
T8	-0.46742014	0.01926170	-0.50527708	-0.42956320

Figure 3c. Example 1 fitted by multivariate least squares, across equation constraints imposed.

and

$$s^2 = \frac{S(\hat{\theta}, \hat{\Sigma})}{nM - p}.$$

Writing $\hat{h} = h(\hat{\theta})$ and $\hat{H} = H(\hat{\theta})$, the Wald test statistic is

$$W = \frac{\hat{h}'(\hat{H}\hat{C}\hat{H}')^{-1}\hat{h}}{qs^2}.$$

One rejects the hypothesis

$$H: h(\theta^0) = 0$$

when W exceeds the upper $\alpha \times 100\%$ critical point of the F-distribution with q numerator degrees of freedom and $nM - p$ denominator degrees of freedom that is, when $W > F^{-1}(1 - \alpha; q, n - p)$.

Recall from Chapter 1 that a convenient method for computing W is to compute a vector of residuals $\hat{e}$ with typical element

$$\hat{e}_s = \text{``}y_s\text{''} - \text{``}f\text{''}(\text{``}x_s\text{''}, \hat{\theta}) = \hat{p}'_{(\alpha)}y_t - \hat{p}'_{(\alpha)}f(x_t, \hat{\theta})$$

compute a design matrix $\hat{F}$ with typical row

$$\hat{f}'_s = \frac{\partial}{\partial\theta'}\text{``}f\text{''}(\text{``}x_s\text{''}, \hat{\theta}) = \hat{p}'_{(\alpha)}\frac{\partial}{\partial\theta'}f(x_t, \hat{\theta})$$

fit the linear model

$$\hat{e} = \hat{F}\beta + u$$

by least squares, and test the hypothesis

$$H: \hat{H}\beta = \hat{h} \quad \text{against} \quad A: \hat{H}\beta \neq \hat{h}.$$

We illustrate.

EXAMPLE 1 (Continued). We wish to test the hypothesis of homogeneity,

$$H: h(\theta^0) = 0 \quad \text{against} \quad A: h(\theta^0) \neq 0$$

$$h(\theta) = \begin{pmatrix} \theta_2 + \theta_3 + \theta_4 \\ \theta_3 + \theta_6 + \theta_7 \\ \theta_4 + \theta_7 + \theta_8 \end{pmatrix}$$

in the model with bivariate response function

$$f(x, \theta) = \begin{pmatrix} \ln \dfrac{\theta_1 + \theta_2 x_1 + \theta_3 x_2 + \theta_4 x_3}{-1 + \theta_4 x_1 + \theta_7 x_2 + \theta_8 x_3} \\[2ex] \ln \dfrac{\theta_5 + \theta_3 x_1 + \theta_6 x_2 + \theta_7 x_3}{-1 + \theta_4 x_1 + \theta_7 x_2 + \theta_8 x_3} \end{pmatrix}$$

using the Wald test. To this end, the multivariate observations (y_t, x_t) are transformed to the univariate entities

$$\text{``}y_s\text{''} = p'_{(\alpha)} y_t \qquad \text{``}x_s\text{''} = (p'_{(\alpha)}, x'_t)'$$

which are then stored in the data set named WORK01 as shown in Figure 4. Using parameter values taken from Figure 3c, the entities

$$\hat{e}_s = \text{``}y_s\text{''} - \text{``}f\text{''}(\text{``}x_s\text{''}, \hat{\theta}) \qquad \hat{f}'_s = \frac{\partial}{\partial \theta'} \text{``}f\text{''}(\text{``}x_s\text{''}, \hat{\theta})$$

are computed and stored in the data set named WORK02. We are now in a position to compute

$$W = \frac{\hat{h}'(\hat{H}\hat{C}\hat{H}')^{-1}\hat{h}}{q s^2}$$

by fitting the model

$$\hat{e}_s = \hat{f}'_s \beta + u_s$$

using least squares and testing

$$H: \hat{H}\beta = \hat{h} \quad \text{against} \quad A: \hat{H}\beta \neq \hat{h}.$$

We have

$$\hat{h} = \begin{pmatrix} -1.28674630 + 0.81856986 + 0.36115784 \\ 0.81856986 - 1.04895916 + 0.03008670 \\ 0.36115784 + 0.03008670 - 0.46742014 \end{pmatrix} \quad \text{(from Fig. 3c)}$$

$$= \begin{pmatrix} -0.10701860 \\ -0.20030260 \\ -0.07617560 \end{pmatrix}$$

$$\hat{H} = \begin{pmatrix} 0 & 1 & 1 & 1 & 0 & 0 & 0 & 0 \\ 0 & 0 & 1 & 0 & 0 & 1 & 1 & 0 \\ 0 & 0 & 0 & 1 & 0 & 0 & 1 & 1 \end{pmatrix}$$

SAS Statements:

```
DATA WORK01;  SET EXAMPLE1;
P1=3.76639099219;  P2=-3.865677509;  Y=P1*Y1+P2*Y2;  OUTPUT;
P1=0.0;  P2=3.339970820524;  Y=P1*Y1+P2*Y2;  OUTPUT;  DELETE;
DATA WORK02;  SET WORK01;
T1=-2.92458126;  T2=-1.28674630;  T3=0.81856986;  T4=0.36115784;
T5=-1.53758854;  T6=-1.04895916;  T7=0.03008670;  T8=-0.46742014;
PEAK=T1+T2*X1+T3*X2+T4*X3;  INTER=T5+T3*X1+T6*X2+T7*X3;
BASE=-1+T4*X1+T7*X2+T8*X3;
E=Y-(P1*LOG(PEAK/BASE)+P2*LOG(INTER/BASE));
DER_T1=P1/PEAK;  DER_T2=P1/PEAK*X1;  DER_T3=P1/PEAK*X2+P2/INTER*X1;
DER_T4=P1/PEAK*X3+(-P1-P2)/BASE*X1;  DER_T5=P2/INTER;
DER_T6=P2/INTER*X2;  DER_T7=P2/INTER*X3+(-P1-P2)/BASE*X2;
DER_T8=(-P1-P2)/BASE*X3;
PROC REG DATA=WORK02;
MODEL E = DER_T1 DER_T2 DER_T3 DER_T4 DER_T5 DER_T6 DER_T7 DER_T8 / NOINT;
HOMOGENE: TEST DER_T2+DER_T3+DER_T4=-0.10701860,
               DER_T3+DER_T6+DER_T7=-0.20030260,
               DER_T4+DER_T7+DER_T8=-0.07617560;
```

Output:

DEP VARIABLE: E

SOURCE	DF	SUM OF SQUARES	MEAN SQUARE	F VALUE	PROB>F
MODEL	8	4.32010E-12	5.40012E-13	0.000	1.0000
ERROR	440	446.857	1.015584		
U TOTAL	448	446.857			

ROOT MSE	1.007762	R-SQUARE	0.0000	
DEP MEAN	0.001628355	ADJ R-SQ	-0.0159	
C.V.	61888.34			

NOTE: NO INTERCEPT TERM IS USED. R-SQUARE IS REDEFINED.

VARIABLE	DF	PARAMETER ESTIMATE	STANDARD ERROR	T FOR H0: PARAMETER=0	PROB > |T|
DER_T1	1	-2.37028E-07	0.277909	-0.000	1.0000
DER_T2	1	3.17717E-07	0.226712	0.000	1.0000
DER_T3	1	5.36973E-08	0.080882	0.000	1.0000
DER_T4	1	1.64816E-08	0.030291	0.000	1.0000
DER_T5	1	-5.10589E-08	0.091930	-0.000	1.0000
DER_T6	1	7.81229E-08	0.083677	0.000	1.0000
DER_T7	1	5.65637E-10	0.036141	0.000	1.0000
DER_T8	1	2.78288E-08	0.019262	0.000	1.0000

TEST: HOMOGENE	NUMERATOR:	7.31205	DF:	3	F VALUE:	7.1998
	DENOMINATOR:	1.01558	DF:	440	PROB >F :	0.0001

Figure 4. Illustration of Wald test computations with Example 1.

$$\hat{h}'(\hat{H}\hat{C}\hat{H}')^{-1}\hat{h}/3 = 7.31205 \qquad \text{(from Fig. 4)}$$

$$s^2 = 1.015584 \qquad \text{(from Fig. 3c or 4)}$$

$$W = 7.1998 \qquad \text{(from Fig. 4 or by division).}$$

Since $F^{-1}(.95; 3, 440) = 2.61$, one rejects at the 5% level. The p-value is smaller than 0.001, as shown in Figure 4. $\qquad\square$

Again following the ideas in Chapter 1, the Wald test statistic is approximately distributed as the noncentral F-distribution, with q numerator degrees of freedom, $nM - p$ denominator degrees of freedom, and noncentrality parameter

$$\lambda = \frac{h'(\theta^0)[H(\theta^0)C(\theta^0)H'(\theta^0)]^{-1}h(\theta^0)}{2}$$

$$C(\theta) = \left[\sum_{t=1}^{n}\left(\frac{\partial}{\partial\theta'}f(x_t,\theta)\right)'\Sigma^{-1}\left(\frac{\partial}{\partial\theta'}f(x_t,\theta)\right)\right]^{-1};$$

written more compactly as $W \doteq F'(q, nM - p, \lambda)$. As noted in Chapter 1, the computation of λ is little different from the computation of W itself; we illustrate.

EXAMPLE 1 (Continued). Consider finding the probability that a 5% level Wald test rejects the hypothesis of homogeneity

$$H: h(\theta) = \begin{pmatrix} \theta_2 + \theta_3 + \theta_4 \\ \theta_3 + \theta_6 + \theta_7 \\ \theta_4 + \theta_7 + \theta_8 \end{pmatrix} = 0$$

at the parameter settings

$$\theta^0 = \begin{pmatrix} -2.82625314 \\ -1.25765338 \\ 0.83822896 \\ 0.36759231 \\ -1.56498719 \\ -0.98193861 \\ 0.04422702 \\ -0.44971643 \end{pmatrix},$$

$$\Sigma = \begin{pmatrix} 0.16492462883510 & 0.09200572942276 \\ 0.09200572942276 & 0.08964264342294 \end{pmatrix},$$

$$n = 224$$

SAS Statements:

```
DATA WORK01;  SET EXAMPLE1;
P1=3.76639099219;  P2=-3.865677509;  Y=P1*Y1+P2*Y2;  OUTPUT;
P1=0.0;  P2=3.339970820524;  Y=P1*Y1+P2*Y2;  OUTPUT;  DELETE;
DATA WORK02;  SET WORK01;
T1=  -2.82625314;  T2=  -1.25765338;  T3=   0.83822896;  T4=   0.36759231;
T5=  -1.56498719;  T6=  -0.98193861;  T7=   0.04422702;  T8=  -0.44971643;
PEAK=T1+T2*X1+T3*X2+T4*X3;  INTER=T5+T3*X1+T6*X2+T7*X3;
BASE=-1+T4*X1+T7*X2+T8*X3;
DER_T1=P1/PEAK;  DER_T2=P1/PEAK*X1;  DER_T3=P1/PEAK*X2+P2/INTER*X1;
DER_T4=P1/PEAK*X3+(-P1-P2)/BASE*X1;  DER_T5=P2/INTER;
DER_T6=P2/INTER*X2;  DER_T7=P2/INTER*X3+(-P1-P2)/BASE*X2;
DER_T8=(-P1-P2)/BASE*X3;
PROC MATRIX;  FETCH F DATA=WORK02(KEEP=DER_T1-DER_T8);  C=INV(F'*F);  FREE F;
FETCH T 1 DATA=WORK02(KEEP=T1-T8);
H = 0 1 1 1 0 0 0 0 / 0 0 1 0 0 1 1 0 / 0 0 0 1 0 0 1 1 ;  HO=H*T';
LAMBDA=HO'*INV(H*C*H')*HO#/2;  PRINT LAMBDA;
```

Output:

	LAMBDA	COL1
ROW1		3.29906

Figure 5. Illustration of Wald test power computations with Example 1.

for data with bivariate response function

$$
f(x, \theta) = \begin{pmatrix} \ln \dfrac{\theta_1 + \theta_2 x_1 + \theta_3 x_2 + \theta_4 x_3}{-1 + \theta_4 x_1 + \theta_7 x_2 + \theta_8 x_3} \\[2ex] \ln \dfrac{\theta_5 + \theta_3 x_1 + \theta_6 x_2 + \theta_7 x_3}{-1 + \theta_4 x_1 + \theta_7 x_2 + \theta_8 x_3} \end{pmatrix}
$$

the value of θ^0 chosen is midway on the line segment joining the last two columns of Table 2.

Recall (Figure 3b) that Σ^{-1} factors as $\Sigma^{-1} = P'P$ with

$$
P = \begin{pmatrix} 3.76639099219 & -3.865677509 \\ 0 & 3.339970820524 \end{pmatrix}.
$$

Exactly as in Figure 4, the multivariate model is transformed in Figure 5 to a univariate model, and the Jacobian of the univariate model evaluated at θ^0—denote it as F—is stored in the data set named WORK02. Next

$$
\lambda = \frac{h'[HCH']h}{2} = 3.29906 \qquad \text{(Fig. 5)}
$$

with $h = h(\theta^0)$, $H = (\partial/\partial\theta')h(\theta^0)$, and

$$C = (F'F)^{-1} = \left[\sum_{t=1}^{n} \left(\frac{\partial}{\partial\theta'} f(x_t, \theta^0) \right)' \Sigma^{-1} \left(\frac{\partial}{\partial\theta'} f(x_t, \theta^0) \right) \right]^{-1}$$

is computed using straightforward matrix algebra. From the Pearson-Hartley charts of the noncentral F-distribution in Scheffé (1959) we obtain

$$1 - F'(2.61; 3, 440, 3.29906) = .55$$

as the approximation to the probability that a 5% level Wald test rejects the hypothesis of homogeneity if the true values of θ^0 and Σ are as above. $\quad\square$

A derivation of the "likelihood ratio" test of the hypothesis

$$H : h(\theta^0) = 0 \quad \text{against} \quad A : h(\theta^0) \neq 0$$

using the ideas of Chapter 1 is straightforward. Recall that $\hat{\theta}$ is the unconstrained minimum of $S(\theta, \hat{\Sigma})$, that $\tilde{\theta}$ minimizes $S(\theta, \hat{\Sigma})$ subject to $h(\theta) = 0$, and that $h(\theta)$ maps $\mathbb{R}^p$ into $\mathbb{R}^q$. As we have seen, an alternative method of computing $\tilde{\theta}$ makes use of the equivalent form of the hypothesis

$$H : \theta^0 = g(\rho^0) \text{ for some } \rho^0 \quad \text{against} \quad A : \theta^0 \neq g(\rho) \text{ for any } \rho.$$

One computes the unconstrained minimum $\hat{\rho}$ of $S[g(\rho), \hat{\Sigma}]$ and puts $\tilde{\theta} = g(\hat{\rho})$. Using the formula given in Chapter 1,

$$L = \frac{(\text{SSE}_{\text{reduced}} - \text{SSE}_{\text{full}})/q}{(\text{SSE}_{\text{full}})/("n" - p)}$$

and using

$$S(\theta, \hat{\Sigma}) = \sum_{s=1}^{nM} ["y_s" - "f"("x_s", \theta)]^2$$

one obtains the statistic

$$L = \frac{[S(\tilde{\theta}, \hat{\Sigma}) - S(\hat{\theta}, \hat{\Sigma})]/q}{S(\hat{\theta}, \hat{\Sigma})/(nM - p)}.$$

One rejects $H : h(\theta^0) = 0$ when L exceeds the $\alpha \times 100\%$ critical point F_α

of the F-distribution with q numerator degrees of freedom and $nM - p$ denominator degrees of freedom, $F_\alpha = F^{-1}(1 - \alpha; q, nM - p)$.

We illustrate the computations with the example. In reading it, recall from Chapter 1 that one can exploit the structure of a composite function in writing code as follows. Suppose code is at hand to compute $f(x, \theta)$ and $F(x, \theta) = (\partial/\partial \theta')f(x, \theta)$. Given the value $\hat{\rho}$, compute $\tilde{\theta} = g(\hat{\rho})$ and $\tilde{G} = (\partial/\partial \rho')g(\hat{\rho})$. Obtain the value $f[x, g(\hat{\rho})]$ from the function evaluation $f(x, \tilde{\theta})$. Obtain $(\partial/\partial \rho')f[x, g(\hat{\rho})]$ by evaluating $F(x, \tilde{\theta})$ and performing the matrix multiplication $F(x, \tilde{\theta})\tilde{G}$.

EXAMPLE 1 (Continued). Consider retesting the hypothesis of homogeneity, expressed as the functional dependence

$$H: \theta^0 = g(\rho^0) \text{ for some } \rho^0 \quad \text{against} \quad A: \theta^0 \neq g(\rho) \text{ for any } \rho$$

with

$$
g(\rho) = \begin{pmatrix}
\rho_1 \\
-\rho_2 - \rho_3 \\
\rho_2 \\
\rho_3 \\
\rho_4 \\
-\rho_5 - \rho_2 \\
\rho_5 \\
-\rho_5 - \rho_3
\end{pmatrix}
$$

in the model with response function

$$
f(x, \theta) = \begin{pmatrix}
\ln \dfrac{\theta_1 + \theta_2 x_1 + \theta_3 x_2 + \theta_4 x_3}{-1 + \theta_4 x_1 + \theta_7 x_2 + \theta_8 x_3} \\
\ln \dfrac{\theta_5 + \theta_3 x_1 + \theta_6 x_2 + \theta_7 x_3}{-1 + \theta_4 x_1 + \theta_7 x_2 + \theta_8 x_3}
\end{pmatrix}
$$

using the "likelihood ratio" test; θ has length $p = 8$ and ρ has length $r = 5$, whence $q = p - r = 3$. The model is bivariate, so $M = 2$, and there are $n = 224$ observations. We adopt the expedient discussed immediately above, reusing the code of Figure 3c; the Jacobian of $g(\rho)$ was displayed earlier on in this section. The result is the SAS code shown in Figure 6. We obtain

$$\text{SSE}(\tilde{\theta}, \hat{\Sigma}) = 474.68221082 \quad \text{(from Fig. 6)}.$$

SAS Statements:

```
DATA WORK01;  SET EXAMPLE1;
P1=3.76639099219;   P2=-3.865677509;   Y=P1*Y1+P2*Y2;   OUTPUT;
P1=0.0;   P2=3.339970820524;   Y=P1*Y1+P2*Y2;   OUTPUT;   DELETE;
PROC NLIN DATA=WORK01 METHOD=GAUSS ITER=50 CONVERGENCE=1.E-13;
PARMS R1=-3 R2=.8 R3=.4 R4=-1.5 R5=.03;
T1=R1;  T2=-R2-R3;  T3=R2;  T4=R3;  T5=R4;  T6=-R5-R2;  T7=R5;  T8=-R5-R3;
PEAK=T1+T2*X1+T3*X2+T4*X3;   INTER=T5+T3*X1+T6*X2+T7*X3;
BASE=-1+T4*X1+T7*X2+T8*X3;
MODEL Y=P1*LOG(PEAK/BASE)+P2*LOG(INTER/BASE);
DER_T1=P1/PEAK;   DER_T2=P1/PEAK*X1;   DER_T3=P1/PEAK*X2+P2/INTER*X1;
DER_T4=P1/PEAK*X3+(-P1-P2)/BASE*X1;   DER_T5=P2/INTER;
DER_T6=P2/INTER*X2;   DER_T7=P2/INTER*X3+(-P1-P2)/BASE*X2;
DER_T8=(-P1-P2)/BASE*X3;
DER.R1=DER_T1;  DER.R2=-DER_T2+DER_T3-DER_T6;  DER.R3=-DER_T2+DER_T4-DER_T8;
DER.R4=DER_T5;  DER.R5=-DER_T6+DER_T7-DER_T8;
```

Output:

NON-LINEAR LEAST SQUARES ITERATIVE PHASE

DEPENDENT VARIABLE: Y   METHOD: GAUSS-NEWTON

ITERATION	R1 R4	R2 R5	R3	RESIDUAL SS
0	-3.00000000 -1.50000000	0.80000000 0.03000000	0.40000000	556.82802354
. . .				
6	-2.72482606 -1.59239423	0.85773951 0.05768367	0.37430609	474.68221082

NOTE: CONVERGENCE CRITERION MET.

NON-LINEAR LEAST SQUARES SUMMARY STATISTICS   DEPENDENT VARIABLE Y

SOURCE	DF	SUM OF SQUARES	MEAN SQUARE
REGRESSION	5	6441.02294156	1288.20458831
RESIDUAL	443	474.68221082	1.07151741
UNCORRECTED TOTAL	448	6915.70515239	
(CORRECTED TOTAL)	447	866.32697265	

PARAMETER	ESTIMATE	ASYMPTOTIC STD. ERROR	ASYMPTOTIC 95 % CONFIDENCE INTERVAL	
			LOWER	UPPER
R1	-2.72482606	0.17837791	-3.07540344	-2.37424867
R2	0.85773951	0.06707057	0.72592147	0.98955755
R3	0.37430609	0.02713134	0.32098315	0.42762902
R4	-1.59239423	0.07748868	-1.74468762	-1.44010083
R5	0.05768367	0.03407531	-0.00928668	0.12465402

Figure 6. Example 1 fitted by multivariate least squares, across equation constraints imposed, homogeneity imposed.

Previously we computed

$$\text{SSE}(\hat{\theta}, \hat{\Sigma}) = 446.85695247 \qquad \text{(from Fig. 3}c\text{)}.$$

The "likelihood ratio" test statistic is

$$L = \frac{[S(\tilde{\theta}, \hat{\Sigma}) - S(\hat{\theta}, \hat{\Sigma})]/q}{S(\hat{\theta}, \hat{\Sigma})/(nM - p)}$$

$$= \frac{(474.68221082 - 446.85695247)/3}{446.85695247/(448 - 8)}$$

$$= 9.133.$$

Comparing with the critical point

$$F^{-1}(.95; 3,440) = 2.61$$

one rejects the hypothesis of homogeneity at the 5% level. This is, by and large, a repetition of the computation displayed in Table 3; the slight change in $\hat{\Sigma}$ has made little difference.   □

In order to approximate the power of the "likelihood ratio" test we proceed as before. We formally treat the transformed model

$$\text{``}y_s\text{''} = \text{``}f\text{''}(\text{``}x_s\text{''}, \theta) + \text{``}e_s\text{''} \qquad s = 1, 2, \ldots, nM$$

as if it were a univariate nonlinear regression model and apply the results of Chapter 1. In a power computation, one is given an expression for the response function $f(x, \theta)$ (with range in $\mathbb{R}^M$), values for the parameters θ^0 and Σ, a sequence of independent variables $\{x_t\}_{t=1}^n$, and the hypothesis

$$H: \theta^0 = g(\rho^0) \text{ for some } \rho^0 \quad \text{against} \quad A: \theta^0 \neq g(\rho) \text{ for any } \rho.$$

Recall that the univariate response function is computed by factoring Σ^{-1} as $\Sigma^{-1} = P'P$ and putting

$$\text{``}f\text{''}(\text{``}x_s\text{''}, \theta) = p'_{(\alpha)} f(x_t, \theta)$$

for

$$s = M(t - 1) + \alpha \qquad \alpha = 1, 2, \ldots, M \quad t = 1, 2, \ldots, n.$$

Applying the ideas of Chapter 1, the null hypothesis induces the location parameter

$$\theta_n^* = g(\rho_n^0)$$

where ρ_n^0 is computed by minimizing

$$\sum_{s=1}^{nM} \left\{ "f"("x_s", \theta^0) - "f"["x_s", g(\rho)] \right\}^2$$

$$= \sum_{t=1}^{n} \left\{ f(x_t, \theta^0) - f[x_t, g(\rho)] \right\}'$$

$$\times \sum_{\alpha=1}^{M} p_{(\alpha)} p'_{(\alpha)} \left\{ f(x_t, \theta^0) - f[x_t, g(\rho)] \right\}$$

$$= \sum_{t=1}^{n} \left\{ f(x_t, \theta^0) - f[x_t, g(\rho)] \right\}' \Sigma^{-1} \left\{ f(x_t, \theta^0) - f[x_t, g(\rho)] \right\}.$$

Let

$$\delta_t = f(x_t, \theta^0) - f[x_t, g(\rho_n^0)]$$

$$F_t = \frac{\partial}{\partial \theta'} f(x_t, \theta^0).$$

Similar algebra results in the following expressions for the noncentrality parameters of Section 5 of Chapter 1:

$$\lambda_1 = \frac{\delta' P_F \delta - \delta' P_{FG} \delta}{2}$$

$$\lambda_2 = \frac{\delta'\delta - \delta' P_F \delta}{2}$$

$$\delta'\delta = \sum_{t=1}^{n} \delta_t' \Sigma^{-1} \delta_t$$

$$\delta' P_F \delta = \left(\sum_{t=1}^{n} \delta_t' \Sigma^{-1} F_t \right) \left(\sum_{t=1}^{n} F_t' \Sigma^{-1} F_t \right)^{-1} \left(\sum_{t=1}^{n} F_t' \Sigma^{-1} \delta_t \right)$$

$$\delta' P_{FG} \delta = \left(\sum_{t=1}^{n} \delta_t' \Sigma^{-1} F_t G \right) \left(G' \sum_{t=1}^{n} F_t' \Sigma^{-1} F_t G \right)^{-1} \left(G' \sum_{t=1}^{n} F_t' \Sigma^{-1} \delta_t \right).$$

One approximates the probability that the "likelihood ratio" rejects H by

$$P(L > F_\alpha) \doteq 1 - H(c_\alpha; q, nM - p, \lambda_1, \lambda_2)$$

where

$$c_\alpha = 1 + \frac{q F_\alpha}{nM - p};$$

$H(x; v_1, v_2, \lambda_1, \lambda_2)$ is the distribution defined and partially tabled in Section 5 of Chapter 1. Recall that if λ_2 is small, the approximation

$$P(L > F_\alpha) \doteq 1 - F'(F_\alpha; q, nM - p, \lambda_1)$$

is adequate, where $F'(x; v_1, v_2, \lambda)$ denotes the noncentral F-distribution. We illustrate with the example.

EXAMPLE 1 (Continued). Consider finding the probability that a 5% level "likelihood ratio" test rejects the hypothesis of homogeneity

$$H: \theta^0 = g(\rho^0) \text{ for some } \rho^0 \quad \text{against} \quad A: \theta^0 \neq g(\rho) \text{ for any } \rho$$

with

$$g(\rho) = \begin{pmatrix} \rho_1 \\ -\rho_2 - \rho_3 \\ \rho_2 \\ \rho_3 \\ \rho_4 \\ -\rho_5 - \rho_2 \\ \rho_5 \\ -\rho_5 - \rho_3 \end{pmatrix}$$

at the parameter settings

$$\theta^0 = \begin{pmatrix} -2.82625314 \\ -1.25765338 \\ 0.83822896 \\ 0.36759231 \\ -1.56498719 \\ -0.98193861 \\ 0.04422702 \\ -0.44971643 \end{pmatrix},$$

$$\Sigma = \begin{pmatrix} 0.16492462883510 & 0.09200572942276 \\ 0.09200572942276 & 0.08964264342294 \end{pmatrix},$$

$$n = 224$$

for data with bivariate response function

$$f(x, \theta) = \begin{pmatrix} \ln \dfrac{\theta_1 + \theta_2 x_1 + \theta_3 x_2 + \theta_4 x_3}{-1 + \theta_4 x_1 + \theta_7 x_2 + \theta_8 x_3} \\[2ex] \ln \dfrac{\theta_5 + \theta_3 x_1 + \theta_6 x_2 + \theta_7 x_3}{-1 + \theta_4 x_1 + \theta_7 x_2 + \theta_8 x_3} \end{pmatrix}.$$

The value of θ^0 chosen is midway on the line segment joining the last two

SAS Statements:

```
DATA WORK01;  SET EXAMPLE1;
P1=3.76639099219;  P2=-3.865677509;  Y=P1*Y1+P2*Y2;  OUTPUT;
P1=0.0;  P2=3.339970820524;  Y=P1*Y1+P2*Y2;  OUTPUT;  DELETE;
DATA WORK02;  SET WORK01;
T1= -2.82625314;  T2= -1.25765338;  T3=  0.83822896;  T4=  0.36759231;
T5= -1.56498719;  T6= -0.98193861;  T7=  0.04422702;  T8= -0.44971643;
PEAK=T1+T2*X1+T3*X2+T4*X3;  INTER=T5+T3*X1+T6*X2+T7*X3;
BASE=-1+T4*X1+T7*X2+T8*X3;
F1=P1/PEAK;  F2=P1/PEAK*X1;  F3=P1/PEAK*X2+P2/INTER*X1;
F4=P1/PEAK*X3+(-P1-P2)/BASE*X1;  F5=P2/INTER;
F6=P2/INTER*X2;  F7=P2/INTER*X3+(-P1-P2)/BASE*X2;  F8=(-P1-P2)/BASE*X3;
YDUMMY=P1*LOG(PEAK/BASE)+P2*LOG(INTER/BASE);  DROP T1-T8;
PROC NLIN DATA=WORK02 METHOD=GAUSS ITER=50 CONVERGENCE=1.E-13;
PARMS R1=-3 R2=.8 R3=.4 R4=-1.5 R5=.03;
T1=R1;  T2=-R2-R3;  T3=R2;  T4=R3;  T5=R4;  T6=-R5-R2;  T7=R5;  T8=-R5-R3;
PEAK=T1+T2*X1+T3*X2+T4*X3;  INTER=T5+T3*X1+T6*X2+T7*X3;
BASE=-1+T4*X1+T7*X2+T8*X3;
MODEL YDUMMY=P1*LOG(PEAK/BASE)+P2*LOG(INTER/BASE);
DER_T1=P1/PEAK;  DER_T2=P1/PEAK*X1;  DER_T3=P1/PEAK*X2+P2/INTER*X1;
DER_T4=P1/PEAK*X3+(-P1-P2)/BASE*X1;  DER_T5=P2/INTER;
DER_T6=P2/INTER*X2;  DER_T7=P2/INTER*X3+(-P1-P2)/BASE*X2;
DER_T8=(-P1-P2)/BASE*X3;
DER.R1=DER_T1; DER.R2=-DER_T2+DER_T3-DER_T6; DER.R3=-DER_T2+DER_T4-DER_T8;
DER.R4=DER_T5; DER.R5=-DER_T6+DER_T7-DER_T8;
```

Output:

SAS   1

NON-LINEAR LEAST SQUARES ITERATIVE PHASE

DEPENDENT VARIABLE: Y   METHOD: GAUSS-NEWTON

ITERATION	R1 R4	R2 R5	R3	RESIDUAL SS
0	-3.00000000 -1.50000000	0.80000000 0.03000000	0.40000000	90.74281456
.				
.				
.				
4	-2.73217450 -1.59899262	0.85819875 0.05540598	0.37461886	7.43156658

Figure 7. Illustration of likelihood ratio test power computations with Example 1.

columns of Table 2. Recall (Fig. 3b) that Σ^{-1} factors as $\Sigma^{-1} = P'P$ with

$$P = \begin{pmatrix} 3.76639099219 & -3.865677509 \\ 0 & 3.339970820524 \end{pmatrix}.$$

Referring to Figure 7, the multivariate model is converted to a univariate model, and the entities $f''(``x_s", \theta^0)$ and $(\partial/\partial\theta')f''(``x_s", \theta)$ are computed and stored in the data set named WORK02. Reusing the code of

SAS Statements:

```
DATA WORK03;  SET WORK02;
R1=-2.73217450;  R2=0.85819875;  R3=0.37461886;  R4=-1.59899262;  R5=0.05540598;
T1=R1;  T2=-R2-R3;  T3=R2;  T4=R3;  T5=R4;  T6=-R5-R2;  T7=R5;  T8=-R5-R3;
PEAK=T1+T2*X1+T3*X2+T4*X3;   INTER=T5+T3*X1+T6*X2+T7*X3;
BASE=-1+T4*X1+T7*X2+T8*X3;
DELTA=P1*LOG(PEAK/BASE)+P2*LOG(INTER/BASE)-YDUMMY;
FG1=F1; FG2=-F2+F3-F6; FG3=-F2+F4-F8;FG4=F5; FG5=-F6+F7-F8;
PROC REG DATA=WORK03;  MODEL DELTA = F1-F8 / NOINT;
PROC REG DATA=WORK03;  MODEL DELTA = FG1-FG5 / NOINT;
```

Output:

SAS   3

DEP VARIABLE: DELTA

SOURCE	DF	SUM OF SQUARES	MEAN SQUARE	F VALUE	PROB>F
MODEL	8	7.401094	0.925137	13358.456	0.0001
ERROR	440	0.030472	.00006925477		
U TOTAL	448	7.431567			

SAS   4

DEP VARIABLE: DELTA

SOURCE	DF	SUM OF SQUARES	MEAN SQUARE	F VALUE	PROB>F
MODEL	5	0.134951	0.026990	1.639	0.1472
ERROR	443	7.296616	0.016471		
U TOTAL	448	7.431567			

Figure 7. (Continued).

Figure 6, ρ_n^0 to minimize

$$\sum_{s=1}^{nM} \left\{ ``f\,"\left(``x_s\,", \theta^0\right) - ``f\,"\left[``x_s\,", g(\rho)\right] \right\}^2$$

is computed using PROC NLIN. From this value and setting $\theta_n^* = g(\rho_n^0)$, the entities

$$``\delta_s\,"=``f\,"\left(``x_s\,", \theta^0\right) - ``f\,"\left(``x_s\,", \theta_n^*\right)$$

$$\frac{\partial}{\partial \theta'}``f\,"\left(``x_s\,", \theta^0\right)\frac{\partial}{\partial \rho'}g\left(\rho_n^0\right)$$

are computed, adjoined to the data in WORK02, and stored in the data set named **WORK03**. Then, as explained in connection with Figure 11a of Chapter 1, one can regress "δ_s" on $(\partial/\partial\theta')$ "f "($"x_s"$, θ^0) to obtain $\delta'\delta$ and $\delta'P_F\delta$ from the analysis of variance table and can regress "δ_s" on $(\partial/\partial\theta')$"$f$ "($"x_s"$, θ^0)($\partial/\partial\rho'$)$g(\rho_n^0)$ to obtain $\delta'P_{FG}\delta$. We have

$$\delta'\delta = 7.431567 \qquad \text{(from Fig. 7)}$$
$$\delta'P_F\delta = 7.401094 \qquad \text{(from Fig. 7)}$$
$$\delta'P_{FG}\delta = 0.134951 \qquad \text{(from Fig. 7)}$$

whence

$$\begin{aligned}
\lambda_1 &= \frac{\delta'P_F\delta - \delta'P_{FG}\delta}{2} \\
&= \frac{7.401094 - 0.134951}{2} \\
&= 3.63307 \\
\lambda_2 &= \frac{\delta'\delta - \delta'P_F\delta}{2} \\
&= \frac{7.431567 - 7.401094}{2} \\
&= 0.01524 \\
c_\alpha &= 1 + \frac{qF_\alpha}{nM - p} \\
&= 1 + 3\frac{2.61}{448 - 8} \\
&= 1.01780.
\end{aligned}$$

Direct computation (Gallant, 1975b) yields

$$\begin{aligned}
P(L > 2.61) &\doteq 1 - H(1.01780; 3, 440, 3.63307, 0.01524) \\
&= 0.610.
\end{aligned}$$

From the Pearson-Hartley charts of the noncentral F-distribution (Scheffé, 1959) one has

$$\begin{aligned}
P(L > 2.61) &\doteq 1 - F'(2.61; 3, 440, 3.63307) \\
&= 0.60. \qquad\qquad \square
\end{aligned}$$

In Chapter 1 we noted that the Lagrange multiplier test had rather bizarre structural characteristics. Take the simple case of testing $H: \theta^0 = \theta^*$ against $A: \theta^0 \neq \theta^*$. If θ^* is near a local minimum or a local maximum of the sum of squares surface, then the test will accept H no matter how large is the distance between $\hat{\theta}$ and θ^*. Also we saw some indications that the Lagrange multiplier test had poorer power than the likelihood ratio test. Thus, it would seem that one would not use the Lagrange multiplier test unless the computation of the unconstrained estimator $\hat{\theta}$ is inordinately burdensome for some reason. We shall assume that this is the case.

If $\hat{\theta}$ is inordinately burdensome to compute, then $\hat{\theta}^{\#}$ will be as well. Thus, it is unreasonable to assume that one has available an estimator $\hat{\Sigma}$ with $\sqrt{n}\,(\hat{\Sigma} - \Sigma)$ bounded in probability when $h(\theta^0) = 0$ is false, since such an estimator will almost always have to be computed from residuals from an unconstrained fit. The exception is when one has replicates at some settings of the independent variable. Accordingly, we shall base the Lagrange multiplier statistic on an estimator $\tilde{\Sigma}_n$ computed as follows.

If the hypothesis is written as a parametric restriction

$$H : h(\theta^0) = 0 \quad \text{against} \quad A : h(\theta^0) \neq 0$$

then let $\tilde{\theta}^{\#}$ minimize $S(\theta, I)$ subject to $h(\theta) = 0$, and put

$$\tilde{\Sigma} = \frac{1}{n} \sum_{t=1}^{n} \left[y_t - f(x_t, \tilde{\theta}^{\#}) \right] \left[y_t - f(x_t, \tilde{\theta}^{\#}) \right]'.$$

If the hypothesis is written as a functional dependence

$$H : \theta^0 = g(\rho^0) \text{ for some } \rho^0 \quad \text{against} \quad A : \theta^0 \neq g(\rho) \text{ for any } \rho$$

then let $\hat{\rho}^{\#}$ minimize $S[g(\rho), I]$ and put

$$\tilde{\Sigma} = \frac{1}{n} \sum_{t=1}^{n} \left\{ y_t - f[x_t, g(\hat{\rho}^{\#})] \right\} \left\{ y_t - f[x_t, g(\hat{\rho}^{\#})] \right\}'.$$

The constrained estimator corresponding to this estimator of scale is $\tilde{\tilde{\theta}}$ that minimizes $S(\theta, \tilde{\Sigma})$ subject to $h(\theta) = 0$. Equivalently, let $\hat{\rho}$ minimize $S[g(\rho), \tilde{\Sigma}]$, whence $\tilde{\tilde{\theta}} = g(\hat{\rho})$.

Factoring $\tilde{\Sigma}^{-1}$ as $\tilde{\Sigma}^{-1} = \tilde{P}'\tilde{P}$, denoting a typical row of $\tilde{P}$ by $\tilde{p}'_{(\alpha)}$ and formally treating the transformed model

$$\text{``}y_s\text{''} = \text{``}f\,\text{''}(\text{``}x_s\text{''}, \theta) + \text{``}e_s\text{''} \qquad s = 1, 2, \ldots, nM$$

with $s = M(t - 1) + \alpha$

$$\text{``}y_s\text{''} = \tilde{p}'_\alpha y_t$$
$$\text{``}x_s\text{''} = (\tilde{p}'_{(\alpha)}, x'_t)'$$
$$\text{``}f\,\text{''}(\text{``}x_s\text{''}, \theta) = \tilde{p}'_{(\alpha)} f(x_t, \theta)$$

as a univariate model, one obtains as the second version of the Lagrange multiplier test given in Chapter 1 the statistic

$$
\tilde{R} = \frac{nM}{S(\tilde{\theta}, \tilde{\Sigma})} \left[\sum_{t=1}^{n} \left[y_t - f(x_t, \tilde{\theta}) \right]' \tilde{\Sigma}^{-1} \left(\frac{\partial}{\partial \theta'} f(x_t, \tilde{\theta}) \right) \right]
$$
$$
\times \left[\sum_{t=1}^{n} \left(\frac{\partial}{\partial \theta'} f(x_t, \tilde{\theta}) \right)' \tilde{\Sigma}^{-1} \left(\frac{\partial}{\partial \theta'} f(x_t, \tilde{\theta}) \right) \right]^{-1}
$$
$$
\times \left[\sum_{t=1}^{n} \left(\frac{\partial}{\partial \theta'} f(x_t, \tilde{\theta}) \right)' \tilde{\Sigma}^{-1} \left[y_t - f(x_t, \tilde{\theta}) \right] \right].
$$

One rejects $H : h(\theta^0) = 0$ if $\tilde{R} > d_\alpha$, where

$$
d_\alpha = \frac{nMF_\alpha}{\dfrac{nM - p}{q} + F_\alpha}
$$

and F_α denotes the $\alpha \times (100\%)$ critical point of the F-distribution with q numerator degrees of freedom and $nM - p$ denominator degrees of freedom that is, $\alpha = 1 - F(F_\alpha; q, nM - p)$.

One can use the same approach used in Chapter 1 to compute $\tilde{R}$. Create a data set with observations

$$
\text{``}\tilde{e}_s\text{''} = \text{``}y_s\text{''} - \text{``}f\text{''}(\text{``}x_s\text{''}, \tilde{\theta}) = \tilde{P}'_{(\alpha)} y_t - P'_{(\alpha)} f(x_t, \tilde{\theta})
$$
$$
\tilde{f}'_s = \frac{\partial}{\partial \theta'} \text{``}f\text{''}(\text{``}x_s\text{''}, \tilde{\theta}) = P'_{(\alpha)} \frac{\partial}{\partial \theta'} f(x_t, \tilde{\theta}).
$$

Let $\tilde{e}$ be the nM-vector with "$\tilde{e}_s$" as elements, and let $\tilde{F}$ be the nM by p matrix with $\tilde{f}'_s$ as a typical row. A linear regression of $\tilde{e}$ on $\tilde{F}$ with no intercept term yields the analysis of variance table

Source	d.f.	Sum of Squares
Regression	p	$\tilde{e}'\tilde{F}(\tilde{F}'\tilde{F})^{-1}\tilde{F}'\tilde{e}$
Error	$nM - p$	$\tilde{e}'\tilde{e} - \tilde{e}'\tilde{F}(\tilde{F}'\tilde{F})^{-1}\tilde{F}'\tilde{e}$
Total	nM	$\tilde{e}'\tilde{e}$

From this table $\tilde{R}$ is computed as

$$
\tilde{R} = nM \frac{\tilde{e}'\tilde{F}(\tilde{F}'\tilde{F})^{-1}\tilde{F}'\tilde{e}}{\tilde{e}'\tilde{e}}.
$$

Let us illustrate.

EXAMPLE 1 (Continued). Consider retesting the hypothesis of homo-geneity, expressed as the functional dependence

$$H: \theta^0 = g(\rho^0) \text{ for some } \rho^0 \quad \text{against} \quad A: \theta^0 \neq g(\rho) \text{ for any } \rho$$

with

$$g(\rho) = \begin{pmatrix} \rho_1 \\ -\rho_2 - \rho_3 \\ \rho_2 \\ \rho_3 \\ \rho_4 \\ -\rho_5 - \rho_2 \\ \rho_5 \\ -\rho_5 - \rho_3 \end{pmatrix}$$

SAS Statements:

```
DATA WORK01;  SET EXAMPLE1;
P1=1.0;  P2=0.0;  Y=P1*Y1+P2*Y2;  OUTPUT;
P1=0.0;  P2=1.0;  Y=P1*Y1+P2*Y2;  OUTPUT;  DELETE;
PROC NLIN DATA=WORK01 METHOD=GAUSS ITER=50 CONVERGENCE=1.E-13;
PARMS R1=-3 R2=.8 R3=.4 R4=-1.5 R5=.03;
T1=R1; T2=-R2-R3; T3=R2; T4=R3; T5=R4; T6=-R5-R2; T7=R5; T8=-R5-R3;
PEAK=T1+T2*X1+T3*X2+T4*X3;  INTER=T5+T3*X1+T6*X2+T7*X3;
BASE=-1+T4*X1+T7*X2+T8*X3;
MODEL Y=P1*LOG(PEAK/BASE)+P2*LOG(INTER/BASE);
DER_T1=P1/PEAK;  DER_T2=P1/PEAK*X1;  DER_T3=P1/PEAK*X2+P2/INTER*X1;
DER_T4=P1/PEAK*X3+(-P1-P2)/BASE*X1;  DER_T5=P2/INTER;
DER_T6=P2/INTER*X2;  DER_T7=P2/INTER*X3+(-P1-P2)/BASE*X2;
DER_T8=(-P1-P2)/BASE*X3;
DER.R1=DER_T1; DER.R2=-DER_T2+DER_T3-DER_T6; DER.R3=-DER_T2+DER_T4-DER_T8;
DER.R4=DER_T5; DER.R5=-DER_T6+DER_T7-DER_T8;
OUTPUT OUT=WORK02 RESIDUAL=E;
```

Output:

SAS   1

NON-LINEAR LEAST SQUARES ITERATIVE PHASE

DEPENDENT VARIABLE: Y   METHOD: GAUSS-NEWTON

ITERATION	R1 R4	R2 R5	R3	RESIDUAL SS
0	-3.00000000 -1.50000000	0.80000000 0.03000000	0.40000000	63.33812691
. . .				
6	-2.71995278 -1.53399610	0.80870662 0.08112412	0.36225861	60.25116542

Figure 8a. Example 1 fitted by least squares, across equation constraints imposed, homogeneity imposed.

SAS Statements:

```
DATA WORK03;  SET WORK02;    E1=E;  IF MOD(_N_,2)=0 THEN DELETE;
DATA WORK04;  SET WORK02;    E2=E;  IF MOD(_N_,2)=1 THEN DELETE;
DATA WORK05;  MERGE WORK03 WORK04;  KEEP E1 E2;
PROC MATRIX FW=20; FETCH E DATA=WORK05(KEEP=E1 E2);
SIGMA=E'*E#/224;  PRINT SIGMA;  P=HALF(INV(SIGMA));  PRINT P;
```

Output:

                                    SAS   3

SIGMA	COL1	COL2
ROW1	0.178738689442	0.09503630224405
ROW2	0.09503630224405	0.09023972761352

P	COL1	COL2
ROW1	3.565728486712	-3.75526011819
ROW2	0	3.328902782166

Figure 8b. Contemporaneous variance-covariance matrix of Example 1 estimated from least squares residuals, across equation constraints imposed, homogeneity imposed.

in the model with response function

$$f(x, \theta) = \begin{pmatrix} \ln \dfrac{\theta_1 + \theta_2 x_1 + \theta_3 x_2 + \theta_4 x_3}{-1 + \theta_4 x_1 + \theta_7 x_2 + \theta_8 x_3} \\ \ln \dfrac{\theta_5 + \theta_3 x_1 + \theta_6 x_2 + \theta_7 x_3}{-1 + \theta_4 x_1 + \theta_7 x_2 + \theta_8 x_3} \end{pmatrix}$$

using the Lagrange multiplier test. Note that θ is a p-vector with $p = 8$, ρ is an r-vector with $r = 5$, $q = p - r = 3$, and there are M equations with $M = 2$ and n observations with $n = 224$.

Before computing the Lagrange multiplier statistic $\tilde{R}$ one must first compute $\hat{\theta}^\#$ as shown in Figure 8a, $\tilde{\Sigma}$ as shown in Figure 8b, and $\tilde{\theta}$ as shown in Figure 8c. The SAS code shown in Figures 8a through 8c is simply the same code shown in Figures 3a through 3c modified by substitutions from Figure 6 so that $S[g(\rho), \Sigma]$ is minimized instead of $S(\theta, \Sigma)$. This substitution is so obvious that the discussion associated with Figures 3a, 3b, 3c, and 6 ought to suffice as a discussion of Figures 8a, 8b, 8c.

We have

$$\tilde{P} = \begin{pmatrix} 3.565728486712 & -3.75526011819 \\ 0 & 3.328902782166 \end{pmatrix} \qquad \text{(from Fig. 8}b\text{)}$$

$$\hat{\rho} = \begin{pmatrix} -2.73001786 \\ 0.85800567 \\ 0.37332245 \\ -1.59315750 \\ 0.05863267 \end{pmatrix} \qquad \text{(from Fig. 8}c\text{)}$$

$$\tilde{\theta} = g(\hat{\rho}) = \begin{pmatrix} \hat{\rho}_1 \\ -\hat{\rho}_2 - \hat{\rho}_3 \\ \hat{\rho}_2 \\ \hat{\rho}_3 \\ \hat{\rho}_4 \\ -\hat{\rho}_5 - \hat{\rho}_2 \\ \hat{\rho}_5 \\ -\hat{\rho}_5 - \hat{\rho}_3 \end{pmatrix} = \begin{pmatrix} -2.7300179 \\ -1.2313281 \\ 0.8580057 \\ 0.3733224 \\ -1.5931575 \\ -0.9166373 \\ 0.0586317 \\ -0.4319541 \end{pmatrix}$$

and

$$S(\tilde{\theta}, \tilde{\Sigma}) = 447.09568448 \qquad \text{(from Fig. 8}c\text{)}.$$

As shown in Figure 9, from these values the entities

$$\text{``}\tilde{e}_s\text{''} = \text{``}y_s\text{''} - \text{``}f\text{''}(\text{``}x_s\text{''}, \tilde{\theta}) = \tilde{P}'_{(\alpha)} y_t - \tilde{P}'_{(\alpha)} f(x_t, \tilde{\theta})$$

and

$$\tilde{f}'_s = \frac{\partial}{\partial \theta'} \text{``}f\text{''}(\text{``}x_s\text{''}, \tilde{\theta}) = \tilde{P}'_{(\alpha)} \frac{\partial}{\partial \theta'} f(x_t, \tilde{\theta})$$

are computed and stored in the data set named WORK02 as

$$\text{``}\tilde{e}_s\text{''} = \text{ETILDE}$$

$$\tilde{f}'_s = (\text{DER\_T1}, \text{DER\_T2}, \dots, \text{DER\_T8}).$$

SAS Statements:

```
DATA WORK01;  SET EXAMPLE1;
P1=3.565728486712;  P2=-3.75526011819;  Y=P1*Y1+P2*Y2;  OUTPUT;
P1=0.0;  P2=3.328902782166;  Y=P1*Y1+P2*Y2;  OUTPUT;  DELETE;
PROC NLIN DATA=WORK01 METHOD=GAUSS ITER=50 CONVERGENCE=1.E-13;
PARMS R1=-3 R2=.8 R3=.4 R4=-1.5 R5=.03;
T1=R1; T2=-R2-R3; T3=R2; T4=R3; T5=R4; T6=-R5-R2; T7=R5; T8=-R5-R3;
PEAK=T1+T2*X1+T3*X2+T4*X3;  INTER=T5+T3*X1+T6*X2+T7*X3;
BASE=-1+T4*X1+T7*X2+T8*X3;
MODEL Y=P1*LOG(PEAK/BASE)+P2*LOG(INTER/BASE);
DER_T1=P1/PEAK;  DER_T2=P1/PEAK*X1;  DER_T3=P1/PEAK*X2+P2/INTER*X1;
DER_T4=P1/PEAK*X3+(-P1-P2)/BASE*X1;  DER_T5=P2/INTER;
DER_T6=P2/INTER*X2;  DER_T7=P2/INTER*X3+(-P1-P2)/BASE*X2;
DER_T8=(-P1-P2)/BASE*X3;
```

Output:

SAS   4

NON-LINEAR LEAST SQUARES ITERATIVE PHASE

DEPENDENT VARIABLE: Y   METHOD: GAUSS-NEWTON

ITERATION	R1 R4	R2 R5	R3	RESIDUAL SS
0	-3.00000000 -1.50000000	0.80000000 0.03000000	0.40000000	522.75679658
.				
.				
.				
6	-2.73001786 -1.59315750	0.85800567 0.05863167	0.37332245	447.09568448

NOTE: CONVERGENCE CRITERION MET.

SAS   5

NON-LINEAR LEAST SQUARES SUMMARY STATISTICS   DEPENDENT VARIABLE Y

SOURCE	DF	SUM OF SQUARES	MEAN SQUARE
REGRESSION	5	5899.23816229	1179.84763246
RESIDUAL	443	447.09568448	1.00924534
UNCORRECTED TOTAL	448	6346.33384677	
(CORRECTED TOTAL)	447	806.65977490	

PARAMETER	ESTIMATE	ASYMPTOTIC STD. ERROR	ASYMPTOTIC 95 % CONFIDENCE INTERVAL LOWER	UPPER
R1	-2.73001786	0.17961271	-3.08302208	-2.37701364
R2	0.85800567	0.06701564	0.72629559	0.98971574
R3	0.37332245	0.02732102	0.31962671	0.42701818
R4	-1.59315750	0.07560672	-1.74175216	-1.44456284
R5	0.05863167	0.03396211	-0.00811621	0.12537956

Figure 8c. Example 1 fitted by multivariate least squares, across equation constraints imposed, homogeneity imposed.

SAS Statements:

```
DATA WORK01;  SET EXAMPLE1;
P1=3.565728486712;  P2=-3.75526011819;  Y=P1*Y1+P2*Y2;  OUTPUT;
P1=0.0;  P2=3.328902782166;  Y=P1*Y1+P2*Y2;  OUTPUT;  DELETE;
DATA WORK02;  SET WORK01;
R1=-2.73001786; R2=0.85800567; R3=0.37332245; R4=-1.59315750; R5=0.05863167;
T1=R1;  T2=-R2-R3;  T3=R2;  T4=R3;  T5=R4;  T6=-R5-R2;  T7=R5;  T8=-R5-R3;
PEAK=T1+T2*X1+T3*X2+T4*X3;   INTER=T5+T3*X1+T6*X2+T7*X3;
BASE=-1+T4*X1+T7*X2+T8*X3;
YTILDE=P1*LOG(PEAK/BASE)+P2*LOG(INTER/BASE);   ETILDE=Y-YTILDE;
DER_T1=P1/PEAK;  DER_T2=P1/PEAK*X1;   DER_T3=P1/PEAK*X2+P2/INTER*X1;
DER_T4=P1/PEAK*X3+(-P1-P2)/BASE*X1;  DER_T5=P2/INTER;
DER_T6=P2/INTER*X2;  DER_T7=P2/INTER*X3+(-P1-P2)/BASE*X2;
DER_T8=(-P1-P2)/BASE*X3;
PROC REG DATA=WORK02;  MODEL ETILDE=DER_T1-DER_T8 / NOINT;
```

Output:

DEP VARIABLE: ETILDE

SOURCE	DF	SUM OF SQUARES	MEAN SQUARE	F VALUE	PROB>F
MODEL	8	24.696058	3.087007	3.216	0.0015
ERROR	440	422.400	0.959999		
U TOTAL	448	447.096			

ROOT MSE	0.979795	R-SQUARE	0.0552
DEP MEAN	0.001515266	ADJ R-SQ	0.0402
C.V.	64661.62		

NOTE: NO INTERCEPT TERM IS USED. R-SQUARE IS REDEFINED.

VARIABLE	DF	PARAMETER ESTIMATE	STANDARD ERROR	T FOR HO: PARAMETER=0	PROB > |T|
DER_T1	1	-0.182493	0.284464	-0.642	0.5215
DER_T2	1	-0.045933	0.228965	-0.201	0.8411
DER_T3	1	-0.029078	0.075294	-0.386	0.6995
DER_T4	1	-0.012788	0.027678	-0.462	0.6443
DER_T5	1	0.055117	0.091764	0.601	0.5484
DER_T6	1	-0.125929	0.076238	-1.652	0.0993
DER_T7	1	-0.019402	0.033583	-0.578	0.5637
DER_T8	1	-0.039163	0.021008	-1.864	0.0630

Figure 9. Illustration of Lagrange multipler test computations with Example 1.

From the regression of $\tilde{e}_s$ on $\tilde{f}'_s$ we obtain

$$\tilde{e}'\tilde{F}(\tilde{F}'\tilde{F})^{-1}\tilde{F}'\tilde{e} = 24.696058 \qquad \text{(from Fig. 9)}.$$

Recall that the parameter estimates shown in Figure 9 are a full Gauss-Newton step from $\tilde{\theta}$ to (hopefully) the minimizer of $S(\theta, \tilde{\Sigma})$. It is interesting to note that if these parameter estimates are added to the last column of Table 2, then the adjacent column is nearly reproduced, as one might expect; replacing $\hat{\Sigma}$ by $\tilde{\Sigma}$ is apparently only a small perturbation.

From the computations above, we can compute

$$\tilde{R} = nM \frac{\tilde{e}'\tilde{F}(\tilde{F}'\tilde{F})^{-1}\tilde{F}'\tilde{e}}{s(\tilde{\theta}, \tilde{\Sigma})}$$

$$= 448 \frac{24.696058}{447.09568448}$$

$$= 24.746$$

which we compare with

$$d_\alpha = nM \frac{F_\alpha}{\dfrac{nM - p}{q} + F_\alpha}$$

$$= 448 \frac{2.61}{\dfrac{440}{3} + 2.61}$$

$$= 7.83.$$

The null hypothesis of homogeneity is rejected.   □

Power computations for the Lagrange multiplier test are rather onerous, as seen from formulas given at the end of Section 6. The worst of it is the annoyance of having to evaluate the distribution function of a general quadratic form in normal variates rather than being able to use readily available tables. If one does not want to go to this bother, then the power of the likelihood ratio test can be used as an approximation to the power of the Lagrange multiplier test.

We saw in Chapter 1 that, for univariate models, inferences based on the asymptotic theory of Chapter 3 are reasonably reliable in samples of moderate size, save in the case of the Wald test statistic, provided one takes the precaution of making degree of freedom corrections and using tables of the F-distribution. This observation would carry over to the present situation if the matrix P with $P'P = \Sigma^{-1}$ used to rotate the model were known. It is the fact that one must use random $\hat{P}$ instead of known P that gives one pause in asserting that what is true in the univariate case is true in the multivariate case as well.

Below we report some simulations that confirm what intuition would lead one to expect. Dividing the Wald and "likelihood ratio" statistics by $S(\hat{\theta}, \hat{\Sigma})/(nM - p)$ and using tables of F instead of tables of the χ^2-distribution does improve accuracy. The Wald test is unreliable. The sampling

Table 4. Accuracy of Null Case Probability Statements.

Variable	Sample Size	Asymptotic Approximation	Monte Carlo	
			Estimate	Standard Error
$P(W > F)$	46	.05	.084	.0051
$P(L > F)$	46	.05	.067	.0046
$P(R > d)$	46	.05	.047	.0039
$P(W' > \chi^2)$	46	.05	.094	.0092
$P(L' > \chi^2)$	46	.05	.072	.0082
$E(s^2)$	46	1.00	1.063	.00083
$P(W > F)$	224	.05	.067	.0056
$P(L > F)$	224	.05	.045	.0046
$R(R > d)$	224	.05	.045	.0046

variation in $\hat{P}$ is deleterious and leads to the need for larger sample sizes before results can be trusted in the multivariate case than in the univariate case. Since $\tilde{P}$ has less sampling variation than $\hat{P}$, the null case Lagrange test probability statements are more reliable than "likelihood ratio" test probability statements. These interpretations of the simulations are subject to all the usual caveats associated with inductive inference. The details are as follows.

EXAMPLE 1 (Continued). The simulations reported in Table 4 were computed as follows. The data in Table 1a were randomly re-sorted and the first $n = 46$ entries were used to form the variables

$$x_t = \ln\left(\frac{(\text{peak price, intermediate price, base price})}{\text{expenditure}}\right)'$$

for $t = 1, 2, \ldots, 46$. For $n = 224$, Table 1a was used in its entirety. At the null case parameter settings

$$\theta^0 = \begin{pmatrix} -2.72482606 \\ -1.23204560 \\ 0.85773951 \\ 0.37430609 \\ -1.59239423 \\ -0.91542318 \\ 0.05768367 \\ -0.43198976 \end{pmatrix}$$

$$\Sigma = \begin{pmatrix} 0.1649246288351 & 0.09200572942276 \\ 0.09200572942276 & 0.08964264342294 \end{pmatrix}$$

independent, normally distributed errors e_t each with mean zero and variance-covariance matrix Σ were generated and used to compute y_t according to

$$y_t = f(x_t, \theta^0) + e_t \qquad t = 1, 2, \ldots, n$$

with

$$f(x, \theta) = \begin{pmatrix} \ln \dfrac{\theta_1 + \theta_2 x_1 + \theta_3 x_2 + \theta_4 x_3}{-1 + \theta_4 x_1 + \theta_7 x_2 + \theta_8 x_3} \\[2ex] \ln \dfrac{\theta_5 + \theta_3 x_1 + \theta_6 x_2 + \theta_7 x_3}{-1 + \theta_4 x_1 + \theta_7 x_2 + \theta_8 x_3} \end{pmatrix}.$$

From each generated sample, the test statistics W, L, R discussed in this section and the statistics W', L' of Section 6 were computed for the hypothesis

$$H: \begin{pmatrix} \theta_2 + \theta_3 + \theta_4 \\ \theta_3 + \theta_6 + \theta_7 \\ \theta_4 + \theta_7 + \theta_8 \end{pmatrix} = 0.$$

This process was replicated N times. The Monte Carlo estimate of, say, $P(L > F)$ is $\hat{p}$ equal to the number of times L exceeded F in the N Monte Carlo replicates divided by N; the reported standard error is $\sqrt{\hat{p}(1 - \hat{p})/N}$. The value of F is computed as $.95 = F(F; 3, nM - p)$. $\mathscr{E}(s^2)$ is the average of $s_i^2 = S(\hat{\theta}, \hat{\Sigma})/(nM - p)$ over the N Monte Carlo trials with standard error computed as

$$\sqrt{\frac{1}{N-1} \frac{\sum_{i=1}^{N}(s_i^2 - \bar{s}^2)}{N}}. \qquad \square$$

The formulas which follow for test statistics that result from a vector notation using the grouped by equation data arrangement are aesthetically more appealing than the formulas presented thus far as noted earlier. Aside from aesthetics, they also serve nicely as mnemonics to the foregoing, because in appearance they are just the obvious modifications of the formulas of Chapter 1 to account for the correlation structure of the errors. Verification that these formulas are correct is left as an exercise.

Recall that in the grouped by equation data arrangement we have M separate regressions of the sort studied in Chapter 1:

$$y_\alpha = f_\alpha(\theta_\alpha^0) + e_\alpha \qquad \alpha = 1, 2, \ldots, M$$

with y_α, $f_\alpha(\theta_\alpha)$, and e_α being n-vectors. These are "stacked" into a single regression

$$y = f(\theta^0) + e$$

by writing

$$y = \underset{nM}{\left(\begin{array}{c} y_1 \\ y_2 \\ \vdots \\ y_M \end{array}\right)_1}$$

$$f(\theta) = \underset{nM}{\left(\begin{array}{c} f_1(\theta_1) \\ f_2(\theta_2) \\ \vdots \\ f_M(\theta_M) \end{array}\right)_1}$$

$$e = \underset{nM}{\left(\begin{array}{c} e_1 \\ e_2 \\ \vdots \\ e_M \end{array}\right)_1}$$

$$\theta = \underset{p}{\left(\begin{array}{c} \theta_1 \\ \theta_2 \\ \vdots \\ \theta_M \end{array}\right)_1}$$

with

$$\mathcal{E}(e) = 0 \qquad \mathcal{C}(e, e') = \Sigma \otimes I.$$

We have available some estimator $\hat{\Sigma}$ of Σ, typically that obtained by finding $\hat{\theta}^{\#}$ to minimize

$$S(\theta, \Sigma) = [y - f(\theta)]'(\Sigma^{-1} \otimes I)[y - f(\theta)]$$

with $\Sigma = I$ and taking as the estimate the matrix $\hat{\Sigma}$ with typical element

$$\hat{\sigma}_{\alpha\beta} = \frac{1}{n}[y_\alpha - f_\alpha(\hat{\theta}_\alpha^{\#})]'[y_\beta - f_\beta(\hat{\theta}_\beta^{\#})].$$

The estimator $\hat{\theta}$ minimizes $S(\theta, \hat{\Sigma})$. Recall that the task at hand is to test a hypothesis that can be expressed either as a parametric restriction

$$H : h(\theta^0) = 0 \quad \text{against} \quad A : h(\theta^0) \neq 0$$

or as a functional dependence

$$H : \theta^0 = g(\rho^0) \text{ for some } \rho^0 \quad \text{against} \quad A : \theta^0 \neq g(\rho) \text{ for any } \rho$$

where ρ is an r-vector, $h(\theta)$ is a q-vector, and $p = r + q$. The various Jacobians required are

$$H(\theta) = \frac{\partial}{\partial \theta'} h(\theta)$$

$$G(\rho) = \frac{\partial}{\partial \rho'} g(\rho)$$

$$F(\theta) = \frac{\partial}{\partial \theta'} f(\theta)$$

being q by p, p by r, and nM by p respectively.
    The Wald test statistic is

$$W = \frac{\hat{h}'(\hat{H}\hat{C}\hat{H})^{-1}\hat{h}}{qs^2}$$

with

$$\hat{C} = \left[F'(\hat{\theta})(\hat{\Sigma}^{-1} \otimes I)F(\hat{\theta}) \right]^{-1}$$

$$s^2 = \frac{S(\hat{\theta}, \hat{\Sigma})}{nM - p}$$

$$\hat{h} = h(\hat{\theta}), \qquad \hat{H} = H(\hat{\theta}).$$

One rejects $H : h(\theta^0) = 0$ when W exceeds $F^{-1}(1 - \alpha, q, nM - p)$.
    The form of the "likelihood ratio" test is unaltered

$$L = \frac{[S(\tilde{\theta}, \hat{\Sigma}) - S(\hat{\theta}, \hat{\Sigma})]/q}{S(\hat{\theta}, \hat{\Sigma})/(nM - p)}$$

where $\tilde{\theta} = g(\hat{\rho})$ and $\hat{\rho}$ minimizes $S[g(\rho), \hat{\Sigma}]$. One rejects when L exceeds $F^{-1}(1 - \alpha; q, nM - p)$.
    As noted above, one is unlikely to use the Lagrange multiplier test unless $S(\theta, \Sigma)$ is difficult to minimize while minimization of $S[g(\rho), \Sigma]$ is relatively easy. In this instance one is apt to use the estimate $\tilde{\Sigma}$ with typical element

$$\tilde{\sigma}_{\alpha\beta} = \frac{1}{n} \left[y_\alpha - f_\alpha(\tilde{\theta}_\alpha^\#) \right]' \left[y_\beta - f_\beta(\tilde{\theta}_\beta^\#) \right]$$

where $\tilde{\theta}^\# = g(\hat{\rho}^\#)$ and $\hat{\rho}^\#$ minimizes $S[g(\rho), I]$. Let $\tilde{\tilde{\theta}} = g(\hat{\hat{\rho}})$, where $\hat{\hat{\rho}}$ minimizes $S[g(\rho), \tilde{\Sigma}]$. The Gauss-Newton step away from $\tilde{\tilde{\theta}}$ (presumably) toward $\hat{\theta}$ is

$$\tilde{D} = \left[\tilde{F}'(\tilde{\Sigma}^{-1} \otimes I)\tilde{F}\right]^{-1}\tilde{F}'(\tilde{\Sigma}^{-1} \otimes I)[y - \tilde{f}]$$

where $\tilde{F} = F(\tilde{\tilde{\theta}})$, and $\tilde{f} = f(\tilde{\tilde{\theta}})$. The Lagrange multiplier test statistic is

$$\tilde{R} = nM\tilde{D}'\left[\tilde{F}'(\tilde{\Sigma}^{-1} \otimes I)\tilde{F}\right]^{-1}\tilde{D}/S(\tilde{\tilde{\theta}}, \tilde{\Sigma}).$$

One rejects when $\tilde{R}$ exceeds

$$d_\alpha = \frac{nMF_\alpha}{\dfrac{nM - p}{q} + F_\alpha}$$

with $F_\alpha = F^{-1}(1 - \alpha; q, nM - p)$.

PROBLEMS

1. Show that if $\theta^\#$ minimizes $S(\theta, I)$ and $\hat{\Sigma} = (1/n)\Sigma_{t=1}^n[y_t - f(x_t, \theta^\#)][y_t - f(x_t, \theta^\#)]'$, then $S(\hat{\theta}, \hat{\Sigma}) = nM$.

2. Show that the matrix $\hat{C}$ of Chapter 1 can be written as

$$\hat{C} = \left(\sum_{s=1}^{nM} \frac{\partial}{\partial\theta}\text{``}f\text{''}(\text{``}x_s\text{''}, \theta)\frac{\partial}{\partial\theta'}\text{``}f\text{''}(\text{``}x_s\text{''}, \theta)\right)^{-1}$$

using the notation of Section 2. Show that $(\partial/\partial\theta')\text{``}f\text{''}(\text{``}x_s\text{''}, \theta) = P'_{(\alpha)}(\partial/\partial\theta')f(x_t, \theta)$, whence

$$\hat{C} = \left[\sum_{t=1}^n \left(\frac{\partial}{\partial\theta'}f(x_t, \theta)\right)'\hat{\Sigma}^{-1}\left(\frac{\partial}{\partial\theta'}f(x_t, \theta)\right)\right]^{-1}.$$

3. Show that the equation $s = M(t - 1) + \alpha$ uniquely defines t and α as a function of s provided that $1 \le \alpha \le M$ and s, t, α are positive integers.

4. Verify that the formulas given for W and $\tilde{R}$ at the end of this section agree with the formulas that precede them.

4. CONFIDENCE INTERVALS

As discussed in Section 6 of Chapter 1, a confidence interval on any (twice continuously differentiable) parametric function $\gamma(\theta)$ can be obtained by inverting any one of the tests of

$$H : h(\theta^0) = 0 \quad \text{against} \quad A : h(\theta^0) \neq 0$$

that were discussed in the previous section. Letting

$$h(\theta) = \gamma(\theta) - \gamma^0$$

one puts in the interval all those γ^0 for which the hypothesis $H : h(\theta^0) = 0$ is accepted at the α level of significance. The same approach applies to confidence regions, the only difference is that $\gamma(\theta)$ and γ^0 will be q-vectors instead of being univariate.

There is really nothing to add to the discussion in Section 6 of Chapter 3. The methods discussed there transfer directly to multivariate nonlinear regression. The only difference is that the test statistics W, L, and $\tilde{R}$ are computed according to the formulas of the previous section. The rest is the same.

5. MAXIMUM LIKELIHOOD ESTIMATION

Given some estimator of scale $\hat{\Sigma}_0$, the corresponding least squares estimator $\hat{\theta}_0$ minimizes $S(\theta, \hat{\Sigma}_0)$, where (recall)

$$S(\theta, \Sigma) = \sum_{t=1}^{n} [y_t - f(x_t, \theta)]' \Sigma^{-1} [y_t - f(x_t, \theta)].$$

A natural tendency is to iterate by putting

$$\hat{\Sigma}_{i+1} = \frac{1}{n} \sum_{t=1}^{n} [y_t - f(x_t, \hat{\theta}_i)][y_t - f(x_t, \hat{\theta}_i)]'$$

and

$$\hat{\theta}_{i+1} = \underset{\theta}{\text{argmin}}\, S(\theta, \hat{\Sigma}_{i+1})$$

where $\text{argmin}_\theta S(\theta, \Sigma)$ means that value of θ which minimizes $S(\theta, \Sigma)$. Continuing this process generates a sequence of estimators

$$\hat{\Sigma}_0 \to \hat{\theta}_0 \to \hat{\Sigma}_1 \to \hat{\theta}_1 \to \hat{\Sigma}_2 \to \hat{\theta}_2 \to \cdots.$$

If the sequence is terminated at any finite step I, then $\hat{\Sigma}_I$ is a consistent estimator of scale with $\sqrt{n}\,(\hat{\Sigma}_I - \Sigma_n^0)$ bounded in probability under the regularity conditions listed in Section 6 (Problem 1). Thus, $\hat{\theta}_I$ is just a least squares estimator, and the theory and methods discussed in Sections 1 through 4 apply. If one iterates until the sequence $\{(\hat{\theta}_i, \hat{\Sigma}_i)\}_{i=1}^{\infty}$ converges, then the limits

$$\hat{\Sigma}_{\infty} = \lim_{i \to \infty} \hat{\Sigma}_i$$

$$\hat{\theta}_{\infty} = \lim_{i \to \infty} \hat{\theta}_i$$

will be a local maximum of a normal errors likelihood surface provided that regularity conditions similar to those listed in Problem 4, Section 4, Chapter 1 are imposed. To see intuitively that this claim is correct, observe that under a normality assumption the random variables $\{y_t\}_{t=1}^{n}$ are independent each with density

$$n_M\big[y_t|f(x_t, \theta), \Sigma\big] = (2\pi)^{-M/2}(\det \Sigma)^{-1/2} e^{-\frac{1}{2}[y_t - f(x_t, \theta)]'\Sigma^{-1}[y_t - f(x_t, \theta)]}.$$

The log-likelihood is

$$\ln \prod_{t=1}^{n} n_M\big[y_t|f(x_t, \theta), \Sigma\big]$$

$$= \text{const} - \sum_{t=1}^{n} \tfrac{1}{2}\big\{\ln \det \Sigma + [y_t - f(x_t, \theta)]'\Sigma^{-1}[y_t - f(x_t, \theta)]\big\}$$

so the maximum likelihood estimator can be characterized as that value of (θ, Σ) which minimizes

$$s_n(\theta, \Sigma) = \frac{1}{n} \sum_{t=1}^{n} \tfrac{1}{2}\big\{\ln \det \Sigma + [y_t - f(x_t, \theta)]'\Sigma^{-1}[y_t - f(x_t, \theta)]\big\}$$

$$= \frac{1}{2}\Big(\ln \det \Sigma + \frac{1}{n}S(\theta, \Sigma)\Big).$$

Further, $s_n(\theta, \Sigma)$ will have a local minimum at each local maximum of the likelihood surface, and conversely. By Problem 11 of Section 6 we have that

$$s_n(\hat{\theta}_i, \hat{\Sigma}_{i+1}) < s_n(\hat{\theta}_i, \hat{\Sigma}_i)$$

provided that $\hat{\Sigma}_{i+1} \neq \hat{\Sigma}_i$. By definition $S(\hat{\theta}_{i+1}, \hat{\Sigma}_{i+1}) \leq S(\hat{\theta}_i, \hat{\Sigma}_{i+1})$. Provided $\hat{\theta}_{i+1} \neq \hat{\theta}_i$, arguments similar to those of Problem 4, Section 4, Chapter 1 can be employed to strengthen the weak inequality to a strict inequality. Thus we have

$$s_n(\hat{\theta}_{i+1}, \hat{\Sigma}_{i+1}) < s_n(\hat{\theta}_i, \hat{\Sigma}_i)$$

unless $(\hat{\theta}_{i+1}, \hat{\Sigma}_{i+1}) = (\hat{\theta}_i, \hat{\Sigma}_i)$, and can conclude that $(\hat{\theta}_{i+1}, \hat{\Sigma}_{i+1})$ is down-hill from $(\hat{\theta}_i, \hat{\Sigma}_i)$. By attending to a few extra details, one can conclude that the limit $(\hat{\theta}_\infty, \hat{\Sigma}_\infty)$ must exist and be a local minimum of $s_n(\theta, \Sigma)$.

One can set forth regularity conditions such that the uniform almost sure limit of $s_n(\theta, \Sigma)$ exists and has a unique minimum (θ^*, Σ^*) (Gallant and Holly, 1980). This fact coupled with the fact that $(\hat{\theta}_0, \hat{\Sigma}_0)$ has almost sure limit (θ^*, Σ^*) under the regularity conditions listed in Section 6 is enough to conclude that $(\hat{\theta}_\infty, \hat{\Sigma}_\infty)$ is tail equivalent to the maximum likelihood estimator and thus for any theoretical purpose can be regarded as if it were the maximum likelihood estimator. As a practical matter one may prefer some other algorithm to iterated least squares as a means to compute the maximum likelihood estimator. In a direct computation, the number of arguments of the objective function that must be minimized can be reduced by "concentrating" the likelihood as follows. Let

$$\hat{\Sigma}(\theta) = \frac{1}{n} \sum_{t=1}^{n} [y_t - f(x_t, \theta)][y_t - f(x_t, \theta)]'$$

and observe that by Problems 8 and 11 of Section 6

$$\min_{\Sigma} s_n(\theta, \Sigma) = s_n[\theta, \hat{\Sigma}(\theta)] = \tfrac{1}{2}[\ln \det \hat{\Sigma}(\theta) + M].$$

Thus it suffices to compute

$$\hat{\theta}_\infty = \operatorname{argmin}_\theta \ln \det \hat{\Sigma}(\theta)$$

and put

$$\hat{\Sigma}_\infty = \hat{\Sigma}(\hat{\theta}_\infty)$$

to have the minimizer $(\hat{\theta}_\infty, \hat{\Sigma}_\infty)$ of $s_n(\theta, \Sigma)$. As before, the reader is referred to Gill, Murray, and Wright (1981) for guidance in the choice of algorithms for minimizing $\ln \det \hat{\Sigma}(\theta)$.

If Assumptions 1 through 7 of Chapter 3 are specialized to the multivariate regression model with normally distributed errors and sample objective function

$$s_n(\theta, \Sigma) = \frac{1}{n} \sum_{t=1}^{n} s(y_t, x_t, \theta, \Sigma)$$

$$s(y, x, \theta, \Sigma) = \tfrac{1}{2}\{\ln \det \Sigma + [y - f(x, \theta)]'\Sigma^{-1}[y - f(x, \theta)]\}$$

one obtains a list of regularity conditions that do not differ in any essential respect from the list given in Section 6; see Gallant and Holly (1980).

Under these regularity conditions the following facts hold: $\hat{\Sigma}_{\infty}$ is consistent for Σ^*, $\sqrt{n}(\hat{\Sigma}_{\infty} - \Sigma_n^0)$ is bounded in probability, and $\hat{\theta}_{\infty}$ minimizes $S(\theta, \hat{\Sigma}_{\infty})$. It follows that $\hat{\theta}_{\infty}$ is a least squares estimator, so that one can apply the theory and methods of Section 3 to have a methodology for inference regarding θ using maximum likelihood estimates. We shall have more to say on this later. However, for joint inference regarding (θ, Σ) or marginal inference regarding Σ one needs the joint asymptotics of $(\hat{\theta}_{\infty}, \hat{\Sigma}_{\infty})$. This is provided by specializing the results of Chapter 3 to the present instance.

In order to develop an asymptotic theory suited to inference regarding Σ it is necessary to subject Σ_n^0 to a Pitman drift. For this, we need some additional notation. Let σ, a vector of length $M(M + 1)/2$, denote the upper triangle of Σ arranged as follows:

$$\sigma = (\sigma_{11}, \sigma_{12}, \sigma_{22}, \sigma_{13}, \sigma_{23}, \sigma_{33}, \ldots, \sigma_{1M}, \sigma_{2M}, \ldots, \sigma_{MM})'.$$

The mapping of σ into the elements of Σ is denoted as $\Sigma(\sigma)$. Let $\text{vec}\,\Sigma$ denote the M^2-vector obtained by stacking the columns of $\Sigma = [\Sigma_{(1)}, \Sigma_{(2)}, \ldots, \Sigma_{(M)}]$ according to

$$\text{vec}\,\Sigma = \begin{pmatrix} \Sigma_{(1)} \\ \Sigma_{(2)} \\ \vdots \\ \Sigma_{(M)} \end{pmatrix}.$$

The mapping of σ into $\text{vec}\,\Sigma(\sigma)$ is a linear map and can be written as

$$\text{vec}\,\Sigma(\sigma) = K\sigma$$

where K is an M^2 by $M(M + 1)/2$ matrix of zeros and ones. Perhaps it is best to illustrate these notations with a 3 by 3 example:

$$\Sigma = \begin{pmatrix} \sigma_{11} & \sigma_{12} & \sigma_{13} \\ \sigma_{21} & \sigma_{22} & \sigma_{23} \\ \sigma_{31} & \sigma_{32} & \sigma_{33} \end{pmatrix}$$

$$\sigma = (\sigma_{11}, \sigma_{12}, \sigma_{22}, \sigma_{13}, \sigma_{23}, \sigma_{33})'$$

$$\Sigma(\sigma) = \begin{pmatrix} \sigma_{11} & \sigma_{12} & \sigma_{13} \\ \sigma_{12} & \sigma_{22} & \sigma_{23} \\ \sigma_{13} & \sigma_{23} & \sigma_{33} \end{pmatrix}$$

$$\underbrace{\begin{pmatrix} \sigma_{11} \\ \sigma_{12} \\ \sigma_{13} \\ \sigma_{12} \\ \sigma_{22} \\ \sigma_{23} \\ \sigma_{13} \\ \sigma_{23} \\ \sigma_{33} \end{pmatrix}}_{\text{vec }\Sigma(\sigma)} = \underbrace{\begin{pmatrix} 1 & 0 & 0 & 0 & 0 & 0 \\ 0 & 1 & 0 & 0 & 0 & 0 \\ 0 & 0 & 0 & 1 & 0 & 0 \\ 0 & 1 & 0 & 0 & 0 & 0 \\ 0 & 0 & 1 & 0 & 0 & 0 \\ 0 & 0 & 0 & 0 & 1 & 0 \\ 0 & 0 & 0 & 1 & 0 & 0 \\ 0 & 0 & 0 & 0 & 1 & 0 \\ 0 & 0 & 0 & 0 & 0 & 1 \end{pmatrix} \begin{pmatrix} \sigma_{11} \\ \sigma_{12} \\ \sigma_{22} \\ \sigma_{13} \\ \sigma_{23} \\ \sigma_{33} \end{pmatrix}}_{K\sigma}.$$

The notation $\Sigma^{1/2}$ denotes a matrix such that $\Sigma = (\Sigma^{1/2})(\Sigma^{1/2})'$, and the notation $\Sigma^{-1/2}$ denotes a matrix such that $\Sigma^{-1} = (\Sigma^{-1/2})'(\Sigma^{-1/2})$. We shall always assume that the factorization algorithm used to compute $\Sigma^{1/2}$, and $\Sigma^{-1/2}$ satisfies $\Sigma^{1/2}\Sigma^{-1/2} = I$.

The data generating model is

$$y_t = f(x_t, \theta_n^0) + [\Sigma(\sigma_n^0)]^{1/2} e_t$$

with θ_n^0 known to lie in some compact set Θ^* and σ_n^0 known to lie in some compact set $\mathcal{S}^*$ over which $\Sigma(\sigma)$ is a positive definite matrix; see Section 6 for a construction of such an $\mathcal{S}^*$. The functional form of $f(x, \theta)$ is known, x is k-dimensional, θ is p-dimensional, and $f(x, \theta)$ takes its values in $\mathbf{R}^M$; y_t and e_t are M-vectors. The errors e_t are independently and identically distributed each with mean zero and variance-covariance matrix the identity matrix of order M. Note that normality is not assumed in deriving the asymptotics. The parameter to be estimated is

$$\lambda_n^0 = (\theta_n^0, \sigma_n^0).$$

Drift is imposed, so that

$$\lim_{n \to \infty} \lambda_n^0 = (\theta^*, \sigma^*) \in \Theta^* \times \mathcal{S}^*.$$

The correspondences with the notations of Chapter 3 are given in Notation 1.

NOTATION 1.

General (Chapter 3)	Specific (Chapter 6)
$e_t = q(y_t, x_t, \gamma_n^0)$	$e_t = [\Sigma(\sigma_n^0)]^{-1/2}[y_t - f(x_t, \theta_n^0)]$
$\gamma \in \Gamma^*$	$\gamma = (\theta', \sigma')', \ \theta \in \Theta^*, \ \sigma \in \mathscr{S}^*$
$y = Y(e, x, \gamma)$	$y = f(x, \theta) + [\Sigma(\sigma)]^{1/2}e$
$s(y, x, \hat{\tau}_n, \lambda)$	$s(y, x, \lambda) = \frac{1}{2}\ln \det \Sigma(\sigma)$ $+ \frac{1}{2}[y - f(x, \theta)]'\Sigma^{-1}(\sigma)[y - f(x, \theta)]$
$\hat{\tau}_n \in T$	No preliminary estimator
$\lambda \in \Lambda^*$	$\lambda = (\theta', \sigma')', \ \theta \in \Theta^*, \ \sigma \in \mathscr{S}^*$
$s_n(\lambda)$ $= \frac{1}{n}\sum_{t=1}^{n} s(y_t, x_t, \hat{\tau}_n, \lambda)$	$s_n(\lambda) = \frac{1}{2}\ln \det \Sigma(\sigma)$ $+ \frac{1}{n}\sum_{t=1}^{n}\frac{1}{2}[y_t - f(x_t, \theta)]'$ $\times \Sigma^{-1}(\sigma)[y_t - f(x_t, \theta)]$
$s_n^0(\lambda)$ $= \frac{1}{n}\sum_{t=1}^{n}\int_{\mathscr{E}} s[Y(e, x_t, \gamma_n^0), x_t, \tau_n^0, \lambda]$ $\times dP(e)$	$s_n^0(\lambda) = \frac{1}{2}\ln \det \Sigma(\sigma_n^0) + \frac{1}{2}M$ $+ \frac{1}{n}\sum_{t=1}^{n}\frac{1}{2}[f(x_t, \theta_n^0) - f(x_t, \theta)]'$ $\times \Sigma^{-1}(\sigma_n^0)[f(x_t, \theta_n^0) - f(x_t, \theta)]$
$s^*(\lambda)$ $= \int_{\mathscr{X}}\int_{\mathscr{E}} s[Y(e, x, \gamma^*), x, \tau^*, \lambda]$ $\times dP(e)\, d\mu(x)$	$s^*(\lambda) = \frac{1}{2}\ln \det \Sigma(\sigma) + \frac{1}{2}M$ $+ \int_{\mathscr{X}}\frac{1}{2}[f(x, \theta^*) - f(x, \theta)]'$ $\times \Sigma^{-1}(\sigma^*)[f(x, \theta^*) - f(x, \theta)]\, d\mu(x)$
$\hat{\lambda}_n$ minimizes $s_n(\lambda)$	$\hat{\lambda}_n$ minimizes $s_n(\lambda)$
$\tilde{\lambda}_n$ minimizes $s_n(\lambda)$ subject to $h(\lambda) = 0$	$\tilde{\lambda}_n = g(\hat{\rho}_n)$ minimizes $s_n(\lambda)$ subject to $h(\lambda) = 0$
λ_n^0 minimizes $s_n^0(\lambda)$	λ_n^0 minimizes $s_n^0(\lambda)$
λ_n^* minimizes $s_n^0(\lambda)$ subject to $h(\lambda) = 0$	$\lambda_n^* = g(\rho_n^0)$ minimizes $s_n^0(\lambda)$ subject to $h(\lambda) = 0$
λ^* minimizes $s^*(\lambda)$	λ^* minimizes $s^*(\lambda)$

In order to use the formulas for the parameters of the asymptotic distributions set forth in Chapter 3 it is necessary to compute $(\partial/\partial\lambda)s[Y(e, x, \gamma^0), x, \lambda]$ and $(\partial^2/\partial\lambda\,\partial\lambda')s[Y(e, x, \gamma^0), x, \lambda]$. To this end, write

$$s\big[Y(e, x, \gamma^0), x, \lambda\big]$$
$$= \tfrac{1}{2}\ln\det\Sigma(\sigma) + \tfrac{1}{2}[u + \delta(x, \theta)]'\Sigma^{-1}(\sigma)[u + \delta(x, \theta)]$$

where $u = \Sigma^{1/2}(\sigma^0)e$ and $\delta(x, \theta) = f(x, \theta^0) - f(x, \theta)$. Note that u has mean zero and variance-covariance matrix Σ_n^0. Letting ξ_i denote a vector with a one in the ith position and zeros elsewhere, we have (Problem 2)

$$\frac{\partial}{\partial\theta_i}s\big[Y(e, x, \gamma^0), x, \lambda\big]$$
$$= -[u + \delta(x, \theta)]'\Sigma^{-1}(\sigma)\frac{\partial}{\partial\theta_i}f(x, \theta)$$

$$\frac{\partial}{\partial\sigma_i}s\big[Y(e, x, \gamma^0), x, \lambda\big]$$
$$= \tfrac{1}{2}\mathrm{tr}\big(\Sigma^{-1}(\sigma)\Sigma(\xi_i)\{I - \Sigma^{-1}(\sigma)[u - \delta(x, \theta)][u - \delta(x, \theta)]'\}\big)$$

$$\frac{\partial^2}{\partial\theta_i\,\partial\theta_j}s\big[Y(e, x, \gamma^0), x, \lambda\big]$$
$$= \left(\frac{\partial}{\partial\theta_i}f(x, \theta)\right)'\Sigma^{-1}(\sigma)\left(\frac{\partial}{\partial\theta_j}f(x, \theta)\right)$$
$$- \sum_{\alpha=1}^{M}\sum_{\beta=1}^{M}\big[\xi_\alpha'\Sigma^{-1}(\sigma)\xi_\beta\big][u_\alpha + \delta_\alpha(x, \theta)]\frac{\partial^2}{\partial\theta_i\,\partial\theta_j}f_\beta(x, \theta)$$

$$\frac{\partial^2}{\partial\sigma_i\,\partial\theta_j}s\big[Y(e, x, \gamma^0), x, \lambda\big]$$
$$= [u + \delta(x, \theta)]'\Sigma^{-1}(\sigma)\Sigma(\xi_i)\Sigma^{-1}(\sigma)\frac{\partial}{\partial\theta_j}f(x, \theta)$$

$$\frac{\partial^2}{\partial\sigma_i\,\partial\sigma_j}s\big[Y(e, x, \gamma^0), x, \lambda\big]$$
$$= -\tfrac{1}{2}\mathrm{tr}\big(\Sigma^{-1}(\sigma)\Sigma(\xi_j)\Sigma^{-1}(\sigma)\Sigma(\xi_i)$$
$$\times\{I - 2\Sigma^{-1}(\sigma)[u - \delta(x, \theta)][u - \delta(x, \theta)]'\}\big).$$

In order to write $(\partial/\partial\sigma)s[Y(e, x, \gamma^0), x, \lambda]$ as a vector we use the fact (Problem 3) that for comformable matrices A, B, C

$$\text{vec}(ABC) = (C' \otimes A)\text{vec } B$$
$$\text{tr}(ABC) = (\text{vec } A')'(I \otimes B)\text{vec } C$$

where (recall) vec A denotes the columns of A stacked into a column vector as defined and illustrated a few paragraphs earlier, and $A \otimes B$ denotes the matrix with typical block $a_{ij}B$ as defined as illustrated in Section 2 of this chapter. Recalling that vec $\Sigma(\sigma) = K\sigma$, we have

$$\frac{\partial}{\partial\sigma_i}s\left[Y(e, x, \gamma^0), x, \lambda\right]$$

$$= \tfrac{1}{2}\text{tr}\left(\Sigma^{-1}(\sigma)\Sigma(\xi_i)\Sigma^{-1}(\sigma)\{\Sigma(\sigma) - [u - \delta(x, \theta)][u - \delta(x, \theta)]'\}\right)$$

$$= \tfrac{1}{2}\text{vec}'\left[\Sigma(\xi_i)\Sigma^{-1}(\sigma)\right]\left[I \otimes \Sigma^{-1}(\sigma)\right]$$
$$\times \text{vec}\{\Sigma(\sigma) - [u - \delta(x, \theta)][u - \delta(x, \theta)]'\}$$

$$= \tfrac{1}{2}\text{vec}'\Sigma(\xi_i)\left[\Sigma^{-1}(\sigma) \otimes I\right]\left[I \otimes \Sigma^{-1}(\sigma)\right]$$
$$\times \text{vec}\{\Sigma(\sigma) - [u - \delta(x, \theta)][u - \delta(x, \theta)]'\}$$

$$= \tfrac{1}{2}\xi_i'K'\left[\Sigma^{-1}(\sigma) \otimes \Sigma^{-1}(\sigma)\right]$$
$$\times \text{vec}\{\Sigma(\sigma) - [u - \delta(x, \theta)][u - \delta(x, \theta)]'\}.$$

From this expression we deduce that

$$(-1)(\partial/\partial\lambda)s\left[Y(e, x, \gamma_n^0), x, \lambda_n^0\right]$$

$$= \begin{pmatrix} \left(\dfrac{\partial}{\partial\theta'}f(x, \theta)\right)'\left(\Sigma_n^0\right)^{-1}u \\[2ex] \tfrac{1}{2}K'\left(\Sigma_n^0 \otimes \Sigma_n^0\right)^{-1}\text{vec}\left[uu' - \left(\Sigma_n^0\right)^{-1}\right] \end{pmatrix}.$$

We now define

NOTATION 2.

$$\Omega = \int_{\mathscr{X}}\left(\frac{\partial}{\partial\theta'}f(x, \theta^*)\right)'\Sigma^{-1}(\sigma^*)\left(\frac{\partial}{\partial\theta'}f(x, \theta^*)\right)d\mu(x)$$

$$\mathscr{F} = \int_{\mathscr{X}}\frac{\partial}{\partial\theta'}f(x, \theta^*)\,d\mu(x)$$

$$\Omega_n^0 = \frac{1}{n}\sum_{t=1}^{n}\left(\frac{\partial}{\partial\theta'}f(x_t, \theta_n^0)\right)'\Sigma^{-1}(\sigma_n^0)\left(\frac{\partial}{\partial\theta'}f(x_t, \theta_n^0)\right)$$

$$\mathscr{F}_n^0 = \frac{1}{n}\sum_{t=1}^{n}\frac{\partial}{\partial\theta'}f(x, \theta_n^0).$$

Then we have

$$\mathscr{U}_n^0 = 0$$

$$\mathscr{I}_n^0 = \begin{pmatrix} \Omega_n^0 & \left(\mathscr{F}_n^0\right)'\left(\Sigma_n^0\right)^{-1}\mathscr{E}\left[u\ \text{vec}'(uu')\right]\left(\Sigma_n^0 \otimes \Sigma_n^0\right)^{-1}K \\ \text{sym} & \frac{1}{4}K'\left(\Sigma_n^0 \otimes \Sigma_n^0\right)^{-1}\text{Var}\left[\text{vec}(uu')\right]\left(\Sigma_n^0 \otimes \Sigma_n^0\right)^{-1}K \end{pmatrix}.$$

The third moment of a normality distributed random variable is zero, whence $\mathscr{E}[u\ \text{vec}'(uu')] = 0$ under normality. From Henderson and Searle (1979) we have that under normality

$$\text{Var}\left[\text{vec}(uu')\right] = \left(\Sigma_n^0 \otimes \Sigma_n^0\right)\left(I + I_{(M,M)}\right)$$

where $I_{(M,M)}$ is a matrix whose entries are zeros and ones defined for a p by q matrix A as

$$\text{vec}\,A = I_{(p,q)}\text{vec}(A').$$

Since for any σ

$$K\sigma = \text{vec}\,\Sigma = I_{(M,M)}\text{vec}\,\Sigma' = I_{(M,M)}\text{vec}\,\Sigma = I_{(M,M)}K\sigma$$

we must have $K = I_{(M,M)}K$, whence, under normality,

$$\mathscr{I}_n^0 = \begin{pmatrix} \Omega_n^0 & 0 \\ 0 & \frac{1}{2}K'\left(\Sigma_n^0 \otimes \Sigma_n^0\right)^{-1}K \end{pmatrix}.$$

Using

$$\text{vec}(uu') = \left(\Sigma_n^0 \otimes \Sigma_n^0\right)^{1/2}\text{vec}\,ee',$$

we have that

$$\mathscr{I}_n^0 = \begin{pmatrix} \Omega_n^0 & \left(\mathscr{F}_n^0\right)'\left(\Sigma_n^{0\,-1/2}\right)'\mathscr{E}\left[e\ \text{vec}'(ee')\right]\left(\Sigma_n^0 \otimes \Sigma_n^0\right)^{-1/2}K \\ \text{sym} & \frac{1}{4}K'\left[\left(\Sigma_n^0 \otimes \Sigma_n^0\right)^{-1/2}\right]'\text{Var}\left[\text{vec}(ee')\right]\left(\Sigma_n^0 \otimes \Sigma_n^0\right)^{-1/2}K \end{pmatrix}$$

in general.

The form of $\mathscr{E}(\partial^2/\partial\sigma\,\partial\sigma')s[Y(e, x, \gamma_n^0), x, \lambda_n^0]$ can be deduced as follows:

$$\mathscr{E}\frac{\partial^2}{\partial\sigma_i\,\partial\sigma_j}s\big[Y(e, x, \gamma_n^0), x, \lambda_n^0\big]$$

$$= -\tfrac{1}{2}\mathrm{tr}\big(\Sigma_n^0\big)^{-1}\Sigma(\xi_j)\big(\Sigma_n^0\big)^{-1}\Sigma(\xi_i)\big[I - 2\big(\Sigma_n^0\big)^{-1}\big(\Sigma_n^0\big)\big]$$

$$= \tfrac{1}{2}\mathrm{vec}'\big[\Sigma(\xi_j)\big(\Sigma_n^0\big)^{-1}\big]\big[I \otimes \big(\Sigma_n^0\big)^{-1}\big]\mathrm{vec}\,\Sigma(\xi_i)$$

$$= \tfrac{1}{2}\xi_j' K'\big(\Sigma_n^0 \otimes \Sigma_n^0\big)^{-1}K\xi_i.$$

Since $\mathscr{E}(\partial^2/\partial\sigma_i\,\partial\theta_j)s[Y(e, x, \gamma_n^0), x, \lambda_n^0] = 0$, we have

$$\mathscr{I}_n^0 = \begin{pmatrix} \Omega_n^0 & 0 \\ 0 & \tfrac{1}{2}K'\big(\Sigma_n^0 \otimes \Sigma_n^0\big)^{-1}K \end{pmatrix}.$$

Normality plays no role in the form of $\mathscr{I}_n^0$.
    In summary we have

NOTATION 3a (In general).

$$\mathscr{I}_n^0 = \begin{pmatrix} \Omega_n^0 & \big(\mathscr{F}_n^0\big)'\big[\big(\Sigma_n^0\big)^{-1/2}\big]'\mathscr{E}\big[e\,\mathrm{vec}'(ee')\big]\big(\Sigma_n^0 \otimes \Sigma_n^0\big)^{-1/2}K \\ \mathrm{sym} & \tfrac{1}{4}K'\big[\big(\Sigma_n^0 \otimes \Sigma_n^0\big)^{-1/2}\big]'\mathrm{Var}\big[\mathrm{vec}(ee')\big]\big(\Sigma_n^0 \otimes \Sigma_n^0\big)^{-1/2}K \end{pmatrix}$$

$$\mathscr{I}_n^0 = \begin{pmatrix} \Omega_n^0 & 0 \\ 0 & \tfrac{1}{2}K'\big(\Sigma_n^0 \otimes \Sigma_n^0\big)^{-1}K \end{pmatrix}$$

$$\mathscr{U}_n^0 = 0.$$

NOTATION 3b (Under normality).

$$\mathscr{I}_n^0 = \begin{pmatrix} \Omega_n^0 & 0 \\ 0 & \tfrac{1}{2}K'\big(\Sigma_n^0 \otimes \Sigma_n^0\big)^{-1}K \end{pmatrix}$$

$$\mathscr{I}_n^0 = \mathscr{I}_n^0$$

$$\mathscr{U}_n^0 = 0.$$

The expressions for $\mathscr{I}*$, $\mathscr{J}*$, and $\mathscr{U}*$ have the same form as above with $(\Omega, \mathscr{F}, \Sigma*)$ replacing $(\Omega_n^0, \mathscr{F}_n^0, \Sigma_n^0)$ throughout.

Let $\hat{\lambda}_n = (\hat{\theta}_\infty, \hat{\Sigma}_\infty)$ denote the minimum of $s_n(\theta, \Sigma)$, and let $\tilde{\lambda}_n = (\tilde{\theta}_\infty, \tilde{\Sigma}_\infty)$ denote the minimum of $s_n(\theta, \Sigma)$ subject to $h(\lambda) = 0$. Define:

NOTATION 4.

$$\hat{\Omega} = \frac{1}{n} \sum_{t=1}^{n} \left(\frac{\partial}{\partial \theta'} f(x_t, \hat{\theta}_\infty) \right)' \hat{\Sigma}_\infty^{-1} \left(\frac{\partial}{\partial \theta'} f(x_t, \hat{\theta}_\infty) \right)$$

$$\hat{S}_t = \begin{pmatrix} \left(\frac{\partial}{\partial \theta'} f(x_t, \hat{\theta}_\infty) \right)' \hat{\Sigma}_\infty^{-1} \hat{u}_t \\ \frac{1}{2} K'(\hat{\Sigma}_\infty \otimes \hat{\Sigma}_\infty)^{-1} \text{vec}\left[\hat{u}_t \hat{u}_t' - \hat{\Sigma}_\infty^{-1} \right] \end{pmatrix}$$

$$\hat{u}_t = y_t - f(x_t, \hat{\theta}_\infty).$$

The expressions for $\tilde{\Omega}$, $\tilde{S}_t$, and $\tilde{u}_t$ are the same with $(\tilde{\theta}_\infty, \tilde{\Sigma}_\infty)$ replacing $(\hat{\theta}_\infty, \hat{\Sigma}_\infty)$ throughout.

We propose the following as estimators of $\mathscr{I}*$ and $\mathscr{J}*$.

NOTATION 5a   (In general).

$$\hat{\mathscr{I}} = \frac{1}{n} \sum_{t=1}^{n} \hat{S}_t \hat{S}_t'$$

$$\hat{\mathscr{J}} = \begin{pmatrix} \hat{\Omega} & 0 \\ 0 & \frac{1}{2} K'(\hat{\Sigma}_\infty \otimes \hat{\Sigma}_\infty)^{-1} K \end{pmatrix}.$$

NOTATION 5b   (Under normality).

$$\hat{\mathscr{I}} = \hat{\mathscr{J}} = \begin{pmatrix} \hat{\Omega} & 0 \\ 0 & \frac{1}{2} K'(\hat{\Sigma}_\infty \otimes \hat{\Sigma}_\infty)^{-1} K \end{pmatrix}.$$

The expressions for $\tilde{\mathscr{I}}$ and $\tilde{\mathscr{J}}$ have the same form with $(\tilde{\Omega}, \tilde{\Sigma}, \tilde{S}_t)$ replacing $(\hat{\Omega}, \hat{\Sigma}, \hat{S}_t)$ throughout.

A test of the marginal hypothesis

$$H : h(\theta^0) = 0 \quad \text{against} \quad A : h(\theta^0) \neq 0$$

where $h(\theta)$ maps $\mathbb{R}^p$ into $\mathbb{R}^q$ is most often of interest in applications. As mentioned earlier, maximum likelihood estimators are least squares estimators, so that, as regards the Wald and the Lagrange multiplier tests, the theory and methods set forth in Section 3 can be applied directly with the maximum likelihood estimators

$$\hat{\theta}_\infty, \hat{\Sigma}_\infty, \tilde{\theta}_\infty, \tilde{\Sigma}_\infty$$

replacing, respectively, the estimators

$$\hat{\theta}, \hat{\Sigma}, \tilde{\theta}, \tilde{\Sigma}$$

in the formulas for the Wald and Lagrange multiplier test statistics. The likelihood ratio test needs modification due to the following considerations.

Direct application of Theorem 15 of Chapter 3 would give

$$L1 = 2n\left[s_n(\tilde{\theta}_\infty, \tilde{\Sigma}_\infty) - s_n(\hat{\theta}_\infty, \hat{\Sigma}_\infty)\right]$$
$$= n\left(\ln \det \tilde{\Sigma}_\infty - \ln \det \hat{\Sigma}_\infty\right)$$

as the likelihood ratio test statistic, whereas application of the results in Section 3 would give

$$L2 = \frac{\left[S(\tilde{\theta}, \hat{\Sigma}_\infty) - S(\hat{\theta}_\infty, \hat{\Sigma}_\infty)\right]/q}{S(\hat{\theta}_\infty, \hat{\Sigma}_\infty)/(nM - p)}$$
$$= \frac{\left[S(\tilde{\theta}, \hat{\Sigma}_\infty) - nM\right]/q}{nM/(nM - p)}$$

where $\tilde{\theta}$ minimizes $S(\theta, \hat{\Sigma}_\infty)$ subject to $h(\theta) = 0$. These two formulas can be reconciled using the equation

$$d \ln \det \Sigma = \mathrm{tr}(\Sigma^{-1} d\Sigma)$$

or

$$\ln \det(\Sigma + \Delta) - \ln \det \Sigma = \mathrm{tr}(\Sigma^{-1}\Delta) + o(\Delta)$$

derived in Problem 4. To within a differential approximation

$$S(\tilde{\theta}, \hat{\Sigma}_\infty) - S(\hat{\theta}_\infty, \hat{\Sigma}_\infty) = n \, \mathrm{tr}\left[\hat{\Sigma}_\infty^{-1}\hat{\Sigma}(\tilde{\theta})\right] - n \, \mathrm{tr}(\hat{\Sigma}_\infty^{-1}\hat{\Sigma}_\infty)$$
$$= n \, \mathrm{tr} \, \hat{\Sigma}_\infty^{-1}\left[\hat{\Sigma}(\tilde{\theta}) - \hat{\Sigma}_\infty\right]$$
$$= n \, \mathrm{tr} \, \hat{\Sigma}_\infty^{-1}(\tilde{\Sigma}_\infty - \hat{\Sigma}_\infty) + n \, \mathrm{tr} \, \hat{\Sigma}_\infty^{-1}\left[\hat{\Sigma}(\tilde{\theta}) - \tilde{\Sigma}_\infty\right]$$
$$\doteq L1 + n \, \mathrm{tr} \, \hat{\Sigma}_\infty^{-1}\left[\hat{\Sigma}(\tilde{\theta}) - \tilde{\Sigma}_\infty\right]$$
$$= L1 + \left[S(\tilde{\theta}, \hat{\Sigma}_\infty) - S(\hat{\theta}_\infty, \hat{\Sigma}_\infty)\right].$$

Thus one can expect that there will be a negligible difference between an inference based on either $L1$ or $L2$ in most applications. Our recommendation is to use $L1 = n(\ln \det \tilde{\Sigma}_\infty - \ln \det \hat{\Sigma}_\infty)$ to avoid the confusion that would result from the use of something other than the classical likelihood ratio test in connection with maximum likelihood estimators. But we do recommend the use of degree of freedom corrections to improve the accuracy of probability statements.

To summarize this discussion, the likelihood ratio test rejects the hypothesis

$$H : h(\theta^0) = 0$$

where $h(\theta)$ maps $\mathbb{R}^p$ into $\mathbb{R}^q$, when the statistic

$$L = n\left(\ln \det \tilde{\Sigma}_\infty - \ln \det \hat{\Sigma}_\infty\right)$$

exceeds qF_α, where F_α denotes the upper $\alpha \times 100\%$ critical point of the F-distribution with q numerator degrees of freedom and $nM - p$ denominator degrees of freedom; $F_\alpha = F^{-1}(1 - \alpha; q, nM - p)$.

We illustrate.

EXAMPLE 1 (Continued). Consider retesting the hypothesis of homogeneity, expressed as the functional dependence

$$H : \theta^0 = g(\rho^0) \text{ for some } \rho^0 \quad \text{against} \quad A : \theta^0 \neq g(\rho) \text{ for any } \rho$$

with

$$g(\rho) = \begin{pmatrix} \rho_1 \\ -\rho_2 - \rho_3 \\ \rho_2 \\ \rho_3 \\ \rho_4 \\ -\rho_5 - \rho_2 \\ \rho_5 \\ -\rho_5 - \rho_3 \end{pmatrix}$$

in the model with response function

$$f(x, \theta) = \begin{pmatrix} \ln \dfrac{\theta_1 + \theta_2 x_1 + \theta_3 x_2 + \theta_4 x_3}{-1 + \theta_4 x_1 + \theta_7 x_2 + \theta_8 x_3} \\ \ln \dfrac{\theta_5 + \theta_3 x_1 + \theta_6 x_2 + \theta_7 x_3}{-1 + \theta_4 x_1 + \theta_7 x_2 + \theta_8 x_3} \end{pmatrix}$$

using the likelihood ratio test; θ has length $p = 8$ and ρ has length $r = 5$,

368   MULTIVARIATE NONLINEAR REGRESSION

SAS Statements:

```
PROC MODEL OUT=MODELO1;
ENDOGENOUS Y1 Y2;
EXOGENOUS  X1 X2 X3;
PARMS T1 -2.98  T2 -1.16  T3 0.787  T4 0.353  T5 -1.51  T6 -1.00
     T7 0.054  T8 -0.474;
PEAK=T1+T2*X1+T3*X2+T4*X3;  INTER=T5+T3*X1+T6*X2+T7*X3;
BASE=-1+T4*X1+T7*X2+T8*X3;
Y1=LOG(PEAK/BASE);  Y2=LOG(INTER/BASE);
PROC SYSNLIN DATA=EXAMPLE1  MODEL=MODELO1 ITSUR NESTIT METHOD=GAUSS OUTS=SHAT;
```

Output:

NONLINEAR ITSUR PARAMETER ESTIMATES

PARAMETER	ESTIMATE	APPROX. STD ERROR	'T' RATIO	APPROX. PROB>|T|
T1	-2.92345	0.2781988	-10.51	0.0001
T2	-1.28826	0.226682	-5.68	0.0001
T3	0.8184883	0.08079815	10.13	0.0001
T4	0.3612072	0.03033416	11.91	0.0001
T5	-1.53759	0.09204265	-16.71	0.0001
T6	-1.04926	0.08368577	-12.54	0.0001
T7	0.02986769	0.03617161	0.83	0.4099
T8	-0.467411	0.01927753	-24.25	0.0001

SYSTEM STATISTICS: SSE =   447.9999   MSE =   2   OBS= 224

COVARIANCE OF RESIDUALS

	Y1	Y2
Y1	0.165141	0.0925046
Y2	0.0925046	0.0898862

Figure 10a. Example 1 fitted by maximum likelihood, across equation constraints imposed.

whence $q = p - r = 3$. The model is bivariate, so $M = 2$ and there are $n = 224$ observations.

In Figure 10a the maximum likelihood estimators are computed by iterating the least squares estimator to convergence, obtaining

$$\hat{\theta}_\infty = \begin{pmatrix} -2.92345 \\ -1.28826 \\ 0.81849 \\ 0.36121 \\ -1.53759 \\ -1.04926 \\ 0.02987 \\ -0.46741 \end{pmatrix} \qquad \text{(from Fig. 10a)}$$

$$\hat{\Sigma}_\infty = \begin{pmatrix} 0.165141 & 0.92505 \\ 0.092505 & 0.08989 \end{pmatrix} \qquad \text{(from Fig. 10a).}$$

SAS Statements:

```
PROC MODEL OUT=MODELO2;
ENDOGENOUS Y1 Y2;
EXOGENOUS  X1 X2 X3;
PARMS R1 -2.72  R2 0.858  R3 0.374  R4 -1.59  R5 0.057;
T1=R1; T2=-R2-R3; T3=R2; T4=R3; T5=R4; T6=-R5-R2; T7=R5; T8=-R5-R3;
PEAK=T1+T2*X1+T3*X2+T4*X3;   INTER=T5+T3*X1+T6*X2+T7*X3;
BASE=-1+T4*X1+T7*X2+T8*X3;
Y1=LOG(PEAK/BASE);  Y2=LOG(INTER/BASE);
PROC SYSNLIN DATA=EXAMPLE1  MODEL=MODELO2 ITSUR NESTIT METHOD=GAUSS OUTS=STILDE;
```

Output:

SAS   10

NONLINEAR ITSUR PARAMETER ESTIMATES

PARAMETER	ESTIMATE	APPROX. STD ERROR	'T' RATIO	APPROX. PROB>|T|
R1	-2.7303	0.1800188	-15.17	0.0001
R2	0.8581672	0.06691972	12.82	0.0001
R3	0.3733482	0.02736511	13.64	0.0001
R4	-1.59345	0.07560367	-21.08	0.0001
R5	0.05854239	0.0339787	1.72	0.0863

SYSTEM STATISTICS: SSE =   448   MSE =   2   OBS= 224

COVARIANCE OF RESIDUALS

	Y1	Y2
Y1	0.179194	0.0953651
Y2	0.0953651	0.0901989

Figure 10b. Example 1 fitted by maximum likelihood, across equation constraints imposed, homogeneity imposed.

Compare these values with those shown in Figures 3b and 3c; the difference is slight.

In Figure 10b the estimator $\hat{\rho}_\infty$ minimizing $\ln \det \Sigma[g(\rho)]$ is obtained by iterated least squares; put $\tilde{\theta}_\infty = g(\hat{\rho}_\infty)$ to obtain

$$\tilde{\theta}_\infty = g(\hat{\rho}_\infty) = \begin{pmatrix} -2.7303 \\ -1.2315 \\ 0.8582 \\ 0.3733 \\ -1.5935 \\ -0.9167 \\ 0.0585 \\ -0.4319 \end{pmatrix} \quad \text{(from Fig. 10b)}$$

$$\tilde{\Sigma}_\infty = \begin{pmatrix} 0.179194 & 0.095365 \\ 0.095365 & 0.090199 \end{pmatrix} \quad \text{(from Fig. 10b)}.$$

```
SAS Statements:

PROC MATRIX;
FETCH SHAT DATA=SHAT(KEEP=Y1 Y2);
FETCH STILDE DATA=STILDE(KEEP=Y1 Y2);
N=224;  L=N#(LOG(DET(STILDE))-LOG(DET(SHAT)));  PRINT L;

Output:
```

SAS   11

L   COL1

ROW1   26.2573

Figure 10c. Illustration of likelihood ratio test computations with Example 1.

Compare these values with those shown in Figure 6; again, the difference is slight.

In Figure 10c, the likelihood ratio test statistic is computed as

$$L = n\left(\ln \det \tilde{\Sigma}_\infty - \ln \det \hat{\Sigma}_\infty\right)$$
$$= 26.2573 \qquad (\text{from Fig. } 10c).$$

$F_\alpha = F^{-1}(.95; 3, 440) = 2.61$, so that

$$qF_\alpha = (3)(2.61) = 7.83.$$

One rejects

$$H : \theta^0 = g(\rho^0) \text{ for some } \rho^0$$

at the 5% level. With this many denominator degrees of freedom, the difference between qF_α and the three degree of freedom chi-square critical value of 7.81 is negligible. In smaller sized samples this will not be the case.

It is of interest to compare

$$n\left(\ln \det \tilde{\Sigma}_\infty - \ln \det \hat{\Sigma}_\infty\right) = 26.2573 \qquad (\text{from Fig. } 10c)$$

with

$$S(\tilde{\theta}, \hat{\Sigma}) - S(\hat{\theta}, \hat{\Sigma}) = 474.6822 - 446.8570 \qquad (\text{from Figs. 6 and } 3c)$$
$$= 27.8252.$$

The differential approximation $d \ln \det \Sigma = \operatorname{tr} \Sigma^{-1} d\Sigma$ seems to be reasonably accurate in this instance.   □

A marginal hypothesis of the form

$$H : h(\sigma^0) = 0 \quad \text{against} \quad A : h(\sigma^0) \neq 0$$

is sometimes of interest in applications. We shall proceed under the assumption that the computation of $(\hat{\theta}_\infty, \hat{\sigma}_\infty)$ is fairly straightforward but that the minimization of $s_n(\theta, \sigma)$ subject to $h(\sigma) = 0$ is inordinately burdensome, as is quite often the case. This assumption compels the use of the Wald test statistic. We shall also assume that the errors are normally distributed.

Under normality, the Wald test statistic for the hypothesis

$$H : h(\sigma^0) = 0 \quad \text{against} \quad A : h(\sigma^0) \neq 0$$

where $h(\sigma)$ maps $\mathbb{R}^{M(M+1)/2}$ into $\mathbb{R}^q$, has the form

$$W = n\hat{h}'(\hat{H}\hat{V}\hat{H})^{-1}\hat{h}$$

where

$$\hat{h} = h(\hat{\sigma}_\infty)$$

$$\hat{H} = \frac{\partial}{\partial \sigma'} h(\hat{\sigma}_\infty)$$

$$\hat{V} = \left[\tfrac{1}{2} K'(\hat{\Sigma}_\infty \otimes \hat{\Sigma}_\infty)^{-1} K \right]^{-1}.$$

The test rejects when W exceeds the upper $\alpha \times 100\%$ critical point of a chi-square random variable with q degrees of freedom.

In performing the computations, explicit construction of the matrix K can be avoided as follows. Consider w defined by

$$\text{vec } uu' = \Sigma(w) = Kw$$

where u is an M-vector. The subscripts are related as follows:

$$\text{vec } uu' = \begin{bmatrix} u_1 u_1 \\ u_1 u_2 \\ \vdots \\ u_\alpha u_\beta \\ \vdots \\ u_M u_{M-1} \\ u_M u_M \end{bmatrix} \quad \leftarrow i = \beta(\beta - 1)/2 + \alpha \rightarrow \quad \begin{bmatrix} w_1 \\ w_2 \\ \vdots \\ w_i \\ \vdots \\ w_{M(M+1)/2-1} \\ w_{M(M+1)/2} \end{bmatrix}.$$

If $u \sim N_M(0, \Sigma)$, then for

$$i = \frac{\beta(\beta - 1)}{2} + \alpha$$

$$j = \frac{\beta'(\beta' - 1)}{2} + \alpha'$$

we have (Anderson, 1984, p. 49) that

$$\mathscr{C}(w_i, w_j) = \mathscr{E}(u_\alpha u_\beta - \sigma_{\alpha\beta})(u_{\alpha'} u_{\beta'} - \sigma_{\alpha'\beta'})$$
$$= \sigma_{\alpha\alpha'}\sigma_{\beta\beta'} + \sigma_{\alpha\beta'}\sigma_{\beta\alpha'}.$$

Thus, the variance-covariance matrix $\mathscr{C}(w, w')$ of the random variable w can be computed easily. Now consider the asymptotics for the model $y_t = u_t = \Sigma^{1/2} e_t$ with e_t independent $N_M(0, I)$. The previous asymptotic results imply

$$\sqrt{n}(\hat{\sigma}_\infty - \sigma) \xrightarrow{\mathscr{L}} N_{M(M+1)/2}\left\{0, \left[\tfrac{1}{2}K(\Sigma \otimes \Sigma)^{-1}K\right]^{-1}\right\}$$

but in this case $\hat{\sigma}_\infty = (1/n)\Sigma_{t=1}^n w_t$ and the central limit theorem implies that

$$\sqrt{n}(\hat{\sigma}_\infty - \sigma) \xrightarrow{\mathscr{L}} N_{M(M+1)/2}[0, \mathscr{C}(w, w')].$$

We conclude that

$$V = \left[\tfrac{1}{2}K(\Sigma \otimes \Sigma)^{-1}K'\right]^{-1} = \mathscr{C}(w, w')$$

and have the following algorithm for computing the elements v_{ij} of V:

DO for $\beta = 1$ to M;
DO for $\alpha = 1$ to β;
$i = \beta(\beta - 1)/2 + \alpha$;
DO for $\beta' = 1$ to M;
DO for $\alpha' = 1$ to β';
$j = \beta'(\beta' - 1)/2 + \alpha'$;
$v_{ij} = \sigma_{\alpha\alpha'}\sigma_{\beta\beta'} + \sigma_{\alpha\beta'}\sigma_{\beta\alpha'}$;
END;
END;
END;
END;

We illustrate with an example.

EXAMPLE 2 (Split plot design).   The split plot experimental design can be viewed as a two way design with multivariate observations in each cell, which is written as

$$y_{ij} = u + \rho_i + \tau_j + e_{ij}$$

where y_{ij}, u, etc. are M-vectors and

$$i = 1, 2, \ldots, I = \text{no. of blocks}$$
$$j = 1, 2, \ldots, J = \text{no. of treatments}$$
$$\mathscr{C}(e_{ij}, e'_{ij}) = \Sigma.$$

In the corresponding univariate split plot analysis, the data are assumed to follow the model

$$y_{kij} = m + r_i + t_j + \eta_{ij} + s_k + (rs)_{ki} + (ts)_{kj} + \epsilon_{kij}$$

where $k = 1, 2, \ldots, M$, Latin letters denote parameters, and Greek letters denote random variables, $\mathrm{Var}(\eta_{ij}) = \sigma_\eta^2$, $\mathrm{Var}(\epsilon_{kij}) = \sigma_\epsilon^2$, and random variables with different subscripts are assumed independent. It is not difficult to show (Problem 5) that the only difference between the two models is that

Table 5. Yields of Three Varieties of Alfalfa (Tons Per Acre) in 1944 Following Four Dates of Final Cutting in 1943.

			Date			
t	Variety	Block	A	B	C	D
1	Ladak	1	2.17	1.58	2.29	2.23
2	Cossac	1	2.33	1.38	1.86	2.27
3	Ranger	1	1.75	1.52	1.55	1.56
4	Ladak	2	1.88	1.26	1.60	2.01
5	Cossac	2	2.01	1.30	1.70	1.81
6	Ranger	2	1.95	1.47	1.61	1.72
7	Ladak	3	1.62	1.22	1.67	1.82
8	Cossac	3	1.70	1.85	1.81	2.01
9	Ranger	3	2.13	1.80	1.82	1.99
10	Ladak	4	2.34	1.59	1.91	2.10
11	Cossac	4	1.78	1.09	1.54	1.40
12	Ranger	4	1.78	1.37	1.56	1.55
13	Ladak	5	1.58	1.25	1.39	1.66
14	Cossac	5	1.42	1.13	1.67	1.31
15	Ranger	5	1.31	1.01	1.23	1.51
16	Ladak	6	1.66	0.94	1.12	1.10
17	Cossac	6	1.35	1.06	0.88	1.06
18	Ranger	6	1.30	1.31	1.13	1.33

Source: Snedecor and Cochran (1980, Table 16.15.1).

the univariate analysis imposes the restriction

$$\Sigma = \sigma_\epsilon^2 I + \sigma_\eta^2 J$$

on the variance-covariance matrix of the multivariate analysis; I is the identity matrix of order M, and J is an M by M matrix whose entries are all ones. Such an assumption is somewhat suspect when the observations

$$y_{ij}' = (y_{1ij}, y_{2ij}, \dots, y_{Mij})$$

represent successive observations on the same plot at different points in time (Gill and Hafs, 1971). An instance is the data shown in Table 5. For these data, $M = 4$, $n = 18$, and the hypothesis to be tested is $H : h(\sigma^0) = 0$, where

$$h(\sigma) = \underbrace{\begin{pmatrix} 1 & 0 & -1 & 0 & 0 & 0 & 0 & 0 & 0 & 0 \\ 1 & 0 & 0 & 0 & 0 & -1 & 0 & 0 & 0 & 0 \\ 1 & 0 & 0 & 0 & 0 & 0 & 0 & 0 & 0 & -1 \\ 0 & 1 & 0 & -1 & 0 & 0 & 0 & 0 & 0 & 0 \\ 0 & 1 & 0 & 0 & -1 & 0 & 0 & 0 & 0 & 0 \\ 0 & 1 & 0 & 0 & 0 & 0 & -1 & 0 & 0 & 0 \\ 0 & 1 & 0 & 0 & 0 & 0 & 0 & -1 & 0 & 0 \\ 0 & 1 & 0 & 0 & 0 & 0 & 0 & 0 & -1 & 0 \end{pmatrix}}_{H} \underbrace{\begin{pmatrix} \sigma_{11} \\ \sigma_{12} \\ \sigma_{22} \\ \sigma_{13} \\ \sigma_{23} \\ \sigma_{33} \\ \sigma_{14} \\ \sigma_{24} \\ \sigma_{34} \\ \sigma_{44} \end{pmatrix}}_{\sigma}$$

```
SAS Statements:

PROC ANOVA DATA=EXAMPLE2;
CLASSES VARIETY BLOCK;
MODEL A B C D = VARIETY BLOCK;
MANOVA / PRINTE;

Output:
```

                                                    SAS   6

                                  ANALYSIS OF VARIANCE PROCEDURE

                                     E = ERROR SS&CP MATRIX

DF=10	A	B	C	D
A	0.56965556	0.23726111	0.25468889	0.36578889
B	0.23726111	0.46912222	0.26341111	0.31137778
C	0.25468889	0.26341111	0.42495556	0.25678889
D	0.36578889	0.31137778	0.25678889	0.60702222

Figure 11. Maximum likelihood estimation of the variance-covariance matrix of Example 2.

The maximum likelihood estimate of Σ is computed in Figure 11 as

$$\hat{\Sigma}_\infty = \begin{pmatrix} 0.0316475 & 0.0131812 & 0.0141494 & 0.0203216 \\ 0.0131812 & 0.0260623 & 0.0146340 & 0.0172988 \\ 0.0141494 & 0.0146340 & 0.0236086 & 0.0142660 \\ 0.0203216 & 0.0172988 & 0.0142660 & 0.0337235 \end{pmatrix} = \Sigma(\hat{\sigma}_\infty)$$

whence

$$h(\hat{\sigma}_\infty) = \begin{pmatrix} 0.00558519 \\ 0.00803889 \\ -0.00207593 \\ -0.00096821 \\ -0.00145278 \\ -0.00714043 \\ -0.00411759 \\ -0.00108488 \end{pmatrix}.$$

```
SAS Statements:

PROC MATRIX;
SSCP =  0.56965556  0.23726111  0.25468889  0.36578889/
        0.23726111  0.46912222  0.26341111  0.31137778/
        0.25468889  0.26341111  0.42495556  0.25678889/
        0.36578889  0.31137778  0.25678889  0.60702222;
S    = (0.56965556  0.23726111  0.46912222  0.25468889  0.26341111
        0.42495556  0.36578889  0.31137778  0.25678889  0.60702222)';
HH   = 1  0  -1  0  0  0  0  0  0  0/
       1  0   0  0  0 -1  0  0  0  0/
       1  0   0  0  0  0  0  0  0 -1/
       0  1   0 -1  0  0  0  0  0  0/
       0  1   0  0 -1  0  0  0  0  0/
       0  1   0  0  0  0 -1  0  0  0/
       0  1   0  0  0  0  0 -1  0  0/
       0  1   0  0  0  0  0  0 -1  0;
N=18;  SIGMA=SSCP#/N;  H=HH*S#/N;
M=4;  V=(O*(1:M*(M+1)#/2))'*(O*(1:M*(M+1)#/2));
DO B = 1 TO M;  DO A = 1 TO B;  I =  B*( B-1)#/2+ A;
DO BB = 1 TO M;  DO AA = 1 TO BB;  J = BB*(BB-1)#/2+AA;
V(I,J)=SIGMA(A,AA)*SIGMA(B,BB)+SIGMA(A,BB)*SIGMA(B,AA);
END;  END;  END;  END;
WALD=N*(H'*INV(HH*V*HH')*H);  PRINT WALD;

Output:
```

SAS   1

	WALD	COL1
ROW1		2.26972

Figure 12. Wald test of a restriction on the variance-covariance matrix of Example 2.

```
SAS Statements:

PROC VARCOMP DATA=EXAMPLE2 METHOD=ML;
CLASSES VARIETY DATE BLOCK;
MODEL
YIELD = BLOCK VARIETY DATE DATE*BLOCK DATE*VARIETY BLOCK*VARIETY /
FIXED = 5;

Output:
```

SAS   2

MAXIMUM LIKELIHOOD VARIANCE COMPONENT ESTIMATION PROCEDURE

```
DEPENDENT VARIABLE: YIELD
```

ITERATION	OBJECTIVE	VAR(VARIETY*BLOCK)	VAR(ERROR)
0	-280.48173508	0.01564182	0.01311867
1	-280.48173508	0.01564182	0.01311867

Figure 13. Maximum likelihood estimation of the variance-covariance matrix of Example 2 under the ANOVA restriction.

Figure 12 illustrates the algorithm for computing $\hat{V}$ discussed above, and we obtain

$$W = 2.26972 \qquad \text{(from Fig. 12).}$$

Entering a table of the chi-square distribution at 8 degrees of freedom, one finds that

$$p = P(W > 2.26972) \doteq 0.97.$$

A univariate analysis of the data seems reasonable.

This happens to be an instance where it is easy to compute the maximum likelihood estimate subject to $h(\sigma) = 0$. From Figure 13 we obtain

$$\tilde{\Sigma}_\infty = \begin{pmatrix} 0.0287605 & 0.0156418 & 0.0156418 & 0.0156418 \\ 0.0156418 & 0.0287605 & 0.0156418 & 0.0156418 \\ 0.0156418 & 0.0156418 & 0.0287605 & 0.0156418 \\ 0.0156418 & 0.0156418 & 0.0156418 & 0.0287605 \end{pmatrix}.$$

For a linear model of this form we have (Problem 6)

$$\begin{aligned} L &= 2n\left[s_n(\tilde{\theta}_\infty, \tilde{\Sigma}_\infty) - s_n(\hat{\theta}_\infty, \hat{\Sigma}_\infty) \right] \\ &= n\left[\ln \det \tilde{\Sigma}_\infty + \operatorname{tr} \tilde{\Sigma}_\infty^{-1} \hat{\Sigma}_\infty - \ln \det \hat{\Sigma}_\infty - M \right] \\ &= 2.61168 \quad \text{(from Fig. 14)} \end{aligned}$$

SAS Statements:

```
PROC MATRIX;
SSCP  = 0.56965556  0.23726111  0.25468889  0.36578889/
        0.23726111  0.46912222  0.26341111  0.31137778/
        0.25468889  0.26341111  0.42495556  0.25678889/
        0.36578889  0.31137778  0.25678889  0.60702222;
N=18;  M=4;  SHAT=SSCP#/N;  STILDE=0.01311867#I(M)+J(M,M,0.01564182);
L=N#(LOG(DET(STILDE))+TRACE(INV(STILDE)*SHAT)-LOG(DET(SHAT))-M);  PRINT L;
```

Output:

                                   SAS   1

                    L   COL1

                   ROW1   2.61168

Figure 14. Likelihood ratio test of a restriction on the variance-covariance matrix of Example 2.

which agrees well with the Wald test statistic.   □

A test of joint hypothesis

$$H : h(\theta^0, \Sigma^0) = 0 \quad \text{against} \quad A : h(\theta^0, \Sigma^0) \neq 0$$

is not encountered very often in applications. In the event that it is, application of Theorem 11, 14, or 15 of Section 5, Chapter 3 is reasonably straightforward.

PROBLEMS

1. Show that $\sqrt{n}\,(\hat{\Sigma}_I - \Sigma_n^0)$ is bounded in probability. Hint: See Problem 1, Section 6.

2. Show that $(\partial/\partial\sigma_i)\Sigma(\sigma) = \Sigma(\xi_i)$. In Problem 4 the expressions $(\partial/\partial\sigma_i)\Sigma^{-1}(\sigma) = -\Sigma^{-1}(\sigma)[(\partial/\partial\sigma_i)\Sigma(\sigma)]\Sigma^{-1}(\sigma)$ and $(\partial/\partial\sigma_i)\ln\det\Sigma(\sigma) = \operatorname{tr}[\Sigma^{-1}(\sigma)(\partial/\partial\sigma_i)\Sigma(\sigma)]$ are derived. Use them to derive the first and second partial derivatives of $s[Y(e, x, \gamma^0), x, \lambda]$ given in the text.

3. Denote the jth column of a matrix A by $A_{(j)}$ and a typical element by a_{ij}. Show that

$$\begin{aligned}
(ABC)_{(j)} &= (AB)C_{(j)} \\
&= \sum_i (c_{ij}A)(B_{(i)}) \\
&= [(C_{(j)})' \otimes A]\operatorname{vec} B.
\end{aligned}$$

Then stack the columns $(ABC)_{(j)}$ to obtain

$$\text{vec}(ABC) = (C' \otimes A)\text{vec } B.$$

Show that

$$\text{tr}(AB) = \sum_i \sum_k a_{ik} b_{ki} = \text{vec}'(A')\text{vec } B$$

whence

$$\text{tr}(ABC) = \text{vec}'(A')\text{vec}(BCI) = \text{vec}'(A')(I \otimes B)\text{vec } C.$$

4. Show that $(\partial/\partial\sigma_i)\Sigma(\sigma) = \Sigma(\xi_i)$. Use $I = \Sigma^{-1}(\sigma)\Sigma(\sigma)$ to obtain $0 = [(\partial/\partial\sigma_i)\Sigma^{-1}(\sigma)]\Sigma(\sigma) + \Sigma^{-1}(\sigma)[(\partial/\partial\sigma_i)\Sigma(\sigma)]$, whence $(\partial/\partial\sigma_i)\Sigma^{-1}(\sigma) = -\Sigma^{-1}(\sigma)[(\partial/\partial\sigma_i)\Sigma(\sigma)]\Sigma^{-1}(\sigma)$. Let a square matrix A have elements a_{ij}, let c_{ij} denote the cofactors of A, and let a^{ij} denote the elements of A^{-1}. From $\det A = \sum_k a_{ik}c_{ik}$ show that $(\partial/\partial a_{ij})\det A = c_{ij} = a^{ji}\det A$. This implies that

$$\frac{\partial}{\partial \text{ vec}'A}\det A = \det A \text{ vec}'(A^{-1})'.$$

Use this fact and the previous problem to show

$$\frac{\partial}{\partial\sigma_i}\det \Sigma(\sigma) = \det \Sigma(\sigma)\text{vec}'\left[\Sigma^{-1}(\sigma)\right]\text{vec}\left(\frac{\partial}{\partial\sigma_i}\Sigma(\sigma)\right)$$

$$= \det \Sigma(\sigma)\text{tr}\left(\Sigma^{-1}(\sigma)\frac{\partial}{\partial\sigma_i}\Sigma(\sigma)\right).$$

Show that $(\partial/\partial\sigma_i)\ln\det \Sigma(\sigma) = \text{tr}[\Sigma^{-1}(\sigma)(\partial/\partial\sigma_i)\Sigma(\sigma)]$.

5. Referring to Example 2, a two way multivariate design has fixed part

$$\mathscr{E}y_{ij} = \mu + \rho_i + \tau_j$$

whereas the split plot ANOVA has fixed part

$$\mathscr{E}y_{ijk} = m + r_i + t_j + s_k + (rs)_{ki} + (ts)_{kj}.$$

Use the following correspondences to show that the fixed part of the designs is the same:

$$
\mu' = (\mu_1, \ldots, \mu_k, \ldots, \mu_M) = (m, \ldots, m, \ldots, m)
$$
$$
\rho_i' = (\rho_{1i}, \ldots, \rho_{ki}, \ldots, \rho_{Mi})
$$
$$
= (r_i + (rs)_{1i}, \ldots, r_i + (rs)_{ki}, \ldots, r_i + (rs)_{Mi})
$$
$$
\tau_j' = (\tau_{1j}, \ldots, \tau_{kj}, \ldots, \tau_{Mj})
$$
$$
= (t_j + (ts)_{1j}, \ldots, t_j + (ts)_{kj}, \ldots, t_j + (ts)_{Mj}).
$$

Show that under ANOVA assumption $\Sigma = \sigma_\epsilon^2 I + \sigma_\eta^2 J$.

6. Suppose that one has the multivariate linear model

$$
y_t' = x_t' B + e_t' \qquad t = 1, 2, \ldots, n
$$

where B is k by p and y_t' is 1 by M. Write

$$
Y = \begin{pmatrix} y_1' \\ y_2' \\ \vdots \\ y_n' \end{pmatrix} \qquad X = \begin{pmatrix} x_1' \\ x_2' \\ \vdots \\ x_n' \end{pmatrix}
$$

and show that

$$
2s_n(B, \Sigma) = \ln \det \Sigma + \frac{\operatorname{tr} \Sigma^{-1}[Y - P_X Y]'[Y - P_X Y]}{n}
$$
$$
+ \frac{\operatorname{tr} \Sigma^{-1}[P_X Y - XB]'[P_X Y - XB]}{n}
$$

where $P_X = X(X'X)^{-1}X'$. One observes from this equation that $\hat{B}$ will be computed as $\hat{B} = (X'X)^{-1}X'Y$ no matter what value is assigned to Σ. Thus, if $(\tilde{B}_\infty, \tilde{\Sigma}_\infty)$ minimizes $s_n(B, \Sigma)$ subject to $\Sigma = \Sigma(\sigma)$, $h(\sigma) = 0$, and $(\hat{B}_\infty, \hat{\Sigma}_\infty)$ is the unconstrained minimizer, then

$$
2s_n(\tilde{B}_\infty, \tilde{\Sigma}_\infty) = \ln \det \tilde{\Sigma}_\infty + \operatorname{tr} \tilde{\Sigma}_\infty^{-1}\hat{\Sigma}_\infty + 0.
$$

6. ASYMPTOTIC THEORY

As in Chapter 4, an asymptotic theory for least squares estimators of the parameters of a multivariate nonlinear regression model obtains by restating Assumptions 1 through 6 of Chapter 3 in context and then applying

Theorems 3 and 5 of Chapter 3. Similarly, the asymptotic distribution of test statistics based on least squares estimators obtains by appending restatements of Assumptions 7 and 13 to the list and applying Theorems 11, 14, and 15. That is what we shall do here.

Recall that, using the grouped by subject data arrangement, the multivariate nonlinear regression model is written as

$$y_t = f(x_t, \theta^0) + e_t \qquad t = 1, 2, \ldots, n$$

with θ^0 known to lie in some compact set Θ^*. The functional form of $f(x, \theta)$ is known, x is k-dimensional, θ is p-dimensional, and $f(x, \theta)$ takes its values in $\mathbb{R}^M$; y_t and e_t are M-vectors. The errors e_t are independently and identically distributed, each with mean zero and nonsingular variance-covariance matrix Σ; viz.

$$\mathscr{E} e_t = 0, \qquad \mathscr{C}(e_t, e_s') = \begin{cases} \Sigma & t = s \\ 0 & t \neq s. \end{cases}$$

The parameter θ^0 is estimated by $\hat{\theta}_n$ that minimizes

$$s_n(\theta, \hat{\Sigma}_n) = \frac{1}{n} \sum_{t=1}^{n} [y_t - f(x_t, \theta)]' \hat{\Sigma}_n^{-1} [y_t - f(x_t, \theta)].$$

Here we shall let $\hat{\Sigma}_n$ by any random variable that converges almost surely to Σ and has $\sqrt{n}(\hat{\Sigma}_n - \Sigma)$ bounded in probability; that is, given $\delta > 0$, there is a bound b and a sample size N such that

$$P\left(\sqrt{n} |\hat{\sigma}_{\alpha\beta n} - \sigma_{\alpha\beta}| < b\right) > 1 - \delta$$

for all $n > N$, $\sigma_{\alpha\beta}$ being a typical element of Σ. Verification that the estimator of Σ proposed in Section 2 satisfies this requirement is left to Problem 1.

Construction of the set T which is presumed to contain $\hat{\Sigma}_n$ requires a little care. Denote the upper triangle of Σ by

$$\tau = (\sigma_{11}, \sigma_{12}, \sigma_{22}, \sigma_{13}, \sigma_{23}, \sigma_{33}, \ldots, \sigma_{1M}, \sigma_{2M}, \ldots, \sigma_{MM})'$$

which is a column vector of length $M(M + 1)/2$. Let $\Sigma(\tau)$ denote the mapping of τ into the elements of Σ, and set $\Sigma(\hat{\tau}_n) = \hat{\Sigma}_n$, $\Sigma(\tau^*) = \mathscr{C}(e_t, e_t')$. Now $\det \Sigma(\tau)$ is a polynomial of degree M in τ and is therefore continuous; moreover, for some $\delta > 0$ we have $\det \Sigma(\tau^*) - \delta > 0$ by assumption. Therefore the set

$$\{\tau : \det \Sigma(\tau) > \det \Sigma(\tau^*) - \delta\}$$

is an open set containing τ^*. Then this set must contain a bounded open ball with center τ^*, and the closure of this ball can be taken as T. The assumption that $\sqrt{n}\,(\hat{\Sigma}_n - \Sigma)$ is bounded in probability means that we have implicitly taken $\tau_n^0 \equiv \tau^* \in T$, and without loss of generality (Problem 2) we can assume that $\hat{\tau}_n$ is in T for all n. Note that $\det \Sigma(\tau) \geq \det \Sigma(\tau^*) - \delta$ for all τ in T, which implies that $\Sigma^{-1}(\tau)$ is continuous and differentiable over T (Problem 3). Put

$$B = \sup\{\sigma^{\alpha\beta}(\tau) : \tau \in T, \alpha, \beta = 1, 2, \ldots, M\}$$

where $\sigma^{\alpha\beta}(\tau)$ denotes a typical element of $\Sigma^{-1}(\tau)$. Since $\Sigma^{-1}(\tau)$ is continuous over the compact set T, we must have $B < \infty$.

We are interested in testing the hypothesis

$$H: h(\theta^0) = 0 \quad \text{against} \quad A: h(\theta^0) \neq 0$$

which we assume can be given the equivalent representation

$$H: \theta^0 = g(\rho^0) \text{ for some } \rho^0 \quad \text{against} \quad A: \theta^0 \neq g(\rho) \text{ for any } \rho$$

where $h: \mathbb{R}^p \to \mathbb{R}^q$, $g: \mathbb{R}^r \to \mathbb{R}^p$, and $p = r + q$. The correspondence with the notation of Chapter 3 is given in Notation 6.

NOTATION 6.

General (Chapter 3)	Specific (Chapter 6)
$e_t = q(y_t, x_t, \gamma_n^0)$	$e_t = y_t - f(x_t, \theta_n^0)$
$\gamma \in \Gamma^*$	$\theta \in \Theta^*$
$y = Y(e, x, \gamma)$	$y = f(x, \theta) + e$
$s(y_t, x_t, \hat{\tau}_n, \lambda)$	$[y_t - f(x_t, \theta)]'\Sigma^{-1}(\hat{\tau}_n)[y_t - f(x_t, \theta)]$
$\hat{\tau}_n \in T$	$\hat{\tau}_n \in T, \tau_n^0 \equiv \tau^*$
$\lambda \in \Lambda^*$	$\theta \in \Theta^*$
$s_n(\lambda)$	$s_n(\theta) = \dfrac{1}{n}\displaystyle\sum_{t=1}^{n}[y_t - f(x_t, \theta)]'$
$\quad = \dfrac{1}{n}\displaystyle\sum_{t=1}^{n} s(y_t, x_t, \hat{\tau}_n, \lambda)$	$\times \Sigma^{-1}(\hat{\tau}_n)[y_t - f(x_t, \theta)]$

General (Chapter 3)	Specific (Chapter 6)

$s_n^0(\lambda)$

$$= \frac{1}{n} \sum_{t=1}^{n} \int_{\mathscr{E}} s[Y(e, x_t, \gamma_n^0), x_t, \tau_n^0, \lambda]$$

$$\times dP(e)$$

$s_n^0(\theta) = M$

$$+ \frac{1}{n} \sum_{t=1}^{n} [f(x_t, \theta_n^0) - f(x_t, \theta)]'$$

$$\times \Sigma^{-1}(\tau^*)[f(x_t, \theta_n^0) - f(x_t, \theta)]$$

$s^*(\lambda)$

$$= \int_{\mathscr{X}} \int_{\mathscr{E}} s[Y(e, x, \gamma^*), x, \tau^*, \lambda]$$

$$\times dP(e) \, d\mu(x)$$

$s^*(\theta)M$

$$+ \int_{\mathscr{X}} [f(x, \theta^*) - f(x, \theta)]' \Sigma^{-1}(\tau^*)$$

$$\times [f(x, \theta^*) - f(x, \theta)] \, d\mu(x)$$

$\hat{\lambda}_n$ minimizes $s_n(\lambda)$

$\hat{\theta}_n$ minimizes $s_n(\theta)$

$\tilde{\lambda}_n$ minimizes $s_n(\lambda)$
  subject to $h(\lambda) = 0$

$\tilde{\theta}_n = g(\hat{\rho}_n)$ minimizes $s_n(\theta)$
  subject to $h(\theta) = 0$

λ_n^0 minimizes $s_n^0(\lambda)$

θ_n^0 minimizes $s_n^0(\theta)$

λ_n^* minimizes $s_n^0(\lambda)$
  subject to $h(\lambda) = 0$

$\theta_n^* = g(\rho_n^0)$ minimizes $s_n^0(\theta)$
  subject to $h(\theta) = 0$

λ^* minimizes $s^*(\lambda)$

θ^* minimizes $s^*(\theta)$

Assumptions 1 through 6 of Chapter 3 read as follows in the present case.

ASSUMPTION 1'. The errors are independently and identically distributed with common distribution $P(e)$.

ASSUMPTION 2'. $f(x, \theta)$ is continuous on $\mathscr{X} \times \Theta^*$, and Θ^* is compact.

ASSUMPTION 3' (Gallant and Holly, 1980). Almost every realization of $\{v_t\}$ with $v_t = (e_t, x_t)$ is a Cesaro sum generator with respect to the product measure

$$\nu(A) = \int_{\mathscr{X}} \int_{\mathscr{E}} I_A(e, x) \, dP(e) \, d\mu(x)$$

and dominating function $b(e, x)$. The sequence $\{x_t\}$ is a Cesaro sum generator with respect to μ and $b(x) = \int_{\mathscr{E}} b(e, x) \, dP(e)$. For each $x \in \mathscr{X}$ there is a neighborhood N_x such that $\int_{\mathscr{E}} \sup_{N_x} b(e, x) \, dP(e) < \infty$.

ASSUMPTION 4′ (Identification). The parameter θ^0 is indexed by n, and the sequence $\{\theta_n^0\}$ converges to θ^*; $\tau_n^0 = \tau^*$, $\sqrt{n}\,(\hat{\tau}_n - \tau^*)$ is bounded in probability, and $\hat{\tau}_n$ converges almost surely to τ^*.

$$s^*(\theta) = M + \int_{\mathscr{X}} [f(x, \theta^*) - f(x, \theta)]' \Sigma^{-1}(\tau^*)$$

$$\times [f(x, \theta^*) - f(x, \theta)]\, d\mu(x)$$

has a unique minimum over Θ^* at θ^*.

ASSUMPTION 5′. Θ^* is compact; $\{\hat{\tau}_n\}$, T, and B are as described in the first few paragraphs of this section. The functions

$$\big[e_\alpha + f_\alpha(x, \theta^0) - f_\alpha(x, \theta)\big]\big[e_\beta + f_\beta(x, \theta^0) - f_\beta(x, \theta)\big]$$

are dominated by $b(e, x)/M^2 B$ over $\mathscr{E} \times \mathscr{X} \times \Theta^* \times \Theta^*$; $b(e, x)$ is that of Assumption 3′.

This is enough to satisfy Assumption 5 of Chapter 3, since

$$s\big[Y(e, x, \theta^0), x, \tau, \theta\big]$$

$$= \big[e + f(x, \theta^0) - f(x, \theta)\big]' \Sigma^{-1}(\tau)\big[e + f(x, \theta^0) - f(x, \theta)\big]$$

$$\leq B \sum_\alpha \sum_\beta \big|e_\alpha + f_\alpha(x, \theta^0) - f_\beta(x, \theta)\big|\big|e_\beta + f_\beta(x, \theta^0) - f_\beta(x, \theta)\big|$$

$$\leq \frac{BM^2 b(e, x)}{M^2 B}$$

$$= b(e, x).$$

The sample objective function is

$$s_n(\theta) = \frac{1}{n} \sum_{t=1}^{n} [y_t - f(x_t, \theta)]' \Sigma^{-1}(\hat{\tau}_n)[y_t - f(x_t, \theta)].$$

Replacing $\hat{\tau}_n$ by $\tau_n^0 \equiv \tau^*$, its expectation is

$$s_n^0(\theta) = \frac{1}{n} \sum_{t=1}^n \mathscr{E}\big[e_t + f(x_t, \theta_n^0) - f(x_t, \theta)\big]'$$
$$\times \Sigma^{-1}(\tau^*)\big[e_t + f(x_t, \theta_n^0) - f(x_t, \theta)\big]$$
$$= \frac{1}{n} \sum_{t=1}^n \mathscr{E} e_t' \Sigma^{-1}(\tau^*) e_t$$
$$+ \frac{1}{n} \sum_{t=1}^n \big[f(x_t, \theta_n^0) - f(x_t, \theta)\big]' \Sigma^{-1}(\tau^*)\big[f(x_t, \theta_n^0) - f(x_t, \theta)\big]$$
$$= M + \frac{1}{n} \sum_{t=1}^n \big[f(x_t, \theta_n^0) - f(x_t, \theta)\big]'$$
$$\times \Sigma^{-1}(\tau^*)\big[f(x_t, \theta_n^0) - f(x_t, \theta)\big];$$

the last equality obtains from

$$\frac{1}{n} \sum_{t=1}^n \mathscr{E} e_t' \Sigma^{-1}(\tau^*) e_t$$
$$= \frac{1}{n} \sum_{t=1}^n \operatorname{tr} \Sigma^{-1}(\tau^*) \mathscr{E} e_t e_t'$$
$$= \frac{1}{n} \sum_{t=1}^n \operatorname{tr} I$$
$$= M.$$

By Lemma 1 of Chapter 3, both $s_n(\theta)$ and $s_n^0(\theta)$ have uniform almost sure limit

$$s^*(\theta) = M + \int_{\mathscr{X}} [f(x, \theta^*) - f(x, \theta)]'$$
$$\times \Sigma^{-1}(\tau^*)[f(x, \theta^*) - f(x, \theta)] \, d\mu(x).$$

Note that the true value θ_n^0 of the unknown parameter is also a minimizer of $s_n^0(\theta)$, so that the use of θ_n^0 to denote them both is not ambiguous. By Theorem 3 of Chapter 3 we have

$$\lim_{n \to \infty} \hat{\theta}_n = \theta^* \qquad \text{almost surely.}$$

Continuing, we have one last assumption in order to be able to claim asymptotic normality.

ASSUMPTION 6′. Θ^* contains a closed ball Θ centered at θ^* with finite, nonzero radius such that

$$\left[e_\alpha + f_\alpha(x, \theta^0) - f_\alpha(x, \theta)\right]\left(\frac{\partial}{\partial \theta_i} f_\beta(x, \theta)\right)$$

$$\left(\frac{\partial}{\partial \theta_i} f_\alpha(x, \theta)\right)\left(\frac{\partial}{\partial \theta_j} f_\beta(x, \theta)\right)$$

$$\left[e_\alpha + f_\alpha(x, \theta^0) - f_\alpha(x, \theta)\right]\left(\frac{\partial^2}{\partial \theta_i \, \partial \theta_j} f(x, \theta)\right)$$

and

$$\left[\left[e_\alpha + f_\alpha(x, \theta^0) - f_\alpha(x, \theta)\right]\left(\frac{\partial}{\partial \theta_i} f_\beta(x, \theta)\right)\right]^2$$

are dominated by $b(e, x)/BM^2$ over $\mathscr{E} \times \mathscr{X} \times \Theta \times \Theta$ for $i, j = 1, 2, \ldots, p$ and $\alpha, \beta = 1, 2, \ldots, M$. Moreover,

$$\mathscr{I}^* = 2\int_\mathscr{X}\left(\frac{\partial}{\partial \theta'} f(x, \theta^*)\right)'\Sigma^{-1}(\tau^*)\left(\frac{\partial}{\partial \theta'} f(x, \theta^*)\right) d\mu(x)$$

is nonsingular.

One can verify that this is enough to dominate

$$\frac{\partial}{\partial \theta_i} s\left[Y(e, x, \theta^0), x, \tau, \theta\right]$$

$$= -2\left[e + f(x, \theta^0) - f(x, \theta)\right]'\Sigma^{-1}(\tau)\frac{\partial}{\partial \theta_i} f(x, \theta)$$

$$\frac{\partial^2}{\partial \theta_i \partial \theta_j} s\left[Y(e, x, \theta^0), x, \tau, \theta\right]$$

$$= 2\left(\frac{\partial}{\partial \theta_i} f(x, \theta)\right)'\Sigma^{-1}(\tau)\left(\frac{\partial}{\partial \theta_j} f(x, \theta)\right)$$

$$-2\sum_{\alpha=1}^{M}\sum_{\beta=1}^{M}\left[e_\alpha + f_\alpha(x, \theta^0) - f_\alpha(x, \theta)\right]\sigma^{\alpha\beta}(\tau)\frac{\partial^2}{\partial \theta_i \, \partial \theta_j} f_\beta(x, \theta)$$

$$\left(\frac{\partial}{\partial \theta_i} s\left[Y(e, x, \theta^0), x, \tau, \theta\right]\right)\left(\frac{\partial}{\partial \theta_j} s\left[Y(e, x, \theta^0), x, \tau, \theta\right]\right)$$

$$= 4\left[\frac{\partial}{\partial \theta_i} f(x, \theta)\right]'\Sigma^{-1}(\tau)\left[e + f(x, \theta^0) - f(x, \theta)\right]$$

$$\times\left[e + f(x, \theta^0) - f(x, \theta)\right]'\Sigma^{-1}(\tau)\left(\frac{\partial}{\partial \theta_j} f(x, \theta)\right)$$

to within a multiplicative constant. Since (Problem 4, Section 5)

$$\frac{\partial}{\partial \tau_i} \Sigma^{-1}(\tau) = -\Sigma^{-1}(\tau)\left(\frac{\partial}{\partial \tau_i}\Sigma(\tau)\right)\Sigma^{-1}(\tau)$$

we have

$$\frac{\partial^2}{\partial \tau_i \partial \theta_j} s\left[Y(e, x, \theta^0), x, \tau, \theta\right]$$

$$= 2\left[e + f(x, \theta^0) - f(x, \theta)\right]'\Sigma^{-1}(\tau)\left(\frac{\partial}{\partial \tau_i}\Sigma(\tau)\right)\Sigma^{-1}(\tau)\frac{\partial}{\partial \theta_j}f(x, \theta).$$

Evaluating at $\theta = \theta^0 = \theta^*$ and integrating, we have

$$\int_{\mathcal{X}}\int_{\mathcal{E}} \frac{\partial^2}{\partial \tau_i \partial \theta_j} s\left[Y(e, x, \theta^*), x, \tau^*, \theta^*\right] dP(e)\, d\mu(x)$$

$$= \int_{\mathcal{X}} - 2\int_{\mathcal{E}} e\, dP(e)\, \Sigma^{-1}(\tau^*)\left(\frac{\partial}{\partial \tau_i}\Sigma(\tau^*)\right)\Sigma^{-1}(\tau^*)\frac{\partial}{\partial \theta_j}f(x, \theta^*)\, d\mu(x)$$

$$= 0$$

because $\int_{\mathcal{E}} e\, dP(e) = 0$. Thus, Assumption 6' is enough to imply Assumption 6 of Chapter 3.

The parameters of the asymptotic distribution of $\hat{\theta}_n$ and various test statistics are defined in terms of the following.

NOTATION 7.

$$\Omega = \int_{\mathcal{X}}\left[\frac{\partial}{\partial \theta'}f(x, \theta^*)\right]'\Sigma^{-1}\left[\frac{\partial}{\partial \theta'}f(x, \theta^*)\right] d\mu(x)$$

$$\Omega_n^0 = \frac{1}{n}\sum_{t=1}^{n}\left(\frac{\partial}{\partial \theta'}f(x_t, \theta_n^0)\right)'\Sigma^{-1}\left(\frac{\partial}{\partial \theta'}f(x_t, \theta_n^0)\right)$$

$$\Omega_n^* = \frac{1}{n}\sum_{t=1}^{n}\left(\frac{\partial}{\partial \theta'}f(x_t, \theta_n^*)\right)'\Sigma^{-1}\left(\frac{\partial}{\partial \theta'}f(x_t, \theta_n^*)\right).$$

NOTATION 8.

$$\mathscr{I}^* = 4\Omega$$
$$\mathscr{J}^* = 2\Omega$$
$$\mathscr{U}^* = 0$$
$$\mathscr{I}_n^0 = 4\Omega_n^0$$

$$\mathscr{I}_n^0 = 2\Omega_n^0$$

$$\mathscr{U}_n^0 = 0$$

$$\mathscr{I}_n^* = 4\Omega_n^*$$

$$\mathscr{J}_n^* = 2\Omega_n^* - \frac{2}{n} \sum_{t=1}^{n} \sum_{\alpha=1}^{n} \sum_{\beta=1}^{n} \left[f_\alpha(x_t, \theta_n^0) - f_\alpha(x_t, \theta_n^*) \right]$$

$$\times \sigma^{\alpha\beta} \frac{\partial^2}{\partial\theta\,\partial\theta'} f_\beta(x_t, \theta_n^*)$$

$$\mathscr{U}_n^* = \frac{4}{n} \sum_{t=1}^{n} \left(\frac{\partial}{\partial\theta'} f(x_t, \theta_n^*) \right)' \Sigma^{-1} \left[f(x_t, \theta_n^0) - f(x_t, \theta_n^*) \right]$$

$$\times \left[f(x_t, \theta_n^0) - f(x_t, \theta_n^*) \right]' \Sigma^{-1} \left(\frac{\partial}{\partial\theta'} f(x_t, \theta_n^*) \right).$$

One can see from Notation 8 that it would be enough to have an estimator of Ω to be able to estimate $\mathscr{I}^*$ and $\mathscr{J}^*$. Accordingly we propose the following.

NOTATION 9.

$$\hat{\Omega} = \frac{1}{n} \sum_{t=1}^{n} \left(\frac{\partial}{\partial\theta'} f(x_t, \hat{\theta}_n) \right)' \hat{\Sigma}^{-1} \left(\frac{\partial}{\partial\theta'} f(x_t, \hat{\theta}_n) \right)$$

$$\tilde{\Omega} = \frac{1}{n} \sum_{t=1}^{n} \left(\frac{\partial}{\partial\theta'} f(x_t, \tilde{\theta}_n) \right)' \hat{\Sigma}^{-1} \left(\frac{\partial}{\partial\theta'} f(x_t, \tilde{\theta}_n) \right).$$

Since $(\mathscr{J}^*)^{-1} \mathscr{I}^* (\mathscr{J}^*)^{-1} = \Omega^{-1}$, we have from Theorem 5 of Chapter 3 that

$$\sqrt{n}\left(\hat{\theta}_n - \theta_n^0 \right) \xrightarrow{\mathscr{L}} N_p(0, \Omega^{-1})$$

$\hat{\Omega}$ converges almost surely to Ω.

Assumptions 7 and 13 of Chapter 3, restated in context, read as follows.

ASSUMPTION 7' (Pitman drift). The sequence θ_n^0 is chosen such that $\lim_{n \to \infty} \sqrt{n}\,(\theta_n^0 - \theta_n^*) = \Delta$. Moreover, $h(\theta^*) = 0$.

ASSUMPTION 13'. The function $h(\theta)$ is a once continuously differentiable mapping of Θ into $\mathbb{R}^q$. Its Jacobian $H(\theta) = (\partial/\partial\theta')h(\theta)$ has full rank $(= q)$ at $\theta = \theta^*$.

From these last two assumptions we obtain a variety of ancillary facts, notably that $\tilde{\theta}_n$ converges almost surely to θ^*, that $\tilde{\Omega}$ converges almost surely to Ω, and that (Problem 6)

$$\mathscr{J}_n^* = 2\Omega_n^* + O\left(\frac{1}{\sqrt{n}}\right).$$

The next task is to apply Theorems 11, 14, and 15 of Chapter 3 to obtain Wald, "likelihood ratio," and Lagrange multiplier test statistics as well as noncentral chi-square approximations to their distributions. With some extra effort, we could derive characterizations theorems as in Chapter 4 to formally justify the degrees of freedom corrections used in Section 3. But we shall not, letting the analogy with univariate nonlinear regression developed in Sections 2 and 3 and the simulations reported in Table 4 serve as justification.

Consider testing

$$H : h\left(\theta_n^0\right) = 0 \quad \text{against} \quad A : h\left(\theta_n^0\right) \neq 0$$

where (recall) $h(\theta)$ is a q-vector with Jacobian $H(\theta) = (\partial/\partial\theta')h(\theta)$, $H(\theta)$ being a q by p matrix. Writing $\hat{h} = h(\hat{\theta}_n)$ and $\hat{H} = H(\hat{\theta}_n)$ and applying Theorem 11 of Chapter 3, we have that the Wald test statistic is

$$W' = n\hat{h}'(\hat{H}\hat{\Omega}^{-1}\hat{H}')^{-1}\hat{h},$$

and that the distribution of W' can be approximated by the noncentral chi-square distribution with q degrees of freedom and noncentrality parameter

$$\alpha = n\frac{h'\left(\theta_n^0\right)\left[H\left(\theta_n^0\right)\left(\Omega_n^0\right)^{-1}H'\left(\theta_n^0\right)\right]^{-1}h\left(\theta_n^0\right)}{2}.$$

Multivariate nonlinear least squares is an instance where $\mathscr{J}_n^* \neq \mathscr{I}_n^*$ but $\mathscr{J}_n^* = (1/2)\mathscr{I}_n^* + O(1/\sqrt{n})$ (Problem 6), whence the likelihood ratio test statistic is

$$L' = n\left[s_n\left(\tilde{\theta}_n\right) - s_n\left(\hat{\theta}_n\right)\right].$$

In the notation of Section 3,

$$L' = S\left(\tilde{\theta}_n, \hat{\Sigma}_n\right) - S\left(\hat{\theta}_n, \hat{\Sigma}_n\right).$$

It is critical that $\hat{\Sigma}_n$ be the same in both terms on the right hand side of this equation. If they differ, then the distributional results that follow are invalid (Problem 8). This seems a bit strange, because it is usually the case in asymptotic theory that any $\sqrt{n}$ -consistent estimator of a nuisance parameter can be substituted in an expression without changing the result. The source of the difficulty is that the first equation in the proof of Theorem 15 is not true if $\hat{\Sigma}_n$ is not the same in both terms.

Applying Theorem 15 of Chapter 3 and the remarks that follow it, we have that the distribution of L' can be approximated by the noncentral chi-square with q degrees of freedom and noncentrality parameter (Problem 7)

$$
\alpha = \left(\sum_{t=1}^{n} \left[f(x_t, \theta_n^0) - f(x_t, \theta_n^*) \right]' \Sigma^{-1} \frac{\partial}{\partial \theta'} f(x_t, \theta_n^*) \right)
$$

$$
\times \frac{\Omega_n^{*-1} \left\{ \sum_{t=1}^{n} \left[f(x_t, \theta_n^0) - f(x_t, \theta_n^*) \right]' \Sigma^{-1} \frac{\partial}{\partial \theta'} f(x_t, \theta_n^*) \right\}'}{2n}.
$$

Up to this point, we have assumed that a correctly centered estimator of the variance-covariance matrix of the errors is available. That is, we have assumed that the estimator $\hat{\Sigma}_n$ has $\sqrt{n}(\hat{\Sigma}_n - \Sigma)$ bounded in probability whether $h(\theta_n^0) = 0$ is true or not. This assumption is unrealistic with respect to the Lagrange multiplier test. Accordingly, we base the Lagrange multiplier test on an estimate of scale $\tilde{\Sigma}_n$, for which we assume

$$
\sqrt{n}(\tilde{\Sigma}_n - \Sigma_n^*) \quad \text{bounded in probability}
$$

$$
\lim_{n \to \infty} \Sigma_n^* = \Sigma.
$$

Of such estimators, that which is mostly likely to be used in applications is obtained by computing $\tilde{\theta}_n^\#$ to minimize $s_n(\theta, I)$ subject to $h(\theta) = 0$, where

$$
s_n(\theta, V) = \frac{1}{n} \sum_{t=1}^{n} \left[y_t - f(x_t, \theta) \right]' V^{-1} \left[y_t - f(x_t, \theta) \right]
$$

and then putting

$$
\tilde{\Sigma}_n = \frac{1}{n} \sum_{t=1}^{n} \left[y_t - f(x_t, \tilde{\theta}_n^\#) \right] \left[y_t - f(x_t, \tilde{\theta}_n^\#) \right]'.
$$

The center is found (Problem 10) by computing $\theta_n^\#$ to minimize $s_n^0(\theta, I)$

subject to $h(\theta) = 0$, where

$$s_n^0(\theta, V) = \operatorname{tr} V^{-1}\Sigma$$
$$+ \frac{1}{n} \sum_{t=1}^{n} \left[f(x_t, \theta_n^0) - f(x_t, \theta) \right]' V^{-1} \left[f(x_t, \theta_n^0) - f(x_t, \theta) \right]$$

and putting

$$\Sigma_n^* = \Sigma + \frac{1}{n} \sum_{t=1}^{n} \left[f(x_t, \theta_n^0) - f(x_t, \theta_n^\#) \right] \left[f(x_t, \theta_n^0) - f(x_t, \theta_n^\#) \right]'.$$

Using the estimator $\tilde{\Sigma}_n$, the formulas for the constrained estimators are revised to read: $\tilde{\theta}_n$ minimizes $s_n(\theta, \tilde{\Sigma}_n)$ subject to $h(\theta) = 0$ and

$$\tilde{\Omega}_n = \frac{1}{n} \sum_{t=1}^{n} \left(\frac{\partial}{\partial \theta'} f(x_t, \tilde{\theta}_n) \right)' \tilde{\Sigma}_n^{-1} \left(\frac{\partial}{\partial \theta'} f(x_t, \tilde{\theta}_n) \right).$$

The form of the efficient score or Lagrange multiplier test depends on how one goes about estimating V^* and $\mathscr{J}^*$ having the estimator $\tilde{\theta}_n$ in hand. In view of the remarks following Theorem 14 of Chapter 3, the choices

$$\tilde{V} = \tilde{\Omega}^{-1} \qquad \tilde{\mathscr{J}} = 2\tilde{\Omega}$$

lead to considerable simplifications in the computations, because it is not necessary to obtain second derivatives of $s_n(\theta)$ to estimate $\mathscr{J}^*$ and one is in the situation where $\tilde{\mathscr{J}}^{-1} = a\tilde{V}$ for $a = \frac{1}{2}$. With these choices, the Lagrange multiplier test becomes (Problem 9)

$$R' = \frac{1}{n} \left(\sum_{t=1}^{n} \left[y_t - f(x_t, \tilde{\theta}_n) \right]' \tilde{\Sigma}_n^{-1} \frac{\partial}{\partial \theta'} f(x_t, \tilde{\theta}_n) \right)$$
$$\times \tilde{\Omega}^{-1} \left(\sum_{t=1}^{n} \left[y_t - f(x_t, \tilde{\theta}_n) \right]' \tilde{\Sigma}_n^{-1} \frac{\partial}{\partial \theta'} f(x_t, \tilde{\theta}_n) \right)'.$$

Let θ_n^{**} minimize $s^0(\theta, \Sigma_n^*)$ subject to $h(\theta) = 0$, and put

$$\Omega_n^{**} = \frac{1}{n} \sum_{t=1}^{n} \left(\frac{\partial}{\partial \theta'} f(x_t, \theta_n^{**}) \right)' (\Sigma_n^*)^{-1} \left(\frac{\partial}{\partial \theta'} f(x_t, \theta_n^{**}) \right).$$

Then the distribution of $\tilde{R}'$ can be characterized as (Problem 9)

$$\tilde{R}' = \tilde{Y} + o_p(1)$$

where

$$\tilde{Y} = \tilde{Z}'(\Omega_n^{**})^{-1}H_n^{**\prime}\left[H_n^{**}(\Omega_n^{**})^{-1}H_n^{**\prime}\right]^{-1}H_n^{**}(\Omega_n^{**})^{-1}\tilde{Z}$$

and

$$\tilde{Z} \sim N_p\left[-\frac{1}{n}\sum_{t=1}^{n}\left(\frac{\partial}{\partial\theta'}f(x_t,\theta_n^{**})\right)'\right.$$

$$\times(\Sigma_n^*)^{-1}\left[f(x_t,\theta_n^0)-f(x_t,\theta_n^{**})\right],\left.\frac{\mathscr{I}_n^{**}}{4}\right]$$

$$\frac{\mathscr{I}_n^{**}}{4} = \frac{1}{n}\sum_{t=1}^{n}\left(\frac{\partial}{\partial\theta'}f(x_t,\theta_n^{**})\right)'$$

$$\times(\Sigma_n^*)^{-1}\Sigma(\Sigma_n^*)^{-1}\left(\frac{\partial}{\partial\theta'}f(x_t,\theta_n^{**})\right)$$

$$H_n^{**} = \frac{\partial}{\partial\theta'}h(\theta_n^{**}).$$

The random variable $\tilde{Y}$ is a general quadratic form in the multivariate normal random variable $\tilde{Z}$, and one can use the methods discussed in Imhof (1961) to compute its distribution. Comparing with the result derived in Problem 5, one sees that the unfortunate consequence of the use of $\tilde{\Sigma}_n$ instead of $\hat{\Sigma}_n$ to compute $\tilde{R}'$ is that one can not use tables of the noncentral chi-square to approximate its nonnull distribution. The null distribution of $\tilde{Y}$ is a chi-square with q degrees of freedom, since if the null is true, $\theta_n^{**} = \theta_n^0$ and $\Sigma_n^* = \Sigma$.

PROBLEMS

1. Show that the regularity conditions listed in this section are sufficient to imply the regularity conditions listed in Section 2 of Chapter 4 for each of the models

$$y_\alpha = f_\alpha(\theta_\alpha) + e_\alpha \qquad \alpha = 1, 2, \ldots, M.$$

Show that this implies that $\hat{\theta}_\alpha^{\#}$ converges almost surely to θ_α^* and that

$\sqrt{n}\,(\hat{\theta}_\alpha^{\#} - \theta_{\alpha n}^0)$ is bounded in probability, where $\overset{\bullet}{\theta}_\alpha^{*}$ and $\theta_{\alpha n}^0$ are defined by Assumption 4' using

$$\theta' = (\theta'_1, \theta'_2, \ldots, \theta'_\alpha, \ldots, \theta'_M).$$

Let

$$\sigma_{\alpha\beta}(\theta) = \frac{1}{n}[y_\alpha - f_\alpha(\theta_\alpha)]'[y_\beta - f_\beta(\theta_\beta)].$$

Show that $\sigma_{\alpha\beta}(\hat{\theta}^{\#})$ converges almost surely to $\sigma_{\alpha\beta}(\theta^{*})$ and that $\sqrt{n}\,[\sigma_{\alpha\beta}(\hat{\theta}^{\#}) - \sigma_{\alpha\beta}(\theta_n^0)]$ is bounded in probability. Hint: Use Taylor's theorem to write

$$\sqrt{n}\left[\sigma_{\alpha\beta}(\hat{\theta}^{\#}) - \sigma_{\alpha\beta}(\theta_n^0)\right]$$
$$= \left(\frac{1}{n}\sum_{t=1}^{n}[y_{\alpha t} - f(x_t, \bar{\theta})]\frac{\partial}{\partial\theta'}f_\beta(x_t, \bar{\theta})\right)\sqrt{n}\,(\hat{\theta} - \theta_n^0)$$
$$+ \left(-\frac{1}{n}\sum_{t=1}^{n}[y_{\beta t} - f(x_t, \bar{\theta})]\frac{\partial}{\partial\theta'}f_\alpha(x_t, \bar{\theta})\right)\sqrt{n}\,(\hat{\theta} - \theta_n^0).$$

2. Apply Lemma 2 of Chapter 3 to conclude that one can assume that $\hat{\tau}_n$ is in T without loss of generality.

3. Use $\Sigma^{-1}(\tau) = $ adjoint $\Sigma(\tau)/\det\Sigma(\tau)$ to show that $\Sigma^{-1}(\tau)$ is continuous and differentiable over T.

4. Verify the computations of Notation 8.

5. (Lagrange multiplier test with a correctly centered estimator of scale.) Assume that an estimator $\hat{\Sigma}$ is available with $\sqrt{n}\,(\hat{\Sigma}_n - \Sigma)$ bounded in probability. Apply Theorem 14, with modifications as necessary to reflect the choices $\tilde{V} = \tilde{\Omega}^{-1}$ and $\tilde{\mathscr{J}} = 2\tilde{\Omega}$, to conclude that the Lagrange multiplier test statistic is

$$R' = \frac{1}{n}\left(\sum_{t=1}^{n}[y_t - f(x_t, \tilde{\theta}_n)]'\hat{\Sigma}^{-1}\frac{\partial}{\partial\theta'}f(x_t, \tilde{\theta}_n)\right)$$
$$\times \tilde{\Omega}^{-1}\tilde{H}'(\tilde{H}\tilde{\Omega}^{-1}\tilde{H}')^{-1}\tilde{H}\tilde{\Omega}^{-1}$$
$$\times \left(\sum_{t=1}^{n}[y_t - f(x_t, \tilde{\theta}_n)]'\hat{\Sigma}^{-1}\frac{\partial}{\partial\theta'}f(x_t, \tilde{\theta}_n)\right)'$$

where $\tilde{H} = H(\tilde{\theta}_n)$ with a distribution that can be characterized as

$$R' = Y + o_p(1)$$

where

$$Y = Z'\Omega_n^{*-1}H_n^{*\prime}\left[H_n^*\Omega_n^{*-1}H_n^{*\prime}\right]^{-1}H_n^*\Omega_n^{*-1}Z$$

and

$$Z \sim N_p\left[-\frac{1}{\sqrt{n}}\sum_{t=1}^{n}\left(\frac{\partial}{\partial\theta'}f(x_t, \theta_n^*)\right)'\Sigma^{-1}\left[f(x_t, \theta_n^0) - f(x_t, \theta_n^*)\right], \Omega_n^*\right]$$

with $H_n^* = H(\theta_n^*)$. Show that the random variable Y has the non-central chi-square distribution with q degrees of freedom and non-centrality parameter

$$\alpha = \left(\sum_{t=1}^{n}\left[f(x_t, \theta_n^0) - f(x_t, \theta_n^*)\right]'\Sigma^{-1}\frac{\partial}{\partial\theta'}f(x_t, \theta_n^*)\right)$$

$$\times \Omega_n^{*-1}H_n^{*\prime}\left[H_n^*\Omega_n^{*-1}H_n^{*\prime}\right]^{-1}H_n^*\Omega_n^{*-1}$$

$$\times \frac{\left(\sum_{t=1}^{n}\left[f(x_t, \theta_n^0) - f(x_t, \theta_n^*)\right]'\Sigma^{-1}\frac{\partial}{\partial\theta'}f(x_t, \theta_n^*)\right)'}{2n}$$

Now use the fact that $\tilde{V} = (1/2)\tilde{\mathcal{J}}^{-1}$ to obtain the simpler form

$$R' = \frac{1}{n}\left(\sum_{t=1}^{n}\left[y_t - f(x_t, \tilde{\theta}_n)\right]'\hat{\Sigma}^{-1}\frac{\partial}{\partial\theta'}f(x_t, \tilde{\theta}_n)\right)$$

$$\times \tilde{\Omega}^{-1}\left(\sum_{t=1}^{n}\left[y_t - f(x_t, \tilde{\theta}_n)\right]'\hat{\Sigma}^{-1}\frac{\partial}{\partial\theta'}f(x_t, \tilde{\theta}_n)\right)'.$$

Use the same sort of argument to show that

$$\alpha = \left(\sum_{t=1}^{n}\left[f(x_t, \theta_n^0) - f(x_t, \theta_n^*)\right]'\Sigma^{-1}\frac{\partial}{\partial\theta'}f(x_t, \theta_n^*)\right)$$

$$\times \frac{\Omega_n^{*-1}\left(\sum_{t=1}^{n}\left[f(x_t, \theta_n^0) - f(x_t, \theta_n^*)\right]'\Sigma^{-1}\frac{\partial}{\partial\theta'}f(x_t, \theta_n^*)\right)'}{2n}.$$

Hint: See the remarks and the example which follow the proof of Theorem 14 of Chapter 3.

6. Verify that $\mathcal{J}_n^* = (1/2)\mathcal{I}_n^* + O(1/\sqrt{n})$. Hint: See the example following Theorem 12 of Chapter 3.

7. Show that

$$\left(\frac{\partial}{\partial\theta}s_n^0(\theta_n^*)\right)'\Omega_n^{*-1}H_n^{*\prime}\left(H_n^*\Omega_n^{*-1}H_n^{*\prime}\right)^{-1}H_n^*\Omega_n^{*-1}\left(\frac{\partial}{\partial\theta}s_n^0(\theta_n^*)\right)$$
$$= \left(\frac{\partial}{\partial\theta}s_n^0(\theta_n^*)\right)'\Omega_n^{*-1}\left(\frac{\partial}{\partial\theta}s_n^0(\theta_n^*)\right).$$

8. Suppose that $\hat{\Sigma}_n$ is computed as

$$\hat{\Sigma}_n = \frac{1}{n}\sum_{t=1}^n [y_t - f(x_t, \hat{\theta}_n)][y_t - f(x_t, \hat{\theta}_n)]'$$

and that $\tilde{\Sigma}_n$ is computed as

$$\tilde{\Sigma}_n = \frac{1}{n}\sum_{t=1}^n [y_t - f(x_t, \tilde{\theta}_n)][y_t - f(x_t, \tilde{\theta}_n)]'.$$

Take it as given that both $\hat{\Sigma}_n$ and $\tilde{\Sigma}_n$ converge almost surely to Σ and that both $\sqrt{n}(\hat{\Sigma}_n - \Sigma)$ and $\sqrt{n}(\tilde{\Sigma}_n - \Sigma)$ are bounded in probability. Show that both $S(\hat{\theta}_n, \hat{\Sigma}_n) \equiv M$ and $S(\hat{\theta}_n, \tilde{\Sigma}_n) \equiv M$, so that

$$S(\tilde{\theta}_n, \tilde{\Sigma}_n) - S(\tilde{\theta}_n, \hat{\Sigma}_n) \equiv 0$$

and cannot be asymptotically distributed as a chi-square random variable. However, both

$$L = S(\tilde{\theta}_n, \hat{\Sigma}_n) - M$$

and

$$L = M - S(\hat{\theta}_n, \tilde{\Sigma}_n)$$

are asymptotically distributed as a chi-square random variable by the results of this section.

9. (Lagrange multiplier test with a miscentered estimator of scale.) Suppose that one uses an estimator of scale $\tilde{\Sigma}_n$ with $\sqrt{n}(\tilde{\Sigma}_n - \Sigma_n^*)$ bounded in probability and $\lim_{n\to\infty}\Sigma_n^* = \Sigma$, as in the text. Use the same argument as in Problem 5 to show that the choices $\tilde{V} = \tilde{\Omega}^{-1}$ and $\tilde{\mathscr{J}} = 2\tilde{\Omega}$ allow the Lagrange multiplier test to be written as

$$\tilde{R}' = \frac{1}{n}\left(\sum_{t=1}^n [y_t - f(x_t, \tilde{\theta}_n)]'\tilde{\Sigma}_n^{-1}\frac{\partial}{\partial\theta'}f(x_t, \tilde{\theta}_n)\right)$$
$$\times\tilde{\Omega}_n^{-1}\left(\sum_{t=1}^n [y_t - f(x_t, \tilde{\theta}_n)]'\tilde{\Sigma}_n^{-1}\frac{\partial}{\partial\theta'}f(x_t, \tilde{\theta}_n)\right)'.$$

Show that the distribution of $\tilde{R}'$ can be characterized as $\tilde{R}' = \tilde{Y} +$ $o_p(1)$ with $\tilde{Y}$ as given in the text. Hint: Let $H = H_n^{**}$, $\mathscr{I} = \mathscr{I}_n^{**}$, $\mathscr{J} = (\partial^2/\partial\theta\,\partial\theta')s_n^0(\theta_n^{**}, \Sigma_n^*)$, and $\Omega = \Omega_n^{**}$. Note that $\mathscr{J} = 2\Omega_n^{**} + o(1)$. Use Theorems 12 and 13 of Chapter 3 to show that

$$
\begin{aligned}
R' &= \frac{1}{4}\left(\sqrt{n}\,\frac{\partial}{\partial\theta}s_n(\tilde{\theta}_n)\right)'\tilde{\Omega}_n^{-1}\left(\sqrt{n}\,\frac{\partial}{\partial\theta}s_n(\tilde{\theta}_n)\right) \\
&= \frac{1}{4}\left(\sqrt{n}\,\frac{\partial}{\partial\theta}s_n(\tilde{\theta}_n)\right)'\Omega^{-1}\left(\sqrt{n}\,\frac{\partial}{\partial\theta}s_n(\tilde{\theta}_n)\right) + o_p(1) \\
&= \frac{1}{4}\left(\sqrt{n}\,\frac{\partial}{\partial\theta}s_n(\theta_n^{**})\right)'\mathscr{J}^{-1}H'(H\mathscr{J}^{-1}H')^{-1}H\Omega^{-1} \\
&\quad \times H'(H\mathscr{J}^{-1}H')^{-1}H\mathscr{J}^{-1}\left(\sqrt{n}\,\frac{\partial}{\partial\theta}s_n(\theta^{**})\right) + o_p(1) \\
&= \frac{1}{4}\left(\sqrt{n}\,\frac{\partial}{\partial\theta}s_n(\theta_n^{**})\right)'\Omega^{-1}H'(H\Omega^{-1}H')^{-1} \\
&\quad \times H\Omega^{-1}\left(\sqrt{n}\,\frac{\partial}{\partial\theta}s_n(\theta^{**})\right) + o_p(1) \\
&= \left(\frac{X}{2}\right)'\Omega^{-1}H'(H\Omega^{-1}H')^{-1}H\Omega^{-1}\left(\frac{X}{2}\right) + o_p(1)
\end{aligned}
$$

where

$$
\frac{X}{2} \sim N_p\left(\frac{\sqrt{n}\,(\partial/\partial\theta)s_n^0(\theta_n^{**})}{2}, \frac{\mathscr{I}_n^{**}}{4}\right).
$$

10. (Computation of the value Σ_n^* that centers $\tilde{\Sigma}_n$.) Assume that $h(\theta) = 0$ can be written equivalently as $\theta = g(\rho)$ for some ρ. Use Theorem 5 of Chapter 3 to show that if one computes $\hat{\rho}_n$ to minimize

$$
s_n(\rho) = \frac{1}{n}\sum_{t=1}^{n}\{y_t - f[x_t, g(\rho)]\}'\{y_t - f[x_t, g(\rho)]\}
$$

then the appropriate centering value is computed by finding ρ_n^0 to minimize

$$
\begin{aligned}
s_n^0(\rho) &= \frac{1}{n}\int_{\mathscr{E}}\{e + f(x_t, \theta_n^0) - f[x_t, g(\rho)]\}' \\
&\quad \times \{e + f(x_t, \theta_n^0) - f[x_t, g(\rho)]\}\,dP(e) \\
&= \operatorname{tr}\Sigma + \frac{1}{n}\sum_{t=1}^{n}\{f(x_t, \theta_n^0) - f[x_t, g(\rho)]\}' \\
&\quad \times \{f(x_t, \theta_n^0) - f[x_t, g(\rho)]\}.
\end{aligned}
$$

Now

$$\tilde{\Sigma}_n = \frac{1}{n} \sum_{t=1}^{n} \{ y_t - f[x_t, g(\hat{\rho}_n)] \} \{ y_t - f[x_t, g(\hat{\rho}_n)] \}'$$

is the solution of the following minimization problem (Problem 11):

minimize

$$s_n(V, \hat{\rho}_n) = \frac{1}{n} \sum_{t=1}^{n} \ln \det(V)$$
$$+ \{ y_t - f[x_t, g(\hat{\rho}_n)] \}' V^{-1} \{ y_t - f[x_t, g(\hat{\rho}_n)] \}$$

subject to

$$V \text{ positive definite, symmetric.}$$

Use Theorem 5 of Chapter 3 to show that the value Σ_n^* that centers $\tilde{\Sigma}_n$ is computed as the solution of the problem

minimize

$$s_n^0(V, \rho_n^0)$$

subject to

$$V \text{ positive definite, symmetric}$$

where

$$s_n^0(V, \rho) = \ln \det(V)$$
$$+ \frac{1}{n} \sum_{t=1}^{n} \int_{\mathscr{E}} \{ e + f(x_t, \theta_n^0) - f[x_t, g(\rho)] \}' V^{-1}$$
$$\times \{ e + f(x_t, \theta_n^0) - f[x_t, g(\rho)] \} \, dP(e)$$
$$= \ln \det(V) + \operatorname{tr}(V^{-1}\Sigma)$$
$$+ \frac{1}{n} \sum_{t=1}^{n} \{ f(x_t, \theta_n^0) - f[x_t, g(\rho)] \}' V^{-1}$$
$$\times \{ f(x_t, \theta_n^0) - f[x_t, g(\rho)] \}.$$

The solution of this minimization problem is (Problem 11)

$$\Sigma_n^0 = \Sigma + \frac{1}{n} \sum_{t=1}^{n} \{ f(x_t, \theta_n^0) - f[x_t, g(\rho_n^0)] \}$$
$$\times \{ f(x_t, \theta_n^0) - f[x_t, g(\rho_n^0)] \}'.$$

11. Let

$$f(V) = \ln \det V + \text{tr}(V^{-1}A)$$

where A is an M by M positive definite symmetric matrix. Show that the minimum of $f(V)$ over the (open) set of all positive definite matrices V is attained at the value $V = A$. Hint:

$$f(V) - f(A) = -\ln \det(V^{-1}A) + \text{tr}(V^{-1}A) - M.$$

Let λ_i be the eigenvalues of $V^{-1}A$. Then

$$f(V) - f(A) = \sum_{i=1}^{M} (\lambda_i - \ln \lambda_i - 1).$$

Since the line $y = x$ plots above the line $y = \ln x + 1$, one has

$$f(V) - f(A) > 0 \qquad \text{if any} \quad \lambda_i \neq 1$$
$$f(V) - f(A) = 0 \qquad \text{if all} \quad \lambda_i = 1.$$

12. (Efficiency of least squares estimators). Define $\hat{\theta}^{\#}$ as the minimizer of $s_n(\theta, \hat{V}_n)$ where $\sqrt{n}\,(\hat{V}_n - V)$ is bounded in probability, $\lim_{n \to \infty} \hat{V}_n = V$ almost surely, and V is positive definite. Show that under Assumptions 1' through 7'

$$\sqrt{n}\left(\hat{\theta}^{\#} - \theta_n^0\right) \xrightarrow{\mathcal{L}} N_p\left(0, \mathcal{J}_V^{-1}\mathcal{I}_V\mathcal{J}_V^{-1}\right)$$

with

$$\mathcal{J}_V = 2\int_{\mathcal{X}}\left(\frac{\partial}{\partial\theta'}f(x,\theta*)\right)'V^{-1}\left(\frac{\partial}{\partial\theta'}f(x,\theta*)\right)d\mu(x)$$

$$\mathcal{I}_V = 4\int_{\mathcal{X}}\left(\frac{\partial}{\partial\theta'}f(x,\theta*)\right)'V^{-1}\Sigma V^{-1}\left(\frac{\partial}{\partial\theta'}f(x,\theta*)\right)d\mu(x).$$

Show that $a'\mathcal{J}_V^{-1}\mathcal{I}_V\mathcal{J}_V^{-1}a$ is minimized when $V = \Sigma$. Note that the equation by equation estimator has $V = I$.

7. AN ILLUSTRATION OF THE BIAS IN INFERENCE CAUSED BY MISSPECIFICATION

The asymptotic theory in Chapter 3 was developed in sufficient generality to permit the analysis of inference procedures under conditions where the data generating model and the model fitted to the data are not the same.

The following example is an instance where a second order polynomial approximation can lead to considerable error. The underlying ideas are similar to those of Example 1.

EXAMPLE 3 (Power curve of a translog test of additivity). The theory of demand states that of the bundles of goods and services that the consumer can afford, he or she will choose that bundle which pleases him or her the most. Mathematically this proposition is stated as follows: Let there be N different goods and services in a bundle, let $q = (q_1, q_2, \ldots, q_N)'$ be an N-vector giving the quantities of each in a bundle, let $p = (p_1, p_2, \ldots, p_N)'$ be the N-vector of corresponding prices, let $u^*(q)$ be the pleasure or utility derived from the bundle q, and let Y be the consumer's total income. The consumer's problem is

$$\text{maximize} \quad u^*(q) \quad \text{subject to} \quad p'q \le Y.$$

The solution has the form $q(x)$, where $x = p/Y$. If one sets

$$g^*(x) = u^*[q(x)]$$

then the demand system $q(x)$ can be recovered by differentiation:

$$q(x) = \left(x' \frac{\partial}{\partial x} g^*(x) \right)^{-1} \frac{\partial}{\partial x} g^*(x).$$

The recovery formula is called Roy's identity, and the function $g^*(x)$ is called the consumer's indirect utility function. See Varian (1978) for regularity conditions and details.

These ideas may be adapted to empirical work by setting forth an indirect utility function $g(x, \lambda)$ which is thought to approximate $g^*(x)$ adequately over a region of interest $\mathscr{X}$. Then Roy's identity is applied to obtain an approximating demand system. Usually one fits to consumer expenditure share data $q_i p_i/Y$, $i = 1, 2, \ldots, N$, although, as we have seen from Example 1, fitting $\ln(q_i p_i/Y) - \ln(q_N p_N/Y)$ to $\ln[x_i q(x)/x_N q(x)]$ is preferable. The result is the expenditure system

$$\frac{q_i p_i}{Y} = f_i(x, \lambda) + e_i \qquad i = 1, 2, \ldots, N - 1$$

with

$$f_i(x, \lambda) = \left(x' \frac{\partial}{\partial x} g(x, \lambda) \right)^{-1} x_i \frac{\partial}{\partial x_i} g(x, \lambda).$$

The index i ranges to $N - 1$ rather than N because expenditure shares sum to one for each consumer and the last may be obtained by subtracting the rest. Converting to a vector notation, write

$$y = f(x, \lambda) + e$$

where y, $f(x, \lambda)$, and e are $(N - 1)$-vectors. Measurements on n consumers yield the regression equations

$$y_t = f(x_t, \lambda) + e_t \qquad t = 1, 2, \ldots, n.$$

Multivariate nonlinear least squares is often used to fit the data, whence, referring to Notation 1, Chapter 3, the sample objective function is

$$s_n(\lambda) = \frac{1}{n} \sum_{t=1}^{n} \tfrac{1}{2} [y_t - f(x_t, \lambda)]'(\hat{S}_n)^{-1}[y_t - f(x_t, \lambda)]$$

and

$$s(y, x, S, \lambda) = \tfrac{1}{2}[y - f(x, \lambda)]'S^{-1}[y - f(x, \lambda)]$$

where $\hat{S}$ is a preliminary estimator of $\mathscr{C}(e, e')$.

Suppose that the consumer's true indirect utility function is additive:

$$g^*(x) = \sum_{i=1}^{N} g_i^*(x_i).$$

Christensen, Jorgenson, and Lau (1975) have proposed that this supposition be tested by using a translog indirect utility function

$$g(x, \lambda) = \sum_{i=1}^{N} \alpha_i \ln x_i + \sum_{i=1}^{N} \sum_{j=1}^{N} \beta_{ij} \ln x_i \ln x_j$$

to obtain the approximating expenditure system

$$f_i(x, \lambda) = \frac{\alpha_i + \sum_{j=1}^{N} \beta_{ij} \ln x_j}{-1 + \sum_{j=1}^{N} \beta_{Mj} \ln x_j}$$

with

$$\lambda = (\alpha_1, \alpha_2, \ldots, \alpha_{N-1}, \beta_{11}, \beta_{12}, \beta_{22}, \beta_{13}, \beta_{23}, \beta_{33}, \ldots, \beta_{1N}, \beta_{2N}, \ldots, \beta_{NN})'$$

and

$$\alpha_N = -1 - \sum_{j=1}^{N-1} \alpha_j \qquad \beta_{ji} = \beta_{ij} \text{ for } i < j \qquad \beta_{Mj} = \sum_{i=1}^{N} \beta_{ij}$$

then testing

$$H: \beta_{ij} = 0 \text{ for all } i \neq j \quad \text{against} \quad A: \beta_{ij} \neq 0 \text{ for some } i \neq j.$$

This is a linear hypothesis of the form

$$h(\lambda) = H\lambda = 0$$

with

$$W = n\hat{\lambda}'_n H'(H\hat{V}H')^{-1} H\hat{\lambda}_n$$

as a possible test statistic where $\hat{\lambda}_n$ minimizes $s_n(\lambda)$ and $\hat{V}$ is as defined in Section 5, Chapter 3.

The validity of this inference depends on whether a quadratic in logarithms is an adequate approximation to the consumer's true indirect utility function. For plausible alternative specifications of $g^*(x)$, it should be true that

$$P(W > c) \doteq \alpha \qquad \text{if } g^* \text{ is additive}$$

$$P(W > c) > \alpha \qquad \text{if } g^* \text{ is not additive}$$

if the translog specification is to be accepted as adequate. In this section we shall obtain an asymptotic approximation to $P(W > c)$ in order to shed some light on the quality of the approximation.

For an appropriately chosen sequence of N-vectors k_α, $\alpha = 1, 2, 3, \ldots$, the consumer's indirect utility function must be of the Fourier form

$$g(x, \gamma) = u_0 + b'x + \tfrac{1}{2}x'Cx$$

$$+ \sum_{\alpha=1}^{\infty} \left\{ u_{0\alpha} + \sum_{j=1}^{\infty} \left[u_{j\alpha}\cos(jk'_\alpha x) - v_{j\alpha}\sin(jk'_\alpha x) \right] \right\}$$

where γ is a vector of infinite length whose entries are b and some triangular arrangement of the $u_{j\alpha}$ and $v_{j\alpha}$; $C = -\sum_{\alpha=1}^{\infty} u_{0\alpha} k_\alpha k'_\alpha$. In consequence, the consumer's expenditure system $f(x, \gamma)$ is that which results by applying Roy's identity to $g(x, \gamma)$. The indirect utility function is additive if and only if the elementary N-vectors are the only vectors k_α which enter $g(x, \gamma)$ with nonzero coefficients—that is, if and only if

$$g(x, \gamma) = u_0 + b'x - \frac{1}{2} \sum_{\alpha=1}^{N} u_{0\alpha} x_\alpha^2$$

$$+ \sum_{\alpha=1}^{N} \left(u_{0\alpha} + \sum_{j=1}^{\infty} \left[u_{j\alpha} \cos(jx_\alpha) - v_{j\alpha} \sin(jx_\alpha) \right] \right).$$

See Gallant (1981) for regularity conditions and details.

The situation is, then, as follows. The data are generated according to

$$y_t = f(x_t, \gamma^0) + e_t \qquad t = 1, 2, \ldots, n.$$

The fitted model is

$$y_t = f(x_t, \lambda) + e_t \qquad t = 1, 2, \ldots, n$$

with λ estimated by $\hat{\lambda}$ minimizing

$$s_n(\lambda) = \frac{1}{n} \sum_{t=1}^{n} s(y_t, x_t, \hat{S}_n, \lambda)$$

where

$$s(y, x, S, \lambda) = \frac{1}{2} [y - f(x, \lambda)]' S^{-1} [y - f(x, \lambda)].$$

The probability $P(W > c)$ is to be approximated for plausible settings of the parameter γ^0 where

$$W = n \hat{\lambda}' H' (H \hat{V} H')^{-1} H \hat{\lambda}.$$

For simplicity, we shall compute the power assuming that (y_t, x_t) are independently and identically distributed. Thus, $\mathscr{U}^*$ and $\mathscr{U}^0$ of Notations 2 and 3, Chapter 3, are zero, and the asymptotic distribution of W is the noncentral chi-square. We assume that $\mathscr{E}(e) = 0$, $\mathscr{C}(ee') = \Sigma$, that $\hat{S}_n$ converges almost surely to Σ, and that $\sqrt{n}(\hat{S}_n - \Sigma)$ is bounded in probability. Direct computation using Notations 1 and 3 of Chapter 3 yields

$$\lambda^0 \quad \text{minimizes} \quad s_n^0(\lambda) = \frac{1}{n} \sum_{t=1}^{n} \tfrac{1}{2}\delta'(x_t, \lambda, \gamma^0)\Sigma^{-1}\delta(x_t, \lambda, \gamma^0) + \frac{N-1}{2}$$

$$\mathscr{J}^0 = \frac{1}{n} \sum_{t=1}^{n} \left(\frac{\partial}{\partial\lambda'}f(x_t, \lambda^0)\right)'\Sigma^{-1}$$

$$\times \left[\Sigma + \delta(x_t, \lambda^0, \gamma^0)\delta'(x_t, \lambda^0, \gamma^0)\right]\Sigma^{-1}$$

$$\times \left(\frac{\partial}{\partial\lambda'}f(x_t, \lambda^0)\right)$$

$$\mathscr{J}^0 = \frac{1}{n} \sum_{t=1}^{n} \left(\frac{\partial}{\partial\lambda'}f(x_t, \lambda^0)\right)'\Sigma^{-1}\left(\frac{\partial}{\partial\lambda'}f(x_t, \lambda^0)\right)$$

$$- \frac{1}{n} \sum_{t=1}^{n} \sum_{i=1}^{N-1} \sum_{j=1}^{N-1} \delta_i(x_t, \lambda^0, \gamma^0)\Sigma^{ij}\frac{\partial^2}{\partial\lambda\,\partial\lambda'}f_j(x_t, \lambda^0)$$

$$\alpha^0 = \frac{n\lambda^{0\prime}H'(HV^0H')^{-1}H\lambda^0}{2}$$

$$V^0 = (\mathscr{J}^0)^{-1}\mathscr{J}^0(\mathscr{J}^0)^{-1}$$

where

$$\delta(x_t, \lambda^0, \gamma) = f(x_t, \gamma^0) - f(x_t, \lambda^0)$$

and Σ^{ij} denotes the elements of Σ^{-1}.

Values of γ^0 were chosen as follows. The parameter γ^0 was truncated to vector of finite length by using only the multiindices

$$k_\alpha = \begin{pmatrix}1\\0\\0\end{pmatrix}, \begin{pmatrix}0\\1\\0\end{pmatrix}, \begin{pmatrix}0\\0\\1\end{pmatrix}, \begin{pmatrix}1\\0\\1\end{pmatrix}, \begin{pmatrix}1\\1\\0\end{pmatrix}, \begin{pmatrix}0\\1\\1\end{pmatrix}, \begin{pmatrix}1\\1\\1\end{pmatrix}$$

and discarding the rest of the infinite sequence $\{k_\alpha\}_{\alpha=1}^{\infty}$. Let K denote the root sum of squares of the parameters of $g(x, \gamma)$ which are not associated with $k_\alpha = k_1, k_2,$ or k_3. For specified values of K, the parameters γ^0 were obtained by fitting $f(x, \gamma^0)$, subject to specified K, to the data used by Christensen, Jorgenson, and Lau (1975), which are shown in Table 6. This provides a sequence of indirect utility functions $g(x, \gamma^0)$ which increase in the degree of departure from additivity. When $K = 0$, $g(x, \gamma^0)$ is additive, and when K is unconstrained, the parameter γ^0 is free to adjust to the data as best it can.

Table 6. Data of Christensen, Jorgenson, and Lau (1975).

Year	Durables Quantity	Durables Price	Non-durables Quantity	Non-durables Price	Services Quantity	Services Price
1929	28.9645	33.9	98.1	38.4	96.1	31.6
1930	29.8164	32.2	93.5	36.4	89.5	32.1
1931	28.9645	31.4	93.1	31.1	84.3	30.9
1932	26.8821	23.9	85.9	26.5	77.1	28.8
1933	25.3676	31.3	82.9	26.8	76.8	26.1
1934	24.6104	27.7	88.5	30.2	76.3	26.8
1935	22.3387	28.8	93.2	31.5	79.5	26.8
1936	24.1371	32.9	103.8	31.6	83.8	27.2
1937	24.1371	29.0	107.7	32.7	86.5	28.3
1938	26.6928	28.4	109.3	31.1	83.7	29.1
1939	26.4088	30.5	115.1	30.5	86.1	29.2
1940	27.0714	29.4	119.9	30.9	88.7	29.5
1941	28.4912	28.9	127.6	33.6	91.8	30.8
1942	29.5325	31.7	129.9	39.1	95.5	32.4
1943	28.6806	38.0	134.0	43.7	100.1	34.2
1944	28.8699	37.7	139.4	46.2	102.7	36.1
1945	28.3966	39.0	150.3	47.8	106.3	37.3
1946	26.6928	44.0	158.9	52.1	116.7	38.9
1947	28.3966	65.3	154.8	58.7	120.8	41.7
1948	31.6149	60.4	155.0	62.3	124.6	44.4
1949	35.8744	50.4	157.4	60.3	126.4	46.1
1950	38.9980	59.2	161.8	60.7	132.8	47.4
1951	43.5414	60.0	165.3	65.8	137.1	49.9
1952	48.0849	64.2	171.2	66.6	140.8	52.6
1953	49.8833	57.5	175.7	66.3	145.5	55.4
1954	53.1016	68.3	177.0	66.6	150.4	57.2
1955	55.4680	63.5	185.4	66.3	157.5	58.5
1956	58.8756	62.2	191.5	67.3	164.8	60.2
1957	61.6206	56.5	194.8	69.4	170.3	62.2
1958	65.3122	66.7	196.8	71.0	175.8	64.2
1959	65.7854	63.3	205.0	71.4	184.7	66.0
1960	68.6251	73.1	208.2	72.6	192.3	68.0
1961	70.6129	72.1	211.9	73.3	200.0	69.1
1962	71.5594	72.4	218.5	73.9	208.7	70.4
1963	73.5472	72.5	223.0	74.9	217.6	71.7
1964	77.2387	76.3	233.3	75.8	229.7	72.8
1965	81.9715	82.3	244.0	77.3	240.7	74.3
1966	87.4615	84.3	255.5	80.1	251.6	76.5
1967	93.8981	81.0	259.5	81.9	264.0	78.8
1968	99.5774	81.0	270.2	85.3	275.0	82.0
1969	106.7710	94.4	276.4	89.4	287.2	86.1
1970	109.1380	85.0	282.7	93.6	297.3	90.5
1971	115.2900	88.5	287.5	96.6	306.3	95.8
1972	122.2000	100.0	299.3	100.0	322.4	100.0

Source: Gallant (1981).

403

404   MULTIVARIATE NONLINEAR REGRESSION

Table 7.   Tests for an Additive Indirect Utility Function.

	Fourier		Translog	
K	Noncentrality	Power	Noncentrality	Power
0.0	0.0	0.010	8.9439	0.872
0.00046	0.0011935	0.010	8.9919	0.874
0.0021	0.029616	0.011	9.2014	0.884
0.0091	0.63795	0.023	10.287	0.924
0.033	4.6689	0.260	14.268	0.987
0.059	7.8947	0.552	15.710	0.993
0.084	82.875	1.000	13.875	0.984
Unconstrained	328.61	1.000	10.230	0.922

Source: Gallant (1981).

The asymptotic approximation to $P(W \geq c)$ with c chosen to give a nominal .01 level test are shown in Table 7. For comparison, the power curve for W computed from the correct model—the Fourier expenditure system—is included in the table.

We see from Table 7 that the translog of explicit additivity is seriously flawed. The actual size of the test is much larger than the nominal significance level of .01, and the power curve is relatively flat. The power does increase near the null hypothesis, as one might expect, but it falls off again as departures from additivity become more extreme.   □

CHAPTER 6

Nonlinear Simultaneous Equations Models

In this chapter, we shall consider nonlinear, simultaneous equations models. These are multivariate models which cannot be written with the dependent variables equal to a vector valued function of the explanatory variables plus an additive error, because it is either impossible or unnatural to do so; in short, the model is expressed in an implicit form $e = q(y, x, \theta)$ where e and y are vector valued. This is as much generality as is needed in applications. The model $q(e, y, x, \theta) = 0$ offers no more generality, since the sort of regularity conditions that permit an asymptotic theory of inference also permit application of the implicit function theorem so that the form $e = q(y, x, \theta)$ must exist; the application where it cannot actually be produced is rare. In this rare instance, one substitutes numerical methods for the computation of e and its derivatives in the formulas that we shall derive.

There are two basic sets of statistical methods customarily employed with these models, those based on a method of moments approach with instrumental variables used to form the moment equations and those based on a maximum likelihood approach with some specific distribution specified for e. We shall discuss both approaches.

Frequently, these models are applied in situations where time indexes the observations and the vector of explanatory variables x_t has lagged values y_{t-1}, y_{t-2}, etc. of the dependent variable y_t as elements: models with a dynamic structure. In these situations, statistical methods cannot be derived from the asymptotic theory set forth in Chapter 3. But it is fairly easy to see intuitively, working by analogy with the statistical methods developed thus far from the asymptotic theory of Chapter 3, what the correct statistical procedures ought to be in dynamic models. Accordingly, we shall lay down

405

this intuitive base and develop the statistical methods for dynamic models in this chapter, deferring consideration of an asymptotic theory that will justify them to the next.

1. INTRODUCTION

In this chapter, the multivariate nonlinear regression model (Chapter 5) will be generalized in two ways.

First, we shall not insist that the model be written in explicit form where the dependent variables $y_{\alpha t}$ are solved out in terms of the independent variables x_t, the parameters θ_α, and the errors $e_{\alpha t}$. Rather, the model may be expressed in implicit form

$$q_\alpha\left(y_t, x_t, \theta_\alpha^0\right) = e_{\alpha t} \qquad t = 1, 2, \ldots, n \quad \alpha = 1, 2, \ldots, M$$

where each $q_\alpha(y, x, \theta_\alpha)$ is a real valued function, y_t is an L-vector, x_t is a k-vector, each θ_α^0 is a p_α-dimensional vector of unknown parameters, and the $e_{\alpha t}$ represent unobservable observational or experimental errors. Note specifically that the number of equations (M) is not of necessity equal to the number of dependent variables (L), although in many applications this will be the case.

Secondly, the model can be dynamic, which is to say that t indexes observations that are ordered in time and that the vector of independent variables x_t can include lagged values of the dependent variable (y_{t-1}, y_{t-2}, etc.) as elements. There is nothing in the theory (Chapter 7) that would preclude consideration of models of the form

$$q_{\alpha t}\left(y_t, \ldots, y_0, x_t, \ldots, x_0, \theta_\alpha^0\right) = e_{\alpha t} \qquad t = 1, 2, \ldots, n \quad \alpha = 1, 2, \ldots, M$$

or similar schemes where the number of arguments of $q_{\alpha t}(\cdot)$ depends on t, but they do not seem to arise in applications, so we shall not consider them. If such a model is encountered, simply replace $q_\alpha(y_t, x_t, \theta_\alpha)$ by $q_{\alpha t}(y_t, \ldots, y_0, x_t, \ldots, x_0, \theta_\alpha)$ at every occurrence in Section 3 and thereafter. Dynamic models frequently will have serially correlated errors $[\mathscr{C}(e_{\alpha s}, e_{\beta t}) \neq 0$ for $s \neq t]$, and this fact will need to be taken into account in the analysis.

Two examples follow. The first has the classical regression structure (no lagged dependent variables or serially correlated errors); the second is dynamic.

EXAMPLE 1 (Consumer demand). This is a reformulation of Example 1 of Chapter 5. In Chapter 5, the analysis was conditional on prices and observed electricity expenditure, whereas in theory it is preferable to condition on prices, income, and consumer demographic characteristics; that is, it is preferable to condition on the data in Table 1b and Table 1c of Chapter 5 rather than condition on Table 1b alone. In practice it is not clear that this is the case, because the data of Tables 1a and 1b are of much higher quality than the data of Table 1c; there are several obvious errors in Table 1c, such as a household with a dryer and no washer or a freezer and no refrigerator. Thus, it is not clear that we are not merely trading an errors in variables problem that arises from theory for a worse one that arises in practice.

To obtain the reformulated model, the data of Tables 1a, 1b, and 1c of Chapter 5 are transformed as follows:

$$y_1 = \ln(\text{peak expenditure share}) - \ln(\text{base expenditure share})$$

$$y_2 = \ln(\text{intermediate expenditure share}) - \ln(\text{base expenditure share})$$

$$y_3 = \ln(\text{expenditure})$$

$$r_1 = \ln(\text{peak price})$$

$$r_2 = \ln(\text{intermediate price})$$

$$r_3 = \ln(\text{base price})$$

$$d_0 = 1$$

$$d_1 = \ln \frac{10 \times \text{peak price} + 6 \times \text{intermediate price} + 8 \times \text{base price}}{24}$$

$$d_2 = \ln(\text{income})$$

$$d_3 = \ln(\text{residence size in sq ft})$$

$$d_4 = \begin{cases} 1 & \text{if the residence is a duplex or apartment} \\ 0 & \text{otherwise} \end{cases}$$

$$d_5 = \begin{cases} 1 & \text{if the residence is a mobile home} \\ 0 & \text{otherwise} \end{cases}$$

$$d_6 = \begin{cases} \ln(\text{heat loss in Btuh}) & \text{if the residence has} \\ & \text{central air conditioning} \\ 0 & \text{otherwise} \end{cases}$$

$$d_7 = \begin{cases} \ln(\text{window air Btuh}) & \text{if the residence has} \\ & \text{window air conditioning} \\ 0 & \text{otherwise} \end{cases}$$

$$
d_8 = \begin{cases} \ln(\text{number of household members} + 1) \\ \quad \text{if the residence has an} \\ \quad \text{electric water heater} \\ 0 \quad \text{otherwise} \end{cases}
$$

$$
d_9 = \begin{cases} 1 \quad \text{if the residence has both an} \\ \quad \text{electric water heater and a washing machine} \\ 0 \quad \text{otherwise} \end{cases}
$$

$$
d_{10} = \begin{cases} \ln(\text{number of household members} + 1) \\ \quad \text{if residence has an electric dryer} \\ 0 \quad \text{otherwise} \end{cases}
$$

$$
d_{11} = \begin{cases} \ln(\text{refrigerator kW}) \quad \text{if the residence has a refrigerator} \\ 0 \quad\quad\quad\quad\quad\quad \text{otherwise} \end{cases}
$$

$$
d_{12} = \begin{cases} \ln(\text{freezer kW}) \quad \text{if the residence has a freezer} \\ 0 \quad\quad\quad\quad\quad \text{otherwise} \end{cases}
$$

$$
d_{13} = \begin{cases} 1 \quad \text{if the residence has an electric range} \\ 0 \quad \text{otherwise.} \end{cases}
$$

As notation, set

$$
y = \begin{pmatrix} y_1 \\ y_2 \\ y_3 \end{pmatrix} \quad r = \begin{pmatrix} r_1 \\ r_2 \\ r_3 \end{pmatrix} \quad d = \begin{pmatrix} d_0 \\ d_1 \\ \vdots \\ d_{13} \end{pmatrix} \quad x = \begin{pmatrix} r \\ d \end{pmatrix}
$$

$$
y_t = \begin{pmatrix} y_{1t} \\ y_{2t} \\ y_{3t} \end{pmatrix} \quad r_t = \begin{pmatrix} r_{1t} \\ r_{2t} \\ r_{3t} \end{pmatrix} \quad d_t = \begin{pmatrix} d_{0t} \\ d_{1t} \\ \vdots \\ d_{13,t} \end{pmatrix} \quad x_t = \begin{pmatrix} r_t \\ d_t \end{pmatrix}.
$$

These data are presumed to follow the model

$$
y_{1t} = \ln \frac{a_1 + r_t' b_{(1)} - y_{3t} 1' b_{(1)}}{a_3 + r_t' b_{(3)} - y_{3t} 1' b_{(3)}} + e_{1t}
$$

$$
y_{2t} = \ln \frac{a_2 + r_t' b_{(2)} - y_{3t} 1' b_{(2)}}{a_3 + r_t' b_{(3)} - y_{3t} 1' b_{(3)}} + e_{2t}
$$

$$
y_{3t} = d_t' c + e_{3t}
$$

where 1 denotes a vector of ones and

$$a = \begin{pmatrix} a_1 \\ a_2 \\ a_3 \end{pmatrix}, \qquad B = \begin{pmatrix} b_{11} & b_{12} & b_{13} \\ b_{21} & b_{22} & b_{23} \\ b_{31} & b_{32} & b_{33} \end{pmatrix} = \begin{pmatrix} b'_{(1)} \\ b'_{(2)} \\ b'_{(3)} \end{pmatrix}, \qquad c = \begin{pmatrix} c_0 \\ c_1 \\ \vdots \\ c_{13} \end{pmatrix}.$$

The matrix B is symmetric and $a_3 = -1$. With these conventions, the nonredundant parameters are

$$a_1, b_{11}, b_{12}, b_{13}, a_2, b_{22}, b_{23}, b_{33}, c_0, c_1, \ldots, c_{13}.$$

The errors

$$e_t = \begin{pmatrix} e_{1t} \\ e_{2t} \\ e_{3t} \end{pmatrix}$$

are taken to be independently and identically distributed each with zero mean.

The theory supporting this model was given in Chapter 5. The functional form of the third equation

$$y_{3t} = d'_t c + e_{3t}$$

and the variables entering into the equation were determined empirically from a data set of which Table 1 of Chapter 5 is a small subset; see Gallant and Koenker (1984).   □

EXAMPLE 2 (Intertemporal consumption and investment). The data shown in Table 1a below are transformed as follows:

$$y_t = \frac{(\text{consumption at time } t)/(\text{population at time } t)}{(\text{consumption at time } t - 1)/(\text{population at time } t - 1)}$$

$$x_t = (1 + \text{stock returns at time } t)\frac{\text{deflator at time } t - 1}{\text{deflator at time } t}.$$

These data are presumed to follow the model

$$\beta^0 (y_t)^{\alpha^0} x_t - 1 = e_t \qquad t = 1, 2, \ldots, 239.$$

Put

$$z_t = (1, y_{t-1}, x_{t-1})'.$$

What should be true of these data in theory is that

$$\mathscr{E}(e_t \otimes z_t) = 0 \qquad t = 2, 3, \ldots, 239$$

$$\mathscr{E}(e_t \otimes z_t)(e_t \otimes z_t)' = \Sigma \qquad t = 2, 3, \ldots, 239$$

$$\mathscr{E}(e_t \otimes z_t)(e_s \otimes z_s)' = 0 \qquad t \neq s.$$

Even though e_t is a scalar, we use the Kronecker product notation to keep the notation consistent with the later sections.

The theory supporting this model specification follows; the reader who has no interest in the theory can skip over the rest of the example.

The consumer's problem is to allocate consumption and investment over time, given a stream $\bar{w}_0, \bar{w}_1, \bar{w}_2, \ldots$ of incomes; $\bar{w}_t$ is the income that the consumer receives at time t. We suppose that the various consumption bundles available at time t can be mapped into a scalar quantity index c_t and that the consumer ranks various consumption streams $c_0, c_1, c_2, \ldots$ according to the utility indicator

$$U(\{c_t\}) = \sum_{t=0}^{\infty} \beta^t u(c_t)$$

where β is a discount factor, $0 < \beta < 1$, and $u(c)$ is a strictly concave increasing function. We suppose that there is a corresponding price index $\bar{p}_{0t}$, so that expenditure on consumption in each period can be computed according to $\bar{p}_{0t} c_t$. Above, we took c_t to be an index of consumption per capita on nondurables plus services and $\bar{p}_{0t}$ to be the corresponding implicit price deflator.

Further, suppose that the consumer has the choice of investing in a collection of N assets with maturities m_j, $j = 1, 2, \ldots, N$; asset j bought at time t cannot be sold until time $t + m_j$, or equivalently, asset j bought at time $t - m_j$ cannot be sold until time t. Let q_{jt} denote the quantity of asset j held at time t, $\bar{p}_{jt}$ the price per unit of that asset at time t; and let $\bar{r}_{jt}$ denote the payoff at time t of asset j bought at time $t - m_j$. If, for example, the jth asset is a default free, zero coupon bond with term to maturity m_j then $\bar{r}_{j, t+m_j}$ is the par value of the bond at time $t + m_j$; if the jth asset is a common stock, then by definition $m_j = 1$ and $\bar{r}_{jt} = \bar{p}_{jt} + \bar{d}_{jt}$, where $\bar{d}_{jt}$ is the dividend per share of the stock paid at time t, if any. Above, we took the first asset to be NYSE stocks weighted by value.

In Tables 1a and 1b, t is interpreted as the instant of time at the end of the month in which recorded. Nondurables and services, population, and the implicit deflator are assumed to be measured at the end of the month in which recorded, and assets are assumed to be purchased at the beginning of the month in which recorded. Thus, for a given row, nondurables and services divided by population is interpreted as c_t, the implicit price deflator as $\bar{p}_{0t}$, the return on stock j as $(\bar{p}_{jt} + \bar{d}_{jt} - \bar{p}_{j,t-1})/\bar{p}_{j,t-1}$, and the return on bill j as $(\bar{p}_{j,t+m_j-1} - \bar{p}_{j,t-1})/\bar{p}_{j,t-1}$.

As an example, putting $r_{jt} = \bar{r}_{jt}/\bar{p}_{0t}$ and $p_{jt} = \bar{p}_{jt}/\bar{p}_{0t}$, if a three month bill is bought February 1 and sold April 30, its return is recorded in the row for February. If t refers to midnight April 30, then the value of t is recorded in the row for April and

$$\frac{c_t}{c_{t-m_j}} = \frac{(\text{April nondurables and services})/(\text{April population})}{(\text{January nondurables and services})/(\text{January population})}$$

$$\frac{r_{jt}}{p_{t-m_j}} = (\text{February return} + 1)\frac{\text{January implicit deflator}}{\text{April implicit deflator}}.$$

As another, if a one month bill is bought April 1 and sold April 30, it is recorded in the row for April. If t refers to midnight April 30, then

$$\frac{c_t}{c_{t-m_j}} = \frac{(\text{April nondurables and services})/(\text{April population})}{(\text{March nondurables and services})/(\text{March population})}$$

$$\frac{r_{jt}}{p_{t-m_j}} = (\text{April return} + 1)\frac{\text{March implicit deflator}}{\text{April implicit deflator}}.$$

With these assumptions, the feasible consumption and investment plans must satisfy the sequence of budget constraints

$$\bar{p}_{0t}c_t + \sum_{t=1}^{N} \bar{p}_{jt}q_{jt} \leq \bar{w}_t + \sum_{t=1}^{N} r_{jt}\bar{q}_{j,t-m_j}$$

$$0 \leq q_{jt}.$$

The consumer seeks to maximize utility, so the budget constraint is effectively

$$c_t = w_t + \sum_{t=1}^{N} r_{jt}q_{j,t-m_j} - \sum_{t=1}^{N} p_{jt}q_{jt}$$

$$0 \leq q_{jt}$$

where $w_t = \bar{w}_t/\bar{p}_{0t}$, $r_{jt} = \bar{r}_{jt}/\bar{p}_{0t}$, and $p_{jt} = \bar{p}_{jt}/\bar{p}_{0t}$, or, in an obvious vector notation,

$$c_t = w_t + r_t' q_{t-m} - p_t' q_t$$
$$0 \le q_t.$$

The sequences $\{w_t\}$, $\{r_t\}$, and $\{p_t\}$ are taken to be outside the consumer's control, or exogenously determined. Therefore the sequence $\{c_t\}$ is determined by the budget constraint once $\{q_t\}$ is chosen. Thus, the only sequence that the consumer can control in attempting to maximize utility is the sequence $\{q_t\}$. The utility associated to some sequence $\{q_t\}$ is

$$V(\{q_t\}) = U(\{w_t + r_t' q_{t-m} - p_t' q_t\})$$
$$= \sum_{t=0}^{\infty} \beta^t u(w_t + r_t' q_{t-m} - p_t' q_t).$$

We shall take the sequences $\{w_t\}$, $\{r_t\}$, and $\{p_t\}$ to be stochastic processes and shall assume that the consumer solves his optimization problem by choosing a sequence of functions Q_t of the form

$$Q_t(w_0, \ldots, w_t, r_0, \ldots, r_t, p_0, \ldots, p_t) \ge 0$$

which maximizes

$$\mathscr{E}_0 V(\{Q_t\}) = \mathscr{E}_0 \sum_{t=0}^{\infty} \beta^t u(w_t + r_t' Q_{t-m} + p_t' Q_t)$$

where

$$\mathscr{E}_t(X) = \mathscr{E}\left[X | (w_0, \ldots, w_t, r_0, \ldots, r_t, p_0, \ldots, p_t)\right].$$

It may be that the consumer can achieve a higher expected utility by having regard to some sequence of vector valued variables $\{v_t\}$ in addition to $\{w_t\}$, $\{r_t\}$, and $\{p_t\}$, in which case the argument list of Q and $\mathscr{E}$ above is replaced by the augmented argument list

$$(w_0, \ldots, w_t, r_0, \ldots, r_t, p_0, \ldots, p_t, v_0, \ldots, v_t).$$

Conceptually, v_t can be infinite dimensional, because anything that is observable or knowable is admissible as additional information in improving the optimum. If one wishes to accommodate this possibility, let $\mathscr{B}_t$

be the smallest σ-algebra such that the random variables

$$\{w_s, r_s, p_s, v_{sj} : s = 0, 1, \ldots, t, \ j = 1, 2, \ldots\}$$

are measurable, let Q_t be $\mathscr{B}_t$-measurable, and let $\mathscr{E}_t(X) = \mathscr{E}(X \mid \mathscr{B}_t)$. Either the augmented argument list or the σ-algebra $\mathscr{B}_t$ is called the consumer's information set, depending on which approach is adopted. For our purposes, the augmented argument list provides enough generality.

Let Q_{0t} denote the solution to the consumer's optimization problem, and consider the related problem

$$\text{maximize} \quad \mathscr{E}_s V(\{Q_t\}_{t=s}^\infty) = \mathscr{E}_s \sum_{t=s}^\infty \beta^t u(w_t + r_t' Q_{t-m} - p_t' Q_t)$$

$$\text{subject to} \quad Q_t = Q_t(w_0, \ldots, w_t, r_0, \ldots, r_t, p_0, \ldots, p_t, v_0, \ldots, v_t) \geq 0$$

with solution Q_{st}. Suppose we piece the two solutions together:

$$\tilde{Q}_t = \begin{cases} Q_{0t} & 0 \leq t \leq s-1 \\ Q_{st} & s \leq t \leq \infty. \end{cases}$$

Then

$$\mathscr{E}_0 V(\{\tilde{Q}_t\}) = \mathscr{E}_0 \sum_{t=0}^\infty \beta^t u(w_t + r_t' \tilde{Q}_{t-m} - p_t' \tilde{Q}_t)$$

$$= \mathscr{E}_0 \sum_{t=0}^{s-1} \beta^t u(w_t + r_t' Q_{0t-m} - p_t' Q_{0t})$$

$$+ \mathscr{E}_0 \mathscr{E}_s \sum_{t=s}^\infty \beta^t u(w_t + r_t' Q_{st-m} - p_t' Q_{st})$$

$$\geq \mathscr{E}_0 \sum_{t=0}^{s-1} \beta^t u(w_t + r_t' Q_{0t-m} - p_t' Q_{0t})$$

$$+ \mathscr{E}_0 \mathscr{E}_s \sum_{t=s}^\infty \beta^t u(w_t + r_t' Q_{0t-m} - p_t' Q_{0t})$$

$$= \mathscr{E}_0 \sum_{t=0}^\infty \beta^t u(w_t + r_t' Q_{0t-m} - p_t' Q_{0t})$$

$$= \mathscr{E}_0 V(\{Q_{0t}\}).$$

This inequality shows that Q_{0t} cannot be a solution to the consumer's optimization problem unless it is also a solution to the related problem for each s.

If we mechanically apply the Kuhn-Tucker theorem (Fiacco and Mc-Cormick, 1968) to the related problem

$$\text{maximize} \quad \mathscr{E}_s V(\{Q_t\}_{t=s}^{\infty}) = \mathscr{E}_s \sum_{t=s}^{\infty} \beta^t u(w_t + r_t' Q_{t-m} - p_t' Q_t)$$

$$\text{subject to} \quad Q_t \geq 0$$

we obtain the first order conditions

$$\frac{\partial}{\partial Q_{js}}\left[\mathscr{E}_0 V(\{Q_t\}) + \lambda_{js} Q_{js}\right] = 0$$

$$\lambda_{js} Q_{js} = 0$$

$$\lambda_{js} \leq 0$$

where the λ_{js} are Lagrange multipliers. If a positive quantity of asset j is observed at time s—if $Q_{js} > 0$—then we must have $\lambda_{js} = 0$ and

$$0 = \frac{\partial}{\partial Q_{js}} \mathscr{E}_s \sum_{t=s}^{\infty} \beta^t u(w_t + r_t' Q_{t-m} - p_t' Q_t)$$

$$= \frac{\partial}{\partial Q_{js}} \mathscr{E}_s \beta^s u(w_s + r_s' Q_{s-m} - p_s' Q_s)$$

$$+ \frac{\partial}{\partial Q_{js}} \mathscr{E}_s \beta^{s+m_j} u\left(w_{s+m_j} + r_{s+m_j}' Q_{(s+m_j)-m} - p_{s+m_j}' Q_{s+m_j}\right)$$

$$= -\beta^s \frac{\partial}{\partial c} u(c_s) p_{js} + \beta^{s+m_j} \mathscr{E}_s \frac{\partial}{\partial c} u\left(C_{s+m_j}\right) r_{j,s+m_j}.$$

A little algebra reduces the first order conditions to

$$\beta^{m_j} \mathscr{E}_t \frac{(\partial/\partial c) u\left(C_{t+m_j}\right)}{(\partial/\partial c) u(c_t)} \frac{r_{j,t+m_j}}{p_{jt}} - 1 = 0$$

for $t = 0, 1, 2, \ldots$ and $j = 1, 2, \ldots, N$. See Lucas (1978) for a more rigorous derivation of these first order conditions.

Suppose that the consumer's preferences are given by some parametric function $u(c, \alpha)$ with $u(c) = u(c, \alpha^0)$. Let $\gamma = (\alpha, \beta)$, and denote the true but unknown value of γ by γ^0. Let z_t be any vector valued random variable whose value is known at time t, and let the index j correspond to some sequence of securities $\{q_{jt}\}$ whose observed values are positive. Put

$$m_t(\gamma) = \left(\beta^{m_j} \frac{(\partial/\partial c) u(c_{t+m_j}, \alpha)}{(\partial/\partial c) u(c_t, \alpha)} \frac{r_{j,t+m_j}}{p_{jt}} - 1\right) \otimes z_t.$$

If we could compute the moments of the sequence of random variables

$\{m_t(\gamma^0)\}$, we could put

$$m_n(\gamma) = \frac{1}{n}\sum_{t=1}^{n} m_t(\gamma)$$

and estimate γ^0 by setting $m_n(\gamma) = \mathscr{E}m_n(\gamma^0)$ and solving for γ.

To this point, the operator $\mathscr{E}(\cdot)$ has represented expectations computed according to the consumer's subjective probability distribution. We shall now impose the rational expectations hypothesis, which states that the consumer's subjective probability distribution is the same distribution as the probability law governing the random variables $\{w_t\}$, $\{r_t\}$, $\{p_t\}$, and $\{v_t\}$. Under this assumption, the random variable $m_t(\gamma^0)$ will have first moment

$$\mathscr{E}m_t(\gamma^0) = \mathscr{E}\mathscr{E}_t\left[\left((\beta^0)^{m_j}\frac{(\partial/\partial c)u(C_{t+m_j},\alpha^0)}{(\partial/\partial c)u(c_t,\alpha^0)}\frac{r_{j,t+m_j}}{p_{jt}}-1\right)\otimes z_t\right]$$

$$= \mathscr{E}\left[\mathscr{E}_t\left((\beta^0)^{m_j}\frac{(\partial/\partial c)u(C_{t+m_j},\alpha^0)}{(\partial/\partial c)u(c_t,\alpha^0)}\frac{r_{j,t+m_j}}{p_{jt}}-1\right)\otimes z_t\right]$$

$$= \mathscr{E}[0\otimes z_t] = 0.$$

For $s+m_j \leq t$ the random variables $m_t(\gamma^0)$ and $m_s(\gamma^0)$ are uncorrelated, since

$$\mathscr{E}m_t(\gamma^0)m_s'(\gamma^0)$$
$$= \mathscr{E}\mathscr{E}_t\left[\left((\beta^0)^{m_j}\frac{(\partial/\partial c)u(C_{t+m_j},\alpha^0)}{(\partial/\partial c)u(c_t,\alpha^0)}\frac{r_{j,t+m_j}}{p_{jt}}-1\right)\otimes z_t\right]$$
$$\times\left[\left((\beta^0)^{m_j}\frac{(\partial/\partial c)u(C_{s+m_j},\alpha^0)}{(\partial/\partial c)u(c_s,\alpha^0)}\frac{r_{j,s+m_j}}{p_{js}}-1\right)\otimes z_s\right]'$$
$$= \mathscr{E}\left[\mathscr{E}_t\left((\beta^0)^{m_j}\frac{(\partial/\partial c)u(C_{t+m_j},\alpha^0)}{(\partial/\partial c)u(c_t,\alpha^0)}\frac{r_{j,t+m_j}}{p_{jt}}-1\right)\otimes z_t\right]$$
$$\times\left[\left((\beta^0)^{m_j}\frac{(\partial/\partial c)u(c_{s+m_j},\alpha^0)}{(\partial/\partial c)u(c_s,\alpha^0)}\frac{r_{j,s+m_j}}{p_{js}}-1\right)\otimes z_s\right]'$$
$$= \mathscr{E}[0\otimes z_t]\left[\left((\beta^0)^{m_j}\frac{(\partial/\partial c)u(c_{s+m_j},\alpha^0)}{(\partial/\partial c)u(c_s,\alpha^0)}\frac{r_{j,s+m_j}}{p_{js}}-1\right)\otimes z_s\right]'$$
$$= 0.$$

Table 1a. Consumption and Stock Returns.

t	Year	Month	Nondurables and Services	Population	Value Weighted NYSE Returns	Implicit Deflator
0	1959	1	381.9	176.6850	0.0093695102	0.6818539
1	.	2	383.7	176.9050	0.0093310997	0.6823039
2	.	3	388.3	177.1460	0.0049904501	0.6814319
3	.	4	385.5	177.3650	0.0383739690	0.6830091
4	.	5	389.7	177.5910	0.0204769890	0.6846292
5	.	6	390.0	177.8300	0.0007165600	0.6876923
6	.	7	389.2	178.1010	0.0371922290	0.6893628
7	.	8	390.7	178.3760	-0.0113433900	0.6910673
8	.	9	393.6	178.6570	-0.0472779090	0.6930894
9	.	10	394.2	178.9210	0.0164727200	0.6945713
10	.	11	394.1	179.1530	0.0194594210	0.6950013
11	.	12	396.5	179.3860	0.0296911900	0.6958386
12	1960	1	396.8	179.5970	-0.0664901060	0.6960685
13	.	2	395.4	179.7880	0.0114439700	0.6967628
14	.	3	399.1	180.0070	-0.0114419700	0.6983212
15	.	4	404.2	180.2220	-0.0163223000	0.7013855
16	.	5	399.8	180.4440	0.0328373610	0.7016008
17	.	6	401.3	180.6710	0.0231378990	0.7024670
18	.	7	402.0	180.9450	-0.0210754290	0.7034826
19	.	8	400.4	181.2380	0.0296860300	0.7047952
20	.	9	400.2	181.5280	-0.0568203400	0.7061469
21	.	10	402.9	181.7960	-0.0045937700	0.7078680
22	.	11	403.8	182.0420	0.0472565590	0.7100049
23	.	12	401.6	182.2870	0.0478186380	0.7109064
24	1961	1	404.0	182.5200	0.0654135870	0.7106436
25	.	2	405.7	182.7420	0.0364446900	0.7108701
26	.	3	409.4	182.9920	0.0318523910	0.7105520
27	.	4	410.1	183.2170	0.0058811302	0.7100707
28	.	5	412.1	183.4520	0.0251120610	0.7100218
29	.	6	412.4	183.6910	-0.0296279510	0.7109602
30	.	7	410.4	183.9580	0.0303546690	0.7127193
31	.	8	411.5	184.2430	0.0251584590	0.7132442
32	.	9	413.7	184.5240	-0.0189705600	0.7138023
33	.	10	415.9	184.7830	0.0265103900	0.7136331
34	.	11	419.0	185.0160	0.0470347110	0.7136038
35	.	12	420.5	185.2420	0.0006585300	0.7143876
36	1962	1	420.8	185.4520	-0.0358958100	0.7155418
37	.	2	420.6	185.6500	0.0197925490	0.7177841
38	.	3	423.4	185.8740	-0.0055647301	0.7191781
39	.	4	424.8	186.0870	-0.0615112410	0.7203390
40	.	5	427.0	186.3140	-0.0834698080	0.7206089
41	.	6	425.2	186.5380	-0.0809235570	0.7208373
42	.	7	427.0	186.7900	0.0659098630	0.7203747
43	.	8	428.5	187.0580	0.0226466300	0.7218203
44	.	9	431.8	187.3230	-0.0487761200	0.7264938
45	.	10	431.0	187.5740	0.0039394898	0.7262181
46	.	11	433.6	187.7960	0.1114552000	0.7269373
47	.	12	434.1	188.0130	0.0139081300	0.7270214

Table 1a. (Continued).

t	Year	Month	Nondurables and Services	Population	Value Weighted NYSE Returns	Implicit Deflator
48	1963	1	434.7	188.2130	0.0508059190	0.7292386
49	.	2	433.7	188.3870	-0.0226500000	0.7299977
50	.	3	436.2	188.5800	0.0347222910	0.7297111
51	.	4	437.0	188.7900	0.0479895880	0.7295194
52	.	5	436.9	189.0180	0.0206833590	0.7308308
53	.	6	440.2	189.2420	-0.0178168900	0.7319400
54	.	7	442.1	189.4960	-0.0018435300	0.7335444
55	.	8	445.6	189.7610	0.0536292610	0.7349641
56	.	9	443.8	190.0280	-0.0126571200	0.7347905
57	.	10	444.2	190.2650	0.0286585090	0.7361549
58	.	11	445.8	190.4720	-0.0047020698	0.7375505
59	.	12	449.5	190.6680	0.0221940800	0.7385984
60	1964	1	450.1	190.8580	0.0256042290	0.7398356
61	.	2	453.7	191.0470	0.0181333610	0.7399162
62	.	3	456.6	191.2450	0.0173465290	0.7402540
63	.	4	456.7	191.4470	0.0051271599	0.7407488
64	.	5	462.1	191.6660	0.0166149310	0.7405324
65	.	6	463.8	191.8890	0.0158939310	0.7416990
66	.	7	466.0	192.1310	0.0202899800	0.7429185
67	.	8	468.5	192.3760	-0.0109651800	0.7432231
68	.	9	468.0	192.6310	0.0315313120	0.7448718
69	.	10	470.0	192.8470	0.0100951200	0.7457447
70	.	11	468.0	193.0390	0.0013465700	0.7465812
71	.	12	474.4	193.2230	0.0034312201	0.7476813
72	1965	1	474.5	193.3930	0.0377800500	0.7481560
73	.	2	477.4	193.5400	0.0066147698	0.7488479
74	.	3	474.5	193.7090	-0.0107356600	0.7506849
75	.	4	479.6	193.8880	0.0345496910	0.7525021
76	.	5	481.2	194.0870	-0.0047443998	0.7554032
77	.	6	479.5	194.3030	-0.0505878400	0.7593326
78	.	7	484.3	194.5280	0.0169978100	0.7602726
79	.	8	485.3	194.7610	0.0299301090	0.7601484
80	.	9	488.7	194.9970	0.0323472920	0.7605893
81	.	10	497.2	195.1950	0.0293272190	0.7626710
82	.	11	497.1	195.3720	0.0008636100	0.7648361
83	.	12	499.0	195.5390	0.0121703600	0.7671343
84	1966	1	500.1	195.6880	0.0100357400	0.7696461
85	.	2	501.5	195.8310	-0.0102875900	0.7730808
86	.	3	502.9	195.9990	-0.0215729900	0.7757009
87	.	4	505.8	196.1780	0.0233628400	0.7785686
88	.	5	504.8	196.3720	-0.0509349700	0.7793185
89	.	6	507.5	196.5600	-0.0109703900	0.7812808
90	.	7	510.9	196.7620	-0.0118703500	0.7827363
91	.	8	508.3	196.9840	-0.0748946070	0.7867401
92	.	9	510.2	197.2070	-0.0066132201	0.7894943
93	.	10	509.8	197.3980	0.0464050400	0.7910945
94	.	11	512.1	197.5720	0.0138342800	0.7922281
95	.	12	513.5	197.7360	0.0047225100	0.7933788

Table 1*a*.   (Continued).

t	Year	Month	Nondurables and Services	Population	Value Weighted NYSE Returns	Implicit Deflator
96	1967	1	516.0	197.8920	0.0838221310	0.7941861
97	.	2	517.7	198.0370	0.0098125497	0.7948619
98	.	3	519.0	198.2060	0.0433843100	0.7959538
99	.	4	521.1	198.3630	0.0420965220	0.7965842
100	.	5	521.0	198.5370	-0.0415207000	0.7988484
101	.	6	523.1	198.7120	0.0232013710	0.8015676
102	.	7	522.1	198.9110	0.0482556600	0.8038690
103	.	8	525.5	199.1130	-0.0056581302	0.8058991
104	.	9	528.2	199.3110	0.0336121990	0.8076486
105	.	10	524.9	199.4980	-0.0276739710	0.8094875
106	.	11	527.9	199.6570	0.0078005102	0.8124645
107	.	12	531.9	199.8080	0.0307225010	0.8155668
108	1968	1	533.0	199.9200	-0.0389530290	0.8200750
109	.	2	533.9	200.0560	-0.0311505910	0.8231879
110	.	3	539.8	200.2080	0.0068653398	0.8262319
111	.	4	540.0	200.3610	0.0898963210	0.8285185
112	.	5	541.2	200.5360	0.0230161700	0.8318551
113	.	6	547.8	200.7060	0.0118528200	0.8335159
114	.	7	550.9	200.8980	-0.0211507900	0.8359049
115	.	8	552.4	201.0950	0.0163246690	0.8388849
116	.	9	551.0	201.2900	0.0422958400	0.8421053
117	.	10	552.1	201.4660	0.0111537600	0.8462235
118	.	11	556.7	201.6210	0.0562853110	0.8492905
119	.	12	554.1	201.7600	-0.0372401590	0.8521928
120	1969	1	557.0	201.8810	-0.0072337599	0.8560144
121	.	2	561.2	202.0230	-0.0502119700	0.8578047
122	.	3	560.6	202.1610	0.0314719300	0.8619336
123	.	4	561.9	202.3310	0.0213753300	0.8667023
124	.	5	566.5	202.5070	0.0029275999	0.8704325
125	.	6	563.9	202.6770	-0.0623450020	0.8751552
126	.	7	565.9	202.8770	-0.0630705430	0.8784238
127	.	8	569.4	203.0900	0.0504970810	0.8816298
128	.	9	568.2	203.3020	-0.0220447110	0.8856037
129	.	10	573.1	203.5000	0.0547974710	0.8888501
130	.	11	572.5	203.6750	-0.0314589110	0.8939738
131	.	12	572.4	203.8490	-0.0180749090	0.8981481
132	1970	1	577.2	204.0080	-0.0763489600	0.9019404
133	.	2	578.1	204.1560	0.0597185420	0.9058986
134	.	3	577.7	204.3350	-0.0023485899	0.9077376
135	.	4	577.1	204.5050	-0.0995834470	0.9123202
136	.	5	580.3	204.6920	-0.0611347710	0.9155609
137	.	6	582.0	204.8780	-0.0502832790	0.9176976
138	.	7	582.8	205.0860	0.0746088620	0.9208991
139	.	8	584.7	205.2940	0.0502020900	0.9235505
140	.	9	588.5	205.5070	0.0426676610	0.9276126
141	.	10	587.3	205.7070	-0.0160981810	0.9324025
142	.	11	587.6	205.8840	0.0521828200	0.9361811
143	.	12	592.6	206.0760	0.0617985200	0.9399258

Table 1a. (Continued).

t	Year	Month	Nondurables and Services	Population	Value Weighted NYSE Returns	Implicit Deflator
144	1971	1	592.2	206.2420	0.0492740680	0.9414049
145	.	2	594.5	206.3930	0.0149685600	0.9434819
146	.	3	592.4	206.5670	0.0441647210	0.9469953
147	.	4	596.1	206.7260	0.0341992900	0.9506794
148	.	5	596.3	206.8910	-0.0365711710	0.9548885
149	.	6	598.5	207.0530	0.0043891501	0.9597327
150	.	7	597.3	207.2370	-0.0398038400	0.9630002
151	.	8	599.1	207.4330	0.0409017500	0.9679519
152	.	9	601.1	207.6270	-0.0056930701	0.9698885
153	.	10	601.7	207.8000	-0.0395274310	0.9729101
154	.	11	604.9	207.9490	-0.0000956400	0.9752025
155	.	12	608.8	208.0880	0.0907427070	0.9797963
156	1972	1	607.9	208.1960	0.0241155400	0.9825629
157	.	2	610.3	208.3100	0.0308808110	0.9875471
158	.	3	618.9	208.4470	0.0091922097	0.9891743
159	.	4	620.6	208.5690	0.0066767102	0.9911376
160	.	5	622.3	208.7120	0.0176741590	0.9942150
161	.	6	623.7	208.8460	-0.0221355410	0.9961520
162	.	7	627.6	208.9880	-0.0018799200	0.9996813
163	.	8	629.7	209.1530	0.0382015890	1.0031760
164	.	9	631.7	209.3170	-0.0064313002	1.0079150
165	.	10	638.2	209.4570	0.0099495398	1.0117520
166	.	11	639.8	209.5840	0.0497901590	1.0151610
167	.	12	640.7	209.7110	0.0116306100	1.0190420
168	1973	1	643.4	209.8090	-0.0253177100	1.0247120
169	.	2	645.3	209.9050	-0.0398146990	1.0309930
170	.	3	643.3	210.0340	-0.0054550399	1.0399500
171	.	4	642.1	210.1540	-0.0464594700	1.0478120
172	.	5	643.2	210.2860	-0.0183557910	1.0541040
173	.	6	646.0	210.4100	-0.0088413004	1.0603720
174	.	7	651.9	210.5560	0.0521348010	1.0632000
175	.	8	643.4	210.7150	-0.0302029100	1.0792660
176	.	9	651.3	210.8630	0.0522540810	1.0815290
177	.	10	649.5	210.9840	-0.0018884100	1.0896070
178	.	11	651.3	211.0970	-0.1165516000	1.0993400
179	.	12	647.7	211.2070	0.0153318600	1.1093100
180	1974	1	648.4	211.3110	-0.0013036400	1.1215300
181	.	2	646.2	211.4110	0.0038444500	1.1363350
182	.	3	645.9	211.5220	-0.0243075400	1.1489390
183	.	4	648.6	211.6370	-0.0433935780	1.1558740
184	.	5	649.3	211.7720	-0.0352610800	1.1667950
185	.	6	650.3	211.9010	-0.0193944290	1.1737660
186	.	7	653.5	212.0510	-0.0730255170	1.1802600
187	.	8	654.5	212.2160	-0.0852853730	1.1926660
188	.	9	652.7	212.3830	-0.1098341000	1.2043820
189	.	10	654.5	212.5180	0.1671594000	1.2122230
190	.	11	651.2	212.6370	-0.0397416390	1.2205160
191	.	12	650.3	212.7480	-0.0234328400	1.2278950

Table 1a.   (Continued).

t	Year	Month	Nondurables and Services	Population	Value Weighted NYSE Returns	Implicit Deflator
192	1975	1	653.7	212.8440	0.1358016000	1.2337460
193	.	2	657.4	212.9390	0.0607054380	1.2376030
194	.	3	659.4	213.0560	0.0293416310	1.2406730
195	.	4	659.7	213.1870	0.0470072290	1.2457180
196	.	5	670.4	213.3930	0.0546782990	1.2502980
197	.	6	669.7	213.5590	0.0517648310	1.2593700
198	.	7	668.3	213.7410	-0.0637501480	1.2721830
199	.	8	670.1	213.9000	-0.0203062710	1.2786150
200	.	9	670.2	214.0550	-0.0366309580	1.2821550
201	.	10	670.8	214.2000	0.0609995690	1.2904000
202	.	11	674.1	214.3210	0.0314961600	1.2966920
203	.	12	677.4	214.4460	-0.0105694800	1.3039560
204	1976	1	684.3	214.5610	0.1251743000	1.3081980
205	.	2	682.9	214.6550	0.0012425600	1.3069260
206	.	3	687.1	214.7620	0.0300192200	1.3092710
207	.	4	690.6	214.8810	-0.0108725300	1.3132060
208	.	5	688.7	215.0180	-0.0088088503	1.3206040
209	.	6	695.0	215.1520	0.0472505990	1.3256120
210	.	7	696.8	215.3110	-0.0073758499	1.3307980
211	.	8	699.6	215.4780	0.0005799900	1.3381930
212	.	9	702.5	215.6420	0.0261333100	1.3449110
213	.	10	705.6	215.7920	-0.0214380810	1.3520410
214	.	11	709.7	215.9240	0.0046152598	1.3587430
215	.	12	715.8	216.0570	0.0585772800	1.3657450
216	1977	1	717.6	216.1860	-0.0398427810	1.3720740
217	.	2	719.3	216.3000	-0.0162227190	1.3831500
218	.	3	716.5	216.4360	-0.0106509200	1.3884160
219	.	4	719.1	216.5650	0.0038957901	1.3953550
220	.	5	722.6	216.7120	-0.0126387400	1.4014670
221	.	6	721.5	216.8630	0.0509454310	1.4105340
222	.	7	728.3	217.0300	-0.0156951400	1.4159000
223	.	8	727.0	217.2070	-0.0140849800	1.4244840
224	.	9	729.1	217.3740	0.0006794800	1.4295710
225	.	10	735.7	217.5230	-0.0394544790	1.4350960
226	.	11	739.4	217.6550	0.0419719890	1.4442790
227	.	12	740.1	217.7850	0.0052549900	1.4508850
228	1978	1	738.0	217.8810	-0.0568409600	1.4581300
229	.	2	744.8	217.9870	-0.0121069800	1.4663000
230	.	3	750.5	218.1310	0.0318689010	1.4743500
231	.	4	750.4	218.2610	0.0833722430	1.4862740
232	.	5	750.3	218.4040	0.0186665390	1.5033990
233	.	6	753.1	218.5480	-0.0129163500	1.5146730
234	.	7	755.6	218.7200	0.0564879100	1.5199840
235	.	8	761.1	218.9090	0.0372171590	1.5284460
236	.	9	765.4	219.0780	-0.0063229799	1.5412860
237	.	10	765.2	219.2360	-0.1017461000	1.5541030
238	.	11	768.0	219.3840	0.0313147900	1.5640620
239	.	12	774.1	219.5300	0.0166718100	1.5694350

Source: Courtesy of the authors, Hansen and Singleton (1982, 1984).

Table 1*b*. **Treasury Bill Returns.**

t	Year	Month	Holding Period		
			1 Month	3 Months	6 Months
0	1959	1	0.0021	0.0067620277	0.0149464610
1	.	2	0.0019	0.0067054033	0.0153553490
2	.	3	0.0022	0.0069413185	0.0156610010
3	.	4	0.0020	0.0071977377	0.0164365770
4	.	5	0.0022	0.0072308779	0.0162872080
5	.	6	0.0024	0.0076633692	0.0175679920
6	.	7	0.0025	0.0080889463	0.0190058950
7	.	8	0.0019	0.0075789690	0.0191299920
8	.	9	0.0031	0.0097180605	0.0230187180
9	.	10	0.0030	0.0103986260	0.0247714520
10	.	11	0.0026	0.0101703410	0.0219579940
11	.	12	0.0034	0.0112402440	0.0246732230
12	1960	1	0.0053	0.0111309290	0.0253483060
13	.	2	0.0029	0.0101277830	0.0230150220
14	.	3	0.0035	0.0106238130	0.0225163700
15	.	4	0.0019	0.0076926947	0.0172802210
16	.	5	0.0027	0.0076853037	0.0174299480
17	.	6	0.0024	0.0079696178	0.0172280070
18	.	7	0.0013	0.0055410862	0.0131897930
19	.	8	0.0017	0.0055702925	0.0127488370
20	.	9	0.0016	0.0057605505	0.0142772200
21	.	10	0.0022	0.0058845282	0.0143817660
22	.	11	0.0013	0.0053367615	0.0121047500
23	.	12	0.0016	0.0060379505	0.0138461590
24	1961	1	0.0019	0.0055921078	0.0121918920
25	.	2	0.0014	0.0058113337	0.0127416850
26	.	3	0.0020	0.0065517426	0.0141991380
27	.	4	0.0017	0.0060653687	0.0131639240
28	.	5	0.0018	0.0057235956	0.0120902060
29	.	6	0.0020	0.0059211254	0.0130860810
30	.	7	0.0018	0.0057165623	0.0123114590
31	.	8	0.0014	0.0056213140	0.0128748420
32	.	9	0.0017	0.0059502125	0.0135532620
33	.	10	0.0019	0.0056616068	0.0136462450
34	.	11	0.0015	0.0057950020	0.0132193570
35	.	12	0.0019	0.0064492226	0.0142288210
36	1962	1	0.0024	0.0067474842	0.0148699280
37	.	2	0.0020	0.0068224669	0.0148422720
38	.	3	0.0020	0.0068684816	0.0148569350
39	.	4	0.0022	0.0069818497	0.0147231820
40	.	5	0.0024	0.0068957806	0.0145040750
41	.	6	0.0020	0.0068334341	0.0141508580
42	.	7	0.0027	0.0073847771	0.0149892570
43	.	8	0.0023	0.0072803497	0.0155196190
44	.	9	0.0021	0.0071101139	0.0151845220
45	.	10	0.0025	0.0069708824	0.0148054360
46	.	11	0.0020	0.0068755150	0.0143437390
47	.	12	0.0023	0.0072675943	0.0150616170

Table 1*b*.   (Continued).

t	Year	Month	1 Month	3 Months	6 Months
				Holding Period	
48	1963	1	0.0025	0.0073993206	0.0151528120
49	.	2	0.0023	0.0074235201	0.0153222080
50	.	3	0.0023	0.0073300600	0.0150877240
51	.	4	0.0025	0.0073734522	0.0152205230
52	.	5	0.0024	0.0073573589	0.0153070690
53	.	6	0.0023	0.0076377392	0.0158174040
54	.	7	0.0027	0.0075825453	0.0157427790
55	.	8	0.0025	0.0083107948	0.0174001460
56	.	9	0.0027	0.0086047649	0.0178933140
57	.	10	0.0029	0.0085899830	0.0179922580
58	.	11	0.0027	0.0088618994	0.0184588430
59	.	12	0.0029	0.0088895559	0.0187247990
60	1964	1	0.0030	0.0089538097	0.0187361240
61	.	2	0.0026	0.0088821650	0.0185148720
62	.	3	0.0031	0.0091063976	0.0192459820
63	.	4	0.0029	0.0089648962	0.0188522340
64	.	5	0.0026	0.0087850094	0.0184588430
65	.	6	0.0030	0.0088362694	0.0184062720
66	.	7	0.0030	0.0087610483	0.0180916790
67	.	8	0.0028	0.0088040829	0.0182579760
68	.	9	0.0028	0.0087461472	0.0181269650
69	.	10	0.0029	0.0090366602	0.0189046860
70	.	11	0.0029	0.0089949369	0.0189571380
71	.	12	0.0031	0.0096279383	0.0203764440
72	1965	1	0.0028	0.0097374916	0.0201922660
73	.	2	0.0030	0.0097503662	0.0203632120
74	.	3	0.0036	0.0101563930	0.0207239390
75	.	4	0.0031	0.0098274946	0.0205006600
76	.	5	0.0031	0.0099694729	0.0204553600
77	.	6	0.0035	0.0098533630	0.0202448370
78	.	7	0.0031	0.0096805096	0.0197677610
79	.	8	0.0033	0.0096729994	0.0199424030
80	.	9	0.0031	0.0096987486	0.0202448370
81	.	10	0.0031	0.0102273230	0.0216147900
82	.	11	0.0035	0.0102949140	0.0214961770
83	.	12	0.0033	0.0104292630	0.0217467550
84	1966	1	0.0038	0.0114099980	0.0241273640
85	.	2	0.0035	0.0117748980	0.0235883000
86	.	3	0.0038	0.0116740470	0.0245782140
87	.	4	0.0034	0.0115190740	0.0244190690
88	.	5	0.0041	0.0118079190	0.0243724580
89	.	6	0.0038	0.0116353030	0.0239523650
90	.	7	0.0035	0.0116611720	0.0240478520
91	.	8	0.0041	0.0120524170	0.0255184170
92	.	9	0.0040	0.0125415330	0.0282398460
93	.	10	0.0045	0.0136051180	0.0286009310
94	.	11	0.0040	0.0133793350	0.0278792380
95	.	12	0.0040	0.0131639240	0.0271854400

Table 1*b*. (Continued).

t	Year	Month	1 Month	3 Months	6 Months
				Holding Period	
96	1967	1	0.0043	0.0122823720	0.0254547600
97	.	2	0.0036	0.0114994050	0.0231827500
98	.	3	0.0039	0.0115102530	0.0232950450
99	.	4	0.0032	0.0102145670	0.0208503010
100	.	5	0.0033	0.0094506741	0.0197293760
101	.	6	0.0027	0.0087846518	0.0192024710
102	.	7	0.0031	0.0100209710	0.0223804710
103	.	8	0.0031	0.0105757710	0.0241363050
104	.	9	0.0032	0.0111052990	0.0248435740
105	.	10	0.0039	0.0111955400	0.0259439950
106	.	11	0.0036	0.0115144250	0.0258800980
107	.	12	0.0033	0.0126451250	0.0286339520
108	1968	1	0.0040	0.0127922300	0.0286719800
109	.	2	0.0039	0.0123600960	0.0252685550
110	.	3	0.0038	0.0128250120	0.0269453530
111	.	4	0.0043	0.0130600930	0.0273784400
112	.	5	0.0045	0.0138714310	0.0293196440
113	.	6	0.0043	0.0146449800	0.0306105610
114	.	7	0.0048	0.0135078430	0.0283122060
115	.	8	0.0042	0.0131379370	0.0273113250
116	.	9	0.0043	0.0132676360	0.0269986390
117	.	10	0.0044	0.0130808350	0.0272654290
118	.	11	0.0042	0.0140209200	0.0285472870
119	.	12	0.0043	0.0139169690	0.0287613870
120	1969	1	0.0053	0.0156394240	0.0327807660
121	.	2	0.0046	0.0158472060	0.0329054590
122	.	3	0.0046	0.0159218310	0.0330774780
123	.	4	0.0053	0.0152308940	0.0315673350
124	.	5	0.0048	0.0150020120	0.0310289860
125	.	6	0.0051	0.0155957940	0.0331374410
126	.	7	0.0053	0.0159739260	0.0351030830
127	.	8	0.0050	0.0181180240	0.0373669860
128	.	9	0.0062	0.0177775620	0.0368773940
129	.	10	0.0060	0.0182124380	0.0373669860
130	.	11	0.0052	0.0178167820	0.0382032390
131	.	12	0.0064	0.0192197560	0.0406687260
132	1970	1	0.0060	0.0201528070	0.0415455100
133	.	2	0.0062	0.0201845170	0.0399575230
134	.	3	0.0057	0.0175420050	0.0353578330
135	.	4	0.0050	0.0160522460	0.0327900650
136	.	5	0.0053	0.0176727770	0.0373940470
137	.	6	0.0058	0.0176465510	0.0366872550
138	.	7	0.0052	0.0163348910	0.0338231330
139	.	8	0.0053	0.0161305670	0.0333294870
140	.	9	0.0054	0.0157260890	0.0328800680
141	.	10	0.0046	0.0148160460	0.0328979490
142	.	11	0.0046	0.0148533580	0.0315511230
143	.	12	0.0042	0.0125738380	0.0254547600

Table 1*b***.   (Continued).**

| t | Year | Month | Holding Period | | |
			1 Month	3 Months	6 Months
144	1971	1	0.0038	0.0123342280	0.0249764920
145	.	2	0.0033	0.0104724170	0.0215225220
146	.	3	0.0030	0.0085597038	0.0180971620
147	.	4	0.0028	0.0086690187	0.0190058950
148	.	5	0.0029	0.0100854640	0.0215620990
149	.	6	0.0037	0.0109888320	0.0232796670
150	.	7	0.0040	0.0131714340	0.0277613400
151	.	8	0.0047	0.0134450200	0.0293101070
152	.	9	0.0037	0.0109888320	0.0238093140
153	.	10	0.0037	0.0116611720	0.0252420900
154	.	11	0.0037	0.0109502080	0.0227504970
155	.	12	0.0037	0.0108597280	0.0224862100
156	1972	1	0.0029	0.0091836452	0.0203764440
157	.	2	0.0025	0.0084762573	0.0186685320
158	.	3	0.0027	0.0085406303	0.0192197560
159	.	4	0.0029	0.0096729994	0.0224862100
160	.	5	0.0030	0.0091643333	0.0205764770
161	.	6	0.0029	0.0096213818	0.0212192540
162	.	7	0.0031	0.0102015730	0.0230324270
163	.	8	0.0029	0.0093638897	0.0218169690
164	.	9	0.0034	0.0115317110	0.0257205960
165	.	10	0.0040	0.0114973780	0.0267322060
166	.	11	0.0037	0.0118292570	0.0260753630
167	.	12	0.0037	0.0123860840	0.0266788010
168	1973	1	0.0044	0.0129822490	0.0277190210
169	.	2	0.0041	0.0144283770	0.0300337080
170	.	3	0.0046	0.0148830410	0.0311747790
171	.	4	0.0052	0.0163263080	0.0353578330
172	.	5	0.0051	0.0159218310	0.0385923390
173	.	6	0.0051	0.0176727770	0.0363070960
174	.	7	0.0064	0.0192872290	0.0398482080
175	.	8	0.0070	0.0212606190	0.0439169410
176	.	9	0.0068	0.0221503970	0.0455567840
177	.	10	0.0065	0.0179196600	0.0393607620
178	.	11	0.0056	0.0188183780	0.0385221240
179	.	12	0.0064	0.0187247990	0.0404498580
180	1974	1	0.0063	0.0191034080	0.0378948450
181	.	2	0.0058	0.0192459820	0.0377205610
182	.	3	0.0056	0.0191277270	0.0383199450
183	.	4	0.0075	0.0215357540	0.0431109670
184	.	5	0.0075	0.0226182940	0.0460462570
185	.	6	0.0060	0.0207901000	0.0436338190
186	.	7	0.0070	0.0189310310	0.0415712590
187	.	8	0.0060	0.0196549890	0.0430125000
188	.	9	0.0081	0.0232796670	0.0497723820
189	.	10	0.0051	0.0157541040	0.0382287500
190	.	11	0.0054	0.0199817420	0.0407782790
191	.	12	0.0070	0.0190396310	0.0393673180

Table 1b. (Continued).

t	Year	Month	Holding Period		
			1 Month	3 Months	6 Months
192	1975	1	0.0058	0.0180318360	0.0361442570
193	.	2	0.0043	0.0144213440	0.0303153990
194	.	3	0.0041	0.0137765410	0.0290288930
195	.	4	0.0044	0.0141247510	0.0309060810
196	.	5	0.0044	0.0138258930	0.0305682420
197	.	6	0.0041	0.0132120850	0.0280662780
198	.	7	0.0048	0.0151025060	0.0319100620
199	.	8	0.0048	0.0159218310	0.0347890850
200	.	9	0.0053	0.0162349940	0.0358686450
201	.	10	0.0056	0.0166865590	0.0402698520
202	.	11	0.0041	0.0141117570	0.0294467210
203	.	12	0.0048	0.0141172410	0.0310940740
204	1976	1	0.0047	0.0131934880	0.0284404750
205	.	2	0.0034	0.0120286940	0.0256940130
206	.	3	0.0040	0.0127384660	0.0284404750
207	.	4	0.0042	0.0125932690	0.0277762410
208	.	5	0.0037	0.0125379560	0.0275911470
209	.	6	0.0043	0.0140357020	0.0306911470
210	.	7	0.0047	0.0136423110	0.0298480990
211	.	8	0.0042	0.0132082700	0.0285162930
212	.	9	0.0044	0.0129737850	0.0276085140
213	.	10	0.0041	0.0129562620	0.0273720030
214	.	11	0.0040	0.0124713180	0.0261569020
215	.	12	0.0040	0.0112657550	0.0235443120
216	1977	1	0.0036	0.0110250710	0.0232796670
217	.	2	0.0035	0.0120649340	0.0258269310
218	.	3	0.0038	0.0119923350	0.0256673100
219	.	4	0.0038	0.0115835670	0.0246578450
220	.	5	0.0037	0.0119640830	0.0254743100
221	.	6	0.0040	0.0127894880	0.0269986390
222	.	7	0.0042	0.0127229690	0.0268921850
223	.	8	0.0044	0.0137610440	0.0295736790
224	.	9	0.0043	0.0142158270	0.0304151770
225	.	10	0.0049	0.0150877240	0.0321971180
226	.	11	0.0050	0.0159218310	0.0339560510
227	.	12	0.0049	0.0154412980	0.0331405400
228	1978	1	0.0049	0.0156871080	0.0336534980
229	.	2	0.0046	0.0164700750	0.0349206920
230	.	3	0.0053	0.0164960620	0.0349899530
231	.	4	0.0054	0.0166230200	0.0351681710
232	.	5	0.0051	0.0163058040	0.0359042880
233	.	6	0.0054	0.0170561080	0.0374757050
234	.	7	0.0056	0.0180656910	0.0388967990
235	.	8	0.0056	0.0175942180	0.0384653810
236	.	9	0.0062	0.0193772320	0.0403403040
237	.	10	0.0068	0.0206396580	0.0438541170
238	.	11	0.0070	0.0223728420	0.0483353140
239	.	12	0.0078	0.0232532020	0.0490025280

Source: Courtesy of the authors, Hansen and Singleton (1982, 1984).

If we specialize these results to the utility function $u(c, \alpha) = c^\alpha/\alpha$ for $\alpha < 1$ and take q_{jt} to be common stocks, we shall have $u'(c, \alpha) = c^\alpha$ and $m_j = 1$, which gives the equations listed at the beginning of this discussion.

A final point: in the derivations we treated $\{w_t\}$, $\{r_t\}$, and $\{p_t\}$ as exogenous to each individual; each individual takes account of the actions of others through the conditional expectation $\mathscr{E}_t$. This determines the consumer's demand schedule. Aggregation over individuals—hold $\{w_t\}$, $\{r_t\}$, and $\{p_t\}$ fixed and add up the $\{q_t\}$ and $\{c_t\}$—determines the demand schedule for the economy. This demand schedule interacts with the aggregate supply schedule (or whatever) to determine (the distribution of) $\{w_t\}$, $\{r_t\}$, and $\{p_t\}$. Thus, when the model is applied to aggregate data, as here, $\{w_t\}$, $\{r_t\}$, and $\{p_t\}$ are to be regarded as endogenous.   □

2. THREE STAGE LEAST SQUARES

Multivariate responses y_t, L-vectors, are assumed to be determined by k-dimensional independent variables x_t according to the system of simultaneous equations

$$ q_\alpha(y_t, x_t, \theta_\alpha^0) = e_{\alpha t} \qquad \alpha = 1, 2, \ldots, L \quad t = 1, 2, \ldots, n $$

where each $q_\alpha(y, x, \theta_\alpha)$ is a real valued function, each θ_α^0 is a p_α-dimensional vector of unknown parameters, and the $e_{\alpha t}$ represent unobservable observational or experimental errors. The analysis is conditional on the sequence of independent variables $\{x_t\}$ as described in Section 2 of Chapter 3, and the x_t do not contain lagged values of the y_t as elements. See the next section for the case when they do.

For any set of values $e_1, e_2, \ldots, e_L$ of the errors, any admissible value for the vector x of independent variables, and any admissible value of the parameters $\theta_1, \theta_2, \ldots, \theta_L$, the system of equations

$$ q_\alpha(y, x, \theta_\alpha) = e_\alpha \qquad \alpha = 1, 2, \ldots, L $$

is assumed to determine y uniquely; if the equations have multiple roots, there is some rule for determining which solution is meant. Moreover, the solution must be a continuous function of the errors, the parameters, and the independent variables. However, one is not obligated to actually be able to compute y given these variables, or even to have complete knowledge of the system, in order to use the methods described below: it is just that the theory (Chapter 3) on which the methods are based relies on this assumption for its validity.

There is one feature of implicit models that one should be aware of. If an equation of the system

$$q_\alpha(y, x, \theta_\alpha) = e_\alpha$$

is transformed using some one to one function $\psi(e)$ to obtain

$$\psi[q_\alpha(y, x, \theta_\alpha)] = u_\alpha$$

where

$$u_\alpha = \psi[e_\alpha]$$

the result is still a nonlinear equation in implicit form, which is equivalent to the original equation for the purpose of determining y from knowledge of the independent variables, parameters, and errors. Setting aside the identity transformation, the distribution of the random variable u_α will differ from that of e_α. Thus one has complete freedom to use transformations of this sort in applications in an attempt to make the error distribution more nearly normally distributed.

However, one must realize that transformations of this sort can either destroy consistency or redefine the population quantity θ_α^0, depending on one's point of view. If one takes the view that the model $q_\alpha(y, x, \theta_\alpha^0) = e$ is correct, then nonlinear transformations will destroy consistency in typical cases. If, as is often true, the equations $q_\alpha(y, x, \theta_\alpha^0) = 0$ obtain from a deterministic theory and are interpreted to mean that some measure of central tendency of the random variables $q_\alpha(y_t, x_t, \theta_\alpha^0)$, $t = 1, 2, \ldots, n$, is zero, then it would seem that one model $q_\alpha(y, x, \theta_\alpha^0) = 0$ has as much standing as the next $\psi[q_\alpha(y, x, \theta_\alpha^0)] = 0$ provided that $\psi(0) = 0$. The choice of various $\psi(\cdot)$ is roughly equivalent to the choice of mean, median, or mode as the measure of central tendency to be employed in the analysis. See Section 4 of Chapter 3 for the theoretical considerations behind these remarks; in particular see Theorem 7 and Problem 6 of that section.

In an application it may be the case that not all the equations of the system are known, or it may be that one is simply not interested in some of them. Reorder the equations as necessary so that it is the first M of the L equations above that are of interest, let θ be a p-vector that consists of the nonredundant parameters in the set $\theta_1, \theta_2, \ldots, \theta_M$, and let

$$
q(y, x, \theta) = \begin{pmatrix} q_1(y, x, \theta_1) \\ q_2(y, x, \theta_2) \\ \vdots \\ q_M(y, x, \theta_M) \end{pmatrix} \qquad e = \begin{pmatrix} e_1 \\ e_2 \\ \vdots \\ e_M \end{pmatrix}
$$

$$
q(y_t, x_t, \theta) = \begin{pmatrix} q_1(y_t, x_t, \theta_1) \\ q_2(y_t, x_t, \theta_2) \\ \vdots \\ q_M(y_t, x_t, \theta_M) \end{pmatrix} \qquad e_t = \begin{pmatrix} e_{1t} \\ e_{2t} \\ \vdots \\ e_{Mt} \end{pmatrix}.
$$

We assume that the error vectors e_t are independently and identically distributed with mean zero and unknown variance-covariance matrix Σ,

$$
\Sigma = \mathscr{C}(e_t, e_t') \qquad t = 1, 2, \ldots, n.
$$

Independence implies a lack of correlation, viz.

$$
\mathscr{C}(e_t, e_s') = 0 \qquad t \neq s.
$$

This is the grouped by subject or multivariate arrangement of the data with the equation index α thought of as the fastest-moving index and the observation index t the slowest. The alternative arrangement is the grouped by equation ordering with t the fastest-moving index and α the slowest. As we saw in Chapter 5, the multivariate scheme has two advantages: it facilitates writing code, and it meshes better with the underlying theory (Chapter 3). However, the grouped by equation formulation is more prevalent in the literature because it was the dominant form in the linear simultaneous equations literature and got carried over when the nonlinear literature developed. We shall develop the ideas using the multivariate scheme and then conclude with a summary in the alternative notation. Let us illustrate with the first example.

EXAMPLE 1 (Continued).   Recall that the model is

$$
y_{1t} = \ln \frac{a_1 + r_t'b_{(1)} - y_{3t}1'b_{(1)}}{a_3 + r_t'b_{(3)} - y_{3t}1'b_{(3)}} + e_{1t}
$$

$$
y_{2t} = \ln \frac{a_2 + r_t'b_{(2)} - y_{3t}1'b_{(2)}}{a_3 + r_t'b_{(3)} - y_{3t}1'b_{(3)}} + e_{2t}
$$

$$
y_{3t} = d_t'c + e_{3t}
$$

where 1 denotes a vector of ones,

$$y_t = \begin{pmatrix} y_{1t} \\ y_{2t} \\ y_{3t} \end{pmatrix}, \quad r_t = \begin{pmatrix} r_{1t} \\ r_{2t} \\ r_{3t} \end{pmatrix}, \quad d_t = \begin{pmatrix} d_{0t} \\ d_{1t} \\ \vdots \\ d_{13,t} \end{pmatrix}, \quad x_t = \begin{pmatrix} r_t \\ d_t \end{pmatrix}$$

and

$$a = \begin{pmatrix} a_1 \\ a_2 \\ a_3 \end{pmatrix}, \quad B = \begin{pmatrix} b_{11} & b_{12} & b_{13} \\ b_{21} & b_{22} & b_{23} \\ b_{31} & b_{32} & b_{33} \end{pmatrix} = \begin{pmatrix} b'_{(1)} \\ b'_{(2)} \\ b'_{(3)} \end{pmatrix}, \quad c = \begin{pmatrix} c_0 \\ c_1 \\ \vdots \\ c_{13} \end{pmatrix}.$$

The matrix B is symmetric and $a_3 = -1$.

Our interest centers in the first and second equations, so we write

$$q(y,x,\theta) = \begin{pmatrix} y_1 - \ln \dfrac{\theta_1 + \theta_2 r_1 + \theta_3 r_2 + \theta_4 r_3 - (\theta_2 + \theta_3 + \theta_4) y_3}{-1 + \theta_4 r_1 + \theta_7 r_2 + \theta_8 r_3 - (\theta_4 + \theta_7 + \theta_8) y_3} \\ y_2 - \ln \dfrac{\theta_5 + \theta_3 r_1 + \theta_6 r_2 + \theta_7 r_3 - (\theta_3 + \theta_6 + \theta_7) y_3}{-1 + \theta_4 r_1 + \theta_7 r_2 + \theta_8 r_3 - (\theta_4 + \theta_7 + \theta_8) y_3} \end{pmatrix}$$

$$\theta = (\theta_1, \theta_2, \theta_3, \theta_4, \theta_5, \theta_6, \theta_7, \theta_8)'$$
$$= (a_1, b_{11}, b_{12}, b_{13}, a_2, b_{22}, b_{23}, b_{33})'.$$

Taking values from Table 1 of Chapter 5, we have

$$x_1 = (r'_1 | d'_1)'$$
$$= (1.36098, 1.05082, 0.058269 |$$
$$1, 0.99078, 9.7410, 6.39693, 0, 1, 0.0000,$$
$$9.4727, 1.09861, 1, 0.00000, -0.35667, 0.00000, 0)'$$
$$x_2 = (1.36098, 1.05082, 0.058269 |$$
$$1, 0.99078, 9.5104, 6.80239, 0, 0, 0.0000,$$
$$0.0000, 1.94591, 1, 0.00000, -0.35667, 0.27763, 1)'$$
$$\vdots$$
$$x_{19} = (1.36098, 1.05082, 0.058269 |$$
$$1, 0.99078, 8.7903, 6.86693, 0, 0, 0.0000,$$
$$10.0858, 1.38629, 1, 0.00000, 0.58501, 0.27763, 1)'$$

$$x_{20} = (1.36098, 1.05082, 0.576613 \mid$$
$$1, 1.07614, 9.1050, 6.64379, 0, 0, 0.0000,$$
$$0.0000, 1.60944, 1, 1.60944, 0.58501, 0.00000, 1)'$$

$$\vdots$$

$$x_{40} = (1.36098, 1.05082, 0.576613 \mid$$
$$1, 1.07614, 11.1664, 7.67415, 0, 0, 9.7143,$$
$$0.0000, 1.60944, 1, 1.60944, 0.33647, 0.27763, 1)'$$

$$x_{41} = (1.36098, 1.36098, 0.058269 \mid$$
$$1, 1.08293, 8.8537, 6.72383, 0, 0, 8.3701,$$
$$0.0000, 1.09861, 1, 1.09861, 0.58501, 0.68562, 1)'$$

$$\vdots$$

$$x_{224} = (1.88099, 1.36098, 0.576613 \mid$$
$$1, 1.45900, 8.8537, 6.88653, 0, 0, 0.0000,$$
$$9.2103, 0.69315, 1, 0.69315, 0.58501, 0.00000, 1)'$$

for the independent or exogenous variables, and we have

$$y_1 = (2.45829, 1.59783, -0.7565)'$$
$$y_2 = (1.82933, 0.89091, -0.2289)'$$

$$\vdots$$

$$y_{19} = (2.33247, 1.31287, 0.3160)'$$
$$y_{20} = (1.84809, 0.86533, -0.0751)'$$

$$\vdots$$

$$y_{40} = (1.32811, 0.72482, 0.9282)'$$
$$y_{41} = (2.18752, 0.90133, 0.1375)'$$

$$\vdots$$

$$y_{224} = (1.06851, 0.51366, 0.1475)'$$

for the dependent or endogenous variables.

One might ask why we are handling the model in this way rather than simply substituting $d'c$ for y_3 above and then applying the methods of Chapter 5. After all, the theory on which we rely is nonstochastic, and we just tacked on an error term at a convenient moment in the discussion. As to the theory, it would have been just as defensible to substitute $d'c$ for y_3

in the nonstochastic phase of the analysis and then tack on the error term. By way of reply, the approach we are taking seems to follow the natural progression of ideas. Throughout Chapter 5, the variable y_3 was regarded as being a potentially error ridden proxy for what we really had in mind. Now, a direct remedy seems more in order than a complete reformulation of the model. Moreover, the specification $\mathscr{E}y_3 = d'c$ was data determined and is rather *ad hoc*. It is probably best just to rely on it for the purpose of suggesting instrumental variables and not to risk the specification error a substitution of $d'c$ for y_3 might entail.   □

Three stage least squares is a method of moments type estimator where instrumental variables are used to form the moment equations. That is, letting z_t denote some K-vector of random variables, one forms sample moments

$$m_n(\theta) = \frac{1}{n}\sum_{t=1}^{n} m(y_t, x_t, \theta)$$

where

$$m(y_t, x_t, \theta) = q(y_t, x_t, \theta) \otimes z_t = \underset{MK}{\begin{pmatrix} q_1(y_t, x_t, \theta_1) \cdot z_t \\ q_2(y_t, x_t, \theta_2) \cdot z_t \\ \vdots \\ q_M(y_t, x_t, \theta_M) \cdot z_t \end{pmatrix}}_1$$

equates them to population moments

$$m_n(\theta) = \mathscr{E}\big[m_n(\theta^0)\big]$$

and uses the solution $\hat{\theta}$ as the estimate of θ^0. If, as is usually the case, the dimension MK of $m_n(\theta)$ exceeds the dimension p of θ, these equations will not have a solution. In this case, one applies the generalized least squares heuristic and estimates θ^0 by that value $\hat{\theta}$ that minimizes

$$S(\theta, V) = \big[nm_n(\theta)\big]'V^{-1}\big[nm_n(\theta)\big]$$

with

$$V = \mathscr{C}\big\{\big[nm_n(\theta^0)\big], \big[nm_n(\theta^0)\big]'\big\}.$$

To apply these ideas, one must compute $\mathscr{E}[m_n(\theta^0)]$ and $\mathscr{C}[m_n(\theta^0), m'_n(\theta^0)]$. Obviously there is an incentive to make this computation as easy as possible. Since

$$m_n(\theta^0) = \frac{1}{n} \sum_{t=1}^{n} e_t \otimes z_t$$

we shall have $\mathcal{E}[m_n(\theta^0)] = 0$ if z_t is uncorrelated with e_t, and

$$\mathcal{C}\{[nm_n(\theta^0)], [nm_n(\theta^0)]'\} = \sum_{t=1}^{n} (\Sigma \otimes z_t z_t') = \Sigma \otimes \sum_{t=1}^{n} z_t z_t'$$

if $\{z_t\}$ is independent of $\{e_t\}$. These conditions will obtain (Problem 1) if we impose the requirement that

$$z_t = Z(x_t)$$

where $Z(x)$ is some (possibly nonlinear) function of the independent variables.

We shall also want z_t to be correlated with $q(y_t, x_t, \theta)$ for values of θ other than θ^0, or the method will be vacuous (Problem 2). This last condition is made plausible by the requirement that $z_t = Z(x_t)$, but, strictly speaking, direct verification of the identification condition (Chapter 3, Section 4)

$$\lim_{n \to \infty} m_n(\theta) = 0 \quad \Rightarrow \quad \theta^0 = \theta$$

is required. This is an almost sure limit. Its computation is discussed in Section 2 of Chapter 3, but it usually suffices to check that

$$\mathcal{E} m_n(\theta) = 0 \quad \Rightarrow \quad \theta^0 = \theta.$$

As we remarked in Chapter 1, few are going to take the trouble to verify this condition in an application, but it is prudent to be on guard (or violations that are easily detected (Problem 3).

The matrix V is unknown, so we adopt the same strategy that was used in multivariate least squares: Form a preliminary estimate $\hat{\theta}^\#$ of θ^0, and then estimate V from residuals. Let

$$\hat{\theta}^\# = \operatorname*{argmin}_{\theta} S\left(\theta, I \otimes \sum_{t=1}^{n} z_t z_t'\right)$$

and put

$$\hat{V} = \left(\frac{1}{n} \sum_{t=1}^{n} q(y_t, x_t, \hat{\theta}^\#) q'(y_t, x_t, \hat{\theta}^\#)\right) \otimes \left(\sum_{t=1}^{n} z_t z_t'\right).$$

There are two alternative estimators of V in the literature. The first of these affords some protection against heteroscedasticity:

$$\bar{V} = \sum_{t=1}^{n} \left[q(y_t, x_t, \hat{\theta}^{\#}) \otimes z_t \right] \left[q(y_t, x_t, \hat{\theta}^{\#}) \otimes z_t \right]'.$$

The second uses two stage least squares residuals to estimate Σ. The two stage least squares estimate of the parameters of the single equation

$$q_\alpha(y_t, x_t, \theta_\alpha) = e_{\alpha t} \qquad t = 1, 2, \ldots, n$$

is

$$\hat{\theta}_\alpha^{\#} = \underset{\theta}{\operatorname{argmin}} \left(\sum_{t=1}^{n} q_\alpha(y_t, x_t, \theta_\alpha) z_t \right)'$$

$$\times \left(\sum_{t=1}^{n} z_t z_t' \right)^{-1} \left(\sum_{t=1}^{n} q_\alpha(y_t, x_t, \theta_\alpha) z_t \right).$$

Two stage least squares is vestigial terminology left over from the linear case (Problem 4). Letting

$$\bar{\sigma}_{\alpha\beta} = \frac{1}{n} \sum_{t=1}^{n} q_\alpha(y_t, x_t, \hat{\theta}_\alpha^{\#}) q_\beta(y_t, x_t, \hat{\theta}_\beta^{\#})$$

the estimate of Σ is the matrix $\bar{\Sigma}$ with typical element $\bar{\sigma}_{\alpha\beta}$, viz.

$$\bar{\Sigma} = [\bar{\sigma}_{\alpha\beta}]$$

and the estimate of V is

$$\bar{V} = \bar{\Sigma} \otimes \sum_{t=1}^{n} z_t z_t'.$$

Suppose that one worked by analogy with the generalized least squares approach used in Chapter 5 and viewed

$$y - f(\theta) = \sum_{t=1}^{n} q(y_t, x_t, \theta) \otimes z_t$$

as a nonlinear regression in vector form, and

$$S(\theta, \hat{V}) = [y - f(\theta)]' \hat{V}^{-1} [y - f(\theta)]$$

$$= \left(\sum_{t=1}^{n} q(y_t, x_t, \theta) \otimes z_t \right)' \hat{V}^{-1} \left(\sum_{t=1}^{n} q(y_t, x_t, \theta) \otimes z_t \right)$$

as the objective function for the generalized least squares estimator of θ. One would conclude that the estimated variance-covariance matrix of $\hat{\theta}$ was

$$
\begin{aligned}
\hat{C} &= \left[\left(\frac{\partial}{\partial \theta'} [y - f(\hat{\theta})] \right)' \hat{V}^{-1} \left(\frac{\partial}{\partial \theta'} [y - f(\hat{\theta})] \right) \right]^{-1} \\
&= \left[\left(\sum_{t=1}^{n} \frac{\partial}{\partial \theta'} q(y_t, x_t, \hat{\theta}) \otimes z_t \right)' \hat{V}^{-1} \left(\sum_{t=1}^{n} \frac{\partial}{\partial \theta'} q(y_t, x_t, \hat{\theta}) \otimes z_t \right) \right]^{-1} \\
&= \left[\left(\sum_{t=1}^{n} Q(y_t, x_t, \hat{\theta}) \otimes z_t \right)' \hat{V}^{-1} \left(\sum_{t=1}^{n} Q(y_t, x_t, \hat{\theta}) \otimes z_t \right) \right]^{-1}.
\end{aligned}
$$

This intuitive approach does lead to the correct answer (Problem 5).

The Gauss-Newton correction vector can be deduced in this way as well (Problem 6):

$$
\begin{aligned}
D(\theta, V) = &-\left[\left(\sum_{t=1}^{n} Q(y_t, x_t, \theta) \otimes z_t \right)' V^{-1} \left(\sum_{t=1}^{n} Q(y_t, x_t, \theta) \otimes z_t \right) \right]^{-1} \\
&\times \left[\left(\sum_{t=1}^{n} Q(y_t, x_t, \theta) \otimes z_t \right)' V^{-1} \left(\sum_{t=1}^{n} q(y_t, x_t, \theta) \otimes z_t \right) \right].
\end{aligned}
$$

The modified Gauss-Newton algorithm for minimizing $S(\theta, V)$ is:

0. Choose a starting estimate θ_0. Compute $D_0 = D(\theta_0, V)$, and find a λ_0 between zero and one such that $S(\theta_0 + \lambda_0 D_0, V) < S(\theta_0, V)$.
1. Let $\theta_1 = \theta_0 + \lambda_0 D_0$. Compute $D_1 = D(\theta_1, V)$, and find a λ_1 between zero and one such that $S(\theta_1 + \lambda_1 D_1, V) < S(\theta_1, V)$.
2. Let $\theta_2 = \theta_1 + \lambda_1 D_1$. ...

The comments in Section 4 of Chapter 1 regarding starting rules, stopping rules, and alternative algorithms apply directly.

In summary, the three stage least squares estimator is computed as follows. The set of equations of interest are

$$
q(y_t, x_t, \theta) = \begin{pmatrix} q_1(y_t, x_t, \theta_1) \\ q_2(y_t, x_t, \theta_2) \\ \vdots \\ q_M(y_t, x_t, \theta_M) \end{pmatrix} \qquad t = 1, 2, \ldots, n
$$

and instrumental variables of the form

$$z_t = Z(x_t)$$

are selected. The objective function that defines the estimator is

$$S(\theta, V) = \left(\sum_{t=1}^{n} q(y_t, x_t, \theta) \otimes z_t \right)' V^{-1} \left(\sum_{t=1}^{n} q(y_t, x_t, \theta) \otimes z_t \right).$$

One minimizes $S(\theta, I \otimes \sum_{t=1}^{n} z_t z_t')$ to obtain a preliminary estimate $\hat{\theta}^{\#}$, viz.

$$\hat{\theta}^{\#} = \underset{\theta}{\operatorname{argmin}} \, S\left(\theta, I \otimes \sum_{t=1}^{n} z_t z_t' \right),$$

and puts

$$\hat{V} = \left(\frac{1}{n} \sum_{t=1}^{n} q(y_t, x_t, \hat{\theta}^{\#}) q'(y_t, x_t, \hat{\theta}^{\#}) \right) \otimes \left(\sum_{t=1}^{n} z_t z_t' \right).$$

The estimate of θ^0 is the minimizer $\hat{\theta}$ of $S(\theta, \hat{V})$, viz.

$$\hat{\theta} = \underset{\theta}{\operatorname{argmin}} \, S(\theta, \hat{V}).$$

The estimated variance-covariance matrix of $\hat{\theta}$ is

$$\hat{C} = \left[\left(\sum_{t=1}^{n} Q(y_t, x_t, \hat{\theta}) \otimes z_t \right)' \hat{V}^{-1} \left(\sum_{t=1}^{n} Q(y_t, x_t, \hat{\theta}) \otimes z_t \right) \right]^{-1}$$

where

$$Q(y_t, x_t, \theta) = \frac{\partial}{\partial \theta'} q(y_t, x_t, \theta).$$

We illustrate with the example.

EXAMPLE 1 (Continued).   A three stage least squares fit of the model

$$q(y, x, \theta) = \left| \begin{array}{l} y_1 - \ln \dfrac{\theta_1 + \theta_2 r_1 + \theta_3 r_2 + \theta_4 r_3 - (\theta_2 + \theta_3 + \theta_4) y_3}{-1 + \theta_4 r_1 + \theta_7 r_2 + \theta_8 r_3 - (\theta_4 + \theta_7 + \theta_8) y_3} \\[3mm] y_2 - \ln \dfrac{\theta_5 + \theta_3 r_1 + \theta_6 r_2 + \theta_7 r_3 - (\theta_3 + \theta_6 + \theta_7) y_3}{-1 + \theta_4 r_1 + \theta_7 r_2 + \theta_8 r_3 - (\theta_4 + \theta_7 + \theta_8) y_3} \end{array} \right|$$

to the data of Table 1, Chapter 5, is shown as Figure 1.

SAS Statements:

```
PROC MODEL OUT=MOD01;
ENDOGENOUS Y1 Y2 Y3;
EXOGENOUS  R1 R2 R3 D0 D1 D2 D3 D4 D5 D6 D7 D8 D9 D10 D11 D12 D13;
PARMS T1 T2 T3 T4 T5 T6 T7 T8 C0 C1 C2 C3 C4 C5 C6 C7 C8 C9 C10 C11 C12 C13;
PEAK= T1+T2*R1+T3*R2+T4*R3-(T2+T3+T4)*Y3;
INTER=T5+T3*R1+T6*R2+T7*R3-(T3+T6+T7)*Y3;
BASE= -1+T4*R1+T7*R2+T8*R3-(T4+T7+T8)*Y3;
Y1=LOG(PEAK/BASE);  Y2=LOG(INTER/BASE);
Y3=D0*C0+D1*C1+D2*C2+D3*C3+D4*C4+D5*C5+D6*C6+D7*C7+D8*C8+D9*C9+D10*C10+D11*C11
   +D12*C12+D13*C13;
PROC SYSNLIN DATA=EG01 MODEL=MOD01 N3SLS METHOD=GAUSS MAXIT=50 CONVERGE=1.E-8
            SDATA=IDENTITY OUTS=SHAT OUTEST=THAT;
INSTRUMENTS  R1 R2 R3 D0 D1 D2 D3 D4 D5 D6 D7 D8 D9 D10 D11 D12 D13 / NOINT;
FIT Y1 Y2  START=(T1 -2.98  T2 -1.16  T3 0.787  T4 0.353  T5 -1.51  T6 -1.00
           T7 0.054  T8 -0.474);
PROC SYSNLIN DATA=EG01 MODEL=MOD01 N3SLS METHOD=GAUSS MAXIT=50 CONVERGE=1.E-8
            SDATA=SHAT ESTDATA=THAT;
INSTRUMENTS  R1 R2 R3 D0 D1 D2 D3 D4 D5 D6 D7 D8 D9 D10 D11 D12 D13 / NOINT;
```

Output:

SYSNLIN PROCEDURE

NONLINEAR 3SLS PARAMETER ESTIMATES

PARAMETER	ESTIMATE	APPROX. STD ERROR	'T' RATIO	APPROX. PROB>!T!	1ST STAGE R-SQUARE
T1	-2.13788	0.58954	-3.63	0.0004	0.6274
T2	-1.98939	0.75921	-2.62	0.0094	0.5473
T3	0.70939	0.15657	4.53	0.0001	0.7405
T4	0.33663	0.05095	6.61	0.0001	0.7127
T5	-1.40200	0.15226	-9.21	0.0001	0.7005
T6	-1.13890	0.18429	-6.18	0.0001	0.5225
T7	0.02913	0.04560	0.64	0.5236	0.5468
T8	-0.50050	0.04517	-11.08	0.0001	0.4646

NUMBER OF OBSERVATIONS		STATISTICS FOR SYSTEM	
USED	220	OBJECTIVE	0.15893
MISSING	4	OBJECTIVE*N	34.96403

COVARIANCE OF RESIDUALS MATRIX USED FOR ESTIMATION

S	Y1	Y2
Y1	0.17159	0.09675
Y2	0.09675	0.09545

Figure 1. Example 1 fitted by nonlinear three stage least squares.

We obtain

$$
\hat{\theta} = \begin{pmatrix} \hat{\theta}_1 \\ \hat{\theta}_2 \\ \hat{\theta}_3 \\ \hat{\theta}_4 \\ \hat{\theta}_5 \\ \hat{\theta}_6 \\ \hat{\theta}_7 \\ \hat{\theta}_8 \end{pmatrix} = \begin{pmatrix} -2.13788 \\ -1.98939 \\ 0.70939 \\ 0.33663 \\ -1.40200 \\ -1.13890 \\ 0.02913 \\ -0.50050 \end{pmatrix} \qquad \text{(from Fig. 1)}
$$

$$
\hat{\Sigma} = \begin{pmatrix} 0.17159 & 0.09675 \\ 0.09675 & 0.09545 \end{pmatrix} \qquad \text{(from Fig. 1)}
$$

$$
S(\hat{\theta}, \hat{\Sigma}) = 34.96403 \qquad \text{(from Fig. 1)}.
$$

These estimates are little changed from the multivariate least squares estimates

$$
\hat{\theta} = \begin{pmatrix} \hat{\theta}_1 \\ \hat{\theta}_2 \\ \hat{\theta}_3 \\ \hat{\theta}_4 \\ \hat{\theta}_5 \\ \hat{\theta}_6 \\ \hat{\theta}_7 \\ \hat{\theta}_8 \end{pmatrix} = \begin{pmatrix} -2.92458 \\ -1.28675 \\ 0.81857 \\ 0.36116 \\ -1.53759 \\ -1.04896 \\ 0.03009 \\ -0.46742 \end{pmatrix} \qquad \text{(from Fig. 3}c\text{, Chapter 5)}.
$$

The main effect of the use of three stage least squares has been to inflate the estimated standard errors.   □

The alternative notational convention is obtained by combining all the observations pertaining to a single equation into an n-vector

$$
q_\alpha(\theta_\alpha) = \underset{n}{\overset{}{\begin{pmatrix} q_\alpha(y_1, x_1, \theta_\alpha) \\ q_\alpha(y_2, x_2, \theta_\alpha) \\ \vdots \\ q_\alpha(y_n, x_n, \theta_\alpha) \end{pmatrix}}}_1 \qquad (\alpha = 1, 2, \ldots, M)
$$

and then stacking these vectors equation by equation to obtain

$$q(\theta) = \begin{pmatrix} q_1(\theta_1) \\ q_2(\theta_2) \\ \vdots \\ q_M(\theta_M) \end{pmatrix}_1^{nM}$$

with

$$\theta = \begin{pmatrix} \theta_1 \\ \theta_2 \\ \vdots \\ \theta_M \end{pmatrix}_1^{\Sigma_{t=1}^M p_\alpha}.$$

If desired, one can impose across equations restrictions by deleting the redundant entries of θ. Arrange the instrumental variables into a matrix Z as follows:

$$Z = \begin{pmatrix} z_1' \\ z_2' \\ \vdots \\ z_n' \end{pmatrix}_n^{K}$$

and put

$$P_Z = Z(Z'Z)^{-1}Z'.$$

With these conventions the three stage least squares objective function is

$$S[\theta, (\Sigma \otimes Z'Z)] = q'(\theta)(\Sigma^{-1} \otimes P_Z)q(\theta).$$

An estimate of Σ can be obtained by computing either

$$\hat{\theta}^{\#} = \underset{\theta}{\text{argmin}}\, S[\theta, (I \otimes Z'Z)]$$

or

$$\hat{\theta}_\alpha^{\#} = \underset{\theta_\alpha}{\text{argmin}}\, q_\alpha'(\theta_\alpha) P_Z q_\alpha(\theta_\alpha) \qquad (\alpha = 1, 2, \ldots, M)$$

and letting $\hat{\Sigma}$ be the matrix with typical element

$$\hat{\sigma}_{\alpha\beta} = \frac{1}{n} q_\alpha'(\hat{\theta}_\alpha^\#) q_\beta(\hat{\theta}_\beta^\#).$$

The estimate of θ^0 is

$$\hat{\theta} = \underset{\theta}{\text{argmin}}\, S[\theta, (\hat{\Sigma} \otimes Z'Z)]$$

with estimated variance-covariance matrix

$$\hat{C} = \left[\left(\frac{\partial}{\partial\theta'}q(\hat{\theta})\right)'(\hat{\Sigma}^{-1} \otimes P_Z)\left(\frac{\partial}{\partial\theta'}q(\hat{\theta})\right)\right]^{-1}.$$

These expressions are svelte by comparison with the summation notation used above. But the price of beauty is an obligation to assume that the errors e_t each have the same variance Σ and are uncorrelated with the sequence $\{z_t\}$. Neither the correction for heteroscedasticity suggested above nor the correction for autocorrelation discussed in the next section can be accommodated within this notational framework.

Amemiya (1977) considered the question of the optimal choice of instrumental variables and found that the optimal choice is obtained if the columns of Z span the same space as the union of the spaces spanned by the columns of $\mathscr{E}(\partial/\partial\theta_\alpha')q(\theta_\alpha^0)$. This can necessitate a large number of columns of Z, which adds to the small sample bias of the estimator and reduces the small sample variance, leading to very misleading confidence intervals. The intuition is: As instruments are added, three stage least squares estimates approach least squares estimates because P_Z approaches the identity matrix. Least squares estimates have small variance but are biased (Tauchen, 1986). Amemiya (1977) proposes some alternative three stage least squares type estimators obtained by replacing $\Sigma^{-1} \otimes P_Z$ with a matrix that has smaller rank but achieves the same asymptotic variance. He also shows that the three stage least squares is not as efficient asymptotically as the maximum likelihood estimator, discussed in Section 5.

The most disturbing aspect of three stage least squares estimators is that they are not invariant to the choice of instrumental variables. Various sets of instrumental variables can lead to quite different parameter estimates even though the model specification and data remain the same. A dramatic illustration of this point can be had by looking at the estimates published by Hansen and Singleton (1982, 1984). Bear in mind when looking at their results that their maximum likelihood estimator is obtained by assuming a distribution for the data and then imposing parametric restrictions implied by the model rather than deriving the likelihood implied by the model and an assumed error distribution; Section 5 takes the latter approach, as does Amemiya's comparison.

One would look to results on the optimal choice of instrumental variables for some guidance that would lead to a resolution of this lack of invariance problem. But they do not provide it. Leaving aside the issue of having either to know the parameter values or to estimate them, one would have to specify the error distribution in order to compute $\mathscr{E}(\partial/\partial\theta_\alpha')q(\theta_\alpha^0)$. But if the error distribution is known, maximum likelihood is the method of choice.

In practice, the most common approach is to use the independent variables x_{it} and low order monomials in x_{it} such as $(x_{it})^2$ or $x_{it}x_{jt}$ as instrumental variables, making no attempt to find the most efficient set using the results on efficiency. We shall return to this issue at the end of the next section.

PROBLEMS

1. Consider the system of nonlinear equations

$$a_0^0 + a_1^0 \ln y_{1t} + a_2^0 x_t = e_{1t}$$

$$b_0^0 + b_1^0 y_{1t} + y_{2t} + b_2^0 x_t = e_{2t}$$

$$t = 1, 2, \ldots$$

where the errors $e_t = (e_{1t}, e_{2t})$ are normally distributed and the independent variable x_t follows the replication pattern

$$x_t = 0, 1, 2, 3, 0, 1, 2, 3, \ldots.$$

Put

$$\theta = (b_0, b_1, b_2)'$$

$$z_t = (1, x_t, x_t^2)'$$

$$m(y_t, x_t, \theta) = (b_0 + b_1 y_{1t} + y_{2t} + b_2 x_t) \otimes z_t$$

$$m_n(\theta) = \frac{1}{n} \sum_{t=1}^{n} m(y_t, x_t, \theta).$$

Show that

$$\lim_{n \to \infty} m_n(\theta^0) = \mathscr{E} m_n(\theta^0) = 0,$$

$$\lim_{n \to \infty} m_n(\theta) = \begin{pmatrix} 1 & c\Sigma_{x=0}^3 \exp(-a_2^0 x) & 1.5 \\ 1.5 & c\Sigma_{x=0}^3 x \exp(-a_2^0 x) & 3.5 \\ 3.5 & c\Sigma_{x=0}^3 x^2 \exp(-a_2^0 x) & 9 \end{pmatrix} \begin{pmatrix} b_0 - b_0^0 \\ b_1 - b_1^0 \\ b_2 - b_2^0 \end{pmatrix}$$

almost surely, where $c = (1/4)\exp[\mathrm{Var}(e_1)/2 - a_0^0]$.

2. Referring to Problem 1, show that if $a_2 \neq 0$ then

$$\lim_{n \to \infty} m_n(\theta) = 0 \quad \Rightarrow \quad \theta^0 = \theta.$$

3. Referring to Problem 1, show that the model is not identified if either $a_2 = 0$ or $z_t = (1, x_t)$.

4. Consider the linear system

$$y_t' \Gamma = x_t' B + e_t' \qquad t = 1, 2, \dots, n$$

where Γ is a square, nonsingular matrix. We shall presume that the elements of y_t and x_t are ordered so that the first column of Γ has $L' + 1$ leading nonzero entries and the first column of B has k' leading nonzero entries; we shall also presume that $\gamma_{11} = 1$. With these conventions the first equation of the system may be written as

$$y_{1t} = (y_{2t}, y_{3t}, \dots, y_{L'}, x_{1t}, x_{2t}, \dots, x_{K'}) \begin{pmatrix} -\gamma_{12} \\ \vdots \\ -\gamma_{1L'} \\ \beta_{11} \\ \vdots \\ \beta_{1K'} \end{pmatrix} + e_{1t}$$

$$= w_t' \delta + e_{1t}.$$

The linear two stage least squares estimator is obtained by putting $z_t = x_t$; that is, the linear two stage least squares estimator is the minimizer $\hat{\delta}$ of

$$S(\delta) = \left(\sum_{t=1}^n (y_{1t} - w_t'\delta)x_t \right)' \left(\sum_{t=1}^n x_t x_t' \right)^{-1} \left(\sum_{t=1}^n (y_{1t} - w_t'\delta)x_t \right).$$

Let $\hat{w}_t$ denote the predicted values from a regression of w_t on x_t, viz.

$$\hat{w}'_t = x'_t \left(\sum_{s=1}^{n} x_s x'_s \right)^{-1} \sum_{s=1}^{n} x_s w'_s \qquad t = 1, 2, \ldots, n.$$

Show that a regression of y_{1t} on $\hat{w}_t$ yields $\hat{\delta}$; that is, show that

$$\hat{\delta} = \left(\sum_{t=1}^{n} \hat{w}_t \hat{w}'_t \right)^{-1} \sum_{t=1}^{n} \hat{w}_t y_{1t}.$$

It is from this fact that the name two stage least squares derives; the first stage is the regression of w_t on x_t, and the second is the regression of y_{1t} on $\hat{w}_t$.

5. Use Notation 8 and Theorem 9 of Chapter 3 to deduce the estimated variance-covariance matrix of the three stage least squares estimator.

6. Use the form of the Gauss-Newton correction vector given in Section 2 of Chapter 5 to deduce the correction vector of the three stage least squares estimator.

3. THE DYNAMIC CASE: GENERALIZED METHOD OF MOMENTS

Although there is a substantial difference in theory between the dynamic case, where errors may be serially correlated and lagged dependent variables may be used as explanatory variables, and the regression case, where errors are independent and lagged dependent variables are disallowed, there is little difference in applications. All that changes is that the variance V of $nm_n(\theta^0)$ is estimated differently.

The underlying system is

$$q_\alpha(y_t, x_t, \theta^0_\alpha) = e_{\alpha t} \qquad t = 0, \pm 1, \pm 2, \ldots \qquad \alpha = 1, 2, \ldots, L$$

where t indexes observations that are ordered in time, $q_\alpha(y, x, \theta_\alpha)$ is a real valued function, y_t is an L-vector, x_t is a k-vector, θ^0_α is a p_α-vector of unknown parameters, and $e_{\alpha t}$ is an unobservable observational or experimental error. The vector x_t can include lagged values of the dependent variable (y_{t-1}, y_{t-}, etc.) as elements. Because of these lagged values, x_t is called the vector of predetermined variables rather than the independent variables. The errors $e_{\alpha t}$ will usually be serially correlated:

$$\mathscr{C}(e_{\alpha s}, e_{\beta t}) = \sigma_{\alpha\beta st} \neq 0 \qquad \alpha, \beta = 1, 2, \ldots, M \quad s, t = 1, 2, \ldots .$$

We do not assume that the errors are stationary, which accounts for the indices st; if the errors were stationary, we would have $\sigma_{\alpha\beta}$.

Attention is restricted to the first M equations and a sample of size n:

$$q_\alpha(y_t, x_t, \theta_\alpha^0) = e_{\alpha t} \qquad t = 1, 2, \ldots, n \quad \alpha = 1, 2, \ldots, M.$$

As in the regression case, let θ be a p-vector containing the nonredundant parameters in the set $\theta_1, \theta_2, \ldots, \theta_M$, and let

$$q(y, x, \theta) = \begin{pmatrix} q_1(y, x, \theta_1) \\ q_2(y, x, \theta_2) \\ \vdots \\ q_M(y, x, \theta_M) \end{pmatrix} \qquad e = \begin{pmatrix} e_1 \\ e_2 \\ \vdots \\ e_M \end{pmatrix}$$

$$q(y_t, x_t, \theta) = \begin{pmatrix} q_1(y_t, x_t, \theta_1) \\ q_2(y_t, x_t, \theta_2) \\ \vdots \\ q_M(y_t, x_t, \theta_M) \end{pmatrix} \qquad e_t = \begin{pmatrix} e_{1t} \\ e_{2t} \\ \vdots \\ e_{Mt} \end{pmatrix}.$$

The analysis is unconditional; indeed, the presence of lagged values of y_t as components of x_t precludes a conditional analysis. The theory on which the analysis is based (Chapter 7) does not rely explicitly on the existence of a smooth reduced form as was the case in the previous section. Let r_t denote those components of x_t that are ancillary; lagged dependent variables are excluded. What is required is the existence of measurable functions $W_t(\cdot)$ that depend on the doubly infinite sequence

$$v_\infty = (\ldots, v_{-1}, v_0, v_1, \ldots)$$

where $v_t = (e_t, r_t)$ such that

$$(y_t, x_t) = W_t(v_\infty)$$

and mixing conditions that limit the dependence between v_s and v_t for $t \neq s$. The details are spelled out in Sections 3 and 5 of Chapter 7.

The estimation strategy is the same as nonlinear three stage least squares (Section 2). One chooses a K-vector of instrumental variables z_t of the form

$$z_t = Z(x_t)$$

(recall that x_t can contain lagged exogeneous and endogenous variables), one forms sample moments

$$m_n(\theta) = \frac{1}{n} \sum_{t=1}^n m(y_t, x_t, \theta)$$

where

$$m(y_t, x_t, \theta) = q(y_t, x_t, \theta) \otimes z_t = \left. \begin{pmatrix} q_1(y_t, x_t, \theta_1) z_t \\ q_2(y_t, x_t, \theta_2) z_t \\ \vdots \\ {}_{MK} q_M(y_t, x_t, \theta_M) z_t \end{pmatrix} \right|_1$$

and one estimates θ^0 by that value $\hat{\theta}$ that minimizes

$$S(\theta, V) = [nm_n(\theta)]' V^{-1} [nm_n(\theta)]$$

with

$$V = \mathscr{C}\{[nm_n(\theta^0)], [nm_n(\theta^0)]'\}.$$

In this case, the random variables

$$m(y_t, x_t, \theta^0) = e_t \otimes z_t \qquad t = 1, 2, \ldots, n$$

are correlated and we have

$$V = \mathscr{E}\left(\sum_{t=1}^n e_t \otimes z_t \right)\left(\sum_{s=1}^n e_s \otimes z_s \right)'$$

$$= \sum_{t=1}^n \sum_{s=1}^n \mathscr{E}(e_t \otimes z_t)(e_s \otimes z_s)'$$

$$= n \sum_{\tau = -(n-1)}^{n-1} S_{n\tau}^0$$

$$= nS_n^0$$

where

$$
S_{n\tau}^0 = \begin{cases} (1/n) \sum_{t=1+\tau}^{n} \mathscr{E}(e_t \otimes z_t)(e_{t-\tau} \otimes z_{t-\tau})' & \tau \geq 0 \\ (S_{n,-\tau}^0)' & \tau < 0. \end{cases}
$$

To estimate V, we shall need a consistent estimator of S_n^0, that is, an estimator $\hat{S}_n$ with

$$
\lim_{n \to \infty} P\left(|S_{n,\alpha\beta}^0 - \hat{S}_{n,\alpha\beta}| > \epsilon\right) = 0 \qquad \alpha, \beta = 1, 2, \ldots, M
$$

for any $\epsilon > 0$. This is basically a matter of guaranteeing that

$$
\lim_{n \to \infty} \mathrm{Var}\left(\hat{S}_{n,\alpha\beta}\right) = 0;
$$

see Theorem 3 in Section 2 of Chapter 7.

A consistent estimate $\hat{S}_{n\tau}$ of $S_{n\tau}^0$ can be obtained in the obvious way by putting

$$
\hat{S}_{n\tau} = \begin{cases} (1/n) \sum_{t=1+\tau}^{n} \left[q(y_t, x_t, \hat{\theta}^{\#}) \otimes z_t\right]\left[\left(q(y_{t-\tau}, x_{t-\tau}, \hat{\theta}^{\#}) \otimes z_{t-\tau}\right]' & \\ & \tau \geq 0 \\ (\hat{S}_{n,-\tau})' & \\ & \tau < 0 \end{cases}
$$

where

$$
\hat{\theta}^{\#} = \operatorname*{argmin}_{\theta} S\left(\theta, I \otimes \sum_{t=1}^{n} z_t z_t'\right).
$$

However, one cannot simply add the $\hat{S}_{n\tau}$ for τ ranging from $-(n-1)$ to $n-1$ as suggested by the definition of S_n^0 and obtain a consistent estimator, because $\mathrm{Var}(\hat{S}_n)$ will not decrease with increasing n. The variance will decrease if a smaller number of summands is used, namely the sum for τ ranging from $-l(n)$ to $l(n)$, where $l(n)$ is the integer nearest $n^{1/5}$.

Consistency will obtain with this modification, but the unweighted sum will not be positive definite in general. As we propose to minimize $S(\theta, \hat{V})$ in order to compute $\hat{\theta}$, the matrix $\hat{V} = n\hat{S}_n$ must be positive definite. The weighted sum

$$
\hat{S}_n = \sum_{\tau = -l(n)}^{l(n)} w\left(\frac{\tau}{l(n)}\right) \hat{S}_{n\tau}
$$

constructed from Parzen weights

$$w(x) = \begin{cases} 1 - 6|x|^2 + 6|x|^3 & 0 \leq x \leq \frac{1}{2} \\ 2(1 - |x|)^3 & \frac{1}{2} \leq x \leq 1 \end{cases}$$

is consistent and positive definite (Theorem 3, Section 2, Chapter 7). The motivation for this particular choice of weights derives from the observation that if $\{e_t \otimes z_t\}$ were a stationary time series then S_n^0 would be the spectral density of the process evaluated at zero; the estimator with Parzen weights is the best estimator of the spectral density in an asymptotic mean square error sense (Anderson, 1971, Chapter 9, or Bloomfield, 1976, Chapter 7).

The generalized method of moments estimator differs from the three stage least squares estimator only in the computation of $\hat{V}$. The rest is the same: the estimate of θ^0 is the minimizer $\hat{\theta}$ of $S(\theta, \hat{V})$, viz.

$$\hat{\theta} = \operatorname*{argmin}_{\theta} S(\theta, \hat{V});$$

the estimated variance-covariance of $\hat{\theta}$ is

$$\hat{C} = \left[\left(\sum_{t=1}^{n} Q(y_t, x_t, \hat{\theta}) \otimes z_t \right)' \hat{V}^{-1} \left(\sum_{t=1}^{n} Q(y_t, x_t, \hat{\theta}) \otimes z_t \right) \right]^{-1}$$

where

$$Q(y_t, x_t, \theta) = (\partial/\partial\theta')q(y_t, x_t, \theta);$$

and the Gauss-Newton correction vector is

$$D(\theta, V) = -\left[\left(\sum_{t=1}^{n} Q(y_t, x_t, \theta) \otimes z_t \right)' V^{-1} \left(\sum_{t=1}^{n} Q(y_t, x_t, \theta) \otimes z_t \right) \right]^{-1}$$

$$\times \left[\left(\sum_{t=1}^{n} Q(y_t, x_t, \theta) \otimes z_t \right)' V^{-1} \left(\sum_{t=1}^{n} q(y_t, x_t, \theta) \otimes z_t \right) \right].$$

We illustrate.

EXAMPLE 2 (Continued). Recall that

$$q(y_t, x_t, \theta) = \beta(y_t)^{\alpha} x_t - 1 \qquad t = 1, 2, \ldots, 239.$$

where, taking values from Table 1a,

$$y_t = \frac{(\text{consumption at time } t)/(\text{population at time } t)}{(\text{consumption at time } t - 1)/(\text{population at time } t - 1)}$$

$$x_t = (1 + \text{stock returns at time } t)\frac{\text{deflator at time } t - 1}{\text{deflator at time } t}.$$

The instrumental variables employed in the estimation are

$$z_t = (1, y_{t-1}, x_{t-1})'.$$

For instance, we have

$$y_2 = 1.01061 \qquad x_2 = 1.00628 \qquad z_2 = (1, 1.00346, 1.008677)'.$$

Recall also that, in theory,

$$\mathcal{E}(e_t \otimes z_t) = 0 \qquad t = 2, 3, \dots, 239$$
$$\mathcal{E}(e_t \otimes z_t)(e_t \otimes z_t)' = \Sigma \qquad t = 2, 3, \dots, 239$$
$$\mathcal{E}(e_t \otimes z_t)(e_s \otimes z_s)' = 0 \qquad t \neq s.$$

Because the variance estimator has the form

$$\hat{V} = n \sum_{\tau = -l(n)}^{l(n)} w\left(\frac{\tau}{l(n)}\right) \hat{S}_{n\tau}$$

where

$$l(n) = n^{1/5},$$

$$w(x) = \begin{cases} 1 - 6|x|^2 + 6|x|^3 & 0 \le x \le \tfrac{1}{2} \\ 2(1 - |x|)^3 & \tfrac{1}{2} \le x \le 1 \end{cases}$$

and

$$\hat{S}_{n\tau} = \begin{cases} \frac{1}{n} \sum_{t=1+\tau}^{n} [q(y_t, x_t, \hat{\theta}^{\#}) \otimes z_t][q(y_{t-\tau}, x_{t-\tau}, \hat{\theta}^{\#}) \otimes z_{t-\tau}]' & \tau \ge 0 \\ (\hat{S}_{n, -\tau})' & \tau < 0. \end{cases}$$

whereas PROC SYSNLIN can only compute a variance estimate of the form

$$\hat{V} = \left(\frac{1}{n} \sum_{t=1}^{n} q(y_t, x_t, \hat{\theta}^{\#}) q'(y_t, x_t, \hat{\theta}^{\#})\right) \otimes \left(\sum_{t=1}^{n} z_t z_t'\right)$$

we are on our own as far as writing code is concerned. Our strategy will be to use PROC MATRIX as follows:

```
DATA WORK01;  SET EG02;
NDSPER = NDS / PEOPLE;  Y = NDSPER / LAG(NDSPER);  X = (1 + STOCKS)*LAG(DEFLATOR) /
DEFLATOR;
DATA WORK02;  SET WORK01;  Z0 = 1;  Z1 = LAG(Y);  Z2 = LAG(X);  IF _N_ = 1 THEN
DELETE;
PROC MATRIX;  FETCH Y DATA = WORK02(KEEP = Y);  FETCH X DATA = WORK02(KEEP = X);
              FETCH Z DATA = WORK02(KEEP = Z0 Z1 Z2);  Z(1,) = 0 0 0;
A = -.4;  B = .9;  V = Z'*Z;  %GAUSS DO WHILE (S>OBJ#(1 + 1.E-5));  %GAUSS END;
TSHARP = A // B;  PRINT TSHARP;
%VARIANCE  V = VHAT;  %GAUSS DO WHILE (S>OBJ#(1 + 1.E-5));  %GAUSS END;
THAT = A // B;  PRINT VHAT THAT CHAT S;
```

where **%GAUSS** is a MACRO which computes a modified (line searched)
Gauss-Newton iterative step:

```
%MACRO GAUSS;
 M = 0 / 0 / 0;  DELM = 0 0 / 0 0 / 0 0;  ONE = 1;
 DO T = 2 TO 239;
  QT = B#Y(T,1)##A#X(T,1) - ONE;
  DELQTA = B#LOG(Y(T,1))#Y(T,1)##A#X(T,1);  DELQTB = Y(T,1)##A#X(T,1);
  MT = QT @ Z(T,)%STR(%');  DELMT = (DELQTA || DELQTB) @ Z(T,)%STR(%');
  M = M + MT;  DELM = DELM + DELMT;
 END;
 CHAT = INV(DELM%STR(%')*INV(V)*DELM);  D = -CHAT*DELM%STR(%')*INV(V)*M;
 S = M%STR(%')*INV(V)*M;  OBJ = S;  L = 2;  COUNT = 0;  A0 = A;  B0 = B;
 DO WHILE (OBJ> = S & COUNT< = 40);
  COUNT = COUNT + ONE;  L = L#.5;  A = A0 + L#D(1,1);  B = B0 + L#D(2,1);  M = 0;
  DO T = 2 TO 239;  M = M + (B#Y(T,1)##A#X(T,1) - ONE)@Z(T,)%STR(%');  END;
  OBJ = M%STR(%')*INV(V)*M;
 END;
%MEND GAUSS;
```

and **%VARIANCE** is a MACRO which computes a variance estimate:

```
%MACRO VARIANCE;
 S0 = 0 0 0 / 0 0 0 / 0 0 0;  S1 = 0 0 0 / 0 0 0 / 0 0 0;  S2 = 0 0 0 / 0 0 0 / 0 0 0;
 ONE = 1;
 DO T = 2 TO 239;
  MT0 = (B#Y(T,1)##A#X(T,1) - ONE) @ Z(T,)%STR(%');  S0 = S0 + MT0*MT0%STR(%');
  IF T>3 THEN DO;
```

```
MT1 = (B#Y(T - 1,1)##A#X(T - 1,1) - ONE)ƏZ(T - 1,)%STR(%'); S1 = S1 +
MT0*MT1%STR(%'); END;
IF T>4 THEN DO;
MT2 = (B#Y(T - 2,1)##A#X(T - 2,1) - ONE)ƏZ(T - 2,)%STR(%'); S2 = S2 +
MT0*MT2%STR(%'); END;
END;
W0 = 1;  W1 = ONE - 6#(1# / 3)##2 + 6#(1# / 3)##3;  W2 = 2#(ONE - (2# / 3))##3;  W3 = 0;
VHAT = (W0#S0 + W1#S1 + W1#S1%STR(%') + W2#S2 + W2#S2%STR(%'));
%MEND VARIANCE;
```

The code is fairly transparent if one takes a one by one matrix to be a scalar and reads ′ for %STR(%'), * for #, * * for ##, / for # / , and ⊗ for Ə.

SAS Statements:

```
%MACRO GAUSS;
 M=0/0/0; DELM=0 0/0 0/0 0; ONE=1;
 DO T=2 TO 239;
  QT = B#Y(T,1)##A#X(T,1)-ONE;
  DELQTA = B#LOG(Y(T,1))#Y(T,1)##A#X(T,1);  DELQTB = Y(T,1)##A#X(T,1);
  MT = QT @ Z(T,)%STR(%');  DELMT = (DELQTA || DELQTB) @ Z(T,)%STR(%');
  M=M+MT;  DELM=DELM+DELMT;
 END;
 CHAT=INV(DELM%STR(%')*INV(V)*DELM);  D=-CHAT*DELM%STR(%')*INV(V)*M;
 S=M%STR(%')*INV(V)*M;  OBJ=S;  L=2;  COUNT=0;  A0=A;  B0=B;
 DO WHILE (OBJ>=S & COUNT<=40);
  COUNT=COUNT+ONE;  L=L#.5;  A=A0+L#D(1,1);  B=B0+L#D(2,1);  M=0;
  DO T=2 TO 239;  M=M+(B#Y(T,1)##A#X(T,1)-ONE)@Z(T,)%STR(%');  END;
  OBJ=M%STR(%')*INV(V)*M;
 END;
%MEND GAUSS;
%MACRO VARIANCE;
 S0=0 0 0/0 0 0/0 0 0;  S1=0 0 0/0 0 0/0 0 0; S2=0 0 0/0 0 0/0 0 0; ONE=1;
 DO T=2 TO 239;
  MT0=(B#Y(T,1)##A#X(T,1)-ONE) @ Z(T,)%STR(%'); S0=S0+MT0*MT0%STR(%');
  IF T>3 THEN DO;
  MT1=(B#Y(T-1,1)##A#X(T-1,1)-ONE)@Z(T-1,)%STR(%'); S1=S1+MT0*MT1%STR(%'); END;
  IF T>4 THEN DO;
  MT2=(B#Y(T-2,1)##A#X(T-2,1)-ONE)@Z(T-2,)%STR(%'); S2=S2+MT0*MT2%STR(%'); END;
 END;
 W0=1;  W1=0;  W2=0;  W3=0;
 VHAT=(W0#S0+W1#S1+W1#S1%STR(%')+W2#S2+W2#S2%STR(%'));
%MEND VARIANCE;
DATA WORK01;  SET EG02;
NDSPER=NDS/PEOPLE; Y=NDSPER/LAG(NDSPER); X=(1+STOCKS)*LAG(DEFLATOR)/DEFLATOR;
DATA WORK02;  SET WORK01; Z0=1; Z1=LAG(Y); Z2=LAG(X);  IF _N_=1 THEN DELETE;
PROC MATRIX;  FETCH Y DATA=WORK02(KEEP=Y);  FETCH X DATA=WORK02(KEEP=X);
              FETCH Z DATA=WORK02(KEEP=Z0 Z1 Z2); Z(1,)=0 0 0;
A=-.4; B=.9; V=Z'*Z; %GAUSS DO WHILE (S>OBJ#(1+1.E-5)); %GAUSS END;
TSHARP=A // B;  PRINT TSHARP;
%VARIANCE V=VHAT;  %GAUSS DO WHILE (S>OBJ#(1+1.E-5)); %GAUSS END;
THAT=A // B;  PRINT VHAT THAT CHAT S;
```

Figure 2. The generalized method of moments estimator for Example 2.

Output:

SAS   1

```
                TSHARP              COL 1

                ROW1          -0.848852
                ROW2           0.998929


        VHAT            COL 1           COL 2           COL 3

        ROW1         0.405822        0.406434        0.398737
        ROW2         0.406434        0.407055        0.399363
        ROW3         0.398737        0.399363        0.392723


                THAT                COL 1

                ROW1           -1.03352
                ROW2            0.998256


                CHAT                COL 1           COL 2

                ROW1             3.58009  -0.00721267
                ROW2         -0.00721267    .0000206032


                S                   COL 1

                ROW1            1.05692
```

Figure 2. (Continued).

While in general this code is correct, for this particular problem

$$\mathscr{E}(e_t \otimes z_t)(e_s \otimes z_s)' = 0 \qquad t \neq s$$

so we shall replace the line

```
W0 = 1;   W1 = ONE - 6#(1# / 3)##2 + 6#(1# / 3)##3;   W2 = 2#(ONE - (2# / 3))##3;   W3 = 0;
```

in %VARIANCE, which computes the weights $w[\tau/l(n)]$, with the line

```
W0 = 1;   W1 = 0;   W2 = 0;   W3 = 0;
```

The computations are shown in Figure 2.   □

Tauchen (1986) considers the question of how instruments ought to be chosen for generalized method of moments estimators in the case where the errors are uncorrelated. In this case the optimal choice of instrumental variables is (Hansen, 1985)

$$z_t = \mathscr{E}_t \frac{\partial}{\partial \theta'} q\left(y_t, x_t, \theta^0 \right) \left[\mathscr{E}_t \left(e_t e_t' \right) \right]^{-1}$$

where $\mathscr{E}_t(\cdot)$ denotes the conditional expectation with respect to all variables (information) relevant to the problem from the present time t to as far into the past as is relevant; see the discussion of this point in Example 2 of Section 2. Tauchen, using the same sort of model as Example 2, obtains the small sample bias and variance for various choices of instrumental variables, which he compares with the optimal choice. He finds that, when short lag lengths are used in forming instrumental variables, nearly asymptotically optimal parameter estimates obtain, and that as the lag length increases, estimates become increasingly concentrated around biased values and confidence intervals become increasingly inaccurate. He also finds that the test of overidentifying restrictions (Section 4) performs reasonably well in finite samples.

The more interesting aspect of Tauchen's work is that he obtains a computational strategy for generating data that follow a nonlinear, dynamic model—one that can be used to formulate a bootstrap strategy to find the optimal instrumental variables in a given application.

PROBLEMS

1. Use the data of Tables 1a and 1b of Section 1 to reproduce the results of Hansen and Singleton (1984).

2. Verify that if one uses the first order conditions for three month treasury bills, $z_t = (1, y_{t-s}, x_{t-s})$ with s chosen as the smallest value that will insure that $\mathscr{E}(q(y_t, x_t, \theta^0) \otimes z_t) = 0$, and Parzen weights, then one obtains

TSHARP	COL1
ROW1	−4.38322
ROW2	1.02499

VHAT	COL1	COL2	COL3
ROW1	0.24656	0.248259	0.246906
ROW2	0.248259	0.249976	0.248609
ROW3	0.246906	0.248609	0.247256

THAT	COL1	
ROW1	-4.37803	
ROW2	1.02505	

CHAT	COL1	COL2
ROW1	22.8898	-0.140163
ROW2	-0.140163	0.00086282

Should Parzen weights be used in this instance?

4. HYPOTHESIS TESTING

As seen in the last two sections, the effect of various assumptions regarding lagged dependent variables, heteroscedasticity, or autocorrelated errors is to alter the form of the variance estimator $\hat{V}$ without affecting the form of the estimator $\hat{\theta}$. Thus, each estimator can be regarded as, at most, a simplified version of the general estimator proposed in Section 5 of Chapter 7, and in consequence the theory of hypothesis testing presented in Section 6 of Chapter 7 applies to all of them. This being the case, here we can lump the preceding estimation procedures together and accept the following as the generic description of the hypothesis testing problem.

Attention is restricted to the first M equations

$$q_{\alpha t}\left(y_t, x_t, \theta_\alpha^0\right) = e_{\alpha t} \qquad t = 1, 2, \ldots, n \quad \alpha = 1, 2, \ldots, M$$

of some system. Let θ be a p-vector containing the nonredundant parameters in the set $\theta_1, \theta_2, \ldots, \theta_M$, and let

$$q(y, x, \theta) = \begin{pmatrix} q_1(y, x, \theta_1) \\ q_2(y, x, \theta_2) \\ \vdots \\ q_M(y, x, \theta_M) \end{pmatrix} \qquad e = \begin{pmatrix} e_1 \\ e_2 \\ \vdots \\ e_M \end{pmatrix}.$$

To estimate θ^0, one chooses a K-vector of instrumental variables z_t of the form

$$z_t = Z(x_t)$$

(recall that x_t can include lagged endogenous and exogenous variables), one constructs the sample moments

$$m_n(\theta) = \frac{1}{n} \sum_{t=1}^{n} m(y_t, x_t, \theta)$$

with

$$m(y_t, x_t, \theta) = q(y_t, x_t, \theta) \otimes z_t = \begin{vmatrix} q_1(y_t, x_t, \theta_1)z_t \\ q_2(y_t, x_t, \theta_2)z_t \\ \vdots \\ q_M(y_t, x_t, \theta_M)z_t \end{vmatrix}_{MK \qquad 1}$$

and one estimates θ^0 by the value $\hat{\theta}$ that minimizes

$$S(\theta, \hat{V}) = [nm_n(\theta)]'\hat{V}^{-1}[nm_n(\theta)]$$

where $\hat{V}$ is some consistent estimate of

$$V = \mathscr{C}\{[nm_n(\theta^0)], [nm_n(\theta^0)]'\}.$$

The estimated variance-covariance matrix of $\hat{\theta}$ is

$$\hat{C} = \left[\left(\sum_{t=1}^{n} Q(y_t, x_t, \hat{\theta}) \otimes z_t\right)' \hat{V}^{-1}\left(\sum_{t=1}^{n} Q(y_t, x_t, \hat{\theta}) \otimes z_t\right)\right]^{-1}$$

where

$$Q(y_t, x_t, \theta) = \frac{\partial}{\partial \theta'} q(y_t, x_t, \theta).$$

The Gauss-Newton correction vector is

$$D(\theta, V) = -\left[\left(\sum_{t=1}^{n} Q(y_t, x_t, \theta) \otimes z_t\right)' V^{-1}\left(\sum_{t=1}^{n} Q(y_t, x_t, \theta) \otimes z_t\right)\right]^{-1}$$

$$\times \left[\left(\sum_{t=1}^{n} Q(y_t, x_t, \theta) \otimes z_t\right)' V^{-1}\left(\sum_{t=1}^{n} q(y_t, x_t, \theta) \otimes z_t\right)\right].$$

With this as a backdrop, interest centers in testing a hypothesis that can be expressed either as a parametric restriction

$$H: h(\theta^0) = 0 \quad \text{against} \quad A: h(\theta^0) \neq 0$$

or as a functional dependence

$$H: \theta^0 = g(\rho^0) \text{ for some } \rho^0 \quad \text{against} \quad A: \theta^0 \neq g(\rho) \text{ for any } \rho.$$

Here, $h(\theta)$ maps $\mathbb{R}^p$ into $\mathbb{R}^q$ with Jacobian

$$H(\theta) = \frac{\partial}{\partial \theta'} h(\theta)$$

which is assumed to be continuous with rank q; $g(\rho)$ maps $\mathbb{R}^r$ into $\mathbb{R}^p$ and has Jacobian

$$G(\rho) = \frac{\partial}{\partial \rho'} g(\rho).$$

The Jacobians are of order q by p for $H(\theta)$ and p by r for $G(\rho)$; we assume that $p = r + q$, and from $h[g(\rho)] = 0$ we have $H[g(\rho)]G(\rho) = 0$. For complete details, see Section 6 of Chapter 3. Let us illustrate with the example.

EXAMPLE 1 (Continued). Recall that

$$q(y, x, \theta) = \begin{vmatrix} y_1 - \ln \dfrac{\theta_1 + \theta_2 r_1 + \theta_3 r_2 + \theta_4 r_3 - (\theta_2 + \theta_3 + \theta_4) y_3}{-1 + \theta_4 r_1 + \theta_7 r_2 + \theta_8 r_3 - (\theta_4 + \theta_7 + \theta_8) y_3} \\ y_2 - \ln \dfrac{\theta_5 + \theta_3 r_1 + \theta_6 r_2 + \theta_7 r_3 - (\theta_3 + \theta_6 + \theta_7) y_3}{-1 + \theta_4 r_1 + \theta_7 r_2 + \theta_8 r_3 - (\theta_4 + \theta_7 + \theta_8) y_3} \end{vmatrix}$$

with

$$\theta = (\theta_1, \theta_2, \theta_3, \theta_4, \theta_5, \theta_6, \theta_7, \theta_8)'.$$

The hypothesis of homogeneity (see Section 4 of Chapter 5) may be written as the parametric restriction

$$h(\theta) = \begin{pmatrix} \theta_2 + \theta_3 + \theta_4 \\ \theta_3 + \theta_6 + \theta_7 \\ \theta_4 + \theta_7 + \theta_8 \end{pmatrix} = 0$$

with Jacobian

$$H(\theta) = \begin{pmatrix} 0 & 1 & 1 & 1 & 0 & 0 & 0 & 0 \\ 0 & 0 & 1 & 0 & 0 & 1 & 1 & 0 \\ 0 & 0 & 0 & 1 & 0 & 0 & 1 & 1 \end{pmatrix}$$

or, equivalently, as the functional dependence

$$
\theta = \begin{pmatrix} \theta_1 \\ \theta_2 \\ \theta_3 \\ \theta_4 \\ \theta_5 \\ \theta_6 \\ \theta_7 \\ \theta_8 \end{pmatrix} = \begin{pmatrix} \theta_1 \\ -\theta_3 - \theta_4 \\ \theta_3 \\ \theta_4 \\ \theta_5 \\ -\theta_7 - \theta_3 \\ \theta_7 \\ -\theta_4 - \theta_7 \end{pmatrix} = \begin{pmatrix} \rho_1 \\ -\rho_2 - \rho_3 \\ \rho_2 \\ \rho_3 \\ \rho_4 \\ -\rho_5 - \rho_2 \\ \rho_5 \\ -\rho_5 - \rho_3 \end{pmatrix} = g(\rho)
$$

with Jacobian

$$
G(\rho) = \begin{pmatrix} 1 & 0 & 0 & 0 & 0 \\ 0 & -1 & -1 & 0 & 0 \\ 0 & 1 & 0 & 0 & 0 \\ 0 & 0 & 1 & 0 & 0 \\ 0 & 0 & 0 & 1 & 0 \\ 0 & -1 & 0 & 0 & -1 \\ 0 & 0 & 0 & 0 & 1 \\ 0 & 0 & -1 & 0 & -1 \end{pmatrix}. \qquad \square
$$

The Wald test statistic for the hypothesis

$$
H : h(\theta^0) = 0 \quad \text{against} \quad A : h(\theta^0) \neq 0
$$

is

$$
W = \hat{h}'(\hat{H}\hat{C}\hat{H}')^{-1}\hat{h}
$$

where $\hat{h} = h(\hat{\theta})$, $H(\theta) = (\partial/\partial\theta')h(\theta)$, and $\hat{H} = H(\hat{\theta})$. One rejects the hypothesis

$$
H : h(\theta^0) = 0
$$

when W exceeds the upper $\alpha \times 100\%$ critical point χ_α^2 of the chi-square distribution with q degrees of freedom; $\chi_\alpha^2 = (\chi^2)^{-1}(1 - \alpha, q)$.

Under the alternative $A : h(\theta^0) \neq 0$, the Wald test statistic is approximately distributed as the noncentral chi-square with q degrees of freedom and noncentrality parameter

$$\lambda = \frac{h'(\theta^0)\left[H(\theta^0)C(\theta^0)H'(\theta^0)\right]^{-1}h(\theta^0)}{2}$$

where

$$C = \left(\sum_{t=1}^{n} \mathscr{E}\left[Q(y_t, x_t, \theta^0) \otimes Z(x_t)\right]' V^{-1} \right.$$

$$\left. \times \sum_{t=1}^{n} \mathscr{E}\left[Q(y_t, x_t, \theta^0) \otimes Z(x_t)\right] \right)^{-1}$$

$$m_n(\theta) = \frac{1}{n} \sum_{t=1}^{n} q(y_t, x_t, \theta) \otimes Z(x_t)$$

$$V = \mathscr{E}\left\{ \left[nm_n(\theta^0)\right], \left[nm_n(\theta^0)\right]' \right\}$$

$$Q(y_t, x_t, \theta) = \frac{\partial}{\partial \theta'} q(y_t, x_t, \theta).$$

Note, in the formulas above, that if x_t is random, then the expectation is $\mathscr{E}[Q(y_t, x_t, \theta^0) \otimes Z(x_t)]$, not $\mathscr{E}[Q(y_t, x_t, \theta^0)] \otimes Z(x_t)$. If there are no lagged dependent variables (and the analysis is conditional), these two expectations will be the same.

EXAMPLE 1 (Continued). Code to compute the Wald test statistic

$$W = \hat{h}'(\hat{H}\hat{C}\hat{H}')^{-1}\hat{h}$$

for the hypothesis of homogeneity

$$h(\theta) = \begin{pmatrix} \theta_2 + \theta_3 + \theta_4 \\ \theta_3 + \theta_6 + \theta_7 \\ \theta_4 + \theta_7 + \theta_8 \end{pmatrix} = 0$$

is shown in Figure 3. The nonlinear three stage least squares estimators $\hat{\theta}$ and $\hat{C}$ are computed using the same code as in Figure 1 of Section 3. The computed values are passed to PROC MATRIX, where the value

$$W = 3.01278 \qquad \text{(from Fig 3)}$$

is computed using straightforward algebra. Since $(\chi^2)^{-1}(.95, 3) = 7.815$, the hypothesis is accepted at the 5% level.

SAS Statements:

```
PROC MODEL OUT=MOD01;
ENDOGENOUS Y1 Y2 Y3;
EXOGENOUS  R1 R2 R3 D0 D1 D2 D3 D4 D5 D6 D7 D8 D9 D10 D11 D12 D13;
PARMS T1 T2 T3 T4 T5 T6 T7 T8 C0 C1 C2 C3 C4 C5 C6 C7 C8 C9 C10 C11 C12 C13;
PEAK= T1+T2*R1+T3*R2+T4*R3-(T2+T3+T4)*Y3;
INTER=T5+T3*R1+T6*R2+T7*R3-(T3+T6+T7)*Y3;
BASE= -1+T4*R1+T7*R2+T8*R3-(T4+T7+T8)*Y3;
Y1=LOG(PEAK/BASE);   Y2=LOG(INTER/BASE);
Y3=D0*C0+D1*C1+D2*C2+D3*C3+D4*C4+D5*C5+D6*C6+D7*C7+D8*C8+D9*C9+D10*C10+D11*C11
   +D12*C12+D13*C13;

PROC SYSLIN DATA=EG01 MODEL=MOD01 N3SLS METHOD=GAUSS MAXIT=50 CONVERGE=1.E-8
            SDATA=IDENTITY OUTS=SHAT OUTEST=TSHARP;
INSTRUMENTS  R1 R2 R3 D0 D1 D2 D3 D4 D5 D6 D7 D8 D9 D10 D11 D12 D13 / NOINT;
FIT Y1 Y2 START = (T1 -2.98   T2 -1.16   T3 0.787   T4 0.353   T5 -1.51   T6 -1.00
                   T7 0.054   T8 -0.474);
PROC SYSLIN DATA=EG01 MODEL=MOD01 N3SLS METHOD=GAUSS MAXIT=50 CONVERGE=1.E-8
            SDATA=SHAT ESTDATA=TSHARP OUTEST=WORK01 COVOUT;
INSTRUMENTS  R1 R2 R3 D0 D1 D2 D3 D4 D5 D6 D7 D8 D9 D10 D11 D12 D13 / NOINT;
FIT Y1 Y2;

PROC MATRIX; FETCH W DATA=WORK01(KEEP = T1 T2 T3 T4 T5 T6 T7 T8);
THAT=W(1,)';   CHAT=W(2:9,);
H=0 1 1 1 0 0 0 0 / 0 0 1 0 0 1 1 0 / 0 0 0 1 0 0 1 1;
W=THAT'*H'*INV(H*CHAT*H')*H*THAT;   PRINT W;
```

Output:

```
                              SAS                                          8


              W              COL1

            ROW1          3.01278
```

Figure 3. Illustration of Wald test computations with Example 1.

In Section 4 of Chapter 5, using the multivariate least squares estimator, the hypothesis was rejected. The conflicting results are due to the larger estimated variance with which the three stage least squares estimator is computed with these data, as one would expect from Tauchen's (1986) work. As remarked earlier, the multivariate least squares estimator is computed from higher quality data than the three stage least squares estimator in this instance, so that the multivariate least squares results are more credible.   □

Let $\tilde{\theta}$ denote the value of θ that minimizes $S(\theta, \hat{V})$ subject to $h(\theta) = 0$. Equivalently, let $\hat{\rho}$ denote the value of ρ that achieves the unconstrained minimum of $S[g(\rho), \hat{V}]$, and put $\tilde{\theta} = g(\hat{\rho})$. The "likelihood ratio" test statistic for the hypothesis

$$H: h(\theta^0) = 0 \quad \text{against} \quad A: h(\theta^0) \neq 0$$

is

$$L = S(\tilde{\theta}, \hat{V}) - S(\hat{\theta}, \hat{V}).$$

It is essential that $\hat{V}$ be the same matrix in both terms on the right hand side; they must be exactly the same, not just "asymptotically equivalent."

One rejects $H: h(\theta^0) = 0$ when L exceeds the upper $\alpha \times 100\%$ critical point χ_α^2 of the chi-square distribution with q degrees of freedom; $\chi_\alpha^2 = (\chi^2)^{-1}(1 - \alpha, q)$.

Let θ^* denote the value of θ that minimizes

$$S^0(\theta, V) = [n\mathscr{E}m_n(\theta)]'V^{-1}[\mathscr{E}m_n(\theta)]$$

subject to

$$h(\theta) = 0.$$

Equivalently, let ρ^0 denote the value of ρ that achieves the unconstrained minimum of $S^0[g(\rho), V]$, and put $\theta^* = g(\rho^0)$.

Under the alternative, $A: h(\theta^0) \neq 0$, the "likelihood ratio" test statistic L is approximately distributed as the noncentral chi-square with q degrees of freedom and noncentrality parameter

$$\lambda = \frac{1}{2}\left(\sum_{t=1}^n \mathscr{E}[q(y_t, x_t, \theta^*) \otimes Z(x_t)]\right)'V^{-1}\left(\sum_{t=1}^n \mathscr{E}[Q(y_t, x_t, \theta^*) \otimes Z(x_t)]\right)$$

$$\times J^{-1}H'(HJ^{-1}H')HJ^{-1}$$

$$\times \left(\sum_{t=1}^n \mathscr{E}[Q(y_t, x_t, \theta^*) \otimes Z(x_t)]\right)'V^{-1}\left(\sum_{t=1}^n \mathscr{E}[q(y_t, x_t, \theta^*) \otimes Z(x_t)]\right)$$

where

$$V = \mathscr{C}\{[nm_n(\theta^0)], [nm_n(\theta^0)]'\}$$

$$Q(y_t, x_t, \theta) = \frac{\partial}{\partial\theta'}q(y_t, x_t, \theta)$$

$$J = \left(\sum_{t=1}^n \mathscr{E}[Q(y_t, x_t, \theta^*) \otimes Z(x_t)]\right)'$$

$$\times V^{-1}\left(\sum_{t=1}^n \mathscr{E}[Q(y_t, x_t, \theta^*) \otimes Z(x_t)]\right)$$

$$H = H(\theta^*) = \frac{\partial}{\partial\theta'}h(\theta^*).$$

Alternative expressions for λ can be obtained using Taylor's theorem and the relationship $H'(HJ^{-1}H')^{-1}H = J - JG(G'JG)^{-1}G'J$ from Section 6 of Chapter 3; see Gallant and Jorgenson (1979).

EXAMPLE 1 (Continued). The hypothesis of homogeneity in the model

$$q(y, x, \theta) = \begin{vmatrix} y_1 - \ln \dfrac{\theta_1 + \theta_2 r_1 + \theta_3 r_2 + \theta_4 r_3 - (\theta_2 + \theta_3 + \theta_4) y_3}{-1 + \theta_4 r_1 + \theta_7 r_2 + \theta_8 r_3 - (\theta_4 + \theta_7 + \theta_8) y_3} \\ y_2 - \ln \dfrac{\theta_5 + \theta_3 r_1 + \theta_6 r_2 + \theta_7 r_3 - (\theta_3 + \theta_6 + \theta_7) y_3}{-1 + \theta_4 r_1 + \theta_7 r_2 + \theta_8 r_3 - (\theta_4 + \theta_7 + \theta_8) y_3} \end{vmatrix}$$

can be expressed as the functional dependence

$$\theta = \begin{pmatrix} \theta_1 \\ \theta_2 \\ \theta_3 \\ \theta_4 \\ \theta_5 \\ \theta_6 \\ \theta_7 \\ \theta_8 \end{pmatrix} = \begin{pmatrix} \theta_1 \\ -\theta_3 - \theta_4 \\ \theta_3 \\ \theta_4 \\ \theta_5 \\ -\theta_7 - \theta_3 \\ \theta_7 \\ -\theta_4 - \theta_7 \end{pmatrix} = \begin{pmatrix} \rho_1 \\ -\rho_2 - \rho_3 \\ \rho_2 \\ \rho_3 \\ \rho_4 \\ -\rho_5 - \rho_2 \\ \rho_5 \\ -\rho_5 - \rho_3 \end{pmatrix} = g(\rho).$$

Minimization of $S[g(\rho), \hat{V}]$ as shown in Figure 4 gives

$$S(\tilde{\theta}, \hat{V}) = 38.34820 \qquad \text{(from Fig. 4)}$$

and we have

$$S(\hat{\theta}, \hat{V}) = 34.96403 \qquad \text{(from Fig. 1)}$$

from Figure 1 of Section 2. Thus

$$\begin{aligned} L &= S(\tilde{\theta}, \hat{V}) - S(\hat{\theta}, \hat{V}) \\ &= 38.34820 - 34.96403 \\ &= 3.38417. \end{aligned}$$

Since $(\chi^2)^{-1}(.95; 3) = 7.815$, the hypothesis is accepted at the 5% level. Notes in Figures 1 and 4 that $\hat{V}$ is computed the same. As mentioned several times, the test is invalid if care is not taken to be certain that this is so. $\square$

SAS Statements:

```
PROC MODEL OUT=MODO2;
ENDOGENOUS Y1 Y2 Y3;
EXOGENOUS  R1 R2 R3 D0 D1 D2 D3 D4 D5 D6 D7 D8 D9 D10 D11 D12 D13;
PARMS RO1 RO2 RO3 RO4 RO5 CO C1 C2 C3 C4 C5 C6 C7 C8 C9 C10 C11 C12 C13;
T1=RO1; T2=-RO2-RO3; T3=RO2; T4=RO3; T5=RO4; T6=-RO5-RO2; T7=RO5; T8=-RO5-RO3;
PEAK= T1+T2*R1+T3*R2+T4*R3-(T2+T3+T4)*Y3;
INTER=T5+T3*R1+T6*R2+T7*R3-(T3+T6+T7)*Y3;
BASE= -1+T4*R1+T7*R2+T8*R3-(T4+T7+T8)*Y3;
Y1=LOG(PEAK/BASE);  Y2=LOG(INTER/BASE);
Y3=D0*C0+D1*C1+D2*C2+D3*C3+D4*C4+D5*C5+D6*C6+D7*C7+D8*C8+D9*C9+D10*C10+D11*C11
   +D12*C12+D13*C13;
PROC SYSNLIN DATA=EGO1 MODEL=MODO2 N3SLS METHOD=GAUSS MAXIT=50 CONVERGE=1.E-8
            SDATA=SHAT;
INSTRUMENTS  R1 R2 R3 D0 D1 D2 D3 D4 D5 D6 D7 D8 D9 D10 D11 D12 D13 / NOINT;
FIT Y1 Y2 START = (RO1 -3 RO2 .8 RO3 .4 RO4 -1.5 RO5 .03);
```

Output:

SAS   8

NONLINEAR 3SLS PARAMETER ESTIMATES

PARAMETER	ESTIMATE	APPROX. STD ERROR	'T' RATIO	APPROX. PROB>!T!
RO1	-2.66573	0.17608	-15.14	0.0001
RO2	0.84953	0.06641	12.79	0.0001
RO3	0.37591	0.02686	13.99	0.0001
RO4	-1.56635	0.07770	-20.16	0.0001
RO5	0.06129	0.03408	1.80	0.0735

NUMBER OF OBSERVATIONS		STATISTICS FOR SYSTEM	
USED	220	OBJECTIVE	0.17431
MISSING	4	OBJECTIVE*N	38.34820

COVARIANCE OF RESIDUALS MATRIX USED FOR ESTIMATION

S	Y1	Y2
Y1	0.17159	0.09675
Y2	0.09675	0.09545

Figure 4. Example 1 fitted by nonlinear three stage least squares, homogeneity imposed.

The Lagrange multiplier test is most apt to be used when the constrained estimator $\tilde{\theta}$ is much easier to compute than the unconstrained estimator $\hat{\theta}$, so it is somewhat unreasonable to expect that a variance estimate $\hat{V}$ computed from unconstrained residuals will be available. Accordingly, let

$$\tilde{\theta}^{\#} = \underset{h(\theta)=0}{\operatorname{argmin}} S\left(\theta, I \otimes \sum_{t=1}^{n} z_t z_t'\right).$$

If the model is a pure regression situation, put

$$\tilde{V} = \left(\frac{1}{n} \sum_{t=1}^{n} q(y_t, x_t, \tilde{\theta}^{\#}) q'(y_t, x_t, \tilde{\theta}^{\#}) \right) \otimes \sum_{t=1}^{n} z_t z_t';$$

if the model is the regression situation with heteroscedastic errors, put

$$\tilde{V} = \sum_{t=1}^{n} \left[q(y_t, x_t, \tilde{\theta}^{\#}) \otimes z_t \right] \left[q(y_t, x_t, \tilde{\theta}^{\#}) \otimes z_t \right]';$$

or if the model is dynamic, put

$$\tilde{V} = n\tilde{S}_n$$

where, letting $l(n)$ denote the integer nearest $n^{1/5}$,

$$\tilde{S}_n = \sum_{\tau = -l(n)}^{l(n)} w\left(\frac{\tau}{l(n)} \right) \tilde{S}_{n\tau}$$

$$w(x) = \begin{cases} 1 - 6|x|^2 + 6|x|^3 & 0 \le x \le \frac{1}{2} \\ 2(1 - |x|)^3 & \frac{1}{2} \le x \le 1 \end{cases}$$

$$\tilde{S}_{n\tau} = \begin{cases} \frac{1}{n} \sum_{t=1+\tau}^{n} \left[q(y_t, x_t, \tilde{\theta}^{\#}) \otimes z_t \right] \left[q(y_{t-\tau}, x_{t-\tau}, \tilde{\theta}^{\#}) \otimes z_{t-\tau} \right]' \\ \hspace{10cm} \tau \ge 0 \\ (\tilde{S}_{n, -\tau})' \\ \hspace{10cm} \tau < 0. \end{cases}$$

Let

$$\tilde{\tilde{\theta}} = \operatorname*{argmin}_{h(\theta)=0} S(\theta, \tilde{V}).$$

The Gauss-Newton step away from $\tilde{\tilde{\theta}}$ (presumably) toward $\hat{\theta}$ is

$$\tilde{D} = D(\tilde{\tilde{\theta}}, \tilde{V}).$$

The Lagrange multiplier test statistic for the hypothesis

$$H: h(\theta^0) = 0 \quad \text{against} \quad A: h(\theta^0) \ne 0$$

is

$$R = \tilde{D}'\tilde{J}\tilde{D}$$

where

$$\tilde{J} = \left(\sum_{t=1}^{n} Q(y_t, x_t, \tilde{\theta}) \otimes z_t \right)' \tilde{V}^{-1} \left(\sum_{t=1}^{n} Q(y_t, x_t, \tilde{\theta}) \otimes z_t \right).$$

One rejects $H: h(\theta^0) = 0$ when R exceeds the upper $\alpha \times 100\%$ critical point χ_{α}^2 of the chi-square distribution with q degrees of freedom; $\chi_{\alpha}^2 = (\chi^2)^{-1}(1 - \alpha, q)$.

The approximate nonnull distribution of the Lagrange multiplier test statistic is the same as the nonnull distribution of the "likelihood ratio" test statistic.

EXAMPLE 1 (Continued). The computations for the Lagrange multiplier test of homogeneity are shown in Figure 5. As in Figure 4, PROC MODEL defines the model $q[y, x, g(\rho)]$. In Figure 5, the first use of PROC SYSNLIN computes

$$\hat{\rho}^{\#} = \operatorname*{argmin}_{\rho} S\left(g(\rho), I \otimes \sum_{t=1}^{n} z_t z_t' \right)$$

$$\tilde{V} = \left(\frac{1}{n} \sum_{t=1}^{n} q(y_t, x_t, \tilde{\theta}^{\#}) q'(y_t, x_t, \tilde{\theta}^{\#}) \right) \otimes \sum_{t=1}^{n} z_t z_t'$$

where

$$\tilde{\theta}^{\#} = g(\hat{\rho}^{\#}).$$

The second use of PROC SYSNLIN computes

$$\hat{\rho} = \operatorname*{argmin}_{\rho} S[g(\rho), \tilde{V}].$$

The subsequence **DATA WO1** statement computes

$$\tilde{q}_t = q(y_t, x_t, \tilde{\theta})$$

$$\tilde{Q}_t = \frac{\partial}{\partial \theta'} q(y_t, x_t, \tilde{\theta})$$

where

$$\tilde{\theta} = g(\hat{\rho}).$$

```
SAS Statements:

PROC MODEL OUT=MOD02;
ENDOGENOUS Y1 Y2 Y3;
EXOGENOUS  R1 R2 R3 D0 D1 D2 D3 D4 D5 D6 D7 D8 D9 D10 D11 D12 D13;
PARMS RO1 RO2 RO3 RO4 RO5 C0 C1 C2 C3 C4 C5 C6 C7 C8 C9 C10 C11 C12 C13;
T1=RO1; T2=-RO2-RO3; T3=RO2; T4=RO3; T5=RO4; T6=-RO5-RO2; T7=RO5; T8=-RO5-RO3;
PEAK= T1+T2*R1+T3*R2+T4*R3-(T2+T3+T4)*Y3;
INTER=T5+T3*R1+T6*R2+T7*R3-(T3+T6+T7)*Y3;
BASE= -1+T4*R1+T7*R2+T8*R3-(T4+T7+T8)*Y3;
Y1=LOG(PEAK/BASE);   Y2=LOG(INTER/BASE);
Y3=D0*C0+D1*C1+D2*C2+D3*C3+D4*C4+D5*C5+D6*C6+D7*C7+D8*C8+D9*C9+D10*C10+D11*C11
   +D12*C12+D13*C13;

PROC SYSNLIN DATA=EG01 MODEL=MOD02 N3SLS METHOD=GAUSS MAXIT=50 CONVERGE=1.E-8
            SDATA=IDENTITY OUTS=STILDE OUTEST=TSHARP;
INSTRUMENTS  R1 R2 R3 D0 D1 D2 D3 D4 D5 D6 D7 D8 D9 D10 D11 D12 D13 / NOINT;
FIT Y1 Y2 START = (RO1 -3 RO2 .8 RO3 .4 RO4 -1.5 RO5 .03);
PROC SYSNLIN DATA=EG01 MODEL=MOD02 N3SLS METHOD=GAUSS MAXIT=50 CONVERGE=1.E-7
            SDATA=STILDE ESTDATA=TSHARP OUTEST=RHOHAT;
INSTRUMENTS  R1 R2 R3 D0 D1 D2 D3 D4 D5 D6 D7 D8 D9 D10 D11 D12 D13 / NOINT;
FIT Y1 Y2;

DATA W01;  IF _N_=1 THEN SET RHOHAT;  SET EG01;  RETAIN RO1-RO5;
T1=RO1; T2=-RO2-RO3; T3=RO2; T4=RO3; T5=RO4; T6=-RO5-RO2; T7=RO5; T8=-RO5-RO3;
PEAK= T1+T2*R1+T3*R2+T4*R3-(T2+T3+T4)*Y3;
INTER=T5+T3*R1+T6*R2+T7*R3-(T3+T6+T7)*Y3;
BASE= -1+T4*R1+T7*R2+T8*R3-(T4+T7+T8)*Y3;
Q1=Y1-LOG(PEAK/BASE);  Q2=Y2-LOG(INTER/BASE);
DQ1T1=-1/PEAK;                   DQ2T1=0;
DQ1T2=-(R1-Y3)/PEAK;             DQ2T2=0;
DQ1T3=-(R2-Y3)/PEAK;             DQ2T3=-(R1-Y3)/INTER;
DQ1T4=-(R3-Y3)/PEAK+(R1-Y3)/BASE; DQ2T4=(R1-Y3)/BASE;
DQ1T5=0;                         DQ2T5=-1/INTER;
DQ1T6=0;                         DQ2T6=-(R2-Y3)/INTER;
DQ1T7=(R2-Y3)/BASE;              DQ2T7=-(R3-Y3)/INTER+(R2-Y3)/BASE;
DQ1T8=(R3-Y3)/BASE;              DQ2T8=(R3-Y3)/BASE;
IF NMISS(OF D0-D13) > 0 THEN DELETE;
KEEP Q1 Q2 DQ1T1-DQ1T8 DQ2T1-DQ2T8 R1-R3 D0-D13;
```

Figure 5. Illustration of Lagrange multiplier test computations with Example 1.

Finally PROC MATRIX is used to compute

$$\tilde{J} = \left(\sum_{t=1}^{n} \tilde{Q}_t \otimes z_t \right)' \tilde{V}^{-1} \left(\sum_{t=1}^{n} \tilde{Q}_t \otimes z_t \right)$$

$$\tilde{D} = \tilde{J}^{-1} \left(\sum_{t=1}^{n} \tilde{Q}_t \otimes z_t \right)' \tilde{V}^{-1} \left(\sum_{t=1}^{n} \tilde{q}_t \otimes z_t \right)$$

and

$$R = \tilde{D}'\tilde{J}\tilde{D}$$
$$= 3.36375 \quad \text{(from Fig. 5).}$$

Since $(\chi^2)^{-1}(.95, 3) = 7.815$, the hypothesis is accepted at the 5% level. □

```
PROC MATRIX;
FETCH Q1 DATA=W01(KEEP=Q1);   FETCH DQ1 DATA=W01(KEEP=DQ1T1-DQ1T8);
FETCH Q2 DATA=W01(KEEP=Q2);   FETCH DQ2 DATA=W01(KEEP=DQ2T1-DQ2T8);
FETCH Z DATA=W01(KEEP=R1-R3 D0-D13);   FETCH STILDE DATA=STILDE(KEEP=Y1 Y2);
M=J(34,1,0);  DELM=J(34,8,0);  V=J(34,34,0);
DO T=1 TO 220;
 QT=Q1(T,)//Q2(T,);   DELQT=DQ1(T,)//DQ2(T,);
 MT = QT @ Z(T,)';   DELMT = DELQT @ Z(T,)';
 M=M+MT;  DELM=DELM+DELMT;   V=V+STILDE @ (Z(T,)'*Z(T,));
END;
CHAT=INV(DELM'*INV(V)*DELM);   D=-CHAT*DELM'*INV(V)*M;
R=D'*INV(CHAT)*D;  PRINT R;
```

Output:

SAS   8

R   COL1

ROW1   3.36375

Figure 5. (Continued).

There is one other test that is commonly used in connection with three stage least squares and generalized method of moments estimation, called the test of the overidentifying restrictions. The terminology is a holdover from the linear case; the test is a model specification test. The idea is that certain linear combinations of the rows of $\sqrt{n}\,m_n(\hat{\theta})$ are asymptotically normally distributed with zero mean if the model is correctly specified. The estimator $\hat{\theta}$ is the minimizer of $S(\theta, \hat{V})$, so it must satisfy the restriction that $[(\partial/\partial\theta)m_n(\theta)]'\hat{V}^{-1}m_n(\theta) = 0$. This is equivalent to a statement that

$$\left[\sqrt{n}\,m_n(\hat{\theta})\right]' - \hat{r}'\hat{H} = 0$$

for some full rank matrix $\hat{H}$ of order $MK - p$ by MK that has rows which are orthogonal to the rows of $[(\partial/\partial\theta)m_n(\theta)]'\hat{V}^{-1}$. This fact and arguments similar to either Theorem 13 of Chapter 3 or Theorem 14 of Chapter 7 lead to the conclusion that $\sqrt{n}\,m_n(\hat{\theta})$ is asymptotically distributed as the singular normal with a rank $MK - p$ variance-covariance matrix. This fact and arguments similar to the proof of Theorem 14 of Chapter 3 or Theorem 16 of Chapter 7 lead to the conclusion that $S(\hat{\theta}, \hat{V})$ is asymptotically distributed as a chi-square random variable with $MK - p$ degrees of freedom under the hypothesis that the model is correctly specified.

One rejects the hypothesis that the model is correctly specified when $S(\hat{\theta}, \hat{V})$ exceeds the upper $\alpha \times 100\%$ critical point χ_α^2 of the chi-square distribution with $MK - p$ degrees of freedom; $\chi_\alpha^2 = (\chi^2)^{-1}(1 - \alpha, MK - p)$.

EXAMPLES 1, 2 (Continued). In Example 1 we have

$$S(\hat{\theta}, \hat{V}) = 34.96403 \qquad \text{(from Fig. 1)}$$
$$MK - p = 2 \times 17 - 8 = 26 \qquad \text{(from Fig. 1)}$$
$$\left(\chi^2\right)^{-1}(.95, 26) = 38.885$$

and the model specification is accepted. In Example 2 we have

$$S(\hat{\theta}, \hat{V}) = 1.05692 \qquad \text{(from Fig. 2)}$$
$$MK - p = 1 \times 3 - 2 = 1 \qquad \text{(from Fig. 2)}$$
$$\left(\chi^2\right)^{-1}(.95, 1) = 3.841$$

and the model specification is accepted. □

PROBLEMS

1. Use Theorem 12 of Chapter 7 and the expressions given in the example of Section 5 of Chapter 7 to derive the Wald test statistic.
2. Use Theorem 15 of Chapter 7 and the expressions given in the example of Section 5 of Chapter 7 to derive the "likelihood ratio" test statistic.
3. Use Theorem 16 of Chapter 7 and the expressions given in the example of Section 5 of Chapter 7 to derive the Lagrange multiplier test statistic in the form

$$R = \tilde{D}'\tilde{H}'\left(\tilde{H}\tilde{J}^{-1}\tilde{H}'\right)^{-1}\tilde{H}\tilde{D}.$$

Use

$$\tilde{H}'\left(\tilde{H}\tilde{J}^{-1}\tilde{H}'\right)^{-1}\tilde{H} = \tilde{J} - \tilde{J}\tilde{G}\left(\tilde{G}'\tilde{J}\tilde{G}\right)^{-1}\tilde{G}'\tilde{J}$$

$$\frac{\partial}{\partial\theta}\left[S(\tilde{\theta}, \tilde{V}) + \tilde{\mu}'h(\tilde{\theta})\right] = 0 \qquad \text{for some Lagrange multiplier } \tilde{\mu}$$

and $\tilde{H}\tilde{G} = 0$ to put the statistic in the form

$$R = \tilde{D}'\tilde{J}\tilde{D}.$$

5. MAXIMUM LIKELIHOOD ESTIMATION

The simplest case, and the one that we consider first, is the regression case where the errors are independently distributed, no lagged dependent variables are used as explanatory variables, and the analysis is conditional on the explanatory variables.

The setup is the same as in Section 2. Multivariate responses y_t, L-vectors, are assumed to be determined by k-dimensional independent variables x_t according to the system of simultaneous equations

$$q_\alpha(y_t, x_t, \theta_\alpha^0) = e_{\alpha t} \qquad \alpha = 1, 2, \ldots, L \quad t = 1, 2, \ldots, n$$

where each $q_\alpha(y, x, \theta_\alpha)$ is a real valued function, each θ_α^0 is a p_α-dimensional vector of unknown parameters, and the $e_{\alpha t}$ represent unobservable observational or experimental errors.

All the equations of the system are used in estimation, so that, according to the notational conventions adopted in Section 2, $M = L$ and

$$q(y, x, \theta) = \begin{pmatrix} q_1(y, x, \theta_1) \\ q_2(y, x, \theta_2) \\ \vdots \\ q_M(y, x, \theta_M) \end{pmatrix} \qquad e = \begin{pmatrix} e_1 \\ e_2 \\ \vdots \\ e_M \end{pmatrix}$$

$$q(y_t, x_t, \theta) = \begin{pmatrix} q_1(y_t, x_t, \theta_1) \\ q_2(y_t, x_t, \theta_2) \\ \vdots \\ q_M(y_t, x_t, \theta_M) \end{pmatrix} \qquad e_t = \begin{pmatrix} e_{1t} \\ e_{2t} \\ \vdots \\ e_{Mt} \end{pmatrix}$$

where θ is a p-vector containing the nonredundant elements of the parameter vectors θ_α, $\alpha = 1, 2, \ldots, M$. The error vectors e_t are independently and identically distributed with common density $p(e \mid \sigma^0)$, where σ is an r-vector. The functional form $p(e \mid \sigma)$ of the error distribution is assumed to be known. In the event that something such as $Q(y_t, x_t, \beta) = u_t$ with $p_t(u_t) = p(u_t \mid x_t, \tau, \sigma)$ is envisaged, one often can find a transformation $\psi(Q, x_t, \tau)$ which will put the model

$$q(y_t, x_t, \theta) = \psi[Q(y_t, x_t, \beta), x_t, \tau] = \psi(u_t, \tau) = e_t$$

into a form that has $p_t(e_t) = p(e_t \mid \sigma)$ where $\theta = (\beta, \tau)$.

Usually normality is assumed in applications, which we indicate by writing $n(e \mid \sigma)$ for $p(e \mid \sigma)$, where σ denotes the unique elements of

$$\Sigma = \mathscr{C}(e_t, e_t')$$

viz.

$$\sigma = \left(\sigma_{11}, \sigma_{12}, \sigma_{22}, \sigma_{13}, \sigma_{23}, \sigma_{33}, \ldots, \sigma_{1M}, \sigma_{2M}, \ldots, \sigma_{MM}\right)'.$$

Let $\Sigma = \Sigma(\sigma)$ denote the mapping of this vector back to the original matrix Σ. With these conventions, the functional form of $n(e \mid \sigma)$ is

$$n(e \mid \sigma) = (2\pi)^{-M/2} \det[\Sigma(\sigma)]^{-1/2} \exp\left\{-\tfrac{1}{2}e'[\Sigma(\sigma)]^{-1}e\right\}.$$

The assumption that the functional form of $p(e \mid \sigma)$ must be known is the main impediment to the use of maximum likelihood methods. Unlike the multivariate least squares case, where a normality assumption does not upset the robustness of validity of the asymptotics, with an implicit model as considered here an error in specifying $p(e \mid \sigma)$ can induce serious bias. The formula for computing the bias is given below, and the issue is discussed in some detail in the papers by Amemiya (1977, 1982) and Phillips (1982b).

Given any value of the error vector e from the set of admissible values $\mathscr{E} \subset \mathbf{R}^M$, any value of the vector of independent variables x from the set of admissible values $\mathscr{X} \subset \mathbf{R}^k$, and any value of the parameter vector θ from the set of admissible values $\Theta \subseteq \mathbf{R}^p$, the model

$$q(y, x, \theta) = e$$

is assumed to determine y uniquely; if the equations have multiple roots, there is some rule for determining which solution is meant. This is the same as stating the model determines a reduced form

$$y = Y(e, x, \theta)$$

mapping $\mathscr{E} \times \mathscr{X} \times \Theta$ onto $\mathscr{Y} \subseteq \mathbf{R}^M$; for each (x, θ) in $\mathscr{X} \times \Theta$ the mapping

$$Y(\cdot, x, \theta): \mathscr{E} \to \mathscr{Y}$$

is assumed to be one to one, onto. It is to be emphasized that while a reduced form must exist, it is not necessary to find it analytically or even to be able to compute it numerically in applications.

These assumptions imply the existence of a conditional density on $\mathscr{Y}$

$$p(y \mid x, \theta, \sigma) = \left|\det \frac{\partial}{\partial y'} q(y, x, \theta)\right| p[q(y, x, \theta) \mid \sigma].$$

A conditional expectation is computed as either

$$\mathscr{E}(T \mid x) = \int_{\mathscr{Y}} T(y) p(y \mid x, \theta, \sigma) \, dy$$

or

$$\mathscr{E}(T \mid x) = \int_{\mathscr{E}} T[Y(e, x, \theta)] p(e \mid \sigma) \, de$$

whichever is the more convenient.

EXAMPLE 3. Consider the model

$$q(y, x, \theta) = \begin{pmatrix} \theta_1 + \ln y_1 + \theta_2 x \\ \theta_3 + \theta_4 y_1 + y_2 + \theta_5 x \end{pmatrix}.$$

The reduced form is

$$Y(e, x, \theta) = \begin{pmatrix} \exp(e_1 - \theta_1 - \theta_2 x) \\ e_2 - \theta_3 - \theta_4 \exp(e_1 - \theta_1 - \theta_2 x) - \theta_5 x \end{pmatrix}.$$

Under normality, the conditional density defined on $\mathscr{Y} = (0, \infty) \times (-\infty, \infty)$ has the form

$$p(y \mid x, \theta, \sigma) = (2\pi)^{-1} (\det \Sigma)^{-1/2} \frac{1}{y_1} \exp\left[-\tfrac{1}{2} q'(y, x, \theta) \Sigma^{-1} q(y, x, \theta)\right]$$

where $\Sigma = \Sigma(\sigma)$. □

The normalized, negative log-likelihood is

$$s_n(\theta, \sigma) = \frac{1}{n} \sum_{t=1}^{n} (-1) \ln p(y \mid x, \theta, \sigma)$$

and the maximum likelihood estimator is the value $(\hat{\theta}, \hat{\sigma})$ that minimizes $s_n(\theta, \sigma)$; that is,

$$(\hat{\theta}, \hat{\sigma}) = \operatorname*{argmin}_{(\theta, \sigma)} s_n(\theta, \sigma)$$

Asymptotically,

$$\sqrt{n} \begin{pmatrix} \hat{\theta} - \theta^0 \\ \hat{\sigma} - \sigma^0 \end{pmatrix} \xrightarrow{\mathscr{L}} N_{p+r}(0, V).$$

Put $\lambda = (\theta, \sigma)$. Either the inverse of

$$\hat{\mathscr{J}} = \frac{1}{n} \sum_{t=1}^{n} \left(\frac{\partial}{\partial \lambda} (-1) \ln p(y_t \mid x_t, \hat{\theta}, \hat{\sigma}) \right) \left(\frac{\partial}{\partial \lambda} (-1) \ln p(y_t \mid x_t, \hat{\theta}, \hat{\sigma}) \right)'$$

or the inverse of

$$\hat{\mathscr{J}} = \frac{\partial^2}{\partial \lambda \, \partial \lambda'} s_n(\hat{\theta}, \hat{\sigma})$$

$$= \frac{1}{n} \sum_{t=1}^{n} \frac{\partial^2}{\partial \lambda \, \partial \lambda'} (-1) \ln p(y_t \mid x_t, \hat{\theta}, \hat{\sigma})$$

will estimate V consistently. Suppose that $\hat{V} = \hat{\mathscr{J}}^{-1}$ is used to estimate V. With these normalization conventions, a 95% confidence interval on the ith element of θ is computed as

$$\hat{\theta}_i \pm z_{.025} \frac{\sqrt{\hat{\mathscr{J}}^{ii}}}{\sqrt{n}}$$

and a 95% confidence interval on the ith element of σ is computed as

$$\hat{\sigma}_i \pm z_{.025} \frac{\sqrt{\hat{\mathscr{J}}^{p+i, \, p+i}}}{\sqrt{n}}$$

where $\mathscr{J}^{ij}$ denotes the ijth element of the inverse of a matrix $\mathscr{J}$, and $z_{.025} = N^{-1}(.025; 0, 1)$.

Under normality

$$s_n(\theta, \sigma) = \text{const} - \frac{1}{n} \sum_{t=1}^{n} \ln \left| \det \frac{\partial}{\partial y'} q(y_t, x, \theta) \right| + \tfrac{1}{2} \ln \det \Sigma(\sigma)$$

$$+ \tfrac{1}{2} \text{tr} \left([\Sigma(\sigma)]^{-1} \frac{1}{n} \sum_{t=1}^{n} q(y_t, x_t, \theta) q'(y_t, x_t, \theta) \right).$$

Define

$$J(y, x, \theta) = \frac{\partial}{\partial y'} q(y, x, \theta).$$

Using the relations (Problem 4, Section 5, Chapter 5)

$$\frac{\partial}{\partial \theta_i} \ln |\det A(\theta)| = \text{tr}[A(\theta)]^{-1} \frac{\partial}{\partial \theta_i} A(\theta)$$

$$\frac{\partial}{\partial \sigma_i} [\Sigma(\sigma)]^{-1} = -[\Sigma(\sigma)]^{-1} [\Sigma(\xi_i)] [\Sigma(\sigma)]^{-1}$$

where ξ_i is the ith elementary $M(M + 1)/2$-vector, and using

$$(-1)\ln p[y \mid x, \theta, \sigma] = \text{const} - \ln |\det J(y, x, \theta)| + \tfrac{1}{2}\ln \det \Sigma(\sigma)$$
$$+ \tfrac{1}{2}q'(y, x, \theta)[\Sigma(\sigma)]^{-1}q(y, x, \theta)$$

we have

$$\frac{\partial}{\partial\theta_i}(-1)\ln p[y \mid x, \theta, \sigma] = -\text{tr}\left([J(y, x, \theta)]^{-1}\frac{\partial}{\partial\theta_i}J(y, x, \theta)\right)$$
$$+ q'(y, x, \theta)[\Sigma(\sigma)]^{-1}\frac{\partial}{\partial\theta_i}q(y, x, \theta)$$

$$\frac{\partial}{\partial\sigma_i}(-1)\ln p[y \mid x, \theta, \sigma] = \tfrac{1}{2}\text{tr}\left\{[\Sigma(\sigma)]^{-1}\Sigma(\xi_i)\right\}$$
$$- \tfrac{1}{2}q'(y, x, \theta)[\Sigma(\sigma)]^{-1}[\Sigma(\xi_i)]$$
$$\times [\Sigma(\sigma)]^{-1}q(y, x, \theta).$$

Interest usually centers in the parameter θ, with σ regarded as a nuisance parameter. If

$$\sigma(\theta) = \underset{\sigma}{\text{argmin}}\, s_n(\theta, \sigma)$$

is easy to compute and the "concentrated likelihood"

$$s_n(\theta) = s_n[\theta, \sigma(\theta)]$$

has a tractable analytic form, then alternative formulas may be used. They are as follows.

The maximum likelihood estimators are computed as

$$\hat{\theta} = \underset{\theta}{\text{argmin}}\, s_n(\theta)$$
$$\hat{\sigma} = \sigma(\hat{\theta})$$

and these will, of course, be the same numerical values that would obtain from a direct minimization of $s_n(\theta, \sigma)$. Partition $\mathcal{I}$ as

$$\mathcal{I} = \begin{pmatrix} \mathcal{I}_{\theta\theta} & \mathcal{I}_{\theta\sigma} \\ \mathcal{I}_{\sigma\theta} & \mathcal{I}_{\sigma\sigma} \end{pmatrix} \begin{matrix} p \text{ rows} \\ r \text{ rows} \end{matrix}$$
$$\begin{matrix} p & r \\ \text{cols} & \text{cols} \end{matrix}$$

Partition V similarly, With these partitionings, the following relationship holds (Rao, 1973, p. 33):

$$V_{\theta\theta} = \left(\mathscr{I}_{\theta\theta} - \mathscr{I}_{\theta\sigma}\mathscr{I}_{\sigma\sigma}^{-1}\mathscr{I}_{\sigma\theta} \right)^{-1}.$$

We have from above that, asymptotically,

$$\sqrt{n}\,(\hat{\theta} - \theta^0) \xrightarrow{\mathscr{L}} N_p(0, V_{\theta\theta}).$$

One can show (Problem 6) that either

$$\hat{\mathscr{X}} = \frac{1}{n} \sum_{t=1}^{n} \left(\frac{\partial}{\partial\theta}\ln p\left[y_t|x_t, \hat{\theta}, \sigma(\hat{\theta})\right]\right)\left(\frac{\partial}{\partial\theta}\ln p\left[y_t|x_t, \hat{\theta}, \sigma(\hat{\theta})\right]\right)'$$

or

$$\hat{\mathscr{L}} = \frac{\partial^2}{\partial\theta\,\partial\theta'}s_n(\hat{\theta})$$

$$= \frac{1}{n} \sum_{t=1}^{n} \frac{\partial^2}{\partial\theta\,\partial\theta'}(-1)\ln p\left[y_t|x_t, \hat{\theta}, \sigma(\hat{\theta})\right]$$

will estimate

$$\mathscr{I}_{\theta\theta} - \mathscr{I}_{\theta\sigma}\mathscr{I}_{\sigma\sigma}^{-1}\mathscr{I}_{\sigma\theta}$$

consistently. Thus, either $\hat{\mathscr{X}}^{-1}$ or $\hat{\mathscr{L}}^{-1}$ may be used to estimate $V_{\theta\theta}$. Note that it is necessary to compute a total derivative in the formulas above; for example,

$$\frac{\partial}{\partial\theta'}(-1)\ln p\left[y|x, \theta, \sigma(\theta)\right]$$

$$= \frac{-1}{p(y|x,\theta,\sigma)}\left(\frac{\partial}{\partial\theta'}p(y|x,\theta,\sigma) + \frac{\partial}{\partial\sigma'}p(y|x,\theta,\sigma)\frac{\partial}{\partial\theta'}\sigma(\theta)\right)\Bigg|_{\sigma=\sigma(\theta)}.$$

Suppose that $\hat{V}_{\theta\theta} = \hat{\mathscr{X}}^{-1}$ is used to estimate $V_{\theta\theta}$. With these normalization conventions, a 95% confidence interval on the ith element of θ is computed as

$$\hat{\theta}_i \pm z_{.025}\frac{\sqrt{\hat{\mathscr{X}}^{ii}}}{\sqrt{n}}$$

where $\mathcal{X}^{ij}$ denotes the ijth element of the inverse of the matrix $\mathcal{X}$, and $z_{.025} = N^{-1}(.025; 0, 1)$.

Under normality,

$$s_n(\theta, \sigma) = \text{const} - \frac{1}{n} \sum_{t=1}^{n} \ln \left| \det \frac{\partial}{\partial y'} q(y_t, x_t, \theta) \right| + \tfrac{1}{2} \ln \det \Sigma(\sigma)$$

$$+ \tfrac{1}{2} \text{tr} \left([\Sigma(\sigma)]^{-1} \frac{1}{n} \sum_{t=1}^{n} q(y_t, x_t, \theta) q'(y_t, x_t, \theta) \right)$$

which implies that

$$\Sigma(\theta) = \Sigma[\sigma(\theta)] = \frac{1}{n} \sum_{t=1}^{n} q(y_t, x_t, \theta) q'(y_t, x_t, \theta)$$

whence

$$s_n(\theta) = \text{const} - \frac{1}{n} \sum_{t=1}^{n} \ln \left| \det \frac{\partial}{\partial y'} q(y_t, x_t, \theta) \right|$$

$$+ \tfrac{1}{2} \ln \det \frac{1}{n} \sum_{t=1}^{n} q(y_t, x_t, \theta) q'(y_t, x_t, \theta).$$

Using the relations (Problem 4, Section 5, Chapter 5)

$$\frac{\partial}{\partial \theta_i} \ln |\det A(\theta)| = \text{tr} \, [A(\theta)]^{-1} \frac{\partial}{\partial \theta_i} A(\theta),$$

$$\frac{\partial}{\partial \theta_i} [\Sigma(\theta)]^{-1} = -[\Sigma(\theta)]^{-1} \left(\frac{\partial}{\partial \theta_i} \Sigma(\theta) \right) [\Sigma(\theta)]^{-1}$$

and

$$(-1)\ln p[y|x, \theta, \sigma(\theta)] = \text{const} - \ln |\det J(y, x, \theta)|$$

$$+ \tfrac{1}{2} \ln \det \Sigma(\theta) + \tfrac{1}{2} q'(y, x, \theta)$$

$$\times [\Sigma(\theta)]^{-1} q(y, x, \theta)$$

we have

$$\frac{\partial}{\partial \theta_i} (-1)\ln p[y|x, \theta, \sigma(\theta)]$$

$$= -\text{tr} \left([J(y, x, \theta)]^{-1} \frac{\partial}{\partial \theta_i} J(y, x, \theta) \right)$$

$$+ \tfrac{1}{2} \text{tr} \left[[\Sigma(\theta)]^{-1} \left(\frac{\partial}{\partial \theta_i} \Sigma(\theta) \right) \{ I - [\Sigma(\theta)]^{-1} q(y, x, \theta) q'(y, x, \theta) \} \right]$$

$$+ q'(y, x, \theta) [\Sigma(\theta)]^{-1} \frac{\partial}{\partial \theta} q(y, x, \theta)$$

where

$$\frac{\partial}{\partial \theta_i} \Sigma(\theta) = \frac{2}{n} \sum_{t=1}^{n} q(y, x, \theta) \frac{\partial}{\partial \theta_i} q'(y, x, \theta).$$

In summary, there are four ways that one might compute an estimate $\hat{V}_{\theta\theta}$ of $V_{\theta\theta}$, the asymptotic variance-covariance matrix of $\sqrt{n}\,(\hat{\theta} - \theta^0)$. Either $\hat{\mathscr{I}}$ or $\hat{\mathscr{J}}$ may be inverted and then partitioned to obtain $\hat{V}_{\theta\theta}$, or either $\hat{\mathscr{X}}$ or $\hat{\mathscr{L}}$ may be inverted to obtain $\hat{V}_{\theta\theta}$.

As to computations, the full Newton downhill direction is obtained by expanding $s_n(\theta, \sigma)$ in a Taylor's expansion about some trial value of the parameter $\lambda_T = (\theta'_T, \sigma'_T)'$:

$$s_n(\theta, \sigma) \doteq s_n(\theta_T, \sigma_T) + \left(\frac{\partial}{\partial \lambda'} s_n(\theta_T, \sigma_T) \right)(\lambda - \lambda_T)$$

$$+ \tfrac{1}{2}(\lambda - \lambda_T)' \left(\frac{\partial^2}{\partial \lambda\, \partial \lambda'} s_n(\theta_T, \sigma_T) \right)(\lambda - \lambda_T).$$

The minimum of this quadratic equation in λ is

$$\lambda_M = \lambda_T - \left(\frac{\partial^2}{\partial \lambda\, \partial \lambda'} s_n(\theta_T, \sigma_T) \right)^{-1} \left(\frac{\partial}{\partial \lambda} s_n(\theta_T, \sigma_T) \right)$$

whence a full Newton step away from the point (θ, σ) and, hopefully, toward the point $\hat{\lambda} = (\hat{\theta}', \hat{\sigma}')'$ is

$$D(\theta, \sigma) = -\left(\frac{\partial^2}{\partial \lambda\, \partial \lambda'} s_n(\theta, \sigma) \right)^{-1} \left(\frac{\partial}{\partial \lambda} s_n(\theta, \sigma) \right).$$

A minimization algorithm incorporating partial step lengths is constructed along the same lines as the modified Gauss-Newton algorithm, which is discussed in Section 4 of Chapter 1. Often,

$$\mathscr{I}(\theta, \sigma) = \frac{1}{n} \sum_{t=1}^{n} \left(\frac{\partial}{\partial \lambda} (-1) \ln p(y_t \mid x_t, \theta, \sigma) \right)$$

$$\times \left(\frac{\partial}{\partial \lambda} (-1) \ln p(y_t \mid x_t, \theta, \sigma) \right)'$$

can be accepted as an adequate approximation to $[(\partial^2/\partial \lambda\, \partial \lambda') s_n(\theta, \sigma)]$; see Problem 5.

To minimize the concentrated likelihood, the same approach leads to

$$D(\theta) = -\left(\frac{\partial^2}{\partial \theta\, \partial \theta'} s_n(\theta) \right)^{-1} \left(\frac{\partial}{\partial \theta} s_n(\theta) \right)$$

as the correction vector and

$$\mathscr{K}(\theta) = \frac{1}{n} \sum_{t=1}^{n} \left(\frac{\partial}{\partial \theta} \ln p[y_t | x_t, \theta, \sigma(\theta)] \right) \left(\frac{\partial}{\partial \theta} \ln p[y_t | x_t, \theta, \sigma(\theta)] \right)'$$

as an approximation to $(\partial^2 / \partial\theta \, \partial\theta') s_n(\theta)$.

As remarked earlier, the three stage least squares estimator only relies on moment assumptions for consistent estimation of the model parameters, whereas maximum likelihood relies on a correct specification of the error density. To be precise, the maximum likelihood estimator estimates the minimum of

$$\bar{s}(\theta, \sigma, \gamma^0) = \frac{1}{n} \sum_{t=1}^{n} \int_{\mathscr{Y}} (-1) \ln p(y | x_t, \theta, \sigma) p(y | x_t, \gamma^0) \, dy$$

where $p(y | x, \gamma^0)$ is the conditional density function of the true data generating process by Theorem 5 of Chapter 3. When the error density $p(e | \sigma)$ is correctly specified, the model parameters θ and the variance σ are estimated consistently by the information inequality (Problem 3). If not, the model parameters may be estimated consistently in some circumstances (Phillips, 1982b), but in general they will not be.

Consider testing the hypothesis

$$H : h(\theta^0) = 0 \quad \text{against} \quad A : h(\theta^0) \neq 0$$

where $h(\theta)$ maps $\mathbb{R}^p$ into $\mathbb{R}^q$ with Jacobian

$$H(\theta) = \frac{\partial}{\partial \theta'} h(\theta)$$

of order q by p, which is assumed to have rank q. The Wald test statistic is

$$W = n h'(\hat{\theta}) [H(\hat{\theta}) \hat{V}_{\theta\theta} H'(\hat{\theta})]^{-1} h(\hat{\theta})$$

where $\hat{V}_{\theta\theta}$ denotes any of the estimators of $V_{\theta\theta}$ described above.

Let $\tilde{\theta}$ denote the minimizer of $s_n(\theta, \sigma)$ or $s_n(\theta) = s_n[\theta, \sigma(\theta)]$ subject to the restriction that $h(\theta) = 0$, whichever is the easier to compute. Let $\tilde{V}_{\theta\theta}$ denote any one of the four formulas for estimating $V_{\theta\theta}$ described above, but with $\tilde{\theta}$ replacing $\hat{\theta}$ throughout. The Lagrange multiplier test statistic is

$$R = n \left(\frac{\partial}{\partial \theta} s_n(\tilde{\theta}) \right)' \tilde{V}_{\theta\theta} \left(\frac{\partial}{\partial \theta} s_n(\tilde{\theta}) \right).$$

The likelihood ratio test statistic is

$$L = 2n[s_n(\tilde{\theta}) - s_n(\hat{\theta})].$$

In each case, the null hypothesis $H: h(\theta^0) = 0$ is rejected in favor of the alternative hypothesis $A: h(\theta^0) \neq 0$ when the test statistic exceeds the upper $\alpha \times 100$ percentage point χ_α^2 of a chi-square random variable with q degrees of freedom; $\chi_\alpha^2 = (\chi^2)^{-1}(1 - \alpha; q)$. Under the alternative hypothesis, each test statistic is approximately distributed as a noncentral chi-square random variable with q degrees of freedom and noncentrality parameter

$$\lambda = n\frac{h'(\theta^0)[H(\theta^0)V_{\theta\theta}H'(\theta^0)]^{-1}h(\theta^0)}{2}$$

where $V_{\theta\theta}$ is computed by inverting and partitioning the matrix

$$\mathscr{I} = \frac{1}{n}\sum_{t=1}^{n}\int_{\mathscr{Y}}\left(\frac{\partial}{\partial\lambda}(-1)\ln p(y\,|\,x_t, \theta^0, \sigma^0)\right)$$

$$\times\left(\frac{\partial}{\partial\lambda}(-1)\ln p(y\,|\,x_t, \theta^0, \sigma^0)\right)' p(y\,|\,x_t, \theta^0, \sigma^0)\,dy.$$

The results of Chapter 3 justify these statistical methods. The algebraic relationships needed to reduce the general results of Chapter 3 to the formulas above are given in the problems. A set of specific regularity conditions that imply the assumptions of Chapter 3, and a detailed verification that this is so, are in Gallant and Holly (1980).

In the dynamic case, the structural model has the same form as above:

$$q(y_t, x_t, \theta^0) = e_t \qquad (t = 1, 2, \ldots, n)$$

with $q(y, x, \theta)$ mapping $\mathscr{Y} \times \mathscr{X} \times \Theta \subset \mathbb{R}^M \times \mathbb{R}^k \times \mathbb{R}^p$ onto $\mathscr{E} \subset \mathbb{R}^M$ and determining the one to one mapping $y = Y(e, x, \theta)$ of $\mathscr{E}$ onto $\mathscr{Y}$. Unlike the regression case, lagged endogenous variables

$$y_{t-1}, y_{t-2}, \ldots, y_{t-l}$$

may be included as components of x_t, and the errors e_t may be correlated. When lagged values are included, we shall assume that the data

$$y_0, y_{-1}, \ldots, y_{1-l}$$

are available, so that $q(y_t, x_t, \theta)$ can be evaluated for $t = 1, 2, \ldots, l$. Let r_t denote the elements of x_t other than lagged endogenous variables.

The leading special case is that of independently and identically distributed errors where the joint density function of the errors and exogenous variables,

$$p(e_n, e_{n-1}, \ldots, e_1, y_0, y_{-1}, \ldots, y_{1-l}, r_n, r_{n-1}, \ldots, r_1)$$

has the form

$$\prod_{t=1}^{n} p(e_t \mid \sigma) p(y_0, \ldots, y_{1-l}) p(r_n, \ldots, r_1).$$

In this case the likelihood is (Problem 1)

$$\prod_{t=1}^{n} \left| \det \frac{\partial}{\partial y'} q(y_t, x_t, \theta) \right| p[q(y_t, x_t, \theta) \mid \sigma] \, p(y_0, \ldots, y_{1-l}) p(r_n, \ldots, r_1)$$

and the conditional likelihood is

$$\prod_{t=1}^{n} \left| \det \frac{\partial}{\partial y'} q(y_t, x_t, \theta) \right| p[q(y_t, x_t, \theta) \mid \sigma].$$

One would rather avoid conditioning on the variables $y_0, y_{-1}, \ldots, y_{1-l}$ because they are not ancillary—their distribution involves θ. However, in most applications it will not be possible to obtain the density $p(y_0, \ldots, y_{1-l})$. Taking logarithms, changing sign, and normalizing leads to the same sample objective function as above:

$$s_n(\theta, \sigma) = \frac{1}{n} \sum_{t=1}^{n} (-1) \ln[p(y_t \mid x_t, \theta)]$$

$$= \frac{1}{n} \sum_{t=1}^{n} (-1) \ln \left(\left| \det \frac{\partial}{\partial y} q(y_t, x_t, \theta) \right| p[q(y_t, x_t, \theta) \mid \sigma] \right).$$

An application of the results of Sections 4 and 6 of Chapter 7 yields the same statistical methods as above. The algebraic relationships required in their derivation are sketched out in Problems 3 through 6.

Sometimes models can be transformed to have identically and independently distributed errors. One example was given above. As another, if the errors from $u_t = Q(y_t, r_t, \beta)$ appear to be serially correlated, a plausible model might be

$$q(y_t, x_t, \theta) = Q(y_t, r_t, \beta) - \tau Q(y_{t-1}, r_{t-1}, \beta) = u_t - \tau u_{t-1} = e_t$$

with $x_t = (y_{t-1}, r_t, r_{t-1})$ and $\theta = (\beta, \tau)$.

As noted, these statistical methods obtain from an application of the results listed in Sections 4 and 6 of Chapter 7. Of the assumptions listed in Chapter 7 that need to be satisfied by a dynamic model, the most suspect is the assumption of near epoch dependence. Some results in this direction are given in Problem 2. A detailed discussion of regularity conditions for the case when $q(y, x, \theta) = y - f(x, \theta)$ and the errors are normally distributed is given in Section 4 of Chapter 7. The general flavor of the regularity conditions is that in addition to the sort of conditions that are required in the regression case, the model must damp lagged y's in the sense that for given r_t and θ one should have $\|y_{t-1}\| > \|Y(0, y_{t-1}, \ldots, y_{t-l}, r_t, \theta)\|$.

In the general dynamic case, the joint density function

$$p(e_n, e_{n-1}, \ldots, e_1, y_0, \ldots, y_{1-l}, r_n, r_{n-1}, \ldots, r_1 \mid \sigma)$$

can be factored as (Problem 1)

$$\prod_{t=1}^{n} p(e_t \mid e_{t-1}, \ldots, e_1, r_t, \ldots, r_1, y_0, \ldots, y_{1-l}, \sigma)$$

$$\times p(y_0, \ldots, y_{1-l}, r_n, \ldots, r_1, \sigma).$$

Letting x_t contain r_t and as many lagged values of y_t as necessary, put

$$p(y_t \mid x_t, x_{t-1}, \ldots, x_1, y_0, \ldots, y_{1-l}, \theta, \sigma)$$

$$= \left| \det \frac{\partial}{\partial y'} q(y_t, x_t, \theta) \right|$$

$$\times p\big[q(y_t, x_t, \theta) \mid q(y_{t-1}, x_{t-1}, \theta), \ldots, q(y_1, x_1, \theta),$$

$$r_t, \ldots, r_1, y_0, \ldots, y_{1-l}, \sigma\big].$$

Thus, the conditional likelihood is

$$p(y_n, \ldots, y_1 \mid r_n, \ldots, r_1, y_0, \ldots, y_{1-l}, \theta, \sigma)$$

$$= \prod_{t=1}^{n} p(y_t \mid x_t, x_{t-1}, \ldots, x_1, y_0, \ldots, y_{1-l}, \theta, \sigma);$$

the sample objective function is

$$s_n(\theta, \sigma) = \frac{1}{n} \sum_{t=1}^{n} (-1)\ln\big[p(y_t \mid x_t, x_{t-1}, \ldots, x_1, y_0, \ldots, y_{1-l}, \theta, \sigma)\big];$$

and the maximum likelihood estimator is

$$(\hat{\theta}, \hat{\sigma}) = \operatorname*{argmin}_{(\theta, \sigma)} s_n(\theta, \sigma).$$

A formal application of the results of Section 4 of Chapter 7 yields that approximately (see Theorem 6 of Chapter 7 for an exact statement)

$$\sqrt{n}\begin{pmatrix} \hat{\theta} - \theta^0 \\ \hat{\sigma} - \sigma^0 \end{pmatrix} \sim N_{p+r}(0, V)$$

with

$$V = \mathscr{I}^{-1}\mathscr{J}\mathscr{I}^{-1}.$$

$\mathscr{I}$ and $\mathscr{J}$ can be estimated using

$$\hat{\mathscr{J}} = \sum_{\tau = -l(n)}^{l(n)} w\left(\frac{\tau}{l(n)}\right)\hat{\mathscr{I}}_{n\tau}$$

and

$$\hat{\mathscr{I}} = \frac{\partial^2}{\partial\lambda\,\partial\lambda'} s_n(\hat{\theta}, \hat{\sigma})$$

where $l(n)$ denotes the integer nearest $n^{1/5}$, and

$$w(x) = \begin{cases} 1 - 6|x|^2 + 6|x|^3 & 0 \le x \le \tfrac{1}{2} \\ 2(1 - |x|)^3 & \tfrac{1}{2} \le x \le 1 \end{cases}$$

$$\hat{\mathscr{I}}_{n\tau} = \begin{cases} \dfrac{1}{n}\displaystyle\sum_{t=1+\tau}^{n}\left(\dfrac{\partial}{\partial\lambda}\ln[p(y_t \,|\, x_t, \ldots, x_1, y_0, \ldots, y_{1-l}, \hat{\theta}, \hat{\sigma})]\right) \\ \qquad \times\left(\dfrac{\partial}{\partial\lambda}\ln[p(y_{t-\tau} \,|\, x_{t-\tau}, \ldots, x_1, y_0, \ldots, y_{1-l}, \hat{\theta}, \hat{\sigma})]\right)' & \tau \ge 0 \\ (\hat{\mathscr{I}}_{n,-\tau})' & \tau < 0. \end{cases}$$

If the model is correctly specified, one can usually show that the conditional expectation of $(\partial/\partial\lambda)\ln p(y_t|x_t, \ldots, x_1, y_0, \ldots, y_{1-l}, \theta^0, \sigma^0)$ given $(y_{t-\tau}, x_{t-\tau}, \ldots, x_1, y_0, \ldots, y_{1-l})$ is zero whence one can take $\hat{\mathscr{J}} = \hat{\mathscr{I}}_{n0}$; see Problem 3. Also, using an argument similar to Problem 5, one can usually take $\hat{V} = \hat{\mathscr{I}}_{n0}^{-1}$ or $\mathscr{J}^{-1}$ if the model is correctly specified.

For testing

$$H: h(\lambda^0) = 0 \quad \text{against} \quad A: h(\lambda^0) \ne 0$$

where $\lambda = (\theta, \sigma)$, the Wald test statistic is (Theorem 12, Chapter 7)

$$W = nh'(\hat{\lambda})[\hat{H}\hat{V}\hat{H}']^{-1}h(\hat{\lambda})$$

with $\hat{H} = (\partial/\partial\lambda')h(\hat{\lambda})$. The null hypothesis $H: h(\lambda^0) = 0$ is rejected in favor of the alternative hypothesis $A: h(\lambda^0) \neq 0$ when the test statistic exceeds the upper $\alpha \times 100$ percentage point χ_α^2 of a chi-square random variable with q degrees of freedom; $\chi_\alpha^2 = (\chi^2)^{-1}(1 - \alpha; q)$.

As a consequence of this result, a 95% confidence interval on the ith element of θ is computed as

$$\hat{\theta}_i \pm z_{.025}\frac{\sqrt{\hat{V}_{ii}}}{\sqrt{n}}$$

and a 95% confidence interval on the ith element of σ is computed as

$$\hat{\sigma}_i \pm z_{.025}\frac{\sqrt{\hat{V}_{p+i,\,p+i}}}{\sqrt{n}}$$

where $z_{.025} = -\sqrt{(\chi^2)^{-1}(.95; 1)} = N^{-1}(.025; 0, 1)$.

Let $\tilde{\lambda} = (\tilde{\theta}, \tilde{\sigma})$ denote the minimizer of $s_n(\lambda)$, subject to the restriction that $h(\lambda) = 0$. Let $\tilde{H}$, $\tilde{\mathscr{I}}$, and $\tilde{\mathscr{J}}$ denote the formulas for $\hat{H}$, $\hat{\mathscr{I}}$, and $\hat{\mathscr{J}}$ above, but with $\tilde{\lambda}$ replacing $\hat{\lambda}$ throughout; put

$$\tilde{V} = \tilde{\mathscr{J}}^{-1}\tilde{\mathscr{I}}\tilde{\mathscr{J}}^{-1}.$$

The Lagrange multiplier test statistic is (Theorem 16, Chapter 7)

$$R = n\left(\frac{\partial}{\partial\lambda}s_n(\tilde{\lambda})\right)'\tilde{\mathscr{J}}^{-1}\tilde{H}'(\tilde{H}\tilde{V}\tilde{H}')^{-1}\tilde{H}\tilde{\mathscr{J}}^{-1}\left(\frac{\partial}{\partial\lambda}s_n(\tilde{\lambda})\right).$$

Again, the null hypothesis $H: h(\lambda^0) = 0$ is rejected in favor of the alternative hypothesis $A: h(\lambda^0) \neq 0$ when the test statistic exceeds the upper $\alpha \times 100$ percentage point χ_α^2 of a chi-square random variable with q degrees of freedom; $\chi_\alpha^2 = (\chi^2)^{-1}(1 - \alpha; q)$.

The likelihood ratio test cannot be used, unless one can show that $\mathscr{I} = \mathscr{J}$; see Theorem 17 of Chapter 7. Formulas for computing the power of the Wald and Lagrange multiplier tests are given in Theorems 14 and 16 of Chapter 7, respectively.

PROBLEMS

This problem set requires a reading of Sections 1 through 3 of Chapter 7 before the problems can be worked.

1. (Derivation of the likelihood in the dynamic case.) Consider the model

$$q(y_t, x_t, \theta^0) = e_t \qquad t = 1, 2, \ldots, n$$

where

$$x_t = (y_{t-1}, y_{t-2}, \ldots, y_{t-l}, r_t).$$

Define the $(n + l)M$-vectors ζ and $e(\zeta)$ by

$$e(\zeta) = \begin{pmatrix} e_n \\ e_{n-1} \\ \vdots \\ e_1 \\ e_0 \\ \vdots \\ e_{1-l} \end{pmatrix} = \begin{pmatrix} q(y_n, x_n, \theta) \\ q(y_{n-1}, x_{n-1}, \theta) \\ \vdots \\ q(y_1, x_1, \theta) \\ y_0 \\ \vdots \\ y_{1-l} \end{pmatrix}, \qquad \zeta = \begin{pmatrix} y_n \\ y_{n-1} \\ \vdots \\ y_1 \\ y_0 \\ \vdots \\ y_{1-l} \end{pmatrix}.$$

Show that $(\partial/\partial\zeta')e(\zeta)$ has a block upper triangular form, so that

$$\det \frac{\partial}{\partial\zeta'} e(\zeta) = \prod_{t=1}^{n} \det \frac{\partial}{\partial y'} q(y_t, x_t, \theta).$$

Show that the joint density function

$$p(e_n, e_{n-1}, \ldots, e_1, y_0, \ldots, y_{1-l}, r_n, r_{n-1}, \ldots, r_1)$$

can be factored as

$$\prod_{t=1}^{n} p(e_t | e_{t-1}, \ldots, e_1, r_t, \ldots, r_1, y_0, \ldots, y_{1-l})$$

$$\times p(y_0, \ldots, y_{1-l}, r_n, \ldots, r_1)$$

and hence that the conditional density

$$p(y_n, \ldots, y_1 | r_n, \ldots, r_1, y_0, \ldots, y_{1-l})$$

can be put in the form

$$\prod_{t=1}^{n} p(y_t | x_t, x_{t-1}, \ldots, x_1, y_0, \ldots, y_{1-l}, \theta).$$

2. (Near epoch dependence.) Consider data generated according to the nonlinear, implicit model

$$q(y_t, y_{t-1}, r_t, \theta^0) = e_t \qquad t = 1, 2, \ldots$$
$$y_t = 0 \qquad t \leq 0$$

where y_t and e_t are univariate. If such a model is well posed, then it must define y_t as a function of y_{t-1}, r_t, θ^0, and e_t. That is, there must exist a reduced form

$$y_t = Y(e_t, y_{t-1}, r_t, \theta^0).$$

Assume that $Y(e, y, r, \theta)$ has a bounded derivative in its first argument:

$$\left| \frac{\partial}{\partial e} Y(e, y, r, \theta) \right| \leq \Delta$$

and is a contraction mapping in its second:

$$\left| \frac{\partial}{\partial y} Y(e, y, r, \theta) \right| \leq d < 1.$$

Let the errors $\{e_t\}$ be independently and identically distributed, and set $e_t = 0$ for $t \leq 0$. With this structure, the underlying data generating sequence $\{V_t\}$ described in Section 2 of Chapter 7 is $V_t = (0, 0)$ for $t \leq 0$ and $V_t = (e_t, r_t)$ for $t = 1, 2 \ldots$. Suppose that θ^0 is estimated by maximum likelihood; that is, $\hat{\theta}$ minimizes

$$s_n(\theta) = \frac{1}{n} \sum_{t=1}^{n} s(y_t, y_{t-1}, r_t, \theta)$$

$$= \frac{1}{n} \sum_{t=1}^{n} -\ln[p(y_t | y_{t-1}, r_t, \theta)]$$

$$= \frac{1}{n} \sum_{t=1}^{n} -\ln\left(\left| \det \frac{\partial}{\partial y_t} q(y_t, y_{t-1}, r_t, \theta) \right| p[q(y_t, y_{t-1}, r_t, \theta)] \right)$$

where $p(e)$ is the density of the errors. We have implicitly absorbed the location and scale parameters of the error density into the definition of $q(y_t, y_{t-1}, r_t, \theta)$.

Let

$$Z_t = \Delta \sum_{j=1}^{\infty} d^j |e_{t-j}|,$$

$$R_t = \sup \frac{|\ln p(y_t | y_{t-1}, r_t, \theta) - \ln p(y_t + h_0 | y_{t-1} + h_1, r_t, \theta)|}{|h_0 + h_1|}$$

where the supremum is taken over the set

$$A_t = \{(h_0, h_1, \theta) : |h_i| \le Z_{t-i}, \theta \in \Theta\}.$$

Assume that for some $p > 4$

$$\|e_1\|_p \le B < \infty$$

$$\|R_t\|_p \le B < \infty.$$

Show that this situation satisfies the hypothesis of Proposition 1 of Chapter 7 by supplying the missing details in the following argument. Define a predictor of y_s of the form

$$\hat{y}_{t,m}^s = \hat{y}_s(V_t, V_{t-1}, \dots, V_{t-m})$$

as follows:

$$\bar{y}_t = 0 \qquad\qquad t \le 0$$

$$\bar{y}_t = Y(0, \bar{y}_{t-1}, r_t, \theta^0) \qquad 0 < t$$

$$\hat{y}_{t,m}^s = \bar{y}_s \qquad\qquad s \le \max(t - m, 0)$$

$$\hat{y}_{t,m}^s = Y(e_s, \hat{y}_{t,m}^{s-1}, r_s, \theta^0) \qquad \max(t - m, 0) < s \le t.$$

By Taylor's theorem, there are intermediate points such that for $t \ge 0$

$$|y_t - \bar{y}_t| = |Y(e_t, y_{t-1}, r_t, \theta^0) - Y(0, \bar{y}_{t-1}, r_t, \theta^0)|$$

$$\le \left| \frac{\partial}{\partial e} Y(\bar{\bar{e}}_t, \bar{\bar{y}}_{t-1}, r_t, \theta^0) e_t \right.$$

$$\left. + \frac{\partial}{\partial y} Y(\bar{\bar{e}}_t, \bar{\bar{y}}_{t-1}, r_t, \theta^0)(y_{t-1} - \bar{y}_{t-1}) \right|$$

$$\le d|y_{t-1} - \bar{y}_{t-1}| + \Delta|e_t|$$

$$\le d^2|y_{t-2} - \bar{y}_{t-2}| + d\Delta|e_{t-1}| + \Delta|e_t|$$

$$\vdots$$

$$\le d^t|y_0 - \bar{y}_0| + \Delta \sum_{j=0}^{t-1} d^j |e_{t-j}|$$

$$= \Delta \sum_{j=0}^{t-1} d^j |e_{t-j}|.$$

For $m > 0$ and $t - m > 0$ the same type of argument yields

$$\left| y_t - \hat{y}_{t,m}^t \right| = \left| Y\left(e_t, y_{t-1}, r_t, \theta^0 \right) - Y\left(e_t, \hat{y}_{t,m}^{t-1}, r_t, \theta^0 \right) \right|$$

$$\leq \left| \frac{\partial}{\partial y} Y\left(e_t, \bar{\bar{y}}_{t-1}, r_t, \theta^0 \right) \left(y_{t-1} - \hat{y}_{t,m}^{t-1} \right) \right|$$

$$\leq d \left| y_{t-1} - \hat{y}_{t,m}^{t-1} \right|$$

$$\vdots$$

$$\leq d^m \left| y_{t-m} - \hat{y}_{t,m}^{t-m} \right|$$

$$= d^m \left| y_{t-m} - \bar{y}_{t-m} \right|$$

$$= \Delta d^m \sum_{j=0}^{t-m-1} d^j \left| e_{t-m-j} \right|$$

where the last inequality obtains by substituting the bound for $|y_t - \bar{y}_t|$ obtained previously. For $t - m < 0$ we have

$$\left| y_t - \hat{y}_{t,m}^t \right| = d^{m-t} |y_0 - y_0| = 0.$$

In either event,

$$\left\| y_t - \hat{y}_{t,m}^t \right\|_p \leq \Delta d^m \sum_{j=0}^{t-m-1} d^j \left\| e_{t-m-j} \right\|_p \leq \frac{B \Delta d^m}{1 - d}.$$

Letting

$$W_t = \left(y_t, y_{t-1}, r_t \right)$$
$$\hat{W}_{t-m}^t = \left(\hat{y}_{t,m}^t, \hat{y}_{t,m}^{t-1}, r_t \right)$$
$$|a| = \sum_{i=1}^k |a_i| \quad \text{(for } a \text{ in } \mathbb{R}^k\text{)}$$

we have

$$s_n(\theta) = \frac{1}{n} \sum_{t=1}^n g_t(W_t, \theta)$$

with

$$g_t(W_t, \theta) = -\ln p\left(y_t \mid y_{t-1}, r_t, \theta \right).$$

For $t \geq 1$

$$\left| g_t(W_t, \theta) - g_t\left(\hat{W}_{t-m}^t, \theta \right) \right| \leq R_t \left| W_t - \hat{W}_{t-m}^t \right|.$$

Letting $q = p/(p + 1) < p$, $r = p/2 > 4$, and

$$B\big(W_t, \hat{W}_{t-m}^t\big) = R_t$$

we have

$$\big\| B\big(W_t, \hat{W}_{t-m}^t\big) \big\|_q \le \big(1 + \|R_t\|_p^p\big)^{1/q} \le \big(1 + B^p\big)^{1/q} < \infty$$

$$\big\| B\big(W_t, \hat{W}_{t-m}^t\big) \big| W_t - \hat{W}_{t-m}^t \big| \big\|_r \le \big\| B\big(W_t, \hat{W}_{t-m}^t\big) \big\|_{2r} \big\| \big| W_t - \hat{W}_{t-m}^t \big| \big\|_{2r}$$

$$\le \frac{B2B\Delta d^m}{1 - d} < \infty$$

$$\eta_m = \sup_t \big\| \big| W_t - \hat{W}_{t-m}^t \big| \big\|_p \le \frac{2B\Delta d^m}{1 - d} < \infty.$$

The rate at which η_m falls off with m is exponential, since $d < 1$, whence η_m is of size $-q(r - 1)/(r - 2)$. Thus all the conditions of Proposition 1 are satisfied.

3. (Information inequality.) Consider the case where the joint density function of the errors and exogenous variables

$$p\big(e_n, e_{n-1}, \ldots, e_1, y_0, y_{-1}, \ldots, y_{1-l}, r_n, r_{n-1}, \ldots, r_1\big)$$

has the form

$$\prod_{t=1}^{n} p(e_t \mid \sigma) p(y_0, \ldots, y_{1-l}) p(r_n, \ldots, r_1)$$

and let

$$p(y_t \mid x_t, \lambda) = \left| \det \frac{\partial}{\partial y'} q(y_t, x_t, \theta) \right| p\big[q(y_t, x_t, \theta) \mid \sigma\big].$$

Assume that $p(y \mid x, \lambda)$ is strictly positive and continuous on $\mathcal{Y}$. Put

$$u(y, x) = \ln p(y \mid x, \lambda) - \ln p(y \mid x, \lambda^0)$$

and supply the missing details in the following argument. By Jensen's inequality

$$\exp\left[\int_{\mathcal{Y}} u(y, x) p(y \mid x, \lambda^0) \, dy \right] \le \int_{\mathcal{Y}} e^{u(y, x)} p(y \mid x, \lambda^0) \, dy$$

$$= \int_{\mathcal{Y}} p(y \mid x, \lambda) \, dy = 1.$$

The inequality is strict unless $u(y, x) = 0$ for every y. This implies that

$$\int_{\mathcal{Y}}(-1)\ln p(y\,|\,x, \lambda)p(y\,|\,x, \lambda^0)\,dy$$

$$> \int_{\mathcal{Y}}(-1)\ln p(y\,|\,x, \lambda^0)p(y\,|\,x, \lambda^0)\,dy$$

for every λ for which $p(y\,|\,x, \lambda) \neq p(y\,|\,x, \lambda^0)$ for some y.

4. (Expectation of the score.) Use Problem 3 to show that if $(\partial/\partial\lambda)\int(-1)\ln p(y\,|\,x, \lambda)p(y\,|\,x, \lambda^0)\,dy$ exists, then it is zero at $\lambda = \lambda^0$. Show that if Assumptions 1 though 6 of Chapter 7 are satisfied, then

$$0 = \int_{\mathcal{Y}}\left(\frac{\partial}{\partial\lambda}(-1)\ln p(y\,|\,x, \lambda)\right)p(y\,|\,x, \lambda^0)\,dy\bigg|_{\lambda=\lambda^0}.$$

5. (Equality of $\mathcal{I}$ and $\mathcal{J}$.) Under the same setup as Problem 3, derive the identity

$$p^{-1}(y\,|\,x, \lambda)\frac{\partial^2}{\partial\lambda_i\,\partial\lambda_j}p(y\,|\,x, \lambda)$$

$$= \frac{\partial^2}{\partial\lambda_i\,\partial\lambda_j}\ln p(y\,|\,x, \lambda)$$

$$+ \left(\frac{\partial}{\partial\lambda_i}\ln p(y\,|\,x, \lambda)\right)\left(\frac{\partial}{\partial\lambda_j}\ln p(y\,|\,x, \lambda)\right).$$

Let ξ_i denote the ith elementary vector. Justify the following steps:

$$\int_{\mathcal{E}}p^{-1}[Y(e, x, \theta)\,|\,x, \lambda]\left(\frac{\partial^2}{\partial\lambda_i\partial\lambda_j}p[Y(e, x, \theta)\,|\,x, \lambda]\right)p(e\,|\,\sigma)\,de$$

$$= \lim_{h\to0}h^{-1}\int_{\mathcal{Y}}p^{-1}(y\,|\,x, \lambda)[(\partial/\partial\lambda_j)p(y\,|\,x, \lambda + h\xi_i)$$

$$- (\partial/\partial\lambda_j)p(y\,|\,x, \lambda)]\,p(y\,|\,x, \lambda)\,dy$$

$$= \lim_{h\to0}h^{-1}\int_{\mathcal{Y}}p^{-1}(y\,|\,x, \lambda)[(\partial/\partial\lambda_j)p(y\,|\,x, \lambda + h\xi_i)]$$

$$\times p(y\,|\,x, \lambda)\,dy$$

$$= \lim_{h\to0}h^{-1}\int_{\mathcal{Y}}\frac{\partial}{\partial\lambda_j}p(y\,|\,x, \lambda + h\xi_i)\,dy$$

$$= \lim_{h\to0}h^{-1}\int_{\mathcal{Y}}\frac{\partial}{\partial\lambda_j}\ln[p(y\,|\,x, \lambda + h\xi_i)]$$

$$\times p(y\,|\,x, \lambda + h\xi_i)\,dy$$

$$= \lim_{h\to0}h^{-1}\cdot 0 = 0.$$

This implies that

$$\mathcal{E}\left(p^{-1}(y_t \mid x_t, \lambda) \frac{\partial^2}{\partial \lambda_i \partial \lambda_j} p(y_t \mid x_t, \lambda) \middle| y_0, \dots, y_{1-l}, r_n, \dots, r_1 \right) = 0.$$

6. (Derivation of $\mathcal{K}$ and $\mathcal{L}$.) Under the same setup as Problem 3, obtain the identity

$$0 = \frac{\partial^2}{\partial\sigma\,\partial\theta'} s_n(\theta, \sigma) \bigg|_{\sigma=\sigma(\theta)} + \frac{\partial^2}{\partial\sigma\,\partial\sigma'} s_n[\theta, \sigma(\theta)] \frac{\partial}{\partial\theta'} \sigma(\theta)$$

and use it to show that

$$\frac{\partial^2}{\partial\theta\,\partial\theta'} s_n[\theta, \sigma(\theta)] = \frac{\partial^2}{\partial\theta\,\partial\theta'} s_n(\theta, \sigma) \bigg|_{\sigma=\sigma(\theta)}$$

$$- \left(\frac{\partial^2}{\partial\theta\,\partial\sigma'} s_n(\theta, \sigma) \right) \left(\frac{\partial^2}{\partial\sigma\,\partial\sigma'} s_n(\theta, \sigma) \right)^{-1}$$

$$\times \left(\frac{\partial^2}{\partial\sigma\,\partial\theta'} s_n(\theta, \sigma) \right) \bigg|_{\sigma=\sigma(\theta)}$$

$$= \mathcal{I}_{\theta\theta} - \mathcal{I}_{\theta\sigma}(\mathcal{I}_{\sigma\sigma})^{-1}\mathcal{I}_{\sigma\theta}.$$

Obtain the identity

$$\frac{\partial}{\partial\theta} \ln p[y \mid x, \theta, \sigma(\theta)] = \frac{\partial}{\partial\theta} \ln p(y \mid x, \theta, \sigma) \bigg|_{\sigma=\sigma(\theta)}$$

$$- \mathcal{I}_{\theta\sigma}(\mathcal{I}_{\sigma\sigma})^{-1} \frac{\partial}{\partial\sigma} \ln p(y \mid x, \theta, \sigma) \bigg|_{\sigma=\sigma(\theta)}.$$

Approximate $\mathcal{I}_{\theta\sigma}$ and $\mathcal{I}_{\sigma\sigma}$ by $\mathcal{J}_{\theta\sigma}$ and $\mathcal{J}_{\sigma\sigma}$, and use the resulting expression to show that

$$\frac{1}{n}\sum_{t=1}^{n}\left(\frac{\partial}{\partial\theta} \ln p[y_t \mid x_t, \theta, \sigma(\theta)] \right)\left(\frac{\partial}{\partial\theta} \ln p[y_t \mid x_t, \theta, \sigma(\theta)] \right)'$$

$$= \mathcal{J}_{\theta\theta} - \mathcal{J}_{\theta\sigma}(\mathcal{J}_{\sigma\sigma})^{-1}\mathcal{J}_{\sigma\theta}.$$

CHAPTER 7

A Unified Asymptotic Theory
for Dynamic Nonlinear Models

The statistical analysis of dynamic nonlinear models, for instance a model such as

$$y_t = f\left(y_{t-1}, x_t, \theta^0\right) + e_t \qquad t = 0, \pm 1, \pm 2, \ldots$$

with serially correlated errors, is little different from the analysis of models with regression structure (Chapter 3) as far as applications are concerned. Effectively all that changes is the formula for estimating the variance $\mathscr{I}_n^0$ of an average of scores. Thus, as far as applications are concerned, the previous intuition and methodology carry over directly to the dynamic situation.

The main theoretical difficulty is to establish regularity conditions that permit a uniform strong law and a continuously convergent central limit theorem that are both plausible (reasonably easy to verify) and resilient to nonlinear transformation. The time series literature is heavily oriented toward linear models and thus is not of much use. The more recent martingale central limit theorems and strong laws are not of much use either, because martingales are essentially a linear concept—a nonlinear transformation of a martingale is not a martingale of necessity. In a series of four papers McLeish (1974, 1975a, 1975b, 1977) developed a notion of asymptotic martingales which he termed mixingales. This is a concept that does extend to nonlinear situations, and the bulk of this chapter is a verification of this claim. The flavor of the extension is this: Conceptually y_t in the model above is a function of all previous errors $e_t, e_{t-1}, \ldots$. But if y_t can be approximated by $\hat{y}_t$ that is a function of $e_t, \ldots, e_{t-m}$ and if the error of approximation $\|y_t - \hat{y}_t\|$ falls off at a polynomial rate in m, then smooth transformations of the form $g(y_t, \ldots, y_{t-l}, x_t, \ldots, x_{t-l}, \gamma)$ follow a uniform strong law and a continuously convergent central limit theorem provided that the error process is strong mixing. The rest of the analysis follows along the lines laid down in Chapter 3 with a reverification made necessary by a weaker form of the uniform strong law: an average of random variables becomes close to the average of the expectations, but the average of the expectations does not necessarily converge. These results were obtained in collaborative research with Halbert White and Jeffrey M. Wooldridge while they visited Raleigh in the summer of 1984. Benedikt M. Pötscher provided helpful comments. National Science Foundation and North Carolina Agricultural Experiment Station support for this work is gratefully acknowledged.

The reader who is applications oriented is invited to scan the regularity conditions to become aware of various pitfalls, isolate the formula for $\hat{\mathscr{I}}$ relevant to the application, and then apply the methods of the previous chapters forthwith. A detailed reading of this chapter is not essential to applications.

The material in this chapter is intended to be accessible to readers familiar with an introductory, measure theoretic probability text such as Ash (1972), Billingsley (1979), Chung (1974), or Tucker (1967). In those instances where the proof in an original source was too terse to be read at that level, proofs with the missing details are supplied here. Proofs of new results or significant modifications to existing results are, of course, given as well. Proofs by citation occur only in those instances when the argument in the original source was reasonably self-contained and readable at the intended level.

1. INTRODUCTION

This chapter is concerned with models which have lagged dependent variables as explanatory variables and (possibly) serially correlated errors. Something such as

$$q\left(y_t, y_{t-1}, x_t, \gamma_1^0\right) = u_t$$

$$u_t = e_t + \gamma_2^0 e_{t-1}$$

$$t = 0, \pm 1, \pm 2, \ldots$$

might be envisaged as the data generating process, with $\{e_t\}$ a sequence of, say, independently and identically distributed random variables. As in Chapter 3, one presumes that the model is well posed, so that in principle, given y_{t-1}, x_t, γ_1^0, u_t, one could solve for y_t. Thus an equivalent representation of the model is

$$y_t = Y\left(u_t, y_{t-1}, x_t, \gamma_1^0\right)$$
$$u_t = e_t + \gamma_2^0 e_{t-1}$$
$$t = 0, \pm 1, \pm 2, \ldots .$$

Substitution yields

$$y_t = Y\left[e_t + \gamma_2 e_{t-1}, Y\left(e_{t-1} + \gamma_2 e_{t-2}, y_{t-2}, x_{t-2}, \gamma_1^0\right), x_t, \gamma_1^0\right]$$

and if this substitution process is continued indefinitely, the data generating process is seen to be of the form

$$y_t = Y\left(t, e_\infty, x_\infty, \gamma^0\right) \qquad t = 0, \pm 1, \pm 2, \ldots$$

with

$$e_\infty = (\ldots, e_{-1}, e_0, e_1, \ldots)$$
$$x_\infty = (\ldots, x_{-1}, x_0, x_1, \ldots).$$

Throughout, we shall accommodate models with a finite past by setting y_t, x_t, e_t equal to zero for negative t; the values of y_0, x_0, and e_0 are the initial conditions in this case.

If one has this sort of data generating process in mind, then a least mean distance estimator could assume the form

$$\hat{\lambda}_n = \operatorname*{argmin}_{\Lambda} s_n(\lambda)$$

with sample objective function

$$s_n(\lambda) = \frac{1}{n} \sum_{t=1}^{n} s\left(t, y_t, y_{t-1}, \ldots, y_{t-l}, x_t, x_{t-1}, \ldots, x_{t-l'}, \hat{\tau}_n, \lambda\right)$$

or

$$s_n(\lambda) = \frac{1}{n} \sum_{t=1}^{n} s_t\left(y_t, y_{t-1}, \ldots, y_0, x_t, x_{t-1}, \ldots, x_0, \hat{\tau}_n, \lambda\right)$$

—the distinction between the two being that one distance function has a finite number of arguments and the number of arguments in the other grows with t. Writing

$$s_n(\lambda) = \frac{1}{n} \sum_{t=1}^{n} s_t\left(y_t, y_{t-1}, \ldots, y_{t-l_t}, x_t, x_{t-1}, \ldots, x_{t-l'_t}, \hat{\tau}_n, \lambda\right)$$

with l_t depending on t accommodates either situation. Similarly, a method of moments estimator can assume the form

$$\hat{\lambda}_n = \underset{\Lambda}{\arg\min}\, s_n(\lambda) = d\left[m_n(\lambda), \hat{\tau}_n\right]$$

with moment equations

$$m_n(\lambda) = \frac{1}{n} \sum_{t=1}^{n} m_t\left(y_t, y_{t-1}, \ldots, y_{t-l_t}, x_t, x_{t-1}, \ldots, x_{t-l'_t}, \hat{\tau}_n, \lambda\right).$$

In the literature, the analysis of dynamic models is unconditional for the most part, and we shall follow that tradition here. Fixed (nonrandom) variables amongst the components of x_t are accommodated by viewing them as random variables that take on a single value with probability one. Under these conventions there is no mathematical distinction between the error process $\{e_t\}_{t=-\infty}^{\infty}$ and the process $\{x_t\}_{t=-\infty}^{\infty}$ describing the independent variables. The conceptual distinction is that the independent variables $\{x_t\}_{t=-\infty}^{\infty}$ are viewed as being determined externally to the model and independently of the error process $\{e_t\}_{t=-\infty}^{\infty}$; that is, the process $\{x_t\}$ is ancillary. In Chapter 6 we permitted x_t to contain lagged dependent variables and used r_t to denote the ancillary components. Here all components of x_t are ancillary unless specifically stated otherwise. Usually, we shall account explicitly for lagged dependent variables as a separate argument of the function they enter. In an unconditional analysis of a dynamic setting, we must permit the process $\{x_t\}_{t=-\infty}^{\infty}$ to be dependent, and, since fixed (nonrandom) variables are permitted, we must rule out stationarity. We shall also permit the error process $\{e_t\}_{t=-\infty}^{\infty}$ to be dependent and nonstationary, primarily because nothing is gained by assuming the contrary. Since there is no mathematical distinction between the errors and the independent variables, we can economize on notation by collecting them into the process $\{v_t\}_{t=-\infty}^{\infty}$ with

$$v_t = (e_t, x_t);$$

denote a realization of the process by

$$v_\infty = (\ldots, v_{-1}, v_0, v_1, \ldots).$$

Recall that if the process has a finite past, then we set $v_t = 0$ for $t < 0$ and take the value of v_0 as the initial condition.

Previously, we induced a Pitman drift by considering data generating processes of the form

$$y_t = Y\left(e_t, x_t, \gamma_n^0\right)$$

and letting γ_n^0 tend to a point γ^*. In the present context it is very difficult technically to handle drift in this way, so instead of moving the data generating model to the hypothesis as in Chapter 3, we shall move the hypothesis to the model by considering

$$H : h\left(\lambda_n^0\right) = h_n^* \quad \text{against} \quad A : h\left(\lambda_n^0\right) \neq h_n^*$$

and letting $h(\lambda_n^0) - h_n^*$ drift toward zero at the rate $O(1/\sqrt{n}\,)$. This method of inducing drift is less traditional, but in some respects is philosophically more palatable. It makes more sense to assume that an investigator slowly discovers the truth as more data become available than to assume that nature slowly accommodates to the investigator's pigheadedness. But withal, the drift is only a technical artifice to obtain approximations to the sampling distributions of test statistics that are reasonably accurate in applications, so that philosophical nitpicking of this sort is irrelevant.

If the data generating model is not going to be subject to drift, there is no reason to put up with the cumbersome notation

$$y_t = Y\left(t, e_\infty, x_\infty, \gamma^0\right)$$

$$s_n(\lambda) = \frac{1}{n} \sum_{t=1}^{n} s_t\left(y_t, y_{t-1}, \ldots, y_{t-l_t}, x_t, x_{t-1}, \ldots, x_{t-l_t'}, \hat{\tau}_n, \lambda\right).$$

Much simpler is to stack the variables entering the distance function into a single vector

$$w_t = \begin{pmatrix} y_t \\ \vdots \\ y_{t-l_t} \\ x_t \\ \vdots \\ x_{t-l_t'} \end{pmatrix}$$

and view w_t as obtained from the doubly infinite sequence

$$v_\infty = \left(\ldots, v_{-1}, v_0, v_1, \ldots\right)$$

by a mapping of the form

$$w_t = W_t(v_\infty).$$

Let w_t be k_t-dimensional. Estimators then take the form

$$\hat{\lambda}_n = \underset{\Lambda}{\mathrm{argmin}}\, s_n(\lambda)$$

$$s_n(\lambda) = \frac{1}{n} \sum_{t=1}^{n} s_t(w_t, \hat{\tau}_n, \lambda)$$

in the case of least mean distance estimators, and

$$\hat{\lambda}_n = \underset{\Lambda}{\mathrm{argmin}}\, s_n(\lambda)$$

$$s_n(\lambda) = d\left[m_n(\lambda), \hat{\tau}_n \right]$$

$$m_n(\lambda) = \frac{1}{n} \sum_{t=1}^{n} m_t(w_t, \hat{\tau}_n, \lambda)$$

in the case of method of moments estimators. We are led then to consider limit theorems for composite functions of the form

$$\frac{1}{n} \sum_{t=1}^{n} g_t\left[W_t(v_\infty), \gamma \right]$$

which is the subject of the next section. There γ is treated as a generic parameter which could be variously γ^0, (τ, λ), or an arbitrary infinite dimensional vector.

Two norms that are used repeatedly in the sequel are defined as follows: If X is a vector valued random variable mapping a probability space $(\Omega, \mathscr{A}, P)$ into $\mathbb{R}^k$, then

$$\|X\|_r = \left[\sum_{i=1}^{k} \int_\Omega |X_i(\omega)|^r \, dP(\omega) \right]^{1/r}$$

$$\|X\| = |X| = \left(\sum_{i=1}^{k} X_i^2 \right)^{1/2}.$$

2. A UNIFORM STRONG LAW AND A CENTRAL LIMIT THEOREM FOR DEPENDENT, NONSTATIONARY RANDOM VARIABLES

Consider a sequence of vector valued random variables

$$V_t(\omega) \qquad t = 0, \pm 1, \pm 2, \ldots$$

defined on a complete probability space $(\Omega, \mathscr{A}, P)$ with range in, say, $\mathbb{R}^l$. Let

$$v_\infty = (\ldots, v_{-1}, v_0, v_1, \ldots)$$

where each v_t is in $\mathbb{R}^l$, and consider vector valued, Borel measurable functions of the form $W_t(v_\infty)$ with range in $\mathbb{R}^{k_t}$ for $t = 0, 1, \ldots$. The subscript t serves three functions. It indicates that time may enter as a variable. It indicates that the focus of the function W_t is the component v_t of v_∞ and that other components v_s enter the computation, as a rule, according as the distance of the index s from the index t; for instance

$$W_t(v_\infty) = \sum_{j=0}^{\infty} \lambda_j v_{t-j}.$$

And it indicates that the dimension k_t of the vector $w_t = W_t(v_\infty)$ may depend on t. Put

$$V_\infty(\omega) = (\ldots, V_{-1}(\omega), V_0(\omega), V_1(\omega), \ldots).$$

Then $W_t[V_\infty(\omega)]$ is a k_t-dimensional random variable depending (possibly) on infinitely many of the random variables $V_t(\omega)$. This notation is rather cumbersome, and we shall often write $W_t(\omega)$ or W_t instead. Let (Γ, ρ) be a compact metric space, and let

$$\{ g_{nt}(w_t, \gamma) : n = 1, 2, \ldots; t = 0, 1, \ldots \}$$
$$\{ g_t(w_t, \gamma) : t = 0, 1, \ldots \}$$

be sequences of real valued functions defined over $\mathbb{R}^{k_t} \times \Gamma$. In this section we shall set forth plausible regularity conditions such that

$$\lim_{n \to \infty} \sup_\Gamma \left| \frac{1}{n} \sum_{t=1}^{n} \left[g_t(W_t, \gamma) - \mathscr{E} g_t(W_t, \gamma) \right] \right| = 0$$

almost surely $(\Omega, \mathscr{A}, P)$ and such that

$$\frac{1}{\sqrt{n}} \sum_{t=1}^{n} \left[g_{nt}(W_t, \gamma_n^0) - \mathscr{E}g_{nt}(W_t, \gamma_n^0) \right] \xrightarrow{\mathscr{L}} N(0,1)$$

for any sequence $\{\gamma_n^0\}$ from Γ, convergent or not. We have seen in Chapter 3 that these are the basic tools with which one constructs an asymptotic theory for nonlinear models. As mentioned earlier, these results represent adaptations and extensions of dependent strong laws and central limit theorems obtained in a series of articles by McLeish (1974, 1975a, 1975b, 1977). Additional details and some of the historical development of the ideas may be had by consulting that series of articles.

We begin with a few definitions. The first defines a quantitative measure of the dependence amongst the random variables $\{V_t\}_{t=-\infty}^{\infty}$.

STRONG MIXING. A measure of dependence between two σ-algebras $\mathscr{F}$ and $\mathscr{G}$ is

$$\alpha(\mathscr{F}, \mathscr{G}) = \sup_{F \in \mathscr{F}, G \in \mathscr{G}} |P(FG) - P(F)P(G)|.$$

The measure will be zero if the two σ-algebras are independent and positive otherwise. Let $\{V_t\}_{t=-\infty}^{\infty}$ be the sequence of random variables defined on the complete probability space $(\Omega, \mathscr{A}, P)$ described above, and let

$$\mathscr{F}_m^n = \sigma(V_m, V_{m+1}, \ldots, V_n)$$

denote the smallest complete (with respect to P) sub-σ-algebra such that the random variables V_t for $t = m, m+1, \ldots, n$ are measurable. Define

$$\alpha_m = \sup_t \alpha\left(\mathscr{F}_{-\infty}^t, \mathscr{F}_{t+m}^{\infty}\right).$$

Observe that the faster α_m converges to zero, the less dependence the sequence $\{V_t\}_{t=-\infty}^{\infty}$ exhibits. An independent sequence has $\alpha_m > 0$ for $m = 0$ and $\alpha_m = 0$ for $m > 0$.

Following McLeish (1975b), we shall express the rate at which such a sequence of nonnegative real numbers approaches zero in terms of size.

SIZE. A sequence $\{\alpha_m\}_{m=1}^{\infty}$ of nonnegative real numbers is said to be of size $-q$ if $\alpha_m = O(m^\theta)$ for some $\theta < -q$.

This definition is stronger than that of McLeish. However, the slight sacrifice in generality is irrelevant to our purposes, and the above definition of size is much easier to work with. Recall that $\alpha_m = O(m^\theta)$ means that there is a bound B with $|\alpha_m| \le Bm^\theta$ for all m larger than some M.

Withers (1981, Corollary 4.a) proves the following. Let $\{\epsilon_t : t = 0, \pm 1, \pm 2, \dots\}$ be a sequence of independent and identically distributed random variables each with mean zero, variance one, and a density $p_\epsilon(t)$ which satisfies $\int_{-\infty}^{\infty} |p_\epsilon(t) - p_\epsilon(t + h)|\, dt \le |h| B$ for some finite bound B. If each ϵ_t is normally distributed, then this condition is satisfied. Let

$$e_t = \sum_{j=0}^{\infty} d_j \epsilon_{t-j}$$

where $d_j = O(j^{-\nu})$ for some $\nu > 3/2$ and $\sum_{j=0}^{\infty} d_j z^j \ne 0$ for complex valued z with $|z| \le 1$. Suppose that $\|\epsilon_t\|_\delta \le \text{const} < \infty$ for some δ with $2/(\nu - 1) < \delta < \nu + 1/2$. Then $\{e_t\}$ is strong mixing with $\{\alpha_m\}$ of size $-[\delta(\nu - 1) - 2]/(\delta + 1)$. For normally distributed $\{\epsilon_t\}$ there will always be such a δ for any ν. These conditions are not the weakest possible for a linear process to be strong mixing; see Withers (1981) and his references for weaker conditions.

The most frequently used time series models are stationary autoregressive moving average models, often denoted ARMA(p, q),

$$e_t + a_1 e_{t-1} + \cdots + a_p e_{t-p} = \epsilon_t + b_1 \epsilon_{t-1} + \cdots + b_q \epsilon_{t-q}$$

with the roots of the characteristic polynomials

$$m^p + a_1 m^{p-1} + \cdots + a_p = 0$$
$$m^q + b_1 m^{q-1} + \cdots + b_q = 0$$

less than one in absolute value. Such processes can be put in the form

$$e_t = \sum_{j=0}^{\infty} d_j \epsilon_{t-j}$$

where the d_j fall off exponentially and $\sum_{j=0}^{\infty} d_j z^j \ne 0$ for complex valued z with $|z| \le 1$ (Fuller, 1976, Theorem 2.7.1 and Section 2.4), whence $d_j = O(j^{-\nu})$ for any $\nu > 0$. Thus, a normal ARMA(p, q) process is strong mixing of size $-q$ for q arbitrarily large; the same is true for any innovation process $\{\epsilon_t\}$ that satisfies Withers's conditions for arbitrary δ.

It would seem from these remarks that an assumption made repeatedly in the sequel, "$\{V_t\}_{t=-\infty}^{\infty}$ is strong mixing of size $-r/(r-2)$ for some $r > 2$," is not unreasonable in applications. If the issue is in doubt, it is probably easier to take $\{V_t\}$ to be a sequence of independent random variables or a finite moving average of independent random variables, which will certainly be strong mixing of arbitrary size, and then show that the dependence of observed data W_t on far distant V_s is limited in a sense we make precise below. This will provide access to our results without the need to verify strong mixing. We shall see an example of this approach when we verify that our results apply to a nonlinear autoregression (Example 1).

We shall not make use of the related concept of uniform mixing, because it requires the innovations $\{\epsilon_t\}$ in the process

$$e_t = \sum_{j=0}^{\infty} d_j \epsilon_{t-j}$$

to be bounded (Athreya and Pantula, 1986).

Consider the vector valued function $W_t(V_\infty)$, which, we recall, depends (possibly) on infinitely many of the coordinates of the vector

$$V_\infty = (\ldots, V_{-1}, V_0, V_1, \ldots).$$

If the dependence of $W_t(V_\infty)$ on coordinates V_s far removed from the position occupied by V_t is too strong, the sequence of random variables

$$W_t = W_t(V_\infty) \qquad t = 0, 1, \ldots$$

will not inherit any limits on dependence from limits placed on $\{V_t\}_{t=-\infty}^{\infty}$. In order to insure that limits placed on the dependence exhibited by $\{V_t\}_{t=-\infty}^{\infty}$ carry over to $\{W_t\}_{t=0}^{\infty}$, we shall limit the influence of V_s on the values taken on by $W_t(V_\infty)$ for values of s far removed from the current epoch t. A quantative measure of this notion is as follows.

NEAR EPOCH DEPENDENCE. Let $\{V_t\}_{t=-\infty}^{\infty}$ be a sequence of vector valued random variables defined on the complete probability space $(\Omega, \mathscr{A}, P)$, and let $\mathscr{F}_m^n$ denote the smallest complete sub-σ-algebra such that the random variables V_t for $t = m, m+1, \ldots, n$ are measurable. Let $W_t = W_t(V_\infty)$ for $t = 0, 1, \ldots$ denote a sequence of Borel measurable functions with range in $\mathbb{R}^{k_t}$ that depends (possibly) on infinitely many of the coordinates of the vector

$$V_\infty = (\ldots, V_{-1}, V_0, V_1, \ldots).$$

Let $\{g_{nt}(w_t)\}$ for $n = 1, 2, \ldots$ and $t = 0, 1, 2, \ldots$ be a doubly indexed sequence of real valued, Borel measurable functions each of which is defined over $\mathbb{R}^{k_t}$. The doubly indexed sequence $\{g_{nt}(W_t)\}$ is said to be near epoch dependent of size $-q$ if

$$\nu_m = \sup_n \sup_t \left\| g_{nt}(W_t) - \mathscr{E}\left[g_{nt}(W_t) \middle| \mathscr{F}_{t-m}^{t+m}\right] \right\|_2$$

is of size $-q$.

Let (Γ, ρ) be a separable metric space, and let $\{g_{nt}(w_t, \gamma)\}$ be a doubly indexed family of real valued functions each of which is continuous in γ for each fixed w_t and Borel measurable in $w_t \in \mathbb{R}^{k_t}$ for fixed γ. The family $\{g_{nt}(W_t, \gamma)\}$ is said to be near epoch dependent of size $-q$ if:

1. The sequence $\{g_{nt}^0(W_t) = g_{nt}(W_t, \gamma_n^0)\}$ is near epoch dependent of size $-q$ for every sequence γ_n^0 from Γ.
2. The sequence

$$\left\{ \bar{g}_{nt}(W_t) = \sup_{\rho(\gamma, \gamma^0) < \delta} g_{nt}(W_t, \gamma) \right\}$$

and

$$\left\{ \underline{g}_{nt}(W_t) = \inf_{\rho(\gamma, \gamma^0) < \delta} g_{nt}(W_t, \gamma) \right\}$$

are near epoch dependent of size $-q$ for each γ^0 in Γ and all positive δ less than some δ^0 which can depend on γ^0.

The above definition is intended to include singly indexed sequences $\{g_t(W_t)\}_{t=1}^\infty$ as a special case with

$$\nu_m = \sup_t \left\| g_t(W_t) - \mathscr{E}\left[g_t(W_t) \middle| \mathscr{F}_{t-m}^{t+m}\right] \right\|_2$$

in this instance. For singly indexed families $\{g_t(W_t, \gamma)\}_{t=0}^\infty$, the definition retains its doubly indexed flavor, as $\{g_{nt}^0(W_t) = g_t(W_t, \gamma_n^0)\}$ is doubly indexed even if $\{g_t(W_t, \gamma)\}$ is not.

Note that if W_t depends on only finitely many of the V_t, for instance

$$W_t = \sum_{j=0}^l f_j(V_{t-j})$$

then any sequence $\{g_{nt}(w_t)\}$ or any family $\{g_{nt}(w_t, \gamma)\}$ will be near epoch dependent, because

$$\left\| g_{nt}(W_t) - \mathscr{E}\left[g_{nt}(W_t) \big| \mathscr{F}_{t-m}^{t+m} \right] \right\|_2 = 0$$

for m larger than l; similarly for $\{g_{nt}^0(w_t)\}$, $\{\bar{g}_{nt}(w_t)\}$, and $\{\underline{g}_{nt}(w_t)\}$.

The situations of most interest here will have the dimension of W_t fixed at $k_t = k$ for all t, and $g_{nt}(w)$ or $g_{nt}(w, \gamma)$ will be smooth, so that

$$g_{nt}(w, \gamma) - g_{nt}(\hat{w}, \gamma) = \frac{\partial}{\partial w'} g_{nt}(\overline{w}, \gamma)(w - \hat{w})$$

where $\overline{w}$ is on the line segment joining w to $\hat{w}$. Letting

$$B_{nt}(w, \hat{w}, \gamma) = \left| \frac{\partial}{\partial w'} g_{nt}(\overline{w}, \gamma) \right|$$

where $|x| = [\Sigma_{i=1}^{k} x_i^2]^{1/2}$, we have

$$\left| g_{nt}(w, \gamma) - g_{nt}(\hat{w}, \gamma) \right| \le B_{nt}(w, \hat{w}, \gamma) |w - \hat{w}|$$

using $|\Sigma x_i y_i| \le |x| |y|$. For functions $g_{nt}(w)$ or $g_{nt}(w, \gamma)$ that are smooth enough to satisfy this inequality, the following lemma and proposition aid in showing near epoch dependence.

PROPOSITION 1. Let $\{V_t\}_{t=-\infty}^{\infty}$, $\{W_t\}_{t=0}^{\infty}$, and $\{g_{nt}(w, \gamma) : t = 0, 1, 2, \ldots; n = 1, 2, \ldots\}$ be as in the definition of near epoch dependence, but with $k_t = k$ for all t. Let

$$\left| g_{nt}(w, \gamma) - g_{nt}(\hat{w}, \gamma) \right| \le B_{nt}(w, \hat{w}, \gamma) |w - \hat{w}|$$

where $|w - \hat{w}| = [\Sigma_{i=1}^{k}(w_i - \hat{w}_i)^2]^{1/2}$ or any other convenient norm on $\mathbb{R}^k$. Suppose that there exist random variables $\hat{W}_{t-m}^{t+m}$ of the form

$$\hat{W}_{t-m}^{t+m} = \hat{W}(V_{t-m}, \ldots, V_t, \ldots, V_{t+m})$$

such that for some $r > 2$ and some pair p, q with $1 \le p, q \le \infty$, $1/p + 1/q = 1$, we have:

1. $\{B_{nt}(W_t, \hat{W}_{t-m}^{t+m}, \gamma)\}$ dominated by random variables d_{ntm} with $\|d_{ntm}\|_q \le \Delta < \infty$.
2. $\{B_{nt}(W_t, \hat{W}_{t-m}^{t+m}, \gamma) | W_t - \hat{W}_{t-m}^{t+m}|\}$ dominated by random variables d_{ntm} with $\|d_{ntm}\|_r \le \Delta < \infty$.

If

$$\eta_m = \sup_t \left\| |W_t - \hat{W}_{t-m}^{t+m}| \right\|_p$$

is of size $-2q(r - 1)/(r - 2)$, then $\{ g_{nt}(W_t, \gamma) \}$ is near epoch dependent of size $-q$.

First, we prove the following lemma.

LEMMA 1. Let $\{V_t\}_{t=-\infty}^{\infty}$ and $\{W_t\}_{t=0}^{\infty}$ be as in Proposition 1, and let $\{g_{nt}(w)\}$ be a sequence of functions defined over $\mathbb{R}^k$ with

$$|g_{nt}(w) - g_{nt}(\hat{w})| \le B_{nt}(w, \hat{w})|w - \hat{w}|.$$

For $\{\hat{W}_{t-m}^{t+m}\}$, r, and q as in Proposition 1 let $\|B_{nt}(W_t, \hat{W}_{t-m}^{t+m})\|_q \le \Delta < \infty$ and let $\|B_{nt}(W_t, \hat{W}_{t-m}^{t+m})|W_t - \hat{W}_{t-m}^{t+m}|\|_r \le \Delta < \infty$. Then $\{g_{nt}(W_t)\}$ is near epoch dependent of size $-q$.

Proof. Let $g(w) = g_{nt}(w)$, $W = W_t$, $\hat{W} = \hat{W}_{t-m}^{t+m}$, and $\mathscr{F} = \mathscr{F}_{t-m}^{t+m}$. For

$$c = \left\{ \left[\|B(W, \hat{W})\|_q \|W - \hat{W}\|_p \right] \left[\|B(W, \hat{W})|W - \hat{W}|\|_r \right]^{-r} \right\}^{1/(1-r)}$$

let

$$B_1(W, \hat{W}) = \begin{cases} B(W, \hat{W}) & B(W, \hat{W})|W - \hat{W}| \le c \\ 0 & B(W, \hat{W})|W - \hat{W}| > c \end{cases}$$

and let $B_2(W, \hat{W}) = B(W, \hat{W}) - B_1(W, \hat{W})$. Then

$$\left\| gW - \mathscr{E}(gW|\mathscr{F}) \right\|_2 \le \|gW - g\hat{W}\|_2$$

[because $\mathscr{E}(gW|\mathscr{F})$ is the best $\mathscr{F}$-measurable approximation to gW in L_2-norm and $\hat{W}$ is $\mathscr{F}$-measurable]

$$\le \|B(W, \hat{W})|W - \hat{W}|\|_2$$

$$\le \|B_1(W, \hat{W})|W - \hat{W}|\|_2$$

$$+ \|B_2(W, \hat{W})|W - \hat{W}|\|_2$$

[by the triangle inequality]]

$$= \left\{ \int [B_1(W,\hat{W})]^2 |W - \hat{W}|^2 \, dP \right\}^{1/2}$$

$$+ \left\{ \int [B_2(W,\hat{W})]^2 |W - \hat{W}|^2 \, dP \right\}^{1/2}$$

$$\leq c^{1/2} \left\{ \int B_1(W,\hat{W}) |W - \hat{W}| \, dP \right\}^{1/2}$$

$$+ c^{(2-r)/2} \left\{ \int c^{-2+r} [B_2(W,\hat{W})]^2 |W - \hat{W}|^2 \, dP \right\}^{1/2}$$

$$\leq c^{1/2} \| B_1(W,\hat{W}) \|_q^{1/2} \| |W - \hat{W}| \|_p^{1/2}$$

$$+ c^{(2-r)/2} \left\{ \int [B_2(W,\hat{W})]^r |W - \hat{W}|^r \, dP \right\}^{1/2}$$

[by the Hölder inequality]

$$= c^{1/2} \| B_1(W,\hat{W}) \|_q^{1/2} \| |W - \hat{W}| \|_p^{1/2}$$

$$+ c^{(2-r)/2} \| B(W,\hat{W}) |W - \hat{W}| \|_r^{r/2}$$

$$= 2^{1/2} \| |W - \hat{W}| \|_p^{(1/2)(r-2)/(r-1)}$$

$$\times \| B(W,\hat{W}) \|_q^{(1/2)(r-2)/(r-1)} \| B(W,\hat{W}) |W - \hat{W}| \|_r^{(1/2)r/(r-1)}$$

after substituting the above expression for c and some algebra. If $\| B(W,\hat{W}) \|_q \leq \Delta$ and $\| B(W,\hat{W}) |W - \hat{W}| \|_r \leq \Delta$, then we have

$$\| gW - \mathscr{E}(gW | \mathscr{F}) \|_2 \leq 2^{1/2} \Delta \| |W - \hat{W}| \|_p^{(1/2)(r-2)/(r-1)}$$

whence

$$\nu_m = \sup_n \sup_t \| g_{nt} W_t - \mathscr{E} \left(g_{nt} W_t | \mathscr{F}_{t-m}^{t+m} \right) \|_2$$

$$\leq 2^{1/2} \Delta \sup_t \| |W_t - \hat{W}_{t-m}^{t+m}| \|_p^{(1/2)(r-2)/(r-1)}$$

$$= 2^{1/2} \Delta \eta_m^{(1/2)(r-2)/(r-1)}.$$

If η_m is of size $-2q(r-1)/(r-2)$ then ν_m is size $-q$. $\square$

Proof of Proposition 1. Now

$$\left| g_{nt}(w, \gamma) - g_{nt}(\hat{w}, \gamma) \right| \le B_{nt}(w, \hat{w}, \gamma)|w - \hat{w}|$$

implies

$$-g_{nt}(w, \gamma) \le B_{nt}(w, \hat{w}, \gamma)|w - \hat{w}| - g_{nt}(\hat{w}, \gamma)$$
$$-g_{nt}(\hat{w}, \gamma) \le B_{nt}(w, \hat{w}, \gamma)|w - \hat{w}| - g_{nt}(w, \gamma)$$

whence, using $\sup\{-x\} = -\inf\{x\}$, one has

$$\left| \inf_{\rho(\gamma, \gamma^0)<\delta} g_{nt}(w, \gamma) - \inf_{\rho(\gamma, \gamma^0)<\delta} g_{nt}(\hat{w}, \gamma) \right|$$
$$\le \sup_{\rho(\gamma, \gamma^0)<\delta} B_{nt}(w, \hat{w}, \gamma)|w - \hat{w}|.$$

A similar argument applied to

$$g_{nt}(w, \gamma) \le B_{nt}(w, \hat{w}, \gamma)|w - \hat{w}| + g_{nt}(\hat{w}, \gamma)$$
$$g_{nt}(\hat{w}, \gamma) \le B_{nt}(w, \hat{w}, \gamma)|w - \hat{w}| + g_{nt}(w, \gamma)$$

yields

$$\left| \sup_{\rho(\gamma, \gamma^0)<\delta} g_{nt}(w, \gamma) - \sup_{\rho(\gamma, \gamma^0)<\delta} g_{nt}(\hat{w}, \gamma) \right|$$
$$\le \sup_{\rho(\gamma, \gamma^0)<\delta} B_{nt}(w, \hat{w}, \gamma)|w - \hat{w}|.$$

We also have

$$\left| g_{nt}(w, \gamma_n^0) - g_{nt}(\hat{w}, \gamma_n^0) \right| \le B_{nt}(w, \hat{w}, \gamma_n^0)|w - \hat{w}|.$$

All three inequalities have the form

$$\left| g_{nt}(w) - g_{nt}(\hat{w}) \right| \le B_{nt}(w, \hat{w})|w - \hat{w}|$$

with $\| B_{nt}(W_t, \hat{W}_{t-m}^{t+m}) \|_q \le \Delta < \infty$ and $\| B_{nt}(W_t, \hat{W}_{t-m}^{t+m})|W_t - \hat{W}_{t-m}^{t+m}| \|_r \le \Delta$ $< \infty$, whence Lemma 1 applies to all three. Thus part 1 of the definition of near epoch dependence obtains for any sequence $\{\gamma_n^0\}$, and part 2 obtains for all positive δ.   □

The following example illustrates how Proposition 1 may be used in applications.

EXAMPLE 1 (Nonlinear autoregression). Consider data generated according to the model

$$y_t = \begin{cases} f(y_{t-1}, x_t, \theta^0) + e_t & t = 1, 2, \dots \\ 0 & t \leq 0. \end{cases}$$

Assume that $f(y, x, \theta)$ is a contraction mapping in y; viz.

$$\left| \frac{\partial}{\partial y} f(y, x, \theta) \right| \leq d < 1.$$

Let the errors $\{e_t\}$ be strong mixing with $\|e_t\|_p \leq K < \infty$ for some $p > 4$; set $e_t = 0$ for $t \leq 0$. As an instance, let

$$e_t = \sum_{j=0}^{l} \gamma_j \epsilon_{t-j}, \qquad \mathcal{E} \epsilon_t = 0, \quad \mathcal{E} |\epsilon_t|^p \leq K < \infty$$

with l finite. With this structure, $V_t = (0, 0)$ for $t \leq 0$ and $V_t = (e_t, x_t)$ for $t = 1, 2, \dots$. Suppose that θ^0 is estimated by least squares—$\hat{\theta}_n$ minimizes

$$s_n(\theta) = \frac{1}{n} \sum_{t=1}^{n} [y_t - f(y_{t-1}, x_t, \theta)]^2.$$

We shall show that this situation satisfies the hypotheses of Proposition 1. To this end, define a predictor of y_s of the form

$$\hat{y}_{t,m}^s = \hat{y}_s(V_t, V_{t-1}, \dots, V_{t-m})$$

as follows:

$$\bar{y}_t = \begin{cases} 0 & t \leq 0 \\ f(\bar{y}_{t-1}, x_t, \theta^0) & 0 < t \end{cases}$$

$$\hat{y}_{t,m}^s = \begin{cases} \bar{y}_s & s \leq \max(t - m, 0) \\ f(\hat{y}_{t,m}^{s-1}, x_s, \theta^0) + e_s & \max(t - m, 0) < s \leq t. \end{cases}$$

For $t \geq 0$ there is a $\tilde{y}_t$ on the line segment joining y_t to $\bar{y}_t$ such that

$$|y_t - \bar{y}_t| = |f(y_{t-1}, x_t, \theta^0) + e_t - f(\bar{y}_{t-1}, x_t, \theta^0)|$$

$$= \left| \frac{\partial}{\partial y} f(\tilde{y}_t, x_t, \theta^0)(y_{t-1} - \bar{y}_{t-1}) + e_t \right|$$

$$\leq d|y_{t-1} - \bar{y}_{t-1}| + |e_t|$$

$$\leq d^2|y_{t-2} - \bar{y}_{t-2}| + d|e_{t-1}| + |e_t|$$

$$\vdots$$

$$\leq d^t|y_0 - \bar{y}_0| + \sum_{j=0}^{t-1} d^j|e_{t-j}|$$

$$= \sum_{j=0}^{t-1} d^j|e_{t-j}|.$$

For $m > 0$ and $t - m > 0$ the same argument yields

$$|y_t - \hat{y}_{t,m}^t| = |f(y_{t-1}, x_t, \theta^0) + e_t - f(\hat{y}_{t,m}^{t-1}, x_t, \theta^0) - e_t|$$

$$\leq d|y_{t-1} - \hat{y}_{t,m}^{t-1}|$$

$$\vdots$$

$$\leq d^m|y_{t-m} - \bar{y}_{t-m}|$$

$$\leq d^m \sum_{j=0}^{t-m-1} d^j|e_{t-m-j}|$$

where the last inequality obtains by substituting the bound for $|y_t - \bar{y}_t|$ obtained previously. For $t - m < 0$ we have

$$|y_t - \hat{y}_{t,m}^t| \leq d^{m-t}|y_0 - \bar{y}_0| = 0.$$

In either event,

$$\|y_t - \hat{y}_{t,m}^t\|_p \leq d^m \sum_{j=0}^{t-m-1} \|e_{t-m-j}\|_p \leq K \frac{d^m}{1-d}.$$

This construction is due to Bierens (1981, Chapter 5).

Letting

$$W_t = (y_t, y_{t-1}, x_t)$$
$$\hat{W}^t_{t-m} = (\hat{y}^t_{t,m}, \hat{y}^{t-1}_{t,m}, x_t)$$

we have

$$s_n(\theta) = \frac{1}{n} \sum_{t=1}^{n} g_t(W_t, \theta)$$

with

$$g_t(W_t, \theta) = [y_t - f(y_{t-1}, x_t, \theta)]^2.$$

For $t \geq 1$

$$
\begin{aligned}
&\left| g_t(W_t, \theta) - g_t(\hat{W}^t_{t-m}, \theta) \right| \\
&= \left| [y_t - f(y_{t-1}, x_t, \theta)]^2 - [\hat{y}^t_{t,m} - f(\hat{y}^t_{t,m}, x_t, \theta)]^2 \right| \\
&= \left| y_t + \hat{y}^t_{t,m} - f(y_{t-1}, x_t, \theta) - f(\hat{y}^t_{t,m}, x_t, \theta) \right| \\
&\quad \times \left| y_t - \hat{y}^t_{t-1} - f(y_{t-1}, x_t, \theta) + f(\hat{y}^t_{t,m}, x_t, \theta) \right| \\
&= \left| 2e_t + f(y_{t-1}, x_t, \theta^0) - f(y_{t-1}, x_t, \theta) \right. \\
&\quad \left. + f(\hat{y}^t_{t,m}, x_t, \theta^0) - f(\hat{y}^t_{t,m}, x_t, \theta) \right| \\
&\quad \times \left| y_t - \hat{y}^t_{t,m} - f(y_{t-1}, x_t, \theta) + f(\hat{y}^t_{t,m}, x_t, \theta) \right| \\
&\leq 2(|e_t| + d|y_{t-1} - \hat{y}^{t-1}_{t,m}|)(|y_t - \hat{y}^t_{t,m}| + d|y_{t-1} - \hat{y}^{t-1}_{t,m}|) \\
&\leq 2\left(|e_t| + d^m \sum_{j=0}^{t-m-2} d^j |e_{t-m-1-j}| \right)(|y_t - \hat{y}^t_{t,m}| + |y_{t-1} - \hat{y}^{t-1}_{t,m}|) \\
&= B(W_t, \hat{W}^t_{t-m})|W_t - \hat{W}^t_{t-m}|
\end{aligned}
$$

where we take as a convenient norm

$$|W_t - \hat{W}^t_{t-m}| = |y_t - \hat{y}^t_{t,m}| + |y_{t-1} - \hat{y}^{t-1}_{t,m}| + |x_t - x_t|.$$

We have at once

$$\left\| B\left(W_t, \hat{W}^t_{t-m}\right) \right\|_p \le 2K\left[1 + \frac{d^m}{1-d}\right] \le \Delta < \infty \qquad \text{all } m, t$$

$$\left\| |W_t - \hat{W}^t_{t-m}| \right\|_p \le 2K\frac{d^m}{1-d} \le \Delta < \infty \qquad \text{all } m, t.$$

Using the Hölder inequality, we have for $r = p/2$ that

$$\left\| B\left(W_t, \hat{W}^t_{t-m}\right)|W_t - \hat{W}^t_{t-m}| \right\|_r$$

$$\le \left\| B\left(W_t, \hat{W}^t_{t-m}\right) \right\|_{2r}\left\| |W_t - \hat{W}^t_{t-m}| \right\|_{2r}$$

$$\le \Delta^2.$$

Note that $B(W_t, \hat{W}^t_{t-m})$ is not indexed by θ, so the above serve as dominating random variables. Put $q = p/(p + 1) < p$, whence

$$\left\| B\left(W_t, \hat{W}^t_{t-m}\right) \right\|_q \le \left(1 + \left\| B\left(W_t, \hat{W}^t_{t-m}\right) \right\|_p^p\right)^{1/q} \le \Delta^0 < \infty.$$

Thus, the example satisfies conditions 1 and 2 of Proposition 1. Lastly, note that

$$\eta_m = \sup_t \left\| |W_t - \hat{W}^t_{t-m}| \right\|_p \le 2K\frac{d^m}{1-d}.$$

The rate at which η_m falls off with increasing m is exponential, since $d < 1$, whence η_m is of size $-q(r-1)/(r-2)$ for any $r > 2$. Thus all conditions of Proposition 1 are satisfied.

If the starting point of the autoregression is random with $y_0 = Y$ where $\|Y\|_p \le K$, the same conclusion obtains. One can see that this is so as follows. In the case of random initial conditions, the sequence $\{V_t\}$ is taken as $V_t = (0,0)$ for $t < 0$, $V_0 = (Y,0)$, and $V_t = (e_t, x_t)$ for $t > 0$. For $t - m > 0$ the predictor $\hat{y}^t_{t,m}$ has prediction error (Problem 2)

$$|y_t - \hat{y}^t_{t,m}| \le d^t|Y| + d^m \sum_{j=0}^{t-m-1} d^j|e_{t-m-j}|$$

$$\le d^m|Y| + d^m \sum_{j=0}^{t-m-1} d^j|e_{t-m-j}|.$$

For $t - m < 0$ one is permitted knowledge of Y and the errors up to time t, so that y_t can be predicted perfectly for $t - m < 0$. Thus, it is possible to devise a predictor $\tilde{y}^t_{t,m}$ with

$$|y_t - \tilde{y}^t_{t,m}| \le d^m|Y| + d^m \sum_{j=0}^{t-m-1} d^j|e_{t-m-j}|.$$

The remaining details to verify the conditions of Proposition 1 for random initial conditions are as above.   □

McLeish (1975b) introduced the concept of mixingales—asymptotic martingales—on which we rely heavily in our treatment of the subject of dynamic nonlinear models. The definition is as follows

MIXINGALE.   Let

$$\{ X_{nt} : n = 1, 2, \ldots; \ t = 1, 2, \ldots \}$$

be a doubly indexed sequence of real valued random variables in $L_2(\Omega, \mathcal{A}, P)$, and let $\mathcal{F}^t_{-\infty}$ be an increasing sequence of sub-σ-algebras. Then $(X_{nt}, \mathcal{F}^t_{-\infty})$ is a mixingale if for sequences of nonnegative constants $\{c_{nt}\}$ and $\{\psi_m\}$ with $\lim_{m \to \infty} \psi_m = 0$ we have for all $t \ge 1$, $n \ge 1$, and $m \ge 0$ that

1. $\|\mathcal{E}(X_{nt}|\mathcal{F}^{t-m}_{-\infty})\|_2 \le \psi_m c_{nt}$,
2. $\|X_{nt} - \mathcal{E}(X_{nt}|\mathcal{F}^{t+m}_{-\infty})\|_2 \le \psi_{m+1} c_{nt}$.

The intention is to include singly indexed sequences $\{X_t\}^\infty_{t=1}$ as a special case of the definition. Thus $(X_t, \mathcal{F}^t_{-\infty})$ is a mixingale if for nonnegative ψ_m and c_t with $\lim_{m \to \infty} \psi_m = 0$ we have

1. $\|\mathcal{E}(X_t|\mathcal{F}^{t-m}_{-\infty})\|_2 \le \psi_m c_t$,
2. $\|X_t - \mathcal{E}(X_t|\mathcal{F}^{t+m}_{-\infty})\|_2 \le \psi_{m+1} c_t$.

There are some indirect consequences of the definition. We must have (Problem 3)

$$\left\|\mathcal{E}\left(X_{nt}|\mathcal{F}^{t-(m+1)}_{-\infty}\right)\right\|_2 \le \left\|\mathcal{E}\left(X_{nt}|\mathcal{F}^{t-m}_{-\infty}\right)\right\|_2$$

$$\left\|X_{nt} - \mathcal{E}\left(X_{nt}|\mathcal{F}^{t+(m+1)}_{-\infty}\right)\right\|_2 \le \left\|X_{nt} - \mathcal{E}\left(X_{nt}|\mathcal{F}^{t+m}_{-\infty}\right)\right\|_2.$$

Thus, ψ_m appearing in the definition could be replaced by $\psi'_m = \min_{n \le m} \psi_n$, so that one can assume that ψ_m satisfies $\psi_{m+1} \leq \psi_m$ without loss of generality. Letting $\mathscr{F}_{-\infty}^{-\infty} = \bigcap_{t=-\infty}^{\infty} \mathscr{F}_{-\infty}^t$ and letting $\mathscr{F}_{-\infty}^{\infty}$ denote the smallest complete sub-σ-algebra such that all the V_t are measurable, we have from

$$\left\| \mathscr{E}\left(X_{nt} | \mathscr{F}_{-\infty}^{-\infty} \right) \right\|_2 \leq \left\| \mathscr{E}\left(X_{nt} | \mathscr{F}_{-\infty}^{t-m} \right) \right\|_2 \leq \psi_m c_{nt}$$

and $\lim_{m \to \infty} \psi_m = 0$ that $\left\| \mathscr{E}(X_{nt} | \mathscr{F}_{-\infty}^{-\infty}) \right\|_2 = 0$ whence $\mathscr{E}(X_{nt} | \mathscr{F}_{-\infty}^{-\infty}) = 0$ almost surely. Consequently, $\mathscr{E}(X_{nt}) = 0$ for all n, $t \geq 1$. By the same sort of argument $X_{nt} - \mathscr{E}(X_{nt} | \mathscr{F}_{-\infty}^{\infty}) = 0$ almost surely.

Every example that we consider will have X_{nt} a function of the past, not the future, so that X_{nt} will perforce be $\mathscr{F}_{-\infty}^t$-measurable. This being the case, condition 2 in the definition of mixingale will be satisfied trivially and is just excess baggage. Nonetheless, we shall carry it along through Theorem 2 because it is not that much trouble and it keeps us in conformity with the literature.

The concept of a mixingale and the concepts of strong mixing and near epoch dependence are related by the following two propositions. Recall that if X is a random variable with range in $\mathbb{R}^k$, then

$$\|X\|_r = \left(\sum_{i=1}^k \int |X_i|^r \, dP \right)^{1/r}.$$

PROPOSITION 2. Suppose that a random variable X defined over the probability space $(\Omega, \mathscr{A}, P)$ is measurable with respect to the sub-σ-algebra $\mathscr{G}$ and has range in $\mathbb{R}^k$. Let $g(x)$ be a real valued, Borel measurable function defined over $\mathbb{R}^k$ with $\mathscr{E}g(X) = 0$ and $\|g(X)\|_r < \infty$ for some $r > 2$. Then

$$\left\| \mathscr{E}\left(gX | \mathscr{F} \right) \right\|_2 \leq 2(2^{1/2} + 1)[\alpha(\mathscr{F}, \mathscr{G})]^{1/2 - 1/r}$$

for any sub-σ-algebra $\mathscr{F}$.

Proof. (Hall and Heyde, 1980, Theorem A.5; McLeish, 1975b, Lemma 2.1.) Suppose that U and V are univariate random variables, each bounded in absolute value by one, and measurable with respect to $\mathscr{F}$ and $\mathscr{G}$ respectively. Let $\nu = \text{sgn}[\mathscr{E}(V | \mathscr{F}) - \mathscr{E}V]$, which is $\mathscr{F}$ measurable. We have

$$
\begin{aligned}
|\mathscr{E}UV - \mathscr{E}U\mathscr{E}V| &= |\mathscr{E}\{U[\mathscr{E}(V|\mathscr{F}) - \mathscr{E}V]\}| \\
&\leq \mathscr{E}|U||\mathscr{E}(V|\mathscr{F}) - \mathscr{E}V| \\
&\leq \mathscr{E}|\mathscr{E}(V|\mathscr{F}) - \mathscr{E}V| \\
&\leq \mathscr{E}\nu[\mathscr{E}(V|\mathscr{F}) - \mathscr{E}V] \\
&= \mathscr{E}[\mathscr{E}(\nu V|\mathscr{F}) - \mathscr{E}\nu\mathscr{E}V] \\
&= \mathscr{E}(\nu V) - \mathscr{E}\nu\mathscr{E}V.
\end{aligned}
$$

The argument is symmetric, so that for $\mu = \mathrm{sgn}[\mathscr{E}(U|\mathscr{G}) - \mathscr{E}U]$ we have

$$
|\mathscr{E}UV - \mathscr{E}U\mathscr{E}V| \leq \mathscr{E}(\mu U) - \mathscr{E}\mu\mathscr{E}U.
$$

But ν is just a particular instance of an $\mathscr{F}$-measurable function that is bounded by one, so we have from this inequality that

$$
|\mathscr{E}\nu V - \mathscr{E}\nu\mathscr{E}V| \leq \mathscr{E}(\mu\nu) - \mathscr{E}\mu\mathscr{E}\nu.
$$

Combining this inequality with the first, we have

$$
|\mathscr{E}UV - \mathscr{E}U\mathscr{E}V| \leq \mathscr{E}|\mathscr{E}(V|\mathscr{F}) - \mathscr{E}V| \leq \mathscr{E}(\mu\nu) - \mathscr{E}\mu\mathscr{E}\nu.
$$

Put $F_{-1} = \{\omega : \nu = -1\}$, $F_1 = \{\omega : \nu = 1\}$, $G_{-1} = \{\omega : \mu = -1\}$, and $G_1 = \{\omega : \mu = 1\}$. Then

$$
\begin{aligned}
\mathscr{E}(\mu\nu) &- \mathscr{E}\mu\mathscr{E}\nu \\
&= P(F_{-1}G_{-1}) - P(F_{-1})P(G_{-1}) + P(F_1G_1) - P(F_1)P(G_1) \\
&\quad - P(F_{-1}G_1) + P(F_{-1})P(G_1) - P(F_1G_{-1}) + P(F_1)P(G_{-1}) \\
&\leq 4\alpha(\mathscr{F}, \mathscr{G}).
\end{aligned}
$$

We have

$$
|\mathscr{E}UV - \mathscr{E}U\mathscr{E}V| \leq \mathscr{E}|\mathscr{E}(V|\mathscr{F}) - \mathscr{E}V| \leq 4\alpha(\mathscr{F}, \mathscr{G})
$$

of which the second inequality will be used below and the first is of some interest in its own right (Hall and Heyde, 1980, p. 277).

The rest of the proof is much the same as the proof of Lemma 1. Put $\alpha = \alpha(\mathscr{F}, \mathscr{G})$, $c = \alpha^{-1/r}\|gX\|_r$, $X_1 = I(|gX| \leq c)$, $X_2 = gX - X_1$, where $I(|gX| \leq c) = 1$ if $|gX| \leq c$ and zero otherwise. If $\mathscr{F}$ and $\mathscr{G}$ are independent, we have $\alpha = 0$ and $\mathscr{E}(gX|\mathscr{F}) = 0$. For $\alpha > 0$

$$\|\mathscr{E}(gX|\mathscr{F})\|_p = \|\mathscr{E}(X_1|\mathscr{F}) + \mathscr{E}(X_2|\mathscr{F}) - \mathscr{E}X_1 + \mathscr{E}X_1\|_p$$

$$\leq \|\mathscr{E}(X_1|\mathscr{F}) - \mathscr{E}X_1\|_p + \|\mathscr{E}(X_2|\mathscr{F})\|_p + \|\mathscr{E}X_1\|_p$$

[by the triangle inequality]

$$\leq \|\mathscr{E}(X_1|\mathscr{F}) - \mathscr{E}X_1\|_p + \|X_2\|_p + \|X_2\|_p$$

[by the conditional Jensen's inequality (Problem 4) and the fact that $|X_2| \geq c \geq \mathscr{E}X_1$]

$$= \left[\int |\mathscr{E}(X_1|\mathscr{F}) - \mathscr{E}X_1|^{p-1}|\mathscr{E}(X_1|\mathscr{F}) - \mathscr{E}X_1| dP\right]^{1/p}$$

$$+ 2\left[c^{p-r}\int c^{r-p}|X_2|^p dP\right]^{1/p}$$

$$\leq \left[(2c)^{p-1}\int |\mathscr{E}(X_1|\mathscr{F}) - \mathscr{E}X_1| dP\right]^{1/p}$$

$$+ 2\left[c^{p-r}\int |X_2|^2 dP\right]^{1/p}$$

[because $X_1 \leq c \leq X_2$]

$$= (2c)^{(p-1)/p}\left[\mathscr{E}|\mathscr{E}(X_1|\mathscr{F}) - \mathscr{E}X_1|\right]^{1/p}$$

$$+ 2c^{(p-r)/p}\|X_2\|_r^{r/p}$$

$$\leq (2c)^{(p-1)/p}(4c\alpha)^{1/p} + 2c^{(p-r)/p}\|gX\|_r^{r/p}$$

[by the inequality derived above the fact that $|gX| \geq |X_2|$]

$$\leq 2(2^{1/p} + 1)\alpha^{1/p - 1/r}\|gX\|_r$$

after substituting the above expression for c and some algebra. $\square$

PROPOSITION 3. Let $\{V_t\}_{t=-\infty}^{\infty}$ be a sequence of vector valued random variables that is strong mixing of size $-2qr/(r-2)$ for some $r > 2$ and $q > 0$. Let $W_t = W_t(V_\infty)$ denote a sequence of functions with range in $\mathbb{R}^{k_t}$ that depends (possibly) on infinitely many of the coordinates of the vector

$$V_\infty = (\ldots, V_{-1}, V_0, V_1, \ldots).$$

Let $\{g_{nt}(w_t)\}$ for $n = 1, 2, \ldots$ and $t = 0, 1, 2, \ldots$ be a sequence of real valued functions with $\mathscr{E}g_{nt}(W_t) = 0$ and $\|g_{nt}(W_t)\|_r < \infty$ that is near epoch dependent of size $-q$. Let $\mathscr{F}_{-\infty}^n$ denote the smallest complete sub-σ-algebra such that the random variables V_t for $t = n, n - 1, \ldots$ are measurable. Then

1. $\left\| \mathscr{E}(g_{nt}W_t | \mathscr{F}_{-\infty}^{t-m}) \right\|_2 \leq \psi_m c_{nt}$,
2. $\left\| g_{nt}W_t - \mathscr{E}(g_{nt}W_t | \mathscr{F}_{-\infty}^{t+m}) \right\|_2 \leq \psi_{m+1} c_{nt}$,

with $\{\psi_m\}$ of size $-q$ and $c_{nt} = \max\{1, \|g_{nt}W_t\|_r\}$.

Proof. Recall that $\mathscr{F}_m^n$ denotes the smallest complete sub-σ-algebra such that $V_m, V_{m+1}, \ldots, V_n$ are measurable. Let m be even. Then

$$\left\| \mathscr{E}\left(g_{nt}W_t | \mathscr{F}_{-\infty}^{t-(m+1)} \right) \right\|_2 = \left\| \mathscr{E}\left[\mathscr{E}\left(g_{nt}W_t | \mathscr{F}_{-\infty}^{t-m} \right) \middle| \mathscr{F}_{-\infty}^{t-(m+1)} \right] \right\|_2$$

[by the law of iterated expectations]

$$\leq \left\| \mathscr{E}\left(g_{nt}W_t | \mathscr{F}_{-\infty}^{t-m} \right) \right\|_2$$

[by the conditional Jensen's inequality (Problem 4)]

$$\leq \left\| \mathscr{E}\left[\mathscr{E}\left(g_{nt}W_t | \mathscr{F}_{t-m/2}^{t+m/2} \right) \middle| \mathscr{F}_{-\infty}^{t-m} \right] \right\|_2$$
$$+ \left\| \mathscr{E}\left(g_{nt}W_t | \mathscr{F}_{-\infty}^{t-m} \right) \right.$$
$$\left. - \mathscr{E}\left[\mathscr{E}\left(g_{nt}W_t | \mathscr{F}_{t-m/2}^{t+m/2} \right) \middle| \mathscr{F}_{-\infty}^{t-m} \right] \right\|_2$$

[by the triangle inequality]

$$\leq 2(2^{1/2} + 1)$$
$$\times \left[\alpha\left(\mathscr{F}_{-\infty}^{t-m}, \mathscr{F}_{t-m/2}^{t+m/2} \right) \right]^{1/2-1/r} \|g_{nt}W_t\|_r$$
$$+ \left\| g_{nt}W_t - \mathscr{E}\left(g_{nt}W_t | \mathscr{F}_{t-m/2}^{t+m/2} \right) \right\|_2$$

[by Proposition 2 applied to the first term and by the conditional Jensen's inequality (Problem 4) applied to the second]

$$\leq 2(2^{1/2} + 1)(\alpha_{m/2})^{1/2-1/r} \|g_{nt}W_t\|_r + \nu_{m/2}$$

by the definition of near epoch dependence. Put $\psi_{m+1} = \psi_m = 2(2^{1/2} + 1)(\alpha_{m/2})^{1/2-1/r} + \nu_{m/2}$ and $c_t = \max\{1, \|g_{nt}W_t\|_r\}$, whence (part 1)

$$\left\| \mathscr{E}\left(g_{nt}W_t | \mathscr{F}_{-\infty}^{t-m} \right) \right\|_2 \leq \psi_m c_{nt}$$

for all $m \geq 0$, $n \geq 1$, $t \geq 1$, and ψ_m is of size $-q$. Again, let m be even,

whence

$$\left\| g_{nt} - \mathscr{E}\left(g_{nt}W_t | \mathscr{F}_{-\infty}^{t+m+1} \right) \right\|_2 \leq \left\| g_{nt} - \mathscr{E}\left(g_{nt}W_t | \mathscr{F}_{-\infty}^{t+m} \right) \right\|_2$$

[because $\mathscr{E}(g_{nt}W_t | \mathscr{F}_{-\infty}^{t+m+1})$ is the best L_2 approximation to $g_{nt}W_t$ by an $\mathscr{F}_{-\infty}^{t+m+1}$ measurable function and $\mathscr{E}(g_{nt}W_t | \mathscr{F}_{-\infty}^{t+m})$ is $\mathscr{F}_{-\infty}^{t+m+1}$ measurable (Problem 5)]

$$\leq \left\| g_{nt}W_t - \mathscr{E}\left(g_{nt}W_t | \mathscr{F}_{t-m}^{t+m} \right) \right\|_2$$

[by the same best L_2 approximation argument]

$$\leq \nu_m$$

[by the definition of near epoch dependence]

$$\leq \psi_{m+1}$$

[by the best L_2 approximation argument]. We have (part 2)

$$\left\| g_{nt}W_t - \mathscr{E}\left(g_{nt}W_t | \mathscr{F}_{-\infty}^{t+m} \right) \right\|_2 \leq \psi_{m+1} c_{nt}$$

for all $m \geq 0$.   □

A mixingale $(X_{nt}, \mathscr{F}_{-\infty}^t)$ with $\{\psi_m\}$ of size $-1/2$ will obey a strong law of large numbers and a central limit theorem provided that additional regularity conditions are imposed on the sequence $\{c_{nt}\}$. An inequality that is critical in showing both the strong law and the central limit theorem is the following.

LEMMA 2 (McLeish's inequality).   Let $(X_{nt}, \mathscr{F}_{-\infty}^t)$ be a mixingale, and put $S_{nj} = \sum_{t=1}^j X_{nt}$. Let $\{a_k\}_{k=-\infty}^\infty$ be a sequence of constants with $a_k = a_{-k}$ and $\sum_{k=1}^\infty \psi_k^2 |a_k^{-1} - a_{k-1}^{-1}| < \infty$. Then

$$\mathscr{E}\left(\max_{j \leq l} S_{nj}^2 \right) \leq 4 \left(\sum_{t=1}^l c_{nt}^2 \right) \left(\sum_{i=-\infty}^\infty a_i \right) \left(\frac{\psi_0^2 + \psi_1^2}{a_0} + 2 \sum_{k=1}^\infty \psi_k^2 (a_k^{-1} - a_{k-1}^{-1}) \right)$$

Proof.   (McLeish, 1975a.)   We have from Doob (1953, Theorem 4.3) that

$$\lim_{m \to \infty} \mathscr{E}\left(X_{nt} | \mathscr{F}_{-\infty}^{t-m} \right) = \mathscr{E}\left(X_{nt} | \mathscr{F}_{-\infty}^{-\infty} \right) = 0$$

$$\lim_{m \to \infty} \mathscr{E}\left(X_{nt} | \mathscr{F}_{-\infty}^{t+m} \right) = \mathscr{E}\left(X_{nt} | \mathscr{F}_{-\infty}^{\infty} \right) = X_{nt}$$

almost surely. It follows that

$$X_{nt} = \sum_{k=-\infty}^\infty \left[\mathscr{E}\left(X_{nt} | \mathscr{F}_{-\infty}^{t+k} \right) - \mathscr{E}\left(X_{nt} | \mathscr{F}_{-\infty}^{t+k-1} \right) \right]$$

almost surely, since

$$\sum_{k=-l}^{m} \left[\mathscr{E}\left(X_{nt}|\mathscr{F}_{-\infty}^{t+k}\right) - \mathscr{E}\left(X_{nt}|\mathscr{F}_{-\infty}^{t+k-1}\right)\right]$$

$$= \mathscr{E}\left(X_{nt}|\mathscr{F}_{-\infty}^{t+m}\right) - \mathscr{E}\left(X_{nt}|\mathscr{F}_{-\infty}^{t-l-1}\right).$$

Put

$$Y_{jk} = \sum_{t=1}^{j} \mathscr{E}\left(X_{nt}|\mathscr{F}_{-\infty}^{t+k}\right) - \mathscr{E}\left(X_{nt}|\mathscr{F}_{-\infty}^{t+k-1}\right)$$

whence $S_{nj} = \sum_{k=-\infty}^{\infty} Y_{jk}$ almost surely. By the Cauchy-Schwartz inequality

$$S_{nj}^2 = \sum_{k=-\infty}^{\infty} a_k^{1/2} \frac{Y_{jk}}{a_k^{1/2}} \le \left(\sum_{k=-\infty}^{\infty} a_k \right)\left(\sum_{k=-\infty}^{\infty} \frac{Y_{jk}^2}{a_k} \right)$$

whence

$$\max_{j \le l} S_{nj}^2 \le \left(\sum_{k=-\infty}^{\infty} a_k \right)\left(\sum_{k=-\infty}^{\infty} \max_{j \le l} \frac{Y_{jk}^2}{a_k} \right).$$

By the monotone convergence theorem

$$\mathscr{E}\left(\max_{j \le l} S_{nj}^2 \right) \le \left(\sum_{k=-\infty}^{\infty} a_k \right) \sum_{k=-\infty}^{\infty} \frac{\mathscr{E}\left(\max_{j \le l} Y_{jk}^2\right)}{a_k}.$$

For fixed k, $\{(Y_{jk}, \mathscr{F}_{-\infty}^{j+k}) : 1 \le j \le l\}$ is a martingale (a definition precedes Lemma 4), since

$$\mathscr{E}\left(Y_{jk}|\mathscr{F}_{-\infty}^{k+j-1}\right) = Y_{j-1,k} + \mathscr{E}\left[\mathscr{E}\left(X_{nj}|\mathscr{F}_{-\infty}^{j+k}\right) - \mathscr{E}\left(X_{nj}|\mathscr{F}_{-\infty}^{j+k-1}\right)\bigg|\mathscr{F}_{-\infty}^{k+j-1}\right]$$

$$= Y_{j-1,k}.$$

A martingale with a last element l, such as the above, satisfies Doob's inequality

$$\mathscr{E}\left(\max_{j \le l} Y_{jk}^2 \right) \le 4\mathscr{E}\left(Y_{lk}^2\right)$$

(Hall and Heyde, 1980, Theorem 2.2, or Doob, 1953, Theorem 3.4), whence

$$\mathscr{E}\left(\max_{j \le l} S_{nj}^2 \right) \le 4\left(\sum_{k=-\infty}^{\infty} a_k \right) \sum_{k=-\infty}^{\infty} \frac{\mathscr{E}\left(Y_{lk}^2\right)}{a_k}.$$

Now (Problem 6)

$$\frac{\mathcal{E}Y_{lk}^2}{a_k} = \sum_{t=1}^{l} \frac{\mathcal{E}\mathcal{E}^2\left(X_{nt}|\mathcal{F}_{-\infty}^{t+k}\right)}{a_k} - \frac{\mathcal{E}\mathcal{E}^2\left(X_{nt}|\mathcal{F}_{-\infty}^{t+k-1}\right)}{a_k}.$$

Let $Z_{ntk} = X_{nt} - \mathcal{E}(X_{nt}|\mathcal{F}_{-\infty}^{t+k})$, whence $\mathcal{E}[X\mathcal{E}(X|\mathcal{B})|\mathcal{B}] = \mathcal{E}^2(X|\mathcal{B})$ implies

$$\mathcal{E}Z_{ntk}^2 = \mathcal{E}X_{nt}^2 - \mathcal{E}\mathcal{E}^2\left(X_{nt}|\mathcal{F}_{-\infty}^{t+k}\right)$$

and

$$\sum_{k=-\infty}^{\infty} \left(\frac{\mathcal{E}\mathcal{E}^2\left(X_{nt}|\mathcal{F}_{-\infty}^{t+k}\right)}{a_k} - \frac{\mathcal{E}\mathcal{E}^2\left(X_{nt}|\mathcal{F}_{-\infty}^{t+k-1}\right)}{a_k} \right)$$

$$= \frac{\mathcal{E}\mathcal{E}^2\left(X_{nt}|\mathcal{F}_{-\infty}^{t}\right)}{a_0} - \frac{\mathcal{E}\mathcal{E}^2\left(X_{nt}|\mathcal{F}_{-\infty}^{t-1}\right)}{a_0}$$

$$- \sum_{k=1}^{\infty} \left(\frac{\mathcal{E}Z_{ntk}^2}{a_k} + \frac{\mathcal{E}Z_{nt,k-1}^2}{a_k} \right)$$

$$+ \sum_{k=1}^{\infty} \left(\frac{\mathcal{E}\mathcal{E}^2\left(X_{nt}|\mathcal{F}_{-\infty}^{t-k}\right)}{a_k} - \frac{\mathcal{E}\mathcal{E}^2\left(X_{nt}|\mathcal{F}_{-\infty}^{t-k-1}\right)}{a_k} \right)$$

$$= \frac{\mathcal{E}\mathcal{E}^2\left(X_{nt}|\mathcal{F}_{-\infty}^{t}\right)}{a_0} - \frac{\mathcal{E}\mathcal{E}^2\left(X_{nt}|\mathcal{F}_{-\infty}^{t-1}\right)}{a_0}$$

$$+ \frac{\mathcal{E}Z_{nt0}^2}{a_1} - \frac{\mathcal{E}\mathcal{E}^2\left(X_{nt}|\mathcal{F}_{-\infty}^{t}\right)}{a_1}$$

$$+ \sum_{k=1}^{\infty} \mathcal{E}Z_{ntk}^2\left(a_{k+1}^{-1} - a_k^{-1}\right)$$

$$+ \sum_{k=1}^{\infty} \mathcal{E}\mathcal{E}^2\left(X_{nt}|\mathcal{F}_{-\infty}^{t-k}\right)\left(a_k^{-1} - a_{k-1}^{-1}\right) \qquad \text{(Problem 7)}$$

$$\leq \frac{c_{nt}^2\psi_0^2}{a_0} + \frac{c_{nt}^2\psi_1^2}{a_1}$$

$$+ \sum_{k=1}^{\infty} c_{nt}^2\psi_{k+1}^2\left(a_{k+1}^{-1} - a_k^{-1}\right)$$

$$+ \sum_{k=1}^{\infty} c_{nt}^2\psi_k^2\left(a_k^{-1} - a_{k+1}^{-1}\right)$$

$$= c_{nt}^2\left(\frac{\psi_0^2}{a_0} + \frac{\psi_1^2}{a_0} + 2\sum_{k=1}^{\infty} \psi_k^2\left(a_k^{-1} - a_{k-1}^{-1}\right) \right).$$

Thus,

$$\sum_{k=-\infty}^{\infty} \frac{\mathcal{E} Y_{lk}^2}{a_k} \leq \left(\sum_{t=1}^{l} c_{nt}^2 \right) \left(\frac{\psi_0 + \psi_1}{a_0} + 2 \sum_{k=1}^{\infty} \psi_k^2 \left(a_k^{-1} - a_{k-1}^{-1} \right) \right). \qquad \square$$

A consequence of Lemma 2 is the following inequality, from which the strong law of large numbers obtains directly.

LEMMA 3.   Let $(X_{nt}, \mathcal{F}_{-\infty}^t)$ be a mixingale with $\{\psi_m\}$ of size $-1/2$, and let $S_{nj} = \sum_{t=1}^{j} X_{nt}$. Then there is a finite constant K that depends only on $\{\psi_m\}$ such that

$$\mathcal{E} \left(\max_{j \leq l} S_{nj}^2 \right) \leq K \left(\sum_{t=1}^{l} c_{nt}^2 \right).$$

If $\psi_m > 0$ for all m, then

$$K = 16 \left[\sum_{k=0}^{\infty} \left(\sum_{m=0}^{k} \psi_m^{-2} \right)^{-1/2} \right]^2.$$

Proof.   (McLeish, 1977.)   By Lemma 2 the result is trivially true if $\psi_m = 0$ for some m, since $\psi_m \geq \psi_{m+1}$. Then assume that $\psi_m > 0$ for all m, put $a_0 = \psi_0$, and put $a_k = [\psi_k (\psi_k^2 + 4 a_{k-1})^{1/2} - \psi_k^2]/a_{k-1}$ for $k \geq 1$, whence a_k is positive and solves

$$a_k^{-1} - a_{k-1}^{-1} = \frac{a_k}{\psi_k^2}.$$

Then

$$\psi_k^{-2} \leq \left(a_k^{-1} + a_{k-1}^{-1} \right) \left(a_k^{-1} - a_{k-1}^{-1} \right) = a_k^{-2} - a_{k-1}^{-2}$$

so that

$$a_k^{-2} \geq \sum_{m=0}^{k} \psi_m^{-2}$$

$$\sum_{k=0}^{\infty} a_k \leq \sum_{k=0}^{\infty} \left(\sum_{m=0}^{k} \psi_m^{-2} \right)^{-1/2}.$$

Now $\psi_m \leq B m^\theta$ for some $\theta < -1/2$, and using an integral approximation we have $\sum_{m=1}^{k}\psi_m^{-2} \le B'k^{-2\theta+1}$ for some B'. Thus $0 < \sum_{k=1}^{\infty}\psi_k^2(a_k^{-1} - a_{k-1}^{-1}) = \sum_{k=1}^{\infty}a_k \le \sum_{k=1}^{\infty}(\sum_{m=0}^{k}\psi_m^{-2})^{-1/2} \le (B')^{-1/2}\sum_{k=1}^{\infty}k^{\theta-1/2} < \infty$. Further, $(\psi_0^2 + \psi_1^2)/a_0 \le 2a_0$ because ψ_m is a decreasing sequence and $a_0^2 = \psi_0^2$. Putting $a_{-k} = a_k$ and substituting into the inequality given by Lemma 2 yields the result. $\square$

A strong law of large numbers for mixingales follows directly using the same argument used to deduce the classical strong law of large numbers from Kolmogorov's inequality, for details see Theorem 2, Section 5.1, and Theorem 1, Section 5.3, of Tucker (1967).

PROPOSITION 4.   Let $(X_t, \mathscr{F}_{-\infty}^t)$ be a mixingale with ψ_m of size $-1/2$.

1. If $\sum_{t=1}^{\infty}c_t^2 < \infty$ then $\sum_{t=1}^{n}X_t$ converges almost surely.
2. If $\sum_{t=1}^{\infty}c_t^2/t^2 < \infty$ then $\lim_{n\to\infty}(1/n)\sum_{t=1}^{n}X_t = 0$ almost surely.

We are now in a position to state and prove a uniform strong law of large numbers. The approach follows Andrews (1986). First we define a smoothness condition, due to Andrews, and then state and prove the uniform strong law.

A-SMOOTH.   Let $\{W_t\}_{t=0}^{\infty}$ be a sequence of random variables defined on the probability space $(\Omega, \mathscr{A}, P)$, each with range in $\mathbb{R}^{k_t}$. A sequence of functions $\{g_t(W_t, \gamma)\}$ defined over a metric space (Γ, ρ) is *A*-smooth if for each γ in Γ there is a constant $\delta > 0$ such that $\rho(\gamma, \gamma^0) \le \delta$ implies

$$\left| g_t(W_t, \gamma) - g_t(W_t, \gamma^0) \right| \le B_t(W_t)h\left[\rho(\gamma, \gamma^0)\right]$$

except on some $E \subset \Omega$ with $P(E) = 0$ where $B_t : \mathbb{R}^{k_t} \to \mathbb{R}^+$ and $h : \mathbb{R}^+ \to \mathbb{R}^+$ are nonrandom functions such that $B_t(w_t)$ is Borel measurable,

$$\frac{1}{n}\sum_{t=1}^{n}\mathscr{E}B_t(W_t) \le \Delta < \infty \qquad \text{for all } n$$

and $h(x)\downarrow h(0) = 0$ as $x \to 0$; δ, $B_t(\cdot)$, and $h(\cdot)$ may depend on γ^0.

THEOREM 1.   (Uniform strong law.)   Let $\{V_t\}_{t=-\infty}^{\infty}$ be a sequence of vector-valued random variables defined on the complete probability space $(\Omega, \mathscr{A}, P)$ that is strong mixing of size $-r/(r-2)$ for some $r > 2$. Let

(Γ, ρ) be a compact metric space and let $W_t = W_t(V_\infty) \equiv W_t(\omega)$ be a Borel measurable function of

$$V_\infty = (\ldots, V_{-1}, V_0, V_1, \ldots)$$

with range in $\mathbb{R}^{k_t}$. Let $\{g_t(w_t, \gamma)\}_{t=0}^\infty$ be a sequence of real-valued functions, each Borel measurable for fixed γ. Suppose:

1. $\{g_t(W_t, \gamma)\}_{t=0}^\infty$ is a near epoch dependent family of size $-\frac{1}{2}$.
2. $\{g_t(W_t, \gamma)\}_{t=0}^\infty$ is A-smooth.
3. There is a sequence $\{d_t\}$ of random variables with

$$\sup_{\Gamma} |g_t[W_t(\omega), \gamma]| \le d_t(\omega)$$

$$\|d_t\|_r \le \Delta < \infty$$

for $t = 0, 1, 2, \ldots$.

Then

$$\lim_{n \to \infty} \sup_{\Gamma} \left| \frac{1}{n} \sum_{t=1}^n [g_t(W_t, \gamma) - \mathscr{E}g_t(W_t, \gamma)] \right| = 0$$

almost surely and

$$\left\{ \frac{1}{n} \sum_{t=1}^n \mathscr{E}g_t(W_t, \gamma) \right\}_{n=1}^\infty$$

is an equicontinuous family.

Proof. (Andrews, 1986). A compact metric space is separable (Problem 8). Let

$$\bar{h}_t(W_t, \gamma^0, \delta) = \sup\{ g_t(W_t, \gamma) : \rho(\gamma, \gamma^0) < \delta \}$$

$$\underline{h}_t(W_t, \gamma^0, \delta) = \inf\{ g_t(W_t, \gamma) : \rho(\gamma, \gamma^0) < \delta \}$$

The continuity of $g_t(w_t, \gamma)$ in γ implied by the A-smooth condition and the separability of (Γ, ρ) insure measurability.

Using the A-smooth condition,

$$
\begin{aligned}
0 &\leq \lim_{\delta \to 0} \sup_{n \geq 1} \left| \frac{1}{n} \sum_{t=1}^{n} \mathscr{E}\bar{h}_t(W_t, \gamma^0, \delta) - \mathscr{E}g_t(W_t, \gamma^0) \right| \\
&\leq \lim_{\delta \to 0} \sup_{n \geq 1} \frac{1}{n} \sum_{t=1}^{n} \left| \mathscr{E}\bar{h}_t(W_t, \gamma^0, \delta) - \mathscr{E}g_t(W_t, \gamma^0) \right| \\
&\leq \lim_{\delta \to 0} \sup_{n \geq 1} \frac{1}{n} \sum_{t=1}^{n} \mathscr{E} \left| \bar{h}_t(W_t, \gamma^0, \delta) - g_t(W_t, \gamma^0) \right| \\
&\leq \lim_{\delta \to 0} \sup_{n \geq 1} \frac{1}{n} \sum_{t=1}^{n} \mathscr{E} \left| B_t(W_t) \right| h(\delta) \\
&\leq \Delta \lim_{\delta \to 0} h(\delta) = 0
\end{aligned}
$$

A similar argument applies to $\underline{h}_t(W_t, \gamma^0, \delta)$. Then given $\epsilon > 0$ and γ^0 in Γ, δ^0 can be chosen small enough that for all $n \geq 1$ and all γ in $\mathcal{O}_{\gamma^0} = \{\gamma : \rho(\gamma, \gamma^0) < \delta^0\}$

$$
\begin{aligned}
\frac{1}{n} \sum_{t=1}^{n} \mathscr{E}g_t(W_t, \gamma^0) &- \frac{\epsilon}{2} \\
&\leq \frac{1}{n} \sum_{t=1}^{n} \mathscr{E}\underline{h}_t(W_t, \gamma^0, \delta^0) \\
&\leq \frac{1}{n} \sum_{t=1}^{n} \mathscr{E}g_t(W_t, \gamma) \\
&\leq \frac{1}{n} \sum_{t=1}^{n} \mathscr{E}\bar{h}_t(W_t, \gamma^0, \delta^0) \\
&\leq \frac{1}{n} \sum_{t=1}^{n} \mathscr{E}g_t(W_t, \gamma^0) + \frac{\epsilon}{2}
\end{aligned}
$$

whence

$$
\left| \frac{1}{n} \sum_{t=1}^{n} \mathscr{E}g_t(W_t, \gamma) - \frac{1}{n} \sum_{t=1}^{n} \mathscr{E}\bar{h}_t(W_t, \gamma^0, \delta^0) \right| < \epsilon
$$

$$
\left| \frac{1}{n} \sum_{t=1}^{n} \mathscr{E}g_t(W_t, \gamma) - \frac{1}{n} \sum_{t=1}^{n} \mathscr{E}\underline{h}_t(W_t, \gamma^0, \delta^0) \right| < \epsilon
$$

$$
\left| \frac{1}{n} \sum_{t=1}^{n} \mathscr{E}g_t(W_t, \gamma) - \frac{1}{n} \sum_{t=1}^{n} \mathscr{E}g_t(W_t, \gamma^0) \right| < \frac{\epsilon}{2}.
$$

The third inequality establishes the second conclusion of Theorem 1; we will use the first and second inequalities below.

The collection $\{\mathcal{O}_{\gamma^0}\}_{\gamma^0 \in \Gamma}$ is an open covering of the compact set Γ so there is a finite subcovering $\{\mathcal{O}_{\gamma_i^0}\}_{i=1}^K$. For every γ in $\mathcal{O}_{\gamma_i^0}$ we have

$$\frac{1}{n} \sum_{t=1}^n g_t(W_t, \gamma) - \frac{1}{n} \sum_{t=1}^n \mathscr{E}g_t(W_t, \gamma)$$

$$\leq \frac{1}{n} \sum_{t=1}^n \bar{h}_t(W_t, \gamma_i^0, \delta_i^0) - \frac{1}{n} \sum_{t=1}^n \mathscr{E}\bar{h}_t(W_t, \gamma_i^0, \delta_i^0) + \epsilon.$$

Every γ in Γ must be in some $\mathcal{O}_{\gamma_i^0}$ so we have

$$\frac{1}{n} \sum_{t=1}^n g_t(W_t, \gamma) - \frac{1}{n} \sum_{t=1}^n \mathscr{E}g_t(W_t, \gamma)$$

$$\leq \max_{1 \leq i \leq K} \left[\frac{1}{n} \sum_{t=1}^n \bar{h}_t(W_t, \gamma_i^0, \delta_i^0) - \frac{1}{n} \sum_{t=1}^n \mathscr{E}\bar{h}_t(W_t, \gamma_i^0, \delta_i^0) \right] + \epsilon$$

$$= \bar{X}_n + \epsilon$$

for all γ in Γ. A similar argument gives

$$\frac{1}{n} \sum_{t=1}^n g_t(W_t, \gamma) - \frac{1}{n} \sum_{t=1}^n \mathscr{E}g_t(W_t, \gamma)$$

$$\geq \min_{1 \leq i \leq K} \left[\frac{1}{n} \sum_{t=1}^n \underline{h}_t(W_t, \gamma_i^0, \delta_i^0) - \frac{1}{n} \sum_{t=1}^n \mathscr{E}\underline{h}_t(W_t, \gamma_i^0, \delta_i^0) \right] - \epsilon$$

$$= \underline{X}_n - \epsilon$$

for all γ in Γ, whence

$$\sup_{\Gamma} \left| \frac{1}{n} \sum_{t=1}^n g_t(W_t, \gamma) - \frac{1}{n} \sum_{t=1}^n \mathscr{E}g_t(W_t, \gamma) \right| \leq \bar{X}_n - \underline{X}_n + \epsilon.$$

In consequence of the definition of near epoch dependence, the sequence $\{\bar{h}_t(W_t, \gamma_i^0, \delta_i^0) - \mathscr{E}\bar{h}_t(W_t, \gamma_i^0, \delta_i^0)\}$ satisfies the hypotheses of Proposition 3 whence the sequence is a mixingale with $\{\psi_t\}$ of size $-\frac{1}{2}$ and $c_t = \max[1, \|\bar{h}_t(W_t, \gamma_i^0, \delta_i^0) - \mathscr{E}\bar{h}_t(W_t, \gamma_i^0, \delta_i^0)\|_r] \leq 1 + 2\Delta < \infty$. Now $\sum_{t=1}^n c_t^2/t^2 < \infty$ and Proposition 4 applies, whence

$$\lim_{n \to \infty} \frac{1}{n} \sum_{t=1}^n \bar{h}_t(W_t, \gamma_i^0, \delta_i^0) - \mathscr{E}\bar{h}_t(W_t, \gamma_i^0, \delta_i^0) = 0.$$

almost surely for $i = 1, 2, \ldots, K$ which implies $\bar{X}_n$ converges almost surely to zero. Similarly, for $\underline{X}_n$ whence

$$\lim_{n \to \infty} \sup_{\Gamma} \left| \frac{1}{n} \sum_{t=1}^{n} g_t(W_t, \gamma) - \frac{1}{n} \sum_{t=1}^{n} \mathscr{E} g_t(W_t, \gamma) \right| \le \epsilon.$$

Now ϵ is arbitrary, which establishes the first conclusion of Theorem 1. □

As seen in Chapter 3, there are two constituents to an asymptotic theory of inference for nonlinear models: a uniform strong law of large numbers and a continuously convergent central limit theorem.

THEOREM 2 (Central limit theorem).   Let $\{V_t\}_{t=-\infty}^{\infty}$ be a sequence of vector valued random variables that is strong mixing of size $-r/(r-2)$ for some $r > 2$. Let (Γ, ρ) be a separable metric space, and let $W_t = W_t(V_\infty)$ be a function of

$$V_\infty = (\ldots, V_{-1}, V_0, V_1, \ldots)$$

with range in $\mathbb{R}^{k_t}$. Let

$$\{ g_{nt}(W_t, \gamma) : n = 1, 2, \ldots; \ t = 0, 1, 2, \ldots \}$$

be a sequence of real valued functions that is near epoch dependent of size $-\frac{1}{2}$. Given a sequence $\{\gamma_n^0\}_{n=1}^{\infty}$ from Γ, put

$$\sigma_n^2 = \mathrm{Var}\left(\sum_{t=1}^{n} g_{nt}(W_t, \gamma_n^0) \right) \qquad n = 1, 2, \ldots$$

$$w_n(s) = \sum_{t=1}^{[ns]} \frac{g_{nt}(W_t, \gamma_n^0) - \mathscr{E} g_{nt}(W_t, \gamma_n^0)}{\sigma_n} \qquad 0 \le s \le 1$$

where $[ns]$ denotes the integer part of ns—the largest integer that does not exceed ns—and $w_n(0) = 0$. Suppose that:

1. $1/\sigma_n^2 = O(1/n)$.
2. $\lim_{n \to \infty} \mathrm{Var}[w_n(s)] = s, \ 0 \le s \le 1$.
3. $\| g_{nt}(W_t, \gamma_n^0) - \mathscr{E} g_{nt}(W_t, \gamma_n^0) \|_r \le \Delta < \infty, \ 1 \le t \le n, \ t = 1, 2, \ldots$.

Then $w_n(\cdot)$ converges weakly in $D[0, 1]$ to a standard Wiener process. In particular,

$$\sum_{t=1}^{n} \frac{g_{nt}(W_t, \gamma_n^0) - \mathscr{E}g_{nt}(W_t, \gamma_n^0)}{\sigma_n} = w_n(1) \xrightarrow{\mathscr{L}} N(0,1). \qquad \square$$

The terminology appearing in the conclusion of Theorem 2 is defined as follows. $D[0,1]$ is the space of functions x or $x(\cdot)$ on $[0,1]$ that are right continuous and have left hand limits; that is, for $0 \le t < 1$, $x(t+) = \lim_{h\downarrow 0} x(t+h)$ exists and $x(t+) = x(t)$, and for $0 < t \le 1$, $x(t-) = \lim_{h\downarrow 0} x(t-h)$ exists. A metric $d(x,y)$ on $D[0,1]$ may be defined as follows. Let Λ denote the class of strictly increasing, continuous mappings λ of $[0,1]$ onto itself; such a λ will have $\lambda(0) = 0$ and $\lambda(1) = 1$ of necessity. For x and y in $D[0,1]$ define $d(x,y)$ to be the infimum of those positive ϵ for which there is λ in Λ with $\sup_t |\lambda(t) - t| < \epsilon$ and $\sup_t |x[\lambda(t)] - y(t)| < \epsilon$. The idea is that one is permitted to shift points on the time axis by an amount ϵ in an attempt to make x and y coincide to within ϵ; note that the points 0 and 1 cannot be so shifted. A verification that $d(x,y)$ is a metric is given by Billingsley (1968, Section 14). If $\mathscr{D}$ denotes the smallest σ-algebra containing the open sets—sets of the form $\mathscr{O} = \{y : d(x,y) < \delta\}$—then $(D, \mathscr{D})$ is a measurable space. $\mathscr{D}$ is called the Borel subsets of $D[0,1]$. The random variables $w_n(\cdot)$ have range in $D[0,1]$ and, perforce, induce a probability measure on $(D, \mathscr{D})$ defined by $P_n(A) = Pw_n^{-1}(A) = P\{\omega : w_n(\cdot)$ in $A\}$ for each A in $\mathscr{D}$. A standard Wiener process $w(\cdot)$ has two determining properties. For each t, the (real valued) random variable $w(t)$ is normally distributed with mean zero and variance t. For each partition $0 \le t_0 \le t_1 \le \cdots \le t_k \le 1$, the (real valued) random variables

$$w(t_1) - w(t_0), w(t_2) - w(t_1), \ldots, w(t_k) - w(t_{k-1})$$

are independent; this property is known as independent increments. Let W be the probability measure on $(D, \mathscr{D})$ induced by this process; $W(A) = Pw^{-1}(A) = P\{\omega : w(\cdot)$ in $A\}$. It exists and puts mass one on the space $C[0,1]$ of continuous functions defined on $[0,1]$ (Billingsley, 1968, Section 9). Weak convergence of $w_n(\cdot)$ to a standard Wiener process means that

$$\lim_{n\to\infty} \int_D f \, dP_n = \int_D f \, dW$$

for every bounded, continuous function f defined over $D[0,1]$. The term weak convergence derives from the fact that the collection of finite, signed, regular, and finitely additive set functions is the dual (Royden, 1968, Chapter 10) of the space of bounded, continuous functions defined on

$D[0, 1]$, and $\lim_{n \to \infty} \int f\, dP_n = \int f\, dW$ for every such f is weak* convergence (pointwise convergence) in this dual space (Dunford and Schwartz, 1957, Theorem IV.6.2.2, p. 262). If h is a continuous mapping from $D[0, 1]$ into $\mathbb{R}^1$, then weak convergence implies $\lim_{n \to \infty} \int_D gh\, dP_n = \int_D gh\, dW$ for all g bounded and continuous on $\mathbb{R}^1$, whence by the change of variable formula $\lim_{n \to \infty} \int_{\mathbb{R}} g\, dP_n h^{-1} = \int_{\mathbb{R}} g\, dW h^{-1}$. Thus the probability measures $P_n h^{-1}$ defined on the Borel subsets of $\mathbb{R}^1$ converge weakly to Wh^{-1}. On $\mathbb{R}^1$ convergence in distribution and weak convergence are equivalent (Billingsley, 1968, Section 3), so that the distribution of $hw_n(\cdot)$,

$$F_n(x) = P\left[hw_n(\cdot) \leq x \right] = P_n h^{-1}(-\infty, x]$$

converges at every continuity point to the distribution of $hw(\cdot)$. In particular the mapping $\pi_1 x(\cdot) = x(1)$ is continuous because $\lim_{n \to \infty} d(y_n, x) = 0$ implies $\lim_{n \to \infty} y_n(1) = x(1)$; recall that one cannot shift the point 1 by choice of λ in Λ. Thus we have that the random variable $w_n(1)$ converges in distribution to the random variable $w(1)$, which is normally distributed with mean zero and unit variance.

The proof of Theorem 2 is due to Wooldridge (1986) and is an adaptation of the methods of proof used by McLeish (1975b, 1977). We shall need some preliminary definitions and lemmas.

Recall that $\{V_t(\omega)\}_{t=-\infty}^{\infty}$ is the underlying stochastic process on $(\Omega, \mathcal{A}, P)$; that $\mathcal{F}_m^n$ denotes the smallest complete sub-σ-algebra such that $V_m, V_{m+1}, \ldots, V_n$ are measurable, $\mathcal{F}_{-\infty}^{-\infty} = \cap_{t=-\infty}^{\infty} \mathcal{F}_{-\infty}^t$, $\mathcal{F}_{-\infty}^{\infty} = \sigma(\cup_{t=-\infty}^{\infty} \mathcal{F}_{-\infty}^t)$; that $W_t(V_\infty)$ is a function of possibly infinitely many of the V_t with range in $\mathbb{R}^{k_t}$ for $t = 0, 1, \ldots$; and that $g_{nt}(w_t, \gamma_n^0)$ maps $\mathbb{R}^{k_t}$ into the real line for $n = 1, 2, \ldots$ and $t = 0, 1, \ldots$. Set

$$X_{nt} = g_{nt}\left(W_t, \gamma_n^0\right) - \mathcal{E} g_{nt}\left(W_t, \gamma_n^0\right)$$

for $t \geq 0$, and $X_{nt} = 0$ for $t < 0$, whence

$$\sigma_n^2 = \text{Var}\left(\sum_{t=1}^{n} X_{nt} \right)$$

$$w_n(s) = \frac{\sum_{t=1}^{[ns]} X_{nt}}{\sigma_n^2}.$$

By Proposition 4, $(X_{nt}, \mathcal{F}_{-\infty}^t)$ for $n = 1, 2, \ldots$ and $t = 1, 2, \ldots$ is a mixingale with $\{\psi_m\}$ of size $-1/2$ and $c_{nt} = \max\{1, \|X_{nt}\|_r\}$. That is,

1. $\|\mathscr{E}(X_{nt}|\mathscr{F}_{-\infty}^{t-m})\|_2 \leq \psi_m c_{nt}$,
2. $\|X_{nt} - \mathscr{E}(X_{nt}|\mathscr{F}_{-\infty}^{t+m})\|_2 \leq \psi_{m+1} c_{nt}$

for $n \geq 1$, $t \geq 1$, $m \geq 0$. We also have from the definition of near epoch dependence that

$$\nu_m = \sup_n \sup_t \left\| X_{nt} - \mathscr{E}\left(X_{nt}|\mathscr{F}_{t-m}^{t+m}\right) \right\|_2$$

is of size $-1/2$. Define

$$S_n(s) = \sum_{t=1}^{[ns]} X_{nt}$$

and

$$S_{nj} = \sum_{t=1}^{j} X_{nt} = S_n\left(\frac{j}{n}\right).$$

Take $S_n(0) = S_{n0} = 0$.

MARTINGALE. Note that $\mathscr{F}_{-\infty}^0, \mathscr{F}_{-\infty}^1, \ldots$ is an increasing sequence of sub-σ-algebras. Relative to these σ-algebras, a doubly indexed process

$$\left\{\left(Z_{nt}, \mathscr{F}_{-\infty}^t\right) : n = 1, 2, \ldots ; t = 0, 1, \ldots\right\}$$

is said to be a martingale if

1. Z_{nt} is measurable with respect to $\mathscr{F}_{-\infty}^t$,
2. $\mathscr{E}|Z_{nt}| < \infty$,
3. $\mathscr{E}(Z_{nt}|\mathscr{F}_{-\infty}^s) = Z_{ns}$ for $s < t$.

The sequence

$$\left\{\left(Y_{nt}, \mathscr{F}_{-\infty}^t\right) : n = 1, 2, \ldots ; t = 1, 2, \ldots\right\}$$

with $Y_{n0} = Z_{n0}$, $Y_{nt} = Z_{nt} - Z_{n,t-1}$ for $t = 1, 2, \ldots$, and $(Z_{nt}, \mathscr{F}_{-\infty}^t)$ as above is called a martingale difference sequence.

UNIFORMLY INTEGRABLE. A collection $\{X_\lambda : \lambda \in \Lambda\}$ of integrable random variables is uniformly integrable if

$$\lim_{M \to \infty} \sup_{\lambda \in \Lambda} \int_{|X_\lambda| > M} |X_\lambda| \, dP = 0.$$

LEMMA 4. Let $(X_{nt}, \mathscr{F}_{-\infty}^t)$ be a mixingale, and put

$$Y_{jk} = \sum_{t=1}^{j} \mathscr{E}\left(X_{nt} | \mathscr{F}_{-\infty}^{t+k}\right) - \mathscr{E}\left(X_{nt} | \mathscr{F}_{-\infty}^{t+k-1}\right).$$

Then for any $\gamma > 1$ and nonnegative sequence $\{a_i\}$ we have

$$\mathscr{E}\left(\max_{j \leq l} |S_{nj}|^\gamma\right) \leq \left(\frac{\gamma}{\gamma - 1}\right)^\gamma \left(\sum_{i=-\infty}^{\infty} a_i\right)^{\gamma - 1} \sum_{k=-\infty}^{\infty} a_k^{1-\gamma} \mathscr{E}|Y_{lk}|^\gamma.$$

Proof. (McLeish, 1975b, Lemma 6.2.)

LEMMA 5. Let $(Y_{nt}, \mathscr{F}_{-\infty}^t)$ for $n = 1, 2, \ldots$ and $t = 1, 2, \ldots$ be a martingale difference sequence [so $\mathscr{E}(Y_{nt} | \mathscr{F}_{\infty}^{t-1}) = 0$ almost surely for all $t \geq 1$], and assume that $|Y_{nt}| \leq K c_{nt}$ almost surely for some sequence of positive constants $\{c_{nt}\}$. Then

$$\mathscr{E}\left(\sum_{t=1}^{n} Y_{nt}\right)^4 \leq 10 K^4 \left(\sum_{t=1}^{n} c_{nt}^2\right)^2.$$

Proof. (McLeish, 1977, Lemma 3.1.)

LEMMA 6. Let $(X_{nt}, \mathscr{F}_{-\infty}^t)$ be a mixingale with $\{\psi_m\}$ of size $-1/2$ and $c_{nt} = \max\{1, \|X_{nt}\|_r\}$ for $r > 2$. If

$$\left\{ X_{nt}^2 : t = 1, 2, \ldots, n; \ n = 1, 2, \ldots \right\}$$

is uniformly integrable, then

$$\left\{ \frac{\max_{j \leq l} \left(S_{n, j+k} - S_{nk}\right)^2}{\sum_{t=k+1}^{k+l} c_{nt}^2} : 1 \leq k + l \leq n, \ k \geq 0, \ n \geq 1 \right\}$$

is uniformly integrable.

Proof. (McLeish, 1977.) For $c \geq 1$ and m to be determined later, put

$$X_{nt}^c = X_{nt} I(|X_{nt}| \leq c c_{nt})$$
$$Y_{nt} = \mathscr{E}\big(X_{nt}^c | \mathscr{F}_{-\infty}^{t+m}\big) - \mathscr{E}\big(X_{nt}^c | \mathscr{F}_{-\infty}^{t-m}\big)$$
$$U_{nt} = X_{nt} - \mathscr{E}\big(X_{nt} | \mathscr{F}_{-\infty}^{t+m}\big) + \mathscr{E}\big(X_{nt} | \mathscr{F}_{-\infty}^{t-m}\big)$$
$$Z_{nt} = \mathscr{E}\big(X_{nt} - X_{nt}^c | \mathscr{F}_{-\infty}^{t+m}\big) - \mathscr{E}\big(X_{nt} - X_{nt}^c | \mathscr{F}_{-\infty}^{t-m}\big)$$

and note that $X_{nt} = Y_{nt} + Z_{nt} + U_{nt}$. Let $E_\alpha X = \int_{[X \geq \alpha]} X\, dP$, $\overline{Y}_{nj} = \sum_{t=1}^{j} Y_{nt}$, $\overline{Z}_{nj} = \sum_{t=1}^{j} Z_{nt}$, $\overline{U}_{nj} = \sum_{t=1}^{j} U_{nt}$, $\bar{c}_{nl}^2 = \sum_{t=1}^{l} c_{nt}^2$. Jensen's inequality implies that $(\sum p_i x_i)^2 \leq \sum p_i x_i^2$ for any positive p_i with $\sum p_i = 1$, whence

$$S_{nj}^2 \leq 3\big(\overline{U}_{nj}^2 + \overline{Y}_{nj}^2 + \overline{Z}_{nj}^2\big)$$

by taking $p_i = 1/3$. In general

$$(X + Y + Z > \alpha) \subset \Big(X > \frac{\alpha}{3}\Big) \cup \Big(Y > \frac{\alpha}{3}\Big) \cup \Big(Z > \frac{\alpha}{3}\Big)$$

whence

$$(X + Y + Z)I(X + Y + Z > \alpha)$$
$$\leq 3XI\Big(X > \frac{\alpha}{3}\Big) + 3YI\Big(Y > \frac{\alpha}{3}\Big) + 3ZI\Big(Z > \frac{\alpha}{3}\Big)$$

and

$$E_\alpha(X + Y + Z) \leq 3E_{\alpha/3}X + 3E_{\alpha/3}Y + 3E_{\alpha/3}Z.$$

It follows that

$$E_\alpha\left(\frac{\max_{j \leq l} S_{nj}^2}{\bar{c}_{nl}^2}\right) \leq 9(y + z + u)$$

where

$$y = E_{\alpha/3}\left(\frac{\max_{j \leq l}\overline{Y}_{nj}^2}{\bar{c}_{nl}^2}\right),$$

$$z = \mathscr{E}\left(\frac{\max_{j \leq l}\overline{Z}_{nj}^2}{\bar{c}_{nl}^2}\right),$$

$$u = \mathscr{E}\left(\frac{\max_{j \leq l}\overline{U}_{nj}^2}{\bar{c}_{nl}^2}\right).$$

For some $\theta < -1/2$ we have

$$0 \le \psi_k = O(k^\theta) = o\left[k^{-1/2}(\ln k)^{-2}\right].$$

Note that for $k \le m$

$$\left\|U_{nt} - \mathscr{E}\left(U_{nt}|\mathscr{F}_{-\infty}^{t+k}\right)\right\|_2 \le \left\|X_{nt} - \mathscr{E}\left(X_{nt}|\mathscr{F}_{-\infty}^{t+m}\right)\right\|_2 \le c_{nt}\psi_m$$

and for $m > k$

$$\left\|U_{nt} - \mathscr{E}\left(U_{nt}|\mathscr{F}_{-\infty}^{t+k}\right)\right\|_2 = \left\|X_{nt} - \mathscr{E}\left(X_{nt}|\mathscr{F}_{-\infty}^{t+k}\right)\right\|_2 \le c_{nt}\psi_k.$$

Similarly $\|\mathscr{E}(U_{nt}|\mathscr{F}_{-\infty}^{t-k})\|_2$ is less than $c_{nt}\psi_m^2$ for $k \le m$ and is less than $c_{nt}\psi_k$ for $k \ge m$. Therefore $(U_{nt}, \mathscr{F}_{-\infty}^t)$ is a mixingale with $\hat{\psi}_k = \psi_{\max(m, k)}$ of size $-1/2$ and $\psi_k \le B/[k^{1/2}\ln^2(k)]$ for all $k > m$. By Lemma 2 with $a_k = m \ln^2 m$ for $|k| \le m$ and $a_k = 1/(k \ln^2 k)$ for $|k| \ge m$ we have

$$u \le 4\left(\sum_{i=-\infty}^{\infty} a_i\right)\left(\text{const } \psi_m^2 m \ln^2 m + \sum_{k=m+1}^{\infty} \psi_k^2\left(a_k^{-1} - a_{k-1}^{-1}\right)\right).$$

Now $\int_2^\infty x^{-1}(\ln x)^{-2}\, dx = \int_{\ln 2}^\infty u^{-2}\, du < \infty$ implies that $0 \le \sum_{k=-\infty}^\infty a_k < \infty$. Further, by Taylor's theorem $0 \le k \ln^2 k - (k-1)\ln^2(k-1) \le k[\ln^2 k - \ln^2(k-1)] \le 2\ln \bar{k}$ for $k-1 \le \bar{k} \le k$, whence $0 \le \sum_{k=2}^\infty \psi_k^2(a_k^{-1} - a_{k-1}^{-1}) \le 2\sum_{k=2}^\infty (\ln k)/(k \ln^4 k) < \infty$. Thus, for arbitrary $\epsilon > 0$ we may choose and fix m sufficiently large that $u \le \epsilon/27$. Note the choice of m depends only on the sequence $\{\psi_k\}$, not on n. Also note that if some of the leading U_{nt} were set to zero, U_{nt} would be a mixingale with the same $\hat{\psi}_k$, but the leading c_{nt} would be zero. Thus the choice of m does not depend on where the sum starts.

Similarly, for $k \le m$, $\|Z_{nt} - \mathscr{E}(Z_{nt}|\mathscr{F}_{-\infty}^{t+k})\|_2$ and $\|\mathscr{E}(Z_{nt}|\mathscr{F}_{-\infty}^{t-k})\|_2$ are less than $\|Z_{nt}\|_2$ and

$$\|Z_{nt}\|_2 \le \|X_{nt} - X_{nt}^c\|_2 \le \max_{t \le n} \mathscr{E}_c X_{nt}^2.$$

For $k > m$, $\|Z_{nt} - \mathscr{E}(Z_{nt}|\mathscr{F}_{-\infty}^{t+k})\|_2 = \|\mathscr{E}(Z_{nt}|\mathscr{F}_{-\infty}^{t+k})\|_2 = 0$. By Lemma 2, with $a_k = a_{k-1} = 1$ for $k \le m + 1$ and $a_{-k} = a_k = k^2$ for $k > m + 1$ we have

$$z \le 4(2m + 4)\left(\max_{t \le n} \mathscr{E}_c X_{nt}^2\right).$$

For our now fixed value of m we may choose c large enough that $z < \epsilon/27$, since $\{X_{nt}^2\}$ is a uniformly integrable set. Note again that c depends neither on n nor on where the sum starts.

With c and m thus fixed, apply lemma 4 to the sequence $\{Y_{nt}\}$ with $\gamma = 4$ and $a = 1$ for $|i| \le m$ and $a_i = i^2$ for $|i| > m$ to obtain

$$\mathscr{E}\left(\max_{j \le l} \overline{Y}_{nj}^4\right) \le \left(\tfrac{4}{3}\right)^4 (2m + 3)^3 \sum_{k=-m}^{m} \mathscr{E}(Y_{lk})^4$$

where

$$Y_{lk} = \sum_{t=1}^{l} \mathscr{E}\left(Y_{nt} | \mathscr{F}_{-\infty}^{t+k}\right) - \mathscr{E}\left(Y_{nt} | \mathscr{F}_{-\infty}^{t+k-1}\right).$$

By Lemma 5, $\sum_{k=-m}^{m} \mathscr{E}(Y_{lk})^4 \le 10(2c)^4 (\sum_{t=1}^{l} c_{nt}^2)^2$, so

$$\mathscr{E}\left(\frac{\max_{j \le l} \overline{Y}_{nj}^4}{\bar{c}_n^4}\right) \le 10\left(\tfrac{4}{3}\right)^4 (2m + 3)^3 (2c)^4.$$

For fixed m and c as chosen previously, one sees from this inequality that there is an α large enough that $y < \epsilon/27$. Thus

$$\mathscr{E}_\alpha\left(\frac{\max_{j \le l} S_{nj}^2}{\bar{c}_n^2}\right) < \epsilon.$$

Note once again that the choice of α depends neither on n nor on where the sum starts; thus

$$\left\{\frac{\max_{j \le l}(S_{n,j+k} - S_{n,k})^2}{\sum_{t=k+1}^{k+l} c_{nt}^2} : 1 \le k + l \le n, \ k \ge 0, \ n \ge 1\right\}$$

is a uniformly integrable set. $\qquad\square$

TIGHTNESS. A family of probability measures $\{P_n\}$ defined on the Borel subsets of $D[0, 1]$ is tight if for every positive ϵ there exists a compact set K such that $P_n(K) > 1 - \epsilon$ for all n. The importance of tightness derives from the fact that it implies relative compactness: every sequence from $\{P_n\}$ contains a weakly convergent subsequence.

LEMMA 7. Let w_n be a sequence of random variables with range in $D[0,1]$, and suppose that

$$\mathscr{U} = \left\{ \frac{\max_{t \le s \le t+\delta}[w_n(s) - w_n(t)]^2}{\delta} : n > N(t,\delta), 0 \le t \le 1, 0 < \delta < 1 \right\}$$

is a uniformly integrable set, where $N(t,\delta)$ is some nonrandom finite valued function. It is understood that if $t + \delta > 1$, then the maximum above is taken over $[t,1]$. Then $\{P_n\}$ with

$$P_n(A) = P\{\omega : w_n(\cdot) \in A\}$$

is tight, and if P' is the weak limit of a subsequence from P_n, then P' puts mass one on the space $C[0,1]$.

Proof. The proof consists in verifying the conditions of Theorem 15.5 of Billingsly (1968). These are:

1. For each positive η there exists an a such that $P\{\omega : |w_n(0)| > a\} \le \eta$ for $n \ge 1$.

2. For each positive ϵ and η, there exists a δ, $0 < \delta < 1$, and an integer n_0, such that

$$P\left\{\omega : \sup_{|s-t|<\delta} |w_n(s) - w_n(t)| \ge \epsilon\right\} \le \eta$$

for all $n \ge n^0$.

Because $w_n(0) = 0$ for all n, condition 1 is trivially satisfied. To show condition 2, let positive ϵ and η be given. As in the proof of Lemma 6, let $E_\alpha X$ denote the integral of X over the set $\{\omega : X \ge \alpha\}$. Note that

$$E_{\lambda^2}\left(\frac{\max_{t \le s \le t+\delta}|w_n(s) - w_n(t)|^2}{\delta}\right)$$

$$\ge \lambda^2 P\left(\max_{t \le s \le t+\delta} |w_n(s) - w_n(t)| \ge \lambda\sqrt{\delta}\right).$$

By hypothesis λ can be chosen so large that both $\epsilon^2/\lambda^2 < 1$ and the left hand side of the inequality is less than $\eta\epsilon^2$ for members of $\mathscr{U}$. Set $\delta = \epsilon^2/\lambda^2$, and set n^0 equal to the largest of the $N(i\delta, \delta)$ for $i = 0, 1, \dots, [1/\delta]$. If $|s - t| < \delta$, then either both t and s lie in an interval of the form $[i\delta, (i + 1)\delta]$ or they lie in abutting intervals of that form, whence

$$
P\left\{\omega: \sup_{|s-t|<\delta} |w_n(s) - w_n(t)| \geq 3\epsilon\right\}
$$

$$
\leq P \bigcup_{i=0}^{[1/\delta]} \left\{\omega: \max_{i\delta \leq s \leq (i+1)\delta} |w_n(s) - w_n(i\delta)| \geq \epsilon\right\}
$$

$$
\leq \sum_{i=0}^{[1/\delta]} \lambda^{-2} E_{\lambda^2}\left(\frac{\max_{i\delta \leq s \leq (i+1)\delta} |w_n(s) - w_n(i\delta)|^2}{\delta}\right)
$$

and if $n > n^0$ this is

$$
\leq \sum_{i=0}^{[1/\delta]} \lambda^{-2} \eta\epsilon^2
$$

$$
\leq \left(1 + \frac{1}{\delta}\right)\delta\eta
$$

$$
\leq 2\eta. \qquad \Box
$$

CONTINUITY SET. Let Y be a (possibly) vector valued random variable. A Y-continuity set is a Borel set B whose boundary ∂B has $P(Y \in \partial B) = 0$. The boundary ∂B of B consists of those limit points of B that are also limit points of some sequence of points not in B. If $Y_n \xrightarrow{\mathscr{L}} Y$, $P_n'(B) = P(Y_n \in B)$, $P'(B) = P(Y \in B)$, and B is a Y-continuity set, then $\lim_{n \to \infty} P_n'(B) = P'(B)$ (Billingsley, 1968, Theorem 2.1).

LEMMA 8. Let V_{ni}, Y_{ni}, Y_i for $i = 1, 2, \ldots, k$ be random variables defined on a probability space $(\Omega, \mathscr{A}, P)$ such that

1. $V_{ni} - Y_{ni} \xrightarrow{P} 0$ for $i = 1, 2, \ldots, k$,

2. $Y_{ni} \xrightarrow{\mathscr{L}} Y_i$ for $i = 1, 2, \ldots, k$,

3. $\lim_{n \to \infty} \left\{ P[\cap_{i=1}^{k}(V_{ni} \in A_i)] - \prod_{i=1}^{k} P(V_{ni} \in A_i)\right\} = 0$.

Condition 3 is called asymptotic independence; the condition must hold for all possible choices of Borel subsets A_i of the real line. Then for all Y_i-continuity sets B_i

$$
\lim_{n \to \infty} \left[P\left(\bigcap_{i=1}^{k} (Y_{ni} \in B_i)\right) - \prod_{i=1}^{k} P(Y_{ni} \in B_i)\right] = 0.
$$

Proof. Conditions 1 and 2 imply that $V_{ni} \xrightarrow{\mathscr{L}} Y_i$, whence for Y_i-continuity sets B_i we have

$$\lim_{n \to \infty} P\left(\bigcap_{i=1}^{k} (V_{ni} \in B_i) \right) = P\left(\bigcap_{i=1}^{k} (Y_i \in B_i) \right)$$

$$\lim_{n \to \infty} \prod_{i=1}^{k} P(V_{ni} \in B_i) = \prod_{i=1}^{k} P(Y_i \in B_i)$$

since $\times_{i=1}^{k} B_i$ is a Y-continuity set of the random variable $Y = (Y_1, Y_2, \ldots, Y_k)'$ with boundary $\times_{i=1}^{k} \partial B_i$ (Problem 9). Condition 3 implies the result. $\square$

Proof of Theorem 2. Recall that we have set

$$X_{nt} = g_{nt}\left(W_t, \gamma_n^0 \right) - \mathscr{E} g_{nt}\left(W_t, \gamma_n^0 \right)$$

$$\sigma_n^2 = \mathrm{Var}\left(\sum_{t=1}^{n} X_{nt} \right)$$

$$w_n(s) = \sum_{t=1}^{[ns]} \frac{X_{nt}}{\sigma_n} \qquad 0 \le s \le 1$$

and that we have the following conditions in force:

1. $1/\sigma_n^2 = O(1/n)$.
2. $\lim_{n \to \infty} \mathrm{Var}[w_n(s)] = s, 0 \le s \le 1$.
3. $\|X_{nt}\|_r \le \Delta < \infty, r > 2, 1 \le t \le n, n = 1, 2, \ldots$.
4. $(X_{nt}, \mathscr{F}_{-\infty}^t)$ is a mixingale with $\{\psi_m\}$ of size $-1/2$ and $c_{nt} = \max\{1, \|X_{nt}\|_r\}$.
5. $\nu_m = \sup_n \sup_t \|X_{nt} - \mathscr{E}(X_{nt}|\mathscr{F}_{t-m}^{t+m})\|_2$ is of size $-1/2$.
6. $\alpha_m = \sup_t \alpha(\mathscr{F}_{-\infty}^t, \mathscr{F}_{t+m}^\infty)$ is of size $-r/(r-2)$.

Condition 3 implies that

$$\left\{ X_{nt}^2 : t = 1, 2, \ldots, n; \, n = 1, 2, \ldots \right\}$$

is a uniformly integrable set (Billingsley, 1968, p. 32). This taken together with condition 4 implies that

$$\mathcal{V} = \left\{ \frac{\max_{j \leq l} (S_{n,\,j+k} - S_{nk})^2}{\sum_{t=k+1}^{k+l} c_{nt}^2} : 1 \leq k + l \leq n,\ k \geq 0,\ n \geq 1 \right\}$$

is a uniformly integrable set by Lemma 6. Condition 1 implies that for any t, $0 \leq t \leq 1$, and any δ, $0 < \delta \leq 1$, if $t \leq s \leq t + \delta$ then

$$\frac{\sum_{j=[nt]}^{[ns]} c_{nj}^2}{\delta \sigma_n^2} = \frac{\sum_{j=[nt]}^{[ns]} \max\left(1, \|X_{nt}\|_r^2\right)}{\delta \sigma_n^2}$$

$$\leq \frac{([ns] - [nt]) \Delta^2}{\delta \sigma_n^2}$$

$$\leq \frac{(n + 1)\,\delta \Delta^2}{\delta \sigma_n^2}$$

$$\leq (n + 1) \Delta^2 O(n^{-1})$$

$$\leq B \Delta^2$$

for n larger than some n^0. For each t and δ put $N(t, \delta) = n_0$, whence

$$\frac{\max_{t \leq s \leq t + \delta} \left[w_n(s) - w_n(t) \right]^2}{\delta}$$

is dominated by $B \Delta^2$ times some member of $\mathcal{V}$ for $n > N(t, \delta)$. Thus Lemma 7 applies whence $\{P_n\}$ is tight, and if P' is the weak limit of a sequence from $\{P_n\}$, then P' puts mass one on $C[0, 1]$; recall that P_n is defined by $P_n(A) = P\{\omega : w_n(\cdot) \text{ in } A\}$ for every Borel subset A of $D[0, 1]$.

Theorem 19.2 of Billingsly (1968) states that if

i. $w_n(s)$ has asymptotically independent increments,
ii. $\{w_n^2(s)\}_{n=1}^{\infty}$ is uniformly integrable for each s,
iii. $\mathcal{E} w_n(s) \to 0$ and $\mathcal{E} w_n^2(s) \to s$ as $n \to \infty$,
iv.   for each positive ϵ and η there is a positive δ, $0 < \delta < 1$, and an integer n_0 such that $P\{\omega : \sup_{|s-t| < \delta} |w_n(s) - w_n(t)| \geq \epsilon\} \leq \eta$ for all $n > n_0$,

then w_n converges weakly in $D[0, 1]$ to a standard Wiener process. We shall verify these four conditions.

We have condition iii at once from the definition of X_{nt} and condition 2. We have just shown that for given t and δ the set

$$
\left\{ \frac{\max_{t \leq s \leq t+\delta}[w_n(s) - w_n(t)]^2}{\delta} \right\}_{n=N(t,\delta)}^{\infty}
$$

is uniformly integrable, so put $\delta = 1$ and $t = 0$ and condition ii obtains. We verified condition iv as an intermediate step in the proof of Lemma 7. It remains to verify condition i.

Consider two intervals $(0, a)$ and (b, c) with $0 < a < b < c \leq 1$. Define

$$
U_n = \mathscr{E}\left[w_n(a) \big| \mathscr{F}_{-\infty}^{[na]} \right]
$$

$$
V_n = \mathscr{E}\left[w_n(c) - w_n(b) \big| \mathscr{F}_{[nc]}^{\infty} \right].
$$

Thus

$$
w_n(a) - U_n = \frac{\sum_{t=1}^{[na]}\left[X_{nt} - \mathscr{E}\left(X_{nt} \big| \mathscr{F}_{-\infty}^{[na]} \right) \right]}{\sigma_n}.
$$

By Minkowski's inequality and condition 5

$$
\left\| w_n(a) - U_n \right\|_2 \leq \sigma_n^{-1} \sum_{t=1}^{[na]} \left\| X_{nt} - \mathscr{E}\left(X_{nt} \big| \mathscr{F}_{-\infty}^{[na]} \right) \right\|_2
$$

$$
\leq \sigma_n^{-1} \sum_{t=1}^{[na]} \left\| X_{nt} - \mathscr{E}\left(X_{nt} \big| \mathscr{F}_{t-[na]+t}^{t+[na]-t} \right) \right\|
$$

$$
\leq \sigma_n^{-1} \sum_{t=1}^{[na]} \nu_{[na]-t}
$$

$$
\leq \sigma_n^{-1} \sum_{m=0}^{[na]} \nu_m.
$$

Since $\{\nu_m\}$ is of size $-1/2$, $\sum_{m=0}^{\infty}\nu_m m^{-1/2} < \infty$. By Kronecker's lemma (Hall and Heyde, 1980, Section 2.6)

$$
[na]^{-1/2} \sum_{m=1}^{[na]} m^{1/2}\left(\nu_m m^{-1/2} \right) = [na]^{-1/2} \sum_{m=1}^{[na]} \nu_m
$$

converges to zero as n tends to infinity. Since σ_n^{-1} is $O(n^{-1/2})$, we have that $\|w_n(a) - U_n\|_2 \to 0$ as $m \to 0$, whence $w_n(a) - U_n \overset{P}{\to} 0$. A similar argument shows that $[w_n(c) - w_n(b)] - V_n \overset{P}{\to} 0$. For any Borel sets A and B, $U_n^{-1}(A) \in \mathscr{F}_{-\infty}^{[na]}$ and $V_n^{-1}(B) \in \mathscr{F}_{[nb]}^{\infty}$; thus

$$\left| P\big(U_n \in A\big) \cap \big(V_n \in B\big) - P\big(U_n \in A\big)P\big(V_n \in B\big) \right| \leq \alpha\Big(\mathscr{F}_{-\infty}^{[na]}, \mathscr{F}_{[nb]}^{\infty} \Big)$$

which tends to zero as n tends to infinity by condition 6. We have now verified conditions 1 and 3 of Lemma 8. Given an arbitrary sequence from $\{P_n\}$, there is a weakly convergent subsequence $\{P_{n'}\}$ with limit P' by relative compactness. Since, by Lemma 7, P' puts mass one on $C[0,1]$, the finite dimensional distributions of w_n converge to the corresponding finite dimensional distributions of P' by Theorem 5.1 of Billingsley (1968). This implies that condition 2 of Lemma 8 holds for the subsequence, whence the conclusion of Lemma 8 obtains for the subsequence. Since the limit given by Lemma 8 is the single value zero and the choice of a sequence from $\{P_n\}$ was arbitrary, we have that condition i, asymptotically independent increments, holds for the three points $0 < a < b < c$. The same argument can be repeated for more points.   □

Theorem 2 provides a central limit theorem for the sequence of random variables

$$\big\{ g_{nt}\big(W_t, \gamma_n^0\big) : n = 1, 2, \ldots ; t = 0, 1, \ldots \big\}.$$

To make practical use of it, we need some means to estimate the variance of a sum, in particular

$$\sigma_n^2 = \mathrm{Var}\left[\sum_{t=1}^{n} g_{nt}\big(W_t, \gamma_n^0\big) \right].$$

Putting

$$X_{nt} = g_{nt}\big(W_t, \gamma_n^0\big) - \mathscr{E}g_{nt}\big(W_t, \gamma_n^0\big)$$

this variance is

$$\sigma_n^2 = \sum_{\tau = -(n-1)}^{(n-1)} R_{n\tau}$$

with

$$R_{n\tau} = \sum_{t=1+|\tau|}^{n} \mathcal{E}\left(X_{nt}X_{n,\,t-|\tau|}\right) \qquad \tau = 0, \pm 1, \pm 2, \ldots, \pm(n-1).$$

The natural estimator of σ_n^2 is

$$\hat{\sigma}_n^2 = \sum_{\tau=-l(n)}^{l(n)} w_\tau \hat{R}_{n\tau}$$

with

$$\hat{R}_{n\tau} = \sum_{t=1+|\tau|}^{n} X_{nt}X_{n,\,t-|\tau|}$$

where w_τ is some set of weights chosen so that $\hat{\sigma}_n^2$ is guaranteed to be positive. Any sequence of weights of the form (Problem 10)

$$w_\tau = \sum_{j=1+|\tau|}^{l(n)} a_j a_{j-|\tau|}$$

will guarantee positivity; the simplest such sequence is the modified Bartlett sequence

$$w_\tau = 1 - \frac{|\tau|}{l(n)}.$$

The truncation estimator $\hat{\sigma}_n^2 = \Sigma_{\tau=-l(n)}^{l(n)} \hat{R}_{n\tau}$ does not have weights that satisfy the positivity condition and can thus assume negative values. We shall not consider it for that reason.

If $\{X_{nt}\}$ were a stationary time series, then estimating the variance of a sum would be the same problem as estimating the value of the spectral density at zero. There is an extensive literature on the optimal choice of weights for the purpose of estimating a spectral density; see for instance Anderson (1971, Chapter 9) or Bloomfield (1976, Chapter 7). In the theoretical discussion we shall use Bartlett weights because of their analytical tractability, but in applications we recommend Parzen weights

$$w_\tau = \begin{cases} 1 - \dfrac{6|\tau|^2}{l^2(n)} + \dfrac{6|\tau|^3}{l^3(n)} & 0 \le \dfrac{|\tau|}{l(n)} \le \dfrac{1}{2} \\[2ex] 2\left[1 - \dfrac{|\tau|}{l(n)}\right]^3 & \dfrac{1}{2} \le \dfrac{|\tau|}{l(n)} \le 1 \end{cases}$$

with $l(n)$ taken as that integer nearest $n^{1/5}$. See Anderson (1971, Chapter 9) for a verification of the positivity of Parzen weights and for a verification that the choice $l(n) \doteq n^{1/5}$ minimizes the mean squared error of the estimator.

At this point we must assume that W_t is a function of past values of V_t, so that W_t is measurable with respect to $\mathscr{F}^t_{-\infty}$. This is an innocuous assumption in view of the intended applications, while proceeding without it would entail inordinately burdensome regularity conditions. The following describes the properties of $\hat{\sigma}_n^2$ subject to this restriction for Bartlett weights; see Problem 12 for Parzen weights.

THEOREM 3. Let $\{V_t\}^\infty_{t=-\infty}$ be a sequence of vector valued random variables that is strong mixing of size $-2qr/(r-2)$ with $q = 2(r-2)/(r-4)$ for some $r > 4$. Let (Γ, ρ) be a separable metric space, and let $W_t = W_t(V_\infty)$ be a function of the past with range in $\mathbb{R}^{k_t}$; that is, W_t is a function of only

$$(\dots, V_{t-2}, V_{t-1}, V_t).$$

Let

$$\{g_{nt}(W_t, \gamma) : n = 1, 2, \dots; t = 0, 1, \dots\}$$

be a sequence of random variables that is near epcoh dependent of size $-q$. Given a sequence $\{\gamma_n^0\}^\infty_{n=1}$ from Γ, put

$$X_{nt} = g_{nt}(W_t, \gamma_n^0) - \mathscr{E}g_{nt}(W_t, \gamma_n^0)$$

and suppose that $\|X_{nt}\|_r \le \Delta < \infty$ for $1 \le t \le n$; $n, t = 1, 2, \dots$. Define

$$\sigma_n^2 = \mathscr{E}\left(\sum_{t=1}^n X_{nt}\right)^2 \quad n = 1, 2, \dots$$

$$\hat{R}_{n\tau} = \sum_{t=1+|\tau|}^n X_{nt} X_{n, t-|\tau|} \quad \tau = 0, \pm 1, \pm 2, \dots, \pm(n-1)$$

$$\hat{\sigma}_n^2 = \sum_{\tau=-l(n)}^{l(n)} \left(1 - \frac{|\tau|}{l(n)}\right)\hat{R}_{n\tau} \quad 1 \le l(n) \le n-1.$$

Then there is a bound B that does not depend on n such that

$$|\sigma_n^2 - \mathscr{E}\hat{\sigma}_n^2| \le Bnl^{-1}(n)$$

$$P(|\hat{\sigma}_n^2 - \mathscr{E}\hat{\sigma}_n^2| > \epsilon) \le \frac{B}{\epsilon^2}nl^4(n).$$

If $\mathscr{E}|g_{nt}(W_t, \gamma_n^0)| < \Delta$ then

$$P\left[\left|\sum_{\tau=-l(n)}^{l(n)}\left(1 - \frac{|\tau|}{l(n)}\right)\sum_{t=1+|\tau|}^{n} X_{nt}\mathscr{E}g_{n,\,t-|\tau|}(W_t, \gamma_n^0)\right| > \epsilon\right] \le \frac{B}{\epsilon^2}nl^3(n)$$

Proof. To establish the first inequality, note that

$$|\sigma_n^2 - \mathscr{E}\hat\sigma_n^2| \le 2l^{-1}(n)\sum_{\tau=0}^{l(n)}\tau\sum_{t=1+\tau}^{n}|\mathscr{E}(X_{nt}X_{n,\,t-\tau})|$$

$$+\,2\sum_{\tau=l(n)}^{n-1}\sum_{t=1+\tau}^{n}|\mathscr{E}(X_{nt}X_{n,\,t-\tau})|$$

$$\le 2l^{-1}(n)\sum_{\tau=0}^{n-1}\tau\sum_{t=1+\tau}^{n}|\mathscr{E}(X_{nt}X_{n,\,t-\tau})|.$$

Now

$$|\mathscr{E}(X_{nt}X_{n,\,t-\tau})| = \left|\mathscr{E}X_{n,\,t-\tau}\mathscr{E}\left(X_{nt}|\mathscr{F}_{-\infty}^{t-\tau}\right)\right|$$

$$\le \|X_{n,\,t-\tau}\|_2\left\|\mathscr{E}\left(X_{nt}|\mathscr{F}_{-\infty}^{t-\tau}\right)\right\|_2$$

$$\le \left(1 + \|X_{n,\,t-\tau}\|_r^{r/2}\right)c_{nt}\psi_\tau$$

where $0 \le c_{nt} \le \max\{1, \|X_{nt}\|_r\} \le (1 + \Delta^{r/2})$ and $0 \le \sum_{\tau=0}^{\infty}\tau\psi_\tau < \infty$ by Proposition 3; note that $q > 2$. Thus, we have

$$|\sigma_n^2 - \mathscr{E}\hat\sigma_n^2| \le 2(1 + \Delta^{r/2})^2nl^{-1}(n)\sum_{\tau=0}^{\infty}\tau\psi_\tau$$

which establishes the first inequality.

To establish the second inequality note that

$$P\left[\left|\sum_{\tau=-l(n)}^{l(n)}\left(1 - \frac{|\tau|}{l(n)}\right)(\hat R_{n\tau} - \mathscr{E}\hat R_{n\tau})\right| < \epsilon\right]$$

$$\ge P\bigcap_{\tau=-l(n)}^{l(n)}\left(\left|1 - \frac{|\tau|}{l(n)}\right||\hat R_{n\tau} - \mathscr{E}\hat R_{n\tau}| < \frac{\epsilon}{2l(n)+1}\right)$$

$$= 1 - P\bigcup_{\tau=-l(n)}^{l(n)}\left\{\left|1 - \frac{|\tau|}{l(n)}\right||\hat R_{n\tau} - \mathscr{E}\hat R_{n\tau}| > \frac{\epsilon}{2l(n)+1}\right\}$$

$$\ge 1 - \sum_{\tau=-l(n)}^{l(n)}\mathrm{Var}(|\hat R_{n\tau} - \mathscr{E}\hat R_{n\tau}|)\frac{[2l(n)+1]^2[1 - |\tau|/l(n)]^2}{\epsilon^2}$$

$$\ge 1 - \sum_{\tau=-l(n)}^{l(n)}\mathscr{E}|\hat R_{n\tau} - \mathscr{E}\hat R_{n\tau}|^2\frac{[2l(n)+1]^2}{\epsilon^2}$$

so that

$$P\left(|\hat{\sigma}_n^2 - \mathscr{E}\hat{\sigma}_n^2| > \epsilon\right) \leq \sum_{\tau=-l(n)}^{l(n)} \mathscr{E}(\hat{R}_{n\tau})^2 \frac{[2l(n)+1]^2}{\epsilon^2}.$$

Suppress the subscript n, and put $X_t = 0$ for $t \leq 0$. By applying in succession a change of variable formula, the law of iterated expectations, and Hölder's inequality, we have

$$\mathscr{E}(\hat{R}_{n\tau})^2 = \mathscr{E} \sum_{s=1+\tau}^{n} \sum_{t=1+\tau}^{n} X_{s-\tau} X_s X_{t-\tau} X_t$$

$$= \mathscr{E} \sum_{h=-(n-1-\tau)}^{(n-1-\tau)} \sum_{t=1+\tau+|h|}^{n} X_{t-|h|-\tau} X_{t-|h|} X_{t-\tau} X_t$$

$$\leq 2 \sum_{h=0}^{2\tau} \sum_{t=1+\tau+h}^{n} \left| \mathscr{E}(X_{t-h-\tau} X_{t-h} X_{t-\tau} X_t) \right|$$

$$+ 2 \sum_{h=2\tau}^{\infty} \sum_{t=1+\tau+h}^{n} \left| \mathscr{E}(X_{t-h-\tau} X_{t-h} X_{t-\tau} X_t) \right|$$

$$= 2 \sum_{h=0}^{2\tau} \sum_{t=1+\tau+h}^{n} \left| \mathscr{E}(X_{t-h-\tau} X_{t-h} X_{t-\tau} X_t) \right|$$

$$+ 2 \sum_{h=2\tau}^{\infty} \sum_{t=1+\tau+h}^{n} \left| \mathscr{E}\left[X_{t-h-\tau} X_{t-h} \mathscr{E}\left(X_{t-\tau} X_t | \mathscr{F}_{-\infty}^{t-h} \right) \right] \right|$$

$$\leq 2 \sum_{h=0}^{2\tau} \sum_{t=1+\tau+h}^{n} \| X_{t-h-\tau} X_{t-h} \|_2 \| X_{t-\tau} X_t \|_2$$

$$+ 2 \sum_{h=2\tau}^{\infty} \sum_{t=1+\tau+h}^{n} \| X_{t-h-\tau} X_{t-h} \|_2 \left\| \mathscr{E}\left(X_{t-\tau} X_t | \mathscr{F}_{-\infty}^{t-h} \right) \right\|_2$$

$$\leq 4\tau n \left(1 + \Delta^{2/4}\right)$$

$$+ 2 \sum_{h=2\tau}^{\infty} \sum_{t=1+\tau+h}^{n} \| X_{t-h-\tau} \|_4 \| X_{t-h} \|_4 \left\| \mathscr{E}\left(X_{t-\tau} X_t | \mathscr{F}_{-\infty}^{t-h} \right) \right\|_2$$

$$\leq \text{const} \cdot \tau n + \text{const} \sum_{h=2\tau}^{\infty} \sum_{t=1+\tau+h}^{n} \left\| \mathscr{E}\left(X_{t-\tau} X_t | \mathscr{F}_{-\infty}^{t-h} \right) \right\|_2.$$

Write $\hat{X}_{t-\tau} = \mathscr{E}(X_{t-\tau} | \mathscr{F}_{t-h/2}^{t+h/2})$ and $\hat{X}_t + \mathscr{E}(X_t | \mathscr{F}_{t-h/2}^{t+h/2})$. By applying the triangle inequality twice, and the conditional Jensen's inequality (Problem

4), we obtain

$$\left\| \mathcal{E}\left(X_{t-\tau} X_t | \mathcal{F}_{-\infty}^{t-h} \right) \right\|_2$$

$$\leq \left\| \mathcal{E}\left(\hat{X}_{t-\tau} \hat{X}_t | \mathcal{F}_{-\infty}^{t-h} \right) \right\|_2 + \left\| \mathcal{E}\left(X_{t-\tau} X_t - \hat{X}_{t-\tau} \hat{X}_t | \mathcal{F}_{-\infty}^{t-h} \right) \right\|_2$$

$$\leq \left\| \mathcal{E}\left(\hat{X}_{t-\tau} \hat{X}_t | \mathcal{F}_{\infty}^{t-h} \right) \right\|_2 + \left\| \mathcal{E}\left[X_t (X_{t-\tau} - \hat{X}_{t-\tau}) | \mathcal{F}_{-\infty}^{t-h} \right] \right\|_2$$

$$+ \left\| \mathcal{E}\left[\hat{X}_{t-\tau} (X_t - \hat{X}_t) | \mathcal{F}_{-\infty}^{t-h} \right] \right\|_2$$

$$\leq \left\| \mathcal{E}\left(\hat{X}_{t-\tau} \hat{X}_t | \mathcal{F}_{-\infty}^{t-h} \right) \right\|_2 + \left\| X_t (X_{t-\tau} - \hat{X}_{t-\tau}) \right\|_2 + \left\| \hat{X}_{t-\tau} (X_t - \hat{X}_t) \right\|_2.$$

The argument used to prove Lemma 1 can be repeated to obtain the inequality

$$\left\| X_t (X_{t-\tau} - \hat{X}_{t-\tau}) \right\|_2 \leq 2^{1/2} \| X_{t-\tau} - \hat{X}_{t-\tau} \|_2^{(1/2)(s-2)/(s-1)} \| X_t \|_2^{(1/2)(s-2)/(s-1)}$$

$$\times \| X_t (X_{t-\tau} - \hat{X}_{t-\tau}) \|_s^{(1/2)s/(s-1)}$$

for $s = r/2 > 2$. Then we have

$$\left\| X_t (X_{t-\tau} - \hat{X}_{t-\tau}) \right\|_2 \leq 2^{1/2} \| X_{t-\tau} - \hat{X}_{t-\tau} \|^{(1/2)(s-2)/(s-1)}$$

$$\times (1 + \Delta^{r/2})^{(1/2)(s-2)/(s-1)}$$

$$\times \left[\| X_{t-\tau} \|_r \| X_t \|_r + \| \hat{X}_{t-\tau} \|_r \| X_t \|_r \right]^{(1/2)s/(s-1)}$$

$$= \mathrm{const} \left\| X_{n,\,t-\tau} - \mathcal{E}\left(X_{n,\,t-\tau} | \mathcal{F}_{t-h/2}^{t+h/2} \right) \right\|_2^{(1/2)(r-4)/(r-2)}$$

where the constant does not depend on n, h, or τ. By the definition of near epoch dependence we have

$$\left\| X_{n,\,t-\tau} - \mathcal{E}\left(X_{n,\,t-\tau} | \mathcal{F}_{t-h/2}^{t+h/2} \right) \right\|_2 \leq \nu_{h/2-\tau}$$

provided that $\tau \leq h/2$. Thus we have

$$\left\| X_t (X_{t-\tau} - \hat{X}_{t-\tau}) \right\|_2 \leq \mathrm{const}\, (\nu_{h/2-\tau})^{(1/2)(r-4)/(r-2)}$$

and by the same argument

$$\left\| \hat{X}_{t-\tau} (X_t - \hat{X}_t) \right\|_2 \leq \mathrm{const}\, (\nu_{h/2})^{(1/2)(r-4)/(r-2)}$$

$$\leq (\mathrm{const.})\, (\nu_{h/2-\tau})^{(1/2)(r-4)/(r-2)}$$

where the constant does not depend on n, h, or τ. Using Proposition 2, we have

$$\left\| \mathscr{E}\left(\hat{X}_{t-\tau} \hat{X}_t | \mathscr{F}_{-\infty}^{t-h} \right) \right\|_2 \leq 2(2^{1/2} + 1)\left[\alpha\left(\mathscr{F}_{-\infty}^{t-h}, \mathscr{F}_{t-h/2}^{t+h/2} \right) \right]^{1/2-1/r} \| X_{n t} \|_r$$

$$\leq 2(2^{1/2} + 1)\Delta\left(\alpha_{h/2} \right)^{1/2-1/r}$$

$$\leq \text{const} \left(\alpha_{h/2-\tau} \right)^{(r-2)/2r}.$$

Combining the various inequalities, we have

$$\sum_{\tau=-l(n)}^{l(n)} \mathscr{E}\left(\hat{R}_{n\tau} \right)^2$$

$$\leq \text{const } n \sum_{\tau=-l(n)}^{l(n)} \left(\tau + \sum_{h=2\tau}^{\infty} \left[(\nu_{h/2-\tau})^{(1/2)(r-4)/(r-2)} + (\alpha_{h/2-\tau})^{(r-2)/2r} \right] \right)$$

$$= \text{const } n \sum_{\tau=-l(n)}^{l(n)} \left(\tau + \sum_{m=0}^{\infty} m^{-(q/2)(r-4)/(r-2)} + m^{-q} \right)$$

$$\leq \text{const } n \left[l^2(n) + l(n) \right].$$

Thus we have

$$P\left(|\hat{\sigma}_n^2 - \mathscr{E}\hat{\sigma}_n^2| > \epsilon \right) \leq \frac{[2l(n) + 1]^2}{\epsilon^2} \text{const } n \left[l^2(n) + l(n) \right]$$

which establishes the second inequality.

Using the same argument as above,

$$P\left\{ \left| \sum_{\tau=-l}^{l} \left(1 - \frac{|\tau|}{l} \right) \sum_{\tau=1+|\tau|}^{n} X_t \mathscr{E} g_{\tau-|\tau|} \right| > \epsilon \right\}$$

$$\leq \sum_{\tau=-l}^{l} \mathscr{E}\left(\sum_{t=1+|\tau|}^{n} X_t \mathscr{E} g_{t-|\tau|} \right)^2 \frac{(2l+1)^2}{\epsilon^2}$$

$$\leq \max_{-l \leq \tau \leq l} \mathscr{E}\left(\sum_{t=1+|\tau|}^{n} X_t \mathscr{E} g_{t-|\tau|} \right)^2 \frac{(2l+1)^3}{\epsilon^2}.$$

Now

$$\mathcal{E}\left(\sum_{t=1+|\tau|}^{n} X_t \mathcal{E} g_{t-|\tau|}\right)^2$$

$$= \mathcal{E} \sum_{s=1+|\tau|}^{n} \sum_{t=1+|\tau|}^{n} \left(\mathcal{E} g_{t-|\tau|} \mathcal{E} g_{s-|\tau|}\right)(X_t X_s)$$

$$= \mathcal{E} \sum_{h=-(n-1-|\tau|)}^{(n-1-|\tau|)} \sum_{t=1+|\tau|}^{n} \left(\mathcal{E} g_{t-|\tau|} \mathcal{E} g_{t-|h|-|\tau|}\right)(X_t X_{t-|h|})$$

$$\leq \sum_{h=-(n-1-|\tau|)}^{(n-1-|\tau|)} \sum_{t=1+|\tau|}^{n} \left| \mathcal{E} g_{t-|\tau|} \mathcal{E}_{t-|h|-|\tau|} \right| \left\| \mathcal{E} X_{t-|h|} \mathcal{E}\left(X_t \mid \mathscr{F}_{-\infty}^{t-|h|}\right)\right|$$

$$\leq \Delta^2 \sum_{h=-(n-1-|\tau|)}^{(n-1-|\tau|)} \sum_{t=1+|\tau|}^{n} \left\| X_{t-|h|} \right\|_2 \left\| \mathcal{E}\left(X_t \mid \mathscr{F}_{-\infty}^{t-|h|}\right)\right\|_2$$

$$\leq \text{const} \sum_{h=-(n-1-|\tau|)}^{(n-1-|\tau|)} \sum_{t=1+|\tau|}^{n} \psi_{|h|}$$

$$\leq \text{const } n \sum_{h=-\infty}^{\infty} \psi_{|h|}$$

$$\leq \text{const } n$$

which establishes the third inequality. □

PROBLEMS

1. (Nonlinear ARMA.) Consider data generated according to the model

$$y_t = \begin{cases} f(y_{t-1}, x_t, \theta^0) + g(e_t, e_{t-1}, \ldots, e_{t-q}, \beta^0) & t = 1, 2, \ldots \\ 0 & t \leq 0 \end{cases}$$

where $\{e_t\}$ is a sequence of independent random variables. Let $\|\sup_\beta g(e_t, e_{t-1}, \ldots, e_{t-q}, \beta)\|_p \leq K < \infty$ for some $p > 4$, and put

$$g_t(W_t, \theta) = \left[y_t - f(y_{t-1}, x_t, \theta)\right]^2.$$

Show that $\{g_t(W_t, \theta)\}$ is near epoch dependent. Hint: Show that $g_t = g(e_t, e_{t-1}, \ldots, e_{t-q}, \beta^0)$ is strong mixing of size $-q$ for all positive q.

2. Referring to Example 1, show that if $V_0 = (Y, 0)$ then $\hat{y}_{t, m}^t$ has prediction error $\left| y_t - \hat{y}_{t, m}^t \right| \leq d^t |Y| + d^m \sum_{j=0}^{t-m-1} d^j |e_{t-m-j}|$ for $t - m$. Use this to show that $[y_t - f(y_{t-1}, x_t, \theta)]^2$ is near epoch dependent.

3. Show that the definition of a mixingale implies that one can assume that $\psi_{m+1} \le \psi_m$ without loss of generality. Hint: See the proof of Proposition 3.

4. The conditional Jensen's inequality is $g[\mathscr{E}(X \mid \mathscr{F})] \le \mathscr{E}(gX \mid \mathscr{F})$ for convex g. Show that this implies $\mathscr{E}[\mathscr{E}(X \mid \mathscr{F})]^p \le \mathscr{E}X^p$, whence $\|\mathscr{E}(X \mid \mathscr{F})\|_p \le \|X\|_p$ for $p \ge 1$.

5. Show that if X and Y are in $L_2(\Omega, \mathscr{A}, P)$ and Y is $\mathscr{F}$-measurable with $\mathscr{F} \subset \mathscr{A}$, then $\|X - \mathscr{E}(X \mid \mathscr{F})\|_2 \le \|X - Y\|_2$. Hint: Consider $[X - \mathscr{E}(X \mid \mathscr{F}) + \mathscr{E}(X \mid \mathscr{F}) - Y]^2$ and show that $\mathscr{E}\{[X - \mathscr{E}(X \mid \mathscr{F})][\mathscr{E}(X \mid \mathscr{F}) - Y]\} = 0$.

6. Show that the random variables

$$U_{tk} = \mathscr{E}\left(X_{nt} \mid \mathscr{F}_{-\infty}^{t+k}\right) - \mathscr{E}\left(X_{nt} \mid \mathscr{F}_{-\infty}^{t+k-1}\right)$$

appearing in the proof of Lemma 2 form a two dimensional array with uncorrelated rows and columns where t is the row index and k is the column index. Show that

$$\mathrm{Var}\left(\sum_{t=1}^{l} U_{tk}\right) = \sum_{t=1}^{l} \mathscr{E}\mathscr{E}^2\left(X_{nt} \mid \mathscr{F}_{-\infty}^{t+k}\right) - \mathscr{E}\mathscr{E}^2\left(X_{nt} \mid \mathscr{F}_{-\infty}^{t+k-1}\right).$$

7. Show that the hypothesis $\sum \psi_k^2 |a_k^{-1} - a_{k-1}^{-1}| < \infty$ permits the reordering of terms in the proof of Lemma 2.

8. Show that a compact metric space (X, ρ) is separable. Hint: Center a ball of radius $1/n$ at each point in X. Thus, there are points $x_{1n}, \ldots, x_{mn}$ within $\rho(x, x_{jn}) < 1/n$ for each x in X. Show that the triangular array that results by taking $n = 1, 2, \ldots$ is a countable dense subset of X.

9. Show that the boundary of $\times_{i=1}^k B_i$ is $\times_{i=1}^k \partial B_i$, where ∂B_i is the boundary of $B_i \subset \mathbb{R}^1$.

10. Write

$$X = \begin{bmatrix} X_1 & 0 & 0 & \cdots & 0 \\ X_2 & X_1 & 0 & \cdots & 0 \\ X_3 & X_2 & X_1 & & 0 \\ \vdots & \vdots & \vdots & & \vdots \\ X_n & X_{n-1} & X_{n-2} & \cdots & X_{n-l(n)} \\ 0 & X_n & X_{n-1} & \cdots & \vdots \\ 0 & 0 & X_n & \cdots & \vdots \\ 0 & 0 & 0 & \cdots & X_n \end{bmatrix} \begin{matrix} \\ \\ \\ \\ \\ \\ \\ \\ \end{matrix}$$

$_{n+l(n)}$ on the left, $_{l(n)}$ on the right

and show that $\hat{\sigma}_n^2 = a'X'Xa$, where

$$a = \left(a_1, a_2, \ldots, a_{l(n)} \right)$$

and hence that $\hat{\sigma}_n^2 \geq 0$ if $w_\tau = \sum_{j=1+|\tau|}^{l(n)} a_j a_{j-|\tau|}$. Show that the trunca-tion estimator $\hat{\sigma}_n^2 = \sum_{\tau=-l(n)}^{l(n)} \hat{R}_{n\tau}$ can be negative.

11. Prove Theorem 3 for Parzen weights assuming that $q \geq 3$. Hint: Verification of the second inequality only requires that the weights be less than one. As to the first, Parzen weights differ from one by a homogeneous polynomial of degree three for $l(n)/n \leq 1/2$ and are smaller than one for $l(n)/n \geq 1/2$.

3. DATA GENERATING PROCESS

In this section we shall give a formal description of a data generating mechanism that is general enough to accept the intended applications yet sufficiently restrictive to permit application of the results of the previous section, notably the uniform strong law of large numbers and the central limit theorem. As the motivation behind our conventions was set forth in Section 1, we can be brief here.

The independent variables $\{x_t\}_{t=-\infty}^{\infty}$ and the errors $\{e_t\}_{t=-\infty}^{\infty}$ are grouped together into a single process $\{v_t\}_{t=-\infty}^{\infty}$ with $v_t = (e_t, x_t)$, each v_t having range in $\mathbb{R}^l$. In instances where we wish to indicate clearly that v_t is being regarded as a random variable mapping the underlying (complete) probability space $(\Omega, \mathscr{A}, P)$ into $\mathbb{R}^l$, we shall write $V_t(\omega)$ or V_t and write $\{V_t(\omega)\}_{t=-\infty}^{\infty}$ or $\{V_t\}_{t=-\infty}^{\infty}$ for the process itself. But for the most part we shall follow the usual convention in statistical writings and let $\{v_t\}_{t=-\infty}^{\infty}$ denote either a realization of the process or the process itself as determined by context.

Recall that $\mathscr{F}_m^n$ is the smallest sub-σ-algebra of $\mathscr{A}$, complete with respect to $(\Omega, \mathscr{A}, P)$, such that $V_m, V_{m+1}, \ldots, V_n$ are measurable; $\mathscr{F}_{-\infty}^{\infty} = \bigcap_{t=-\infty}^{\infty} \mathscr{F}_{-\infty}^t$. Situations with a finite past are accommodated by putting $V_t = 0$ for $t < 0$ and letting V_0 represent initial conditions, fixed or random, if any. Note that if $\{V_t\}$ has a finite past, then $\mathscr{F}_{-\infty}^t$ will be the trivial σ-algebra $\{\phi, \Omega\}$ plus its completion for $t < 0$.

ASSUMPTION 1. $\{V_t(\omega)\}_{t=-\infty}^{\infty}$ is a sequence of random variables each defined over a complete probability space $(\Omega, \mathscr{A}, P)$ and each with range in $\mathbb{R}^l$.

Let

$$v_\infty = (\dots, v_{-1}, v_0, v_1, \dots)$$

denote a doubly infinite sequence, a point in $\times_{t=-\infty}^\infty \mathbb{R}^l$. Recall (Section 1) that the dependent variables $\{y_t\}_{t=-\infty}^\infty$ are viewed as obtaining from v_∞ via a reduced form such as

$$y_t = Y(t, v_\infty, \gamma^0) \qquad t = 0, \pm 1, \pm 2, \dots$$

but, since we shall be studying the limiting behavior of functions of the form

$$s_n(\lambda) = \frac{1}{n} \sum_{t=1}^n s_t(y_t, y_{t-1}, \dots, y_{t-l_t}, x_t, x_{t-1}, \dots, x_{t-l_t'}, \hat{\tau}_n, \lambda)$$

it is more convenient to group observations into a vector

$$w_t = \begin{pmatrix} y_t \\ \vdots \\ y_{t-l_t} \\ x_t \\ \vdots \\ x_{t-l_t'} \end{pmatrix}$$

dispense with consideration of $Y(t, v_\infty, \gamma^0)$, and put conditions directly on the mapping

$$w_t = W_t(v_\infty)$$

with range in $\mathbb{R}^{k_t}$, $k_t = l_t + l_t'$. The most common choices for k_t are $k_t = \text{const}$, fixed for all t, and $k_t = \text{const } t$. Recall that the subscript t associated to $W_t(V_\infty)$ serves three functions. It indicates that time may enter as a variable, it indicates that $W_t(v_\infty)$ depends primarily on the component v_t of v_∞ and to a lesser extent on components of v_s of v_∞ with $|t - s| > 0$, and it indicates that the dimension k_t of the vector $w_t = W_t(v_\infty)$ may depend on t. As $W_t(v_\infty)$ represents data, it need only be defined for $t = 0, 1, \dots$, with W_0 representing initial conditions, fixed or random, if any. We must also require that W_t depend only on the past to invoke Theorem 3.

ASSUMPTION 2. Each function $W_t(v_\infty)$ in the sequence $\{W_t\}_{t=0}^\infty$ is a Borel measurable mapping of $\mathbb{R}_{-\infty}^\infty = \times_{t=-\infty}^\infty \mathbb{R}^l$ into $\mathbb{R}^{k_t}$. That is, if B is a

Borel subset of $\mathbb{R}^{k_t}$, then the preimage $W_t^{-1}(B)$ is an element of the smallest σ-algebra $\mathscr{B}_{-\infty}^{\infty}$ containing all cylinder sets of the form

$$\cdots \times \mathbb{R}^l \times B_m \times B_{m+1} \times \cdots \times B_n \times \mathbb{R}^l \times \cdots$$

where each B_t is a Borel subset of $\mathbb{R}^l$. Each function $W_t(v_\infty)$ depends only on the past; that is, it depends only on $(\ldots, v_{t-2}, v_{t-1}, v_t)$.

The concern in the previous section was to find conditions such that a sequence of real valued random variables of the form

$$\{ g_t(W_t, \gamma) : \gamma \in \Gamma, t = 0, 1, \ldots \}$$

will obey a uniform strong law and such that a sequence of the form

$$\left\{ g_{nt}\left(W_t, \gamma_n^0\right) : \gamma_n^0 \in \Gamma; t = 0, 1, \ldots; n = 1, 2 \ldots \right\}$$

will follow a central limit theorem. Aside from some technical conditions, the inquiry produced three conditions.

The first condition limits the dependence that $\{V_t\}_{t=-\infty}^{\infty}$ can exhibit.

ASSUMPTION 3. $\{V_t\}_{t=-\infty}^{\infty}$ is strong mixing of size $-4r/(r-4)$ for some $r > 4$.

The second is a bound $\|d_t\|_r \leq \Delta < \infty$ on the rth moment of the dominating functions $d_t \geq |g_t(W_t, \gamma)|$ in the case of the strong law and a similar rth moment condition $\|g_{nt}(W_t, \gamma_n^0) - \mathscr{E}g_{nt}(W_t, \gamma_n^0)\|_r \leq \Delta < \infty$ in the case of the central limit theorem; r above is that of Assumption 3. There is a trade off: the larger the moment r can be so bounded, the more dependence $\{V_t\}$ is allowed to exhibit.

The third condition is a requirement that $g_t(W_t, \gamma)$ or $g_{nt}(W_t, \gamma)$ be nearly a function of the current epoch. In perhaps the majority of applications the condition of near epoch dependence will obtain trivially, because $W_t(V_\infty)$ will be of the form

$$W_t(V_\infty) = W_t(V_{t-m}, \ldots, V_t)$$

for some finite value of m that does not depend on t. In other applications, notably the nonlinear autoregression, the dimension of W_t does not depend on t, $g_t(w, \gamma)$ or $g_{nt}(w, \gamma)$ is smooth in the argument w, and W_t is nearly a function of the current epoch in the sense that $\eta_m = \|W_t - \mathscr{E}(W_t \mid \mathscr{F}_{t-m}^{t+m})\|_2$ falls off at a geometric rate in m, in which case the near epoch dependence condition obtains by Proposition 1. For applications not falling into these two categories, the near epoch dependence condition must be verified directly.

4. LEAST MEAN DISTANCE ESTIMATORS

Recall that a least mean distance estimator $\tilde{\lambda}_n$ is defined as the solution of the optimization problem

$$\text{minimize} \quad s_n(\lambda) = \frac{1}{n} \sum_{t=1}^{n} s_t(w_t, \hat{\tau}_n, \lambda).$$

As with $\{v_t\}_{t=-\infty}^{\infty}$, we shall let $\{w_t\}_{t=0}^{\infty}$ denote either a realization of the process—that is, data—or the process itself, as determined by context. For emphasis, we shall write $W_t(v_\infty)$ when considering it as a function defined on $\mathbb{R}^\infty_{-\infty}$, and write $W_t(V_\infty)$, W_t, $W_t[V_\infty(\omega)]$, or $W_t(\omega)$ when considering it as a random variable. The random variable $\hat{\tau}_n$ corresponds conceptually to a preliminary estimator of nuisance parameters; λ is a p-vector, and each $s_t(w_t, \tau, \lambda)$ is a real valued, Borel measureable function defined on some subset of $\mathbb{R}^{k_t} \times \mathbb{R}^u \times \mathbb{R}^p$. A constrained least mean distance estimator $\tilde{\lambda}_n$ is the solution of the optimization problem

$$\text{minimize} \quad s_n(\lambda) \quad \text{subject to} \quad h(\lambda) = h_n^*$$

where $h(\lambda)$ maps $\mathbb{R}^p$ into $\mathbb{R}^q$.

The objective of this section is to find the asymptotic distribution of the estimator $\tilde{\lambda}_n$ under regularity conditions that do not rule out specification error. Some ancillary facts regarding the asymptotic distribution of the constrained estimator $\tilde{\lambda}_n$ under a Pitman drift are also derived for use in later sections on hypothesis testing. We shall leave the data generating mechanism fixed and impose drift by moving h_n^*; this is the exact converse of the approach taken in Chapter 3. Example 1, least squares estimation of the parameters of a nonlinear autoregression, will be used for illustration throughout this section.

EXAMPLE 1 (Continued).   The data generating model is

$$y_t \begin{cases} f(y_{t-1}, x_t, \gamma^0) + e_t & t = 1, 2, \dots \\ 0 & t \leq 0 \end{cases}$$

with $|(\partial/\partial y)f(y, x, \gamma)| \leq d < 1$ for all relevant x and γ.

The process

$$V_t = \begin{cases} (e_t, x_t) & t = 1, 2, \ldots \\ (0, 0) & t \le 0 \end{cases}$$

generates the underlying sub-σ-algebras $\mathscr{F}^t_{-\infty}$ that appear in the definition of strong mixing and near epoch dependence. The data consist of

$$W_t = (y_t, y_{t-1}, x_t) \qquad t = 0, 1, 2, \ldots .$$

As we saw in Section 2, $\|e_t\|_p \le \Delta < \infty$ for some $p > 4$ is enough to guarantee that for the least squares sample objective function

$$s_n(\lambda) = \frac{1}{n} \sum_{t=1}^{n} [y_t - f(y_{t-1}, x_t, \lambda)]^2$$

the family

$$s_t(W_t, \lambda) = [y_t - f(y_{t-1}, x_t, \lambda)]^2 \qquad t = 0, 1, \ldots$$

is near epoch dependent of size $-q$ for any $q > 0$. The same is true of the family of scores

$$\frac{\partial}{\partial \lambda} s_t(W_t, \lambda) = \frac{\partial}{\partial \lambda} [y_t - f(y_{t-1}, x_t, \lambda)] \qquad t = 0, 1, \ldots$$

assuming suitable smoothness (Problem 2).

If we take $\|V_t\|_r \le \Delta < \infty$ for some $r > 4$ and assume that $\{V_t\}$ is strong mixing of size $-r/(r-2)$, then Theorems 1 and 2 can be applied to the sample objective function and the scores respectively. If $\{V_t\}$ is strong mixing of size $-4r/(r-4)$, then Theorem 3 may be applied to the scores.

As we shall see later, if the parameter λ is to be identified by least squares, it is convenient if the orthogonality condition

$$\mathscr{E} e_t g(y_{t-1}, x_t) = 0$$

holds for all square integrable $g(y_{t-1}, x_t)$. The easiest way to guarantee that the orthogonality condition holds is to assume that $\{e_t\}$ is a sequence of independent random variables and that the process $\{e_t\}$ is independent of $\{x_t\}$, whence e_t and (y_{t-1}, x_t) are independent.   □

In contrast to Chapter 3, $s_n(\lambda)$ and, hence, $\hat{\lambda}_n$ do not, of necessity, possess almost sure limits. To some extent this is a simplification, as the ambiguity as to whether some fixed point λ^* or a point λ_n^0 that varies with n ought to be regarded as the location parameter of $\hat{\lambda}_n$ is removed. Here, λ_n^0 is the only possibility. This situation obtains due to the use of a weaker strong law, Theorem 1 of this chapter, instead of Theorem 1 of Chapter 3. The estimator $\hat{\tau}_n$ is centered at τ_n^0 defined in Assumption 4.

NOTATION 1.

$s_n(\lambda) = (1/n)\Sigma_{t=1}^n s_t(w_t, \hat{\tau}_n, \lambda)$.

$s_n^0(\lambda) = (1/n)\Sigma_{t=1}^n \mathcal{E} s_t(W_t, \tau_n^0, \lambda)$.

$\hat{\lambda}_n$ minimizes $s_n(\lambda)$.

$\tilde{\lambda}_n$ minimizes $s_n(\lambda)$ subject to $h(\lambda) = 0$.

λ_n^0 minimizes $s_n^0(\lambda)$.

λ_n^* minimizes $s_n^0(\lambda)$ subject to $h(\lambda) = 0$.

In the above, the expectation is computed as

$$\mathcal{E} s_t(W_t, \tau, \lambda) = \int_\Omega s_t[W_t(\omega), \tau, \lambda]\, dP(\omega).$$

Identification does not require that the minimum of $s_n^0(\lambda)$ become stable as in Chapter 3, but does require that the curvature near each λ_n^0 become stable for large n.

ASSUMPTION 4 (Identification). The nuisance parameter estimator $\hat{\tau}_n$ is centered at τ_n^0 in the sense that $\lim_{n\to\infty}(\hat{\tau}_n - \tau_n^0) = 0$ almost surely and $\sqrt{n}(\hat{\tau}_n - \tau_n^0)$ is bounded in probability. The estimation space Λ^* is compact, and for each $\epsilon > 0$ there is an N such that

$$\inf_{n > N}\ \inf_{|\lambda - \lambda_n^0| \geq \epsilon}\ [s_n^0(\lambda) - s_n^0(\lambda_n^0)] > 0.$$

In the above, $|\lambda - \lambda^0| = [\Sigma_{i=1}^p (\lambda_i - \lambda_i^0)^2]^{1/2}$ or any other convenient norm and it is understood that the infimum is taken over λ in Λ^* with $|\lambda - \lambda^0| > \epsilon$.

For the example, sufficient conditions such that the identification condition obtains are as follows.

EXAMPLE 1 (Continued). We have

$$s_n^0(\lambda) = \frac{1}{n} \sum_{t=1}^{n} \mathscr{E}\left[e_t + f(y_{t-1}, x_t, \gamma^0) - f(y_{t-1}, x_t, \lambda)\right]^2$$

$$= \frac{1}{n} \sum_{t=1}^{n} \mathscr{E}e_t^2 + \frac{2}{n} \sum_{t=1}^{n} \mathscr{E}e_t\left[f(y_{t-1}, x_t, \gamma^0) - f(y_{t-1}, x_t, \lambda)\right]$$

$$+ \frac{1}{n} \sum_{t=1}^{n} \mathscr{E}\left[f(y_{t-1}, x_t, \gamma^0) - f(y_{t-1}, x_t, \lambda)\right]^2$$

$$= \frac{1}{n} \sum_{t=1}^{n} \sigma_t^2 + \frac{1}{n} \sum_{t=1}^{n} \mathscr{E}\left[f(y_{t-1}, x_t, \gamma^0) - f(y_{t-1}, x_t, \lambda)\right]^2.$$

Using Taylor's theorem and the fact that γ^0 minimizes $s_n^0(\lambda)$,

$$s_n^0(\lambda) - s_n^0(\gamma^0)$$

$$= (\lambda - \gamma^0)'\left[\frac{1}{n} \sum_{t=1}^{n} \mathscr{E}\left(\frac{\partial}{\partial\lambda}f(y_{t-1}, x_t, \bar{\lambda})\right)\left(\frac{\partial}{\partial\lambda}f(y_{t-1}, x_t, \bar{\lambda})\right)'\right](\lambda - \gamma^0).$$

A sufficient condition for identification is that the smallest eigenvalue of

$$S(\lambda) = \frac{1}{n} \sum_{t=1}^{n} \mathscr{E}\left(\frac{\partial}{\partial\lambda}f(y_{t-1}, x_t, \lambda)\right)\left(\frac{\partial}{\partial\lambda}f(y_{t-1}, x_t, \lambda)\right)'$$

is bounded from below for all λ in Λ^* and all n larger than some N. We are obliged to impose this same condition later in Assumption 6. □

We append some additional conditions to the identification condition to permit application of the uniform strong law.

ASSUMPTION 5. The sequences $\{\hat{\tau}_n\}$ and $\{\tau_n^0\}$ are contained in T, which is a closed ball with finite, nonzero radius. On $T \times \Lambda^*$, the family $\{s_t[W_t(\omega), \tau, \lambda]\}_{t=0}^{\infty}$ is near epoch dependent of size $-1/2$, it is A-smooth in (τ, λ) (Problem 1), and there is a sequence of random variables $\{d_t\}$ with $\sup_{T\times\Lambda^*}|s_t[W_t(\omega), \tau, \lambda]| \leq d_t(\omega)$ and $\|d_t\|_r \leq \Delta < \infty$ for all t, where r is that of Assumption 3.

LEMMA 9. Let Assumptions 1 through 5 hold. Then

$$\lim_{n\to\infty} \sup_{\Lambda^*} |s_n(\lambda) - s_n^0(\lambda)| = 0$$

almost surely, and $\{s_n^0(\lambda)\}_{n=0}^{\infty}$ is an equicontinuous family.

Proof. Writing $\mathscr{E}s_t(W_t, \hat{\tau}_n, \lambda)$ to mean $\mathscr{E}s_t(W_t, \tau, \lambda)|_{\tau=\hat{\tau}_n}$ we have

$$\sup_{\Lambda^*} |s_n(\lambda) - s_n^0(\lambda)|$$

$$\leq \sup_{\Lambda^*} \left| \frac{1}{n} \sum_{t=1}^{n} [s_t(W_t, \hat{\tau}_n, \lambda) - \mathscr{E}s_t(W_t, \hat{\tau}_n, \lambda)] \right|$$

$$+ \sup_{\Lambda^*} \left| \frac{1}{n} \sum_{t=1}^{n} \mathscr{E}s_t(W_t, \hat{\tau}_n, \lambda) - \mathscr{E}s_t(W_t, \tau_n^0, \lambda) \right|$$

$$\leq \sup_{\Lambda^* \times T} \left| \frac{1}{n} \sum_{t=1}^{n} [s_t(W_t, \tau, \lambda) - \mathscr{E}s_t(W_t, \tau, \lambda)] \right|$$

$$+ \sup_{\Lambda^*} \left| \frac{1}{n} \sum_{t=1}^{n} \mathscr{E}s_t(W_t, \hat{\tau}_n, \lambda) - \mathscr{E}s_t(W_t, \tau_n^0, \lambda) \right|.$$

Except on an event that occurs with probability zero, we have that the first term on the right hand side of the last inequality converges to zero as n tends to infinity by Theorem 1, and the same for the second term by the equicontinuity of the average guaranteed by Theorem 1 and the almost sure convergence of $\hat{\tau}_n - \tau_n^0$ to zero guaranteed by Assumption 4. $\quad\square$

THEOREM 4 (Consistency). Let Assumptions 1 through 5 hold. Then

$$\lim_{n \to \infty} (\hat{\lambda}_n - \lambda_n^0) = 0$$

almost surely.

Proof. Fix ω not in the exceptional set given by Lemma 9, and let $\epsilon > 0$ be given. For N given by Assumption 4 put

$$\delta = \inf_{n > N} \inf_{|\lambda - \lambda_n^0| \geq \epsilon} \left[s_n^0(\lambda) - s_n^0(\lambda_n^0) \right].$$

Applying Lemma 9, there is an N' such that $\sup_{\Lambda^*}|s_n(\lambda) - s_n^0(\lambda)| < \delta/2$ for all $n > N'$. Since $s_n(\hat{\lambda}_n) \leq s_n(\lambda_n^0)$, we have for all $n > N'$ that

$$s_n^0(\hat{\lambda}_n) - \frac{\delta}{2} \leq s_n(\hat{\lambda}_n) \leq s_n(\lambda_n^0) \leq s_n^0(\lambda_n^0) + \frac{\delta}{2}$$

or $|s_n^0(\hat{\lambda}_n) - s_n^0(\lambda_n^0)| < \delta$. Then for all $n > \max(N, N')$ we must have $|\hat{\lambda}_n - \lambda_n^0| < \epsilon$. $\quad\square$

The asymptotic distribution of $\hat{\lambda}_n$ is characterized in terms of the following notation.

NOTATION 2.

$$\overline{\mathcal{U}}_n(\lambda) = \sum_{\tau=-l(n)}^{l(n)} w\left(\frac{\tau}{l(n)}\right) \overline{\mathcal{U}}_{n\tau}(\lambda)$$

$$w(x) = \begin{cases} 1 - 6|x|^2 + 6|x|^3 & 0 \leq |x| \leq \frac{1}{2} \\ 2(1-|x|)^3 & \frac{1}{2} \leq |x| \leq 1 \end{cases}$$

$l(n) =$ the integer nearest $n^{1/5}$

$$\overline{\mathcal{U}}_{n\tau}(\lambda) = \begin{cases} \dfrac{1}{n} \sum_{t=1+\tau}^{n} \left(\mathscr{E}\dfrac{\partial}{\partial\lambda} s_t(W_t, \tau_n^0, \lambda)\right)\left(\mathscr{E}\dfrac{\partial}{\partial\lambda} s_{t-\tau}(W_{t-\tau}, \tau_n^0, \lambda)\right)' & \tau \geq 0 \\ \mathcal{U}'_{n,-\tau}(\lambda) & \tau < 0 \end{cases}$$

$$\overline{\mathscr{I}}_n(\lambda) = \sum_{\tau=-(n-1)}^{n-1} \overline{\mathscr{I}}_{n\tau}(\lambda)$$

$$\overline{\mathscr{I}}_{n\tau}(\lambda) = \begin{cases} \dfrac{1}{n} \sum_{t=1+\tau}^{n} \mathscr{E}\left(\dfrac{\partial}{\partial\lambda} s_t(W_t, \tau_n^0, \lambda)\right)\left(\dfrac{\partial}{\partial\lambda} s_{t-\tau}(W_{t-\tau}, \tau_n^0, \lambda)\right)' - \overline{\mathcal{U}}_{n\tau}(\lambda) & \\ & \tau \geq 0 \\ \overline{\mathscr{I}}'_{n,-\tau}(\lambda) & \\ & \tau < 0 \end{cases}$$

$$\overline{\mathscr{J}}_n(\lambda) = \frac{1}{n} \sum_{t=1}^{n} \mathscr{E}\frac{\partial^2}{\partial\lambda\,\partial\lambda'} s_t(W_t, \tau_n^0, \lambda)$$

$$\mathscr{J}_n^0 = \overline{\mathscr{J}}_n(\lambda_n^0) \qquad \mathscr{I}_n^0 = \overline{\mathscr{I}}_n(\lambda_n^0) \qquad \mathcal{U}_n^0 = \overline{\mathcal{U}}_n(\lambda_n^0)$$

$$\mathscr{J}_n^* = \overline{\mathscr{J}}_n(\lambda_n^*) \qquad \mathscr{I}_n^* = \overline{\mathscr{I}}_n(\lambda_n^*) \qquad \mathcal{U}_n^* = \overline{\mathcal{U}}_n(\lambda_n^*).$$

We illustrate their computation with the example.

EXAMPLE 1 (Continued).   The first and second partial derivatives of

$$s_t(w_t, \lambda) = [y_t - f(y_{t-1}, x_t, \lambda)]^2$$

are

$$\frac{\partial}{\partial\lambda}s(w_t, \lambda) = -2[y_t - f(y_{t-1}, x_t, \lambda)]\frac{\partial}{\partial\lambda}f(y_{t-1}, x_t, \lambda)$$

$$\frac{\partial^2}{\partial\lambda\,\partial\lambda'}s(w_t, \lambda) = 2\left(\frac{\partial}{\partial\lambda}f(y_{t-1}, x_t, \lambda)\right)\left(\frac{\partial}{\partial\lambda}f(y_{t-1}, x_t, \lambda)\right)'$$

$$-2[y_t - f(y_{t-1}, x_t, \lambda)]\frac{\partial^2}{\partial\lambda\,\partial\lambda'}f(y_{t-1}, x_t, \lambda).$$

Evaluating the first derivative at $\lambda = \gamma^0$ and $y_t = f(y_{t-1}, x_t, \gamma^0) + e_t$, we have, recalling that e_t and (y_{t-1}, x_t) are independent,

$$\mathscr{E}\frac{\partial}{\partial\lambda}s(W_t, \lambda)\Big|_{\lambda=\gamma^0} = -2\mathscr{E}e_t\mathscr{E}\frac{\partial}{\partial\lambda}f(y_{t-1}, x_t, \gamma^0) = 0$$

whence $\mathscr{U}_n^0 = 0$.

Put

$$F_t = \frac{\partial}{\partial\lambda}f(y_{t-1}, x_t, \lambda)\Big|_{\lambda=\gamma^0}$$

$$\sigma_t^2 = \mathscr{E}e_t^2.$$

Then

$$\mathscr{E}\left(\frac{\partial}{\partial\lambda}s(W_t, \lambda)\right)\left(\frac{\partial}{\partial\lambda}s(W_t, \lambda)\right)'\Big|_{\lambda=\gamma^0}$$

$$= \begin{cases} 4\mathscr{E}e_t^2\mathscr{E}F_tF_t' & s = t \\ 4\mathscr{E}e_t\mathscr{E}e_sF_tF_s' & s < t \end{cases}$$

$$= \begin{cases} 4\sigma_t^2\mathscr{E}F_tF_t' & s = t \\ 0 & s < t \end{cases}$$

and

$$\mathscr{E}\frac{\partial^2}{\partial\lambda\,\partial\lambda'}s(W_t, \lambda)\Big|_{\lambda=\gamma^0} = 2\mathscr{E}F_tF_t' - 2\mathscr{E}e_t\mathscr{E}\frac{\partial}{\partial\lambda\,\partial\lambda'}f(y_{t-1}, x_t, \gamma^0).$$

In summary,

$$\mathscr{I}_n^0 = \frac{4}{n}\sum_{t=1}^{n}\sigma_t^2\mathscr{E}F_tF_t'$$

$$\mathscr{J}_n^0 = \frac{2}{n}\sum_{t=1}^{n}\mathscr{E}F_tF_t'. \qquad \square$$

General purpose estimators of $(\mathcal{I}_n^0, \mathcal{J}_n^0)$ and $(\mathcal{I}_n^*, \mathcal{J}_n^*)$—$(\hat{\mathcal{I}}_n, \hat{\mathcal{J}}_n)$ and $(\tilde{\mathcal{I}}_n, \tilde{\mathcal{J}}_n)$ respectively—may be defined as follows.

NOTATION 3.

$$\mathcal{I}_n(\lambda) = \sum_{t=-l(n)}^{l(n)} w\left(\frac{\tau}{l(n)}\right)\mathcal{I}_{n\tau}(\lambda)$$

$$\mathcal{I}_{n\tau}(\lambda) = \begin{cases} \dfrac{1}{n}\displaystyle\sum_{t=1+\tau}^{n}\left(\dfrac{\partial}{\partial\lambda}s_t(w_t,\hat{\tau}_n,\lambda)\right)\left(\dfrac{\partial}{\partial\lambda}s_{t-\tau}(w_{t-\tau},\hat{\tau}_n,\lambda)\right)' & \tau \geq 0 \\[2mm] \mathcal{I}'_{n,-\tau}(\lambda) & \tau < 0 \end{cases}$$

$$w(x) = \begin{cases} 1 - 6|x|^2 + 6|x|^3 & 0 \leq |x| \leq \tfrac{1}{2} \\[1mm] 2(1 - |x|)^3 & \tfrac{1}{2} \leq |x| \leq 1 \end{cases}$$

$l(n) =$ the integer nearest $n^{1/5}$

$$\mathcal{J}_n(\lambda) = \frac{1}{n}\sum_{t=1}^{n}\frac{\partial^2}{\partial\lambda\,\partial\lambda'}s_t(w_t,\hat{\tau}_n,\lambda)$$

$$\hat{\mathcal{I}} = \mathcal{I}_n(\hat{\lambda}), \qquad \hat{\mathcal{J}} = \mathcal{J}_n(\hat{\lambda}), \qquad \tilde{\mathcal{I}} = \mathcal{I}_n(\tilde{\lambda}), \qquad \tilde{\mathcal{J}} = \mathcal{J}_n(\tilde{\lambda}).$$

The special structure of specific applications will suggest alternative estimators. For instance, with Example 1 one would prefer to take $\mathcal{I}_{n\tau}(\lambda) \equiv 0$ for $\tau \neq 0$.

The normalized sum of the scores is asymptotically normally distributed under the following regularity conditions, as we show in Theorem 4 below.

ASSUMPTION 6. The estimation space Λ^* contains a closed ball Λ with finite, nonzero radius. The points $\{\lambda_n^0\}$ are contained in a concentric ball of smaller radius. Let $g_t(W_t, \tau, \lambda)$ be a generic term that denotes an element of $(\partial/\partial\lambda)s_t(W_t, \tau, \lambda)$, $(\partial^2/\partial\lambda\,\partial\lambda')s_t(W_t, \tau, \lambda)$, $(\partial^2/\partial\tau\,\partial\lambda')s_t(W_t, \tau, \lambda)$, or $[(\partial/\partial\lambda)s_t(W_t, \tau, \lambda)][(\partial/\partial\lambda)s_t(W_t, \tau, \lambda)]'$. On $T \times \Lambda$, the family $\{g_t[W_t(\omega), \tau, \lambda]\}$ is near epoch dependent of size $-q$ with $q = 2(r - 2)/(r - 4)$, where r is that of Assumption 3, $g_t[W_t(\omega), \tau, \lambda]$ is A-smooth in (τ, λ), and there is a sequence of random variables $\{d_t\}$ with $\sup_{T\times\Lambda}|g_t[W_t(\omega), \tau, \lambda]| \leq d_t(\omega)$ and $\|d_t\|_r \leq \Delta < \infty$ for all t. There is an

N and constants $c_0 > 0$, $c_1 < \infty$ such that for all δ in $\mathbb{R}^p$ we have

$$c_0\delta'\delta \le \delta'\bar{\mathcal{J}}_n(\lambda)\delta \le c_1\delta'\delta \qquad \text{all} \quad n > N, \quad \text{all} \quad \lambda \in \Lambda$$

$$c_0\delta'\delta \le \delta'\mathcal{J}_n^0\delta \le c_1\delta'\delta \qquad \text{all} \quad n > N$$

$$c_0\delta'\delta \le \delta'\mathcal{J}_n^*\delta \le c_1\delta'\delta \qquad \text{all} \quad n > N$$

$$\lim_{n\to\infty} \delta'\left(\mathcal{J}_n^0\right)^{-1/2}\mathcal{J}_{[ns]}^0\left(\mathcal{J}_n^0\right)^{-1/2'}\delta = \delta'\delta \qquad \text{all} \quad 0 < s \le 1$$

$$\lim_{n\to\infty} \delta'\left(\mathcal{J}_n^*\right)^{-1/2}\mathcal{J}_{[ns]}^*\left(\mathcal{J}_n^*\right)^{-1/2'}\delta = \delta'\delta \qquad \text{all} \quad 0 < s \le 1.$$

Also,

$$\lim_{n\to\infty} \frac{1}{n}\sum_{t=1}^{n} \mathcal{E}\frac{\partial^2}{\partial\tau\partial\lambda'}s_t\left(W_t, \tau_n^0, \lambda_n^0\right) = 0.$$

Recall that $[ns]$ denotes the integer part of ns, that $\mathcal{J}^{-1/2}$ denotes a matrix with $\mathcal{J}^{-1} = (\mathcal{J}^{-1/2})'(\mathcal{J}^{-1/2})$ and $\mathcal{J}^{1/2}$ a matrix with $\mathcal{J} = (\mathcal{J}^{1/2})(\mathcal{J}^{1/2})'$, and that factorizations are always taken to be compatible so that $\mathcal{J}^{1/2}\mathcal{J}^{-1/2} = I$.

As mentioned in Chapter 3, the condition

$$\lim_{n\to\infty} \frac{1}{n}\sum_{t=1}^{n} \mathcal{E}\frac{\partial^2}{\partial\tau\partial\lambda'}s_t\left(W_t, \tau_n^0, \lambda_n^0\right) = 0$$

permits two step (first τ, then λ) estimation. If it is not satisfied, the easiest approach is to estimate τ and λ jointly.

The requirement that

$$\lim_{n\to\infty} \delta'\left(\mathcal{J}_n^0\right)^{-1/2}\mathcal{J}_{[ns]}^0\left(\mathcal{J}_n^0\right)^{-1/2'}\delta = \delta'\delta$$

is particularly unfortunate because it is nearly the same as requiring that

$$\lim_{n\to\infty} \mathcal{J}_n^0 = \mathcal{J}*$$

as in Chapter 3. This has the effect of either restricting the amount of heteroscedasticity that $(\partial/\partial\lambda)s_t(W_t, \tau_n^0, \lambda_n^0)$ can exhibit or requiring the use of a variance stabilizing transformation (see Section 2 of Chapter 3). But the restriction is dictated by the regularity conditions of the central limit theorem and there is no way to get around it, because asymptotic normality cannot obtain if the condition is violated (Ibragimov, 1962). We verify that the condition holds for the example.

EXAMPLE 1 (Continued). For the example,

$$y_t = \begin{cases} f(y_{t-1}, x_t, \gamma^0) + e_t & t = 1, 2, \ldots \\ 0 & t \le 0 \end{cases}$$

with $|(\partial/\partial y)f(y, x, \gamma^0)| \le d < 1$, we shall verify that

$$\mathscr{I}_n^0 = \frac{4}{n} \sum_{t=1}^{n} \mathscr{E} e_t^2 \mathscr{E} G(y_{t-1}, x_t)$$

$$G(y, x) = \left(\frac{\partial}{\partial \lambda} f(y, x, \lambda) \right) \left(\frac{\partial}{\partial \lambda} f(y, x, \lambda) \right)' \Big|_{\lambda = \gamma^0}$$

satisfies the condition

$$\lim_{n \to \infty} \delta' \left(\mathscr{I}_n^0 \right)^{-1/2} \mathscr{I}_{[ns]}^0 \left(\mathscr{I}_n^0 \right)^{-1/2'} \delta = \delta' \delta.$$

To do so, define

$$\bar{y}_t = \begin{cases} 0 & t \le 0 \\ f(\bar{y}_{t-1}, x_t, \gamma^0) & 0 < t \end{cases}$$

$$\hat{y}_{t,m}^s = \begin{cases} 0 & s \le \max(t - m, 0) \\ f(\hat{y}_{t,m}^{s-1}, x_s, \gamma^0) + e_s & \max(t - m, 0) < s \le t \end{cases}$$

$$Y_t = \sum_{j=0}^{t-1} d^j |e_{t-j}|$$

$$g(y, x) = \text{a typical element of } G(y, x)$$

and assume that $\{e_t\}$ and $\{x_t\}$ are sequences of identically distributed random variables, and that $|\bar{y}_t|$ and $|(\partial/\partial y)g(y, x)|$ are bounded by some $\Delta < \infty$. As in Section 2, for $m \ge 0$ and $t \ge 0$ there is a $\tilde{y}_t$ on the line segment joining y_t to $\bar{y}_t$ such that

$$\begin{aligned}
|y_t - \bar{y}_t| &= \left| f(y_{t-1}, x_t, \gamma^0) + e_t - f(\bar{y}_{t-1}, x_t, \gamma^0) \right| \\
&= \left| \frac{\partial}{\partial y} f(\tilde{y}_t, x_t, \gamma^0)(y_{t-1} - \bar{y}_{t-1}) + e_t \right| \\
&\leq d|y_{t-1} - \bar{y}_{t-1}| + |e_t| \\
&\quad \vdots \\
&\leq d^t|y_0 - \bar{y}_0| + \sum_{j=0}^{t-1} d^j |e_{t-j}| \\
&= Y_t.
\end{aligned}$$

For $t - m > 0$ the same argument yields

$$\begin{aligned}
|y_t - \hat{y}_{t,m}^t| &= \left| f(y_{t-1}, x_t, \gamma^0) + e_t - f(\hat{y}_{t,m}^{t-1}, x_t, \gamma^0) - e_t \right| \\
&\leq d|y_{t-1} - \hat{y}_{t,m}^{t-1}| \\
&\quad \vdots \\
&\leq d^m|y_{t-m}| \\
&\leq d^m|\bar{y}_{t-1}| + d^m|y_{t-m} - \bar{y}_{t-m}| \\
&\leq d^m(\Delta + Y_{t-m}).
\end{aligned}$$

The assumption that the sequences of random variables $\{e_t\}$ and $\{x_t\}$ are identically distributed causes the sequence of random variables $\{G(\hat{y}_{t-1,m}^{t-1}, x_t)\}$ to be identically distributed. Thus

$$\mathscr{E} e_t^2 \mathscr{E} G(\hat{y}_{t-1,m}^{t-1}, x_t) = V \qquad \text{all } t.$$

But

$$\begin{aligned}
\mathscr{E}\left| g(y_{t-1}, x_t) - g(\hat{y}_{t-1,m}^{t-1}, x_t) \right| &\leq \mathscr{E}\left| \frac{\partial}{\partial y} g(\bar{\bar{y}}_{t-1}, x_t) \right| |y_{t-1} - \hat{y}_{t-1,m}^{t-1}| \\
&\leq \Delta \mathscr{E}|y_{t-1} - \hat{y}_{t-1,m}^{t-1}| \\
&\leq \Delta d^m \mathscr{E}(\Delta + Y_{t-m}) \\
&\leq \Delta d^m \left(\Delta + \sum_{j=0}^{\infty} d^j \mathscr{E}|e_0| \right) \\
&= \text{const } d^m
\end{aligned}$$

where the constant does not depend on n or t. Thus

$$\lim_{n \to \infty} \delta' \left(\mathscr{I}_n^0 \right)^{-1/2} \left(\mathscr{I}_{[ns]}^0 \right) \left(\mathscr{I}_n^0 \right)^{-1/2'} \delta$$
$$= \delta' [V + O(d^m)]^{-1/2} [V + O(d^m)] [V + O(d^m)]^{-1/2'} \delta.$$

Now m is arbitrary, so we must have

$$\lim_{n \to \infty} \delta' \left(\mathscr{I}_n^0 \right)^{-1/2} \left(\mathscr{I}_{[ns]}^0 \right)^{-1/2'} \left(\mathscr{I}_n^0 \right)^{-1/2} \delta = \delta' \delta. \qquad \square$$

With Assumption 6 one has access to Theorems 1 through 3, and asymptotic normality of the scores and the estimator $\hat{\lambda}_n$ follows directly using basically the same methods of proof as in Chapter 3. The details are as follows.

LEMMA 10. Under Assumptions 1 through 6, interchange of differentiation and integration is permitted in these instances:

$$\frac{\partial}{\partial \lambda} s_n^0(\lambda) = \frac{1}{n} \sum_{t=1}^{n} \mathscr{E} \frac{\partial}{\partial \lambda} s_t (W_t, \tau_n^0, \lambda)$$

$$\frac{\partial^2}{\partial \lambda \, \partial \lambda'} s_n^0(\lambda) = \frac{1}{n} \sum_{t=1}^{n} \mathscr{E} \frac{\partial^2}{\partial \lambda \, \partial \lambda'} s_t (W_t, \tau_n^0, \lambda).$$

Moreover,

$$\lim_{n \to \infty} \sup_{\Lambda} \left| \frac{\partial}{\partial \lambda} s_n(\lambda) - \frac{\partial}{\partial \lambda} s_n^0(\lambda) \right| = 0 \qquad \text{almost surely}$$

$$\lim_{n \to \infty} \sup_{\Lambda} \left| \frac{\partial^2}{\partial \lambda \, \partial \lambda'} s_n(\lambda) - \frac{\partial^2}{\partial \lambda \, \partial \lambda'} s_n^0(\lambda) \right| = 0 \qquad \text{almost surely}$$

and the families

$$\left\{ \frac{\partial}{\partial \lambda} s_n^0(\lambda) \right\}_{n=1}^{\infty} \quad \text{and} \quad \left\{ \frac{\partial^2}{\partial \lambda \, \partial \lambda'} s_n^0(\lambda) \right\}_{n=1}^{\infty}$$

are equicontinuous on Λ.

Proof. The proof that interchange is permitted is the same as in Lemma 3 of Chapter 3. Almost sure convergence and equicontinuity follow directly from Theorem 1 using the same argument as in Lemma 9. $\qquad \square$

THEOREM 5 (Asymptotic normality of the scores). Under Assumptions 1 through 6,

$$\sqrt{n}\left(\mathscr{I}_n^0\right)^{-1/2}\frac{\partial}{\partial\lambda}s_n(\lambda_n^0) \xrightarrow{\mathscr{L}} N_p(0,I)$$

$$\lim_{n\to\infty}\left(\mathscr{I}_n^0 + \mathscr{U}_n^0 - \hat{\mathscr{I}}\right) = 0 \qquad \text{in probability.}$$

Proof. For each i where $i = 1, 2, \ldots, p$ we have

$$\sqrt{n}\,\frac{\partial}{\partial\lambda_i}s_n(\lambda_n^0) = \frac{1}{\sqrt{n}}\sum_{t=1}^n \frac{\partial}{\partial\lambda_i}s_t(W_t, \tau_n^0, \lambda_n^0)$$

$$+ \frac{1}{n}\sum_{t=1}^n \frac{\partial^2}{\partial\lambda_i\partial\tau'}s_t(W_t, \bar{\tau}_n, \lambda_n^0)\sqrt{n}\left(\hat{\tau} - \tau_n^0\right)$$

where $\bar{\tau}_n$ is on the line segment joining $\hat{\tau}_n$ to τ_n^0. By Assumption 4, $\lim_{n\to\infty}\hat{\tau}_n - \tau_n^0 = 0$ almost surely and $\sqrt{n}(\hat{\tau}_n - \tau_n^0) = O_p(1)$, whence

$$\lim_{n\to\infty}\frac{1}{n}\sum_{t=1}^n \frac{\partial^2}{\partial\lambda_i\partial\tau}\left[s_t(W_t, \bar{\tau}_n, \lambda_n^0) - s_t(W_t, \tau_n^0, \lambda_n^0)\right]\sqrt{n}\left(\hat{\tau}_n - \tau_n^0\right) = 0$$

almost surely. By Assumption 6 we have

$$\lim_{n\to\infty}\frac{1}{n}\sum_{t=1}^n \frac{\partial^2}{\partial\lambda_i\partial\tau'}s_t(W_t, \tau_n^0, \lambda_n^0)\sqrt{n}\left(\hat{\tau}_n - \tau_n^0\right) = 0$$

almost surely. As the elements of $(\mathscr{I}_n^0)^{-1/2}$ must be bounded (Problem 3), we have, recalling that $(\partial/\partial\lambda)s_n^0(\lambda_n^0) = 0$,

$$\sqrt{n}\left(\mathscr{I}_n^0\right)^{-1/2}\frac{\partial}{\partial\lambda}s_n(\lambda_n^0)$$

$$= \frac{1}{n}\sum_{t=1}^n \left(\mathscr{I}_n^0\right)^{-1/2}\left(\frac{\partial}{\partial\lambda}s_t(W_t, \tau_n^0, \lambda_n^0) - \mathscr{E}\frac{\partial}{\partial\lambda}s_t(W_t, \tau_n^0, \lambda_n^0)\right) + o_s(1)$$

where the interchange of integration and differentiation is permitted by Lemma 10. Let δ be a nonzero p-vector, and put

$$g_{nt}(W_t, \gamma_n^0) = \delta'\left(\mathscr{I}_n^0\right)^{-1/2}\frac{\partial}{\partial\lambda}s_t(W_t, \tau_n^0, \lambda_n^0).$$

Assumption 3 guarantees that $\{V_t\}_{t=-\infty}^{\infty}$ is strong mixing of size $-4r/(r-4)$ with $r > 4$, so $\{V_t\}_{t=-\infty}^{\infty}$ is strong mixing of size $-\frac{1}{2}$ as required by Theorem 2, and Assumption 6 guarantees that $\{g_{nt}(W_t, \gamma_n^0)\}$ is near epoch dependent of size $-q$ with $q = 2(r-2)/(r-4) > \frac{1}{2}$ (Problem 4). We have

$$
\begin{aligned}
\sigma_{[ns]}^2 &= \mathrm{Var}\left(\sum_{t=1}^{[ns]} g_{nt}(W_t, \gamma_n^0) \right) \\
&= [ns]\,\delta'\big(\mathscr{I}_n^0\big)^{-1/2} \mathscr{I}_{[ns]}^0 \big(\mathscr{I}_n^0\big)^{-1/2'} \delta
\end{aligned}
$$

which, by Assumption 6, satisfies

1. $1/\sigma_n^2 = 1/\sigma_{[n]}^2 = O(1/n)$,
2. $\lim_{n\to\infty} \sigma_{[ns]}^2/\sigma_n^2 = s$.

Further, Assumption 6 and Problem 3 implies

3. $\left\| g_{nt}(W_t, \gamma_n^0) - \mathscr{E} g_{nt}(W_t, \gamma_n^0) \right\|_r$

$$
\leq \big[\delta'\big(\mathscr{I}_n^0\big)\delta\big]^{1/2} \max_{1 \leq i \leq p} \left\| \frac{\partial}{\partial \lambda_i} s_t(W_t, \tau_n^0, \lambda_n^0) - \mathscr{E}\frac{\partial}{\partial \lambda_i} s_t(W_t, \tau_n^0, \lambda_n^0) \right\|_r
$$

$$
\leq \left(\frac{1}{c_0} \delta'\delta \right)^{1/2} \big(\|d_t\|_r + \|d_t\|_1 \big).
$$

Thus,

$$
\begin{aligned}
&\sqrt{n}\,\delta'\big(\mathscr{I}_n^0\big)^{-1/2} \frac{\partial}{\partial\lambda} s_n(\lambda_n^0) \\
&= \frac{1}{\sqrt{n}} \sum_{t=1}^{n} \big[g_{nt}(W_t, \gamma_n^0) - \mathscr{E} g_{nt}(W_t, \gamma_n^0) \big] \\
&= (\delta'\delta)^{1/2} \frac{1}{\sigma_n} \sum_{t=1}^{n} g_{nt}(W_t, \gamma^0) \\
&\overset{\mathscr{L}}{\to} N(0, \delta'\delta)
\end{aligned}
$$

by Theorem 2. This proves the first assertion.

To prove the second, put

$$
X_{nt} = \delta'\left(\frac{\partial}{\partial\lambda} s_t(W_t, \tau_n^0, \lambda_n^0) - \mathscr{E}\frac{\partial}{\partial\lambda} s_t(W_t, \tau_n^0, \lambda_n^0) \right).
$$

Referring to Theorem 3, note that

$$\sigma_n^2 = \mathcal{E}\left(\sum_{t=1}^n X_{nt}\right)^2 = n\delta'\mathcal{I}_n^0\delta$$

$$\hat{R}_{n\tau} = \sum_{t=1+|\tau|}^n X_{nt}X_{n,\,t-|\tau|}$$

$$\hat{\sigma}_n^2 = \sum_{\tau=-l(n)}^{l(n)} w\left(\frac{\tau}{l(n)}\right)\hat{R}_{n\tau}$$

$$= n\delta'\left(\hat{\mathcal{I}} - \mathcal{U}_n^0\right)\delta$$

$$- \sum_{\tau=-l(n)}^{l(n)} w\left(\frac{\tau}{l(n)}\right) \sum_{t=1+|\tau|}^n X_{n,\,t-|\tau|}\mathcal{E}\delta'\frac{\partial}{\partial\lambda}s_t\left(W_t, \tau_n^0, \lambda_n^0\right)$$

$$- \sum_{\tau=-l(n)}^{l(n)} w\left(\frac{\tau}{l(n)}\right) \sum_{t=1+|\tau|}^n X_{n\tau}\mathcal{E}\delta'\frac{\partial}{\partial\lambda}s_{t-|\tau|}\left(W_{t-|\tau|}, \tau_n^0, \lambda_n^0\right)$$

$$= n\delta'\left(\hat{\mathcal{I}} - \mathcal{U}_n^0\right)\delta - Z_n$$

where $w(x)$ denotes Parzen weights. By Assumption 3, $\{V_t\}$ is strong mixing of size $-4r/(r-4)$ for $r > 4$, as required by Theorem 3; by Assumption 2, W_t depends only on the past; by Assumption 6, $\{X_{nt}\}$ is near epoch dependent of size $-q$ with $q = 2(r-2)/(r-4)$; so we have from Theorem 3 that

$$\lim_{n\to\infty} P\left(\frac{1}{n}|Z_n| > \epsilon\right) = \lim_{n\to\infty} \frac{B}{\epsilon^2}\frac{l^3(n)}{n} = 0$$

$$\lim_{n\to\infty} \frac{1}{n}|\sigma_n^2 - \mathcal{E}\hat{\sigma}_n^2| = \lim_{n\to\infty} Bl^{-1}(n) = 0$$

$$\lim_{n\to\infty} P\left[\frac{1}{n}|\hat{\sigma}_n^2 - \mathcal{E}\hat{\sigma}_n^2| > \epsilon\right] = \lim_{n\to\infty} \frac{B}{\epsilon^2}\frac{l^4(n)}{n} = 0$$

whence

$$\lim_{n\to\infty} \delta'\left(\mathcal{I}_n^0 - \hat{\mathcal{I}} + \mathcal{U}_n^0\right)\delta = 0 \qquad \text{in probability}$$

for every $\delta \neq 0$.   □

THEOREM 6 (Asymptotic normality). Let Assumptions 1 through 6 hold. Then

$$\sqrt{n}\left(\mathscr{I}_n^0\right)^{-1/2}\mathscr{I}_n^0\left(\hat{\lambda}_n - \lambda_n^0\right) \overset{\mathscr{L}}{\to} N_p(0, I)$$

$$\lim_{n\to\infty}\left(\mathscr{I}_n^0 - \hat{\mathscr{I}}\right) = 0 \qquad \text{almost surely.}$$

Proof. By Lemma 2 of Chapter 3 we may assume without loss of generality that $\hat{\lambda}_n, \lambda_n^0 \in \Lambda$ and that $(\partial/\partial\lambda)s_n(\hat{\lambda}_n) = o_s(n^{-1/2})$, $(\partial/\partial\lambda)s_n^0(\lambda_n^0) = o(n^{-1/2})$.

By Taylor's theorem

$$\sqrt{n}\frac{\partial}{\partial\lambda}s_n(\lambda_n^0) = \sqrt{n}\frac{\partial}{\partial\lambda}s_n(\hat{\lambda}_n) + \bar{\mathscr{I}}\sqrt{n}\left(\lambda_n^0 - \hat{\lambda}_n\right)$$

where $\bar{\mathscr{I}}$ has rows $(\partial/\partial\lambda')(\partial/\partial\lambda_i)s_n(\bar{\lambda}_{in})$ with $\|\bar{\lambda}_{in} - \lambda_n^0\| \leq \|\hat{\lambda}_n - \lambda_n^0\|$. Lemma 10 permits interchange of differentiation and integration, we have $\lim_{n\to\infty}\|\hat{\lambda}_n - \lambda_n^0\| = 0$ almost surely by Theorem 4, so that application of Theorem 1 yields $\lim_{n\to\infty}\mathscr{I}_n^0 - \bar{\mathscr{I}} = 0$ almost surely (Problem 5). Thus, we may write

$$\sqrt{n}\left(\mathscr{I}_n^0\right)^{-1/2}\left[\mathscr{I}_n^0 + o_s(1)\right]\left(\hat{\lambda}_n - \lambda_n^0\right) = -\sqrt{n}\left(\mathscr{I}_n^0\right)^{-1/2}\frac{\partial}{\partial\lambda}s_n(\lambda_n^0) + o_s(1)$$

recalling that $(\partial/\partial\lambda)s_n(\hat{\lambda}_n) = o_s(n^{-1/2})$ and that $(\mathscr{I}_n^0)^{-1/2} = O(1)$ by Assumption 6 (Problem 3). The right hand side is $O_p(1)$ by Theorem 5, and $(\mathscr{I}_n^0)^{1/2}$ and $(\mathscr{I}_n^0)^{-1}$ are $O(1)$ by Assumption 6, so that $\sqrt{n}(\hat{\lambda}_n - \lambda_n^0) = O_p(1)$ and we can write

$$\sqrt{n}\left(\mathscr{I}_n^0\right)^{-1/2}\mathscr{I}_n^0\left(\hat{\lambda}_n - \lambda_n^0\right) = -\sqrt{n}\left(\mathscr{I}_n^0\right)^{-1/2}\frac{\partial}{\partial\lambda}s_n(\lambda_n^0) + o_p(1)$$

which proves the first result.

The same argument used to show $\lim_{n\to\infty}\mathscr{I}_n^0 - \bar{\mathscr{I}} = 0$ almost surely can be used to show that $\lim_{n\to\infty}\mathscr{I}_n^0 - \hat{\mathscr{I}} = 0$ almost surely. □

Next we shall establish some ancillary facts concerning the estimator $\tilde{\lambda}_n$ that minimizes $s_n(\lambda)$ subject to $H: h(\lambda) = h_n^*$ under the assumption that the elements of the q-vector $\sqrt{n}[h(\lambda_n^0) - h_n^*]$ are bounded. Here h_n^* is a variable quantity chosen to adjust to λ_n^0 so that the elements of the vector are bounded, which contrasts with Chapter 3, where λ_n^0 was taken as the variable quantity and h_n^* was held fixed at zero. As in Chapter 3, these results are for use in deriving asymptotic distributions of test statistics and are not meant to be used as a theory of constrained estimation. See Section 7 of Chapter 3 for a discussion of how a general asymptotic theory of estimation can be adapted to estimation subject to constraints.

ASSUMPTION 7 (Pitman drift). The function $h(\lambda)$ that defines the null hypothesis $H: h(\lambda_n^0) = h_n^*$ is a twice continuously differentiable mapping of Λ as defined by Assumption 6 into $\mathbb{R}^q$ with Jacobian denoted as $H(\lambda) = (\partial/\partial\lambda')h(\lambda)$. The eigenvalues of $H(\lambda)H'(\lambda)$ are bounded below over Λ by $c_0^2 > 0$ and above by $c_1^2 < \infty$. In the case where $p = q$, $h(\lambda)$ is assumed to be a one to one mapping with a continuous inverse. In the case $q < p$, there is a continuous function $\phi(\lambda)$ such that the mapping

$$\begin{pmatrix} \rho \\ \tau \end{pmatrix} = \begin{pmatrix} \phi(\lambda) \\ h(\lambda) \end{pmatrix}$$

has a continuous inverse

$$\lambda = \psi(\rho, \tau)$$

defined over $S = \{(\rho, \tau): \rho = \phi(\lambda), \tau = h(\lambda), \lambda \in \Lambda\}$. Moreover, $\psi(\rho, \tau)$ has a continuous extension to the set

$$R \times T = \{\rho: \rho = \phi(\lambda)\} \times \{\tau: \tau = h(\lambda)\}.$$

The sequence $\{h_n^*\}$ is chosen such that

$$\sqrt{n}\left[h(\lambda_n^0) - h_n^*\right] = O(1).$$

There is an N such that for all δ in $\mathbb{R}^p$

$$\delta'\mathcal{U}_n^0\delta \le c_1\delta'\delta \qquad \text{all} \quad n > N$$
$$\delta'\mathcal{U}_n^*\delta \le c_1\delta'\delta \qquad \text{all} \quad n > N.$$

The purpose of the functions $\phi(\lambda)$ and $\psi(\rho, \tau)$ in Assumption 7 is to insure the existence of a sequence $\{\lambda_n^\#\}$ that satisfies $h(\lambda_n^\#) = 0$ but has $\lim_{n \to \infty}(\lambda_n^\# - \lambda_n^0) = 0$. This is the same as assuming that the distance between λ_n^0 and the projection of λ_n^0 onto $\Lambda_n^* = \{\lambda: h(\lambda) = h_n^*\}$ decreases as $|h_n^* - h(\lambda_n^0)|$ decreases. The existence of the sequence $\{\lambda_n^\#\}$ and the identification condition (Assumption 4) is enough to guarantee that $|\lambda_n^0 - \lambda_n^*|$ decreases as $|h(\lambda_n^0) - h_n^*|$ decreases (Problem 7). The bounds on the eigenvalues of $H(\lambda)H'(\lambda)$ (Assumption 7) and $\bar{\mathcal{J}}_n(\lambda)$ (Assumption 6) guarantee that $|\lambda_n^0 - \lambda_n^*|$ decreases as fast as $|h(\lambda_n^0) - h^*|$ decreases, as we show in the next two lemmas.

LEMMA 11. Let $\mathscr{I}$ be a symmetric p by p matrix, and let H be a matrix of order q by p with $q < p$. Suppose that the eigenvalues of $\mathscr{I}$ are bounded below by $c_0 > 0$ and above by $c_1 < \infty$, and that those of HH' are bounded below by c_0^2 and above by c_1^2. Then there is a matrix G of order p by $p - q$ with orthonormal columns such that $HG = 0$, the elements of

$$A = \begin{pmatrix} G'\mathscr{I} \\ H \end{pmatrix}$$

are bounded above by pc_1, and $|\det A| \geq (c_0)^{2p}$.

Proof. Let

$$H = USV_{(1)}'$$

be the singular value decomposition (Lawson and Hanson, 1974, Chapter 4) of H, where S is a diagonal matrix of order q with positive entries on the diagonal, and $V_{(1)}'$ is of order q by p and $U'U = UU' = V_{(1)}'V_{(1)} = I$ of order q. From $HH' = US^2U'$ we see that $c_0^2 \leq s_{ii}^2 \leq c_1^2$. Choose $V_{(2)}'$ of order $p - q$ by p such that

$$V = \begin{pmatrix} V_{(1)}' \\ V_{(2)}' \end{pmatrix}$$

satisfies

$$I = V'V = V_{(1)}V_{(1)}' + V_{(2)}V_{(2)}'$$
$$= \begin{pmatrix} V_{(1)}'V_{(1)} & V_{(1)}'V_{(2)} \\ V_{(2)}'V_{(1)} & V_{(2)}'V_{(2)} \end{pmatrix} = \begin{pmatrix} I_{(1)} & 0 \\ 0 & I_{(2)} \end{pmatrix}.$$

Put $G' = V_{(2)}'$, note that $HG = 0$, and consider

$$AA' = \begin{pmatrix} V_{(2)}'\mathscr{I} \\ USV_{(1)}' \end{pmatrix} (\mathscr{I}V_{(2)} \,\vdots\, V_{(1)}SU')$$

$$= \begin{pmatrix} V_{(2)}'\mathscr{I}\mathscr{I}V_{(2)} & V_{(2)}'\mathscr{I}V_{(1)}SU' \\ USV_{(1)}'\mathscr{I}V_{(2)} & US^2U' \end{pmatrix}$$

$$= \begin{pmatrix} V_{(2)}' & 0 \\ 0 & U \end{pmatrix} \begin{pmatrix} \mathscr{I} & 0 \\ 0 & S \end{pmatrix} \begin{pmatrix} V_{(1)} & V_{(2)} \\ I & 0 \end{pmatrix} \begin{pmatrix} V_{(1)}' & I \\ V_{(2)}' & 0 \end{pmatrix} \begin{pmatrix} \mathscr{I} & 0 \\ 0 & S \end{pmatrix} \begin{pmatrix} V_{(2)} & 0 \\ 0 & U' \end{pmatrix}$$

$$= BCDD'C'B'.$$

The elements of B and D are bounded by one, so we must have that each element of BCD is bounded above by pc_1. Then each element of AA' is bounded above by $p^2 c_1^2$. Since a diagonal element of AA' has the form $\sum_i a_{ij}^2$, we must have $|a_{ij}| \le pc_1$. Now (Mood and Graybill, 1963, p. 206)

$$
\begin{aligned}
\det AA' &= \det(US^2 U') \det\left[V_{(2)}' \mathscr{I}\mathscr{I} V_{(2)} - V_{(2)}' \mathscr{I} V_{(1)} SU' \right. \\
&\qquad \left. \times (US^2 U')^{-1} USV_{(1)}' \mathscr{I} V_{(2)} \right] \\
&= \det S^2 \det\left(V_{(2)}' \mathscr{I}\mathscr{I} V_{(2)} - V_{(2)}' \mathscr{I} V_{(1)} V_{(1)}' \mathscr{I} V_{(2)} \right) \\
&= \det S^2 \det\left(V_{(2)}' \mathscr{I} V_{(2)} V_{(2)}' \mathscr{I} V_{(2)} \right) \\
&\ge (c_0)^{2p} \det^2\left(V_{(2)}' \mathscr{I} V_{(2)} \right).
\end{aligned}
$$

But

$$
c_0 x'x = c_0 x' V_{(2)}' V_{(2)} x \le x' V_{(2)}' \mathscr{I} V_{(2)} x
$$

whence $(c_0)p \le \det V_{(2)}' \mathscr{I} V_{(2)}$ and

$$
(c_0)^{4p} \le \det A'A = \det^2 A. \qquad \square
$$

LEMMA 12. Under Assumptions 1 through 7 there is a bound B that does not depend on n such that $|\lambda_n^0 - \lambda_n^*| \le B|h(\lambda_n^0) - h_n^*|$, where $|\lambda| = (\sum_{i=1}^p \lambda_i^2)^{1/2}$.

Proof. The proof for the case $q = p$ is immediate, as the one to one mapping $\tau = h(\lambda)$ has a Jacobian whose inverse has bounded elements. Consider the case $q < p$.

Let $\epsilon > 0$ be given. For N_0 given by Assumption 4 put

$$
\delta = \inf_{n > N_0} \inf_{|\lambda - \lambda_n^0| > \epsilon} \left| s_n^0(\lambda) - s_n^0(\lambda_n^0) \right|.
$$

Let $\psi(\rho, \tau)$ be the continuous function defined on $R \times T$ given by Assumption 7. Now $h_n^* = h(\lambda_n^*)$ by definition, and put $\rho_n^* = \phi(\lambda_n^*)$, $h_n^0 = h(\lambda_n^0)$, and $\rho_n^0 = \phi(\lambda_n^0)$. The image of a compact set is compact, and the Cartesian product of two compact sets is compact, so $R \times T$ is compact. A continuous function on a compact set is uniformly continuous so $\lim_{n \to \infty} |h_n^0 - h_n^*| = 0$ implies that

$$
\lim_{n \to \infty} \sup_R \left| \psi(\rho, h_n^0) - \psi(\rho, h_n^*) \right| = 0.
$$

In particular, putting $\lambda_n^\# = \psi(\rho_n^0, h_n^*)$, we have

$$\lim_{n \to \infty} |\lambda_n^\# - \lambda_n^0| = 0.$$

By Assumption 6 the points $\{\lambda_n^0\}$ are in a concentric ball of radius strictly smaller than the radius of Λ, so we must have $\lambda_n^\#$ in Λ for all n greater than some N_1. By Lemma 9, the family $\{s_n^0(\lambda)\}$ is equicontinuous, so that there is an N_2 such that $|\lambda - \lambda_n^0| < \eta$ implies that

$$\left| s_n^0(\lambda) - s_n^0(\lambda_n^0) \right| < \delta$$

for all $n > N_2$. Choose N_3 large enough that $|\lambda_n^\# - \lambda_n^0| < \eta$ for all $n > N_3$. The point $\lambda_n^\#$ satisfies the constraint $h(\lambda_n^\#) = h_n^*$, so we must have $s_n^0(\lambda_n^*) \le s_n^0(\lambda_n^\#)$ for $n > N_1$. For $n > \max(N_0, N_1, N_2, N_3)$ we have

$$s_n^0(\lambda_n^*) \le s_n^0(\lambda_n^\#) < s_n^0(\lambda_n^0) + \delta$$

whence $|s_n^0(\lambda_n^*) - s_n^0(\lambda_n^0)| < \delta$ and we must have $|\lambda_n^* - \lambda_n^0| < \epsilon$. We have shown that $\lambda_n^0 - \lambda_n^* = o(1)$ as $|h_n^0 - h_n^*|$ tends to zero.

The first order conditions for the problem

$$\text{minimize} \quad s_n^0(\lambda) \quad \text{subject to} \quad h(\lambda) = h_n^*$$

are

$$\frac{\partial}{\partial \lambda'} s_n^0(\lambda_n^*) + \theta' H(\lambda_n^*) = 0$$

$$h(\lambda_n^*) = h_n^*.$$

By Taylor's theorem we have

$$\frac{\partial}{\partial \lambda} s_n^0(\lambda_n^0) = \frac{\partial}{\partial \lambda} s_n^0(\lambda_n^*) + \left[\mathscr{J}_n^* + o(1) \right](\lambda_n^0 - \lambda_n^*)$$

$$h(\lambda_n^0) - h_n^* = h(\lambda_n^*) - h_n^* + \left[H_n^* + o(1) \right](\lambda_n^0 - \lambda_n^*).$$

Using $(\partial / \partial \lambda) s_n^0(\lambda_n^0) = 0$ for large n and $h(\lambda_n^*) - h_n^* = 0$, we have upon substitution into the first order conditions that

$$\left[\mathscr{J}_n^* + o(1) \right](\lambda_n^0 - \lambda_n^*) = -H_n^{*\prime}\theta$$

$$\left[H_n^* + o(1) \right](\lambda_n^0 - \lambda_n^*) = h(\lambda_n^0) - h_n^*.$$

Let G_n^* be the matrix given by Lemma 11 with orthonormal columns,

$H_n^* G_n^* = 0$, $0 < (c_0)^{2p} \leq \det A_n^*$, and $\max_{ij} |a_{ijn}^*| \leq pc_1 < \infty$, where

$$A_n^* = \begin{pmatrix} G_n^{*\prime} \mathscr{I}_n^* \\ H^* \end{pmatrix}.$$

Let a_{ij} denote the elements of matrix A, and consider the region

$$\left\{ a_{ij} : 0 < (c_0)^{2p} - \epsilon \leq \det A, \ |a_{ij}| \leq pc_1 + \epsilon \right\}.$$

On this region we must have $|a^{ij}| \leq B < \infty$, where a^{ij} denotes an element of A^{-1}. For large n the matrix A_n^* is in this region by Lemma 11, as is the matrix

$$A_n = \begin{pmatrix} G_n^{*\prime} [\mathscr{I}_n^* + o(1)] \\ H_n^* + o(1) \end{pmatrix}$$

since the elements of G_n^* are bounded by one. In consequence we have

$$\left(\lambda_n^0 - \lambda_n^* \right) = A_n^{-1} \begin{pmatrix} 0 \\ h(\lambda_n^0) - h_n^* \end{pmatrix}$$

where the elements of A_n^{-1} are bounded above by B for all n larger than some N. Thus we have $|\lambda_n^0 - \lambda_n^*| \leq B|h(\lambda_n^0) - h_n^*|$ for large n. □

THEOREM 7. Let Assumptions 1 through 7 hold. Then

$$\sqrt{n} \left(\lambda_n^0 - \lambda_n^* \right) = O(1)$$

$$\lim_{n \to \infty} \tilde{\lambda}_n - \lambda_n^* = 0 \qquad \text{almost surely,}$$

$$\sqrt{n} (\mathscr{I}_n^*)^{-1/2} \frac{\partial}{\partial \lambda} [s_n(\lambda_n^*) - s_n^0(\lambda_n^*)] \xrightarrow{\mathscr{L}} N_p(0, I)$$

$$\lim_{n \to \infty} \left(\mathscr{I}_n^* + \mathscr{U}_n^* - \tilde{\mathscr{I}} \right) = 0 \qquad \text{in probability}$$

$$\lim_{n \to \infty} \mathscr{I}_n^* - \tilde{\mathscr{I}} = 0 \qquad \text{almost surely.}$$

Proof. The first result obtains from Lemma 12, since $\sqrt{n} [h(\lambda_n^0) - h_n^*] = 0(1)$ by Assumption 7.

The proof of the second is nearly word for word same as the first part of the proof of Lemma 12. One puts

$$\delta = \inf_{n > N_0} \inf_{|\lambda - \lambda_n^0| > \epsilon} \left| s_n^0(\lambda) - s_n^0(\lambda_n^0) \right|$$

and for fixed ω has from Lemma 9 that $|\lambda - \lambda_n^0| < \eta$ implies

$$\left| s_n(\lambda) - s_n^0(\lambda_n^0) \right| < \frac{\delta}{2}$$

for all n larger than N_1. For n larger than N_2 one has

$$\left| \lambda_n^\# - \lambda_n^0 \right| < \eta$$

as in the proof of Lemma 12. The critical inequality becomes, for the same fixed ω,

$$s_n^0(\tilde{\lambda}_n) - \frac{\delta}{2} < s_n(\tilde{\lambda}_n) \le s_n(\lambda_n^\#) < s_n^0(\lambda_n^0) + \frac{\delta}{2}$$

whence $s_n^0(\tilde{\lambda}_n) - s_n^0(\lambda_n^0) < \delta$ and we must have $|\tilde{\lambda}_n - \lambda_n^0| < \epsilon$. Combining this with the first result gives the second.

The proof of the third and fourth results is the same as the proof of Theorem 5, recalling that $(\partial/\partial\lambda)s_n^0(\lambda_n^*)$ is the mean of $(\partial/\partial\lambda)s_n(\lambda_n^*)$ by Lemma 10.

The fifth result is an immediate consequence of Lemma 10 and the second result. $\square$

PROBLEMS

1. Suppose that w_t has fixed dimension $k_t = k$ for all t, that the dependence of $s_t(w_t, \tau, \lambda)$ on t is trivial, $s_t(w, \tau, \lambda) = s(w, \tau, \lambda)$, and that the partial derivatives of $s(w, \tau, \lambda)$ with respect to τ and λ are dominated by integrable $\{d_t\}$. Show that $s_t(W_t(\omega), \tau, \lambda)$ is A-smooth.

2. Referring to Example 1, show that the family $\{(\partial/\partial\lambda_i)[y_t - f(y_{t-1}, x_t, \lambda)]^2\}$ is near epoch dependent of size $-q$ for any $q > 0$. List the regularity conditions used.

3. Let c_{0n} be the smallest eigenvalue of $\mathcal{I}_n^0$, and c_{1n} the largest. Prove that Assumption 6 implies that $c_0 \le c_{0n} \le c_{1n} \le c_1$ all $n \ge N$. Prove that $\det \mathcal{I}_n^0 \ge (c_0)^p$ for all $n \ge N$ and that $\delta'(\mathcal{I}_n^0)^{-1}\delta \le (1/c_0)\delta'\delta$ for all $n \ge N$. Show that $(\mathcal{I}_n^0)^{-1}$ can always be factored in such a way that the elements of $(\mathcal{I}_n^0)^{-1/2}$ are bounded.

4. Show that if the elements of $(\partial/\partial\lambda)s_t(W_t, \tau_n^0, \lambda_n^0)$ are near epoch dependent of size $-q$, then so are the elements of $\delta'A_n(\partial/\partial\lambda)$ $s_t(W_t, \tau_n^0, \lambda_n^0)$ if A_n has bounded elements.

5. Let $\lim_{n \to \infty} \sup_{T \times \Lambda} |(1/n)\sum_{t=1}^{n} f_t(\tau, \lambda) - \mathscr{E} f_t(\tau, \lambda)| = 0$ almost surely, let $\{(1/n)\sum_{t=1}^{n} \mathscr{E} f_t(\tau, \lambda)\}_{n=1}^{\infty}$ be an equicontinuous family on $T \times \Lambda$, and let $\lim_{n \to \infty} |(\hat{\tau}_n, \hat{\lambda}_n) - (\tau_n^0, \lambda_n^0)| = 0$ almost surely. Show that $\lim_{n \to \infty} |(1/n)\sum_{t=1}^{n} f_t(\hat{\tau}_n, \hat{\lambda}_n) - \mathscr{E} f_t(\tau_n^0, \lambda_n^0)| = 0$ almost surely.

6. Prove Lemma 11 with $\mathscr{J}$ not necessarily symmetric but with the singular values of $\mathscr{J}$ bounded below by $c_0 > 0$ and above by $c_1 < \infty$.

7. The purpose of the function $\psi(\rho, \tau)$ in Assumption 7 is to guarantee the existence of a sequence $\{\lambda_n^\#\}$ that satisfies $h(\lambda_n^\#) = 0$ and $\lim_{n \to \infty} \lambda_n^\# - \lambda_n^0 = 0$. Prove Lemma 12 using this condition instead of the existence of $\psi(\rho, \tau)$.

5. METHOD OF MOMENTS ESTIMATORS

Recall that a method of moments estimator $\hat{\lambda}_n$ is defined as the solution of the optimization problen

$$\text{minimize} \quad s_n(\lambda) = d\left[m_n(\lambda), \hat{\tau}_n\right]$$

where $d(m, \tau)$ is a measure of the distance of m to zero, $\hat{\tau}_n$ is an estimator of nuisance parameters, and $m_n(\lambda)$ is a vector of sample moments,

$$m_n(\lambda) = \frac{1}{n} \sum_{t=1}^{n} m_t(w_t, \hat{\tau}_n, \lambda).$$

The dimensions involved are as follows: w_t is a k_t-vector, τ is a u-vector, λ is a p-vector, and each $m_t(w_t, \tau, \lambda)$ is a Borel measurable function defined on some subset of $\mathbb{R}^{k_t} \times \mathbb{R}^u \times \mathbb{R}^p$ and with range in $\mathbb{R}^v$. Note that v is a constant; specifically, it does not depend on t. As previously, we use lowercase w_t to mean either a random variable or data as determined by context. For emphasis, we shall write $W_t(v_\infty)$ when considered as a function on $\mathbb{R}^\infty_{-\infty}$, and write $W_t(V_\infty)$, W_t, $W_t[V_\infty(\omega)]$, or $W_t(\omega)$ when considered as a random variable depending on the underlying probability space $(\Omega, \mathscr{A}, P)$ through function composition with the process $\{V_t(\omega)\}_{t=-\infty}^{\infty}$. A constrained method of moments estimator $\tilde{\lambda}_n$ is the solution of the optimization problem

$$\text{minimize} \quad s_n(\lambda) \quad \text{subject to} \quad h(\lambda) = h_n^*$$

where $h(\lambda)$ maps $\mathbb{R}^p$ into $\mathbb{R}^q$.

As in the previous section, the objective is to find the asymptotic distribution of the estimator $\hat{\lambda}_n$ under regularity conditions that do not rule out specification error. Some ancillary facts regarding $\tilde{\lambda}_n$ under a Pitman drift are also derived for use in the next section. As in the previous section, drift is imposed by moving h_n^*.

As the example, we shall consider the estimation procedure that is most commonly used to analyze data that are presumed to follow a nonlinear dynamic model. The estimator is called nonlinear three stage least squares by some authors (Jorgenson and Laffont, 1974; Gallant, 1977b; Amemiya, 1977; Gallant and Jorgenson, 1979) and generalized method of moments by others (Hansen, 1982). Usually, the term three stage least squares refers to a model with regression structure, and generalized method of moments to dynamic models.

EXAMPLE 2 (Generalized method of moments). Data are presumed to follow the model

$$q_t\left(y_t, x_t, \gamma^0\right) = e_t \qquad t = 0, 1, \ldots$$

where y_t is an L-vector of endogenous variables, x_t is a k_t-vector with exogenous variables and (possibly) lagged values of y_t as elements (the elements of x_t are collectively termed predetermined variables rather than exogenous variables due to the presence of lagged values of y_t), γ^0 is a p-vector, and $q_t(y, x, \gamma)$ maps $\mathbb{R}^L \times \mathbb{R}^{k_t} \times \mathbb{R}^p$ into $\mathbb{R}^M$ with $M \le L$. Note that M, L, and p do not depend on t. Instrumental variables—a sequence of K-vectors $\{z_t\}$—are assumed available for estimation. These variables have the form $z_t = Z_t(x_t)$, where $Z_t(x)$ is some (possibly) nonlinear, vector valued function of the predetermined variables that are presumed to satisfy

$$\mathscr{E}e_t \otimes z_t = 0 \qquad t = 0, 1, \ldots$$

where, recall (Chapter 5, Section 2),

$$e \otimes z = \begin{pmatrix} e_{1t}z_t \\ e_{2t}z_t \\ \vdots \\ e_{Mt}z_t \end{pmatrix}.$$

More generally, z_t may be any K-vector that has $\mathscr{E}e_t \otimes z_t = 0$, but since a trivial dependence of $q_t(y_t, x_t, \gamma)$ on elements of x_t is permitted, the form $z_t = Z_t(x_t)$ is not very restrictive. Also, z_t may depend on some preliminary estimator $\hat{\tau}_n$ and be of the form

$$\hat{z}_t = Z_t(x_t, \hat{\tau}_n) \qquad \text{with} \qquad \mathscr{E}e_t \otimes Z_t\left(x_t, \tau_n^0\right) = 0$$

or depend on the parameter γ^0 (Hansen, 1982a) with

$$\mathscr{E}e_t \otimes Z_t(x_t, \gamma^0) = 0.$$

The moment equations are

$$m_n(\lambda) = \frac{1}{n} \sum_{t=1}^n m_t(w_t, \hat{\tau}_n, \lambda)$$

with $w_t' = (y_t', x_t')$ and

$$m_t(w_t, \hat{\tau}_n, \lambda) = q_t(y_t, x_t, \lambda) \otimes Z_t(x_t)$$
$$m_t(w_t, \hat{\tau}_n, \lambda) = q_t(y_t, x_t, \lambda) \otimes Z_t(x_t, \hat{\tau}_n)$$

or

$$m_t(w_t, \hat{\tau}_n, \lambda) = q_t(y_t, x_t, \lambda) \otimes Z_t(x_t, \lambda).$$

Hereafter, we shall consider the case $z_t = Z_t(x_t)$ because it occurs most frequently in practice. Our theory covers the other cases, but application is more tedious because the partial derivatives of $m_t(w_t, \tau, \lambda)$ with respect to τ and λ become more complicated.

If $v = M \times K = p$, one can use method of moments in the classical sense by putting sample moments equal to population moments, viz. $m_n(\lambda) = 0$, and solving for λ to get $\hat{\lambda}_n$. But in most applications $M \times K > p$ and the equations cannot be solved. However, one can view the equation

$$m_n(\gamma^0) = \frac{1}{n} \sum_{t=1}^n e_t \otimes z_t$$

as a nonlinear regression with p parameters and $M \times K$ observations, and apply the principle of generalized least squares to estimate γ^0. Let τ_n^0 denote the upper triangle of $[(1/n)\mathscr{E}(\sum_{t=1}^n e_t \otimes z_t)(\sum_{s=1}^n e_s \otimes z_s)']^{-1}$, and put

$$D(\tau_n^0) = \left[\frac{1}{n}\mathscr{E}\left(\sum_{t=1}^n e_t \otimes z_t \right)\left(\sum_{s=1}^n e_s \otimes z_s \right)' \right]^{-1}.$$

Using the generalized least squares heuristic, one estimates γ^0 by $\hat{\lambda}_n$ that minimizes

$$d[m_n(\lambda), \hat{\tau}_n] = \frac{1}{2}m_n'(\lambda)D(\hat{\tau}_n)m_n(\lambda).$$

We shall assume that the estimator $\hat{\tau}_n$ satisfies $\lim_{n \to \infty} \hat{\tau}_n - \tau_n^0 = 0$ almost surely and that $\sqrt{n}(\hat{\tau}_n - \tau_n^0)$ is bounded in probability. The obvious approach to obtain such an estimate is to find the minimum $\hat{\lambda}^{\#}$ of $m'_n(\lambda)[I \otimes (1/n)\Sigma_{t=1}^n z_t z'_t]^{-1} m_n(\lambda)$, and put

$$D(\hat{\tau}_n) = \left[\sum_{\tau=-l(n)}^{l(n)} w\left(\frac{\tau}{l(n)}\right) S_{n\tau}(\hat{\lambda}^{\#}) \right]^{-1}$$

where

$$S_{n\tau}(\lambda) = \begin{cases} \dfrac{1}{n} \sum_{t=1+\tau}^{n} [q_t(y_t, x_t, \lambda) \otimes z_t][q_{t-\tau}(y_{t-\tau}, x_{t-\tau}, \lambda) \otimes z_{t-\tau}]' & \\ & \tau \geq 0 \cdot \\ S'_{n,-\tau}(\lambda) & \\ & \tau < 0 \end{cases}$$

If $e_t \otimes z_t$ and $e_s \otimes z_s$ are uncorrelated for all time gaps $|s - t|$ larger than some l, as in many applications to financial data (Hansen and Singleton, 1982), then we can obtain the conditions $\lim \hat{\tau}_n - \tau_n^0 = 0$ and $\sqrt{n}(\hat{\tau}_n - \tau_n^0)$ bounded in probability using Taylor's expansions and Theorems 1 and 2 with $l(n) \equiv l$ and $w(x) \equiv 1$. But if $e_t \otimes z_t$ and $e_s \otimes z_s$ are correlated for every s, t pair, then this sort of approach will fail for any $l(n)$ with $\lim_{n \to \infty} l(n) = \infty$, because Theorem 3 is not enough to imply the critical result that $\sqrt{n}\,\delta'[D^{-1}(\hat{\tau}_n) - D^{-1}(\tau_n^0)]\delta$ is bounded in probability. But as noted in the discussion preceding Theorem 3, $\delta'D^{-1}(\hat{\tau}_n)\delta$ is an estimate of a spectral density at zero, so that if $\{e_t \otimes z_t\}$ were stationary we should have the critical result with $w(x)$ taken as Parzen weights and $l(n) = [n^{1/5}]$. It is an open question as to whether $\sqrt{n}\,\delta'[D^{-1}(\hat{\tau}_n) - D^{-1}(\tau_n^0)]\delta$ is bounded in probability under the sort of heteroscedasticity permitted by Theorem 2, or if stationarity is essential. □

We call the reader's attention to some heavily used notation and then state the identification condition.

NOTATION 4.

$m_n(\lambda) = (1/n)\Sigma_{t=1}^n m_t(w_t, \hat{\tau}_n, \lambda),$

$m_n^0(\lambda) = (1/n)\Sigma_{t=1}^n \mathscr{E} m_t(W_t, \tau_n^0, \lambda),$

$s_n(\lambda) = d[m_n(\lambda), \hat{\tau}_n],$

$s_n^0(\lambda) = d[m_n^0(\lambda), \tau_n^0],$

$\hat{\lambda}_n$ minimizes $s_n(\lambda),$

$\tilde{\lambda}_n$ minimizes $s_n(\lambda)$ subject to $h(\lambda) = 0,$

λ_n^0 minimizes $s_n^0(\lambda),$

λ_n^* minimizes $s_n^0(\lambda)$ subject to $h(\lambda) = 0.$

ASSUMPTION 8 (Identification). The nuisance parameter estimator $\hat{\tau}_n$ is centered at τ_n^0 in the sense that $\lim_{n \to \infty} \hat{\tau}_n - \tau_n^0 = 0$ almost surely and $\sqrt{n}(\hat{\tau}_n - \tau_n^0)$ is bounded in probability. Either the solution λ_n^0 of the moment equations $m_n^0(\lambda) = 0$ is unique for each n, or there is one solution that can be regarded as being naturally associated to the data generating process. Put $M_n^0 = (\partial/\partial \lambda')m_n^0(\lambda_n^0)$ and $M_n^* = (\partial/\partial \lambda')m_n^0(\lambda_n^*)$; there is an N and constants $c_0 > 0$, $c_1 < \infty$ such that for all δ in $\mathbb{R}^p$ we have

$$c_0^2 \delta'\delta \leq \delta'M_n^{0\prime}M_n^0\delta \leq c_1^2\delta'\delta$$

$$c_0^2 \delta'\delta \leq \delta'M_n^{*\prime}M_n^*\delta \leq c_1^2\delta'\delta.$$

As mentioned in Section 4 of Chapter 3, the assumption that $m_n^0(\lambda_n^0) = 0$ is implausible in misspecified models when the range of $m_n(\lambda)$ is in a higher dimension than the domain. As the case $m_n^0(\lambda_n^0) \neq 0$ is much more complicated than the case $m_n^0(\lambda_n^0) = 0$ and we have no need of it in the body of the text, consideration of it is deferred to Problem 1. The example has $m_n^0(\lambda_n^0) = 0$ with $\lambda_n^0 \equiv \gamma^0$ for all n by construction.

The following notation defines the parameters of the asymptotic distribution of $\hat{\lambda}_n$.

NOTATION 5.

$$\bar{K}_n(\lambda) = \sum_{\tau = -l(n)}^{l(n)} w\left(\frac{\tau}{l(n)}\right) \bar{K}_{n\tau}(\lambda)$$

$$w(x) = \begin{cases} 1 - 6|x|^2 + 6|x|^3 & 0 \leq |x| \leq \frac{1}{2} \\ 2(1 - |x|)^3 & \frac{1}{2} \leq |x| \leq 1 \end{cases}$$

$l(n) =$ the integer nearest $n^{1/5}$

$$\bar{K}_{n\tau}(\lambda) = \begin{cases} \dfrac{1}{n} \sum_{t=1+\tau}^{n} [\mathscr{E}m_t(W_t, \tau_n^0, \lambda)][\mathscr{E}m_{t-\tau}(W_{t-\tau}, \tau_n^0, \lambda)]' & \tau \geq 0 \\ K'_{n,-\tau}(\lambda) & \tau < 0 \end{cases}$$

$$\bar{S}_n(\lambda) = \sum_{\tau = -(n-1)}^{(n-1)} \bar{S}_{n\tau}(\lambda)$$

$$\bar{S}_{n\tau}(\lambda) = \begin{cases} \dfrac{1}{n} \sum_{t=1+\tau}^{n} \mathscr{E}m_t(W_t, \tau_n^0, \lambda)m'_{t-\tau}(W_{t-\tau}, \tau_n^0, \lambda) - \bar{K}_{n\tau}(\lambda) & \tau \geq 0 \\ \bar{S}'_{n,-\tau}(\lambda) & \tau < 0 \end{cases}$$

$$\overline{M}_n(\lambda) = \frac{1}{n} \sum_{t=1}^{n} \mathscr{E} \frac{\partial}{\partial \lambda'} m_t(W_t, \tau_n^0, \lambda)$$

$$\overline{D}_n(\lambda) = \frac{\partial^2}{\partial m \, \partial m'} d\left[m_n^0(\lambda), \tau_n^0\right]$$

$$\overline{\mathscr{I}}_n(\lambda) = \overline{M}_n'(\lambda) \overline{D}_n(\lambda) \overline{S}_n(\lambda) \overline{D}_n(\lambda) \overline{M}_n(\lambda)$$

$$\overline{\mathscr{J}}_n(\lambda) = \overline{M}_n'(\lambda) \overline{D}_n(\lambda) \overline{M}_n(\lambda)$$

$$\overline{\mathscr{U}}_n(\lambda) = \overline{M}_n'(\lambda) \overline{D}_n(\lambda) \overline{K}_n(\lambda) \overline{D}_n(\lambda) \overline{M}_n(\lambda)$$

$$\mathscr{I}_n^0 = \overline{\mathscr{I}}_n(\lambda_n^0) \quad \mathscr{J}_n^0 = \overline{\mathscr{J}}_n(\lambda_n^0) \quad \mathscr{U}_n^0 = \overline{\mathscr{U}}_n(\lambda_n^0) \quad S_n^0 = \overline{S}(\lambda_n^0)$$

$$\mathscr{I}_n^* = \overline{\mathscr{I}}_n(\lambda_n^*) \quad \mathscr{J}_n^* = \overline{\mathscr{J}}_n(\lambda_n^*) \quad \mathscr{U}_n^* = \overline{\mathscr{U}}_n(\lambda_n^*) \quad S_n^* = S(\lambda_n^*).$$

We shall illustrate the computations with the example.

EXAMPLE 2 (Continued). Recall that the data follow the model

$$q_t(y_t, x_t, \gamma^0) = e_t \qquad t = 1, 2, \ldots, n$$

with

$$m_t(w_t, \lambda) = q_t(y_t, x_t, \lambda) \otimes Z(x_t)$$
$$= q_t(y_t, x_t, \lambda) \otimes z_t$$

and

$$m_n(\lambda) = \frac{1}{n} \sum_{t=1}^{n} m_t(w_t, \lambda).$$

Since

$$m_n^0(\gamma^0) = \mathscr{E} \frac{1}{n} \sum_{t=1}^{n} e_t \otimes z_t = 0$$

we have $\lambda_n^0 = \gamma^0$ for all n, and since, for each t, $\mathscr{E} m_t(w_t, \lambda_n^0) = \mathscr{E} e_t \otimes z_t = 0$, we have $K_{n\tau}^0 = 0$. Further,

$$S_{n\tau}(\lambda_n^0) = \begin{cases} \dfrac{1}{n} \sum_{t=1+\tau}^{n} \mathscr{E} e_t e_{t-\tau}' \otimes z_t z_{t-\tau}' & \tau \geq 0 \\[2mm] S_{n, -\tau}(\lambda_n^0) & \tau < 0 \end{cases}$$

$$S_n^0 = \sum_{\tau = -(n-1)}^{n-1} S_{n\tau}(\lambda_n^0).$$

We have

$$
\begin{aligned}
\overline{M}_n(\lambda) &= \frac{1}{n} \sum_{t=1}^{n} \mathscr{E} \frac{\partial}{\partial \lambda} m(W_t, \lambda) \\
&= \frac{1}{n} \sum_{t=1}^{n} \mathscr{E} \frac{\partial}{\partial \lambda'} [q_t(y_t, x_t, \lambda) \otimes z_t] \\
&= \frac{1}{n} \sum_{t=1}^{n} \mathscr{E} \left[\left(\frac{\partial}{\partial \lambda'} q_t(y_t, x_t, \lambda) \right) \otimes z_t \right] \\
&= \frac{1}{n} \sum_{t=1}^{n} \mathscr{E} [Q_t(\lambda) \otimes z_t].
\end{aligned}
$$

Recall that

$$
d(m, \tau_n^0) = \tfrac{1}{2} m' D(\tau_n^0) m
$$

with

$$
\begin{aligned}
D(\tau_n^0) &= \left[\frac{1}{n} \mathscr{E} \left(\sum_{t=1}^{n} e_t \otimes z_t \right) \left(\sum_{s=1}^{n} e_s \otimes z_s \right)' \right]^{-1} \\
&= (S_n^0)^{-1}.
\end{aligned}
$$

Thus,

$$
\overline{D}_n(\lambda) = \frac{\partial^2}{\partial m \, \partial m'} \tfrac{1}{2} m'(S_n^0)^{-1} m \bigg|_{m = m_n(\lambda)} = (S_n^0)^{-1}
$$

and

$$
\begin{aligned}
\mathscr{I}_n^0 &= \left(\frac{1}{n} \sum_{t=1}^{n} \mathscr{E} Q_t(\lambda_n^0) \otimes z_t \right)' (S_n^0)^{-1} \left(\frac{1}{n} \sum_{t=1}^{n} \mathscr{E} Q_t(\lambda_n^0) \otimes z_t \right) \\
&= \mathscr{I}_n^0.
\end{aligned}
$$

An important special case is the one where the x_t are taken as fixed (random variables with zero variance) and the errors $\{e_t\}$ are taken as independently and identically distributed with $\mathscr{E} e_t e_t' = \Sigma$. In this case

$$
S_n^0 = \Sigma \otimes \frac{1}{n} \sum_{t=1}^{n} z_t z_t'
$$

and

$$
\mathcal{I}_n^0 = \left(\frac{1}{n} \sum_{t=1}^{n} \mathcal{E} Q_t(\lambda_n^0) \otimes z_t \right)'
$$

$$
\times \left(\Sigma \otimes \frac{1}{n} \sum_{t=1}^{n} z_t z_t' \right)^{-1} \left(\frac{1}{n} \sum_{t=1}^{n} \mathcal{E} Q_t(\lambda_n^0) \otimes z_t \right)
$$

$$
= \mathcal{J}_n^0. \qquad \qquad \square
$$

General purpose estimators of $(\mathcal{I}_n^0, \mathcal{J}_n^0)$ and $(\mathcal{I}_n^*, \mathcal{J}_n^*)$, denoted $(\hat{\mathcal{I}}, \hat{\mathcal{J}})$ and $(\tilde{\mathcal{I}}, \tilde{\mathcal{J}})$ respectively, may be defined as follows.

NOTATION 6.

$$
S_n(\lambda) = \sum_{\tau=-l(n)}^{l(n)} w\left(\frac{\tau}{l(n)} \right) S_{n\tau}(\lambda)
$$

$$
S_n(\lambda) = \begin{cases} \dfrac{1}{n} \sum_{t=1+\tau}^{n} m_t(w_t, \hat{\tau}_n, \lambda) m'_{t-\tau}(w_{t-\tau}, \hat{\tau}_n, \lambda) & \tau \ge 0 \\[2mm] S'_{n,-\tau}(\lambda) & \tau < 0 \end{cases}
$$

$$
w(x) = \begin{cases} 1 - 6|x|^2 + 6|x|^3 & 0 \le x \le \frac{1}{2} \\[2mm] 2(1 - |x|)^3 & \frac{1}{2} \le x \le 1 \end{cases}
$$

$l(n) = $ the integer nearest $n^{1/5}$

$$
M_n(\lambda) = \frac{1}{n} \sum_{t=1}^{n} \frac{\partial}{\partial \lambda'} m_t(w_t, \hat{\tau}_n, \lambda)
$$

$$
D_n(\lambda) = \frac{\partial^2}{\partial m \, \partial m'} d\left[m_n(\lambda), \hat{\tau}_n \right]
$$

$$
\mathcal{I}_n(\lambda) = M_n'(\lambda) D_n(\lambda) S_n(\lambda) D_n(\lambda) M_n(\lambda)
$$

$$
\mathcal{J}_n(\lambda) = M_n'(\lambda) D_n(\lambda) M_n(\lambda)
$$

$$
\hat{\mathcal{I}} = \mathcal{I}_n(\hat{\lambda}_n) \qquad \hat{\mathcal{J}} = \mathcal{J}_n(\hat{\lambda}_n)
$$

$$
\tilde{\mathcal{I}} = \mathcal{I}_n(\tilde{\lambda}_n) \qquad \tilde{\mathcal{J}} = \mathcal{J}_n(\tilde{\lambda}_n).
$$

For the example (generalized method of moments) one is presumed to have an estimate $D(\hat{\tau}_n)$ of $(S_n^0)^{-1}$ available in advance of the computations. In applications, it is customary to reuse this estimate to obtain an estimate of $\mathcal{I}_n^0$ rather than try to estimate S_n^0 afresh. We illustrate:

EXAMPLE 2 (Continued). Recall that by assumption $\lim_{n \to \infty} D(\hat{\tau}_n) - (S_n^0)^{-1} = 0$ almost surely and $\sqrt{n}\,[D(\hat{\tau}_n) - (S_n^0)^{-1}]$ is bounded in probability. Thus, for the case

$$m_t(w_t, \lambda) = q_t(y_t, x_t, \lambda) \otimes z_t$$

we have

$$\hat{\mathscr{J}} = \left(\frac{1}{n} \sum_{t=1}^n Q_t(\hat{\lambda}_n) \otimes z_t\right)' D(\hat{\tau}_n)\left(\frac{1}{n} \sum_{t=1}^n Q_t(\hat{\lambda}_n) \otimes z_t\right)$$

$$= \hat{\mathscr{J}}$$

where, recall,

$$Q_t(\lambda) = \frac{\partial}{\partial \lambda'} q_t(y_t, x_t, \lambda).$$

In the special case where $\{e_t\}$ is a sequence of independent and identically distributed random variables with $\mathscr{E}e_t e_t' = \Sigma$ and z_t taken as fixed, we have

$$D(\hat{\tau}_n) = \left(\hat{\Sigma} \otimes \frac{1}{n} \sum_{t=1}^n z_t z_t'\right)^{-1}$$

and

$$\hat{\mathscr{J}} = \left(\frac{1}{n} \sum_{t=1}^n Q_t(\hat{\lambda}_n) \otimes z_t\right)'$$

$$\times \left(\hat{\Sigma} \otimes \frac{1}{n} \sum_{t=1}^n z_t z_t'\right)^{-1}\left(\frac{1}{n} \sum_{t=1}^n Q_t(\hat{\lambda}_n) \otimes z_t\right)$$

$$= \hat{\mathscr{J}}. \qquad \qquad \square$$

The following conditions permit application of the uniform strong law for dependent observations to the moment equations, the Jacobian, and the Hessian of the moment equations.

ASSUMPTION 9. The sequences $\{\hat{\tau}_n\}$ and $\{\tau_n^0\}$ are contained in Υ, which is a closed ball with finite, nonzero radius. The sequence $\{\lambda_n^0\}$ is contained in Λ^*, which is a closed ball with finite, nonzero radius. Let

$g_t(W_t, \tau, \lambda)$ be a generic term that denotes, variously,

$$m_{\alpha t}(W_t, \tau, \lambda) \qquad \frac{\partial}{\partial \lambda_i} m_{\alpha t}(W_t, \tau, \lambda) \qquad \frac{\partial^2}{\partial \lambda_i \, \partial \lambda_j} m_{\alpha t}(W_t, \tau, \lambda)$$

$$\frac{\partial}{\partial \tau_l} m_{\alpha t}(W_t, \tau, \lambda) \quad \text{or} \quad \frac{\partial}{\partial \lambda_i} m_{\alpha t}(W_t, \tau, \lambda) \frac{\partial}{\partial \lambda_j} m_{\beta t}(W_t, \tau, \lambda)$$

for $i, j = 1, 2, \ldots, p$, $l = 1, 2, \ldots, u$, and $\alpha = 1, 2, \ldots, v$. On $\Upsilon \times \Lambda^*$, the family $\{ g_t[W_t(\omega), \tau, \lambda] \}$ is near epoch dependent of size $-q$ with $q = 2(r - 2)/(r - 4)$, where r is that of Assumption 3, $g_t[W_t(\omega), \tau, \lambda]$ is A-smooth in (τ, λ) and there is a sequence of random variables $\{ d_t \}$ with $\sup_{\Upsilon \times \Lambda^*} g_t[W_t(\omega), \tau, \lambda] \leq d_t(\omega)$ and $\|d_t\|_r \leq \Delta < \infty$ for all t.

Observe that the domination condition in Assumption 9 guarantees that $m_n^0(\lambda)$ takes its range in some compact ball, because

$$\max_\alpha \sup_n \sup_{\lambda \in \Lambda^*} \left| m_{\alpha n}^0(\lambda) \right| \leq \sup_n \frac{1}{n} \sum_{t=1}^n \mathscr{E} d_t$$

$$\leq 1 + \|d_t\|_r^r < \infty.$$

We shall need to restrict the behavior of the distance function $d(m, \tau)$ on a slightly larger ball $\mathscr{M}$. The only distance functions used in the text are quadratic:

$$d(m, \tau) = m' D(\tau) m$$

with $D(\tau)$ continuous and positive definite on Υ. Thus, we shall abstract minimally beyond the properties of quadratic functions. See Problem 1 for the more general case.

ASSUMPTION 10. Let $\mathscr{M}$ be a closed ball with a concentric ball of smaller radius that contains $\bigcup_{n=1}^\infty \{ m = m_n^0(\lambda) : \lambda \in \Lambda^* \}$. The distance function $d(m, \tau)$ and derivatives $(\partial/\partial m)d(m, \tau)$, $(\partial^2/\partial m \, \partial m')d(m, \tau)$, $(\partial^2/\partial m \, \partial \tau')d(m, \tau)$ are continuous on $\mathscr{M} \times \Upsilon$. Moreover, $(\partial/\partial m)d(0, t) = 0$ for all τ in Υ [which implies $(\partial^2/\partial m \, \partial \tau')d(0, \tau) = 0$ for all τ in Υ], and $(\partial^2/\partial m \, \partial m')d(m, \tau)$ is positive definite over $\mathscr{M} \times \Upsilon$.

Before proving consistency, we shall collect together a number of facts needed throughout this section as a lemma:

LEMMA 13. Under Assumptions 1 through 3 and 8 through 10, interchange of differentiation and integration is permitted in these instances:

$$\frac{\partial}{\partial \lambda_i} m^0_{\alpha n}(\lambda) = \frac{1}{n} \sum_{t=1}^{n} \mathscr{E} \frac{\partial}{\partial \lambda_i} m_{\alpha t}(W_t, \tau^0_n, \lambda)$$

$$\frac{\partial^2}{\partial \lambda_i \partial \lambda_j} m^0_{\alpha n}(\lambda) = \frac{1}{n} \sum_{t=1}^{n} \mathscr{E} \frac{\partial^2}{\partial \lambda_i \partial \lambda_j} m_{\alpha t}(W_t, \tau^0_n, \lambda).$$

Moreover,

$$\lim_{n \to \infty} \sup_{\Lambda^*} \left| m_{\alpha n}(\lambda) - m^0_{\alpha n}(\lambda) \right| = 0 \qquad \text{almost surely}$$

$$\lim_{n \to \infty} \sup_{\Lambda^*} \left| \frac{\partial}{\partial \lambda_i} [m_{\alpha n}(\lambda) - m^0_{\alpha n}(\lambda)] \right| = 0 \qquad \text{almost surely}$$

$$\lim_{n \to \infty} \sup_{\Lambda^*} \left| \frac{\partial^2}{\partial \lambda_i \partial \lambda_j} [m_{\alpha n}(\lambda) - m^0_{n\alpha}(\lambda)] \right| = 0 \qquad \text{almost surely}$$

$$\lim_{n \to \infty} \sup_{\Lambda^*} \left| s_n(\lambda) - s^0_n(\lambda) \right| = 0 \qquad \text{almost surely}$$

$$\lim_{n \to \infty} \sup_{\Lambda^*} \left| \frac{\partial}{\partial \lambda_i} [s_n(\lambda) - s^0_n(\lambda)] \right| = 0 \qquad \text{almost surely}$$

$$\lim_{n \to \infty} \sup_{\Lambda^*} \left| \frac{\partial^2}{\partial \lambda_i \partial \lambda_j} [s_n(\lambda) - s^0_n(\lambda)] \right| = 0 \qquad \text{almost surely}$$

and the families $\{m^0_{\alpha n}(\lambda)\}$, $\{(\partial/\partial \lambda_i) m^0_{\alpha n}(\lambda)\}$, $\{(\partial^2/\partial \lambda_i \partial \lambda_j) m^0_{\alpha n}(\lambda)\}$, $\{s^0_n(\lambda)\}$, $\{(\partial/\partial \lambda_i) s^0_n(\lambda)\}$, and $\{(\partial^2/\partial \lambda_i \partial \lambda_j) s^0_n(\lambda)\}$ are equicontinuous; indices range over $i, j = 1, 2, \ldots, p$, $\alpha = 1, 2, \ldots, v$, and $n = 1, 2, \ldots, \infty$ in the above.

Proof. The proof for the claims involving $m_n(\lambda)$ and $m^0_n(\lambda)$ is the same as the proof of Lemma 10.

For $m_n(\lambda)$ in $\mathscr{M}$ we have

$$s_n(\lambda) = d[m_n(\lambda), \hat{\tau}_n]$$

$$\frac{\partial}{\partial \lambda_i} s_n(\lambda) = \sum_{\alpha} \frac{\partial}{\partial m_\alpha} d[m_n(\lambda), \hat{\tau}_n] \frac{\partial}{\partial \lambda_i} m_{\alpha n}(\lambda)$$

and

$$\frac{\partial^2}{\partial\lambda_i\,\partial\lambda_j} = \sum_\alpha\sum_\beta \frac{\partial^2}{\partial m_\alpha\,\partial m_\beta} d\left[m_n(\lambda),\hat{\tau}_n\right]\frac{\partial}{\partial\lambda_i}m_{\alpha n}(\lambda)\frac{\partial}{\partial\lambda_j}m_{\beta n}(\lambda)$$

$$+\sum_\alpha \frac{\partial}{\partial m_\alpha} d\left[m_n(\lambda),\hat{\tau}_n\right]\frac{\partial^2}{\partial\lambda_i\,\partial\lambda_j}m_{\alpha n}(\lambda).$$

Consider the second equation. A continuous function on a compact set is uniformly continuous; thus $(\partial/\partial m_\alpha)d(m,\tau)$ is uniformly continuous on $\mathcal{M}\times\mathfrak{T}$. Given $\epsilon>0$, choose δ small enough that $|m-m^0|<\delta$ and $|\hat{\tau}-\tau|<\delta$ imply $|(\partial/\partial m_\alpha)[d(m,\hat{\tau})-d(m^0,\tau^0)]|<\epsilon$. Fix a realization of $\{V_t\}_{t=1}^\infty$ for which $\lim_{n\to\infty}|\hat{\tau}_n-\tau_n^0|=0$ and $\lim_{n\to\infty}\sup_{\Lambda*}|m_n(\lambda)-m_n^0(\lambda)|=0$; almost every realization is such, by Assumption 9 and Theorem 1. Choose N large enough that $n>N$ implies $\sup_{\Lambda*}|m_n(\lambda)-m_n^0(\lambda)|<\delta$ and $|\hat{\tau}_n-\tau_n^0|<\delta$. This implies uniform convergence, since we have $\sup_{\Lambda*}|(\partial/\partial m_\alpha)\{d[m_n(\lambda),\hat{\tau}_n]-d[m_n^0(\lambda),\tau_n^0]\}|<\epsilon$ for $n>N$. By equicontinuity, we can choose η such that $|\lambda-\lambda^0|<\eta$ implies $|m_n^0(\lambda)-m_n^0(\lambda^0)|<\delta$. For $|\lambda-\lambda^0|<\eta$ we have

$$\sup_n\left|\frac{\partial}{\partial m_\alpha}\left\{d\left[m_n^0(\lambda),\tau_n^0\right]-d\left[m_n^0(\lambda_n^0),\tau_n^0\right]\right\}\right|<\epsilon$$

which implies that $\{(\partial/\partial m_\alpha)d_n[m_n^0(\lambda),\tau_n^0]\}$ is an equicontinuous family.

As $(\partial/\partial\lambda_i)s_n(\lambda)$ is a sum of products of uniformly convergent, equicontinuous functions, it has the same properties.

The argument for $s_n(\lambda)$ and $(\partial^2/\partial\lambda_\alpha\,\partial\lambda_\beta)s_n(\lambda)$ is the same. $\qquad\square$

As we have noted earlier, in many applications it is implausible to assume that $m_n^0(\lambda)$ has only one root over Λ^*. Thus, the best consistency result that we can show is that $s_n(\lambda)$ will eventually have a local minimum near λ_n^0 and that all other local minima of $s_n(\lambda)$ must be some fixed distance δ away from λ_n^0, where δ does not depend on λ_n^0 itself. Hereafter, we shall take $\hat{\lambda}_n$ to mean the root given by Theorem 8.

THEOREM 8 (Existence of consistent local minima). Let Assumptions 1 though 3 and 8 through 10 hold. Then there is a $\delta>0$ such that the value of $\hat{\lambda}_n$ which minimizes $s_n(\lambda)$ over $|\lambda-\lambda_n^0|\le\delta$ satisfies

$$\lim_{n\to\infty}\left(\hat{\lambda}_n-\lambda_n^0\right)=0$$

almost surely.

Proof. By Lemma 13 the family $\{\overline{M}_n(\lambda) = (\partial/\partial\lambda')m_n^0(\lambda)\}$ is equicontinuous over Λ^*. Then there is a δ small enough that $|\overline{\lambda} - \lambda_n^0| \le \delta$ implies

$$(\lambda - \lambda_n^0)'\big[\overline{M}_n'(\overline{\lambda})\,\overline{M}_n(\overline{\lambda}) - M_n^{0\prime}M_n^0\big](\lambda - \lambda_n^0) > -\frac{c_0^2}{2}|\lambda - \lambda_n^0|^2$$

where c_0^2 is the eigenvalue defined in Assumption 8. Let ν_0 be the smallest eigenvalue of $(\partial^2/\partial m\,\partial m')d(m,\tau)$ over $\mathcal{M} \times \mathfrak{T}$; it is positive by Assumption 10 and continuity over a compact set. Recalling that $m_n^0(\lambda_n^0) = 0$, $d(0,\tau) = 0$, and $(\partial/\partial m)d(0,\tau) = 0$, we have by Taylor's theorem that for N given by Assumption 8

$$\inf_{n>N}\ \inf_{\epsilon \le |\lambda - \lambda_n^0| \le \delta}\ \big|s_n^0(\lambda) - s_n^0(\lambda_n^0)\big|$$

$$= \inf_{n>N}\ \inf_{\epsilon \le |\lambda - \lambda_n^0| \le \delta}\ m_n^{0\prime}(\lambda)\Big(\frac{\partial^2}{\partial m\,\partial m'}d(\overline{m},\tau_n^0)\Big)m_n^0(\lambda)$$

$$\ge \nu_0 \inf_{n>N}\ \inf_{\epsilon \le |\lambda - \lambda_n^0| \le \delta}\ m_n^{0\prime}(\lambda)m_n^0(\lambda)$$

$$= \nu_0 \inf_{n>N}\ \inf_{\epsilon \le |\lambda - \lambda_n^0| \le \delta}\ (\lambda - \lambda_n^0)'\overline{M}_n'(\overline{\lambda})\,\overline{M}_n(\overline{\lambda})(\lambda - \lambda_n^0)$$

$$\ge \nu_0 \inf_{n>N}\ \inf_{\epsilon \le |\lambda - \lambda_n^0| \le \delta}\ (\lambda - \lambda_n^0)'M_n^{0\prime}(\lambda_n^0)M_n^0(\lambda_n^0)(\lambda - \lambda_n^0)$$

$$-\nu_0\frac{c_0^2}{2}|\lambda - \lambda_n^0|^2$$

$$\ge \nu_0\frac{c_0^2}{2}\epsilon^2$$

where $\overline{m}$ is on the line segment joining the zero vector to m, and $\overline{\lambda}$ is on the line segment joining λ to λ^0.

Fix ω not in the exceptional set given by Lemma 13. Choose N' large enough that $n > N'$ implies that $\sup_{\Lambda^*}|s_n(\lambda) - s_n^0(\lambda)| < \nu_0 c_0^2 \epsilon^2/4$ for all $n > N'$. Since $s_n(\hat\lambda_n) \le s_n(\lambda_n^0)$, we have for all $n > N'$ that

$$s_n^0(\hat\lambda_n) - \frac{\nu_0 c_0^2 \epsilon^2}{4} \le s_n(\hat\lambda_n) \le s_n(\lambda_n^0) \le s_n^0(\lambda_n^0) + \frac{\nu_0 c_0^2 \epsilon^2}{4}$$

or $0 < s_n^0(\hat\lambda_n) - s_n^0(\lambda_n^0) < \nu_0 c_0^2 \epsilon^2/2$. Then for all $n > \max(N, N')$ we must have $|\hat\lambda_n - \lambda_n^0| < \epsilon$. $\qquad\square$

We append some additional conditions needed to prove the asymptotic normality of the score function $(\partial/\partial\lambda)s_n(\lambda_n^0)$.

ASSUMPTION 11. The points $\{\lambda_n^0\}$ are contained in a closed ball Λ that is concentric with Λ^* but with smaller radius. There is an N and constants $c_0 > 0$, $c_1 < \infty$ such that for δ in $\mathbb{R}^p$ we have

$$c_0 \delta'\delta \le \delta' \bar{\mathscr{I}}_n(\lambda)\delta \le c_1\delta'\delta \qquad \text{all } n > N \quad \text{all } \lambda \in \Lambda$$

$$c_0 \delta'\delta \le \delta' \mathscr{I}_n^0 \delta \le c_1 \delta'\delta \qquad \text{all } n > N$$

$$c_0 \delta'\delta \le \delta' \mathscr{I}_n^* \delta \le c_1 \delta'\delta \qquad \text{all } n > N$$

$$\lim_{n \to \infty} \delta'\left(S_n^0\right)^{-1/2} S_{[ns]}^0 \left(S_n^0\right)^{-1/2'} \delta = \delta'\delta \qquad \text{all } 0 < s \le 1$$

$$\lim_{n \to \infty} \delta'\left(S_n^*\right)^{-1/2} S_{[ns]}^* \left(S_n^*\right)^{-1/2'} \delta = \delta'\delta \qquad \text{all } 0 < s \le 1.$$

Also,

$$\lim_{n \to \infty} \frac{1}{n} \sum_{t=1}^{n} \mathscr{E} \frac{\partial}{\partial \tau'} m_t\left(W_t, \tau_n^0, \lambda_n^0\right) = 0.$$

THEOREM 9 (Asymptotic normality of the scores). Under Assumptions 1 through 3 and 8 through 11,

$$\sqrt{n}\left(\mathscr{I}_n^0\right)^{-1/2} \frac{\partial}{\partial \lambda} s_n\left(\lambda_n^0\right) \xrightarrow{\mathscr{L}} N_p(0, I)$$

$$\lim_{n \to \infty} \left(\mathscr{I}_n^0 + \mathscr{U}_n^0 - \hat{\mathscr{I}}\right) = 0 \qquad \text{in probability.}$$

Proof. By the same argument used to prove Theorem 5 we have

$$\sqrt{n}\left(S_n^0\right)^{-1/2}\left[m_n\left(\lambda_n^0\right) - m_n^0\left(\lambda_n^0\right)\right] \xrightarrow{\mathscr{L}} N_v(0, I)$$

$$\lim_{n \to \infty} \left[S_n^0 + K_n^0 - S_n(\hat{\lambda}_n)\right] = 0 \qquad \text{almost surely.}$$

A typical element of the vector $\sqrt{n}\,(\partial/\partial m)d[m_n(\lambda_n^0), \hat{\tau}_n]$ can be expanded about $[m_n^0(\lambda_n^0), \tau_n^0]$ to obtain

$$\sqrt{n}\,\frac{\partial}{\partial m_\alpha}d\left[m_n\left(\lambda_n^0\right), \hat{\tau}_n\right] = \sqrt{n}\,\frac{\partial}{\partial m_\alpha}d\left[m_n^0\left(\lambda_n^0\right), \tau_n^0\right]$$

$$+ \frac{\partial}{\partial \tau'}\frac{\partial}{\partial m_\alpha}d(\bar{m}, \bar{\tau})\sqrt{n}\left(\hat{\tau}_n - \tau_n^0\right)$$

$$+ \frac{\partial}{\partial m'}\frac{\partial}{\partial m_\alpha}d(\bar{m}, \bar{\tau})\sqrt{n}\left[m_n\left(\lambda_n^0\right) - m_n^0\left(\lambda_n^0\right)\right]$$

where $(\bar{m}, \bar{\tau})$ is on the line segment joining $[m_n(\lambda_n^0), \hat{\tau}_n]$ to $[m_n^0(\lambda_n^0), \tau_n^0]$. We have that $\sqrt{n}[m_n(\lambda_n^0) - m_n^0(\lambda_n^0)]$ converges in distribution and so is bounded in probability; we have assumed that $\sqrt{n}(\hat{\tau}_n - \tau_n^0)$ is bounded in probability. Then using the uniform convergence of $m_n(\lambda) - m_n^0(\lambda)$ to zero given by Lemma 13, the convergence of $\hat{\tau}_n - \tau_n^0$ to zero, and the continuity of $d(m, \tau)$ and its derivatives, we can write

$$\sqrt{n}\,\frac{\partial}{\partial m}d\big[m_n(\lambda_n^0), \hat{\tau}_n\big]$$

$$= \sqrt{n}\,\frac{\partial}{\partial m}d\big[m_n^0(\lambda_n^0), \tau_n^0\big]$$

$$+ \frac{\partial^2}{\partial m\,\partial\tau'}d\big[m_n^0(\lambda_n^0), \tau_n^0\big]\sqrt{n}\,(\hat{\tau}_n - \tau_n^0)$$

$$+ \frac{\partial^2}{\partial m\,\partial m'}d\big[m_n^0(\lambda_n^0), \tau_n^0\big]\sqrt{n}\,\big[m_n(\lambda_n^0) - m_n^0(\lambda_n^0)\big]$$

$$+ o_p(1).$$

Since λ_n^0 is an interior point of Λ^* by Assumption 11, we have $\sqrt{n}\,(\partial/\partial\lambda)s_n^0(\lambda_n^0) = O(1)$, whence

$$\sqrt{n}\,\frac{\partial}{\partial\lambda}s_n(\lambda_n^0) = \sqrt{n}\,\frac{\partial}{\partial\lambda}s_n(\lambda_n^0) - \sqrt{n}\,\frac{\partial}{\partial\lambda}s_n^0(\lambda_n^0) + o(1)$$

$$= \sqrt{n}\,M_n'(\lambda_n^0)\frac{\partial}{\partial m}d\big[m_n(\lambda_n^0), \hat{\tau}_n\big]$$

$$- \sqrt{n}\,\overline{M}_n'(\lambda_n^0)\frac{\partial}{\partial m}d\big[m_n^0(\lambda_n^0), \tau_n^0\big] + o(1)$$

$$= \sqrt{n}\,\big[M_n'(\lambda_n^0) - \overline{M}_n'(\lambda_n^0)\big]\frac{\partial}{\partial m}d\big[m_n^0(\lambda_n^0), \tau_n^0\big]$$

$$+ M_n'(\lambda_n^0)\bigg(\frac{\partial^2}{\partial m\,\partial\tau'}d\big[m_n^0(\lambda_n^0), \tau_n^0\big]\bigg)\sqrt{n}\,(\hat{\tau}_n - \tau_n^0)$$

$$+ M_n'(\lambda_n^0)\bigg(\frac{\partial^2}{\partial m\,\partial m'}d\big[m_n^0(\lambda_n^0), \tau_n^0\big]\bigg)$$

$$\times\sqrt{n}\,\big[m_n(\lambda_n^0) - m_n^0(\lambda_n^0)\big] + o_p(1).$$

We have assumed that $m_n^0(\lambda_n^0) = 0$, whence $(\partial/\partial m)d[m_n^0(\lambda_n^0), \tau_n^0] = 0$ and $(\partial^2/\partial m\,\partial\tau')d[m_n^0(\lambda_n^0), \tau_n^0] = 0$, and this equation simplies to

$$\sqrt{n}\,\frac{\partial}{\partial\lambda}s_n(\lambda_n^0)$$

$$= M_n'(\lambda_n^0)\bigg(\frac{\partial^2}{\partial m\,\partial m'}d\big[m_n^0(\lambda_n^0), \tau_n^0\big]\bigg)\sqrt{n}\,\big[m_n(\lambda_n^0) - m_n^0(\lambda_n^0)\big] + o_p(1).$$

In general this simplification will not obtain, and the asymptotic distribution of $\sqrt{n}\,(\partial/\partial\lambda)s_n(\lambda_n^0)$ will be more complicated than the distribution that we shall obtain here (Problem 1).

Now

$$\left(\mathscr{I}_n^0\right)^{-1/2} = \left(S_n^0\right)^{-1/2}\left(D_n^0\right)^{-1}\left(M_n^{0\prime}\right)^{-1}.$$

Assumptions 8, 10, and 11 assure the existence of the various inverses and the existence of a uniform (in n) bound on their elements. Then

$$\sqrt{n}\left(\mathscr{I}_n^0\right)^{-1/2}\frac{\partial}{\partial\lambda}s_n(\lambda_n^0) = \sqrt{n}\left(S_n^0\right)^{-1/2}\left[m_n(\lambda_n^0) - m_n^0(\lambda_n^0)\right] + o_p(1).$$

and the first result obtains. Lemma 13 and Theorem 8 guarantee that $\lim_{n\to\infty}(M_n^0 - \hat{M}_n) = 0$ almost surely, Assumption 10 and Theorem 8 guarantee that $\lim_{n\to\infty}(D_n^0 - \hat{D}_n) = 0$ almost surely, and we have already that $\lim_{n\to\infty}(S_n^0 + K_n^0 - \hat{S}_n) = 0$ almost surely, whence the second result obtains. $\qquad\square$

Asymptotic normality of the unconstrained estimator follows at once.

THEOREM 10 (Asymptotic normality). Let Assumptions 1 through 3 and 8 through 11 hold. Then

$$\sqrt{n}\left(\mathscr{I}_n^0\right)^{-1/2}\mathscr{I}_n^0\left(\hat{\lambda}_n - \lambda_n^0\right) \xrightarrow{\mathscr{L}} N_p(0, I)$$

$$\lim_{n\to\infty}\left(\mathscr{I}_n^0 - \hat{\mathscr{I}}\right) = 0 \qquad \text{almost surely.}$$

Proof. The proof is much the same as the proof of Theorem 6. $\qquad\square$

Next we establish some ancillary facts regarding the constrained estimator subject to a Pitman drift for use in the next section.

ASSUMPTION 12 (Pitman drift). The function $h(\lambda)$ that defines the null hypothesis $H: h(\lambda_n^0) = h_n^*$ is a twice continuously differentiable mapping of Λ as defined by Assumption 11 into $\mathbb{R}^q$ with Jacobian denoted as $H(\lambda) = (\partial/\partial\lambda')h(\lambda)$. The eigenvalues of $H(\lambda)H'(\lambda)$ are bounded below over Λ by $c_0^2 > 0$, and above by $c_1^2 < \infty$. In the case where $p = q$, $h(\lambda)$ is assumed to be a one to one mapping with a continuous inverse. In the case $p < q$, there is a continuous function $\phi(\lambda)$ such that the mapping

$$\begin{pmatrix} \rho \\ \tau \end{pmatrix} = \begin{pmatrix} \phi(\lambda) \\ h(\lambda) \end{pmatrix}$$

has a continuous inverse

$$\lambda = \psi(\rho, \tau)$$

defined over $S = \{(\rho, \tau): \rho = \phi(\lambda), \tau = h(\lambda), \lambda \in \Lambda\}$. Moreover, $\psi(\rho, \tau)$ has a continuous extension to the set

$$R \times T = \{\rho: \rho = \phi(\lambda)\} \times \{\tau: \tau = h(\lambda)\}.$$

The sequence $\{h_n^*\}$ is chosen such that

$$\sqrt{n}\left[h(\lambda_n^0) - h_n^*\right] = O(1).$$

There is an N such that for all δ in $\mathbb{R}^p$

$$\delta' \mathcal{U}_n^0 \delta \leq c_1 \delta' \delta \qquad \text{all} \quad n > N$$

$$\delta' \mathcal{U}_n^* \delta \leq c_1 \delta' \delta \qquad \text{all} \quad n > N.$$

THEOREM 11. Let Assumptions 1 through 3 and 8 through 12 hold. Then there is a $\delta > 0$ such that the value of λ_n which minimizes $s_n(\lambda)$ over $|\lambda - \lambda_n^*| < \delta$ subject to $h(\lambda) = h_n^*$ satisfies

$$\lim_{n \to \infty} \left(\tilde{\lambda}_n - \lambda_n^*\right) = 0 \qquad \text{almost surely.}$$

Moreover,

$$\sqrt{n}\left(\lambda_n^0 - \lambda_n^*\right) = O(1)$$

$$\sqrt{n}\left(\mathcal{I}_n^*\right)^{-1/2} \frac{\partial}{\partial \lambda}\left[s_n(\lambda_n^*) - s_n^0(\lambda_n^*)\right] \xrightarrow{\mathcal{L}} N_p(0, I)$$

$$\lim_{n \to \infty} \left(\mathcal{I}_n^* + \mathcal{U}_n^* - \tilde{\mathcal{I}}\right) = 0 \qquad \text{in probability}$$

$$\lim_{n \to \infty} \mathcal{I}_n^* - \tilde{\mathcal{I}} = 0 \qquad \text{almost surely.}$$

Proof. The proof is much the same as the proof of Theorem 7.   □

PROBLEMS

1. Let Assumptions 1 through 3 and 8 through 11 hold except that $m_n^0(\lambda_n^0) \neq 0$; also $(\partial/\partial m)d(0, \tau)$ and $(\partial^2/\partial m \, \partial m')d(0, \tau)$ need not be zero. Presume that the estimator of the nuisance parameter τ_n^0 can be

put in the form

$$\sqrt{n}\left(\hat{\tau}_n - \tau_n^0\right) = A_n^0 \frac{1}{\sqrt{n}} \sum_{t=1}^n f_t(W_t) + o_p(1)$$

where $\{f_t(W_t)\}$ satisfies the hypotheses of Theorem 2 and $c_0 \delta' \delta \le \delta'(A_n^0)'(A_n^0)\delta \le c_1 \delta' \delta$ for finite, nonzero c_0, c_1 and all n larger than some N. Define

$$Z_t = \begin{pmatrix} m_t\left(W_t, \tau_n^0, \lambda_n^0\right) \\ \text{vec}\, \frac{\partial}{\partial \lambda} m_t'\left(W_t, \tau_n^0, \lambda_n^0\right) \\ f_t(W_t) \end{pmatrix}$$

$$\mathscr{K}_{n\tau}^0 = \begin{cases} \dfrac{1}{n} \displaystyle\sum_{t=1+\tau}^n (\mathscr{E}Z_t)(\mathscr{E}Z_{t-\tau})' & \tau \ge 0 \\[4mm] \mathscr{K}_{n,-\tau}^{0\prime} & \tau < 0 \end{cases}$$

$$\mathscr{S}_n^0 = \sum_{\tau = -(n-1)}^{n-1} \mathscr{S}_{n\tau}^0$$

$$\mathscr{S}_{n\tau}^0 = \begin{cases} \dfrac{1}{n} \displaystyle\sum_{t=1+\tau}^n \mathscr{E}Z_t Z_{t-\tau}' - \mathscr{K}_{n\tau}^0 & \tau \ge 0 \\[4mm] \mathscr{S}_{n,-\tau}^{0\prime} & \tau < 0 \end{cases}$$

$$\mathscr{A}_n^0 = \left(M_n^{0\prime} D_n^0 \;\vdots\; \frac{\partial}{\partial m'} d\left(m_n^0, \tau_n^0\right) \otimes I_p \;\vdots\; M_n^{0\prime} \frac{\partial^2}{\partial m\, \partial \tau'} d\left(m_n^0, \tau_n^0\right) A_n^0 \right)$$

$$\mathscr{I}_n^0 = \mathscr{A}_n^0 \mathscr{S}_n^0 \mathscr{A}_n^{0\prime}$$

$$\mathscr{J}_n^0 = \frac{\partial^2}{\partial \lambda\, \partial \lambda'} s_n^0(\lambda_n^0).$$

Show that

$$\sqrt{n}\left(\mathscr{I}_n^0\right)^{-1/2} \frac{\partial}{\partial \lambda} s_n(\lambda_n^0) \xrightarrow{\mathscr{L}} N_p(0, I)$$

$$\sqrt{n}\left(\mathscr{I}_n^0\right)^{-1/2} \mathscr{J}_n^0\left(\hat{\lambda}_n - \lambda_n^0\right) \xrightarrow{\mathscr{L}} N_p(0, I).$$

6. HYPOTHESIS TESTING

The results obtained thus far may be summarized as follows:

SUMMARY. Let Assumptions 1 through 3 hold, and let either Assumptions 4 through 7 or 8 through 12 hold. Then on a closed ball Λ with finite, nonzero radius

$$s_n(\lambda) - s_n^0(\lambda) \overset{\text{a.s.}}{\to} 0 \qquad \text{uniformly on } \Lambda$$

$$\frac{\partial}{\partial \lambda}\left[s_n(\lambda) - s_n^0(\lambda)\right] \overset{\text{a.s.}}{\to} 0 \qquad \text{uniformly on } \Lambda$$

$$\frac{\partial^2}{\partial \lambda\, \partial \lambda'}\left[s_n(\lambda) - s_n^0(\lambda)\right] \overset{\text{a.s.}}{\to} 0 \qquad \text{uniformly on } \Lambda$$

$$\left\{s_n^0(\lambda)\right\}_{n=1}^{\infty} \text{ is equicontinuous}$$

$$\left\{\frac{\partial}{\partial \lambda}s_n^0(\lambda)\right\}_{n=1}^{\infty} \text{ is equicontinuous}$$

$$\left\{\frac{\partial^2}{\partial \lambda\, \partial \lambda'}s_n^0(\lambda)\right\}_{n=1}^{\infty} \text{ is equicontinuous}$$

$$\sqrt{n}\left(\mathscr{I}_n^0\right)^{-1/2}\left(\frac{\partial}{\partial \lambda}\right)s_n\left(\lambda_n^0\right) \overset{\mathscr{L}}{\to} N_p(0, I)$$

$$\sqrt{n}\left(\mathscr{I}^*\right)^{-1/2}\frac{\partial}{\partial \lambda}\left[s_n(\lambda_n^*) - s_n^0(\lambda_n^*)\right] \overset{\mathscr{L}}{\to} N_p(0, I)$$

$$\sqrt{n}\left(\lambda_n^0 - \lambda_n^*\right) = O(1) \qquad \sqrt{n}\left[h\left(\lambda_n^0\right) - h_n^*\right] = O(1)$$

$$\hat{\lambda}_n - \lambda_n^0 \overset{\text{a.s.}}{\to} 0 \qquad\qquad \tilde{\lambda}_n - \lambda_n^* \overset{\text{a.s.}}{\to} 0$$

$$\mathscr{I}_n^0 + \mathscr{U}_n^0 - \hat{\mathscr{I}} \overset{P}{\to} 0, \qquad \mathscr{I}_n^* + \mathscr{U}_n^* - \tilde{\mathscr{I}} \overset{P}{\to} 0$$

$$\mathscr{I}_n^0 - \hat{\mathscr{I}} \overset{\text{a.s.}}{\to} 0, \qquad\qquad \mathscr{I}_n^* - \tilde{\mathscr{I}} \overset{\text{a.s.}}{\to} 0$$

$$c_0\delta'\delta \le \delta'\bar{\mathscr{I}}(\lambda)\delta \le c_1\delta'\delta \quad \text{all } n > N \quad \text{all } \delta \in \mathbb{R}^p, \quad \text{all } \lambda \in \Lambda$$

$$c_0\delta'\delta \le \delta'\mathscr{I}_n^*\delta \le c_1\delta'\delta \quad \text{all } n > N \quad \text{all } \delta \in \mathbb{R}^p$$

$$c_0\delta'\delta \le \delta'\mathscr{I}_n^*\delta \le c_1\delta'\delta \quad \text{all } n > N \quad \text{all } \delta \in \mathbb{R}^p$$

$$\delta'\mathscr{U}_n^0\delta \le c_1\delta'\delta \quad \delta'\mathscr{U}_n^*\delta \le c_1\delta'\delta \quad \text{all } n > N \quad \text{all } \delta \in \mathbb{R}^p$$

where $0 < c_0 < c_1 < \infty$. Moreover, $\hat{\lambda}_n$ and $\tilde{\lambda}_n$ are tail equivalent to random variables that take their values in the interior of Λ, and λ_n^0 and λ_n^*

are interior to Λ for large n. Thus, in the sequel we may take $\hat{\lambda}_n$, $\tilde{\lambda}_n$, λ_n^0, and λ_n^* interior to Λ without loss of generality. $\square$

Taking the Summary as the point of departure, consider testing

$$H : h\left(\lambda_n^0\right) = h_n^* \quad \text{against} \quad A : h\left(\lambda_n^0\right) \neq h_n^*$$

where $h(\lambda)$ maps $\Lambda \subset \mathbf{R}^p$ into $\mathbf{R}^q$. As in Chapter 3, we shall study three test statistics for this problem: the Wald test, the "likelihood ratio" test, and the Lagrange multiplier test. Each statistic, say T as a generic term, is decomposed into a sum of two random variables

$$T_n = X_n + a_n$$

where a_n converges in probability to zero and X_n has a known, finite sample distribution. Such a decomposition permits the statement

$$\lim_{n \to \infty} \left[P(T_n > t) - P(X_n > t) \right] = 0.$$

Because we allow specification error and nonstationarity, we shall not necessarily have T_n converging in distribution to a random variable X. However, the practical utility of convergence in distribution in applications derives from the statement

$$\lim_{n \to \infty} \left[P(T_n > t) - P(X > t) \right] = 0$$

because $P(X < t)$ is computable and so can be used to approximate $P(T_n > t)$. Since the value $P(X_n > t)$ that we shall provide is computable, we shall capture the full benefits of a classical asymptotic theory.

We introduce some additional notation.

NOTATION 7.

$$V_n^0 = \left(\mathcal{J}_n^0\right)^{-1} \mathcal{J}_n^0 \left(\mathcal{J}_n^0\right)^{-1} \qquad V_n^* = \left(\mathcal{J}_n^*\right)^{-1} \mathcal{J}_n^* \left(\mathcal{J}_n^*\right)^{-1}$$

$$\hat{V} = \hat{\mathcal{J}}^{-1} \hat{\mathcal{J}} \hat{\mathcal{J}}^{-1} \qquad\qquad \tilde{V} = \tilde{\mathcal{J}}^{-1} \tilde{\mathcal{J}} \tilde{\mathcal{J}}^{-1}$$

$$\hat{h} = h(\hat{\lambda}) \qquad\qquad\qquad H(\lambda) = \frac{\partial}{\partial \lambda'} h(\lambda)$$

$$H_n^0 = H\left(\lambda_n^0\right) \qquad\qquad H_n^* = H(\lambda_n^*)$$

$$\hat{H} = H(\hat{\lambda}_n) \qquad\qquad\qquad \tilde{H} = H(\tilde{\lambda}_n).$$

In Theorem 12,

$$V = V_n^0, \qquad \mathcal{I} = \mathcal{I}_n^0, \qquad \mathcal{J} = \mathcal{J}_n^0, \qquad \mathcal{U} = \mathcal{U}_n^0, \qquad H = H_n^0.$$

In Theorems 13, 14, 15, and 16,

$$V = V_n^*, \qquad \mathcal{I} = \mathcal{I}_n^*, \qquad \mathcal{J} = \mathcal{J}_n^*, \qquad \mathcal{U} = \mathcal{U}_n^*, \qquad H = H_n^*.$$

The first test statistic considered is the Wald test statistic

$$W = n(\hat{h} - h_n^*)'(\hat{H}\hat{V}\hat{H}')^{-1}(\hat{h} - h_n^*).$$

As shown below, one rejects the hypothesis $H: h(\lambda_n^0) = h_n^*$ when W exceeds the upper $\alpha \times 100\%$ critical point of a chi-square random variable with q degrees of freedom to achieve an asymptotically level α test in a correctly specified situation. As noted earlier, the principal advantage of the Wald test is that it requires only one unconstrained optimization to compute it. The principal disadvantages are that it is not invariant to reparametrization and its sampling distribution is not as well approximated by our characterizations as are the "likelihood ratio" and Lagrange multiplier tests.

THEOREM 12. Let Assumptions 1 through 3 hold, and let either Assumptions 4 through 7 or 8 through 12 hold. Let

$$W = n(\hat{h} - h_n^*)'(\hat{H}\hat{V}\hat{H}')^{-1}(\hat{h} - h_n^*).$$

Then

$$W \sim Y + o_p(1)$$

where

$$Y = Z'\left[H\mathcal{J}^{-1}(\mathcal{I} + \mathcal{U})\mathcal{J}^{-1}H'\right]^{-1}Z$$

and

$$Z \sim N_q\left\{\sqrt{n}\left[h(\lambda_n^0) - h_n^*\right], HVH'\right\}.$$

(Recall that $V = V_n^0$, $\mathcal{I} = \mathcal{I}_n^0$, $\mathcal{J} = \mathcal{J}_n^0$, $\mathcal{U} = \mathcal{U}_n^0$, and $H = H_n^0$.) If $\mathcal{U} = 0$, then Y has the noncentral chi-square distribution with q degrees of freedom and noncentrality parameter $\alpha = n[h(\lambda_n^0) - h_n^*]'(HVH')^{-1}[h(\lambda_n^0) - h_n^*]/2$. Under the null hypothesis $\alpha = 0$.

Proof. We may assume without loss of generality that $\hat{\lambda}_n, \lambda_n^0 \in \Lambda$ and that $(\partial/\partial\lambda)s_n(\hat{\lambda}_n) = o_s(n^{-1/2})$, $(\partial/\partial\lambda)s_n^0(\lambda_n^0) = o(n^{-1/2})$. By Taylor's theorem

$$\sqrt{n}\left[h_i(\hat{\lambda}_n) - h_i(\lambda_n^0)\right] = \frac{\partial}{\partial\lambda'}h_i(\bar{\lambda}_{in})\sqrt{n}\left(\hat{\lambda}_n - \lambda_n^0\right) \qquad i = 1, 2, \ldots, q$$

where $\|\bar{\lambda}_{in} - \lambda_n^0\| \le \|\hat{\lambda}_n - \lambda_n^0\|$. By the almost sure convergence of $\|\lambda_n^0 - \hat{\lambda}_n\|$ to zero, $\lim_{n\to\infty}\|\bar{\lambda}_{in} - \lambda_n^0\| = 0$ almost surely, whence $\lim_{n\to\infty}[(\partial/\partial\lambda)h_i(\bar{\lambda}_{in}) - (\partial/\partial\lambda)h_i(\lambda_n^0)] = 0$ almost surely. Thus we may write

$$\sqrt{n}\left[h(\hat{\lambda}_n) - h(\lambda_n^0)\right] = [H + o_s(1)]\sqrt{n}\left(\hat{\lambda}_n - \lambda_n^0\right).$$

Again by Taylor's theorem

$$\sqrt{n}\,\mathscr{I}^{-1/2}\frac{\partial}{\partial\lambda}s_n(\lambda_n^0) = \sqrt{n}\,\mathscr{I}^{-1/2}\frac{\partial}{\partial\lambda}s_n(\hat{\lambda}_n)$$

$$+\mathscr{I}^{-1/2}[\mathscr{I} + o_s(1)]\sqrt{n}\left(\hat{\lambda}_n - \lambda_n^0\right).$$

By the Summary, the left hand side is $O_p(1)$, and $\mathscr{I}^{1/2}$ and $\mathscr{I}^{-1}$ are both $O(1)$, whence $\sqrt{n}\,(\hat{\lambda}_n - \lambda_n^0) = O_p(1)$ and

$$\sqrt{n}\,\frac{\partial}{\partial\lambda}s_n(\lambda_n^0) = \mathscr{I}\sqrt{n}\left(\hat{\lambda}_n - \lambda_n^0\right) + o_p(1).$$

Combining these two equations, we have

$$\sqrt{n}\left[h(\hat{\lambda}_n) - h(\lambda_n^0)\right] = [H + o_s(1)]\,\mathscr{I}^{-1}\mathscr{I}^{1/2}\mathscr{I}^{-1/2}\mathscr{I}\sqrt{n}\left(\hat{\lambda}_n - \lambda_n^0\right)$$

$$= [H + o_s(1)]\,\mathscr{I}^{-1}\mathscr{I}^{1/2}\mathscr{I}^{-1/2}\left(\sqrt{n}\,\frac{\partial}{\partial\lambda}s_n(\lambda_n^0) + o_p(1)\right)$$

$$= H\mathscr{I}^{-1}\mathscr{I}^{1/2}\mathscr{I}^{-1/2}\sqrt{n}\,\frac{\partial}{\partial\lambda}s_n(\lambda_n^0) + o_p(1)$$

because all terms are $O_s(1)$ save the $o_s(1)$ and $o_p(1)$ terms. The equicontinuity of $\{\bar{\mathscr{I}}_n(\lambda)\}$, the almost sure convergence of $\|\hat{\lambda}_n - \lambda_n^0\|$ to zero, and $\det \bar{\mathscr{I}}(\lambda) \ge \Delta > 0$ imply that

$$(\hat{H}\hat{V}\hat{H}')^{-1} - \left[H\mathscr{I}^{-1}(\mathscr{I} + \mathscr{U})\mathscr{I}^{-1}H'\right]^{-1} = o_s(1).$$

Since

$$\sqrt{n}\left[h(\hat{\lambda}_n) - h_n^*\right] = \sqrt{n}\left[h(\lambda_n^0) - h_n^*\right]$$
$$+ H\mathscr{J}^{-1}\mathscr{I}^{1/2}\mathscr{I}^{-1/2}\sqrt{n}\,\frac{\partial}{\partial\lambda}s_n(\lambda_n^0) + o_p(1)$$

and all terms on the right are bounded in probability, we have that

$$W = n\left[h(\hat{\lambda}_n) - h_n^*\right]'(\hat{H}\hat{V}\hat{H}')^{-1}\left[h(\hat{\lambda}_n) - h_n^*\right]$$
$$= n\left[h(\hat{\lambda}_n) - h_n^*\right]'\left[H\mathscr{J}^{-1}(\mathscr{I} + \mathscr{U})\mathscr{J}^{-1}H'\right]^{-1}\left[h(\hat{\lambda}_n) - h_n^*\right] + o_p(1).$$

By the Skorokhod representation theorem (Serfling, 1980, Section 1.6), there are random variables X_n with the same distribution as $\mathscr{I}^{-1/2}\sqrt{n}\,(\partial/\partial\lambda)$ $s_n(\lambda_n^0)$ such that $X_n \sim X + o_s(1)$ where $X \sim N_p(0, I)$. Then

$$\sqrt{n}\left[h(\hat{\lambda}_n) - h_n^*\right] \sim \sqrt{n}\left[h(\lambda_n^0) - h_n^*\right] + H\mathscr{J}^{-1}\mathscr{I}^{1/2}X + o_p(1)$$

because H, $\mathscr{J}^{-1}$, $\mathscr{I}^{1/2}$ are bounded. Let $Z = \sqrt{n}\,[h(\lambda_n^0) - h_n^*] + H\mathscr{J}^{-1}\mathscr{I}^{1/2}X$, and the result follows.   □

In order to characterize the distribution of the Lagrange multiplier and "likelihood ratio" test statistics we shall need the following characterization of the distribution of the score vector evaluated at the constrained value λ_n^*.

THEOREM 13.   Let Assumptions 1 through 3 hold and let either Assumptions 4 through 7 or 8 through 12 hold. Then

$$\sqrt{n}\,\frac{\partial}{\partial\lambda}s_n(\lambda^*) \sim X + o_s(1)$$

where

$$X \sim N_p\left(\sqrt{n}\,\frac{\partial}{\partial\lambda}s_n^0(\lambda_n^*), \mathscr{I}_n^*\right).$$

Proof.   By either Theorem 7 or Theorem 11

$$\sqrt{n}\,(\mathscr{I}_n^*)^{-1/2}\frac{\partial}{\partial\lambda}s_n(\lambda_n^*) - \sqrt{n}\,(\mathscr{I}_n^*)^{-1/2}\frac{\partial}{\partial\lambda}s_n^0(\lambda_n^*) \xrightarrow{\mathscr{L}} N_p(0, I).$$

By the Skorokhod representation theorem (Serfling, 1980, Section 1.6) there are random variables Y_n with the same distribution as $\sqrt{n}\,(\mathscr{I}_n^*)^{-1/2}(\partial/\partial\lambda)$

$s_n(\lambda_n^*)$ such that $Y_n - \sqrt{n}\,(\mathscr{I}_n^*)^{-1/2}(\partial/\partial\lambda)s_n^0(\lambda_n^*) \sim Y + o_s(1)$, where $Y \sim N_p(0, I)$. Let

$$X = (\mathscr{I}_n^*)^{1/2}Y - \sqrt{n}\,\frac{\partial}{\partial\lambda}s_n^0(\lambda_n^*)$$

whence

$$X \sim N_p\left(\sqrt{n}\,\frac{\partial}{\partial\lambda}s_n^0(\lambda_n^*),\ \mathscr{I}_n^*\right).$$

Since $(\mathscr{I}_n^*)^{1/2}$ is bounded, $(\mathscr{I}_n^*)^{1/2}o_s(1) = o_s(1)$ and the result follows.   □

Both the "likelihood ratio" and Lagrange multiplier test statistics are effectively functions of the score vector evaluated at $\tilde{\lambda}_n$. The following result gives an essential representation.

THEOREM 14. Let Assumptions 1 through 3 hold, and let either Assumptions 4 through 7 or 8 through 12 hold. Then

$$\sqrt{n}\,\frac{\partial}{\partial\lambda}s_n(\tilde{\lambda}_n) = H'\big(H\mathscr{I}^{-1}H'\big)^{-1}H\mathscr{I}^{-1}\sqrt{n}\,\frac{\partial}{\partial\lambda}s_n(\lambda_n^*) + o_p(1)$$

$$= O_p(1)$$

where $\mathscr{I} = \mathscr{I}_n^*$ and $H = H_n^*$.

Proof. By Taylor's theorem

$$\sqrt{n}\,\frac{\partial}{\partial\lambda}s_n(\tilde{\lambda}_n) = \sqrt{n}\,\frac{\partial}{\partial\lambda}s_n(\lambda_n^*) + \bar{\mathscr{J}}\sqrt{n}\,(\tilde{\lambda}_n - \lambda_n^*)$$

$$\sqrt{n}\,h(\tilde{\lambda}_n) = \sqrt{n}\,h(\lambda_n^*) + \bar{H}\sqrt{n}\,(\tilde{\lambda}_n - \lambda_n^*)$$

where $\bar{\mathscr{J}}$ has rows

$$\frac{\partial}{\partial\lambda'}\frac{\partial}{\partial\lambda_i}s_n(\bar{\lambda}_{in})\qquad i = 1, 2, \ldots, p$$

and $\bar{H}$ has rows

$$\frac{\partial}{\partial\lambda'}h_j(\bar{\lambda}_{jn})\qquad j = 1, 2, \ldots, q$$

with $\bar{\lambda}_{in}$ and $\bar{\lambda}_{jn}$ on the line segment joining $\tilde{\lambda}_n$ to λ_n^*. Now $\sqrt{n}\,[h(\tilde{\lambda}_n) - h_n^*] = o_s(1)$. Recalling that $\sqrt{n}\,[h(\lambda_n^*) - h_n^*] \equiv 0$, we have $\bar{H}\sqrt{n}\,(\tilde{\lambda}_n - \lambda_n^*) = o_s(1)$. Since $\|\tilde{\lambda}_n - \lambda_n^*\|$ converges almost surely to zero, $\bar{\lambda}_{in} - \lambda_n^*$ and $\bar{\lambda}_{jn} - \lambda_n^*$ converge almost surely to zero and $\bar{\mathscr{J}} = \mathscr{J} + o_s(1)$ by the equicontinuity of $\{\mathscr{J}_n(\lambda)\}_{n=1}^{\infty}$; continuity of $H(\lambda)$ on Λ compact implies equicontinuity, whence $\bar{H} = H + o_s(1)$. Moreover, there is an N corresponding to almost every realization of $\{V_t\}$ such that $\det(\bar{\mathscr{J}}) > 0$ for all $n > N$. Defining $\bar{\mathscr{J}}^{-1}$ arbitrarily when $\det(\bar{\mathscr{J}}) = 0$ we have

$$\bar{\mathscr{J}}^{-1}\bar{\mathscr{J}}\sqrt{n}\,(\tilde{\lambda}_n - \lambda_n^*) \equiv \sqrt{n}\,(\tilde{\lambda}_n - \lambda_n^*)$$

for all $n > N$. Thus, $\bar{\mathscr{J}}^{-1}\bar{\mathscr{J}}\sqrt{n}\,(\tilde{\lambda}_n - \lambda_n^*) = \sqrt{n}\,(\tilde{\lambda}_n - \lambda_n^*) + o_s(1)$. Combining these observations, we may write

$$\bar{H}\sqrt{n}\,(\tilde{\lambda}_n - \lambda_n^*) = o_s(1)$$

$$\sqrt{n}\,(\tilde{\lambda}_n - \lambda_n^*) = \bar{\mathscr{J}}^{-1}\sqrt{n}\,\frac{\partial}{\partial\lambda}s_n(\tilde{\lambda}_n) - \bar{\mathscr{J}}^{-1}\sqrt{n}\,\frac{\partial}{\partial\lambda}s_n(\lambda_n^*) + o_s(1)$$

whence

$$\bar{H}\bar{\mathscr{J}}^{-1}\sqrt{n}\,\frac{\partial}{\partial\lambda}s_n(\tilde{\lambda}_n) = \bar{H}\bar{\mathscr{J}}^{-1}\sqrt{n}\,\frac{\partial}{\partial\lambda}s_n(\lambda_n^*) + o_s(1).$$

Now $\sqrt{n}\mathscr{J}^{-1/2}[(\partial/\partial\lambda)s_n(\lambda_n^*) - (\partial/\partial\lambda)s_n^0(\lambda_n^*)]$ converges in distribution, and by Taylor's theorem

$$\sqrt{n}\mathscr{J}^{-1/2}\frac{\partial}{\partial\lambda}s_n^0(\lambda_n^*) = \sqrt{n}\mathscr{J}^{-1/2}\frac{\partial}{\partial\lambda}s_n^0(\lambda_n^0) + \sqrt{n}\mathscr{J}^{-1/2}\bar{\bar{\mathscr{J}}}(\lambda_n^* - \lambda_n^0)$$

$$= o(1) + O(1)$$

so we have that $\sqrt{n}\mathscr{J}^{-1/2}(\partial/\partial\lambda)s_n^0(\lambda_n^*)$ is bounded. Since $\mathscr{J}^{-1/2}$ is bounded, $\sqrt{n}\,(\partial/\partial\lambda)s_n(\lambda_n^*)$ is bounded in probability. By Lemma 2 of Chapter 3, there is a sequence of Lagrange multipliers $\tilde{\theta}_n$ such that

$$\sqrt{n}\,\frac{\partial}{\partial\lambda}s_n(\tilde{\lambda}_n) + \tilde{H}'\sqrt{n}\,\tilde{\theta}_n = o_s(1).$$

By continuity of $H(\lambda)$ and the almost sure convergence of $\|\tilde{\lambda}_n - \lambda_n^*\|$ to zero we have $\tilde{H} = H + o_s(1)$. Defining $(\bar{H}\bar{\mathscr{J}}^{-1}\tilde{H})^{-1}$ similarly to $\bar{\mathscr{J}}^{-1}$ above and recalling that $\sqrt{n}\,(\partial/\partial\lambda)s_n(\lambda_n^*)$ is bounded in probability,

$$H'\big(H\mathscr{J}^{-1}H'\big)^{-1}H\mathscr{J}^{-1}\sqrt{n}\,\frac{\partial}{\partial\lambda}s_n(\lambda_n^*)$$

$$=\tilde{H}'\big(\overline{H}\tilde{\mathscr{J}}^{-1}\tilde{H}'\big)^{-1}\overline{H}\tilde{\mathscr{J}}^{-1}\sqrt{n}\,\frac{\partial}{\partial\lambda}s_n(\lambda_n^*)+o_p(1)$$

$$=\tilde{H}'\big(\overline{H}\tilde{\mathscr{J}}^{-1}\tilde{H}'\big)^{-1}\overline{H}\tilde{\mathscr{J}}^{-1}\sqrt{n}\,\frac{\partial}{\partial\lambda}s_n(\tilde{\lambda}_n)+o_p(1)$$

$$=\tilde{H}'\big(\overline{H}\tilde{\mathscr{J}}^{-1}\tilde{H}'\big)^{-1}\overline{H}\tilde{\mathscr{J}}^{-1}\tilde{H}'\sqrt{n}\,\tilde{\theta}_n+o_p(1)$$

$$=\tilde{H}'\sqrt{n}\,\tilde{\theta}_n+o_p(1)$$

$$=\sqrt{n}\,\frac{\partial}{\partial\lambda}s_n(\tilde{\lambda}_n)+o_p(1).\qquad\square$$

The second test statistic considered is the "likelihood ratio" test statistic

$$L=2n\big[s_n(\tilde{\lambda}_n)-s_n(\hat{\lambda}_n)\big].$$

As shown below, one rejects the hypothesis $H:h(\lambda_n^0)=h_n^*$ when L exceeds the upper $\alpha\times100\%$ critical point of a chi-square random variable with q degrees of freedom to achieve an asymptotically level α test in a correctly specified situation. The principal disadvantages of the "likelihood ratio" test are that it takes two minimizations to compute it and it requires that

$$(H_n^*)(\mathscr{J}_n^*)^{-1}(\mathscr{I}_n^*)(\mathscr{J}_n^*)^{-1}(H_n^*)'=(H_n^*)(\mathscr{J}_n^*)^{-1}(H_n^*)'+o(1)$$

to achieve its null case asymptotic distribution. As seen earlier, when this condition holds, there is Monte Carlo evidence that indicates that the asymptotic approximation is quite accurate if degree of freedom corrections are applied.

THEOREM 15. Let Assumptions 1 through 3 hold, and let either Assumptions 4 through 7 or 8 through 12 hold. Let

$$L=2n\big[s_n(\tilde{\lambda}_n)-s_n(\hat{\lambda}_n)\big].$$

Then

$$L\sim Y+o_p(1)$$

where

$$Y=Z'\mathscr{J}^{-1}H'\big(H\mathscr{J}^{-1}H'\big)^{-1}H\mathscr{J}^{-1}Z$$

and

$$Z \sim N_p\left(\sqrt{n}\,\frac{\partial}{\partial\lambda}s_n^0(\lambda_n^*),\,\mathscr{I}\right).$$

Recall that $V = V_n^*$, $\mathscr{I} = \mathscr{I}_n^*$, $\mathscr{J} = \mathscr{J}_n^*$, $\mathscr{U} = \mathscr{U}_n^*$, and $H = H_n^*$.

If $HVH' = H\mathscr{J}^{-1}H'$, then Y has the noncentral chi-square distribution with q degrees of freedom and noncentrality parameter

$$\alpha = n\frac{(\partial/\partial\lambda')s_n^0(\lambda_n^*)\mathscr{J}^{-1}H'\left(H\mathscr{J}^{-1}H'\right)^{-1}H\mathscr{J}^{-1}(\partial/\partial\lambda)s_n^0(\lambda_n^*)}{2}.$$

Under the null hypothesis, $\alpha = 0$.

Proof. By Taylor's theorem

$$2n\left[s_n(\tilde{\lambda}_n) - s_n(\hat{\lambda}_n)\right] = 2n\left(\frac{\partial}{\partial\lambda}s_n(\hat{\lambda}_n)\right)'(\tilde{\lambda}_n - \hat{\lambda}_n)$$

$$+ n(\tilde{\lambda}_n - \hat{\lambda}_n)'\left(\frac{\partial^2}{\partial\lambda\,\partial\lambda'}s_n(\overline{\lambda}_n)\right)(\tilde{\lambda}_n - \hat{\lambda}_n)$$

where $\|\overline{\lambda}_n - \hat{\lambda}_n\| \le \|\tilde{\lambda}_n - \hat{\lambda}_n\|$. By the Summary, $\|\tilde{\lambda}_n - \lambda_n^*\|$ and $\|\hat{\lambda}_n - \lambda_n^0\|$ converge almost surely to zero and $\{(\partial^2/\partial\lambda\,\partial\lambda')s_n(\lambda)\}_{n=1}^{\infty}$ is equicontinuous, whence $(\partial^2/\partial\lambda\,\partial\lambda')s_n(\overline{\lambda}_n) = \mathscr{J} + o_s(1)$. By Lemma 2 of Chapter 3, $2n[(\partial/\partial\lambda)s_n(\hat{\lambda}_n)]'(\tilde{\lambda}_n - \hat{\lambda}_n) = o_s(1)$, whence

$$2n\left[s_n(\tilde{\lambda}_n) - s_n(\hat{\lambda}_n)\right] = n(\tilde{\lambda}_n - \hat{\lambda}_n)'[\mathscr{J} + o_s(1)](\tilde{\lambda}_n - \hat{\lambda}_n) + o_s(1).$$

Again by Taylor's theorem

$$[\mathscr{J} + o_s(1)]\sqrt{n}\,(\tilde{\lambda}_n - \hat{\lambda}_n) = \sqrt{n}\,\frac{\partial}{\partial\lambda}s_n(\tilde{\lambda}_n)$$

whence, using the same type of argument as in the proof of Theorem 14,

$$\sqrt{n}\,(\tilde{\lambda}_n - \hat{\lambda}_n) = [\mathscr{J} + o_s(1)]^{-1}[\mathscr{J} + o_s(1)]\sqrt{n}\,(\tilde{\lambda}_n - \hat{\lambda}_n) + o_s(1)$$

$$= [\mathscr{J} + o_s(1)]^{-1}\sqrt{n}\,\frac{\partial}{\partial\lambda}s_n(\tilde{\lambda}_n) + o_s(1)$$

which is bounded in probability by Theorem 14. Thus

$$2n\left[s_n(\tilde{\lambda}_n) - s_n(\hat{\lambda}_n)\right] = n(\tilde{\lambda}_n - \hat{\lambda}_n)'\mathscr{J}(\tilde{\lambda}_n - \hat{\lambda}_n) + o_p(1)$$

$$\sqrt{n}\,(\tilde{\lambda}_n - \hat{\lambda}_n) = \mathscr{J}^{-1}\sqrt{n}\,\frac{\partial}{\partial\lambda}s_n(\tilde{\lambda}_n) + o_p(1)$$

whence

$$2n\left[s_n(\tilde{\lambda}_n) - s_n(\hat{\lambda}_n)\right] = n\left(\frac{\partial}{\partial\lambda}s_n(\tilde{\lambda}_n)\right)'\mathscr{J}^{-1}\left(\frac{\partial}{\partial\lambda}s_n(\tilde{\lambda}_n)\right) + o_p(1)$$

and the distributional result follows at once from Theorem 13 and 14. To see that Y is distributed as the noncentral chi-square when $HVH' = H\mathscr{J}^{-1}H'$, note that $\mathscr{J}^{-1}H'(H\mathscr{J}^{-1}H')^{-1}H\mathscr{J}^{-1}\mathscr{J}$ is idempotent under this condition. $\qquad\square$

The last statistic considered is the Lagrange multiplier test statistic

$$R = n\left(\frac{\partial}{\partial\lambda}s_n(\tilde{\lambda}_n)\right)'\tilde{\mathscr{J}}^{-1}\tilde{H}'(\tilde{H}\tilde{\mathscr{J}}^{-1}\tilde{\mathscr{J}}\tilde{\mathscr{J}}^{-1}\tilde{H}')^{-1}\tilde{H}\tilde{\mathscr{J}}^{-1}\left(\frac{\partial}{\partial\lambda}s_n(\tilde{\lambda}_n)\right).$$

As shown below, one rejects the hypothesis $H: h(\lambda_n^0) = h_n^*$ when R exceeds the upper $\alpha \times 100\%$ critical point of a chi-square random variable with q degrees of freedom to achieve an asymptotically level α test in a correctly specified situation. Using the first order condition

$$\frac{\partial}{\partial\lambda}\mathscr{L}(\tilde{\lambda}_n, \tilde{\theta}_n) = \frac{\partial}{\partial\lambda}\left\{s_n(\lambda) + \theta'\left[h(\tilde{\lambda}_n) - h_n^*\right]\right\} = 0$$

for the problem

$$\text{minimize} \quad s_n(\lambda) \quad \text{subject to} \quad h(\lambda) = h_n^*$$

one can replace $(\partial/\partial\lambda)s_n(\tilde{\lambda})$ by $\tilde{\theta}'(\partial/\partial\lambda)h(\tilde{\lambda}_n)$ in the expression for R—whence the term Lagrange multiplier test; it is also called the efficient score test. Its principal advantage is that it requires only one constrained optimization for its computation. If the constraint $h(\lambda) = h_n^*$ completely specifies $\tilde{\lambda}_n$ or results in a linear model, this can be an overwhelming advantage. The test can have rather bizaare structural characteristics. Suppose that $h(\lambda) = h_n^*$ completely specifies $\tilde{\lambda}_n$. Then the test will accept any h_n^* for which $\tilde{\lambda}_n$ is a local minimum, maximum, or saddle point of $s_n(\lambda)$, regardless of how large is $\|h(\hat{\lambda}) - h_n^*\|$. As we have seen, Monte Carlo evidence suggests that the asymptotic approximation is reasonably accurate.

THEOREM 16. Let Assumptions 1 through 3 hold, and let either Assumptions 4 through 7 or 8 through 12 hold. Let

$$R = n\left(\frac{\partial}{\partial\lambda}s_n(\tilde{\lambda}_n)\right)'\tilde{\mathscr{J}}^{-1}\tilde{H}'(\tilde{H}\tilde{V}\tilde{H}')^{-1}\tilde{H}\tilde{\mathscr{J}}^{-1}\left(\frac{\partial}{\partial\lambda}s_n(\tilde{\lambda}_n)\right).$$

Then

$$R \sim Y + o_p(1)$$

where

$$Y = Z' \mathscr{J}^{-1} H' \left[H \mathscr{J}^{-1} (\mathscr{J} + \mathscr{U}) \mathscr{J}^{-1} H' \right]^{-1} H \mathscr{J}^{-1} Z$$

and

$$Z \sim N_p \left(\sqrt{n} \, \frac{\partial}{\partial \lambda} s_n^0(\lambda_n^*), \, \mathscr{J} \right).$$

Recall that $V = V_n^*$, $\mathscr{J} = \mathscr{J}_n^*$, $\mathscr{J} = \mathscr{J}_n^*$, $\mathscr{U} = \mathscr{U}_n^*$, and $H = H_n^*$.

If $\mathscr{U} = 0$, then Y has the noncentral chi-square distribution with q degrees of freedom and noncentrality parameter $\alpha = n[(\partial/\partial\lambda)s_n^0(\lambda_n^*)]' \mathscr{J}^{-1} H'(HVH')^{-1} H \mathscr{J}^{-1}[(\partial/\partial\lambda)s_n^0(\lambda_n^*)]/2$. Under the null hypothesis, $\alpha = 0$.

Proof. By the summary,

$$\tilde{\mathscr{J}}^{-1} \tilde{H}' (\tilde{H} \tilde{V} \tilde{H}')^{-1} \tilde{H} \tilde{\mathscr{J}}^{-1}$$

$$= \mathscr{J}^{-1} H' \left[H \mathscr{J}^{-1} (\mathscr{J} + \mathscr{U}) \mathscr{J}^{-1} H' \right]^{-1} H \mathscr{J}^{-1} + o_s(1).$$

By Theorem 14, $\sqrt{n} \, (\partial/\partial\lambda) s_n(\tilde{\lambda}_n)$ is bounded in probability, whence we have

$$R = n \left(\frac{\partial}{\partial \lambda} s_n(\tilde{\lambda}_n) \right)' \mathscr{J}^{-1} H'$$

$$\times \left[H \mathscr{J}^{-1} (\mathscr{J} + \mathscr{U}) \mathscr{J}^{-1} H' \right]^{-1} H \mathscr{J}^{-1} \left(\frac{\partial}{\partial \lambda} s_n(\tilde{\lambda}_n) \right) + o_p(1)$$

$$= n \left(\frac{\partial}{\partial \lambda} s_n(\lambda_n^*) \right)' \mathscr{J}^{-1} H'$$

$$\times \left[H \mathscr{J}^{-1} (\mathscr{J} + \mathscr{U}) \mathscr{J}^{-1} H' \right]^{-1} H \mathscr{J}^{-1} \left(\frac{\partial}{\partial \lambda} s_n(\lambda_n^*) \right) + o_p(1).$$

The distributional result follows by Theorem 13. The matrix $\mathscr{J}^{-1} H' [H \mathscr{J}^{-1} \mathscr{J} \mathscr{J}^{-1} H']^{-1} H \mathscr{J}^{-1} \mathscr{J}$ is idempotent, so Y follows the noncentral chi-square distribution if $\mathscr{U} = 0$.   □

References

Aitchison, J., and S. M. Shen (1980), Logistic-normal distributions: Some properties and uses, *Biometrika* **67**, 261–272.

Akaike, H. (1969), Fitting autoregressive models for prediction, *Annals of the Institute of Statistical Mathematics* **21**, 243–247.

Amemiya, T. (1974), The nonlinear two-stage least squares estimator, *Journal of Econometrics* **2**, 105–110.

Amemiya, T. (1977), The maximum likelihood and nonlinear three stage least squares estimator in the general nonlinear simultaneous equations model, *Econometrica* **45**, 955–968.

Amemiya, T. (1982), Correction to a lemma, *Econometrica* **50**, 1325–1328.

Anderson, T. W. (1971), *The Statistical Analysis of Time Series*, New York: Wiley.

Anderson, T. W. (1984), *An Introduction to Multivariate Statistical Analysis*, Second Edition. New York: Wiley.

Andrews, D. W. K. (1986), Consistency in nonlinear econometric models: A generic uniform law of large numbers, Cowles Foundation Discussion Paper, Yale University, New Haven, Connecticut.

Athreya, K. B., and S. G. Pantula (1986), Mixing properties of Harris chains and AR processes, *Journal of Applied Probability*, in press.

Ash, R. B. (1972), *Real Analysis and Probability*, New York: Academic.

Balet-Lawrence, S. (1975), Estimation of the parameters in an implicit model by minimizing the sum of absolute value of order p. Ph.D. Dissertation, North Carolina State University, Raleigh, North Carolina.

Barnett, W. A. (1976), Maximum likelihood and iterated Aitken estimation of nonlinear systems of equations, *Journal of the American Statistical Association* **71**, 354–360.

Bartle, R. G. (1964), *The Elements of Real Analysis*, New York: Wiley.

Bates, D. M. and D. G. Watts (1980), Relative curvature measures of nonlinearity, *Journal of the Royal Statistical Society Series B* **42**, 1–25.

Bates, D. M., and D. G. Watts (1981), Parameter transformations for improved approximate confidence regions in nonlinear least squares, *The Annals of Statistics* **9**, 1152–1167.

Beale, E. M. L. (1960), Confidence regions in non-linear estimation, *Journal of the Royal Statistical Society Series B* **22**, 41–76.

Berger, R. L., and N. A. Lanberg (1984), Linear least squares estimates and nonlinear means, *Journal of Statistical Planning and Inference* **10**, 277–288.

Berk, K. N. (1974), Consistent autoregressive spectral estimates, *The Annals of Statistics* **2**, 489–502.

Bierens, H. J. (1981), *Robust Methods and Asymptotic Theory*, Lecture Notes in Economics and Mathematical Systems 192, Berlin: Springer.

Billingsley, P. (1968), *Convergence of Probability Measures*, New York: Wiley.

Billingsley, P. (1979), *Probability and Measure*, New York: Wiley.

Blackwell, D. and M. A. Girshick (1954), *Theory of Games and Statistical Decisions*, New York: Wiley.

Blackorby, C., D. Primont, and R. Russell (1978), *Duality, Separability, and Functional Structure: Theory and Economic Applications*, New York: North-Holland.

Bloomfield, P. (1976), *Fourier Analysis of Time Series: An Introduction*, New York: Wiley.

Box, G. E. P., and H. L. Lucas (1959), The design of experiments in nonlinear situations, *Biometrika* **46**, 77–90.

Burguete, J. F. (1980), Asymptotic theory of instrumental variables in nonlinear regression, Ph.D. Dissertation, North Carolina State University, Raleigh, North Carolina.

Burguete, J. F., A. R. Gallant, and G. Souza (1982), On unification of the asymptotic theory of nonlinear econometric models, *Econometric Reviews* **1**, 151–190.

Bussinger, P. A., and G. H. Golub (1969), Singular value decomposition of a complex matrix, *Communications of the ACM* **12**, 564–565.

Caves, D. W., and L. R. Christensen, (1980), Econometric analysis of residential time-of-use electricity experiments, *Journal of Econometrics* **14**, 285–286.

Christensen, L. R., D. W. Jorgenson, and L. J. Lau (1975), Transcendental logarithmic utility functions, *The American Economic Review* **65**, 367–383.

Chung, K. L. (1974), *A Course in Probability Theory*, Second Edition, New York: Academic.

Cook, R. D., and J. A. Witmer (1985), A note on parameter-effects curvature, *Journal of the American Statistical Association* **80**, 872–878.

Cox, D. R., and D. V. Hinkley (1974), *Theoretical Statistics*, London: Chapman and Hall.

Deaton, A., and J. Muellbauer (1980), *Economics and Consumer Behavior*, Cambridge, England: Cambridge U. P.

Dennis, J. E., D. M. Gay, and R. E. Welch (1977), An adaptive nonlinear least-squares algorithm, Department of Computer Sciences Report No. TR 77-321, Cornell University, Ithaca, New York.

Doob, J. L. (1953), *Stochastic Processes*, New York: Wiley.

Dunford, N., and J. T. Schwartz (1957), *Linear Operators, Part I: General Theory*, New York: Wiley.

Durbin, J. (1970), Testing for serial correlation in least-squares regression when some of the regressors are lagged dependent variables, *Econometrica* **38**, 410–421.

Edmunds, D. E., and V. B. Moscatelli (1977), Fourier approximation and embeddings of Sobolev spaces, *Dissertationes Mathematicae* **CXLV**, 1–46.

Efron, B. (1975), Defining the curvature of a statistical problem (with applications to second order efficiency), *The Annals of Statistics* **3**, 1189–1242.

Elbadawi, I., A. R. Gallant, and G. Souza (1983), An elasticity can be estimated consistently without a priori knowledge of functional form, *Econometrica* **51**, 1731–1751.

Eppright, E. S., H. M. Fox, B. A. Fryer, G. H. Lamkin, V. M. Vivian, and E. S. Fuller (1972), Nutrition of infants and preschool children in the north central region of the United States of America, *World Review of Nutrition and Dietetics* **14**, 269–332.

Fiacco, A. V., and G. P. McCormick (1968), *Nonlinear Programming: Sequential Unconstrained Minimization Techniques*, New York: Wiley.

Fox, M. (1956), Charts on the power of the *T*-Test, *The Annals of Mathematical Statistics* **27**, 484–497.

Fuller, W. A. (1976), *Introduction to Statistical Time Series*, New York: Wiley.

Gallant, A. R. (1973), Inference for nonlinear models, Institute of Statistics Mimeograph Series No. 875, North Carolina State University, Raleigh, North Carolina.

Gallant, A. R. (1975a), The power of the likelihood ratio test of location in nonlinear regression models, *Journal of the American Statistical Association* **70**, 199–203.

Gallant, A. R. (1975b), Testing a subset of the parameters of a nonlinear regression model, *Journal of the American Statistical Association* **70**, 927–932.

Gallant, A. R. (1975c), Seemingly unrelated nonlinear regressions, *Journal of Econometrics* **3**, 35–50.

Gallant, A. R. (1975d), Nonlinear regression, *The American Statistician* **29**, 73–81.

Gallant, A. R. (1976), Confidence regions for the parameters of a nonlinear regression model, Institute of Statistics Mimeograph Series No. 1077, North Carolina State University, Raleigh, North Carolina.

Gallant, A. R. (1977a), Testing a nonlinear regression specification: A nonregular case, *Journal of the American Statistical Association* **72**, 523–530.

Gallant, A. R. (1977b), Three-stage least-squares estimation for a system of simultaneous, nonlinear, implicit equations, *Journal of Econometrics* **5**, 71–88.

Gallant, A. R. (1980), Explicit estimators of parametric functions in nonlinear regression, *Journal of the American Statistical Association* **75**, 182–193.

Gallant, A. R. (1981), On the bias in flexible functional forms and an essentially unbiased form: The Fourier flexible form, *Journal of Econometrics* **15**, 211–245.

Gallant, A. R. (1982), Unbiased determination of production technologies, *Journal of Econometrics* **20**, 285–323.

Gallant, A. R., and W. A. Fuller (1973), Fitting segmented polynomial regression models whose join points have to be estimated, *Journal of the American Statistical Association* **68**, 144–147.

Gallant, A. R., and J. J. Goebel (1975), Nonlinear regression with autoregressive errors, Institute of Statistics Mimeograph Series No. 986, North Carolina State University, Raleigh, North Carolina.

Gallant, A. R., and J. J. Goebel (1976), Nonlinear regression with autocorrelated errors, *Journal of the American Statistical Association* **71**, 961–967.

Gallant, A. R., and A. Holly (1980), Statistical inference in an implicit, nonlinear, simultaneous equation model in the context of maximum likelihood estimation, *Econometrica* **48**, 697–720.

Gallant, A. R., and D. W. Jorgenson (1979), Statistical inference for a system of simultaneous, nonlinear implicit equations in the context of instrumental variable estimation, *Journal of Econometrics* **11**, 275–302.

Gallant, A. R., and R. W. Koenker (1984), Cost and benefits of peak-load pricing of electricity: A continuous-time econometric approach, *Journal of Econometrics* **26**, 83–114.

Gill, J. L., and H. D. Hafs (1971), Analysis of repeated measurements of animals, *Journal of Animal Science* **33**, 331–336.

Gill, P. E., W. Murray, and M. H. Wright (1981), *Practical Optimization*, New York: Academic.

Golub, G. H., and V. Pereyra (1973), The differentiation of pseudo-inverses and nonlinear least-squares problems whose variables separate, *SIAM Journal of Numerical Analysis* **10**, 413–432.

Grossman, W. (1976), Robust nonlinear regression, in Johannes Gordesch and Peter Naeve, editors, *Compstat 1976, Proceedings in Computational Statistics*, Wien, Austria: Physica.

Guttman, I., and D. A. Meeter (1965), On Beale's measures of non-linearity, *Technometrics* **7**, 623–637.

Hall, P. G., and C. C. Heyde (1980), *Martingale Limit Theory and Its Application*, New York: Academic.

Halperin, M. (1963), Confidence interval estimation in nonlinear regression, *Journal of the Royal Statistical Society Series B* **25**, 330–333.

Hammersley, J. M., and D. C. Handscomb (1964), *Monte Carlo Methods*, New York: Wiley.

Hansen, L. P. (1982), Large sample properties of generalized method of moments estimators, *Econometrica* **50**, 1029–1054.

Hansen, L. P. (1985), Using martingale difference approximations to obtain covariance matrix bounds for generalized method of moments estimators, *Journal of Econometrics* **30**, 203–238.

Hansen, L. P., and K. J. Singleton (1982), Generalized instrumental variables estimators of nonlinear rational expectations models, *Econometrica* **50**, 1269–1286.

Hansen, L. P., and K. J. Singleton (1984), Errata, *Econometrica* **52**, 267–268.

Hartley, H. O. (1961), The modified Gauss-Newton method for the fitting of nonlinear regression functions by least squares, *Technometrics* **3**, 269–280.

Hartley, H. O. (1964), Exact confidence regions for the parameters in nonlinear regression laws, *Biometrika* **51**, 347–353.

Hartley, H. O., and A. Booker (1965), Nonlinear least squares estimation, *Annals of Mathematical Statistics* **36**, 638–650.

Henderson, H. V., and S. R. Searle (1979), Vec and vech operators for matrices, with some uses in Jacobians and multivariate statistics, *The Canadian Journal of Statistics* **7**, 65–81.

Hoadley, B. (1971), Asymptotic properties of maximum likelihood estimators for the independent not identically distributed case, *The Annals of Mathematical Statistics* **42**, 1977–1991.

Holly, A. (1978), Tests of nonlinear statistical hypotheses in multiple equation nonlinear models, *Cahiers du Laboratoire d'Econometrie*, Ecole Polytechnique, Paris.

Huber, P. J. (1964), Robust estimation of a location parameter, *Annals of Mathematical Statistics* **35**, 73–101.

Huber, P. J. (1982), Comment on "The unification of the asymptotic theory of nonlinear econometric models," *Econometric Reviews* **1**, 191–192.

Ibragimov, I. A. (1962), Some limit theorems for stationary processes, *Theory of Probability and Its Applications* **7**, 349–382.

Imhof, P. (1961), Computing the distribution of quadratic forms in normal variates, *Biometrika* **48**, 419–426.

Jennrich, R. I. (1969), Asymptotic properties of nonlinear least squares estimation, *The Annals of Mathematical Statistics* **40**, 633–643.

Jensen, D. R. (1981), Power of invariant tests for linear hypotheses under spherical symmetry, *Scandinavian Journal of Statistics* **8**, 169–174.

Jorgenson, D. W., and J.-J. Laffont (1974), Efficient estimation of nonlinear simultaneous equations with additive disturbances, *Annals of Economic and Social Measurement* **3**, 615–640.

Judge, G. G., W. E. Griffiths, R. C. Hill, and T.-C. Lee (1980), *The Theory and Practice of Econometrics*, New York: Wiley.

Lawson, C. L., and R. J. Hanson (1974), *Solving Least Squares Problems*, Englewood Cliffs, New Jersey: Prentice-Hall.

Levenberg, K. (1944), A method for the solution of certain problems in least squares, *Quarterly Journal of Applied Mathematics* **2**, 164–168.

Loeve, M. (1963), *Probability Theory*, Third Edition. Princeton, New Jersey: Van Nostrand.

Lucas, R. E., Jr. (1978), Asset prices in an exchange economy, *Econometrica* **46**, 1429–1445.

Luenberger, D. G. (1969), *Optimization by Vector Space Methods*, New York: Wiley.

Malinvaud, E. (1970a), *Statistical Methods of Econometrics*, Amsterdam: North-Holland.

Malinvaud, E. (1970b), The consistency of nonlinear regressions, *The Annals of Mathematical Statistics* **41**, 956–969.

Marquardt, D. W. (1963), An algorithm for least-squares estimation of nonlinear parameters, *Journal of the Society for Industrial and Applied Mathematics* **11**, 431–441.

McLeish, D. L. (1974), Dependent central limit theorems and invariance principles, *The Annals of Probability* **2**, 630–628.

McLeish, D. L. (1975a), A maximal inequality and dependent strong laws, *The Annals of Probability* **3**, 829–839.

McLeish, D. L. (1975b), Invariance principles of dependent variables, *Zeitschrift für Wahrscheinlichkeitstheorie und verwandte Gebiete* **32**, 165–178.

McLeish, D. L. (1977), On the invariance principle for nonstationary mixingales, *The Annals of Probability* **5**, 616–621.

Mood, A. M., and F. A. Graybill (1963), *Introduction to the Theory of Statistics*, Second Edition, New York: McGraw-Hill.

Neveu, J. (1965), *Mathematical Foundations of the Calculus of Probability*, San Francisco: Holden-Day.

Office of the President (1972), *Economic Report of the President, Transmitted to Congress, January 1972*, Washington, D.C.: U.S. Government Printing Office, 250.

Office of the President (1974), *Economic Report of the President, Transmitted to Congress, January 1974*, Washington, D.C.: U.S. Government Printing Office, 305.

Osborne, M. R. (1972), Some aspects of non-linear least squares calculations, in Lootsma, F. A., editor, *Numerical Methods for Non-linear Optimization*, New York: Academic.

Pantula, S. G. (1985), Autoregressive processes with several unit roots, Institute of Statistics Mimeograph Series No. 1665, North Carolina State University, Raleigh, North Carolina.

Pearson, E. S., and H. O. Hartley (1951), Charts of the power function of the analysis of variance tests, derived from the non-central F-distribution, *Biometrika* **38**, 112–130.

Phillips, P. C. B. (1982a), Comment on "The unification of the asymptotic theory of nonlinear econometric models," *Econometric Reviews* **1**, 193–199.

Phillips, P. C. B. (1982b), On the consistency of nonlinear FIML, *Econometrica* **50**, 1307–1324.

Pratt, J. W. (1961), Length of confidence intervals, *Journal of the American Statistical Association* **56**, 549–567.

Rao, C. R. (1973), *Linear Statistical Inference and Its Applications*, Second Edition, New York: Wiley.

Ratkowsky, D. A. (1983), *Nonlinear Regression Modeling, A Unified Practical Approach*, New York: Marcel Dekker.

Ross, G. J. S. (1970), The efficient use of function minimization in non-linear maximum-likelihood estimation, *Applied Statistics* **19**, 205–221.

Rossi, P. E. (1983), Specification and analysis of econometric production models, Ph.D. Dissertation, University of Chicago, Chicago, Illinois.

Royden, H. L. (1968), *Real Analysis*, Second Edition, New York: Macmillan.

Ruskin, D. M. (1978), M-estimates of nonlinear regression parameters and their jackknife constructed confidence intervals, Ph.D. Dissertation, University of California at Los Angeles, Los Angeles, California.

Scheffé, H. (1959), *The Analysis of Variance*, New York: Wiley.

Searle, S. R. (1971), *Linear Models*, New York: Wiley.

Serfling, R. J. (1980), *Approximation Theorems of Mathematical Statistics*, New York: Wiley.

Snedecor. G. W., and W. G. Cochran (1980), *Statistical Methods*, Seventh Edition, Ames, Iowa: Iowa State U.P.

Souza, G. (1979), Statistical inference in nonlinear models: A pseudo likelihood approach, Ph.D. Dissertation, North Carolina State University, Raleigh, North Carolina.

Tauchen, G. E. (1986), Statistical properties of generalized method of moments estimates of structural parameters using financial market data, *Journal of Business and Economic Statistics*, in press.

Tucker, H. G. (1967), *A Graduate Course in Probability*, New York: Academic.

Turner, M. E., R. J. Monroe, and H. L. Lucas (1961), Generalized asymptotic regression and nonlinear path analysis, *Biometrics* **17**, 120–149.

U.S. Bureau of the Census (1960), *Historical Statistics of the United States*, Washington, D.C.: U.S. Government Printing Office, 116–119.

Varian, H. R. (1978), *Microeconomic Analysis*, New York: Norton.

White, H. L., Jr. (1980), Nonlinear regression on cross section data, *Econometrica* **48**, 721–746.

White, H. L., Jr. (1982), Comment on "The unification of the asymptotic theory of nonlinear econometric models," *Econometric Reviews* **1**, 201–205.

Williams, E. J. (1962), Exact fiducial limits in nonlinear estimation, *Journal of the Royal Statistical Society Series B* **24**, 125–139.

Withers, C. S. (1981), Conditions for linear processes to be strong-mixing, *Zeitschrift für Wahrscheinlichkeitstheorie und verwandte Gebiete* **57**, 477–480.

Wooldridge, J. M. (1986), Nonlinear econometric models with dependent observations. Ph.D. Dissertation, University of California at San Diego, San Diego, California.

Wouk, A. (1979), *A Course of Applied Functional Analysis*, New York: Wiley.

Zellner, A. (1962), An efficient method of estimating seemingly unrelated regressions and tests for aggregation bias, *Journal of the American Statistical Association* **57**, 348–368.

Zellner, A. (1976), Bayesian and non-Bayesian analysis of the regression model with multivariate student-*t* error terms, *Journal of the American Statistical Association* **71**, 400–405.

Author Index

Aitchison, J., 271
Akaike, H., 137
Amemiya, T., 150, 151, 439, 467, 567
Anderson, T.W., 129, 137, 372, 446, 533, 534
Andrews, D.W.K., 515, 516
Ash, R.B., 488
Athreya, K.B., 496

Balet-Lawrence, S., 150
Barnett, W.A., 150
Bartle, R.G., 188, 208
Bates, D.M., 146, 147
Beale, E.M.L., 83, 140, 146, 147
Berger, R.L., 191
Berk, K.N., 130
Bierens, H.J., 503
Billingsley, P., 488, 520, 521, 527, 528, 529, 530, 532
Blackorby, C., 273, 290
Blackwell, D., 45
Bloomfield, P., 446, 533
Booker, A., 29, 151
Box, G.E.P., 5
Burguete, J.F., 149, 157, 197
Bussinger, P.A., 143

Caves, D.W., 290
Christensen, L.R., 272, 290, 399, 402, 403
Chung, K.L., 164, 488
Cochran, W.G., 373
Cook, R.D., 147
Cox, D.R., 166

Deaton, A., 270, 271
Dennis, J.E., 39
Doob, J.L., 511, 512
Dunford, N., 521
Durbin, J., 185

Edmunds, D.E., 173
Efron, B., 146
Elbadawi, I., 272, 318
Eppright, E.S., 143

Fiacco, A.V., 414
Fox, H.M., 143
Fox, M., 52
Fryer, B.A., 143
Fuller, E.S., 143, 144
Fuller, W.A., 129, 495

Gallant, A.R., 4, 5, 20, 25, 40, 55, 73, 81, 82, 84, 128, 136, 141, 142, 144, 146, 149, 150, 151, 157, 158, 159, 171, 185, 255, 272, 278, 289, 290, 318, 341, 357, 382, 401, 403, 404, 409, 459, 475, 567
Gay, D.M., 39
Gill, J.L., 374
Gill, P.E., 28, 29, 38, 57, 357
Girshick, M.A., 45
Goebel, J.J., 128, 136
Golub, G.H., 37, 143, 171
Graybill, F.A., 562
Griffiths, W.E., 124
Grossman, W., 150
Guttman, I., 5, 6, 83

601

Hafs, H.D., 374
Hall, P.G., 507, 508, 512, 531
Halperin, M., 140
Hammersley, J.M., 83
Handscomb, D.C., 83
Hansen, L.P., 420, 425, 439, 451, 567, 568, 569
Hanson, R.J., 561
Hartley, H.O., 26, 28, 29, 52, 140, 151
Henderson, H.V., 363
Heyde, C.C., 507, 508, 512, 531
Hill, R.C., 124
Hinkley, D.V., 166
Holly, A., 150, 157, 158, 159, 171, 185, 255, 357, 382, 475
Huber, P.J., 20, 149, 175, 198

Imhof, P., 391

Jennrich, R.I., 150, 157, 160
Jensen, D.R., 104
Jorgenson, D.W., 151, 272, 399, 402, 403, 459, 567
Judge, G.G., 124

Koenker, R.W., 278, 289, 290, 409

Laffont, J.J., 151, 567
Lamkin, G.H., 143
Lanberg, N.A., 191
Lau, L.J., 272, 399, 402, 403
Lawson, C.L., 561
Lee, T.C., 124
Levenberg, K., 26
Loeve, M., 168
Lucas, H.L., 5, 140
Lucas, R.E., 414
Luenberger, D.G., 195

McCormick, G.P., 414
McLeish, D.L., 487, 494, 506, 507, 511, 514, 521, 523, 524
Malinvaud, E., 12, 150, 157
Marquardt, D.W., 26, 37
Meeter, D.A., 5, 6, 83
Monroe, R.J., 140
Mood, A.M., 562
Moscatelli, V.B., 173
Muellbauer, J., 270, 271
Murray, W., 28, 29, 38, 57, 357

Office of the President, 131
Osborne, M.R., 38

Pantula, S.G., 129, 496
Pearson, E.S., 52
Pereyra, V., 37, 171
Phillips, P.C.B., 149, 467, 474
Pratt, J.W., 118
Primont, D., 273, 290

Rao, C.R., 190, 211, 215, 230, 234, 471
Ratkowsky, D.A., 146
Ross, G.J.S., 147
Rossi, P.E., 271
Royden, H.L., 23, 160, 172, 215, 520
Ruskin, D.M., 150, 151
Russell, R.R., 273, 290

Scheffe, H., 52, 54, 333, 341
Schwartz, J.T., 521
Searle, S.R., 121, 122, 363
Serfling, R.J., 190, 211, 224, 226, 230, 234, 588
Shen, S.M., 271
Singleton, K.J., 420, 425, 439, 569
Snedecor, G.W., 373
Souza, G., 149, 150, 157, 174, 272, 318

Tauchen, G.E., 439, 450, 457
Tucker, H.G., 21, 23, 167, 488
Turner, M.E., 140

U.S. Bureau of the Census, 131

Varian, H.R., 398
Vivian, V.M., 143

Watts, D.G., 146, 147
Welch, R.E., 39
White, H.L., 149, 222
Williams, E.J., 140
Withers, C.S., 495
Witmer, J.A., 147
Wooldridge, J.M., 521
Wouk, A., 195
Wright, M.H., 28, 29, 38, 57, 357

Zellner, A., 104

Subject Index

Across equations restrictions, 323, 438
Akaike's method, 139
Almost sure convergence, defined, 170
Ancillary statistics, 166
Argmin, 355
A-smooth, 515
Asymptotic normality:
  dynamic models:
    of least mean distance estimators, 559,
      564
    of method of moments estimators, 581,
      582
  regression structure:
    consequences of misspecification, 190
    of least mean distance estimators, 189
    of method of moments estimators, 211,
      215
  of scores:
    dynamic models:
      constrained estimators under Pitman
        drift, 588
      least mean distance estimators, 556
      method of moments estimators, 579
    regression structure:
      least mean distance estimators, 186
      method of moments estimators, 209
Autocorrelated errors, 128, 442, 452
Autocovariance function, 127
Autoregressive process, 128

Bartlett weights, 533
Bounded in probability:
  autocorrelations, 128

defined, 215
scale estimator, 213, 380, 392, 395

Centering nuisance parameter estimators,
  213
Central Limit Theorem:
  dynamic models:
    definition of terminology, 520
    stated, 519
    variance estimation, 532
  regression structure:
    definition of terminology, 170
    stated, 163
Cesaro Sum Generator:
  alternative computational formulas, 162
  consequence of accelerated convergence
    in misspecified situations, 190
  defined, 159
  introductory discussion, 157
  probability of conditional failure, 168
Chain rule, 10, 16
Characteristic polynomial, 129
Characterizations of statistics, univariate
    least squares:
  estimator of variance, 260
  Lagrange multiplier test statistic, 265
  least squares estimator, 259
  likelihood ratio test statistic, 264
  nonlinear function of least squares
    estimator, 261
  reciprocal of estimator of variance, 261
  restricted estimator of variance, 263
  Wald test statistic, 262

Chi-square distribution:
  central, 120
  noncentral, 120
Cholesky factorization, 133, 237
Compact parameter space, discussion, 170
Compartment analysis, 6
Composite function rule, 10, 16
Concentrated likelihood, 357, 470, 473
Conditional analysis:
  dynamic models, 466, 475, 477, 489
  regression structure:
    defined, 166
    detailed discussion of probability
      structure, 167
Conditional distribution, regression
    structure:
  of dependent variables, 169
  of errors, 168
Conditional Jensen's inequality, 540
Conditional likelihood, dynamic models,
    467, 475, 477
Confidence regions:
  multivariate models, 355
  univariate least squares:
    correspondence between expected
      length, area, or volume and power
      of test, 116
    Lagrange multiplier, 110
    likelihood ratio, 107
    measures of nonlinearity, 146
    structural characteristics of, 107, 112,
      118, 119
    Wald, 104
Consistency:
  dynamic models:
    of least mean distance estimators, 548
    of method of moments estimators, 577
  regression structure:
    of least mean distance estimators, 180
    of method of moments estimators,
      208
Constrained estimation:
  asymptotic normality, 243, 245
  consistency, 243, 244
  variance formulas, 243, 245
Constraint, equivalent representations, 240,
    242, 560
Consumer demand, 268, 407
Contemporaneous correlation, 268
Continuity set, 528

Continuous convergence, 163, 519
Convergence in distribution, regression
    structure, defined, 170
Covariance stationary, 123, 127
Coverage function, 118
Critical function, 116

Data generating model:
  dynamic models:
    defined, 541
    introductory discussion, 488
  regression structure, defined, 156
Differentiation:
  chain rule, 10, 16
  composite function rule, 10, 16
  gradient, 8
  Jacobian, 9
  Hessian, 8
  matrix derivative, 8
  vector derivative, 8
Disconnected confidence regions, 119
Distance function:
  dynamic models:
    least mean distance estimators, 544
    method of moments estimators, 566
  regression structure:
    least mean distance estimators, 176
    method of moments estimators, 197,
      204
Dynamic models:
  defined, 148, 406
  estimation by generalized method of
    moments, 442
  estimation by maximum likelihood,
    465

Efficiency of method of moments
    estimators, 212
Efficient score test, *see* Lagrange multiplier
    test
Electricity expenditure, 268, 407
Equicontinuity, defined, 161
Equivalent local power of Wald, likelihood
    ratio, and efficient score tests,
    239
Estimated parameter:
  dynamic models, 546
  regression structure, 175
Estimated variance-convariance of method
    of moments estimators, 453

Estimation space, regression structure,
  defined, 154, 176
Explicit form, 406

F-distribution:
  central, 120
  doubly noncentral, 121
  noncentral, 121
Fixed regressors, 247
Fourier form, 318, 400
Functional dependence, 57, 240, 242, 320,
  454

Gauss-Newton method:
  generalized method of moments, 446,
    453
  multivariate least squares, 317
  three stage least squares, 434
  univariate least squares:
    algorithm, 28
    algorithm failure, 39
    convergence proof, 44
    informal discussion, 26
    starting values, 29
    step length determination, 28
    stopping rules, 29
General inference problem, introductory
  discussion, 154
Generalized method of moments estimator:
  bias of, 451
  choice of instrumental variables, 451
  confidence intervals, 451
  consistency of variance estimates,
    446
  defined, 442, 446, 567
  estimated variance-covariance of, 446
  Gauss-Newton correction vector, 446
  insuring positive definite variance
    estimate, 446
  overidentifying restrictions, 451, 464
  sample objective function, 444, 567
  see also Three stage least squares
    estimators
Generalized nonlinear least squares, 127,
  128
Gradient, 8
Grid search, 36
Grouped by equation data arrangement:
  multivariate models:
    covariance matrix for, 300, 318

  defined, 291, 299, 351
  test statistics, 353, 354
  simultaneous equations models, 428, 437
Grouped by subject data arrangement:
  multivariate models:
    defined, 291
    test statistics, 326, 333, 342, 388, 389,
      390
  simultaneous equations models, 428

Hartley-Booker estimator, 29, 151
Hartley's method, see Gauss-Newton
  method
Hessian, 8
Heteroscedastic errors, 123, 124, 156
Heteroscedastic invariant:
  dependence invariant:
    Lagrange multiplier test, 123, 139
    variance estimator, 123, 138
    Wald test, 123, 138
  Lagrange multiplier test, 123, 126
  variance estimator, 123, 125, 433, 438,
    452
  Wald test, 123, 125

Identification condition:
  dynamic models:
    least mean distance estimators:
      defined, 546
      example, 548
    method of moments estimators:
      defined, 570
      discussed, 570
  regression structure:
    least mean distance estimators:
      defined, 176
      example, 176
    method of moments estimators:
      defined, 199
      discussed, 199
      example, 199
  univariate least squares, 20
Implicit form, 406
Independent and identically distributed
  regressors, consequence of in
  misspecified situations, 191, 247
Infinite dimensional parameter space,
  examples, 171, 172
Information inequality, 474, 484
Information set, 412

Instrumental variables, 150, 431, 435, 439, 444, 446, 452
Intertemporal consumption and investment, 409
Intrinsic curvature, 146
Iterated least squares estimators, 355

Jacobian, 9
Join point, 143

Kronecker product:
  defined, 299
  manipulative facts, 299, 362, 377
  relation to vec operator, 362, 377
Kuhn-Tucker theorem, 414

Lack-of-fit test, 140, 146
  choice of regressors, 141
  noncentrality parameters, 141
Lagged dependent variables, 442, 452
Lagrange multiplier test:
  dynamic models:
    asymptotic distribution, 593
    defined, 593
    discussed, 593
  multivariate models:
    least squares:
      computation, 343
      with correctly centered estimator of scale, 392
      defined, 390, 342, 353
      with miscentered estimator of scale, 394
      Monte Carlo simulations, 349
      power computations, 349
    maximum likelihood, 366
  regression structure:
    asymptotic distribution, 230
    defined, 219
    discussed, 219
  simultaneous equations models:
    generalized method of moments estimators, 460
    maximum likelihood estimators:
      dynamic models, 474, 479
      regression structure, 474
    three stage least squares estimators, 460
  univariate models:
    asymptotic distribution, 87
    computation, 90

corresponding confidence region, 110
  defined, 89
  with heteroscedastic errors, 126
  informal discussion, 85, 102
  Monte Carlo simulations, 100
  with nonstationary, serially correlated errors, 139
  power computations, 97
  testing specification, 139
Large residual problem, 39
Least mean distance estimator:
  dynamic models:
    consistency, 548
    constrained, 544
    defined, 544
    identification, 546
    introductory discussion, 489
    summary of results, 584
  regression structure:
    constrained, 176
    defined, 148, 174, 176
    introductory discussion, 148
    summary of results, 217
Least squares estimator:
  multivariate models:
    asymptotic normality, 387
    consistency, 384
    defined, 301, 380
    efficiency, 397
    identification, 383
    iterated least squares estimators, 355
    Pitman drift, 387
    sample objective function, 383
  univariate models:
    asymptotic normality, 258
    characterized as linear function of errors, 16, 259
    computation, see Gauss-Newton method
    consistency, 256
    data generating model, 255
    defined, 1, 3, 253
    distribution of, 17, 18
    first order conditions, 15
    identification, 255
    informal discussion of regularity conditions, 19
    Pitman drift, 258
    sample objective function, 253, 256
    score function, 258

Likelihood ratio test:
  dynamic models:
    asymptotic distribution, 591
    defined, 591
    discussed, 591
  multivariate models:
    least squares estimators:
      computation, 333
      defined with degrees of freedom
        corrections, 333, 353
      defined without degrees of freedom
        corrections, 388
      importance of identical estimators of
        scale, 394
      Monte Carlo simulations, 349
      power computations, 336
    maximum likelihood estimators:
      relationship to least squares
        "likelihood ratio" test, 366
      for test of location, 366
  regression structure:
    alternate form, 235
    asymptotic distribution, 233
    defined, 220
    discussed, 220
  simultaneous equations models:
    generalized method of moments
      estimators, 457
    maximum likelihood estimators:
      dynamic models, 479
      regression structure, 474
    three stage least squares estimators,
      457
  univariate models:
    asymptotic distribution, 70
    computation, 57
    corresponding confidence region, 107
    defined, 56
    with heteroscedastic errors, 127
    informal discussion, 55
    Monte Carlo simulations, 81, 82, 84
    with nonstationary, serially correlated
      errors, 139
    power computations, 69
    testing specification, 139, 141
Linear regression model, *see* Univariate
  nonlinear regression model
Logistic-normal distribution, 271

McLeish's inequality, 511
Martingale, 522

Martingale difference sequence, 522
Matrix derivatives, 8
Maximum likelihood estimators:
  multivariate models:
    computed by iterated least squares,
      355
    concentrated likelihood, 357
    considered as least squares estimator,
      358
    data generating model, 359
    defined, 356
    derivatives of distance function, 361,
      377
    likelihood, 356
    Pitman drift, 358
    sample objective function, 361
    score function, 362
  simultaneous equations models:
    asymptotic variance-covariance matrix
      of, 473
    bias, 467
    concentrated likelihood, 470, 473
    conditional likelihood, 467, 475, 477
    defined:
      for dynamic models with
        independent errors, 475
      for general dynamic models, 477
      for regression structures, 468
    derivation of likelihood in dynamic
      case, 480
    expectation of score, 485
    information inequality, 474, 484
    near epoch dependence, 477, 481
    Newton downhill direction, 473
    normally distributed errors, 466
    sample objective function, 468, 476,
      477
    transformations, 466
Measures of nonlinearity, 146
M-estimator:
  iteratively rescaled M-estimator,
    example, 174, 176, 178, 181, 183,
    184, 186, 205
  scale invariant M-estimator, example,
    198, 199, 201
Method of moments estimator:
  dynamic models:
    consistency, 577
    constrained, 566
    defined, 566
    identification, 570

Method of moments estimator (*Continued*)
  introductory discussion, 489
  summary of results, 584
  under misspecification, 570, 582
  regression structure:
    constrained, 197
    defined, 151, 197
    introductory discussion, 150
    summary of results, 217
  simultaneous equations models,
    431
Misspecification:
  defined, 149
  example, 220, 222, 225, 233, 235
Mixingale:
  defined, 506
  sufficient conditions for, 509
Mixing conditions, 443
Modified Gauss-Newton method, *see*
    Gauss-Newton method
Multivariate nonlinear regression model:
  defined, 290
  identification, 383
  Pitman drift, 387
  rotation to univariate model, 317

Near epoch dependence:
  defined, 496
  examples of, 477, 481, 502
  sufficient conditions for, 498
Newton downhill direction, 473
Nonlinear ARMA, 539
Nonlinear autoregression, 502, 544, 546,
    549, 553
Nonlinear least squares under specification
    error, example, 220, 222, 225,
    233, 235
Nonlinear regression model, *see* Univariate
    nonlinear regression model
Normal distribution:
  multivariate, 120
  univariate, 120
Null hypothesis:
  dynamic models, defined, 585
  regression structure, defined, 218

Optimal choice of instrumental variables,
    439, 451
  bootstrap strategy, 451
Overidentifying restrictions, 451, 464

Parameter effect curvatures, 146
Parameter space, defined, 155
Parametric restriction, 57, 240, 242, 320,
    453
Parzen weights, 446, 533
Pitman drift:
  dynamic models:
    introductory discussion, 490
    least mean distance estimators, 560
    method of moments estimators, 581
  regression structure:
    consequence:
      in constrained estimation, 191, 214
      in misspecified situations, 191
    examples, 171, 172, 195
    introductory discussion, 154
Population moments, 431
Predetermined variables, 442
Principal component vectors, 142

Quadratic forms, 121
Quadratic-linear response function, 143

Rank qualification, 20
Rao's efficient score test, *see* Lagrange
    multiplier test
Rational expectations hypothesis, 415
Reduced form, 157, 443, 467, 542
Regression structure, 148, 166, 167, 406,
    442, 465
Regular conditional probability:
  defined, 168
  example, 168
Reparametrization, 147
Robustness of validity, 467

Sample moments, 431, 444, 453
Sample objective function:
  dynamic models:
    least mean distance estimators, 544
    method of moment estimators, 566
  regression structure:
    least mean distance estimators:
      almost sure limit, 176
      defined, 176
      expectation, 176
    method of moment estimators:
      almost sure limit, 198
      defined, 197